WILLIAM M. BLACKMORE

Laboratory Hamsters

AMERICAN COLLEGE OF LABORATORY ANIMAL MEDICINE SERIES

Steven H. Weisbroth, Ronald E. Flatt, and Alan L. Kraus, eds.:
The Biology of the Laboratory Rabbit, 1974

Joseph E. Wagner and Patrick J. Manning, eds.:
The Biology of the Guinea Pig, 1976

Edwin J. Andrews, Billy C. Ward, and Norman H. Altman, eds.:
Spontaneous Animal Models of Human Disease, Volume I, 1979;
Volume II, 1979

Henry J. Baker, J. Russell Lindsey, and Steven H. Weisbroth, eds.:
The Laboratory Rat, Volume I: Biology and Diseases, 1979;
Volume II: Research Applications, 1980

Henry L. Foster, J. David Small, and James G. Fox, eds.:
The Mouse in Biomedical Research, Volume I: History,
Genetics, and Wild Mice, 1981; Volume II: Diseases, 1982;
Volume III: Normative Biology, Immunology, and Husbandry, 1983;
Volume IV: Experimental Biology and Oncology, 1982

James G. Fox, Bennett J. Cohen, and Franklin M. Loew, eds.:
Laboratory Animal Medicine, 1984

G. L. Van Hoosier, Jr., and Charles W. McPherson, eds.:
Laboratory Hamsters, 1987

Laboratory Hamsters

EDITED BY

G. L. Van Hoosier, Jr.
School of Medicine
Division of Animal Medicine
University of Washington
Seattle, Washington

Charles W. McPherson
School of Veterinary Medicine
North Carolina State University
Raleigh, North Carolina

Academic Press, Inc.
Harcourt Brace Jovanovich, Publishers

Orlando San Diego New York Austin
Boston London Sydney Tokyo Toronto

Copyright © 1987 by Academic Press, Inc.
ALL RIGHTS RESERVED.
NO PART OF THIS PUBLICATION MAY BE REPRODUCED OR
TRANSMITTED IN ANY FORM OR BY ANY MEANS, ELECTRONIC
OR MECHANICAL, INCLUDING PHOTOCOPY, RECORDING, OR
ANY INFORMATION STORAGE AND RETRIEVAL SYSTEM, WITHOUT
PERMISSION IN WRITING FROM THE PUBLISHER.

ACADEMIC PRESS, INC.
Orlando, Florida 32887

United Kingdom Edition published by
ACADEMIC PRESS INC. (LONDON) LTD.
24–28 Oval Road, London NW1 7DX

Library of Congress Cataloging in Publication Data

Laboratory hamsters.

   (American College of Laboratory Animal Medicine series)
   Includes index.
   1. Hamsters as laboratory animals.   2. Hamsters.
I. Van Hoosier, G. L.   II. McPherson, Charles W.
III. Series.
QL737.R638L33   1987   636'.93233   86-28738
ISBN 0–12–714165–0 (alk. paper)

PRINTED IN THE UNITED STATES OF AMERICA

87 88 89 90   9 8 7 6 5 4 3 2 1

Contents

List of Contributors	viii	
Foreword	x	
Preface	xiii	
List of Reviewers for Chapters in This Volume	xv	

Part I   The Golden or Syrian Hamster

Chapter 1   Historical Perspectives and Taxonomy
J. Derrell Clark

I.	Description and Behavior	3
II.	Geographical Distribution and Habitat	3
III.	Taxonomy	4
IV.	Nomenclature	4
V.	Cytogenetics	4
VI.	History of Laboratory Use	5
	References	6

Chapter 2   Morphophysiology
W. Sheldon Bivin, Glenn A. Olsen, and Kathleen A. Murray

I.	Introduction	10
II.	General Features	10
III.	Digestive System	12
IV.	Respiratory System	19
V.	Cardiovascular and Lymphatic Systems	24
VI.	Urinary and Reproductive Systems	27
VII.	Endocrine System	32
VIII.	Nervous System	35
IX.	Special Sensory Systems	35
	References	36

Chapter 3   Clinical Chemistry and Hematology
Farol N. Tomson and K. Jane Wardrop

I.	Introduction	43
II.	Sources of Variation	44
III.	Specimen Handling	44
IV.	Clinical Chemistry	46
V.	Hematology	54
	References	58

Chapter 4   Care and Management
Melvin W. Balk and Gilbert M. Slater

I.	Introduction	61
II.	Housing	61
III.	Sanitation Procedures	62
IV.	Environmental Requirements	62
V.	Nutrition and Feeding	63
VI.	Practical Breeding	63
VII.	Handling, Restraint, Identification, and Sexing	64
	References	66

Chapter 5   Biomethodology
Jerald Silverman

I.	Introduction	70
II.	Clinical Examination	70
III.	Radiographic Techniques	71
IV.	Injections and Intubations	72
V.	Collections	75
VI.	Anesthesia	77
VII.	Surgical Procedures	82
VIII.	Physiological Measurements	87
IX.	Perfusions	88
X.	Laparoscopy	89
XI.	Miscellaneous Techniques	89
XII.	Euthanasia	89
XIII.	Necropsy	90
	References	91

Chapter 6	Viral Diseases		Chapter 11	Drugs Used in Hamsters with a Review	
	John C. Parker, James R. Ganaway,			of Antibiotic-Associated Colitis	
	and Cynthia S. Gillett			*J. David Small*	
I.	Introduction	95	I.	Introduction	179
II.	Viruses	96	II.	Drug Therapy in Hamsters	180
	References	106	III.	Antibiotic-Associated Enterocolitis	180
				References	196
Chapter 7	Bacterial and Mycotic Diseases				
	Craig S. Frisk		Chapter 12	Experimental Biology: Use in Oncologic	
I.	Introduction	112		Research	
II.	Proliferative Ileitis, Hamster Enteritis, Atypical			*John J. Trentin*	
	Ileal Hyperplasia	112	I.	Introduction	201
III.	Cecal Mucosal Hyperplasia	121	II.	The Cheek Pouch as a Privileged Tumor	
IV.	Tyzzer's Disease	121		Transplant Site	202
V.	Salmonellosis	125	III.	Use of Conditioned Hamsters for Growth and	
VI.	Campylobacteriosis	127		Transplantation of Human Tumors	202
VII.	Tularemia	128	IV.	Transmissible (Epizootic) Leukemia of Hamsters	202
VIII.	Pseudotuberculosis	129	V.	Nonrandom Chromosome Abnormalities in	
IX.	Streptococcal Infections	129		Transformed Syrian Hamster Cell Lines	203
X.	*Pasteurella* Infections	129	VI.	Role of Adenovirus Type 2-Induced Early Tumor	
XI.	Bacterial Respiratory Infections	129		Antigens in Natural Cell-Mediated Defense against	
XII.	Actinomycosis	130		Such Tumors in Hamsters	204
XIII.	Cutaneous Bacterial Abscesses	130		References	205
XIV.	Postirradiation Infections	131		Appendix 1. Spontaneous Tumor Development:	
XV.	Miscellaneous Bacterial Infections	131		Benign Tumors	207
XVI.	Mycotic Infections	131		Appendix 2. Spontaneous Tumor Development:	
	References	131		Malignant Tumors	208
				Appendix 3. Induced Tumors: Viral Oncogenesis	210
Chapter 8	Parasitic Diseases			Appendix 4. Induced Tumors: Survival and Tumor	
	Joseph E. Wagner			Incidence after Irradiation	214
I.	Introduction	135			
II.	Parenteral Parasites	136	Chapter 13	Experimental Biology: Use in	
III.	Parasites of the Alimentary System	137		Immunobiology	
IV.	Parasites of the Integument	149		*J. Wayne Streilein*	
	References	153	I.	Introduction	215
			II.	Ontogeny of Immunity	216
Chapter 9	Neoplastic Diseases		III.	Immunochemistry	217
	John D. Strandberg		IV.	Immunogenetics	218
I.	Introduction	157	V.	Lymphoreticular Cells	219
II.	Respiratory System	158	VI.	Recent Advances in Hamster Immunobiology	221
III.	Gastrointestinal System	159		References	223
IV.	Urinary System	161			
V.	Musculoskeletal System	161	Chapter 14	Experimental Biology: Use in Infectious	
VI.	Tumors of Soft Connective Tissues	161		Disease Research	
VII.	Integumentary System	161		*J. K. Frenkel*	
VIII.	Circulatory System	163	I.	Hamsters as Unique Infection Models	228
IX.	Hematopoietic System	163	II.	Infections with Disseminated Lesions	229
X.	Endocrine System	164	III.	Infections with Cerebral and Ocular Lesions	233
XI.	Reproductive System	166	IV.	Infections with Muscle Lesions	235
XII.	Nervous System	167	V.	Infections with Respiratory Tract Lesions	235
	References	167	VI.	Infections with Kidney Lesions	236
			VII.	Infections with Liver Lesions	237
Chapter 10	Noninfectious Diseases		VIII.	Infections with Gastrointestinal Lesions	238
	Gene B. Hubbard		IX.	Infections with Reticuloendothelial and Blood	
	and Robert E. Schmidt			Vascular Lesions	239
I.	Introduction	169	X.	Infections with Lesions in Endocrine Organs	240
II.	Diseases Associated with Aging	169	XI.	Infections with Skin and Other Lesions	241
III.	Nutritional Disease	175	XII.	Hamsters as Sentinels of Infection	242
IV.	Polycystic Disease	175	XIII.	Hamsters as Models of Cellular Immunity against	
V.	Traumatic Disease	175		Infection	242
VI.	Prenatal Mortality	175	XIV.	Conclusions	243
	References	177		References	243

CONTENTS

Chapter 15	Experimental Biology: Genetic Models in Biomedical Research *F. Homburger and Jeanne Peterson*	
I.	Introduction	251
II.	Specific Disease Models	253
	References	261

Chapter 16	Experimental Biology: Other Research Uses of Syrian Hamsters *Christian E. Newcomer, Douglas A. Fitts, Brian D. Goldman, Michael R. Murphy, Ghanta N. Rao, Gerald Shklar, and Joel L. Schwartz*	
I.	Introduction	264
II.	Physiological Regulation of Ingestive Behavior	264
III.	Behavior and Neuroscience Research	267
IV.	Reproductive Physiology	269
V.	Hibernation and Cold Adaptation Research	273
VI.	Teratology	275
VII.	Toxicology	277
VIII.	Radiobiological Research	278
IX.	Oral Pathology: Chemical Carcinogenesis in the Hamster Cheek Pouch	281
X.	Chemical Carcinogenesis at Sites Other than Cheek Pouch	285
XI.	Dental Caries and Periodontal Disease	288
	References	289

Appendix	Selected Normative Data for the Syrian Golden Hamster *Charles W. McPherson*	
	Text	301

Part II   The Striped or Chinese Hamster

Chapter 17	Biology and Care *Albert Chang, Arthur Diani, and Mark Connell*	
I.	History	305
II.	Description and Behavior	306
III.	Biology	306
IV.	Care and Husbandry	313
	References	316

Chapter 18	Diseases *Warren C. Ladiges*	
I.	Infectious Diseases	321
II.	Neoplastic Diseases	324
III.	Metabolic and Genetic Conditions	325
IV.	Traumatic Disorders	326
V.	Reproductive Disorders	326
VI.	Miscellaneous Diseases	327
	References	328

Chapter 19	Use in Research *Arthur Diani and George Gerritsen*	
I.	Introduction	330
II.	Spontaneous Diabetes Mellitus	330
III.	Chemically Induced Diabetes Mellitus	335
IV.	General Toxicology	336
V.	Genetic Toxicology	338
VI.	Cytogenetics of Germ and Somatic Cells	341
VII.	Cell Lines Derived from Chinese Hamster Tissues and Organs	341
	References	341

Part III   The European Hamster

Chapter 20	Biology, Care, and Use in Research *Ulrich Mohr and Heinrich Ernst*	
I.	History and Taxonomy	351
II.	Care and Management	352
III.	Anatomical and Histological Characteristics	353
IV.	Hematological and Clinicochemical Parameters	356
V.	Hibernation	357
VI.	Diseases	360
VII.	Use in Research	361
	References	364

Part IV   Other Hamsters

Chapter 21	Biology, Care, and Use in Research *Connie A. Cantrell and Dennis Padovan*	
	Introduction	369
	Phodopus sungorus (Dzungarian hamster)	370
	References	374
	Mystromys albicaudatus (South African hamster)	376
	References	380
	Mesocricetus brandti (Turkish hamster)	382
	References	383
	Mesocricetus newtoni (Rumanian hamster)	384
	References	385
	Cricetulus migratorius (Armenian hamster)	385
	References	386

Index		389

List of Contributors

Numbers in parentheses indicate the pages on which the authors' contributions begin.

Melvin W. Balk (61), Charles River Laboratories, Wilmington, Massachusetts 01887-0630

W. Sheldon Bivin (9), Laboratory Animal Resources, School of Veterinary Medicine, Louisiana State University, Baton Rouge, Louisiana 70803

Connie A. Cantrell (369), Division of Animal Resources, Virginia Commonwealth University, Richmond, Virginia 23298

Albert Chang (305), Diabetes and Gastrointestinal Diseases Research Unit, The Upjohn Company, Kalamazoo, Michigan 49001

J. Derrell Clark (3), Department of Medical Microbiology, College of Veterinary Medicine, University of Georgia, Athens, Georgia 30602

Mark Connell (305), Diabetes and Gastrointestinal Diseases Research Unit, The Upjohn Company, Kalamazoo, Michigan 49001

Arthur Diani (305, 329), Diabetes and Gastrointestinal Diseases Research Unit, The Upjohn Company, Kalamazoo, Michigan 49001

Heinrich Ernst (351), Institut für Experimentelle Pathologie, Medizinische Hochschule Hannover, D-3000 Hannover 61, Federal Republic of Germany

Douglas A. Fitts (263), Department of Psychology, University of Washington, Seattle, Washington 98195

J. K. Frenkel (228), Department of Pathology and Oncology, University of Kansas School of Medicine, Kansas City, Kansas 66103

Craig S. Frisk (111), Mayo Clinic, Rochester, Minnesota 55905

James R. Ganaway (95), Microbiological Associates, Bethesda, Maryland 20816

George Gerritsen (329), Diabetes and Gastrointestinal Diseases Research Unit, The Upjohn Company, Kalamazoo, Michigan 49001

Cynthia S. Gillett (95), Research Animal Resources Center, University of Wisconsin, Madison, Wisconsin 53706

Brian D. Goldman (263), Worcester Foundation for Experimental Biology, Shrewsbury, Massachusetts 01545

F. Homburger (251), Bio-Research Institute, Cambridge, Massachusetts 02139

Gene B. Hubbard (169), U.S. Army Institute of Surgical Research, Comparative Medicine Branch, Fort Sam Houston, Texas 78234

Warren C. Ladiges (321), Division of Animal Medicine, University of Washington, Seattle, Washington 98195

Charles W. McPherson (301), School of Veterinary Medicine, North Carolina State University, Raleigh, North Carolina 27606

Ulrich Mohr (351), Institut für Experimentelle Pathologie, Medizinische Hochschule Hannover, D-3000 Hannover 61, Federal Republic of Germany

Michael R. Murphy (263), Systems Research Laboratory, Brooks Air Force Base, San Antonio, Texas 78235

Kathleen A. Murray (9), Laboratory Animal Resources, School of Veterinary Medicine, Louisiana State University, Baton Rouge, Louisiana 70803

Christian E. Newcomer (263), Division of Comparative Medicine, Massachusetts Institute of Technology, Cambridge, Massachusetts 02139

Glenn A. Olsen (9), Laboratory Animal Resources, School of Veterinary Medicine, Louisiana State University, Baton Rouge, Louisiana 70803

LIST OF CONTRIBUTORS

Dennis Padovan (369), Division of Animal Resources, Virginia Commonwealth University, Richmond, Virginia 23298

John C. Parker (95), Microbiological Associates, Bethesda, Maryland 20816

Jeanne Peterson (251), Department of Biology, Boston College, Chestnut Hill, Massachusetts 02167

Ghanta N. Rao (263), National Toxicology Program, National Institute of Environmental Health Sciences, Research Triangle Park, North Carolina 27709

Robert E. Schmidt (169), California Veterinary Diagnostics, West Sacramento, California 95691

Joel L. Schwartz (263), Department of Oral Medicine and Oral Pathology, Harvard School of Dental Medicine, Boston Massachusetts 02115

Gerald Shklar (263), Department of Oral Medicine and Oral Pathology, Harvard School of Dental Medicine, Boston, Massachusetts 02115

Jerald Silverman (69), Laboratory Animal Center, The Ohio State University, Columbus, Ohio 43220

Gilbert M. Slater (61), Charles River Laboratories, Wilmington, Massachusetts 01887-0630

J. David Small (179), Comparative Medicine Branch, National Institute of Environmental Health Sciences, Research Triangle Park, North Carolina 27709

John D. Strandberg (157), Division of Comparative Medicine, The John Hopkins University, School of Medicine, Baltimore, Maryland 21205

J. Wayne Streilein (215), Department of Microbiology and Immunology, University of Miami School of Medicine, Miami, Florida 33101

Farol N. Tomson (43), Laboratory Animal Resources Center, Washington State University, Pullman, Washington 99164-1165

John J. Trentin (201), Division of Experimental Biology, Baylor College of Medicine, Texas Medical Center, Houston, Texas 77030

Joseph E. Wagner (135), Research Animal Diagnostic Lab, College of Veterinary Medicine, University of Missouri, Columbia, Missouri 65211

K. Jane Wardrop (43), Department of Veterinary Microbiology and Pathology, College of Veterinary Medicine, Washington State University, Pullman, Washington 99164-7040

Foreword

Common laboratory animals have generally been developed from domesticated species that were closely associated with people since prehistoric times as competitors for food, carriers of disease, sources of food, or companions in the household. Hamsters are a relatively recent introduction as subjects for biomedical research. Rats, guinea pigs, chickens, and dogs were involved in investigations that led to the discovery of vitamins and the prevention and cure of nutritional deficiency diseases, while wild hamsters were scurrying around Syrian wheat fields collecting grain for winter storage.

Syrian hamsters made a successful transition from fields to laboratories in a surprisingly short time. Chinese hamsters were the first to be used in medical research in 1919 when Hsieh described their susceptibility to *Leishmania,* the causative agent of kala-azar. However, Dr. Saul Adler at the Hebrew University in Jerusalem found Chinese animals unsatisfactory in studies of Mediterranean leishmaniasis in 1930. The animals would not breed in captivity, many were lost in *Pasturella* epidemics, and regular shipments of satisfactory animals were difficult to obtain from China. Fortunately Adler persuaded his colleague, I. Aharoni, a zoologist, to organize an expedition to Syria to search for living specimens of an indigenous hamster. Ten young golden hamsters and their mother were excavated from a burrow eight feet deep in a wheat field near Aleppo. After some misfortunes nine animals were delivered to H. Ben-Menahem, director of the animal facilities at the Hebrew University. Of these, five gnawed their way out of a temporary cage the first night, but the remaining animals proved properly prolific under the director's personal care, and a flourishing colony was finally produced. In fact, within sixteen months after animals were collected in Syria, Adler and O. Theodor reported at a meeting of The Royal Society in London their successful use of laboratory-reared Syrian hamsters in the study of Mediterranean kala-azar. The rapid domestication of field-collected animals may have been facilitated by the fact that the laboratory and field environments were similar; the personnel were experienced hamster handlers, familiar not only with Chinese, but also with European and Middle Eastern species; and the collected animals were not severely stressed by an extended period of shipment. Adler was so delighted with his new research animals that he immediately delivered pairs to research institutions in Great Britain and France in 1931 and later made shipments to both Egypt and India. Three shipments reached the United States in 1938. Within the first twenty-five years after their introduction as laboratory animals 2023 scientific studies were published in which golden hamsters were used as subjects, and within the next six years 2089 citations were added to a master bibliography after identification and verification.

At present the actual number of hamsters used in research is somewhat less than half a million a year. The numbers of guinea pigs, rabbits, and hamsters used are quite similar. The United States Department of Agriculture data show that an average of 480,000 rabbits, 450,000 guinea pigs, and 420,000 hamsters were used annually in experimental studies from 1973 to 1984. However, from 1973 to 1975 more hamsters were used than guinea pigs or rabbits, from 1976 to 1981 rabbits were used in the greatest numbers, but from 1982 to 1984 guinea pigs served the most frequently as research subjects. Mice and rats are the animals used in the largest numbers, with annual figures in the millions. The number of rats involved in research studies is about ten times the number of guinea pigs, hamsters, or rabbits, but about three to four times more mice are used than rats. A recent Congressional report estimated that in 1978, 60–68 million animals were involved in experimental studies.

FOREWORD

To provide for such a multitude of animals and cope with the continually increasing costs and complexities of experimental research, it is essential to have accurate, authoritative, comprehensive, and current books to assist in research planning, provision of optimal animal husbandry, refinement of techniques, and acquisition of knowledge of current research in specialized areas. It is appropriate that the American College of Laboratory Animal Medicine is producing a series of books on the biology, husbandry, and experimental uses of laboratory animals in response to this need. Eleven volumes have already appeared in the series: one each on the laboratory rabbit (1974) and the guinea pig (1976), two on the rat (1979, 1980), four on the mouse (1981, 1982, 1983), two on spontaneous animal models of human disease (1979), and one on laboratory animal medicine (1984). A comparable book on laboratory hamsters is needed. The pioneer book edited by Hoffman, Robinson, and Magalhaes published in 1968, "The Golden Hamster: Its Biology and Use in Medical Research," with extensive bibliography and stereotaxic atlas, covered only a single species and is now both out of date and out of print. "The Hamster: Reproduction and Behavior," a book edited by S. Siegel in 1985, is the work of nineteen investigators on the use of Syrian hamsters in two major areas of research. Nine chapters deal with social behavior and its neurological basis. Four chapters summarize recent studies in reproductive endocrinology, while single chapters treat the history of domestication, biological rhythms, and regulation of energy balance. These reviews should be invaluable for those interested in the areas treated, but the book does not provide material on basic biology and husbandry for several species of hamsters.

Before leaving the subject of the number of animals used in research and the need for adequate source books, I would like to pause for a moment of speculation. Is there a possible correlation between the publication of reference books and the number of research animals used a few years later? The hamster, rabbit, and guinea pig data presented above may suggest this possibility.

This book reviews the history, biology, husbandry, spontaneous diseases, and experimental biology of eight species of hamsters—sixteen chapters on the Syrian (golden), three on the Chinese, one on the European, and a final chapter on five different species.

The Syrian hamster, often designated as "the" laboratory hamster, is the most frequently used species, perhaps because of its intermediate size, adaptability to laboratory conditions, relatively cooperative disposition, and, in part, the enthusiasm of its original sponsor. The chapters of particular value are those on clinical chemistry, biomethodology, spontaneous diseases, and drug therapy. The systematic treatment of anatomy, physiology, and husbandry should prove very useful. Comprehensive reviews on the applications of Syrian hamsters in experimental biology cover oncology, immunobiology, infectious diseases, and genetic disease models. A final chapter contains shorter reviews on research with Syrian hamsters in experimental physiology, radiobiology, teratology, toxicology, carcinogenesis, and dental pathology.

Although hamsters are now well established as laboratory animals and are familiar as pets or in zoos, they are native to the Old and not the New World, and thus the variety of species and their interrelationships are not understood by most Americans. In fact, the sudden appearance of these new animals in laboratories gave rise to the notion that like mules, they were hybrids and not natural animals. Even early research publications suggested that some species had arisen from hybridization in nature. One species, named triton after a mythical god that was part man and part fish, looks less like a hamster and more like a rat with a long tail and very pointed nose.

As a group, hamsters are characterized as stout-bodied, stubby-tailed, broad-headed, cheek-pouched, nidificating, fossorial rodents. The "true" hamsters are members of the subfamily Cricetinae, which was originally in the family Muridae, or mouse family. But in 1946, Simpson separated five subfamilies from Muridae and placed them in Cricetidae, or hamster family. Traditionally, taxonomists have been divided into two types, lumpers and splitters, who favor large or small groups in classification. Thus, some taxonomists lump hamsters with Muridae, the largest mammalian family, while others split them and related subfamilies into Cricetidae. But taxonomy is now also being transformed into systematics, a type of classification in which organisms are grouped in such a manner as to show their relationship to one another. As new experimental techniques are developed, evidence is obtained to support changes in classification, and some groups are discontinued and others replace them.

The following tabulation on Hamster Classification Schemes provides a summary of three recent publications. Corbet and Hill of the British Museum of Natural History in "A World List of Mammalian Species" (1980) include scientific and vernacular names and geographical ranges for fifteen species of hamsters. Honacki, Kinman, and Koeppl in "Mammal Species of the World: A Taxonomic and Geographic Reference" (1982) from the Association of Systematic Collections of Lawrence, Kansas, provide similar information for twenty-four species of hamsters. Nowak and Paradiso in "Walker's Mammals of the World" (1983) describe seventeen species of hamsters. Both Corbet and Hill and Nowak and Paradiso have returned the hamster subfamily to Muridae, but Honacki and co-workers have retained the use of Cricetidae. Except for evolutionary biologists, those in experimental research are rarely concerned with the assignment of animals to particular families or larger groups. However, the accurate identification of the species used is essential for proper interpretation of results in all experimental research. Some hamsters, for example, have gallbladders, others do not; some species breed easily in captivity, others have special requirements; some are docile, others extremely savage; the number of mammae range from eight to seventeen, depending on the species.

Fortuitously, the genus of a hamster may provide an indication of its approximate size. Hamsters of the genus *Cricetus* are the largest. *Mesocricetus* species are medium sized. *Calomyscus* and *Phodopus* species are small, mouselike, or dwarf animals. *Cricetulus,* from the derivation of the word, should be small, or diminutive, but within the genus some are small and others are ratlike in size.

Common, or vernacular, names are most specific if they are derived directly from the scientific name by translating the Latin to English. Hence, golden for *auratus*, Newton's for *newtoni*, long-tailed for *longicaudatus*, Dzungurian for *sungorus*, etc., are the most specific common names. The use of geographical names may be misleading or vague at best since at least three hamster species live in Syria and seven or more may be found in China. Reference to Syrian or Chinese hamsters does not indicate which animals were actually used.

It is hoped that a brief digression into classification has served to introduce the wide variety of animals in the hamster subfamily to facilitate the accurate identification of species used as experimental subjects and also to encourage the use of less common species, particularly in comparative studies.

"Laboratory Hamsters" will provide a welcome source for specific information and stimulate productivity in many areas of biomedical research.

Hulda Magalhaes
Professor of Zoology, emerita
Bucknell University
Lewisburg, Pennsylvania

Hamster Classification Schemes

Corbet and Hill (1980)	Honacki *et al.* (1982)	Nowak and Paradiso (1983)	Common names	Geographic distribution
Family Muridae	Family Cricetidae	Family Muridae		
Subfamily Cricetinea	Subfamily Cricetinea	Subfamily Cricetinea		
Calomyscus	*Calomyscus*	*Calomyscus*	Mouse-like hamsters	
C. bailwardi	C. bailwardi	C. bailwardi	Bailward's hamster	Turkestan; Iran; Afghanistan; Pakistan
	C. baluchi			Pakistan; Baluchistan
	C. hotsoni			Pakistan; Baluchistan
	C. mystax			Turkmenistan
	C. urartensis			Iran; Azerbaidzhan
Phodopus	*Phodopus*	*Phodopus*	Dwarf or small desert hamsters	
	P. campbelli		Campbell's hamster	Mongolia
P. roborovskii	P. roborovskii	P. roborovskii	Roborovski's or desert hamster	Mongolia; China
P. sungorus[a]	P. sungorus	P. sungorus	Dzungarian or striped hairy-footed hamster	Siberia; Manchuria
Cricetus	*Cricetus*	*Cricetus*	Common large hamster	
C. cricetus[a]	C. cricetus	C. cricetus	European or black-bellied hamster	Central Europe; USSR
Cricetulus	*Cricetulus*	*Cricetulus*	Small hamsters	
C. alticola	C. alticola	C. alticola	Mountain or Ladak hamster	Pakistan; Kashmir
C. barabensis[a]	C. barabensis	C. barabensis	Striped Chinese hamster	Mongolia; China
C. curtatus	C. curtatus	C. curtatus	Short or Mongolian hamster	Mongolia; China
C. eversmanni	C. eversmanni	C. eversmanni	Eversmann's hamster	Kazakh
	C. griseus[a]		Gray Chinese hamster	China; Mongolia
C. kamensis	C. kamensis	C. kamensis	Cliff or Tibetan Hamster	Tibet
C. longicaudatus	C. longicaudatus	C. longicaudatus	Lesser long-tailed hamster	China; Mongolia
C. migratorius[a]	C. migratorius	C. migratorius	Migratory or Armenian hamster	SE Europe; Iran; Mongolia
	C. obscurus		Dark hamster	China
	C. pseudogriseus		False gray hamster	Transbai Kalia; Mongolia
C. triton	C. triton	C. triton	Chimera, Greater long-tailed or rat-like hamster	China; Korea; Siberia
Mesocricetus	*Mesocricetus*	*Mesocricetus*	Medium-size hamsters	
M. auratus[a]	M. auratus	M. auratus	Golden or Syrian hamster	Asia Minor: Syria
	M. brandti	M. brandti	Brandt's or Turkish hamster	Caucasus; Syria
M. newtoni[a]	M. newtoni	M. newtoni	Newton's or Rumanian hamster	Bulgaria; Rumania
M. raddei	M. raddei	M. raddei	Radde's hamster	Caucasus
Subfamily Nesomyinae	Subfamily Nesomyinae			
Mystromys	*Mystromys*	*Mystromys*		
M. albicaudatus[a]	M. albicaudatus	M. albicaudatus	White tailed rat; South African hamster	South Africa; Lesotho

[a]Laboratory hamsters.

Preface

It is with a sigh of relief and a debt of gratitude that we undertake the final task of composing the Preface for this volume. During the three years in which this book has evolved, there have been numerous highs and lows, but there is no doubt that the high quality of the chapters prepared by the various contributors is a reward that amply justifies the efforts we have expended. With the rapid explosion of information, the compiling, distilling, and collating of existing information from disparate sources are scientific responsibilities equal to that of developing new information. Accordingly, should any of our readers have an opportunity to undertake an analogous effort in the future, we would encourage giving it a high priority.

We appreciate the sponsorship of the American College of Laboratory Animal Medicine (ACLAM) for this work and the advice and suggestions of many of its members. The American College of Laboratory Animal Medicine was founded in 1957 to encourage education, training, and research in laboratory animal medicine. Its primary goals include professional certification of veterinarians in laboratory animal medicine (diplomates) and continuing education. This book represents a part of a program developed to further the educational goals of ACLAM.

Our aim and that of ACLAM has been to produce a work on hamsters useful to the widest possible audience, i.e., the general scientific community, including investigators using or considering the use of hamsters in research, veterinarians, students of veterinary medicine, and those professionally concerned with the care and management of hamsters (supervisory personnel, animal technicians, and technologists concerned with the day-to-day applied care of laboratory hamsters), as well as commercial producers of hamsters.

The contributors were assembled based on their recognition as experts in their specific disciplines. With their detailed presentations, we have assembled an authoritative reference work that will be of interest to almost anyone utilizing hamsters.

In keeping with the aims of this project, we have focused our attention on the Syrian golden hamster, but include sections on Chinese, European, and other hamsters used in the laboratory. Of the sixteen chapters on the Syrian golden hamster, four are devoted to its basic biology and care, five to diseases, five focus on the main uses of hamsters, and two (biomethodology and drug therapy) are relevant to hamster biology and disease as well as to the research use of hamsters. The chapters on other hamsters reflect on the same distribution of material.

We wish to express our supreme gratitude to the authors who contributed their time, expertise, and effort. Some of the contributors are known to us by reputation only and through correspondence and phone calls. We look forward to meeting them personally, trusting that future occasions will present such an opportunity. Chapter reviewers are recognized for their suggestions which contributed to the overall quality of the final product. Special thanks are extended to our wives, Marlene Van Hoosier and Lillian McPherson, for their patience with the long hours we spent on the book when they might have wanted us to be doing something else! The environment and "silent partnership" of the University of Washington and North Carolina State University were essential for this project. We also gratefully acknowledge and sincerely thank Alice Ruff, Brenda Gregory, and Norma Walker for their dedicated and competent secretarial assistance. The assistance and suggestions provided by the staff of Academic Press were especially helpful. Finally, it should be noted that no individual associated with this book received financial remuneration. All royalties go to the American College of Laboratory Animal Medicine for use in the preparation of additional books and its continuing education program in laboratory animal science.

In accordance with current ACLAM policy, there is no official dedication to this book. Before we became aware of this, we had decided to propose Dr. John J. Trentin, the author of Chapter 12, "Experimental Biology: Use in Oncologic Research." It was in association with John in 1962 that one of us (GVH) started to work with hamsters on the oncogenic effects of human viruses. John's dedication and commitment to science, his "sixth sense" for productive areas of research, encouragement, support, and personal friendship have been an inspiration throughout the years.

Gerald L. Van Hoosier, Jr.
Charles W. McPherson

List of Reviewers for Chapters in This Volume

Banks, Keith L.	Washington State University, Pullman
Barthold, Stephen W.	Yale University, New Haven
Berman, Leonard D.	Mallory Institute of Pathology, Boston
Boorman, Gary A.	National Institute of Environmental Health Sciences, Research Triangle Park
Cameron, Thomas P.	National Cancer Institute, Bethesda
Capen, Charles C.	Ohio State University, Columbus
Goldman, Harvey	Beth Israel Hospital, Boston
Harkness, John E.	Mississippi State University, Mississippi State
Hessler, Jack R.	University of Tennessee Center for the Health Sciences, Memphis
Kohn, Dennis F.	Columbia University, New York
Magalhaes, Hulda	Bucknell University, Lewisburg
Mattingly, Steele F.	University of Cincinnati Medical Center, Cincinnati
Montgomery, Charles A.	National Institute of Environmental Health Sciences, Research Triangle Park
Moreland, Alvin F.	University of Florida, Gainesville
Pour, Parviz	Eppley Cancer Research Institute, Omaha
Ringler, Daniel H.	University of Michigan, Ann Arbor
Schneider, Gerald E.	Massachusetts Institute of Technology, Cambridge
Silvers, Willys K.	University of Pennsylvania, Philadelphia
Smith, Abigail L.	Yale University, New Haven
Steffen, Joseph M.	University of Louisville, Louisville
Stuhlman, Robert A.	Wright State University, Dayton
TimmWood, Karen I.	Oregon State University, Corvallis
Wescott, Richard B.	Washington State University, Pullman
Wrightman, Stephen R.	Eli Lilly Co., Indianapolis

Part I

The Golden or Syrian Hamster

Chapter 1

Historical Perspectives and Taxonomy

J. Derrell Clark

    I. Description and Behavior 3
   II. Geographical Distribution and Habitat 3
 III. Taxonomy 4
 IV. Nomenclature 4
  V. Cytogenetics 4
 VI. History of Laboratory Use 5
      A. History 5
      B. Initial Research Use 6
         References 6

I. DESCRIPTION AND BEHAVIOR

The Syrian hamster is a stocky animal with short legs and a very short tail (Fig. 1). Adults are usually between 150 and 170 mm in length from the tip of the nose to the base of the tail, which is about 12 mm long. Young mature male hamsters weigh about 85–110 gm, and adult females weigh about 95–120 gm. As they age the weight will increase somewhat. The fur is short, soft, and smooth. The common coloring of the dorsal surface is a reddish golden brown. Consequently, the Syrian hamster is also commonly referred to as the golden hamster and sometimes as the golden Syrian hamster. The ventral surface of the body is uniformly gray. The skin is extremely loose. The ears are of medium length, gray, and almost naked. The eyes are large and dark. The front feet each have four digits and claws; the rear feet each have five digits. Hamsters have large well-developed cheek pouches in which food may be stored. They are terrestrial and cursorial, adept diggers, and mainly nocturnal, with limited daylight activity. Their natural habitat is dry, rocky steppes or brushy slopes. They live in burrows which they construct. Hamsters are primarily granivorous, but also eat green plant parts and shoots, roots, insects, and fruit. Adults generally live one to a burrow and will readily fight one another (Nowak and Paradiso, 1983; Anderson and Jones, 1984).

II. GEOGRAPHICAL DISTRIBUTION AND HABITAT

Syrian hamsters are indigenous to northwest Syria with a narrowly restricted geographical distribution in the vicinity of Aleppo. They live in deep, chambered burrows in which they store grain and other feedstuffs foraged from the fields at night (Billingham and Silvers, 1963).

Fig. 1. Syrian hamster, *Mesocricetus auratus*.

III. TAXONOMY

The rodent suborder Myomorpha consists of three superfamilies including Muroidea, which includes hamsters and murine rodents.

The classification of hamsters and murine rodents has varied over the past century. Whether they should all be included in a prodigious family Muridae or be distributed in two, Muridae and Cricetidae, or more families constitutes the principal variation (Anderson and Jones, 1984). Some authorities include hamsters in the family Muridae (Nowak and Paradiso, 1983; Anderson and Jones, 1984). Others include them in the family Cricetidae (Honacki *et al.*, 1982). Cricetidae and Muridae are characterized by having four digits and a thumb knob on each of the forefeet, and five digits on each of the rear feet. Hamsters (family Cricetidae) differ from typical rats and mice (family Muridae) in several characteristics. The tail of hamsters is usually covered with short hairs. The cusps of the molar teeth always occur in two parallel longitudinal rows. Members of this family have a variety of body shapes and lengths and are divided into five to seven structurally or geographically discontinuous groups. In Muridae, the molar teeth cusps are arranged in three rows, which may form transverse ridges. The tail ranges from an average length to very long, is generally covered with short bristles, and is scaly. Hamsters are placed in the subfamily Cricetinae. Also based on various taxonomic ranks (Anderson and Jones, 1984; Corbet and Hill, 1980; Honacki *et al.*, 1982), there are five to seven genera in this subfamily: *Cricetulus, Cricetus, Mesocricetus, Phodopus, Tscherskia, Mystromys,* and *Calomyscus*. The Chinese hamster (*Cricetulus griseus,* also known as *Cricetulus barabensis*) is described in Chapters 17, 18, and 19, the European hamster (*Cricetus cricetus*) is covered in Chapter 20, and other hamsters—that is, the Dzungarian hamster (*Phodopus sungorus*), the South African hamster (*Mystromys albicaudatus*), the Turkish hamster (*Mesocricetus brandti*), the Rumanian hamster (*Mesocricetus newtoni*), and the Armenian hamster (*Cricetulus migratorius*)—are described in Chapter 21.

The Syrian hamster, *Mesocricetus auratus,* was first described and originally named *Cricetus auratus* by Waterhouse in 1839. The original description was based on study of the skin and skull of a female. Nehring (1902) was the second known person to study the Syrian hamster. He used a preserved female at the Beirut Museum, and in 1898 established the genus *Mesocricetus* for certain hamsters from southeastern Europe and Asia Minor. In 1902, he renamed the Syrian hamster *Mesocricetus auratus*. However, for several decades, the Syrian hamster continued to be considered a subgenus of *Cricetus* and was termed *Cricetus (Mesocricetus) auratus*. In 1940, Ellerman pointed out that *Mesocricetus* was a well-characterized and distinct genus (Granados, 1951). Animals in the genus *Mesocricetus* differ from those in *Cricetus* by being smaller, having shorter tails, and having a greater number of mammae. *Mesocricetus auratus* are restricted to the vicinity of Aleppo in northwestern Syria. According to Honacki *et al.* (1982), the current taxonomic position of the Syrian hamster is outlined as: order, Rodentia; family, Cricetidae; subfamily, Cricetinae; genus, *Mesocricetus*; and species, *auratus*.

IV. NOMENCLATURE

In the recommended rules for a nomenclatural system, the symbol for *M. auratus* is (SYR) (Institute of Laboratory Animal Resources, 1970). Information regarding inbred strains, genetically defined stocks, and mutants for coat color and other coat characteristics of hamsters has been provided (Yoon and Peterson, 1979; Slater, 1979; Nixon *et al.*, 1979; Nixon, 1979). These variants include white, cream, piebald, albino, hairless, and longhair (teddy bear).

V. CYTOGENETICS

The karyotype of the Syrian hamster has been extensively studied using conventionally stained or banded chromosomes (Awa *et al.*, 1959; Ohno and Weiler, 1961; Ishihara *et al.*, 1962; Lehman *et al.*, 1963; Galton and Holt, 1964; Utsumi *et al.*, 1965; Hsu and Arrighi, 1971; Popescu and DiPaolo, 1972, 1980; Yamamoto *et al.*, 1973; Bigger and Savage, 1976; Pavia *et al.*, 1977). The diploid chromosome number has been determined to be 44. There has been a lack of consistency and agreement among the various studies of the chromosome complement, since all elements are not easily identifiable by staining. On the basis of G-banding patterns, which is more definitive, idiograms for *M. auratus* have been proposed (Popescu and DiPaolo, 1972; Bigger and Savage, 1976; DiPaolo and Popescu, 1977; Pavia *et al.*, 1977). The karyotype of a normal Syrian hamster is presented in Fig. 2.

Using staining techniques, Lehman *et al.* (1963) made detailed karyotypic comparisons of somatic metaphases, especially of the sex chromosomes, of three inbred strains and two noninbred stocks. They reported that the 22 chromosome pairs consist of 5 median metacentrics, 9 submedian metacentrics, 3 subtelocentrics, 4 large acrocentrics, and 1 small acrocentric. Only 8 of these could readily be identified individually.

The five pairs of chromosomes, X and numbers 5, 10, 14, and 20, with median or near-median centromeres are individually recognizable. Of these, the X chromosome is the longest and has a near-median centromere position. The nine submedian metacentrics are numbers 1, 2, 3, 4, 8, 9, 12, 13, and 15. Chromosome pairs 1 and 2 are slightly shorter than the longest chromosomes and frequently can be separately identified by their arm ratios. Chromosomes 3 and 4 have weaker arm ratios and are slightly shorter than 1 and 2, but are very similar to numbers 8 and 9. These latter four pairs are the most difficult to determine. However, there is a slight difference in length. Chromosome pair 13 is only slightly smaller than 12. Chromosome pair 15 is not always clearly distinguishable from pair 13.

There are three pairs of subtelocentrics, 6, 7, and 11. Chromosome pair 6 is easily identified, since it has the strongest arm ratio in the complement. Its length may vary between that of 3 and 9. Usually chromosome pair 7 can be differentiated from number 6 by arm ratio. Chromosome pair 11 may be confused with pair 7 because of similar arm ratios.

Chromosome pairs 16, 17, 18, and 19 are large acrocentrics and are identifiable as a group. These chromosomes are of medium length within the complement. Pair 16 appears to have no detectable short arms. Pair 19 is shorter than the others. Chromosome pair 21 is a single small acrocentric which is unique and easily recognized.

Using the trypsin–Giemsa banding techniques, Pavia *et al.* (1977) arranged autosomes into groups based on total length and arm ratios using the criteria of Lehman *et al.* (1963), namely:

First row: The five largest chromosomes after the X are arranged by descending arm ratios, with the exception of number 1 preceding number 2.

Second row: The next five chromosome pairs (6–10) are arranged by descending arm ratio, number 6 having the highest arm ratio and number 10, the median landmark, placed at the end of this row.

Third row: The next four largest chromosome pairs are arranged by descending arm ratios, with number 14 remaining a median landmark, and number 15 left in its traditional position.

Fourth row: The acrocentrics are arranged in order of descending size, the smallest chromosome pairs (20 and 21) placed at end of karyotype according to size.

Polymorphism has been reported in descriptions of the Y chromosome. Forms of the Y chromosome in the Syrian hamster have been described as medium-sized acrocentric (Ohno and Weiler, 1961), submedian metacentric (Awa *et al.*, 1959; Lehman *et al.*, 1963; Ishihara *et al.*, 1962), and long telocentric (Lehman *et al.*, 1963). In most of the observations of Lehman *et al.* (1963) the Y chromosome consistently was determined to be a medium-sized, submedian metacentric with a length in the range of pairs 11–15. However, in some animals the Y chromosome was telocentric and longer than chromosome pairs 16–19.

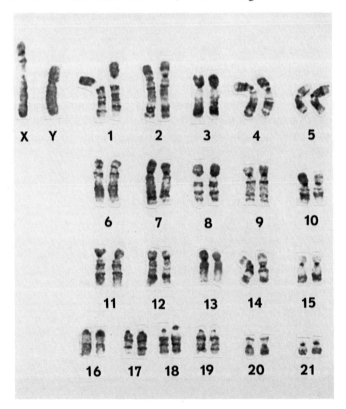

Fig. 2. Karyotype of a normal Syrian hamster showing typical G-banding patterns; chromosomes arranged in karyotypes according to the nomenclature proposed by Popescu and DiPaolo (1972). The contribution of this illustration by Dr. N. C. Popescu (Bethesda, Maryland) is gratefully acknowledged.

VI. HISTORY OF LABORATORY USE

A. History

Hamsters have a recent history of domestication. Laboratory hamsters are believed to have originated from three or four littermates captured near Aleppo in Syria in 1930. Thus, they not only have a very narrow genetic base, but they have been domesticated for a relatively short time in comparison with mice, rats, and guinea pigs. More recently, additional wild hamsters have been trapped and studied (Murphy, 1971).

A comprehensive history of the capture and domestication of the Syrian hamster has been published (Murphy, 1985).

In the late 1920s, Dr. Saul Adler of the Hebrew University, Jerusalem, was studying leishmaniasis. At the time Chinese hamsters were the only suitable experimental animals for these studies. Since they had to be imported from the Far East and did not reproduce in captivity, he was interested in finding an indigenous species for experimental use, preferably *Cricetulus phaeus*, a close relative of the Chinese hamster (Yerganian, 1972).

In 1930, Professor I. Aharoni of the Department of Zoology of the Hebrew University went on a zoological expedition to Syria. Mr. Aharoni agreed to collect and provide specimens of *C. phaeus* for Dr. Adler's work. Mr. Aharoni collected *C. phaeus*, but on April 12, 1930, he also recovered an adult female Syrian hamster with a litter of 11 from a burrow near Aleppo, Syria. Mr. Aharoni was successful in raising most of the young and presented 9 Syrian hamsters to Mr. H. Ben-Menachen, head of the Hebrew University Animal Facilities, in July 1930. Five of the animals promptly escaped, one female was killed, and only three or four animals survived. [There is discrepancy regarding the sex and number of the remaining hamsters. In Adler's account one male and two females remained (Adler, 1948). However, Aharoni states in his memoirs that of four surviving animals, one was a female (Yerganian, 1972).] From these littermates, on August 18, 1930, Mr. Ben-Menachen succeeded in producing the first known litters born in captivity. Thereafter, a laboratory-reared colony of Syrian hamsters was quickly established (Adler, 1948; Yerganian, 1972).

Dr. Adler quickly realized that Syrian hamsters were useful laboratory animals. He decided to distribute the animals as widely as possible and avoid the risk of maintaining a valuable species in one laboratory. In 1931, he carried a few animals to Professor Edward Hindle of the Wellcome Bureau of Scientific Research in England (Bruce and Hindle, 1934). He also distributed them to the Medical Research Council and the College of France. In 1938, animals were sent from Palestine to the U.S. Public Health Service Hospital, Carville, Louisiana and Western Reserve University in Ohio (Bond, 1945). A third colony of Syrian hamsters was established at the Rockefeller Foundation in 1938 (Poiley, 1950).

Therefore, it appears that most of the Syrian hamsters used as laboratory animals or pets are descendants of the original three or four surviving littermates.

In 1971, 12 Syrian hamsters captured by Syrian farmers were imported to the United States. The hamsters were sent to the Massachusetts Institute of Technology, where observations on their natural history and comparisons with domesticated stock were made (Murphy, 1971). Descendants of these animals are maintained at the National Institutes of Health, Bethesda, Maryland (Murphy, 1985) (see also Chapter 13).

B. Initial Research Use

Syrian hamsters were introduced to the scientific community by Adler and Theodor (1931), who reported on their use in kala-azar studies. They were quickly accepted as research animals and were used in a variety of studies (Anonymous, 1969; Granados, 1951). A comprehensive bibliography was published in 1968 on the literature from 1931 to 1963 (Magalhaes, 1968). Some examples of early studies involving hamsters include breeding and growth (Bruce and Hindle, 1934); susceptibility to *Brucella abortus* (Tchernomoretz and Ellenbogen, 1937); susceptibility to human leprosy (Adler, 1937); reproductive physiology (Klein, 1938); reproductive cycle (Deanesly, 1938); susceptibility to tubercle bacilli (Griffith and Pagel, 1939); maintenance (Laidlaw, 1939); host for St. Louis encephalitis (Broun *et al.*, 1941); influenza virus infection (Wheeler and Nungester, 1942); effect of hormones on reproduction (Peczenik, 1942); experimental rickets (Jones, 1945); experimental cholesterol arteriosclerosis (Altschul, 1946); tumor transplantation (Crabb, 1946); inoculation with Newcastle disease virus (Reagan *et al.*, 1947); effect of X-radiation (Kivy, 1947) biological exchange of trace elements (Smith, 1947); and toxic effects of beryllium (Sprague *et al.*, 1948).

Before 1954, over half of the publications regarding use of Syrian hamsters involved microbiology, parasitology, or dental disease (Magalhaes, 1954). By 1963, the relative use of hamsters in these fields of study had declined, but hamsters were being actively used in cancer research and reproductive physiology (Magalhaes, 1968).

ACKNOWLEDGMENTS

Figure 2, Syrian hamster karyotype, was kindly supplied by Dr. Nicholas C. Popescu, Department of Health and Human Services, Public Health Service, National Institutes of Health, National Cancer Institute, Bethesda, Maryland.

REFERENCES

Adler, S. (1937). Inoculation of human leprosy into Syrian hamster. *Lancet* **233**, 714–715.

Adler, S. (1948). Origin of the golden hamster, *Cricetus auratus* as a laboratory animal. *Nature (London)* **162**, 256–257.

Adler, S., and Theodor, O. (1931). Investigations on Mediterranean Kala-azar. II. *Leishmania infantum*. *Proc. Soc. London, Ser. B* **108**, 453–463.

Altschul, R. (1946). Experimental cholesterol arteriosclerosis. *Arch. Pathol.* **42**, 277–284.

Anderson, S., and Jones, J., Jr. (1984). "Order and Families of Recent Mammals of the World." Wiley, New York.

Anonymous (1969). The Syrian hamster: Its utility in research. *Charles River Dig.* **8,** No. 2.

Awa, A., Sasaki, M., and Takayama, S. (1959). An *in vitro* study of the somatic chromosomes in several mammals. *Jpn. J. Zool.* **12,** 257–266.

Bigger, T., and Savage, J. (1976). Location of nucleolar organizing regions on the chromosomes of the Syrian hamster (*Mesocricetus auratus*) and the Djungarian hamster (*Phodopus sungorus*). *Cytogenet. Cell Genet.* **16,** 495–504.

Billingham, R., and Silvers, W. (1963). Skin transplants and the hamster. *Sci. Am.* **208,** 118–127.

Bond, C. (1945). The golden hamster (*Cricetus auratus*): Care, breeding, and growth, *Physiol. Zool.* **18,** 52–59.

Broun, G., Muether, R., Mezera, R., and LeGier, M. (1941). Transmission of St. Louis encephalitis to the hamster. *Proc. Soc. Exp. Biol. Med.* **46,** 601–603.

Bruce, H., and Hindle, E. (1934). The golden hamster, *Cricetus (Mesocricetus) auratus* Waterhouse. Notes on its breeding and growth. *Proc. Zool. Soc. London* Part 1, 361–366.

Corbet, G., and Hill, J. (1980). "A World List of Mammalian Species." Comstock Publ. Assoc., London.

Crabb, E. (1946). A transplantable 9,10-dimethyl-1,2-benzanthracene sarcoma in the Syrian hamster. *Cancer Res.* **6,** 627–636.

Deansely, R. (1938). The reproductive cycle of the golden hamster (*Cricetus auratus*). *Proc. Zool. Soc. London* **108,** 31–37.

DiPaolo, J., and Popescu, N. (1977). Banding pattern analysis of initial structural chromosome alterations induced by *N*-methyl-*N'*-nitro-*N*-nitrosoguanidine in Syrian hamster cells. *Mutat. Res.* **44,** 359–368.

Ellerman, J. R. (1940). "The Families and Genera of Living Rodents." Trustees Brit. Mus., London.

Galton, M., and Holt, S. (1964). DNA replication patterns of the sex chromosomes in somatic cells of the Syrian hamster. *Cytogenetics* **3,** 97–111.

Granados, H. (1951). Nutritional studies on growth and reproduction of the golden hamster (*Mesocricetus auratus auratus*). *Acta Physiol. Scand.* **24,** Suppl. 87, 1–138.

Griffith, A., and Pagel, W. (1939). The susceptibility of the golden hamster (*Cricetus auratus*) to bovine, human, and avian tubercle bacilli and to the vole strain of the acid-fast bacillus (Wells). *J. Hyg.* **39,** 154–160.

Honacki, J., Kinman, K., and Koeppl, J. (1982). "Mammal Species of the World." Allen Press, Inc. and The Assoc. of Systematics Collections, Lawrence, Kansas.

Hsu, T., and Arrighi, F. (1971). Distribution of constitutive heterochromatin in mammalian chromosomes. *Chromosoma* **34,** 243–253.

Institute of Laboratory Animal Resources (Committee on Nomenclature) (1970). A nomenclatural system for outbred animals. *Lab. Anim. Care* **20,** 903–906.

Ishihara, T., Moore, G., and Sandberg, A. (1962). Chromosome constitution of two tumors of the golden hamster. *J. Natl. Cancer Inst. (U.S.)* **29,** 161–195.

Jones, J. (1945). Experimental rickets in the hamster. *J. Nutr.* **30,** 143–146.

Kivy, E. (1947). The effect of roentgen irradiation on the testes of the golden hamster. *Anat. Rec.* **99,** 650–651.

Klein, M. (1938). Relation between the uterus and the ovaries in the pregnant hamster. *Proc. Soc. London, Ser. B* **125,** 348–364.

Laidlaw, P. (1939). Maintenance of the golden hamster. *Int. J. Lepr.* **7,** 513–516.

Lehman, J., MacPherson, I., and Moorhead, P. (1963). Karyotype of the Syrian hamster. *J. Natl. Cancer Inst. (U.S.)* **31,** 639–650.

Magalhaes, H. (1954). The golden hamster as a laboratory animal. *J. Anim. Tech. Assoc.* **5,** 39–44.

Magalhaes, H. (1968). Housing, care, and breeding. *In* "The Golden Hamster—Its Biology and Use in Medical Research" (R. A. Hoffman, P. F. Robinson and H. Magalhaes, eds.), pp. 15–23. Iowa State Univ. Press, Ames.

Murphy, M. (1971). Natural history of the Syrian golden hamster—a reconnaissance expedition. *Am. Zool.* **11,** 632.

Murphy, M. (1985). History of the capture and domestication of the Syrian golden hamster (*Mesocricetus auratus* Waterhouse). *In* "The Hamster" (H. I. Siegel, ed.), pp. 3–20. Plenum, New York.

Nehring, A. (1902). Über *Mesocricetus auratus* Waterhouse. *Zool. Anz.* **26,** 57–60.

Nixon, C. (1979). Origin of inbred strains: Hamsters. Part III. Strains maintained at the Massachusetts Institute of Technology. *In* "Inbred and Genetically Defined Strains of Laboratory Animals" (P. L. Altman and D. D. Katz, eds.), Part 2, p. 434. Fed. Am. Soc. Exp. Biol., Bethesda, Maryland.

Nixon, C., Robinson, R., Yoon, C., and Peterson, J. S. (1979). Mutants for coat color and other coat characteristics: Hamster. *In* "Inbred and Genetically Defined Strains of Laboratory Animals" (P. L. Altman and D. D. Katz, eds.), Part 2, p. 453. Fed. Am. Soc. Exp. Biol., Bethesda, Maryland.

Nowak, R. M., and Paradiso, J. L. (1983). "Walker's Mammals of the World," 4th ed. Johns Hopkins Univ. Press, Baltimore, Maryland.

Ohno, S., and Weiler, C. (1961). Sex chromosome behavior pattern in germ and somatic cells of *Mesocricetus auratus*. *Chromosoma* **12,** 362–373.

Pavia, R., Smith, L., and Goldenberg, D. (1977). An analysis of the G-banded chromosomes of the golden hamster. *Int. J. Cancer* **20,** 460–465.

Peczenik, O. (1942). Actions of sex hormones on oestrous cycle and reproduction of the golden hamster. *J. Endocrinol.* **3,** 157–167.

Poiley, S. M. (1950). Breeding and care of the Syrian hamster, *Cricetus auratus*. *In* "The Care and Breeding of Laboratory Animals" (E. J. Farris, ed.), pp. 118–152. Wiley, New York.

Popescu, N., and DiPaolo, J. (1972). Identification of Syrian hamster chromosomes by acetic–saline–Giemsa (ASG) and trypsin techniques. *Cytogenetics* **11,** 500–507.

Popescu, N., and DiPaolo, J. (1980). Chromosomal interrelationship of hamster species of the genus *Mesocricetus*. *Cytogenet. Cell Genet.* **28,** 10–23.

Reagan, R., Lillie, M., Poelma, L., and Brueckner, A. L. (1947). Transmission of the virus of Newcastle disease to the Syrian hamster. *Am. J. Vet. Res.* **8,** 136–138.

Slater, G. (1979). Origin of inbred strains: Hamster. Part II. Strains maintained by Charles River Breeding Laboratory. *In* "Inbred and Genetically Defined Strains of Laboratory Animals" (P. L. Altman and D. D. Katz, eds.), Part 2, p. 434. Fed. Am. Soc. Exp. Biol., Bethesda, Maryland.

Smith, R. (1947). Studies on the biological exchange of radioantimony in animals and man. *Fed. Proc., Fed. Am. Soc. Exp. Biol.* **6,** 205.

Sprague, G., Pettengill, A., and Stokinger, H. (1948). Initial studies of the inhalation toxicity of beryllium sulfate. *Fed. Proc., Fed. Am. Soc. Exp. Biol.* **7,** 257.

Tchernomoretz, I., and Ellenbogen, V. (1937). A note on the susceptibility of the Syrian hamster, *Cricetus auratus* to *Brucella abortus* Bang. *J. Comp. Pathol. Ther.* **50,** 136–140.

Utsumi, K., Kitamura, I., and Trentin, J. (1965). Karyologic studies of normal cells and of adenovirus-type-12-induced tumor cells of the Syrian hamster. *J. Natl. Cancer Inst. (U.S.)* **35,** 759–769.

Waterhouse, G. R. (1839). Description of a new species of hamster (*Cricetus auratus*). *Proc. Zool. Soc.* **VII,** 57–58.

Wheeler, A., and Nungester, W. (1942). Effect of mucin on influenza virus infection in hamsters. *Science* **96,** 92–93.

Yamamoto, T., Rabinowitz, Z., and Sachs, L. (1973). Identification of the chromosomes that control malignancy. *Nature (London), New Biol.* **243,** 247–250.

Yerganian, G. (1972). History and cytogenetics of hamsters. *Prog. Exp. Tumor Res.* **16,** 2–41.

Yoon, C., and Peterson, J. (1979). Genetically defined but non-inbred strains: Hamsters. *In* "Inbred and Genetically Defined Strains of Laboratory Animals" (P. L. Altman and D. D. Katz, eds.), Part 2, p. 435. Fed. Am. Soc. Exp. Biol., Bethesda, Maryland.

Chapter 2

Morphophysiology

W. Sheldon Bivin, Glenn H. Olsen, and Kathleen A. Murray

I.	Introduction	10
II.	General Features	10
	A. Body Conformation	10
	B. Musculoskeletal Characteristics	10
	C. Vibrissae	11
	D. Hair	11
	E. Brown Adipose Tissue	11
	F. Flank Organ	12
	G. Thoracic Cavity	12
	H. Abdominal and Pelvic Cavities	12
	I. Physiological Data	12
III.	Digestive System	12
	A. Oral Cavity	12
	B. Teeth	13
	C. Tongue	14
	D. Salivary Glands	15
	E. Cheek Pouches	16
	F. Stomach	17
	G. Small Intestine	17
	H. Cecum	18
	I. Large Intestine	18
	J. Pancreas	18
	K. Liver and Gallbladder	19
IV.	Respiratory System	19
	A. Upper Respiratory System	19
	B. Oropharynx and Trachea	20
	C. Lower Respiratory System	20
V.	Cardiovascular and Lymphatic Systems	24
	A. Heart	24
	B. Cheek Pouch	25
	C. Intestine	25
	D. Pancreas and Liver	25
	E. Adrenal Gland	25
	F. Kidney	25
	G. Ovaries and Uterus	26
	H. Orbital Venous Sinus	26
	I. Lymphatic System	26

VI. Urinary and Reproductive Systems 27
    A. Kidney and Ureters 27
    B. Bladder and Urethra 27
    C. Male Reproductive System 28
    D. Female Reproductive System 29
VII. Endocrine System 32
    A. Adrenal Gland 32
    B. Islets of Langerhans 33
    C. Parathyroid Gland 33
    D. Thyroid Gland 33
    E. Pituitary Gland 34
    F. Pineal Gland 34
VIII. Nervous System 35
    A. Central Nervous System 35
    B. Peripheral Nervous System 35
IX. Special Sensory Systems 35
    A. Vomeronasal Organ 35
    B. Vision 36
        References 36

I. INTRODUCTION

This chapter summarizes the anatomical and physiological characteristics of the Syrian golden hamster. Although there have been numerous articles written on regional anatomy and physiology, only one text, the 1968 edition of *The Golden Hamster* by R. A. Hoffman, P. F. Robinson, and H. Magalhaes, has ever attempted to summarize this information. In this chapter, a special effort will be made to provide material which will serve as a primary reference for the biomedical research scientist and laboratory animal medicine specialist who uses or cares for hamsters in their professional activities.

II. GENERAL FEATURES

A. Body Conformation

The golden hamster differs from most rodents in that it has a very short tail, a compact stout body with flank glands, and cheek pouches. Dorsal and ventral views of an adult hamster showing general body regions and the position of the flank glands are outlined in Fig. 1.

The sexually mature male is characterized by an enlarged, rounded posterior end when the testes are present in the scrotal sacs, darkly pigmented flank glands, a long anogenital distance, and a penis. The mature female has a pointed posterior, reduced flank glands, a short anogenital distance, a vaginal opening with a small clitoris, and a separate urethral opening (Fig. 1).

The newborn can be easily sexed by comparing the distance between the genital papilla and the anus. This distance is greater in the males than in the females, and the genital papilla of the male is larger than that of the female.

B. Musculoskeletal Characteristics

Very little descriptive literature is available on the musculoskeletal system of the golden hamster. The only systematic study of the epaxial masculature is by Salih and Kent (1964). They describe the origins, insertions, gross structure, and relationships of 17 muscles or muscle groups located dorsal to the transverse processes of the vertebrae. The vertebral formula

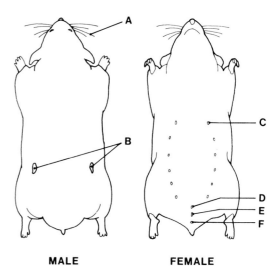

Fig. 1. Dorsal view of a male hamster (left) and ventral view of a female hamster (right). A, Vibrissae; B, flank glands; C, mammary glands; D, urethral opening; E, vaginal opening; F, anus.

for hamsters is 7 cervical, 13 thoracic, 6 lumbar, 4 sacral, and 13–14 caudal (Salih and Kent, 1966).

The golden hamster has been used as an experimental animal in many studies of neuromuscular disorders (Caufield 1966; Homburger 1972; Chokroverty et al., 1977, 1978). Reyes and Chokroverty (1980) concluded that the distribution and size of muscle fibers within the quadriceps muscles should be considered in any study involving denervation by nerve section or muscle inactivity by tenotomy or immobilization. Unfortunately, these are the only references on morphophysiology of the muscles of the hamster, other than those associated with the cheek pouches.

C. Vibrissae

Mammalian vibrissae (sinus hairs, bluthaare, whiskers) are specialized hairs that form a highly refined vibrotactile receptor system in most species. Independent movement of individual vibrissae is important for spatial orientation and communication in numerous species. The hamster is unique in that this species moves its cranial vibrissae in various complex patterns keyed to particular exploratory situations. This movement repertoire suggests that a precisely operating motor system controls the vibrissae. The facial musculature, vibrissal vasculature, and vibrissal connective tissues have been proposed as important components of the vibrissal motor system (Wineski, 1985).

D. Hair

The growth of hair in the golden hamster begins at 9 days of age and continues through 51 days of age. Hair growth ceases during this telogen phase (except in the flank gland areas) for a period of 20–24 days, after which growth resumes. Of further interest is the fact that hair growth always begins in the pigmented areas before it begins in the unpigmented areas. This information is of particular importance in skin carcinogenic studies (Musser and Silverman, 1980). Annual cycles of pelage color which appear to be under neuroendocrine control occur in the hamster. Tyrosinase levels peak in the spring when pigmented hair is produced and in the fall when unpigmented hair is produced. The melanin content of hair follicles is high in summer and low in the winter, but serum tyrosinase levels do not differ during the spring and fall molts. This would suggest that there must be a neuroendocrine mechanism, which is photoperiodically sensitive, to prevent the raised tyrosinase levels of the fall molt being expressed as melanogenesis (Logan and Weatherhead, 1978).

E. Brown Adipose Tissue

The thermogenic role of brown adipose tissue is well documented and has been described as "nonshivering thermogenesis." It is established that in hibernating, newborn, and cold-acclimated animals this tissue has a major role in heat production (Smith and Horwitz, 1969; Bruck, 1970; Jansky, 1973).

A major portion of the brown adipose tissue is ventral to and between the scapulae, extending from the lower cervical spine to the midthoracic level (Fig. 2). Other lobes extend beneath the scapulae and occupy the axillary apices with thin projections rostrally between the caudal cervical muscles. Only minute amounts are found around the thymus and thyroid glands. Only this thoracic brown adipose tissue is associated with the sympathetic nervous chain. In the abdominal region it is common to see the adrenal glands, renal hila, and parts of the ureters sheathed with brown fat.

The main function of the brown adipose tissue is to heat the blood passing through it. Therefore, it is not surprising to find a very rich capillary net, filled with red blood corpuscles, coursing throughout the tissue. It is the density of the red blood

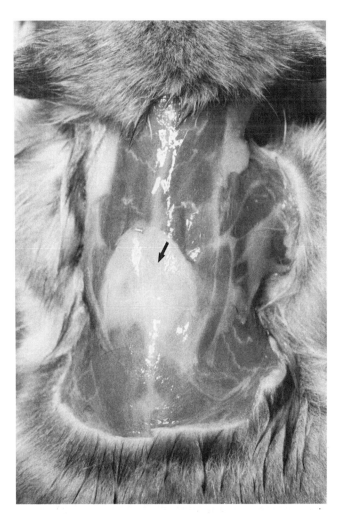

Fig. 2. Interscapular brown adipose tissue (arrow) in an adult hamster.

cells which gives the tissue its typical tawny-orange to brown color. This, of course, is sharply contrasted with the white adipose tissue, which is poorly vascularized and is mainly composed of white to yellow lipids. The vascularity of the brown adipose tissue is four to six times greater than that of white adipose tissue (Afzelius, 1970).

Because the hamster is a hibernating animal, it is interesting to note the changes which occur in brown adipose tissue during periods of cold acclimation. The relative weight of the tissue increases, its color darkens, and tissue protein content increases, whereas fat content decreases, and the multilocular structure of the tissue predominates. During heat acclimation the opposite is true. The color becomes lighter, fat content increases, protein decreases, and the tissue structure is mostly unilocular (Rabi and Cassuto, 1976).

F. Flank Organ

The flank organ, or costovertebral gland, of the Syrian golden hamster is a complex structure composed of sebaceous glands, terminal hair, and pigment cells. In young animals, the position of the flank organs is indicated by a small, lightly pigmented area. These areas become darkly pigmented in mature males, and their location is shown in Fig. 1. For years these flank glands have been considered to be a secondary male sexual characteristic. It is also observed that the male animals investigate the female flank organs and clitorial glands prior to mating. Recent studies have shown that the growth of hair and sebaceous glands are androgen-dependent functions. Apparently testosterone is converted to dihydrotestosterone (DHT) in the flank organ. DHT has been found in the nuclei of peripheral and differentiating sebocytes of the flank organs. These observations suggest that androgen may have a role in lipogenesis, hair growth, and flank organ growth (Lucky et al., 1985).

G. Thoracic Cavity

The thorax is bounded by 13 thoracic vertebrae dorsally, seven sternal and six poststernal rib pairs plus their cartilages laterally, and the sternum ventrally. The cranial aperture of the thorax is bounded dorsally by the seventh cervical vertebra, laterally by the first rib pair, and ventrally by the sternum. The caudal limit of the thorax is at the level of the thirteenth rib and is closed by a muscular diaphragm. The diaphragm originates dorsally on the first lumbar vertebrae and is composed of a very well-developed pars muscularis and a transparent centrum tendineum. The hiatus aorticus is located dorsally between the two crula and ventral to the last three thoracic vertebrae. The esophageal hiatus is located ventrally and penetrates the tendinous center near its junction with the right drus. Ventrally and just to the right of the esophageal hiatus is the foramen vena cavae.

The pleura forms a large right and left pleural sac surrounding the lungs. Interspaced between the right and left pleural cavities is the pericardial cavity surrounding the heart. Two sternopericardial ligaments connect the pericardial sac to the sternum. There is no macroscopically identifiable cavum mediastini serosum (Schwarze and Michel, 1960).

H. Abdominal and Pelvic Cavities

The abdominal cavity is the largest serous cavity and extends from the diaphragm to the entrance of the pelvic cavity. Its dorsal wall is bounded by six lumbar vertebrae. The caudal portion is supported by the wings of the ilium. The ventral abdominal wall is supported cranially by the lower end of the sternum and the xiphoid process. A prominent ventral midline of fused fascial planes forms the linea alba (Schwarze and Michel, 1960).

The walls of the pelvic cavity are supported dorsally by three sacral vertebrae and the first coccygeal vertebra. The ilium and the ischium form the lateral walls, and the ventral float is composed of the pubis and the ischium. The retroperitoneal section of the pelvic cavity is very short (Schwarze and Michel, 1960).

Within the abdominal cavity there are several ligaments which connect the diaphragm with the liver, stomach, and spleen. These ligaments are all continuous with the peritoneum. A large falciform ligament extends on the ventral abdominal floor from the tip of the sternum to the umbilicus. Caudally the hepatogastricum ligament and the hepatoduodenale ligament are continuous with the lesser omentum. The greater omentum is thin and fatty as it extends from the greater curvature of the forestomach, the glandular stomach, and the duodenum to the transverse colon. The foramen epiploicum, which is the opening into the omental bursa, is situated between the caudal vena cava and the portal vein (Schwarze and Michel, 1960). The common mesentery originates from the dorsal body wall and extends craniocaudally to the rectum, becoming increasingly shorter in the pelvic region.

I. Physiological Data

As a source for ready reference, Table I shows acceptable ranges for physiological data of the golden hamster.

III. DIGESTIVE SYSTEM

A. Oral Cavity

The free margins of the lips form a three-cornered flap which completely blocks the mouth opening when the mouth is closed and aids in the filling of the cheek pouches. The max-

Table I
Physiology of the Golden Hamster[a]

Characteristic	Range
Pulse rate	250–500/min
Respiratory rate	33–127/min (mean 74)
Temperature	36.2°–38.8°C renal
Puberty	Male, 30–40 days
	Female, 36–84 days
Breeding age	Male, 60–90 days
	Female 56–90 days
Estrous cycle	Female, 4 days
Duration of heat	20 hr
Ovulation	Spontaneous, 8–12 hr from the beginning of estrus
Gestation	15–18 days
Litter size	4–12 young
Weight at birth	2–3 gm
Weaning	21 days (35–40 gm)
Length of lactation	3–4 weeks
Number of teats	7 pair
Reproductive life	Male, 9–12 months
	Female, 6 months to 2 years
Life span	1–3 years
Chromosomes	44
Blood volume	7.4 ± 3.1% of body weight
Food requirements	10–16 gm hard food daily
	10–20 gm green vegetables daily

[a]Adapted from Klomberg and Wolff (1978) and Van Hoosier and Ladiges (1984).

Table II
Ages at Which Teeth Erupt in the Syrian Hamster[a]

Type	Radiographic evidence of calcification (days)	Eruption begins[b]		Occlusion attained (days)
Incisors		Within	24 hr	2
First molars (M1)	2		7–8 days	8–9
		Upper	13–14 days	
Second molars (M2)	7	Lower	12–13 days	16–17
		Upper	32–35 days	
Third molars (M3)	18	Lower	30–32 days	40–45

[a]Adapted from Keyes and Dale (1944); Hinrichsen (1955), and Gaunt (1961).
[b]These eruption dates are in conflict with more recent data cited by Schwarze and Michel (1960): Incisors, 4–5 days; M1, 10 days; M2, 20 days; M3, 33 days.

imum size of the oral aperture is 12 mm vertically by 17 mm horizontally (Schwarze and Michel, 1960).

The hard palate is 15.0–16.5 mm long and runs from the caudal edge of the incisor teeth to the caudal end of the last molars. The bony frame of the hard palate is composed of the palatine process of the incisive bone, the palatine process of the maxilla, and the horizontal lamina of the palatine bone. The hard-palate mucous membrane forms two groups of seven ridges located on either side of the midline. The ridges slope labially near the molars and in a pharyngeal direction when not near the molars. The nasopalatine duct enters the oral cavity on the labial surface of the first pair of ridges.

The soft palate is 4–6 mm long and begins caudal to the last molar. It is generally pink and smooth.

B. Teeth

Dentition in the hamster, 2(I 1/1, C 0/0, P 0/0, M 3/3) = 16, is monophyodont, bunodont, and brachyodont, with open-rooted incisors and rooted molars. Primary eruption is regular with rapid development (Table II), which may be slightly advanced in animals from small litters compared with animals from large litters (Table II) (Keyes and Dale, 1944).

Incisors grow irregularly depending on factors including age, sex, diet, and season. (Females generally have larger teeth than males.) The short (2.5±0.5 mm) maxillary incisors can replace themselves in 1 week while the longer (10.0±1.8 mm) mandibular incisors can require 2.5–3 weeks for replacement. Maxillary incisors have labiolingual and mesiodistal dimensions which are smaller than the comparable dimensions of the mandibular incisors. The mandibular incisor bevel is concave and longer than the maxillary incisor bevel, and both extend lingually to end in an abrupt notch at the gingival margin. The diastema separating the incisors from the molars is larger in the maxilla (9–11 mm) than in the mandible (4–6 mm). Mandibular incisors can be positioned on either the labial or lingual surface of the maxillary incisors. In addition, the mandibular symphysis may not close throughout life (Schwarze and Michel, 1960).

The molar teeth form a convex arch with maxillary molars pointing outward and mandibular molars pointing inward (Schwarze and Michel, 1960). Molars are characterized as having three pair of cusps in rostrocaudal alignment. Cusps are described as being conical, sharply pointed, separated by broad buccolingual sulci, and having deep occlusal fossae and grooves (Table III). Newly erupted molar crowns are completely covered by enamel, and roots are covered by a thin layer of cell-free cementum. Attrition causes exposed primary dentin to become metamorphosed with secondary dentin deposited on the underlining pulpal wall. Secondary cementum is also deposited at the root apices. Occlusion is by cuspal interdigitation. Attrition is initiated on the mesial slopes of mandibular cusps and distal slopes of maxillary cusps, and spreads over the occlusal surface (Keyes and Dale, 1944).

Morphologically, the crowns allow retention of fine food particles, making the hamster susceptible to caries comparable to humans (Dale *et al.*, 1941; Arnold, 1942; Lovelace *et al.*, 1958). Other factors known to induce caries are a negative calcium balance and diets rich in corn. Male hamsters suffer

Table III
Measurements of the Molar Teeth of Syrian Hamsters[a,b]

Type	Number of roots	Buccolingual dimensions			Mesiodistal dimension	Total tooth length	Root length
		Paracusps	Mesocusps	Distocusps			
Maxilla							
First molar	4	1.13 ± 0.03	1.30 ± 0.02	1.35 ± 0.02	2.33 ± 0.03	3.2	3 × Crown height
Second molar	4		1.39 ± 0.05	1.26 ± 0.05	1.77 ± 0.06	3.2	2.5 × Crown height
Third molar	3		1.35 ± 0.05	1.00 ± 0.04	1.56 ± 0.03	2.7	1.5 × Crown height
Mandible							
First molar	2	0.82 ± 0.05	1.04 ± 0.04	1.15 ± 0.04	2.11 ± 0.08	4.1	3 × Crown height
Second molar	2		1.28 ± 0.02	1.28 ± 0.04	1.89 ± 0.03	3.9	2.5 × Crown height
Third molar	2		1.39 ± 0.04	1.16 ± 0.05	1.96 ± 0.05	2.9	1.5 × Crown height

[a]From Keyes and Dale (1944).
[b]Each datum represents an average measurement of 15–20 teeth. Values given in millimeters.

from caries more frequently than females. In spite of the great amount of tooth decay that is seen in some hamsters, no statistical difference in life span is found on different diets (Lovelace et al., 1958).

C. Tongue

The tongue is a well-developed, spoon-shaped, very flexible organ which is 23–28 mm long and 4.5–7 mm wide. The tongue slopes steeply to the pharynx and is attached only laterally and ventrally. The muscular bulge at the base of the tongue contains the small hyoid bone. The central part of the tongue is composed of many striated muscle fibers that are disposed in three planes (longitudinal, transverse, and vertical). The individual muscle fibers within the bundles are each surrounded by an endomysium with capillaries close to the muscle fibers. Four different types of lingual papillae are identified in the hamster: filiform, fungiform, foliate, and vallate papillae (Fig. 3).

Abundant filiform papillae are found on the rostral, dorsal, and lateral surface of the tongue. These papillae are conical in shape with the vertex pointing caudally. The reader is referred to Fernandez et al. (1978) for a complete histological description.

Occasional fungiform papillae are located diffusely throughout the rostral dorsal surface of the tongue. Each papilla has a taste bud on its dorsal surface. A single vallate papilla is present on the caudal dorsal midline of the tongue. Multiple taste buds are located in epithelial depressions on either side of the vallate papilla. Foliate papillae are found on the lateral caudal tongue surface with taste buds arranged in longitudinal rows in the epithelium of the interpapillary clefts (Miller and Chaudhry, 1976).

Morphologically, all taste buds in the hamster are spindle-shaped and 100–125 µm long by 50–80 µm wide. They have a communication with the oral cavity through the taste pore, and the taste buds extend through the stratified squamous epithelium of the tongue to the basement membrane. In addition, there are hairlike cytoplasmic extensions into the taste pores (Miller and Chaudhry, 1976).

Hamster taste buds are similar ultrastructurally to those of other mammals such as the rat (Farbonann, 1965; Uga, 1969) and rabbit (Murray and Murray, 1967).

The sensory innervation from taste receptor cells in the taste buds in the fungiform papillae is carried by the chorda tympani nerve. Fibers of the glossopharyngeal nerve innervate receptors in the vallate and foliate papillae (Frank, 1973; Hyman and Frank, 1980a).

Each taste bud is responsive to one or two of the basic tastes. If two, these combinations are always either acid–salt or sweet–salt. Sweet and acid responses are negatively correlated. The ordering of stimuli for sweet receptors is sweet > bitter > salt > acid, while the ordering for acid receptors is almost the opposite: acid > salt > bitter > sweet (Frank,

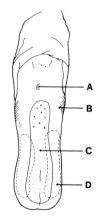

Fig. 3. Dorsal view of the hamster tongue illustrating the location of the vallate (A), foliate (B), fungiform (C), and filliform papillae (D).

1973). Mixtures of sucrose and electrolytes are less effective stimuli than sucrose alone, with the electrolytes causing a reduction in effect of the stimulating chemical (Hyman and Frank, 1980b).

Three classes of nerve receptor fibers are identified (Hyman and Frank, 1980b) as sweet-best, salt-best, and acid-best. Fibers of one class all react in the same manner to mixtures of chemicals. Sweet-best fibers have three characteristic responses: (1) an additive effect in the presence of a second chemical that is also a very effective stimulus, (2) a reduction in stimulus by addition of an electrolyte that is not itself an effective stimulus, and (3) a variability in the effect of electrolytes on two different effective stimuli. The response of salt-best fibers to mixtures is either indistinguishable from or smaller than the response to a single-component stimulus and is nearly equal to the sum of the two stimuli given individually. The effect of electrolyte mixtures on acid-best fibers is nearly equal to their individual effects (Hyman and Frank, 1980b).

D. Salivary Glands

The major salivary glands found in the hamster are the submaxillary, parotid, and sublingual. In addition, there is a retrolingual gland found in association with the submaxillary gland (Boerner-Patzelt, 1956). Hamster salivary glands are compound tubuloacinar structures.

Acinar elements complete their differentiation early in the postnatal development. Convoluted granular tubules have a more gradual and prolonged postnatal development, becoming completely differentiated upon sexual maturity. Syrian hamster submaxillary acinar cells have sulfomucins and sialomucins unlike the Chinese hamster acinar cells, which contain only acid mucopolysaccharides with low levels of sulfated carbohydrates (Devi and Jacoby, 1966).

The submaxillary gland contains a granular tubular segment developing from striated duct tissue at sexual maturity which functions as part of the intralobular duct system, similar to the rat and mouse. Submaxillary acini are described as morphologically and histochemically intermediate between mucous and serous acini (Shackelford, 1961). The sialic acid content of the female hamster submaxillary gland is greater than that found in the male. Staining the submaxillary glands with alcian blue produces a more intense color in the female. Ovariectomy was found to reduce female sialic acid content (and alcian blue staining ability), while estradiol administration elevates male sialic acid levels to near-female levels. Female sex hormones influence this uptake of sialic acid into the submaxillary mucin (Shackelford and Klapper, 1962a).

The organization of the parenchyma of the hamster submaxillary gland is similar to that of the rat and mouse (Flon and Gerstner, 1968). Narrow intercalated ducts connect the acini to the branches of the granular convoluted tubules. These tubules are continuous with the striated duct segments of the excretory system. The major part of the submaxillary gland consists of acinar elements, closely resembling the rat's gland (Devi and Jacoby, 1966). The tubule/acinus ratio, size, and appearance of the cells of the convoluted tubules are similar for both male and female. The hamster cells of the convoluted granular tubules are not as tall as in the rat and mouse, and with a more definite basal cushion containing striations or concentrations of small granules. The nucleus is located just above the cushion and below the supranuclear cytoplasm. Striated duct-type cells are frequently found with the granular cells in the convoluted tubules. Granules found in the acinar and intercalated duct cells are small, refractile, of approximately equal size and stain reaction, and appear evenly distributed. Two types of granules are found in the convoluted tubules. The larger type of granules are found near the nucleus and progressively decrease in size toward the cell border. The smaller granule type is finally dispersed within the cell cytoplasm (Flon and Gerstner, 1968). Most submaxillary acini produce both sialo- and sulfomucin acid mucopolysaccharides (Spicer and Duvenci, 1964).

The parotid gland, located in fatty tissue below the ear, covers the caudal one-fourth of the M. masseter and is structurally similar to other mammals (Shackelford, 1961). Four salivary ducts (two dorsal, one central, and one ventral) emerge from each parotid salivary gland. These ducts run on the lateral margin of the masseter muscle and then cross this muscle. The gland is entirely covered by the cheek pouch and ends near the maxillary edge of the entrance to the cheek pouch, 3 mm labially from the first maxillary molar (Schwarze and Michel, 1960).

Serous acini are connected to striated ducts via long, branching intercalated ducts containing granular formations. Electron micrographs of both parotid and submaxillary glands demonstrate a well-developed endoplasmic reticulum. Granular tubules and striated ducts have a system of basal infoldings closely associated with the mitochondria. In the hamster most proteolytic enzyme activity is confined to the parotid gland (Shackelford, 1961), whereas the rat and mouse show the most proteolytic enzyme activity in the submaxillary gland (Shackelford and Klapper, 1962b). Proteolytic enzyme activity is lowest in the hamster submaxillary gland.

The sublingual gland is composed of mucous acini, serous demilunes, and numerous myoepithelial cells (Shackelford, 1961; Alm et al., 1973). Pilocarpine stimulation, which produces marked vacuolation and nuclear disorientation in the submaxillary gland, produces no marked effect in the sublingual gland (Shackelford and Klapper, 1962b).

The hamster has a plexus of adrenergic nerves surrounding both acini and striated ducts of the salivary glands especially within the sublingual glands. In the periacinar areas, nerves are less abundant. The striated duct cells in the hamster sublingual gland have a rich supply of axons. The submaxillary glands have double innervation of both sympathetic and parasympathetic fibers (Alm et al., 1973; Bloomn and Carlosöö, 1973).

Fig. 4. Gross dissection illustrating the margins of the hamster cheek pouch (dashed lines) and pouch retractor muscle (arrow).

E. Cheek Pouches

Hamsters have two well-developed buccal pouches located beneath the skin on the lateral side of the neck and head (Fig. 4). The pouches are paired, internal muscular sacs opening on each side of the vestibulum oris and extending dorsocaudally to the shoulder (West, 1958). Pouches are used for carrying food, most commonly to carry pelleted foods or whole grains (corn) as opposed to finer milled or ground foods (Keyes and Dale, 1944). Females more frequently carry food in pouches than males, and younger animals more frequently keep food in pouches than older animals (Keyes and Dale, 1944). The storage capacity of these pouches is well illustrated in Fig. 5. A unique characteristic of the cheek pouch is that it can be easily everted, a feature which facilitates many research endeavors.

The cheek pouches are described as resembling the finger of a glove when filled. They cover part of the parotid gland, the masseter muscle, the external orbital gland, the lateral neck muscles, and some of the lateral shoulder muscles. The pouches measure 35–40 mm long and 4–8 mm wide when empty and up to 20 mm wide when filled. The cutaneous mucosa has fine, pale-pink folds covered with tiny wartlike protrusions. These protrusions are most numerous at the pouch entrance (Schwarze and Michel, 1960).

The primordium of the pouch first appears 3 days preceding birth. The primordium forms as a lateral outgrowth of oral epithelium in the form of a double-walled epithelial cup. From the posterior rim of this cup a double sheet of epithelium continues to grow posteriorly. Ten days postpartum the layers of the double epithelial sheet separate to form the pouch lumen. There follows an eightfold increase in pouch area between day 10 and day 60 (from 3 cm^2 to 25 cm^2). Also formed during this 50-day period are over 90 circular epithelial pegs with one to four connective-tissue papillae each. Pegs are most numerous on the medial pouch wall in the area of insertion of antagonistically acting muscles (Gillette, 1957).

The pouch wall is composed of four layers: stratified squamous epithelium on the surface, dense fibrous connective tissue, longitudinal striated muscle fibers, and loose areolar or submuscular connective tissue where the pouch joins underlying structures. Blood vessels and nerves are found in the connective tissue and muscle layers (Fulton *et al.*, 1946; Billingham *et al.*, 1960; Lindenmann and Strauli, 1968). The fine structure of the hamster cheek pouch begins with a thin homogeneous basement membrane separating basal cells from dense, randomly arranged collagen fibers of the adjacent lamina propria. Tonofibrils are found in great numbers in the basal cells and the stratum spinosum. Numerous mitochondria are found in the basal cells, stratum spinosum, and stratum granulosum, while mitochondria are less frequent in the stratum corneum. The upper cell layers of the stratum spinosum and stratum granulosum contain small (0.1 μm diameter), very dense granules. Densely compacted tonofibrils are the principal intracellular component of the stratum corneum. Desmosomes are found throughout the various cell layers (Albright and Listgarten, 1962).

The pouch retractor muscles arise from the lumbodorsal fascia and spinous processes of the last three thoracic vertebrae with insertion in two divisions on the dorsolateral and ventromedial surfaces of the pouch (West, 1958). The longitudinal and sphincterlike muscles are derived from the buccinator muscle. The longitudinal muscles contract the pouch, drawing it toward the aperture, while the sphincterlike muscles constrict the aperture. The retractor muscle draws the pouch dorsocaudally, acting as an antagonist to the longitudinal muscles (Priddy and Brodie, 1948). These muscles are probably not powerful enough to empty the cheek pouches; rather, the pouches are emptied by a massaging action with the front feet and action of the tongue.

Three superficial branches of the facial nerve (r. marginal-

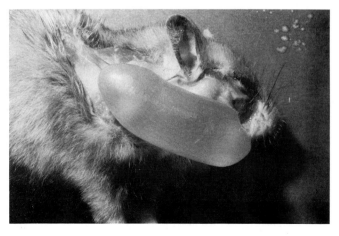

Fig. 5. Hamster cheek pouch distended with water.

mandibular br., dorsal buccolabial br., and a ventral buccolabial br.) supply the pouch (Schwarze and Michel, 1960).

Although the cheek pouches are considered an immunologically privileged site, the absence of lymphatic structures cannot be used to explain long survival times of pouch implants, as lymph vessels are found between the muscular and submuscular connective tissue layers and occasionally between the muscular and dense connective tissue layers. The number and shape are variable, with size somewhat dependent on degree of pouch enlargement. Larger lymphatic vessels frequently accompany arteries, veins, and nerves. The lymph vessels are frequently filled with a precipitatelike material which is stained pink by hematoxylin–eosin preparations. Lymphocytes and granulocytes are occasionally seen in the vessels. Valves occur frequently along the lymph vessels (Lindenmann and Strauli, 1968).

Chaisson (1954) describes four genera of palearctic true hamsters that have internal cheek pouches (*Cricetulus, Cricetus, Phodopus,* and *Mesocricetus*), while the African hamster (*Mystromys*) lacks cheek pouches. Other members of Rodentia with cheek pouches include some squirrel species, three genera of African Muridae, and the African gerbil (Chaisson, 1954).

F. Stomach

The hamster has a distinctly compartmentalized stomach consisting of two parts, the forestomach (pars cardiaca) and the glandular stomach (pars pylorica) (Fig. 6). The separation between the forestomach and the glandular stomach is a constriction with a sphincterlike muscular structure which may regulate movement of ingesta between the sections of the stomach. The stomach is approximately 3.5 cm long by 2.0 cm wide. The esophageal opening is cranial to the constriction separating the forestomach and the glandular stomach. The histological appearance of the glandular stomach is representative of that of a monogastric animal (Hoover *et al.,* 1969). Fewer fundic region argentaffin cells are found in the hamster than in the rat or the rabbit. Distribution of these cells in the basal half of the mucosa is uniform. The argentaffin cells in this region are irregular and often multipolar. Argentaffin cells in the pyloric region are rounded, and more reduced in number than in the fundic region but more numerous than in the corresponding region in the rabbit, mouse, or guinea pig. Argentaffin cells are absent from the esophagus and the stratified squamous epithelium of the forestomach (Dawson, 1945). The forestomach is similar both histologically and ultrastructurally to the rumen (Sakata and Tamate, 1976). However, the keratinized mucosal epithelial lining lacks papillae like those found in true ruminants (Hoover *et al.,* 1969). Microflora and volatile fatty acids are found in the forestomach as are found in the rumen of cattle and sheep (Mangold, 1929; Hoover *et al.,* 1969). The

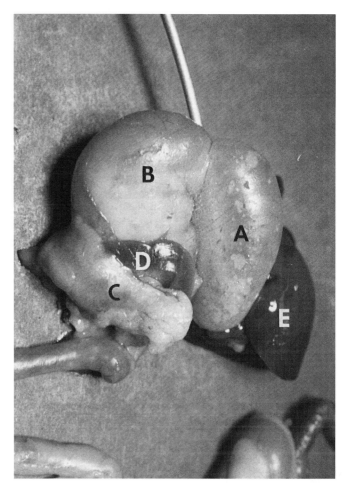

Fig. 6. Compartmentalized hamster stomach. (A) Forestomach; (B) glandular stomach; (C) pancreas; (D) spleen; (E) liver.

functional significance of the forestomach is questionable, since growth and metabolized energy values are not affected by its removal. Resident time of ingested feed in the forestomach is approximately 1 hr, reducing any effectiveness that fermentation in the forestomach might have on the total nutrition of the hamster (Ehle and Warner, 1978).

The average pH of the forestomach is always higher than the glandular stomach. Gram-positive species prevail, with gram-negative species consisting primarily of coliforms. Microorganisms reach numbers of 1×10^8 to 1×10^9/gm of contents in the forestomach and 1×10^7/gm of contents in the glandular stomach, which is comparable to numbers found in the rat (Kunstyr, 1974).

G. Small Intestine

The duodenum, beginning at the pyloris, runs laterally and slightly caudally toward the right before turning to run caudally as the descending duodenal loop. This descending loop passes ventral to the transverse colon and then the initial section of the jejunum. The duodenum now makes a reverse turn to the left to become the ascending duodenal loop. The ascending loop travels about one-half the distance of the descending loop, at which point it turns to cross the descending loop and become the jejunum. The jejunum is about two and one-half times the length of the duodenum and is formed into a number of small looplets. Duodenal villi first appear at 13.5 days of gestation. By birth the villi have changed from a square to a columnar shape. After birth, absorbed colostrum appears as distinct round lipid droplets in the intestinal epithelium. The droplets can be found for up to 6 days (Koruhmaru *et al.*, 1979). A short (<2 cm) ileum is continuous with the jejunum and leads into the cecum (Fig. 7) (Hoffman *et al.*, 1968). In the hamster intestine, calcium-binding activity is greater in the duodenum than in the ileum. The level of calcium-binding activity is directly related to the degree of calcium transport in the intestines (Kallfelz and Wasserman, 1972). The enzymes invertase β-fructofuranosidase) and maltase (α-glucosidase) are localized within the brush borders of the small intestinal epithelium. The brush border is the intracellular locus for disaccharide hydrolysis and the digestive site for carbohydrates in the hamster (Miller and Crane, 1961a,b,c).

H. Cecum

The hamster cecum is slightly sacculated and divided into an apical and a basal portion (Krueger and Rieschel, 1950). A groove in the external cecal wall corresponds to the location of a semilunar valve delineating the division between the two portions. This apicobasal cecal valve is formed by a bending of the cecum back on itself, thus producing an ellipsoidal, crescent-shaped piece of tissue extending into the cecal lumen. The circular muscle of the apical portion of the cecum covers the crescent (Yonce and Krueger, 1952). There are four valves in the hamster ileocecal–colic junctions: the ileocecal valve, the apicobasal semilunar valve, the basal semilunar valve, and the chevron valve. The ileocecal valve has a semilunar lip projecting into the lumen of the ileum. The apicobasal semilunar valve, located on the apical side of the ileocecal junction, functions to shunt the contents of the small intestine directly into the colon. The basal semilunar valve is located 5 mm toward the colon from the apicobasal semilunar valve, while the chevron valve is located at the cecal–colic junction and is formed from the ends of muscular chevrons located in the colon (Krueger and Rieschel, 1950).

I. Large Intestine

The ascending colon moves cranially from the cecal–colic junction, then under the jejunum, and turns, forming a horseshoe-shaped loop to the right before becoming the transverse colon. The transverse colon extends from the right to the left side of the dorsal abdomen just caudal to the liver and stomach. The colon then turns caudally to become the descending colon before emptying through the rectum and anus.

The feeding pattern in adult hamsters, as described by Borer *et al.* (1979), consists of 5-min feeding periods in which 0.9–1.33 gm of food is ingested followed by a 2-hr fast. Hamsters show no circadian rhythm in their feeding. Hamsters will hoard and gather food even when food intake is restricted in order to avoid fasts of longer than 12 hr and to avoid greater than a 20% weight loss (Borer *et al.*, 1979).

J. Pancreas

The hamster pancreas is a well-defined, yellowish white structure in the central abdomen (Fig. 6). It averages 0.46 gm

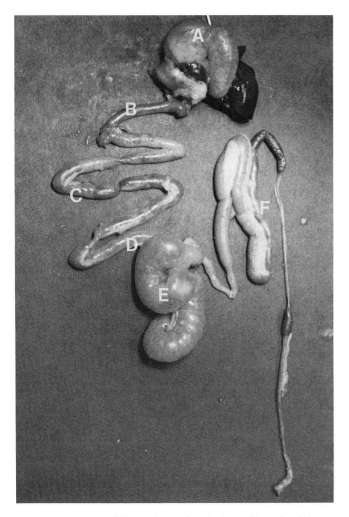

Fig. 7. Photograph of dissected gastrointestinal tract illustrating (A) stomach, (B) duodenum, (C) jejunum, (D) ileum, (E) cecum, and (F) colon.

of weight or approximately 0.4 to 0.5% of total body weight. The general shape of the pancreas is similar to the Greek letter λ, with the irregularly shaped central area lying mediodorsally to the cranial duodenum. It is bordered on the right by the pyloris, duodenum, and ascending colon. The gastric lobe of the pancreas is attached to the stomach and pylorus by the mesogastric membrane formed from the greater omentum. The gastric lobe comprises 25% of the weight of the pancreas. The splenic lobe follows the greater curvature of the glandular stomach and comprises 40% of the pancreatic weight. It is covered by the greater omentum as it runs cranially to the area of the aorta and caudal vena cava. The splenic lobe is caudal to the spleen but not attached to the splenic hilus. Instead, it has a connective tissue attachment to the cranial descending colon. The thin, leaf-shaped duodenal lobe is the smallest, averaging only 12% of the total pancreatic weight. It extends laterally from the duodenal loop to run caudally enclosed in the mesentery, between the duodenum and cranial ascending colon (Takahashi et al., 1977).

Three major pancreatic ducts join the common bile duct as the latter passes through the pancreatic head to enter the dorsal side of the duodenum. In addition, several smaller ducts from the head region drain directly into the common duct. There is a single main duct in both the duodenal and splenic lobes while the number of main ducts in the gastric lobe varies between one and two.

Organogenesis of the pancreas is first seen at 10.5 days of gestation (Boyer, 1968). The islets of Langerhans develop only after birth, during the first 14 postnatal days. The islets of Langerhans then double in size by 4 weeks after birth but do not reach full size until about 1 year of age. The acini of the hamster pancreas reach full size at 4 weeks postpartum (Orgrowsky et al., 1980).

K. Liver and Gallbladder

The four lobes of the hamster liver are the right dorsocaudal, left dorsocaudal, dorsal median, and ventral median. The dorsal median lobe is divided into a large right dorsal part and a smaller left dorsal part. The ventral median lobe is divided by the umbilical fissure into a right cranioventral part and a lateral left cranioventral part (Nettleblad, 1954).

No significant difference is noted in liver weights of male and nonpregnant female hamsters. A significant increase is noted during pregnancy. The increase in liver weight in late pregnancy is proportionately greater than the increase in body weight ($p < 0.25$). This relative hepatomegaly is related to a 37% reduction in mean bile flow in 8-day pregnant hamsters, and a 62% reduction in bile flow in 14-day pregnant hamsters as compared with nonpregnant females. In addition, a reduced secretion of total bile acids due to decreased cholic acid secretion, a decrease in hepatic Na^+,K^+-ATPase activity, and a decrease in the acid-independent bile flow is observed (Reyes and Kern, 1979).

The gallbladder is located in the right cranioventral section of the ventral median lobe of the liver. The gallbladder wall consists of three major layers: mucosa, muscularis, and serosa. Goblet cells are completely absent from the epithelium, while casklike and crayonlike cells are scattered in the epithelial layers. Casklike cells are prominent in the bottom of folds. Mitochondria are granular, rod-shaped, or occasionally filamentous in shape. Glycogen and fat droplets of various sizes occur in the apical and basal areas of the epithelium cells. The fat droplets contain neutral fat cholesterol or P.A.S. positive lipids. Small vacuoles of unknown significance also are found in the apical sections of epithelial cells. The lamina propria is similar to other species, but occasionally contains glandlike formations called Rokitansky–Aschoff sinuses (Yamada, 1959, 1962). Hamster bile acids have a similar composition to human bile (Reyes and Kern, 1979).

IV. RESPIRATORY SYSTEM

A. Upper Respiratory System

The respiratory system of the hamster consists of the nasal cavity, nasal sinuses, pharynx, larynx, trachea, and lungs. The hamster nasal cavity consists of vestibular, nonolfactory, and olfactory portions. Much of the nonolfactory nasal cavity surface is lined by cuboidal and columnar epithelium. Goblet cells and ciliated respiratory epithelium are present over only a small portion of the nasal cavity surface. The hamster has several nasal serous glands which open into the internal ostium of the external nares. These glands include one infraseptal gland, two nasoturbinate glands, five maxilloturbinate glands, one ventromedial nasal gland, four or five dorsal medial nasal glands, and the lateral nasal gland (Stenson's gland). Even though most of these glands are classified as serous structures, the hamster does have some nasal glands, which contain acidic mucopolysaccharide (mucus) components, and the vomeronasal gland, which contains neutral mucopolysaccharides (Adams and McFarland, 1972).

Of particular interest is the keen olfactory capability of most mammals, including hamsters. On the basis of behavioral data, it has been shown that mammals are sensitive to extremely low concentrations and can discriminate a wide range of odors. Unlike the rat, where four endoturbinates and two ectoturbinates are present, the hamster has four endoturbinates and three ectoturbinates. These very intricately folded turbinates project into the lumen of the nasal cavity and thus provide for an increased nasal mucosal surface. This nasal mucosa serves several well-accepted functions in that it filters incoming air, humidifies the air if it is too dry, and warms and/or cools the inspired air before it passes on to the lungs (O'Connell et al., 1978, 1979). Recent work has shown that the hamster has specialized receptive fields within the mucosa of these turbinates for olfactory bulb neurons. The topographical projection of input from localized regions in the epithelium onto second-order neurons in the olfactory bulb provides an explanation for the highly sensitive and discriminating capabilities of the hamster olfactory system (Costanzo and O'Connell, 1980).

A number of physiological measurements have been made by various workers on the temperature of the respiratory mucosa, on respiratory water loss, on streaming patterns of the mucociliary blanket, and on the pathway of air currents (Dawes, 1952; Cole, 1954; Schmidt-Nielsen et al., 1970). However, these measurements lack meaning unless related to an exact area of the respiratory tract. The rostral portion of the nasal cavity has goblet cells on the surface of the dorsal meatus, the septum, the lateral wall of the ventral meatus, and the medial surface of the ventral lamella of the maxilloturbinate; thus mucociliary action is restricted to these areas. Most of the caudal portion of the nasal cavity is lined by ciliated pseudostratified columnar epithelium. Apparently nonciliated stratified squamous epithelium occurs where there is the greatest impact of inspiratory air, and the ciliated columnar epithelium occurs where the greatest amount of air flow passes further back in the nasal cavity. The epithelium may be only two cell layers thick in areas least exposed to air. Thus the functions of waterproofing of the respiratory epithelium and removal of entrapped particulated matter by mucus would appear to be greatest in the area most exposed to air currents (Adams and McFarland, 1972).

B. Oropharynx and Trachea

The larynx is composed of a small epiglottic cartilage surrounded by two plates of thyroid cartilage. Caudal to these is the cricoid cartilage with the small arytenoid cartilage on its cranial dorsal rim. The tracheal bifurcation is situated at the height of the fourth rib pair.

C. Lower Respiratory System

The gross anatomy of the lung (Fig. 8) is characterized by a large single left lobe and several lobes on the right side (cranial, middle, caudal, intermediate, and accessory). The cranial lobe is oval in shape and lies adjacent to the cranial four to five intercostal spaces. A tetrahedron-shaped middle lobe envelopes most of the heart, except for the apex. The caudal lobe is the largest portion of the right lung, and it has the shape of a right triangle. This lobe reaches across the caudal six to seven intercostal spaces. An intermediate lobe is separated from the caudal lobe by the mesentery of the caudal vena cava and lies between the two lungs. It is in contact with the caudal lobe

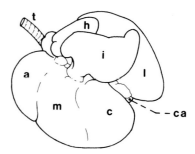

Fig. 8. Lobes of the hamster lung. (a) Cranial lobe; (m) middle lobe; (c) caudal lobe; (i) intermediate lobe; (ca) caudal accessory lobe; (l) left lobe; (h) heart; (t) trachea.

through a hilar connection and by the pulmonary ligament. Dorsally the esophagus runs across the intermediate lobe, whereby it produces an impression. Some authors consider this lobe to be a part of the caudal lobe (Hoffman et al., 1968), whereas others consider it to be a separate lobe (Schwarze and Michel, 1960). A small cone-shaped caudal accessory lobe lies between the caudal vena cava, the caudal lobe, and the intermediate lobe.

Table IV

Pulmonary Measurements Determined from a Lung Cast: Frequencies and Generations of Airways by Lobe and Range of Sizes[a]

	Number of airways[b]		
Lobe	Bronchi	Bronchioles	Terminal bronchioles
Left	18 (2–7)	44 (4–9)	156 (4–9)
Right			
Cranial	9 (3–7)	20 (5–8)	80 (5–9)
Middle	5 (3–5)	45 (4–7)	86 (5–8)
Infracardiac	4 (3–4)	22 (4–7)	48 (5–8)
Caudal	13 (3–6)	77 (5–9)	170 (5–10)
			540 (4–10)
Total	49 (2–7)	208 (4–9)	

	Length (mm)		Diameter (mm)	
Airway[c]	Range	Average ± SE	Range	Average ± SE
Bronchi	1.0–5.5	2.70 ± 0.21	0.5–2.0	0.67 ± 0.04
Bronchioles	0.1–4.0	1.39 ± 0.05	0.1–	0.24 ± 0.01
Terminal bronchioles	0.1–0.2	0.15 ± 0.003	<0.5 0.1–0.2	0.16 ± 0.003

Average number of respiratory units per terminal bronchiole: 2.4 ± 0.1 (100 terminal bronchioles and respiratory units counted)

[a]Reprinted from Kennedy et al. (1978) by permission of Cambridge University Press.
[b]Number of generations given in parentheses.
[c]Airways classified by histological definition described in text.

Table V
Measurements of Airways Determined from Histological Sections[a]

	Airway dimensions[b]			
	Diameter (mm)		Length (mm)	
Airway	Range	Average ± SE	Range	Average ± SE
Large				
Trachea	1.9–2.1	2.0 ± 0.04 (7)		
Bronchi	0.5–2.0	0.70 ± 0.06 (25)		
Bronchioles	0.10–<0.5	0.23 ± 0.02 (25)		
Terminal bronchioles	0.1–0.2	0.17 ± 0.004 (25)		
Respiratory region				
Respiratory bronchioles	0.05–0.20	0.16 ± 0.01 (23)	0.12–0.50	0.23 ± 0.03 (23)
Alveolar ducts	0.09–0.20	0.13 ± 0.01 (23)	0.24–0.55	0.32 ± 0.08 (23)
Alveolar sacs	0.10–0.20	0.13 ± 0.01 (18)	0.20–0.40	0.27 ± 0.01 (18)
Alveoli	0.03–0.10	0.06 ± 0.002 (74)	0.03–0.10	0.06 ± 0.002 (74)

Average number of alveoli per respiratory bronchiole: 4.2 ± 0.23 (one side only) (23)
Average number of alveoli per alveolar duct: 5.1 ± 0.18 (one side only) (23)
Average number of alveoli per alveolar sac: 5.2 ± 0.27 (18)

		Cell type frequencies and cell numbers in airways[d]		
Airway	Cell type[c]	Average total no. per 150 μm epithelial length	SE	Percentage of total
Trachea	Ciliated	13.4	0.54	36.1
Main bronchi	Goblet	14.3	0.54	38.5
	Basal	7.5	0.34	20.2
	Undetermined	1.9	0.23	5.1
	Total	37.1	0.84	
Bronchi	Ciliated	9.5	0.34	35.3
	Goblet	7.3	0.56	27.1
	Basal	4.8	0.33	17.8
	Clara	2.0	0.21	7.4
	Undetermined	3.3	0.37	12.3
	Total	26.9	0.46	
Bronchioles	Ciliated	11.6	0.31	48.7
	Clara	11.5	0.27	48.3
	Undetermined	0.7	0.21	2.9
	Total	23.8	0.29	

Airway	Height of epithelium[e] Average ± SE (μm)
Trachea	20.0 ± 0.29
Bronchi	20.2 ± 0.34
Bronchioles	"Scalloped" appearance: Clara cells, ~ 20 μm; ciliated cells, ~ 10 μm (without cilia)

Alveolar region
Average number of cells per alveolus: 11.4 ± 0.8 (100 alveoli counted in each of five animals; measured on longitudinal sections of alveolar ducts and sacs; does not include macrophages)
Frequency of macrophages in plastic sections: average of 38 ± 0.89 macrophages (1000 alveolar cells, counted in each of five animals
In each of the five animals studied, there were approximately one-third the number of type 2 cells as compared to the total number of type 1 and endothelial cells.

[a] Reprinted from Kennedy et al. (1978) by permission of Cambridge University Press.
[b] Seven animals. Number of units counted given in parentheses.
[c] Undetermined category includes intermediate and brush cells.
[d] Ten fields, two from each of five animals.
[e] Five measurements made from each of five animals.

The airway branching of the hamster is strongly monopodial, and the airway bifurcations are very sharp. The trachea bifurcates into a thicker right and a thinner left main bronchus. The left main bronchus does not divide into lobar bronchi, but rather into four lateral and three medial segmental bronchi at the level of the sixth to tenth rib pairs. In contrast, the right main bronchus forms four lobar bronchi before branching into smaller bronchi. Reisseisen's membrane, a layer of smooth muscle and elastic tissue, lines the lobar bronchi. The lobar bronchi terminate in segmental bronchi of varying size (300–500 μm in diameter). A ciliated columnar epithelium with mucous cells lines all of the bronchi. The segmental bronchi branch into the intralobar bronchi with a cuboidal epithelium and finally into the smaller respiratory bronchioles (100–140

Table VI

Comparison of Human and Hamster Respiratory Systems[a]

Respiratory system and characteristic	Airway				
	Trachea	Bronchus	Bronchiole	Respiratory bronchiole	Alveoli
Human					
Diameter	20–25 mm	≥1.0 mm	0.5–1.0 mm	≤0.5 mm	270 μm
Epithelium	Ciliated pseudostratified columnar with goblet cells; no Clara cells	Same as trachea; height, 30–50 μm	Ciliated simple columnar; goblet cells become infrequent and drop out; Clara cells appear	Low columnar to cuboidal; many Clara cells, few ciliated cells; ciliation drops out along this airway	Types 1 and 2 cells (Rhodin, 1974)
Lamina propria	Prominent (elastic lamina)	Thin (diffuse network)	Thin: loose connective tissue		
Muscularis	Not prominent; only trachealis muscle	Prominent muscularis; muscle fascicles; continuous circular or spiral	Prominent muscularis; muscle fascicles circular and spiral; separated by connective tissue	Layer of smooth muscle cells is thin and incomplete	Narrow bundles of smooth muscle cells encircle the entrance to each alveolus of the alveolar duct
Submucosa	Mucous and seromucous glands	Mucous and seromucous glands	No glands	—	—
Adventitia	About 20 C-shaped cartilage rings; lymph nodes	Cartilage plates; lymph nodes, lymphatics; bundles of collagenous and elastic fibers; blood vessels, nerve bundles	Connective tissue, elastic fibers, blood vessels, lymphatics	Collagenous connective tissue containing elastic fibers	—
Hamster					
Diameter	~2.0 mm	0.5–2.0 mm	0.1–<0.5	0.05–0.2 mm	~76 μm
Epithelium	Ciliated pseudostratified columnar in places to low columnar; many goblet cells, no Clara cells; height, 20 μm	Ciliated low columnar; many goblet cells, some Clara cells; height, 20 μm	Ciliated low columnar; Clara and ciliated cells; cilia drop out in the terminal bronchiole; height of Clara cells 20 μm, of ciliated cells 10 μm	Clara cells	Types 1 and 2
Lamina propria	Prominent	Very thin	—	—	—
Muscularis	Only trachealis muscle	Muscle fascicles continuous, primarily circular	Loose network: primary circular muscle fascicles	Rare single muscle fibers	—
Submucosa	Mucous glands	No glands	No glands	—	—
Adventitia	C-Shaped cartilage rings; mast cells; lymph nodes, nodules; lymphatics	Cartilage plates drop out as the airways enter the lung; mast cells; lymphatics; rare lymph nodules	Connective tissue—elastic, reticular fibers; blood vessels; lymphatics	—	—

[a]Adapted and reprinted from Kennedy et al. (1978) by permission of Cambridge University Press.

μm in diameter). The typical number of branches to the terminal bronchioles is about 10 to 18 (Table IV). The rat, hamster, and guinea pig lungs are essentially free of respiratory bronchioles, although the hamster and guinea pig have a transition to alveolarized airways within a single generation that could be classified as producing one order of respiratory bronchioles (Phalen and Oldham, 1983). Each bronchiole gives rise to two or three alveolar ducts, each 200–300 in length. Pulmonary alveoli, 35–75 μm in diameter, line these alveolar ducts. For a more detailed description of the termination of these bronchi, the reader is referred to the works of Kus and Sawicki (1976), Kennedy et al. (1978), and Phalen et al. (1978).

Table V summarizes the dimensions of the air pathways, identifies the type and number of cells found in each airway section, gives an approximation of cell height of the epithelium at three levels within the airway, and gives an average number of cells per alveolus (Kennedy et al., 1978).

Because the hamster has been used as an animal model for human respiratory research, Table VI is included, which compares the human and hamster respiratory system. Altogether, these data produce a pattern of lung morphometry for the hamster lung which contrasts somewhat with lung models used in particle deposition estimates for human lungs. The main difference lies in the declining surface area of peripheral airways present in the hamster lung (Kennedy et al., 1978).

In relation to human lung carcinogenesis, it is important to note that the histological appearance of the hamster trachea closely resembles the human bronchus (Kendrick et al., 1974), the site of most human lung cancer. One cell type of particular interest is the Clara cell. This cell is found in hamster bronchial epithelium and is thought to provide the major component of the distal "mucous ciliary escalator" (Kilburn, 1967) involved in the removal of foreign particles and carcinogens from the lung. Preliminary studies have suggested that it may be the cell of origin of carcinoid lung tumors and oat cell tumors (Lisco et al., 1974). It is also considered to be the cell of origin of human bronchioloalveolar carcinomas (Kuhn, 1972). For more detailed study of this cell the reader is referred to the works of Plopper et al. (1980a,b).

In hamsters, as in most other mammals at rest, inspiration is active and expiration is passive. The functional residual capacity (FRC) is determined by a static balance of forces between lung and chest wall, their transmural pressures being equal and opposite in sign. Pleural pressure is negative at FRC in an anesthetized hamster, so the lung is subjected to a negative transpulmonary pressure and may be attempting to resist further collapse. Because of this difference, changes in lung volume must be carefully interpreted when hamsters are used as models of human disease (Koo et al., 1976; Lai, 1979). Table VII summarizes the body weight, lung weight, lung volume, and quasi-static compliance of lung and chest wall.

The pulmonary arteries are typical elastic arteries with walls of equal thickness. The lobar arteries and their branches are muscular-type vessels. The venules have one to three layers of myocardial cells on their surface. The myocardium in the pulmonary veins passes into the atrial myocardium without any distinct boundary (Kus and Sawicki, 1976).

Lymphatic nodules, without germinal centers, are located in the walls of lobar and segmental bronchi, in the pulmonary parenchyma and along the venules (Kus and Sawicki, 1976).

Pulmonary and acid–base parameters have been measured with a total-body plethysmograph and indwelling catheters. Measurements for pH and PCO_2 of arterial samples in the euthermic hamster with a body temperature of 37°–38°C showed an arterial blood pH of 7.40 and PCO_2 of 45.3 torr. The hibernating hamster has a body temperature of 9°C, an arterial blood pH of 7.57, and a PCO_2 of 36.1 torr. These data have been confirmed in other hibernating species and suggest that the pH rises more during hibernation than it does during euthermic conditions. This is in contrast to poikilothermic animals, which would show a greater pH increase for the same drop in body temperature. Thus hibernating animals are relatively acidotic (Malen et al., 1973).

Table VII

Body Weight, Lung Weight, Lung Volume, and Quasi-Static Compliance of Lung and Chest Wall[a]

Variable[b]	Observed mean ± SE
Body wt (gm)	122.3 ± 3.0
Lung wt (gm)	0.74 ± 0.02
TLC_{25} (ml)	7.2 ± 0.14
VC (ml)	5.2 ± 0.13
FRC (ml)	2.4 ± 0.06
Cst(L) (ml/cm H_2O)	0.63 ± 0.03
Cst(W) (ml/cm H_2O)	3.39 ± 0.53

[a]Adapted from Koo et al. (1976).
[b]TLC_{25}, Lung volume at transpulmonary pressure 25 cm H_2O; VC, vital capacity; FRC, functional residual capacity; Cst(L), quasi-static compliance of lung; Cst(W), quasi-static compliance of chest wall.

Table VIII

Ventilation and Respiratory Volumes[a]

Weight (gm)	Mean resting respiratory rate/min	Tidal volume (ml)	Mean minute volume (ml)
130	30	1.4	42
100	32	1.03	33
90	32	0.91	30

[a]From Hoffman et al. (1968).

Measurements also show that the resting respiratory rate is inversely proportional to the body weight, whereas tidal volume and mean minute volume are directly related (Table VIII). These figures were collected within a zone of thermal neutrality, 27°–28°C. Any increase or decrease in environmental temperature would increase the resting ventilation.

V. CARDIOVASCULAR AND LYMPHATIC SYSTEMS

A. Heart

As in other mammals, the heart is four-chambered with two atria and two ventricles. The heart is located on the midline in the thoracic cavity and is in contact with the thoracic wall between ribs 3 and 5, which makes the cardiac puncture bleeding technique a relatively simple task. Three leaflets are present on the aortic and pulmonary valves, whereas two major and minor accessory leaflets are present on the left and right atrioventricular valves.

Comprehensive illustrations of the major blood vessels are shown in Figs. 9 and 10. As observed, the heart and vascular systems of the hamster closely resemble that of the rat. One difference of interest is that the media of the coronary arteries of the rat is composed of longitudinal muscle fibers; whereas, in the hamster the media of the coronary arteries is more like that found in other rodents and is composed of circular muscle (Kovalevskii, 1966).

Electrocardiographic studies have shown that the rate of contraction varies between 250 and 500 beats per minute. The average P–Q interval is 48 msec (range 40–60 msec) and the average QRS interval is 15 msec (range 13–20 msec). The amplitude of the T wave is 0.33 ± 0.07 mV, and the P-wave amplitude is 0.19 ± 0.03 mV (Hoffman et al., 1968).

Further work has shown that the blood pressure of the hamster does not increase with age as in most mammals, but may

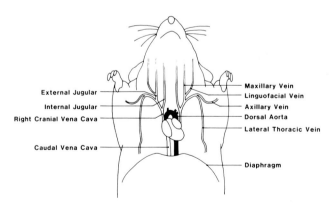

Fig. 9. Schematic of the major veins of the golden hamster.

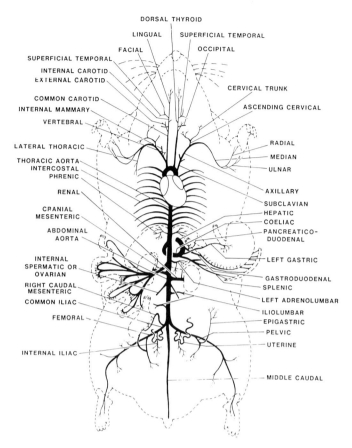

Fig. 10. Schematic of the major arteries of the golden hamster.

in fact decrease. Blood pressure readings vary from 76.3 ± 12.3 mm Hg in young animals, to 79.4 ± 13.5 mm Hg in 1.5- to 2.0-year-old animals, to 77.7 ± 13 mm Hg in older hamsters (Hoffman et al., 1968).

The mean cardiac output is 461 ± 29 ml/min/kg of body weight, which is much higher than the 250 ml/min/kg of body weight seen in the rat. Although no explanation for this high cardiac index is apparent, the hamster has been observed to have an exceptional oxyhemoglobin dissociation curve. Both the half-saturation oxygen tension and the Bohr shift are lower than would be predicted for an animal of this size. Given these observations, it might be expected that the hamster would have a relatively high cardiac output if it is to supply similar amounts of oxygen to the tissues at comparable oxygen tensions (Duling and Weiner, 1972).

The blood volume in the hamster is estimated to be 7.4 ± 3.1% of body weight. This compares to other body fluid compartments as follows: gut water is 7.6–8.5% of the body weight, and total body water is 67.4–68.9% of body weight, with the female showing the higher values (Hoffman et al., 1968).

Another interesting feature of the cardiovascular system in rats, mice, and hamsters is the existence of specialized sheaths of muscle fibers along the walls of the pulmonary veins. Further studies have shown these fibers to be analogous to the fibers of the myocardium. Hence the hamster, rat, and mouse differ significantly from other animals and humans, in whom the myocardium of the left atrium stops at the mouth of the pulmonary vein and does not extend along their walls (Kovalevskii, 1966).

B. Cheek Pouch

Blood supply to the pouch wall is from two branches of the mandibular labial artery (ventral and middle saccular arteries), one branch of the linguofacial (dorsal saccular artery) and one branch of the external carotid (caudal saccular artery). The retractor musculature is supplied by branches of the caudal auricular and thoracodorsal arteries. Venous return from the pouch wall is via tributaries of the rostral and caudal facial veins (Priddy and Brodie, 1948).

C. Intestine

The major blood supply to the intestinal tract comes via the abdominal aorta to the cranial mesenteric artery. The first branch is the cranial pancreaticoduodenal artery to the cranial portion of the descending duodenum and pancreas. The next branch is the colic artery to the transverse colon. This is followed closely by the caudal pancreaticoduodenal artery to the caudal portions of the descending duodenum, the ascending duodenum, and pancreas. The ileocecocolic artery to the ileum, cecum, and ascending colon splits off just ahead of the formation of the jejunal trunk with its 8–10 major branches to the jejunal loops. The caudal portion of the descending colon and the cranial portion of the rectum are supplied by the caudal mesenteric artery. According to unpublished works by Kent (1985) and Orsi *et al.* (1977), the hamster is the only mammal which may have the caudal mesenteric artery arising from the right common iliac artery and not from the dorsal aorta as in other mammals. Whereas Kent (1985) describes this as a common occurrence, Orsi *et al.* (1977) related that this anomaly was rare. However, it seems apparent that in many cases the caudal mesenteric artery arises at or immediately adjacent to the bifurcation of the abdominal aorta into the right and left common iliac arteries. The caudal rectum and anus receive blood from the caudal hemorrhoidal artery, a branch of the left common iliac artery (Schwarze and Michel, 1960; Hoffman *et al.*, 1968).

D. Pancreas and Liver

The blood supply to the pancreas arises from the celiac artery. The splenic artery travels through the splenic lobe of the pancreas to eventually become the left gastroepiploic artery, which in turn anastomoses with the right gastroepiploic artery. The right gastroepiploic artery is also the blood supply to the gastric pancreatic lobe. The pancreatic head and duodenal lobes receive arterial blood from the cranial pancreaticoduodenal artery. Vnous return from the pancreas is via the splenic, right gastroepiploic, and cranial pancreaticoduodenal veins primarily, with a small portion of the tail of the duodenal lobe draining by the caudal pancreaticoduodenal vein (Schwarze and Michel, 1960).

The liver receives its arterial blood from the main hepatic artery. This vessel enters the liver, along with the portal vein, at the hepatic portal (Schwarze and Michel, 1960).

E. Adrenal Gland

The major blood supply to the adrenals is by the cranial and caudal suprarenal arteries. These arteries become medullary arteries which traverse the cortex, usually without division and certainly without evidence of contributing to the supply of the cortical cells, to penetrate the medulla and divide into capillaries. These capillaries form a common medullary plexus which is fed zonally by the medullary arteries. The arrangement of the number of medullary arteries is related more to the size of the adrenal medulla than to the overall size of the gland. Some medullary capillaries anastomose with adjacent primary, secondary, tertiary, or subsequent radicals of the main adrenal vein according to their position, while others join the central vein. Occasionally they are seen joining the sinusoids associated with the zona reticularis. Corticovenous blood is collected into the peripheral radicals of the central vein in the inner part of the zona reticularis or at the corticomedullary junction. It then passes in progressively larger venous channels between the medullary chromaffin cells to reach the main vein. Although medullary capillaries anastomose with these venous channels, no evidence of a true portal circulation between the cortex and medulla has been demonstrated. Therefore, blood will usually flow from the medullary capillary plexus into the venous channels rather than in the opposite direction. The hamster differs from the rabbit in that it apparently has a poorly developed nervous or hormonal mechanism for controlling adrenal medullary arteries and their branches (Coupland and Selby, 1976).

F. Kidney

The renal artery and vein divide into six to eight corresponding pairs of interlobar vessels near the hilum (Lacy and Schmidt-Nielsen, 1979a). The vascular bundles of the renal medulla are cone-shaped structures with their bases in the outer stripe and the apexes near the tip of the papilla. They are most prominent in the inner stripe of the medulla. Two types of vascular bundles exist: simple and complex (Kriz *et al.*, 1976). The simple bundle consists of exclusively arterial and venous vasa recta and is found in the hamster, rabbit, guinea pig, prairie dog, opossum, cat, dog, pig, rhesus monkey, and human (Kriz et al., 1976). The complex type contains the arterial and venous vasa recta as well as the thin descending limbs of short loops of Henle and has been found in the rat (Kriz, 1967), mouse (Kriz and Koepsell, 1974), and fat sand rat *Psammomys obesus* (Kaissling et al., 1975). It is thought that the simple type is more primitive and that the complex type allows for optimal medullary function and greater urine-concentrating capabilities.

G. Ovaries and Uterus

In the hamster, as well as in the guinea pig and rat, the uterine artery provides a major portion of the blood supply to the ovary. On each side, the uterine artery approaches the uterus near its caudal end, runs parallel to the uterine horn, and terminates in a branch to the ovary (Del Campo and Ginther, 1972). The left ovarian artery originates from the aorta approximately 1.0 mm cranial to the origin of the right artery, and both terminate at the ovary, relatively independent of the uterine arteries. Both the uterine and ovarian arteries coil as they approach the ovaries and, in some hamsters, small anastomoses between the ovarian and uterine arteries are seen close to the ovary (Del Campo and Ginther, 1972). This differs from the guinea pig and rat, where prominent anastomoses exist between the ovarian and uterine arteries ahead of the termination at the ovary. The ovarian and uterine veins join to form the uteroovarian vein, which then enters the vena cava. The junction of the ovarian and uterine veins is relatively further from the ovary (1.2 cm) and nearer to the vena cava in the hamster than that in the guinea pig and rat. In the hamster, as well as the guinea pig and rat, there is a direct unilateral uteroluteal pathway (Ginther, 1967). It is thought that the uterine artery is the functional unit of this unilateral pathway (Del Campo and Ginther, 1972).

H. Orbital Venous Sinus

Because of the relatively small size of superficial veins, the absence of an accessible tail vein, and the small size and mobility of the heart, some investigators have avoided using the hamster in hematological studies. However, it is valuable to know that the hamster has an orbital venous sinus similar to the mouse. This space or sinus is an elongated venous channel which lies in a groove of bone formed in the presphenoid and maxillary bones. The venous space represents an orbital or precavernous sinus formed by the confluence of the ophthalmic veins, which then drain posteriorly into the cavernous sinus. Physiologically it is important to remember that the hamster's blood has a rapid coagulation time, which necessitates the use of an anticoagulant (1:1000 heparin) in any blood collection instruments (Pansky et al., 1961; Edwards, 1967; Timm, 1979).

References dealing with the vasculature to other specialized areas of the body are cited by Hoffman et al. (1968).

I. Lymphatic System

The distribution of lymph nodes throughout the body is shown in Fig. 11. There are 12 lymphocenters (parotid, mandibular, retropharyngeal, axillary, mediastinal, dorsal thoracic, celiac, cranial mesenteric, lumbar, iliac, superficial inguinal, and popliteal) which are composed of 16 groups of nodes (Kawashima, 1972). Major concentrations of lymph nodes are located in the jejunal mesentery, near the junction of the jejunum and duodenum, near the ileocecal junction, and in the mesentery of the transverse colon. By region, these nodes occupy approximately the same locations as those described for other rodents (Belisle and Saint-Marie, 1981b).

Histologically the deep cortex of a lymph node consists of one to several basic elements referred to as deep cortex units, some of which are variably fused to one another to form deep complexes. A unit is a semirounded structure which lies next to the peripheral cortex and bulges into the medulla. Each of these units is centered on the opening(s) of an afferent lymphatic vessel, and it is divided into a "center" and a "periphery." The center is nearly devoid of reticular fingers, whereas the periph-

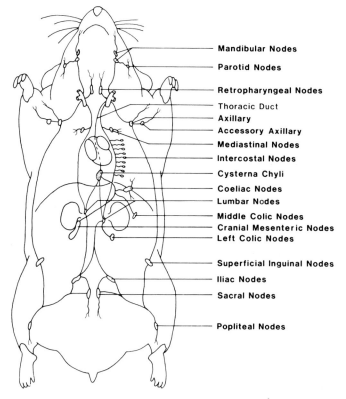

Fig. 11. Schematic of the lymphatic system of the golden hamster.

ery exhibits a dense framework of fibers. The periphery is the site of concentration of most postcapillary venules of a unit and contains lymphatic sinuses which are often loaded with small lymphocytes. Functionally, it is thought that the center is a site of cellular proliferation and retention, whereas the periphery is a site of rapid lymphocyte migration in and out of the unit. Some support has been given to the view that the immune reactivity of a unit is reflected by the lymphocyte content of peripheral sinuses. The full significance of such "sinus loading" is still under investigation (Belisle and Saint-Marie, 1981a).

VI. URINARY AND REPRODUCTIVE SYSTEMS

A. Kidney and Ureters

The reddish brown, bean-shaped kidneys are located to the right and left of the median plane in the lumbar region. The right kidney is located at the level of the second to the fourth lumbar vertebrae and the left kidney at the level of the third to fourth lumbar vertebrae. The dorsal surface of the kidney is flattened while the ventral surface protrudes; both poles are rounded. Generally, the right kidney weighs less than the left and the kidneys of female animals are heavier than those of males. The combined weight of both kidneys is approximately 1.0% of the body weight (Schwarze and Michel, 1960).

The unipapillate hamster kidney is distinguished by a very long papilla which extends out into the ureter. This anatomical arrangement has allowed *in vivo* urine collection from single collecting tubules. A papillectomy decreases the length of the renal medulla, which in turn decreases the ability of the kidney to concentrate urine (Jarausch and Ullrich, 1957).

Studies of the mammalian renal pelvis have led to the classification of two broad categories. The type I pelvis is simple in shape, without extensions and relatively small. This type of pelvis is found in mammals which do not have elevated urine osmolality, such as the pig and beaver. The type II pelvis is larger and more extensive, with a highly folded pelvic wall which allows for high urine-concentrating capacities. The hamster, rat, dog, sheep, and sand rat are examples of mammals with a type II renal pelvis (Pfeiffer, 1968).

The lower pelvis is funnel-shaped, while the upper pelvis is complexly shaped and extends deep into the kidney parenchyma. Approximately 2 to 3 mm of the papilla is visible in the intact kidney, since it projects into the hilus. The upper pelvis begins at the papilla base and is dorsoventrally flattened into two parallel sheets which extend deep into each kidney pole. Each parallel sheet has three scalloped impressions which have six to eight extensions (fornices) that project deep into the

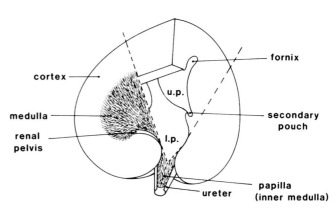

Fig. 12. Schematic of hamster kidney. Removal of the medulla in the wedge-shaped area reveals upper (u.p.) and lower (l.p.) pelvic formation. (Adapted from Lacy and Schmidt-Nielsen, 1979a.)

renal parenchyma from the upper margins. Several smaller extensions (secondary pouches) project toward the hilum from the lower margins (Fig. 12) (Lacy and Schmidt-Nielsen, 1979a).

Ultrastructural examination of the hamster renal pelvis revealed no differences between (1) male and female hamsters, (2) hamsters on high- and low-protein diets, or (3) hamsters on water deprivation and those receiving water *ad libitum*. Three morphologically distinct epithelia line the hamster renal pelvis. A portion of the outer medulla is covered by transitional epithelium, but the majority of surface area is covered by a simple epithelium that ranges from squamous to low cuboidal. The inner medulla is covered by a single layer of tall columnar cells which become progressively shorter toward the inner–outer medullary junction. This epithelium is morphologically similar to collecting-duct epithelium of the inner medulla. The cortical pelvic epithelium is transitional epithelium which is similar to ureter and urinary bladder epithelium. Morphological comparisons with other epithelia suggest that the inner and outer medullary pelvic epithelia are permeable to solutes and/or water, while the transitional epithelium covering the cortex appears relatively impermeable (Lacy and Schmidt-Nelsen, 1979b).

An animal model for human renal cancer may be induced in the male hamster by estrogen administration (Van Hoosier and Ladiges, 1984).

B. Bladder and Urethra

The gross anatomy of the hamster's bladder and urethra is similar to that described for the dog (Schwarze and Michel, 1960). In the female, the urethra is surrounded by connective tissue and runs dorsally on the pelvic symphysis terminating in a 3- to 3.5-mm wide cone-shaped peg ventral to the vagina. Glandular tissue is located in the lateral walls of the caudal urethra. This tissue contains a large number of secretory ducts which are lined with a nonstratified, cuboidal to squamous epithelium. The male has an approximately 10-mm-long intrapelvic urethral section which begins at the neck of the bladder and continues caudally toward the pelvic symphysis. At the caudal end of the pelvic symphysis the urethra curves ventrally and continues, extrapelvically, along the ventral surface of the penis. The intrapelvic urethra is characterized by numerous urethral glands located in the mucosa; the extrapelvic urethral mucosa is aglandular.

The dog and the Syrian hamster are two of the most reliable models for studying the effect of chemical carcinogens on the urinary bladder (Van Hoosier and Ladiges, 1984).

C. Male Reproductive System

In neonatal hamsters, the anogenital distance is longer in the male than in the female. Also, the genital papilla in the male is more prominent, slightly longer, and not as symmetrical as that of the female (Magalhaes, 1968). In sexually mature male hamsters, the testes within the scrotal sacs cause the posterior end of the body to appear round. There is no mediastinum in the testis, and the penis is retracted when animals are not mating. Other external features that are characteristic of the sexually mature male are darkly pigmented flank glands and a longer anogenital distance, compared with the female.

During the first 2 weeks after birth, the gonads and accessory reproductive organs in both sexes grow and develop slowly. At 16 days the testis tubules begin to form lumina, testicles descend at about 26 days of age, and sperm heads first appear by 36 days (Ortiz, 1948). As in most other rodents, the hamster sperm head is sickle-shaped and flattened when seen in surface view (Yanagimachi and Noda, 1970). Hamster spermatozoa do not become fertile until they reach the cauda epididymis (Cummins, 1976). All male accessory glands have reached their adult histological condition by 26 days, and there is evidence of secretion in the seminal vesicle and coagulating gland at this time (Ortiz, 1948).

Figure 13 illustrates the urogenital system of the mature male hamster. A thick fat pad covers the proximal end of the testicle, the caput, and almost half of the body of the epididymis. The body of the epididymis runs along the dorsomedial side of the testicle. The cauda epididymis is very well developed. The ductus deferens is 15–20 mm long and begins at the cauda epididymis. It runs in a craniomedial direction to the level of the prostate, where it turns caudally and unites with the excretory duct of the seminal vesicle (vesicular gland) to enter the urethra. The diameter of the ductus deferens doubles at the level of the prostate due to a large number of ampulla glands in the mucosa. Male accessory sex glands include interior and exterior ampulla glands, seminal vesicles, coagulating glands, a three-lobed prostate (cranial, caudal, and central), and bulbourethral glands (Schwarze and Michel, 1960).

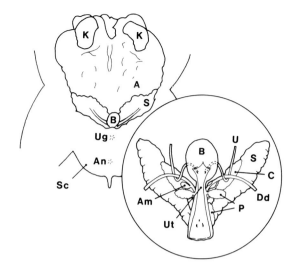

Fig. 13. Ventrodorsal view of the male urogenital system of the hamster illustrating the kidney (K), bladder (B), seminal vesicle (S), urogenital opening (Ug), anus (An), scrotum (Sc), and adipose tissue (A). The insert is a ventral view of the accessory sex organs. The bladder and urethra have been opened to show entrances of ducts. Shown in the insert are the seminal vesicle (S), ureter (U), ductus deferens (Dd), urethra (Ut), coagulating gland (C), prostate (P), and ampullary gland (Am). A bulbourethral gland enters the urethra further caudally and is not illustrated.

Male hamsters reach sexual maturity when they weigh approximately 90 gm at about 12 weeks of age (Van Hoosier and Ladiges, 1984). Neither the seminal vesicles, the prostate, nor the coagulating glands are necessary for normal mating behavior or for reproduction (Pauker, 1948; Weinerth et al., 1961). In animals exposed to natural daylight and temperature, there is a seasonal variation in the size and weight of the reproductive organs, with the lowest weights found during the winter months (Reiter, 1973). Spermatogenesis is reduced or absent during this time. Testicular regression, declining testosterone levels, and elmination of copulatory behavior is seen in male hamsters exposed to less than 12.5 hr of light per day (Gaston and Menaker, 1967; Elliot, 1976; Morin and Zucker, 1978). Some of the light-microscopic change reported in involuted testes include Leydig cells of decreased size and increased nucleocytoplasmic ratio, decreased number of mitochondria, decreased amount of internal and tubular endoplasmic reticulum, increased number of lipid droplets, decreased cell size, and cytoplasmic differentiation of macrophages (Wing and Lin, 1977).

Light-microscopic preparations of the testes reveal small blood vessels and clusters of Leydig cells and macrophages which are located mostly in the angular areas of the interstitium and are surrounded by irregularly shaped lymphatic sinusoids (Wing and Ling, 1977). The general testicular interstitial organization of the hamster is similar to that of the guinea pig (Fawcett et al., 1973).

Intersitial cells of Leydig are the major source of testicular androgen. In the reproductively active hamster, Leydig cells contain abundant endoplasmic reticulum, which is predominantly in the form of flattened cisternae rather than agranular tubular reticulum. The tubular form is the more common type of endoplasmic reticulum seen in other mammalian species. It is also noted that serum cholesterol levels appear to be inversely related to the amount of agranular tubular reticulum (Wing and Lin, 1977). Serum cholesterol level is highest in the hamster, in which the tubular reticulum is least developed, while the guinea pig has a very low serum cholesterol and an especially abundant endoplasmic reticulum (Christensen, 1965). The rat (Christensen and Gillim, 1969) and mouse (Christensen and Fawcett, 1966) are intermediary between the guinea pig and hamster. Tubular reticulum is thought to be the site of cholesterol synthesis in Leydig cells (Christensen, 1965). The greater the volume of of the agranular tubules in the Leydig cells, the lesser is their dependence on exogenous cholesterol for the biosynthesis of androgens (Wing and Lin, 1977).

Hamsters, like rats and mice, have a superficial rete testis which lies immediately beneath the tunica albuginea. The extratesticular rete is particularly well developed in hamsters. In rats, mice, and hamsters, the epithelial lining of the rete testis is generally low cuboidal, with a highly irregular nucleus containing one to three nucleoli. Intraepithelial lymphocytes are found in the epididymis in less than 1% of the epithelial cells (Martin, 1976).

Age-related, light microscopy changes have been described in the hamster testis. With age, the lumina of the seminiferous tubules increase as the thickness of the wall decreases and the tubule diameter remains constant. These changes begin centrally, resulting in eventual arrest of the final stages of spermatogenesis. Other aging changes described include more pronounced collagenous fibers in the epididymis, ductus deferens, blood vessels, basement membrane, and tunica albuginea; connective tissue replacement of the circular muscles and taller epithelium, as well as an increase in the number of hyalinized cells in the epididymis (Soderwall et al., 1950).

There is little change in the reproductive hormone profile in the aging male hamster. There is no significant change between serum testosterone and luteinizing hormone (LH) levels of 4-, 11-, 24- and 31-month-old hamsters. Serum follicle-stimulating hormone (FSH) levels significantly increase between 4- and 11-month-old hamsters but remain stable thereafter. Also, very few differences are observed, with age, in the weight of the testes and accessory sex organs in both sexually active and inactive male hamsters. These results differ from earlier studies done in the rat and mouse, both of which demonstrated age-related decreases in serum testosterone, LH, and FSH, as well as decreased testicular and accessory sex gland weights (Swanson et al., 1982).

The male hamster has a baculum or os penis consisting of two distal lateral prongs and a dorsal prong (Callery, 1951). Adult male hamsters, like other laboratory Muridae and Cavidae, have a urethral plug which is a white, rubberlike cast of the urethra (Kunstyr et al., 1982). The function of the urethral plug is not known; however, it is similar in structure, ultrastructure (Rosenbauer et al., 1980; Campean et al., 1980), and amino acid composition (Kunstyr et al., 1982) to the vaginal plug found after mating in female rats and mice.

D. Female Reproductive System

Figure 14 illustrates the urogenital system of the mature female hamster. In the female, the urinary opening is cranial, the anus is caudal, and the vaginal opening is located between the two. Flank glands are present in the female but are less obvious than those of the male, due to lighter pigmentation. The caudal end of the female is pointed toward the tail. The mean number of mammary glands is 14, with a reported range of 12–17 teats per female hamster. Teats are arranged in two rows of seven, and the glands extend from the thorax to the inguinal region (Fig. 1). In general, when extra teats are present, they are between either the first and second or sixth and seventh pair of teats (Anderson and Sinha, 1972).

The 3- to 4-mm-long, oval-shaped ovaries are located dorsolaterally to the kidneys and are completely enclosed within membranous sacs called bursae ovaricae. Each bursa is composed of three layers: (1) an inner, discontinuous bursal epithelium that faces the ovary; (2) a middle layer of connective tissue that contains fibroblasts, bundles of smooth muscle cells, and blood vessels; and (3) an outer continuous epithelium that faces the peritoneal cavity. One side of the bursa has a thin layer of connective tissue, the "window," through

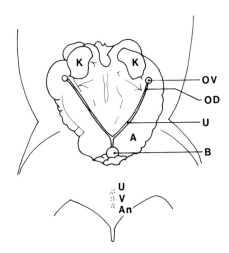

Fig. 14. Female urogenital system of the hamster illustrating the kidney (K), bladder (B), ovary (OV), oviduct (OD), uterine horn (U), adipose tissue (A), urethral opening (U), vaginal opening (V), and anus (An).

which the ovary may be seen. The rest of the bursa is surrounded by a thick layer of fat. The presence of the ovarian bursa restricts exchange of materials between bursal and peritoneal cavities. Fluid is found within the bursa and is thought to be predominantly an ovarian exudate. The bundles of smooth muscle cells within the bursa may serve to regulate fluid volume and pressure within the bursal cavity, creating an environment that may be important for normal ovarian function, egg transport into the oviduct, and possible fertilization. The ultrastructure of the bursal cells and the organization of the bursa does not change during the estrous cycle (Martin et al., 1981).

The ovarian theca folliculi is composed of an inner vascular layer and an outer avascular layer. The theca interna consists of epithelioid cells with large lipid droplets and tubular mitochondria characteristic of steroid-producing cells. The external layer consists of undifferentiated mesenchymal cells or fibroblasts, smooth muscle cells, and cells that appear to be an intermediary stage between the undifferentiated cells and the smooth muscle cells (McReynolds et al., 1973).

Smooth muscle-like cells are found in the theca externa of mature antral follicles, in ovarian stroma surrounding corpora lutea, and in the interstitial ovarian tissue (McReynolds et al., 1973). Smooth muscle cells have also been reported in the theca externa of sheep (O'Shea, 1971), rat (Osvaldo-Decima, 1970), monkey, cat, rabbit, guinea pig (Burden, 1972), and gerbil (McReynolds et al., 1973). The rat (Osvaldo-Decima, 1970; O'Shea, 1971) and gerbil (McReynolds et al., 1973) have smooth muscle cells in the corpora lutea and interstitial tissue as well. Smooth muscle-like cells have also been described in atretic follicles of the rat and monkey (Osvaldo-Decima, 1960). Perifollicular smooth muscle in the rat and monkey is only one to three cell layers thick, while five to six cell layers of smooth muscle are seen in the hamster and gerbil. It has been suggested that smooth muscle contraction may play an important role in the ovulatory process of hamsters and gerbils as well as in the contraction or collapse of the follicle following ovulation and during atresia (McReynolds et al., 1973).

The caudal vagina, immediately cranial to the vulva, is flattened dorsoventrally. A pair of vaginal pouches protrude from the lateral walls of the lower vagina, extend caudally, and are directed ventrally and laterally. These pouches terminate just proximal to the vaginal orifice. The upper vagina is more cylindrical and extends cranial to and beyond the os cervicis as a crescentic cul-de-sac in the dorsal wall of which the cervix is embedded (Kent, 1968). The dorsal lip of the cervix is longer than the ventral lip. The caudal part of the cervix exists as a single cervical canal, while cranially there are dual cervical canals which lead to the two uteri. The uterus is duplex. The undivided part is 7–8 mm long, while each horn is approximately 20 mm in length and 2–3 mm in diameter. The regions of the long, tightly coiled oviduct are described in detail by Oeri (1960) and Strauss (1956).

In general, histology of the ovary, uterus, and vagina of the hamster is similar to that described for other mammals. Histological aging changes in the hamster ovary, uterus, and vagina are similar to those found in the mouse and rat. Aging changes seen in the female tract include an increase in the ratio of atretic to normal follicles; increased fibrous connective tissue in germinal epithelium of the ovary, fallopian tubes, uterus, and vagina; increased uterine diameter; and atrophy of the circular uterine musculature. The reader is referred to Rolle and Charipper (1949) for a complete description.

Reproductive maturity occurs in the female by 3 months of age when the animal reaches a weight of 90–100 gm (Van Hoosier and Ladiges, 1984). The vagina opens at approximately 10 days of age, in advance of sexual maturity, which differs from most rodents where vaginal opening and sexual maturity chronologically coincide (Ortiz, 1948, 1955). Females have a 4-day estrous cycle with rapid development and regression of corpora lutea during a single cycle, unlike the mouse or rat, where there is retention of several sets of corpora lutea from previous cycles. Following ovulation, the corpus luteum organizes rapidly. The differentiation and subsequent regression of the hamster corpora lutea is described by Nakano (1960). Pituitary gonadotropic hormones regulate the maturation and ovulation of ovarian follicles. Cyclic modifications in the reproductive tract as well as the onset and duration of the females' receptivity to the male are regulated by ovarian hormones.

Data on the reproductive cycle of the female Syrian hamster, gathered from several sources, is shown in Table IX. Onset of psychic estrus occurs approximately 8 hr prior to ovulation. Psychic estrus in the female is evidenced by lordosis. When the controlled photoperiod consists of light from 6 AM to 6 PM, ovulation occurs between midnight and 1:00 AM following an evening of heat (Soderwall et al., 1960). The mean number of eggs ovulated per hamster is 10 (Austin, 1955). The approximate size of a mature ovum is 70 μm, and its maximal fertilizable life is 13 hr.

It has been conventional to call day 1 of the estrous cycle the 24 hr prior to and including ovulation. Day 2 is the subsequent 24 hr, including the morning with a copious postovulatory discharge. The reader is referred to Table IX for details of the cycle. Upper vaginal and cervical epithelium exhibits cyclic changes in response to ovarian hormonal activity. However, vaginal pouches constantly accumulate desquamated cells from the vagina as well as leukocytes and noncornified cells from higher levels. For this reason, vaginal smears made occasionally are not always reliable; however, if smears are made daily, cells from earlier stages of the cycle will not accumulate and the smear will more nearly mirror the condition of the tract (Orsini, 1961).

Table IX
Reproductive Cycle of Female Syrian Hamsters[a,b]

Day of cycle	Stage of cycle and duration	Vaginal smear	Histology of upper vagina and cervix	Ovary	Plasma progesterone	Mating response
1		Leukocytes very few to rare	Transmigration of leukocytes occasional	Advanced luteolysis; extensive phagocytic activity and lutea cell autolysis indicative of an advanced involutional state		0
	Proestrus (~3 hr)	Nonnucleated squamous cells	Epithelium foliaceous; elongated cells extending into lumen forming loose reticulum underlaid by thick stratified squamous epithelium		1.9 ± 0.8 ng/ml	+
		No leukocytes	No transmigratory leukocytes			++
	(Estrus (~12 hr)	Nucleated epithelial cells; a few squamous cells; elongated cells rise to abundance; some oval cells	Early: begin exfoliation of columnar and elongated cells	Ovulation between midnight and 1 AM, depending on photoperiod		++
			Late: exfoliation heavy, begin desquamation of underlying layers			+++
	Metestrus (~4 hr)	No leukocytes; heavy influx of oval epithelial cells; other types rapidly decrease	Continuing desquamation of deeper cells	Luteinization: luteinizing granulosa cells at an estimated 6–8 hr after ovulation		+++
			Postovulatory discharge present			++
2	Diestrus (~76 hr)	Oval epithelial cells abundant. Leukocytes appear and rise to abundance.	Early: transmigration of leukocytes commences. Epithelium very low, stratified; post-ovulatory discharge heavy.		1.8 ± 4 ng/ml	
			Late: heavy transmigration of leukocytes; low, stratified epithelium proliferating			0
3	Middle diestrus	Oval epithelial cells sparse; a few squamous cells; leukocytes falling to sparse	Stratified epithelium thicker, surface cells becoming columnar; transmigration of leukocytes decreased	Secretory phase of corpus luteum luteal cell hypertrophy	6.6 ± 1.1 ng/ml	0
4	Late diestrus	Oval epithelial cells very few; squamous cells few	Surface columnar cells elongated toward lumen; vacuolelike spaces in epithelium; rapid proliferation of underlying squamous layer	Lueolysis: reduction of luteal cell size, condensation of agranular endoplasmic reticulum, regressive changes in mitochondria and marked drop in luteal progesterone activity	<0.5 ng/ml	0

[a] Adapted from Kent (1968) and Leavitt et al. (1973).
[b] Material in this table was based on day 2 being defined as the day of postovulatory discharge. The reader is cautioned that some reports in the literature arbitrarily define the day of postovulatory discharge as day 1 of the cycle.

VII. ENDOCRINE SYSTEM

A. Adrenal Gland

The adrenal glands are craniomedial to the kidneys, embedded in adipose tissue, at the level of the last thoracic vertebra or the first lumbar vertebra. They are 1.5–2 mm long and 0.8–1.2 mm wide (Schwarze and Michel, 1960).

The adrenals of the hamster differ in many respects from those of other mammals that have been studied so far. Developmentally the male adrenal begins its increase in size by the fourth week and is significantly heavier and has a lower number of nuclei per constant area in the zona reticularis. In contrast to the female, this zone is well developed in the male and consists of cells that are much larger than the reticularis cells of the female. The sexual dimorphism therefore is due to an enlargement of these reticularis cells. Whether or not the cortex as a whole contains a higher number of cells with proliferative activity remains unsettled (Zieger et al., 1974). This difference in the size of the adrenals in hamsters continues into the adult and is the reverse of that found in most rodents, where the female usually has larger adrenal glands. The greater adrenal mass in the male hamster is apparently due to an increased thickness of the zona reticularis. The sex differences in adrenal cortex zone thickness are not conspicuous in other mammalian species. For example, the cortex of human adrenals have three zones in a zona glomerulosa/zona fasciculata/zona reticularis ratio of about 1:3:2 regardless of sex (Warner, 1971). This ratio was similar to that of female hamsters, while the zone ratio in males was about 1:3:5.

Studies of the adrenal steroids of female rats have shown that the rate of steroid secretion and the steroid plasma level are higher and that the half-life of the steroids is shorter than in males. In hamsters, however, it is the male adrenal gland that is heavier and that has the higher rate of steroid secretion and a shorter steroid half-life. It is also interesting that the sexual dimorphism is particularly evident in the zona reticularis and not in the outer fasciculata, as is the case in the rat, where this zone is made up of spongiocytic cells quite unlike the compact cells found in the adrenal of the hamster. The significance of the different location of the sexual dimorphism is so far unexplained (Zieger et al., 1974). This sexual difference, in addition to being reversed, is larger than that reported for other mammals.

The histological architecture of the hamster adrenal gland is similar to that of most other eutherian species. It has relatively tough connective tissue capsular fibers which pass between the clusters of precamul cells in the outermost zones. There are three main cortical zones which can be recognized: the zona glomerulosa, zona fasciculata, and zona reticularis. These zones are not easily distinguishable until 30 days following birth.

The zona glomerulosa is the narrow shell of basophilic cells arranged in clusters or cords. This layer is usually composed of only a few flattened cells. Between the zona glomerulosa and the zona fasciculata is a narrow band of small flattened cells occasionally referred to as the transitional layer. These cells possess characteristics common to both bordering zones.

The zona fasciculata forms the greatest portion of the adrenal cortex and consists of long parallel columns of cells with steroidal nuclei resting in an eosinophilic cytoplasm. These columns of cells are supported by a fine reticular network and are separated by narrow sinusoids. The area immediately adjacent to the adrenal cortex is the zona reticularis. This is a poorly demarcated area in which the cells rest on a reticular network which is more dense than elsewhere in the gland. It is in this layer that the cells of the male are much larger, with vesicular nuclei, and there is a more intimate association of the reticular and medullary cells.

Several functions have been suggested for the zona reticularis. Due to the increased numbers of degenerating and dead cells found in this zone, it has been postulated that this zone is a zone of increased cell degeneration and cell death. Subsequently, this interpretation has been questioned because dead and degenerating cells are found in other zones as well. Stereological studies show that this zone has a reactivity to adrenocorticotropin (ACTH) similar to that of the zona fasciculata. The zona reticularis has also been postulated to be a source of adrenal sex steroids, and in the gerbil there appears to be an interrelationship between the adrenal cortex and the ovary, since ovariectomy stimulated the adrenal gland and there is a marked increase in plasma cortisol. Lipid droplets found in the zona reticularis may provide cholesterol esters, which are precursors for steroid genesis in the adrenal gland (Nickerson, 1979). Many investigators have observed the immune response of the Syrian hamster as different from other rodents in many respects (Sherman et al., 1963; Adner et al., 1965). A quarter of a century ago it was reported that a seasonal enlargement of the adrenals in male hamsters was accompanied by enhanced susceptibility to experimental poliomyelitis. The susceptibility increased with cortisone treatment and decreased with testicular hypertrophy or testosterone injection (Teodoru and Schwartzman, 1954). More recently, sexual dimorphism in adrenal weight and immune response was observed in hamsters. A suppressive effect of androgen on antibody-mediated responsiveness has been observed (Blazkovec and Orsini, 1976). The respective secretion modes of corticosteroids in male and female hamsters were studied. Correlation of these findings to the zone thickness of the adrenal cortex remains to be elucidated (Ohtaki, 1979).

Gerbils and hamsters are related taxonomically, belonging to the same family. The gerbil and hamster have a remarkably efficient mechanism for water conservation, presumably as an adaptation to an arid environment in their natural habitat. Prolactin is important in some animals for regulation of water balance, although this has not been shown for gerbils and hamsters. However, the adrenal gland is essential for survival in gerbils in as much as adrenalectomized gerbils do not survive despite replacement therapy, whereas the hamster has been known to survive following bilateral adrenalectomy (Coupland and Selby, 1976).

Histologically, the hamster adrenal medulla is similar to that of most other mammals. Two types of precursor cells are found in the medulla of the newborn hamster: a few undifferentiated sympathetic cells and many undifferentiated pheochromocytes. In the adult hamster, single sympathetic ganglion cells are interspersed among the medullary cells. Light microscopy reveals two types of pheochromocytes, which some investigators believe are actually different functional states of the same cell (I. Ito, 1954; T. Ito, 1957).

Catecholamine content and location in the hamster adrenal medulla is similar to that of other species; that is, peripherally located norepinephrine-containing cells and more centrally located epinephrine-containing cells (Camanni and Molinatti, 1958; Eranko, 1955). The mode of medullary hormonal release in the hamster is an "indirect secretion." Adrenal medullary hormones first enter intercellular spaces and then blood vessels, rather than direct release into the bloodstream like other endocrine secretions (Graumann, 1956; Ito, 1957).

Although a special dynamic control system of the adrenal medullary arteries or their branches has neither been demonstrated nor proved to be absent at the present time, it is pertinent to consider what factors might be responsible for controlling epinephrine production. At the present time the most likely candidate for this would seem to be innervational. For example, it is known that a variety of enzymes essential to catecholamine synthesis are at least partly under neural control (Coupland and Selby, 1976).

B. Islets of Langerhans

Gross anatomy of the pancreas is described in the digestive system section in this chapter (Section III, J). The islets of Langerhans are round to oval, compact masses of cells separated from the exocrine acini by a thin layer of reticulum (Jewell and Charipper, 1951). The greatest concentration of islets is found in the tail or caudal portion of the pancreas (Jewell and Charipper, 1951). Two cell types (A and B) are found in a 1:4 ratio (Muller, 1959). This ratio is similar to that described in humans and in other rodents (Muller, 1959). Two subtypes of the A cell have been described: A_1 (silver-positive) and A_2 (silver-negative). A_2 cells are thought to be the source of glucagon (Alm and Hellman, 1964) and B cells the source of insulin. The function of A_1 cells is yet to be determined.

A or alpha (α) cells are large with basally located spherical nuclei and fine eosinophilic granules throughout the cytoplasm when stained with Mallory–Herdenbain Azan stain (Jewell and Charipper, 1951). A cells are always located near small blood vessels. B or beta (β) cells have large cytoplasmic granules and randomly distributed nuclei of variable shapes (Muller, 1959).

The vasculature of individual islets is similar in most mammals, including humans and hamsters. A single afferent vessel enters an islet and branches extensively, which results in 12–20 efferent vessels (Muller, 1959).

C. Parathyroid Gland

The hamster has two parathyroid glands, 0.7–1.0 mm long and 0.3–0.5 mm wide, embedded in the lateral margin of the thyroid gland at the level of the caudal end of the cricoid cartilage (Michel, 1957). Accessory parathyroid glands are rare. Histologically, only one cell type has been described in the hamster parathyroid gland (Weymouth and Baker, 1954). The cells are irregularly shaped with relatively large nuclei, cytoplasmic lipid inclusions, and cytoplasmic argyrophilic granules. Morphologically, the hamster parathyroid parenchymal cell is similar to the principal cell of the human parathyroid gland. The parathyroid has a thin connective tissue capsule, sparse connective tissue stroma, and an extensive vascular system (Michel, 1957).

D. Thyroid Gland

The bilobed hamster thyroid gland is located on the ventrolateral surface of the first two tracheal cartilages. There is a thin, ventrally located isthmus (Michel, 1957). The thyroid gland has a connective tissue capsule as well as an interlobar and perifollicular connective tissue stroma (Sato, 1959). Shortly after weaning, adipose cells develop interfollicularly (Spagnoli and Charipper, 1955). After day 3, single and small clusters of chromophobic parafollicular cells may be seen. A rich anastomotic plexus of capillaries is associated with each follicle, although no true intraepithelial capillaries have been described in the hamster, as has been described in some other mammals (Sato, 1959).

The follicles of the hamster thyroid are uniformly small for the first 2 months, after which there is a gradual size increase. In general, larger follicles are found in the periphery with smaller follicles congregating centrally (Sato, 1959). The cuboidal to tall columnar follicular epithelial cells have basally located nuclei, randomly distributed cytoplasmic mitochondria, and a supranuclear located Golgi body (Knigge, 1957). During the first week of life, abundant mitotic figures are present in follicular cells, after which the number gradually decreases (Sato, 1959). Functional activity cannot be reliably determined by histological evaluation in the hamster (Reiter and Hoffman, 1968).

The sympathetic thyroid innervation is, in general, composed of thyroidal sympathetic, adrenergic nerve terminals which form a network around vessels as well as single terminals between, and sometimes around, follicles. The interfollicular terminals are especially numerous in adult hamsters; they are also numerous in sheep and mice, but few in adult rats and dogs, and even fewer in pigs. The functional importance of this interspecies variation in the number of thyroidal interfollicular sympathetic nerve terminals is presently unclear (Melander et al., 1975).

With advancing age, there is a high prevalence of perifollicular amyloidosis (64%) in the hamster thyroid gland (Spagnoli and Charipper, 1955; McMartin, 1979). Plasma levels of thyroxin (T_4), triiodothyronine (T_3), and growth hormone, as well as the free-T_4 and free-T_3 index are significantly depressed in 20-month-old hamsters compared to values of 2-month-old animals kept under the same photoperiodic conditions. Also, plasma thyroid-stimulating hormone (TSH) cholesterol is elevated in older hamsters compared with young hamsters (Vaughan et al., 1982). Aging (Pittman, 1962; Frolkis et al., 1973) and hypothyroidism (Cornwell et al., 1961; Walton et al., 1965) have been associated with decreased thyroid activity and increased plasma cholesterol in both rats and humans.

Hamsters exposed to prolonged cold exhibit an increase in thyroidal activity (Knigge, 1960; Tashima, 1963). However, in the hamster, cold exposure does not result in pronounced morphological changes as it does in homeotherms (Deane and Lyman, 1953, 1954). The reason for this variation is presently unclear.

E. Pituitary Gland

The female hamster's pituitary gland weighs 3–5 mg and is 25% heavier than that of the male. Percentage weights of the pars distalis, pars intermedia, and neurohypophysis of the female have been reported to be 77, 8, and 15% respectively, while those of the male are 71, 9, and 20% (Knigge, 1954a; Legait, 1962).

General anatomy and cytology of the hamster pituitary gland is similar to that of other mammals. The pituitary gland is somewhat flattened dorsoventrally and is broadest at the rostral end (Hanke and Charipper, 1948). The pars distalis is separated from a prominent pars intermedia by a persistent cleft, and the neurohypophysis surrounds the pars nervosa laterally and ventrally (Hanke and Charipipper, 1948). This positional arrangement of the lobes of the pituitary gland is similar to that of the rat (Hanke and Charipper, 1948), except that the caudal extension of the adenohypophysis is longer in the hamster. The neurohypophysis is similar to that of the guinea pig (Vanderburgh, 1917) and differs from animals such as the cat, in that the infundibular cavity does not extend into the tissue of the neurohypophysis (De Beer, 1926).

The pars distalis, the largest portion of the gland, contains three distinct cell types: basophils, acidophils, and chromophobes. In the mature male hamster, more than 50% of the cells are chromophobes and there are slightly more basophilic cells than acidophilic cells. This differs from the mature female with greater than 50% chromophobes and slightly more acidophilic cells than basophilic cells (Hanke and Charipper, 1948; Knigge, 1954b). The numerous processes of the pars distalis, which extend laterally and anteriorly, make total hypophysectomy difficult in the hamster (Knigge, 1954a).

The pars intermedia is relatively avascular and the connective tissue is less dense compared to that of the pars distalis. Cells of the pars intermedia may interdigitate with neurohypophyseal tissue. The neurohypophysis consists of highly vascular neural tissue which contains glial cells, unmyelinated nerve fibers of the hypothalamo–hypophyseal tract, and pituicytes (Hanke and Charipper, 1948).

F. Pineal Gland

The pineal gland of the hamster is located superficially between the parietal and occipital brain lobes and is approximately 1 mm in diameter. It is encapsulated by a richly vascular layer of pia mater, and it has a supportive connective tissue stroma. Traditionally, two main cell types have been described: pinealocytes and glial cells. Pinealocytes are large cells with large nuclei and prominent nucleoli. Glial cells have oval to triangular-shaped nuclei and a more basophilic cytoplasm, when compared with pinealocytes. Additional cell types in the pineal stroma consist of occasional fibroblasts and ependymal cells (DasGupta, 1968). More recently, it has been proposed that the pineal gland parenchyma may actually be divided into a medulla and cortex (Heidbuchel and Vollrath, 1983). This morphological demarcation is most evident in the hamster. In the hamster, cortical-region pinealocytes are larger and contain more cytoplasm than medullary pinealocytes. Also, there is a 30- to 80-μm zone consisting mainly of interstitial cells, which demarcates the cortex from the medulla (Sheridan and Reiter, 1970; Hewing, 1976). The existence of a demarcated pineal cortex and medulla is not as evident in the rat. In the rat, nuclear size appears to be the distinguishing feature between cortical-type and medullary-type areas. Peripherally located cells have larger nuclei than centrally located cells. Functional significance of these morphological variations is presently undetermined (Heidbuchel and Vollrath, 1983).

A photic information pathway may pass from the retina to the pineal gland via the thoracic cord (Moore, 1978). Retinohypothalamic fibers project to the optic chiasm and into the contralateral hypothalamic suprachiasmatic nuclei (SCN). Efferent fibers of the SCN project to the hypothalamic ventral tuberal area (Swanson and Cowan, 1975), which has reciprocal connections with neurons in the lateral hypothalamic area. Neuronal axons from the lateral hypothalamic area project directly to the intermediolateral cell column of the upper thoracic spinal cord (Saper et al., 1976). From the upper thoracic cord, preganglionic sympathetic fibers pass up the sympathetic trunk and enter the superior cervical ganglion (SCG). The neuronal axons of postganglionic neurons in the SCG terminate in the pineal gland (Kappers, 1960, 1965). These nerve processes end in the perivascular spaces in the vicinity of pinealocyte terminals within the pineal gland (Kappers, 1976). The most abundant neurotransmitter is norepinephrine (NE) (Pellegrino de Iraldi and Zieher, 1966), which is released primarily during the dark phase of the light–dark cycle (Zatz, 1981). NE release in the rat has been shown to exhibit a circadian rhythm (Wurtman and Axelrod, 1966), but this has not been demonstrated in the hamster (Morgan et al., 1976). NE interacts with β-adrenergic receptors on pinealocytes and initiates a series of events, which results in production and release of melatonin and possibly other indole and polypeptide compounds (Reppert and Klein, 1980; Reiter, 1982).

The pineal gland has been implicated in the regulation of seasonal reproductive cycles in several photoperiodic mammalian species (Herbert, 1972; Goldman et al., 1979; Reiter, 1980). Exposure to short photoperiods of 12.5 hr or less daily illumination induces gonadal quiescence in the hamster. Removal of the pineal gland completely prevents this response to short days (Reiter, 1980). The reader is referred to an excellent review of the neuroendocrine effects of the pineal gland and of melatonin by Reiter (1982).

The ability of the pineal gland to produce melatonin is markedly depressed in advanced age in both the hamster and gerbil (Reiter et al., 1980). In the hamster, this change occurs between 14 and 18 months of age. Greenberg and Weiss (1978) have shown that the density of β-adrenergic receptors within the rat pineal gland decreases significantly with age. Assuming a similar depression in pineal levels of β-adrenergic receptors in hamsters, this change accounts for the depressed melatonin production seen with age. NE stimulation of the β-adrenergic receptors on pinealocytes is responsible for the initiation of melatonin production (Reiter et al., 1980).

VIII. NERVOUS SYSTEM

A. Central Nervous System

Only a few reports are available in the literature on the architecture and function of the central nervous system in the hamster. A fairly complete description in the development of the cranial nerves and the spinal cord is presented by Magalhaes (1968). Also, there are two stereotaxic atlases of the hamster brain. The first atlas was by Smith and Bodmer (1963), and the most current one is by Knigge and Joseph (1968). The latter study identifies an isocortex and an allocortex, and suggests that the isocortex is better developed in hamsters than it is in rabbits.

Figure 15 is a sagittal view of the hamster brain. It has an

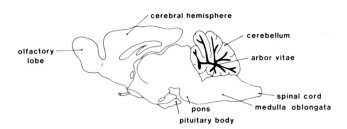

Fig. 15. Sagittal view of the hamster brain.

average total weight of 1.4 gm, of which the hypothalamus is approximately 0.42%. The cerebrum measures 12.2±0.5 mm long by 13.3±0.4 mm wide and 9.5±0.4 mm high. The hypothalamus is 5.2±0.3 mm long by 4.0±0.2 mm wide. The olfactory bulb is 5.8±0.6 mm long. The cerebellum with the paraflocculus is 13.0±0.8 mm long and without the paraflocculus is 10.5±0.6 mm long (Pilleri, 1960).

Dorsally, the cerebral hemispheres consist of large frontal, lateral, and occipital lobes. The sulcus paramediamus is low and flat, while the uvula is lengthened considerably caudally. The paraflocculus and flocculus have only one layer. The paleocortex is very broad, being wider caudally than rostrally. The pons and medulla oblongata are longer than wide. On a median sagittal section the corpus callosum appears short and narrow. The medial region of the cerebellum, near the sigmodon is distinctly differentiated into 10 layers (Pilleri, 1960).

B. Peripheral Nervous System

The hamster has 8 cervical, 13 thoracic, 6 lumbar, 3 sacral, and 2–3 caudal pairs of nerves (Michel, 1963). There are nerve plexuses in the cervical, brachial, and lumbosacral areas. Rhoades (1981) describes the cortical and spinal somatosensory input to the superior colliculus. Pyramidal cells from contralateral lamina IV and from lamina V send axons to the colliculus. Numerous polysynaptic pathways are reported by which the somatosensory information arrives at the colliculus. A limited number of references are recorded on the innervation of organs, skeletal muscle, teeth, and tactile hair (Magalhaes, 1968).

IX. SPECIAL SENSORY SYSTEMS

A. Vomeronasal Organ

This organ was first described by Jacobson in 1811 (as cited in Schwarge and Michel, 1960), and it is still referred to as Jacobson's organ by some scholars. The organ consists of a 7-mm-long tubular structure which was embryologically derived from the olfactory placode. It is located bilaterally in the rostroventral nasal septum and is completely enclosed in a bony capsule. The cranial end communicates with the nasal cavity through a rostral orifice, whereas the caudal portion is closed. The lateral wall is lined with nonsensory respiratory epithelium, devoid of goblet cells, and the medial wall of the lumen is lined with sensory olfactory epithelium. The free surface of these sensory cells is covered with structures resembling cilia and microvilli. The vomeronasal nerve consists of nonmyelinated nerve fibers and lies just beneath the sensory epithelium. Mucous and serous acini are distributed on the dorsolateral and ventromedial sides of the lumen. These glands differ from the alcian blue-positive olfactory glands of Bowman in that they are PAS-positive (Taniguchi and Mochizuki, 1982; Taniguchi et al., 1982).

Although the true functions of this organ are not fully understood, there have been several partially substantiated suggestions. Some reports suggest that it may have an olfactory function, whereas others suggest it is a receptor for some type of pheromone. The male hamster has vomeronasal chemoreceptors located within the organ which are important in sexual behavior. An autonomically controlled vascular pump is used to transport stimulants to the receptors. Any interruption of the efferent nerves controlling the pump results in sexual behavioral deficits similar to those produced by interruption of the afferent nerves carrying information from the vomeronasal organ to the brain (Meredith et al., 1980). Severance of the afferent vomeronasal nerves results in deficits in mounting and ejaculation (Powers and Winans, 1975; Winans and Powers, 1977).

B. Vision

The golden hamster is a predominantly nocturnal rodent with small eyes. Vision is probably less important to it than hearing, somatic sensation, and olfaction. However, the hamster has special characteristics that justify a full description of its visual pathway. First, it is functionally different from passive animals like the rabbit, in that it has a panoramic visual field and long streak-shaped specialization areas in the retina. Second, it differs from the more active mammals such as the cat and primates, which have an enormous binocular visual field and an obvious area centralis or fovea. The hamster has a rather uniform retinal organization, with a crude area centralis whose center does not occupy corresponding points in the two eyes (Tiao and Blakemore, 1976a; Greiner and Weidman, 1978). Third, the hamster, perhaps more than any other animal studied, seems to divide the visual function between two main central structures, the visual cortex and the superior colliculus. This hypothesis of two visual systems within a single brain provides mechanisms for identifying novel stimuli, commanding the orienting response in the colliculus and for pattern recognition in the visual cortex (Tiao, 1976). The response properties of hamster visual cortex cells are less differentiated than those in the tectum when compared to either cat or monkey. Furthermore, there is relatively little physiological evidence to support the theory that the visual cortex of the hamster is more involved with the identification of visual stimuli than the colliculus (Chalupa, 1981).

It is also interesting to note that each divergent area centralis points into the middle of the hamster's view of its own vibrissae. Since the latter are surely the most important channel of somatic sensory input and there is, in the superior colliculus, good correspondence between visual and somatic maps (Tiao and Blakemore, 1976c), perhaps it is not surprising to find a coincidence between the presumed region of maximum resolution in the retina and the images of the vibrissae (Tiao, 1976).

The hamster also has special significance for the developmental biologist. Because of the short gestation period, the young are born extremely immature with a poorly developed visual system. Thus, the newborn hamster is an excellent model for experimental manipulation of the developing visual pathway (Tiao and Blakemore, 1976b).

ACKNOWLEDGMENTS

The authors wish to thank Mike Broussard and Harry Cowgill of the Instructional Resources Center, School of Veterinary Medicine, The Louisiana State University (Baton Rouge), for their assistance in the preparation of the figures and photographs for this manuscript. We also owe a great deal of thanks to Jan Townsley, Beth Wilson, and Diane Gaeke for their patience and clerical support.

REFERENCES

Adams, D. R., and McFarland, L. Z. (1972). Morphology of the nasal fossae and associated structures of the hamster (*Mesocricetus auratus*). *J. Morphol.* **137**(2), 161–179.

Adner, M. M., Sherman, J. D., and Dameshek, W. (1965). The normal development of the lymphoid mass in the golden hamster and its relationship to the effect of thymectomy. *Blood* **25**, 511–521.

Afzelius, B. A. (1970). Brown adipose tissue. Its gross anatomy, histology and cytology. *In* "Brown Adipose Tissue" (O. Lindberg, ed.), pp. 1–31. Am. Elsevier, New York.

Albright, J. T., and Listgarten, M. A. (1962). Observations on the fine structure of the hamster cheek pouch epithelium. *Arch. Oral Biol.* **7**, 613–620.

Alm, A., and Hellman, B. (1964). Distribution of two types of A cells in the pancreatic islets of some mammalian species. *Acta Endocrinol. (Copenhagen)* **46**, 307–316.

Alm, P., Bloom, G. D., and Carlsoo, B. (1973). Adrenergic and cholinergic nerves of bovine, guinea pig and hamster salivary glands. *Z. Zellforsch. Mikrosk. Anat.* **138**, 407–420.

Anderson, R. R., and Sinha, K. N. (1972). Number of mammary glands and litter size in the golden hamster. *J. Mammol.* **53**, 382–384.

Arnold, F. A. (1942). The production of carious lesions in the molar teeth of hamster (*Cricetus auratus*). *Public Health Rep.* **57**, 1599–1604.

2. MORPHOPHYSIOLOGY

Austin, C. R. (1955). Ovulation, fertilization and early cleavage in the hamster (*Mesocricetus auratus*). *J. R. Microsc. Soc.* **73**, 141–154.

Belisle, C., and Sainte-Marie, G. (1981a). Tridimensional study of the deep cortex of the rat lymph node. III. Morphology of the deep cortex units. *Anat. Rec.* **199**, 213–226.

Belisle, C., and Sainte-Marie, G. (1981b). Topography of the deep cortex of the lymph nodes of various mammalian species. *Anat. Rec.* **201**, 553–561.

Billingham, E. R., Ferrigan, L. W., and Silvers, W. K. (1960). Cheek pouch of the Syrian hamster and tissue transplantation immunity. *Science* **132**, 1488.

Blazkovec, A. A., and Orsini, M. W. (1976). Ontogenic aspects of sexual dimorphism and the primary immune response to sheep erythrocytes in hamsters from pre-puberty through senescence. *Int. Arch. Allergy Appl. Immunol.* **50**, 55–67.

Bloomn, G. D., and Carlosöö, B. (1973). Adrenergic and cholinergic nerves of the bovine guinea pig and hamster salivary glands. *Z. Zellforsch. Mikrosk. Anat.* **138**, 407–420.

Boerner-Patzelt, D. (1956). Die Mundspeicheldrusen des Goldhamsters. *Anat. Anz.* **102**, 317–332.

Borer, K. T., Rowland, N., Mirow, A., Borer, R. C., Jr., and Kelch, R. P. (1979). Physiological and behavioral responses to starvation in the golden hamster. *Am. J. Physiol.* **236**, E105–E112.

Boyer, C. C. (1968). Embryology. *In* "The Golden Hamster:—Its Biology and Use in Medical Research" (R. A. Hoffman, P. F. Robinson, and H. Magalhaes, eds.), pp. 73–89. Iowa State Univ. Press, Ames.

Bruck, K. (1970). Nonshivering thermogenesis and brown adipose tissue in relation to age and their integration in the thermoregulatory system. *In* "Brown Adipose Tissue" (O. Lindberg, ed.), pp. 117–154. Am. Elsevier, New York.

Burden, H. W. (1972). Ultrastructural observations on ovarian perifollicular smooth muscle in the cat, guinea pig, and rabbit. *Am. J. Anat.* **133**, 125–142.

Callery, R. (1951). Development of the os genitale in the golden hamster, *Mesocricetus* (*Cricetus*) *auratus*. *J. Mammol.* **32**, 204–207.

Camanni, F., and Molinatti, G. (1958). Selective depletion of noradrenaline in the adrenals of the hamster produced by reserpine. *Acta Endocrinol. (Copenhagen)* **29**, 369–374.

Campean, N., Campean, C., and Rosenbauer, K. A. (1980). Experimentelle untersuchungen zur Frage der vaginalen pfropfentstehung bei Laboratoriumstieren. *Z. Versuchstierkd.* **22**, 50–62.

Caufield, J. B. (1966). Electron microscopic observations on the dystrophic hamster muscle. *Ann. N.Y. Acad. Sci.* **138**, 151–199.

Chaisson, R. B. (1954). The phylogenetic significance of rodent cheek pouches. *J. Mammol.* **35**, 425–427.

Chalupa, L. M. (1981). Some observations on the functional organization of the golden hamster's visual system. *Behav. Brain Res.* **3**, 189–200.

Chokroverty, S., Bernshohn, V., Reyes, M. G., and Chokroverty, M. (1977). Effects of adrenocortico-tropic hormone on muscle acetylcholinesterase and nonspecific esterases. *Acta Neurol. Scand.* **33**, 226–230.

Chokroverty, S., Reyes, M. G., Chokroverty, M., and Kaplan, R. (1978). Effect of prednisone on motor endplate fine structure. A morphometric study in hamsters. *Ann. Neurol.* **3**, 358–365.

Christensen, A. K. (1965). The fine structure of testicular interstitial cells in guinea pigs. *J. Cell Biol.* **26**, 911–935.

Christensen, A. K., and Fawcett, D. W. (1966). The fine structure of testicular interstitial cells in mice. *Am. J. Anat.* **118**, 551–571.

Christensen, A. K., and Gillim, S. W. (1969). The correlation of fine structure and function in steroid-secreting cells, with emphasis on those of the gonads. *In* "The Gonads" (K. W. McKerns, ed.), pp. 415–488. Appleton-Century-Crofts, New York.

Cole, P. (1954). Recordings of respiratory air temperature. *J. Laryngol. Otol.* **68**, 295–307.

Cornwell, D. G., Druger, F. A., Hamwi, G. J., and Brown, J. B. (1961). Studies on the characterization of human serum lipoproteins separated by ultracentrifugation in a density gradient. *Am. J. Clin. Nutr.* **9**, 24–40.

Costanzo, R. M., and O'Connell, R. J. (1980). Receptive fields of second-order neurons in the olfactory bulb of the hamster. *J. Gen. Physiol.* **76**, 53–68.

Coupland, R. E., and Selby, J. E. (1976). The blood supply of the mammalian adrenal medulla: A comparative study. *Acta Anat.* **122**, 539–551.

Cummins, J. M. (1976). Effects of epididymal occlusion on sperm maturation in the hamster. *J. Exp. Zool.* **197**, 187–190.

Dale, P. P., Layansky, T. P., and Keyes, P. H. (1941). Production of dental caries in Syrian hamsters. *J. Dent. Res.* **23**, 445–451.

DasGupta, T. K. (1968). The anatomy of the pineal organ in the Syrian hamster. *Acta Anat.* **71**, 28–35.

Dawes, J. D. (1952). The course of the nasal airstreams. *J. Laryngol. Otol.* **66**, 583–593.

Dawson, A. B. (1945). Argentaffin cells of the gastric mucosa of the rabbit, guinea pig, mouse and hamster. *Anat. Rec.* **91**, 53–63.

Deanne, H. W., and Lyman, C. P. (1953). Comparison of effect of cold environment on adrenal and thyroid glands of a hibernator (hamster) and a nonhibernator (rat). *Anat. Rec.* **115**, 389–390.

Deane, H. W., and Lyman, C. P. (1954). Body temperature, thyroid and adrenal cortex of hamsters during cold exposure and hibernation, with comparisons to rats. *Endocrinology (Baltimore)* **55**, 300–315.

De Beer, G. R. (1926). "The Comparative Anatomy, Histology, and Development of the Pituitary Body." Oliver & Boyd, Edinburgh.

Del Campo, C. H., and Ginther, O. J. (1972). Vascular anatomy of the uterus and ovaries and the unilateral luteolytic effect of the uterus: Guinea pigs, rats, hamsters, and rabbits. *Am. J. Vet. Res.* **33**, 2561–2578.

Devi, N. S., and Jacoby, F. (1966). The submaxillary gland of golden hamster and its post-natal development. *J. Anat.* **100**, 269.

Duling, B. R., and Weiner, D. E. (1972). Measurements of regional blood flow in the golden hamster. *Proc. Soc. Exp. Biol. Med.* **139**, 607–609.

Edwards, A. G. (1967). Modified bleeding technique for rats and hamsters. *J. Am. Vet. Med. Assoc.* **155**, 1217.

Ehle, F. R., and Warner, R. G. (1978). Nutritional implications of the hamster forestomach. *J. Nutr.* **108**, 1047–1053.

Elliot, J. A. (1976). Circadian rhythms and photoperiodic time measurement in mammals. *Fed. Proc., Fed. Am. Soc. Exp. Biol.* **35**, 2339–2346.

Eranko, O. (1955). Distribution of fluorescing islets, adrenaline and noradrenaline in the adrenal medulla of the hamster. *Acta Endocrinol. (Copenhagen)* **18**, 174–179.

Farbonann, A. I. (1965). Fine structure of the taste bud. *J. Ultrastruct. Res.* **12**, 328–350.

Fawcett, D. W., Neaves, W. B., and Flores, M. N. (1973). Comparative observations on intertubular lymphatics and the organization of the interstitial tissue of the mammalian testis. *Biol. Reprod.* **9**, 500–532.

Fernandez, B., Suarez, I., and Zapata, A. (1978). Ultrastructure of the filiform papillae on the tongue of the hamster. *J. Anat.* **126**, 487–494.

Flon, H., and Gerstner, R. (1968). Salivary glands of the hamster. I. Submandibular gland: A histochemical study after preservation with various fixatives. *Acta Histochem.* **31**, 234–253.

Frank, M. (1973). An analysis of hamster afferent taste nerve response functions. *J. Gen. Physiol.* **61**, 588–618.

Frolkis, V. V., Verzhikovskaya, N. V., and Valueva, G. V. (1973). The thyroid and age. *Exp. Gerontol.* **8**, 285–296.

Fulton, G. P., Jackson, R. G., and Lutz, B. R. (1946). Cinephotomicroscopy of normal blood circulation in the cheek pouch of the hamster, *Cricetus auratus*. *Anat. Rec.* **96**, 537 (abstr.).

Gaston, S., and Menaker, M. (1967). Photoperiodic control of hamster testis. *Science* **158**, 925–928.

Gaunt, W. A. (1961). The growth of the teeth and jaws of the golden hamster. *Acta Anat.* **47**, 301–327.

Gillette, R. (1957). Normal growth and differentiation of hamster pouch epithelium. *Anat. Rec.* **128**, 555–556.

Ginther, O. J. (1967). Local utero-ovarian relationships. *J. Anim. Sci.* **26**, 578–585.

Goldman, B., Hall, V., Hollister, C., Roychoudbury, P., Tamarkin, L., and Westrom, W. (1979). Effects of melatonin on the reproductive system in intact and pinealectomized male hamsters maintained under various photoperiods. *Endocrinology (Baltimore)* **104**, 82–88.

Graumann, W. (1956). Formation and secretion of periodic positive material in the adrenal medulla of golden hamsters. *Z. Anat. Entwicklungsgesch.* **119**, 415–430.

Greenberg, L. H., and Weiss, B. (1978). β-adrenergic receptors in aged rat brain. Reduced number and capacity of pineal gland to develop supersensitivity. *Science* **201**, 61–63.

Greiner, J. V., and Weidman, T. A. (1978). Development of the hamster retina: A morphologic study. *Am. J. Vet. Res.* **39**, 665–670.

Hanke, H. H., and Charipper, H. A. (1948). The anatomy and cytology of the pituitary gland of the golden hamster (*Cricetus auratus*). *Anat. Rec.* **102**, 123–139.

Heidbuchel, U., and Vollrath, L. (1983). Morphological findings related to the problem of cortex and medulla in the pineal glands of rat and hamster. *J. Anat.* **136**, 723–734.

Herbert, J. (1972). Initial observations on pinealectomized ferrets kept for long periods in either daylight or artificial illumination. *Endocrinology (Baltimore)* **55**, 591.

Hewing, M. (1976). Die postnatale Entwicklung der Epiphysis cerebri beim Gold-hamster. *Verh. Anat. Ges.* **70**, 85–91.

Hinrichsen, K. (1955). Funktionsentwicklung des Pardontium beim Goldhamstermolaren. *Acta Anat.* **23**, 161–180.

Hoffman, R. A., Robinson, P. F., and Magalhaes, H., eds. (1968). "The Golden Hamster:—Its Biology and Use in Medical Research." Iowa State Univ. Press, Ames.

Homburger, F. (1972). Disease models in Syrian hamsters. *In* "Pathology of the Syrian Hamster" (F. Homburger, ed.), pp. 69–86. Karger, Basel.

Hoover, W. H., Mannings, C. L., and Sheerin, H. W. (1969). Observations on digestion in the golden hamster. *J. Anim. Sci.* **28**, 349–352.

Hyman, A. M., and Frank, M. E. (1980a). Effects of binary taste stimuli on the neural activity of the hamster chorda tympani. *J. Gen. Physiol.* **76**, 125–142.

Hyman, A. M., and Frank, M. E. (1980b). Sensitivities of single nerve fibers in the hamster chorda tympani to mixtures of taste stimuli. *J. Gen. Physiol.* **76**, 143–173.

Ito, I. (1954). Neurosecretory phenomena at the ganglion cells in the adrenal medulla of the golden hamster. *Okajimas Folia Anat. Jpn.* **26**, 221–226.

Ito, T. (1957). Postnatal histogenesis of the adrenal medulla of the hamster with special reference to the functional structure. *Okajimas Folia Anat. Jpn.* **30**, 239–256.

Jansky, L. (1973). Non-shivering thermogenesis and its thermoregulatory significance. *Biol. Rev. Cambridge Philos. Soc.* **48**, 85–132.

Jarausch, K. H., and Ullrich, K. J. (1957). Technique of withdrawal of urine samples from single collecting tubules in mammalian kidney by means of polyethylene capillaries. *Pfluegers Arch. Gesamte Physiol. Menschen Tiere* **264**, 88–94.

Jewell, H. A., and Charipper, H. A. (1951). The morphology of the pancreas of the golden hamster, *Cricetus auratus*, with special reference to the histology and cytology of the islets of Langerhans. *Anat. Rec.* **111**, 401–415.

Kaissling, B., de Rouffignac, C., Barrett, J. M., and Kriz, W. (1975). The structural organization of the kidney of the desert rodent. *Psammomys obesus*. *Anat. Embryol.* **148**, 121.

Kallfelz, F. A., and Wasserman, R. H. (1972). Correlation between ^{47}Ca absorption and intestinal calcium-binding activity in the golden hamster. *Proc. Soc. Exp. Biol. Med.* **139**, 77–79.

Kappers, J. A. (1960). The development, topographical relations and innervation of the epiphysis cerebri in the albino rat. *Z. Zellforsch. Mikrosk. Anat.* **52**, 163–215.

Kappers, J. A. (1965). Survey of the innervation of the epiphysis cerebri and the accessory pineal organs of vertebrates. *Prog. Brain Res.* **10**, 87–151.

Kappers, J. A. (1976). The mammalian pineal gland, a survey. *Acta Neurochir.* **34**, 109–149.

Kawashima, Y. (1972). The lymph system in rodents. *Jpn. J. Vet. Res.* **20**, 35–43.

Kendrick, J., Nettesheim, P., and Hammons, A. S. (1974). Tumor induction in tracheal grafts: A new experimental model for respiratory carcinogenesis studies. *J. Natl. Cancer Inst. (U.S.)* **52**, 1317–1325.

Kennedy, A. R., Desrosiers, A., Terzaghi, M., and Little, J. B. (1978). Morphometric and histological analysis of the lungs of Syria golden hamsters. *J. Anat.* **125**(3), 527–553.

Kent, G. C. (1968). Physiology of reproduction. *In* "The Golden Hamster—Its Biology and Use in Medical Research" (R. A. Hoffman, P. F. Robinson, and H. Magalhaes, eds.), pp. 119–138. Iowa State Univ. Press, Ames.

Kent, C. G. (1985). "Cardiovascular System of the Syrian Golden Hamster." Louisiana State Univ., Baton Rouge (unpublished).

Keyes, D. H., and Dale, P. P. (1944). A preliminary survey of the pouches and dentition of the Syrian hamster. *J. Dent. Res.* **23**, 427–438.

Kilburn, K. H. (1967). Cilia and mucus transport as determinants of the response of the lung to air pollutants. *Arch. Environ. Health* **14**, 77–91.

Klomburg, von S., and Wolff, D. (1978). The Syrian Golden Hamster—physiology, feeding, diseases. *Berl. Muench. Tieraertzl. Wochenschr.* **91**, 150–153.

Knigge, K. M. (1954a). The effect of hypophysectomy on the adrenal gland of the hamster (*Mesocricetus auratus*). *Am. J. Anat.* **94**, 225–272.

Knigge, K. M. (1954b). Response of the hamster's adrenal cortex to desoxycorticosterone acetate, with observations on the kidney, thyroid and hypophysis. *Endocrinology (Baltimore)* **55**, 731–744.

Knigge, K. M. (1957). Influence of cold exposure upon the endocrine glands of the hamster, with an apparent dichotomy between morphological and functional response of the thyroid. *Anat. Rec.* **127**, 75–95.

Knigge, K. M. (1960). Neuroendocrine mechanisms influencing ACTH and TSH secretion and their rode in cold acclimation. *Fed. Proc., Fed. Am. Soc. Exp. Biol.* **19**, 45–51.

Knigge, K. M., and Joseph, S. A. (1968). A stereotaxic atlas of the brain of the golden hamster. *In* "The Golden Hamster—Its Biology and Use in Medical Research (R. A. Hoffman, P. F. Robinson, and H. Magalhaes, eds.), p. 104. Iowa State Univ. Press, Ames.

Koo, K. W., Leith, D. E., Sherter, C. B., and Snider, G. L. (1976). Respiratory mechanics in normal hamsters. *J. Appl. Physiol.* **40**, 936–942.

Korohmaru, M., Takao, N., and Mochizuki, K. (1979). Morphological changes of developing duodenal villi in golden hamsters. *Jpn. J. Vet. Sci.* **41**, 151–165.

Kovalevskii, G. V. (1966). Some special morphological features of the heart and lungs of small laboratory animals. *Dokl. Akad. Nauk SSSR* **168**, 1214–1216.

Kriz, W. (1967). Der architektonische und funktionelle Aufbau der Rattenniere. *Z. Zellforsch. Mikrosk. Anat.* **82**, 495.

Kriz, W., and Koepsell, H. (1974). The structural organization of the mouse kidney. *Z. Anat. Entwicklungsgesch.* **144**, 137.

Kriz, W., Barrett, J. M., and Peter, S. (1976). The renal vasculature: Anatomical-functional aspects. *Int. Rev. Physiol.* **11**, 1–21.

Krueger, H., and Rieschel, R. (1950). Values of the hamster cecum. *Fed. Proc., Fed. Am. Soc. Exp. Biol.* **9**, 72–73 (abstr.).

Kuhn, C. (1972). Fine structure of bronchiolo-alveolar cell carcinoma. *Cancer (Philadelphia)* **30**, 1107–1118.

Kunstyr, I. (1974). Some quantitative and qualitative aspects of the stomach microflora of the conventional rat and hamster. *Zentralbl. Veterinaer med., Reihe A* **21**, 553–561.

Kunstyr, I., Kupper, W., Weisser, H., Haumann, S., and Messow, C. (1982).

Urethral plug—a new secondary male sex characteristic in rat and other rodents. *Lab. Anim.* **16,** 151–155.

Kus, J., and Sawicki, B. (1976). The bronchial tree, blood vessels, and lymphatic nodules in the pulmonary parenchyma of the golden hamster, *Mesocricetus auratus. Folia Biol. (Krakow)* **24**(3), 293–7.

Lacy, E., and Schmidt-Nielsen, B. (1979a). Anatomy of the renal pelvis in the hamster. *Am. J. Anat.* **154,** 291–320.

Lacy, E., and Schmidt-Nielsen, B. (1979b). Ultrastructural organization of the hamster renal pelvis. *Am. J. Anat.* **155,** 403–424.

Lai, Y. L. (1979). Lung volume and pleural pressure in the anesthetized hamster. *J. Appl. Physiol.* **46,** 927–931.

Leavitt, W. W., Basom, C. R., Bagwell, J. N., and Blaka G. C. (1973). Structure and function of the hamster corpus luteum during the estrous cycle. *Am. J. Anat.* **136,** 135–150.

Legait, E. (1962). Recherches morphologiques et histophysiologiques sur le pars intermedia des Ronguers. *Biol. Med. (Paris)* **51,** 190–204.

Lindenmann, R., and Strauli, P. (1968). Lymphatic vessels in the cheek pouch of the golden hamster. *Transplantation* **6,** 557–561.

Lisco, H., Kennedy, A. R., and Little, J. B. (1974). Histologic observations on the pathogenesis of lung cancer in hamsters following administration of polonium-210. *Exp. Lung Cancer: Carcinog. Bioassays, Int. Symp., 1974,* pp. 468–474.

Logan, A., and Weatherhead, B. (1978). Pelage color cycles and hair follicle tyrosinase activity in the Syrian hamster. *J. Invest. Dermatol.* **71,** 295–298.

Lovelace, F., Will, L., Sperling, G., and McCay, C. M. (1958). Teeth, bones and aging of Syrian hamsters. *J. Gerontol.* **13,** 27–31.

Lucky, A. W., Eisenfeld, A. J., and Visintin, A. A. (1985). Autoradiographic localization of tritiated dihydrotestosterone in the flank organ of the albino hamster. *J. Invest. Dermatol.* **84,** 122–125.

McMartin, D. N. (1979). Morphologic lesions in aging Syrian hamsters. *J. Gerontol.* **34,** 502–511.

McReynolds, H. D., Siraki, C. M., Bramson, P. H., and Pollock, R. J. (1973). Smooth muscle-like cells in ovaries of the hamster and gerbil. *Z. Zellforsch. Mikrosk. Anat.* **140,** 1–8.

Magalhaes, H. (1968). Gross anatomy. *In* "The Golden Hamster—Its Biology and Use in Medical Research" (R. A. Hoffman, P. F. Robinson, and H. Magalhaes, eds.), pp. 91–109. Iowa State Univ. Press, Ames.

Malen, A., Arens, H., and Waechter, A. (1973). Pulmonary respiration and acid-base state in hibernating marmots and hamsters. *Respir. Physiol.* **17**(1), 45–61.

Mangold, E. (1929). "Handbuch der Ernährung und des Stoffwechsels der landwirtshaftlichen Nutztiere," Vol. II, p. 321. Springer-Verlag, Berlin and New York.

Martin, D. Y. M. (1976). The mammalian rete testis—A morphological examination. *Anat. Rec.* **186,** 493–524.

Martin, G. G., Sack, M., and Talbot, P. (1981). The structure of bursae ovaricae surrounding the ovaries of the golden hamster. *Anat. Rec.* **201,** 485–498.

Melander, A., Sundler, F., and Westgren, U. (1975). Sympathetic innervation of the thyroid: Variation with species and with age. *Endocrinology (Baltimore)* **96,** 102–106.

Meredith, M., Marques, D. M., O'Connell, R. J., and Stern, F. L. (1980). Vomeronasal pump: Significance of male sexual behavior. *Science* **207,** 1224–1226.

Michel, G. (1957). Anatomy of the thyroid, parathyroids, and adrenals of Syrian golden hamsters (*Mesocricetus auratus* waterhouse). *Zentralbl. Veterinaermed.* **4,** 497–508.

Michel, G. (1963). Zur Ausbildung der Spinalnerven des syrischen Goldhamsters. *Z. Versuchstierkd.* **2,** 91–104.

Miller, D., and Crane, R. K. (1961a). A procedure for the isolation of the epithelial brush border membrane of hamster small intestine. *Anal. Biochem.* **2,** 284–286.

Miller, D., and Crane, R. K. (1961b). The digestive function of the epithelium of the small intestine. I. An intracellular locus of disaccharide and sugar phosphate ester hydrolysis. *Biochim. Biophys. Acta* **52,** 281–293.

Miller, D., and Crane, R. K. (1961c). The digestive function of the epithelium of the small intestine. II. Localization of disaccharide hydrolysis in the isolated brush border portion of intestinal epithelial cells. *Biochim. Biophys. Acta* **52,** 293–298.

Miller, R. L., and Chaudhry, A. P. (1976). Comparative ultrastructure of vallate, foliate and fungiform taste buds of golden Syrian hamster. *Acta Anat.* **95,** 75–92.

Moore, R. Y. (1978). Central neural control of circadian rhythms. *Front. Neuroendocrinol.* **5,** 185–206.

Morgan, W. W., Reiter, R. J., and Pfeil, K. A. (1976). Hamster pineal noradrenaline: Levels over a regulated lighting period and the influence of superior cervical ganglionectomy. *Life Sci.* **19,** 437–440.

Morin, L. P., and Zucker, I. (1978). Photoperiodic regulation of copulatory behavior in the male hamster. *J. Endocrinol.* **77,** 249–258.

Muller, D. (1959). Investigation of the morphology of the islands of Langerhans and of Feyrter's duct of the pancreas of the Syrian golden hamster. *Anat. Anz.* **106,** 369–385.

Murray, R. G., and Murray, A. (1967). Fine structure of taste buds of rabbit foliate papillae. *J. Ultrastruct. Res.* **19,** 327–353.

Musser, T. K., and Silverman, J. (1980). The hair cycle of the Syrian golden hamster (*Mesocricetus auratus*). *Lab. Anim. Sci.* **30,** 681–683.

Nakano, A. (1960). Histological studies of the prenatal and postnatal development of the ovary of the golden hamster (*Cricetus auratus*). *Okajimas Folia Anat. Jpn.* **35,** 183–217.

Nettleblad, S. C. (1954). Die Lobierung und innere Topographie der Saugerleber nebst Beitragen zur Kenntnis der Leberentwicklung beim Gold-Hamster (*Cricetus auratus*). [The lobes and internal topography of the mammalian liver including a contribution to the knowledge of liver development in the golden hamster (*Cricetus auratus*).] *Acta Anat.* **21,** Suppl. 20, 1–251.

Nickerson, P. A. (1979). Adrenal cortex in retired breeder mongolian gerbils (*Meriones unguiculatus*) and golden hamsters (*Mesocricetus auratus*). *Am. J. Pathol.* **95,** 347–358.

O'Connell, R. J., Singer, A. G., Macrides, F., Pfaffmann, C., and Agosta, W. C. (1978). Responses of the male golden hamster to mixtures of odorants identified from vaginal discharge. *Behav. Biol.* **24,** 244–255.

O'Connell, R. J., Singer, A. G., Pfaffmann C., and Agosta, W. C. (1979). Pheromones of hamster vaginal discharge. Attraction to femtogram amounts of dimethyl disulfide and to mixtures of volatile components. *J. Chem. Ecol.* **5,** 575–585.

Oeri, H. U. (1960). The structure of the oviduct of the golden hamster. *Z. Saugetierkd.* **25,** 52–65.

Ohtaki, S. (1979). Conspicuous sex differences in zona reticularis of the adrenal cortex of the Syrian hamster. *Lab. Anim. Sci.* **29,** 765–769.

Orgrowsky, D., Fawcett, J., Althoff, J., Wilson, R. B., and Pour, P. (1980). Structure of the pancreas in Syrian hamsters, scanning electron-microscopic observations. *Acta Anat.* **107,** 121–128.

Orsi, A. M., Silva, P. P. E., Dias, S. M., and Olivera, M. C. (1977). Considerations about the branching of the aorta abdominalis in hamster. *Anat. Anz.* **142,** 507–511.

Orsini, M. W. (1961). The external vaginal phenomena characterizing the stages of the estorus cycle, pregnancy, pseudopregnancy, lactation, and the anestrous hamster. *Mesocricetus auratus* Waterhouse. *Proc. Anim. Care Panel* **11,** 193–206.

Ortiz, E. (1948). The postnatal development of the reproductive system of the golden hamster (*Cricetus auratus*) and its reactivity to hormones. *Physiol. Zool.* **20,** 45–66.

Ortiz, E. (1955). The relation of advancing age to reactivity of the reproductive system in the female hamster. *Anat. Rec.* **122,** 517–537.

O'Shea, J. D. (1971). Smooth muscle-like cells in the theca externa of ovarian follicles in the sheep. *J. Reprod. Fertil.* **24,** 283–285.

Osvaldo-Decima, L. (1970). Smooth muscle in the ovary of the rat and monkey. *J. Ultrastruct. Res.* **29**, 218–237.

Pansky, B., Jacobs, M., House, E. L., and Tassoni, J. P. (1961). The orbital region as a source of blood samples in the golden hamster. *Anat. Rec.* **139**, 409–412.

Paukcr, R. S. (1948). The effects of removing seminal vesicles, prostate and testes on the mating behavior of the golden hamster (*Cricetus auratus*). *J. Comp. Physiol. Psychol.* **41**, 252–257.

Pellegrino de Iraldi, A., and Zieher, L. M. (1966). Noradrenaline and dopamine content of normal, decentralized, and denervated pineal glands of the rat. *Life Sci.* **5**, 149–154.

Pfeiffer, E. W. (1968). Comparative anatomical observations of the mammalian renal pelvis and medulla. *J. Anat.* **102**, 321–331.

Phalen, R. F., and Oldham, M. J. (1983). Tracheobronchial airway structure as revealed by casting techniques. *Am. Rev. Respir. Dis.* **128**, 51–4.

Phalen, R. F., Yeh, H. C., Schum, G. M., and Raabe, O. G. (1978). Application of an idealized model of morphometry of the mammalian tracheobronchial tree. *Anat. Rec.* **190**(2), 167–176.

Pilleri, G. (1960). Materialien zur vergleichenden Anatomie des Gehirns der Myomorpha. *Acta Anat.* **41**, Suppl., 69–88.

Pittman, J. A. (1962). The thyroid and ageing. *J. Am. Geriatr. Soc.* **10**, 10–12.

Plopper, C. G., Mariassy, A. T., and Hill, L. H. (1980a). Ultrastructure of the nonciliated bronchiolar epithelial Clara cell of mammalian lung. 1. A comparison of rabbit, guinea pig, rat, hamster and mouse. *Exp. Lung Res.* **1**(2), 139–154.

Plopper, C. G., Hill, L. H., and Mariassy, A. T. (1980b). Ultrastructure of the nonciliated bronchiolar epithelial Clara cell of mammalian lung. 3. A study of man with comparison of 15 mammalian species. *Exp. Lung Res.* **1**(2), 171–180.

Powers, J. B., and Winans, S. S. (1975). Vomeronasal organ: Critical role in mediating sexual behavior of the male hamster. *Science* **187**, 961.

Priddy, R. B., and Brodie, A. F. (1948). Facial musculature, nerves and blood vessels of the hamster in relation to the cheek pouch. *J. Morphol.* **83**, 149–180.

Rabi, T., and Cassuto, Y. U. (1976). Metabolic adaptations in brown adipose tissue of the hamster in extreme ambient temperatures. *Am. J. Physiol.* **231**, 153–160.

Reiter, A. J., and Hoffman, R. A. (1968). The endocrine system. *In* "The Golden Hamster—Its Biology and Use in Medical Research" (R. A. Hoffman, P. F. Robinson, and H. Magalhaes, eds.). Iowa State Univ. Press, Ames.

Reiter, R. J. (1973). Pineal control of a seasonal reproductive rhythm in male golden hamsters exposed to natural daylight and temperature. *Endocrinology (Baltimore)* **92**, 423–430.

Reiter, R. J. (1980). The pineal and its hormones in the control of reproduction in mammals. *Endocr. Rev.* **1**, 109.

Reiter, R. J. (1982). Neuroendocrine effects of the pineal gland and of melatonin. *Front. Neuroendocrinol.* **7**, 287–317.

Reiter, R. J., Johnson, L. Y., Stegr, R. W., Richardson, B. A., and Petterborg, L. J. (1980). Pineal biosynthetic activity and neuroendocrine physiology in the aging hamster and gerbil. *Peptides* (Fayetteville, N.Y.) **1**, Suppl., pp. 66–77.

Reppert, S. M., and Klein, B. C. (1980). Mammalian pineal gland: Basic and clinical aspects. *In* "The Endocrine Functions of the Brain" (M. Motta, ed.), pp. 327–371. Raven Press, New York.

Reyes, H., and Kern, F., Jr. (1979). Effects of pregnancy on bile flow and biliary lipids in the hamster. *Gastroenterology* **76**, 144–150.

Reyes, M. G., and Chokroverty, S. (1980). Distribution and size of myofibers in quadriceps muscle of Syrian hamster. *Acta Neuropathol.* **51**, 241–243.

Rhoades, R. W. (1981). Cortical and spinal somatosensory input to the superior colliculus in the golden hamster: An anatomical and electrophysiological study. *J. Comp. Neurol.* **195**, 415–432.

Rhodin, J. A. G. (1974). Histology: A Text and Atlas. Oxford Univ. Press, London and New York.

Rolle, G. K., and Charipper, H. A. (1949). The effects of advancing age upon the histology of the ovary, uterus and vagina of the golden hamster (*Cricetus auratus*). *Anat. Rec.* **105**, 281.

Rosenbauer, K. A., Campean, N., and Campean, C. (1980). Licht und rast erelektronmikroskopische Untersuchunge über die Morphologie des vaginal pfropfes bei Laboratoriumstieren. *Anat. Anz.* **47**, 301–316.

Sakata, T., and Tamate, H. (1976). Light and electron microscopic observation of the forestomach mucosa in the golden hamster. *Tohoku J. Agric. Res.* **27**, 26–39.

Salih, M. S., and Kent, G. C., Jr. (1964). The epaxial muscles of the golden hamster. *Anat. Rec.* **150**, 319–334.

Salih, M. S., and Kent, G. C., Jr. (1966). The vertebral column of the golden hamster. *Proc. La. Acad. Sci.* **24**, 161–174.

Saper, C. B., Loewy, A. D., Swanson, L. W., and Cowan, W. M. (1976). Direct hypothalamus-autonomic connections. *Brain Res.* **117**, 305–312.

Sato, T. (1959). The postnatal histogenesis of the thyroid gland of the golden hamster (*Cricetus auratus*). *Okajimas Folia Anat. Jpn.* **33**, 225–253.

Schmidt-Nielsen, L., Bretz, W. L., and Taylor, C. R. (1970). Panting in dogs: Unidirectional air flow over evaporated surfaces. *Science* **169**, 1102–1104.

Schwarze, E., and Michel, G. (1960). Visceral anatomy of the Syrian golden hamster. *Wiss. Z. Karl Marx Univ. Leipzig, Math. Naturwiss. Reihe* **9**, 95–126.

Shackelford, J. M. (1961). The structure and function of the salivary glands with particular emphasis on the hamster (*Cricetus auratus*). Dissertation, University of Alabama, Birmingham (*Diss. Abstr.* **22**, 1347–1348).

Shackleford, J. M., and Klapper, C. E. (1962a). A sexual dimorphism of hamster submaxillary mucin. *Anat. Rec.* **142**, 495–503.

Shackleford, J. M., and Klapper, C. E. (1962b). Casein degrading ability of hamster, rat and mouse salivary glands. *Arch. Oral Biol.* **7**, 337–342.

Sheridan, M. N., and Reiter, R. J. (1970). Observations on the pineal system in the hamster. I. Relation of the superficial and deep pineal to the epithalamus. *J. Morphol.* **131**, 153–161.

Sherman, J. D., Adner, M. M., and Dameshek, W. (1963). Effective thymectomy on the golden Hamster (*Mesocricetus auratus*) I. Wasting disease. *Blood* **22**, 252–271.

Smith, O. A., and Bodmer, C. N. (1963). A stereotaxic atlas of the brain of the golden hamster (*Cricetus auratus*). *J. Comp. Neurol.* **120**, 53–63.

Smith, R. E., and Horwitz, B. A. (1969). Brown fat and thermogenesis. *Physiol. Rev.* **49**, 330–425.

Soderwall, A. L., Van Brocklin, G., and Ramey, R. (1950). Senescent histological appearances in reproductive systems of aged male and female hamsters. *Anat. Rec.* **108**, 115–116, 603–604.

Soderwall, A. L., Kent, H. A., Turbyfill, C. L., and Britenbaker, A. L. (1960). Variation in gestation length and litter size of the golden hamster (*Mesocricetus auratus*). *J. Gerontol.* **15**, 246–248.

Spagnoli, H. H., and Charipper, H. A. (1955). The effects of aging on the histology and cytology of the pituitary gland of the golden hamster (*Cricetus auratus*), with brief reference to simultaneous changes in the thyroid and testis. *Anat. Rec.* **121**, 117–139.

Spicer, S. S., and Duvenci, J. (1964). Histochemical characteristics of mucopolysaccharides in exorbital lacrimal glands. *Anat. Rec.* **149**, 333.

Strauss, F. (1956). The time and place of fertilization of the golden hamster egg. *J. Embryol. Exp. Morphol.* **4**, 42–56.

Swanson, L. J., Desjardins, C., and Turek, F. W. (1982). Aging of the reproductive system in the male hamster: Behavioral and endocrine patterns. *Biol. Reprod.* **26**, 791–799.

Swanson, L. W., and Cowan, W. M. (1975). The efferent connections of the suprachiasmatic nucleus of the hypothalamus. *J. Comp. Neurol.* **16**, 1–12.

Takahashi, M., Pour, P., Althoff, J., and Donnelly, T. (1977). The pancreas of the Syrian hamster (*Mesocricetus auratus*). I. Anatomical study. *Lab. Anim. Sci.* **27,** 336–343.

Taniguchi, K., and Mochizuki, K. (1982). Morphological studies on the vomeronasal organ in the golden hamster. *Jpn. J. Vet. Sci.* **44,** 419–426.

Taniguchi, K., Taniguchi, K., and Mochizuki, K. (1982). Developmental studies on the vomeronasal organ of the golden hamster. *Jpn. J. Vet. Sci.* **44,** 709–716.

Tashima, L. S. (1963). Effect of cold exposure on thyroid gland activity of *Mesocricetus auratus* and *Cetellus tridecemlineatus*. Ph.D. Thesis, Harvard University, Cambridge, Massachusetts.

Teodoru, C. V., and Schwartzman, G. (1954). Relation of certain endocrine disturbances to susceptibility of golden Syrian hamsters to experimental poliomyelitis. *J. Exp. Med.* **100,** 563–574.

Tiao, Y. C. (1976). Functional organization in the visual cortex of the golden hamster. *J. Comp. Neurol.* **168,** 459–481.

Tiao, Y. C., and Blakemore, C. (1976a). Regional specialization in the golden hamster's retina. *J. Comp. Neurol.* **168,** 439–458.

Tiao, Y. C., and Blakemore, C. (1976b). Functional organization in the visual cortex of the golden hamster. *J. Comp. Neurol.* **168,** 459–482.

Tiao, Y. C., and Blakemore, C. (1976c). Functional organization in the superior colliculus of the golden hamster. *J. Comp. Neurol.* **168,** 483–504.

Timm, K. I. (1979). Orbital venous anatomy of the rat. *Lab. Anim. Sci.* **29,** 636–638.

Uga, S. (1969). A study on the cytoarchitecture of taste buds of rat circumvallate papillae. *Arch. Histol. Jpn.* **31,** 59–72.

Vanderburgh, C. M. (1917). The hypophysis of the guinea pig. *Anat. Rec.* **12,** 95–112.

Van Hoosier, G. L., and Ladiges, W. C. (1984). Biology and diseases of hamsters. *In* "Laboratory Animal Medicine" (J. G. Fox, B. J. Cohen, and F. M. Loew, eds.), pp. 121–147. Academic Press, New York.

Vaughan, M. K., Richardson, B. A., Croft, C. M., Powanda, M. C., and Reiter, R. J. (1982). Interaction of ageing, photoperiod and melatonin on plasma thyroid hormones and cholesterol levels in female Syrian hamsters (*Mesocricetus auratus*). *Gerontology* **28,** 345–353.

Walton, K. W., Campbell, D. A., and Tonks, E. L. (1965). The significance of alterations in serum lipids in thyroid dysfunction. *J. Clin. Sci.* **29,** 199–216.

Warner, N. E. (1971). The adrenals. *In* "Basic Endocrine Pathology," pp. 71–96. Year Book Med. Publ., Chicago, Illinois.

Weinerth, J. L., Battaglia, C. R., and Magalhaes, H. (1961). The nonessential nature of seminal vesicles in golden hamsters. *Am. Zool.* **1,** 397.

West, W. T. (1958). Histologic study of living striated muscle fibers *in situ* in the cheek pouch of the golden hamster. *Am. J. Anat.* **103,** 349–373.

Weymouth, R. J., and Baker, B. L. (1954). The presence of argyrophilic granules in the parenchymal cells of the parathyroid glands. *Anat. Rec.* **119,** 519–527.

Winans, S. S., and Powers, J. B. (1977). Olfactory and vomeronasal deafferentation of male hamsters: Histological and behavioral analyses. *Brain Res.* **126,** 325–344.

Wineski, L. E. (1985). Facial morphology and vibrissal movement in the golden hamster. *J. Morphol.* **183,** 199–217.

Wing, T., and Lin, H. (1977). The fine structure of testicular interstitial cells in the adult golden hamster with special reference to seasonal changes. *Cell Tissue Res.* **157,** 385–393.

Wurtman, R. J., and Axelrod, J. (1966). A 24 hour rhythm in the content of norepinephrine in the pineal and salivary glands of the rat. *Life Sci.* **5,** 665–669.

Yamada, K. (1959). The minute structure of the hamster gall bladder with special reference to the functions of the epithelium. *Okajimas Folia Anat. Jpn.* **33,** 321–351.

Yamada, K. (1962). Chemocytological observations on two peculiar epithelial cell types in the gall bladders of laboratory rodents. *Z. Zellsforsch. Mikrosk. Anat.* **56,** 180–187.

Yanagimachi, R., and Noda, Y. D. (1970). Fine structure of the hamster sperm head. *Am. J. Anat.* **128,** 367–388.

Yonce, L., and Krueger, H. (1952). Apico-basal value of the hamster cecum. *Am. J. Physiol.* **171,** 781 (abstr.).

Zatz, M. (1981). Pharmacology of the rat pineal gland. *In* "The Pineal Gland" (R. J. Reiter, ed.), Vol. 1, pp. 229–242. CRC Press, Boca Raton, Florida.

Zieger, G., Lux, B., and Kubatsch, B. (1974). Sex dimorphism in the adrenal of hamsters. *Acta Endocrinol. (Copenhagen)* **75,** 550–560.

Chapter 3

Clinical Chemistry and Hematology

Farol N. Tomson and K. Jane Wardrop

I.	Introduction	43
II.	Sources of Variation	44
III.	Specimen Handling	44
	A. Introduction	44
	B. Specimen Collection	44
	C. Sample Preparation	45
	D. Analytical Technique	45
IV.	Clinical Chemistry	46
	A. Introduction	46
	B. Glucose	46
	C. Urea Nitrogen	46
	D. Creatinine	47
	E. Proteins	47
	F. Bilirubin	48
	G. Lipids	49
	H. Electrolytes and Lactic Acid	49
	I. Enzymes	50
	J. Hormones	53
	K. Urinalysis	54
V.	Hematology	54
	A. General Characteristics	54
	B. Peripheral Blood	55
	C. Coagulation	57
	D. Bone Marrow	58
	References	58

I. INTRODUCTION

The sustained use of the hamster as a laboratory animal has renewed more interest in hamster laboratory data. Studies concerning the special qualities of the hamster (ability to conserve water and to hibernate) produced the earlier published data (Hoffman, 1968; Robinson, 1968). The recent use of the hamster in chronic toxicological studies and the introduction of ultramicrotechniques into laboratories have made more information available.

This chapter was written to summarize these data in a useful manner; values listed are those of the Syrian hamster (*Mesocricetus auratus*) unless otherwise noted. For more information about hamster clinical chemistry and hematology values, including the methodologies used, readers should consult the references or specific texts such as "Clinical Biochemical and Hematological Reference Values in Normal Experimental Animals" (Mitruka and Rawnsley, 1977), "Laboratory Animal Medicine" (Fox et al., 1984), "The Clinical Chemistry of Laboratory Animals" (Loeb and Quimby, 1987), and "Fundamentals of Clinical Chemistry" (Tietz, 1976).

II. SOURCES OF VARIATION

There are many variables that can influence laboratory data. Pakes and associates (1984) presented the salient complicating factors affecting rodent research and categorized the variables as physical, chemical, and microbial components. Ringler and Dabich (1979) also reviewed the more significant variables in rodent research. Both references are summarized in Table I.

Variables within the environment of the hamster are under direct control of the research team. Caging, temperature, ventilation, lighting, noise, and time of sample collection are parameters that investigators can hold constant.

The type of anesthetic agent used to sedate hamsters and the relative concentration of dietary constituents in hamster diets are often overlooked as significant variables. Phenobarbital, for example, was shown to increase serum cholesterol to four times the preanesthetic level (Jones and Armstrong, 1965). Feldman and associates (1982) demonstrated that increasing the protein concentration of the diet also increased the urea nitrogen concentration in the hamster.

An interesting study by Grodsky and colleagues (1974), utilizing diabetic Chinese hamsters, showed how a change in geographical location and a reduction of dietary fat caused ketonuria to disappear from their animals.

The age, sex, and strain of the hamster can contribute to variations in the hemoglobin, hematocrit, alkaline phosphatase, cholesterol, and phosphorus levels in the blood (Fox et al., 1984; Homburger et al., 1966; Maxwell et al., 1985; Volkert and Musacchia, 1970).

The physiological responses that are linked to circadian rhythms remain intriguing variables in hamster research. Cincotta and Meier (1984) demonstrated dramatic differences in the hypoglycemic response to insulin which varied according to the time of day when insulin was injected. Hypoglycemia was observed at 8 and 20 hr after the onset of light, but no hypoglycemic activity was detected at 4 hr after the onset of light.

Other variables such as chemical contaminants and latent infections may be more difficult for the research team to control, yet the presence of some of these contaminants has been shown to have a profound impact on other laboratory animal data (Pakes et al., 1984; Ringler and Dabich, 1979). To reduce the effects of these variables, researchers may find it necessary to use certified feed, chemically treated water, filtered air, and specific pathogen-free hamsters.

III. SPECIMEN HANDLING

A. Introduction

Generating laboratory data from any small laboratory animal poses several practical problems for investigators. Defining and controlling these problems will lead to more comparable data among animals (and laboratories), thereby improving the quality of animal research.

B. Specimen Collection

Obtaining a representative blood sample, free of significant artifact, is difficult. Venous blood should be taken directly from a vein with minimal trauma to surrounding tissues. Small volumes of blood (0.02–0.04 ml) may be obtained from the hamster using jugular, femoral, or lateral thigh veins. (Refer to Chapter 5 for more information on blood collection techniques.)

Table I
Variables Affecting Clinical Laboratory Results

Type of variable	Examples
Physical[a]	Cage design
	Temperature
	Humidity
	Ventilation
	Lighting
	Noise
Chemical[a]	Ambient air quality
	Drinking water quality
	Animal diet quality
	Drugs
Microbial[a]	Viruses
	Mycoplasmas
	Bacteria
	Parasites
Animal factors[b]	Strain
	Age
	Sex
Laboratory factors[b]	Restraint method
	Bleeding site
	Sampling time
	Sample type (serum, plasma)
	Sample storage
	Analytical method

[a] Adapted from Pakes et al. (1984).
[b] Adapted from Ringler and Dabich (1979).

Blood from cardiac punctures may contain tissue fluids that influence certain enzyme determinations. Maxwell and associates (1985) indicated that serum from cardiac samples may be contaminated with various enzymes such as creatine phosphokinase, aspartate aminotransferase, lactate dehydrogenase, and alanine aminotransferase, thereby yielding greater values than when blood is collected from other routes.

By using small amounts of blood and alternating the collection sites weekly, serial sample collecting can routinely be done. However, too frequent serial sampling may interfere with some data. Gerritsen and Blanks (1974) demonstrated that a second sample taken 15 min after the first yielded higher serum glucose values. The stress of restraint and release of epinephrine accounted for these differences.

Unsedated animals must be handled on a regular basis to avoid excitement and the need for excessive restraint. Serum adrenal corticosteroid values and leukograms may change if the animal struggles or becomes excited during blood collection.

Some data vary among animals and laboratories because of different times of the day when the samples are collected. Samples taken in the late afternoon differ from samples taken in the morning. Lighting, estrous cycles, room cleaning schedules, and noise from the cage washing machines have all been shown to influence certain constituents of the blood from other rodents (Pakes *et al.*, 1984; Ringler and Dabich, 1979).

Perhaps the most variable factor to address with the hamster (and other small rodents) is their feeding schedule prior to sampling. Hamsters are usually fed *ad libitum*, which results in variable amounts of food in the stomach. To help eliminate this metabolic digestive variable, some researchers have fasted their animals before sampling. A study by Gerritsen and Blanks (1974) indicated that more consistent insulin values were obtained when hamsters were subjected to an overnight fast followed by a 20-min feed. More attention needs to be given to the feeding behavior of hamsters. The significance of coprophagy in altering hamster clinical chemistry values is not known.

C. Sample Preparation

Routine hematological tests are performed using the anticoagulant ethylenediaminetetraacetic acid (EDTA). Sodium citrate is commonly used for coagulation studies. Serum is used for most of the chemistry tests, as anticoagulants can interfere with these analyses. Anticoagulants such as EDTA, oxalate, and citrate prevent clotting by chelating calcium ions. This is undesirable in enzyme assays, where calcium ions serve as cofactors in these reactions. If an anticoagulant has to be used, heparin interferes least with clinical chemistry tests and is preferred (Tietz, 1976). Fluoride–oxalate, an antiglycolytic anticoagulant mixture, can also be used in the analysis of glucose, especially if the plasma cannot be separated quickly. When serum is needed for the test, the sample should be allowed to stand for 30 min before removing the serum, to allow for clotting and clot retraction. Separation must not be delayed beyond 45 min without refrigeration. The concentrations of many blood constituents (glucose, phosphates, lipids, chlorides) may change if the sample remains at room temperature (Mitruka and Rawnsley, 1977). Certain enzymes (e.g., lactate dehydrogenase) are not stable if frozen. Care should be taken to consult appropriate references to determine the proper method of storage.

Lipemic samples are a common problem in rodent clinical chemistry and should be avoided. The increased turbidity in these samples typically produces erroneously high results for substances measured by end point assays. The bromocresol green procedure for albumin, the cresophthalein method for calcium, and the acid–ammonium molybdate procedure for inorganic phosphate are affected by lipemia (Statland and Winkel, 1979). Lipemic-induced elevations in glucose and bilirubin concentrations have been reported in dogs (Handelman and Blue, 1983). If samples from fasted animals cannot be collected, or if a pathological lipemia is present, ultracentrifugation or precipitation procedures can be used to separate lipoproteins from the remainder of the serum. Polyethylene glycol 6000 is one precipitating agent that has been recently evaluated with good results (Thompson and Kunze, 1984).

Hemolyzed samples will also produce errors in chemistry values. This can be due to either (1) interference with absorbance by the spectrophotometer or (2) release from the erythrocyte of the substance being measured. Almost all commonly assayed serum enzymes are present in high concentrations within erythrocytes, and visible hemolysis will produce elevated serum values. Potassium and inorganic phosphate may also be affected (Williams *et al.*, 1978).

D. Analytical Technique

In microanalytical chemical chemistry, procedures usually require only a few milliliters of sample. Ultramicrotechniques refer to the study of samples in the microliter range. Early in the 1900s the determination of glucose required 25–50 ml of blood (Tietz, 1976). In the mid-1900s, techniques were developed that require only 1 or 2 ml of blood. Currently, clinical microchemistry has developed techniques requiring a few microliters of sample for a single determination. Many of these laboratory tests are now routinely performed on commercially available automated equipment. The advancement of this laboratory equipment along with the development of ultramicrotechniques has made the hamster (as well as other smaller animals) more useful as a research model. For example, 17 different blood chemistry tests can be performed on a 400-μl hamster blood sample (0.40 ml) (Maxwell *et al.*, 1985).

The adoption of the international unit (I.U.) for enzyme determinations has made the comparison of data among laboratories more meaningful, but still these comparisons may not be entirely accurate. Methodologies for enzymes, including substrate used and reaction temperatures, must also be reported.

IV. CLINICAL CHEMISTRY

A. Introduction

In most abnormal biological conditions, animals demonstrate specific biochemical changes caused by cellular dysfunction. These changes can be detected through the use of specific laboratory tests. Researchers and clinicians are then faced with the problem of trying to evaluate a test result as being normal or abnormal. "Reference values" is a term being used to distinguish experimentally controlled test data from what the true "normal" values are in the real population. Researchers should determine baseline or reference values on their own animals through the use of control groups. The clinician can test two or three normal animals from the same group in some situations, but otherwise must rely on the published reference values.

B. Glucose

Glucose (Table II) is the most commonly measured carbohydrate in the blood. Its concentration reflects the nutritional, hormonal, and emotional state of the animal, and can be affected by a number of factors. Feeding schedules, hibernation, restraint, and anesthesia have been shown to alter blood glucose levels.

In one study, animals anesthetized with thiobarbiturates had serum glucose values of 300 mg/dl when compared to 144 mg/dl for the unanesthetized controls (Turner and Howards, 1977). The hyperglycemia lasted up to 5 hr. Hyperglycemia has also been reported in hibernating hamsters (Lyman and Leduc, 1953).

The most significant variable affecting glucose values appears to be whether the sample is taken from a fasted or nonfasted animal. Also, the length of fast seems to affect some values. Bannon and Friedell (1966) demonstrated plasma glucose values of 174 ± 55 mg/dl for nonfasted hamsters, 106 ± 30 mg/dl for hamsters fasted 4–6 hr, and 123 ± 67 mg/dl for hamsters fasted 24 hr. In the same study, tumor-bearing (fibrosarcoma) hamsters averaged 24-hr fasting values of 38 ± 24 mg/dl compared to their nonfasted controls at 109 ± 43 mg/dl.

Hamsters are normally fed *ad libitum,* and when they are fasted, hamsters do not completely empty their stomachs. Rowland's study (1984) into the feeding behavior of hamsters concluded that hamsters do not adapt to feed restriction schedules like the rat. This aspect of hamster husbandry probably accounts for most of the variability in the published data.

Gerritsen and Blanks (1974), working with Chinese hamsters, showed nonfasting serum glucose levels of 96 mg/dl. When they were fasted overnight and refed for 20 min prior to sampling the next day, the glucose values were 148 mg/dl.

If there is going to be a delay between the collection of the blood and the test procedure, a preservative must be added which will inhibit glycolysis. About 7% of the glucose will disappear from the serum during the first hour it sits at room temperature (Tietz, 1976). Thymol and fluoride are commonly used to prevent glycolysis, and collection tubes containing these compounds are commercially available. (Refer to Chapter 19 for a discussion of glucose values in Chinese hamsters.)

C. Urea Nitrogen

Serum urea nitrogen (Table III) is a commonly used indicator for renal dysfunction in most animals. In hamsters, as in other animals, urea nitrogen levels are greatly altered by the type of diet eaten, the amount of diet eaten, and the duration of fasting (Feldman *et al.,* 1982). Therefore, the fasting status and the

Table II
Blood Glucose

Sample	Mean ± SD (mg/dl)	Range (mg/dl)	Number	Reference
Plasma, nonfasting	174 ± 55	—	45	Bannon and Friedell (1966)
Plasma, 4- to 6-hr fast	106 ± 31	—	20	Bannon and Friedell (1966)
Serum, 12-hr fast	121 ± 34	73–173	31 Males	Maxwell *et al.* (1985)
	135 ± 38	37–198	32 Females	Maxwell *et al.* (1985)
Plasma, 24-hr fast	123 ± 63	—	25	Bannon and Friedell (1966)
Serum	73 ± 13	—	84 Males	Mitruka and Rawnsley (1977)
	65 ± 10	—	80 Females	Mitruka and Rawnsley (1977)

Table III
Urea Nitrogen

Sample	Mean ± SD (mg/dl)	Range (mg/dl)	Number	Reference
Plasma, nonfasting	22 ± 3	—	46	Bannon and Friedell (1966)
Serum, 12-hr fast, 3 months old	19 ± 4	12–26	30 Males	Maxwell et al. (1985)
	18 ± 2	15–33	32 Females	Maxwell et al. (1985)
Plasma, 24-hr fast	20 ± 4	—	33	Bannon and Friedell (1966)
Serum, nonfasting, 10 months old				D. B. Feldman (personal communication, 1984)
12% Protein	18 ± 2	—	3 Males	
	17 ± 1	—	3 Females	
18% Protein	23 ± 1	—	3 Males	
	23 ± 1	—	3 Females	
24% Protein	22 ± 1	—	3 Males	
	27 ± 1	—	3 Females	

percentage of dietary protein in the feed are highly significant factors to consider when evaluating urea nitrogen levels.

Feldman and associates (1982) reported significantly higher urea nitrogen levels in hamsters fed diets containing 18 and 24% protein when compared to hamsters fed diets with 12% protein. They also cited higher urea nitrogen levels in female hamsters.

D. Creatinine

Creatinine (Table IV), a nonprotein nitrogen compound produced during muscle metabolism, is not affected as much by the diet as urea nitrogen. The amount produced depends on protein catabolism and is found in elevated levels in the serum if there is a deficiency in the glomerular filtration.

Feldman and associates (1982) emphasized the significance of using serum creatinine as a measure of renal function, since urea nitrogen levels varied widely even in hamsters with nephritis.

In the evaluation of creatinine results from other rodents, one should recognize that certain drugs or compounds (barbiturates, glucose, protein, acetone, ketones, ascorbic acid, and sulfobromophthalein) may affect the true values (Mitruka and Rawnsley, 1977; Ringler and Dabich, 1979).

E. Proteins

The liver synthesizes many kinds of proteins (glycoproteins, lipoproteins, mucoproteins, fibrinogen, albumin, and globulins; see Tables V–VII) in response to the normal demands of

Table IV
Serum Creatinine

Sample	Mean ± SD (mg/dl)	Range (mg/dl)	Number	Reference
Serum	1.04 ± 0.28	—	84	Mitruka and Rawnsley (1977)
Serum, nonfasting, 10 months old				D. B. Feldman (personal communication, 1984)
12% Protein	0.33 ± 0.70	—	3 Males	
	0.50 ± 0.06	—	3 Females	
18% Protein	0.30 ± 0.15	—	3 Males	
	0.47 ± 0.03	—	3 Females	
24% Protein	0.33 ± 0.09	—	3 Males	
	0.43 ± 0.03	—	3 Females	
Serum, 12-hr fast, 3 months old	0.56 ± 0.08	0.4–0.7	30 Males	Maxwell et al. (1985)
	0.59 ± 0.15	0.4–1.0	32 Females	Maxwell et al. (1985)

Table V
Total Proteins

Sample	Mean ± SD (gm/dl)	Range (gm/dl)	Number	Reference
Serum	6.9 ± 0.3	—	84 Males	Mitruka and Rawnsley (1977)
	7.3 ± 0.5	—	80 Females	Mitruka and Rawnsley (1977)
Serum, nonfasted	5.2 ± 0.5	—	96	Dodds et al. (1977)
Serum, 12-hr fast	6.3 ± 0.3	5.8–7.0	31 Males	Maxwell et al. (1985)
	5.9 ± 0.3	5.2–6.7	32 Females	Maxwell et al. (1985)
Plasma, nonfasted	6.3 ± 0.3	—	—	Dent (1977)

homeostasis. These proteins are usually measured in serum because the plasma component contains significant fibrinogen levels. If plasma is used, House and associates (1961) have indicated that the hamster fibrinogen is linked to the α-globulin rather than to the β-globulin.

Coe and Ross (1983) described IgG$_1$, IgG$_2$, IgA, and IgM in immunized hamsters. Copeland and Ginsberg (1982) reported electrophoretic variation in the plasma glycoproteins between the diabetic and nondiabetic Chinese hamster, and concluded that the difference is genetic rather than metabolic.

Acute-phase proteins (i.e., those protein components that change after injury) have also been investigated in the hamster. Coe and Ross (1983) described an acute-phase protein that was 50- to 100-fold higher in female than in male hamsters. This female protein (FP) was structurally homologous with human C-reactive protein and amyloid P component. They found this female protein consistently in hamster amyloid, and further suggested an equilibrium between FP in the serum and amyloid.

The most extensive study on hamster plasma proteins, using 500 male hamsters (House et al., 1961), revealed the following significant changes in proteins with age: (1) a decrease in albumin during the first year, (2) an increase in α$_2$-globulin by 6 months, (3) a decrease in β-globulin at 8 weeks, (4) varying γ-globulin and fibrinogen levels, and (5) a decrease in albumin/globulin ratio with age.

Pregnancy and hibernation have been shown to alter serum proteins. Pregnancy decreased the albumin and increased the α-globulin (Peterson et al., 1961). Hibernation increased albumin, total protein, and β-globulin, while decreasing the γ-globulin fraction (South and Jeffay, 1958).

Amyloidosis, common in aging hamsters, is accompanied by lowered albumin and elevated globulin levels (Gleister et al., 1971). Experimentally induced inflammations will elevate γ-globulins in hamsters (Betts et al., 1964).

F. Bilirubin

Bilirubin (Table VIII) is formed from the breakdown of hemoglobin. It is carried in the plasma to the hepatocytes, where it is conjugated to form bilirubin diglucuronide, and excreted in the bile. Measurement of total bilirubin includes both the conjugated and unconjugated (free) forms, and is a useful

Table VI
Albumin

Sample	Mean ± SD (gm/dl)	Range (gm/dl)	Number	Reference
Serum	2.3 ± 0.4	—	84 Males	Mitruka and Rawnsley (1977)
	3.5 ± 0.3	—	80 Females	Mitruka and Rawnsley (1977)
Serum, nonfasted, 168–169 days of age	3.9 ± 0.3	—	10 Males	Thomas et al. (1979)
	3.5 ± 0.3	—	5 Males	Thomas et al. (1979)
Serum, 12-hr fast	4.3 ± 0.22	4.0–4.9	31 Males	Maxwell et al. (1985)
	4.1 ± 0.28	3.5–4.5	31 Females	Maxwell et al. (1985)
Plasma, nonfasted	3.7 ± 0.30	—	Males	Dent (1977)
	3.6 ± 0.24	—	Females	Dent (1977)

Table VII
Globulins

Sample	Percentage total protein (means ± SD)	Number	Reference
α_1-Globulin			
Plasma	8.0 ± 1.7	21 Males	House et al. (1961)
Serum	9.3 ± 1.9	84 Males	Mitruka and Rawnsley (1977)
	7.6 ± 1.5	80 Females	Mitruka and Rawnsley (1977)
	11.6 ± 1.7	21	Dodds et al. (1977)
α_2-Globulin			
Plasma	24.7 ± 8.6	21 Males	House et al. (1961)
Serum	26.7 ± 6.9	84 Males	Mitruka and Rawnsley (1977)
	23.5 ± 5.5	80 Females	Mitruka and Rawnsley (1977)
	8.6 ± 2.6	21	Dodds et al. (1977)
β-Globulin			
Plasma	13.1 ± 3.8	21 Males	House et al. (1961)
Serum	8.1 ± 3.6	84 Males	Mitruka and Rawnsley (1977)
	11.4 ± 2.7	80 Females	Mitruka and Rawnsley (1977)
	9.6 ± 4.4	21	Dodds et al. (1977)
γ-Globulin			
Serum	10.2 ± 3.1	84 Males	Mitruka and Rawnsley (1977)
	9.2 ± 2.8	80 Females	Mitruka and Rawnsley (1977)
	4.0 ± 1.8	21	Dodds et al. (1977)

screening test for liver damage or hemolysis, although abnormal serum bilirubin levels have not usually been reported from hamsters with hepatic dysfunction (Wardrop and Van Hoosier, 1987).

Compared to corresponding values in humans, hamsters have lower bilirubin values (Mitruka and Rawnsley, 1977). Hemolysis in serum will account for large errors in bilirubin tests using nitrous acid (Benjamin and McKelvie, 1978). If the sample is exposed to direct sunlight, losses of up to 50% of the available bilirubin will occur in 1 hr (Benjamin and McKelvie, 1978). Freezing has also been shown to increase bilirubin levels in serum samples (Bayard, 1974).

G. Lipids

The principal lipids (Tables IX,X) found in serum are cholesterol, phospholipids, triglycerides, and fatty acids. Cholesterol, because of its association with human atherosclerosis, has been investigated in most laboratory animals. In the hamster, circulating lipids in general are much lower than human levels (Mitruka and Rawnsley, 1977), but higher than the other rodents (Cox and Gökcen, 1974). Vaughan and associates (1982) demonstrated a reduction in plasma cholesterol due to a shortened photoperiod (from 14 hr light to 10 hrs light per 24 hr) and a reduction in cholesterol due to lower environmental temperatures. Plasma triglycerides were not affected in this study. Serum lipids were found to be increased during hamster hibernation (Denyes and Baumber, 1965).

Circulating hamster lipids contain three different lipoproteins which have been identified by Sicart and associates (1984) as very low-density lipoproteins (VLDL), low-density lipoproteins (LDL), and high-density lipoproteins (HDL). They described the LDL as exceptionally rich in triglycerides (32%), with the plasma cholesterol being equally distributed between the α- (HDL) and β- (LDL,VLDL) lipoproteins. Cox and Gökcen (1974) claimed that lipid metabolism in hamsters differs from that in humans because of the four times higher free fatty acid levels and a greater lecithin–cholesterol acyltransferase activity in hamsters. Their analysis of hamster urine indicated that cholesterol was the main lipid present.

H. Electrolytes and Lactic Acid

The range of normal values of electrolytes (Table XI) in hamster serum show little deviation from the accepted normal values of other rodents. Laboratory methods, sample handling,

Table VIII
Total Bilirubin

Sample	Mean ± SD (mg/dl)	Range (mg/dl)	Number	Reference
Serum, nonfasted, 168–169 days of age	0.82 ± 0.17	—	10 Males	Thomas et al. (1979)
	0.83 ± 0.31	—	5 Females	Thomas et al. (1979)
Serum, 12-hr fast	0.4 ± 0.18	0.1–0.8	31 Males	Maxwell et al. (1985)
	0.3 ± 0.16	0.1–0.9	31 Females	Maxwell et al. (1985)
Serum	0.42 ± 0.12	—	84 Males	Mitruka and Rawnsley (1977)
	0.36 ± 0.11	—	80 Females	Mitruka and Rawnsley (1977)

Table IX
Cholesterol

Sample	Mean ± SD (mg/dl)	Range (mg/dl)	Number	Reference
Serum	55 ± 12	—	84 Males	Mitruka and Rawnsley (1977)
	52 ± 11	—	80 Females	Mitruka and Rawnsley (1977)
Plasma, nonfasted	237 ± 29	—	Males	Dent (1977)
	182 ± 49	—	Females	Dent (1977)
Plasma, nonfasted, 10-hr light cycle	160 (graph reading)		40 Males	Vaughan et al. (1984)
Serum, 12-hr fast	94 ± 23	55–145	31 Males	Maxwell et al. (1985)
	136 ± 20	89–188	32 Females	Maxwell et al. (1985)
	89 ± 17	—	25 Males	Cox and Gökcen (1974)

and animal ages probably account for the range of values reported.

Younger hamsters have higher serum phosphorus levels than older animals (Dent, 1977). Hemolysis of the samples can elevate potassium concentrations to 8 mEq/liter in the serum (Bannon and Friedell, 1966).

I. Enzymes

The determination of serum enzyme activity (Table XII) is a useful diagnostic aid. The interpretation of these enzyme values requires an awareness not only of the specific enzyme and its reactions, but also of the laboratory method used to measure its activity. Readers are encouraged to review the individual enzymes and their current analytical methods in recent biochemical texts and laboratory journals.

Enzyme concentrations are influenced by various diets and nutritional deficiencies in all animals. As is the case with other clinical chemistry values, investigators should establish their own laboratory reference values before assessing experimental alterations.

1. Alkaline Phosphatase

The alkaline phosphatase in hamster circulation is composed of the isoenzymes from bone, liver, and intestine (Cox and Gökcen, 1973). Alkaline phosphatase from bone appears to be the main serum isoenzyme, while the intestinal isoenzyme comprises 10–15% of the total activity. The hamster bone and intestinal isoenzymes are more heat-stable, while the liver isoenzyme is heat-sensitive. Liver alkaline phosphatase, though present at weaning, was not detectable in older normal hamsters.

Higher alkaline phosphatase values were found among immature hamsters, as in other animals, since growing bone releases more enzyme than mature bone (Dent, 1977; Eugster et al., 1966).

Elevated alkaline phosphatase levels may be seen in any disease of the liver, but dramatic increases are usually seen in animals with bile duct obstructions. Increased concentrations of alkaline phosphatase in hamsters have been most commonly associated with leukemias and tumors of the prostate gland, bone, and liver (Eugster et al., 1966).

The specific cause of the extreme variability shown in Table

Table X
Serum Lipids

Sample	Mean ± SD (gm/dl)	Range (gm/dl)	Number	Reference
Triglycerides, 12-hr fast	123 ± 42.7[a]	72–227	30 Males	Maxwell et al. (1985)
	129 ± 27.0[a]	79–186	32 Females	Maxwell et al. (1985)
Total lipids, 12-hr fast	345 ± 121	—	35 Males	Cox and Gökcen (1974)
Phospholipids, 12-hr fast	161 ± 29	—	25 Males	Cox and Gökcen (1974)

[a]Median values.

Table XI
Electrolytes and Lactic Acid

Electrolyte; Lactic Acid	Mean ± SD	Range	Number	Reference
Sodium (mEq/liter)				
Plasma, nonfasting	147 ± 5	—	34	Bannon and Friedell (1966)
	144 ± 3	—		Dent (1977)
Serum, nonfasting	129 ± 9	—	199	Burns and de Lannoy (1966)
Serum	128 ± 2	—	84 Males	Mitruka and Rawnsley (1977)
	134 ± 2	—	80 Females	Mitruka and Rawnsley (1977)
Potassium (mEq/liter)				
Plasma, nonfasting	4.8 ± 1.9	—	33	Bannon and Friedell (1966)
Serum, nonfasting	4.6 ± 1.3	2.3–9.8	199	Burns and de Lannoy (1966)
Serum	4.7 ± 0.5	—	84 Males	Mitruka and Rawnsley (1977)
	5.3 ± 0.5	—	80 Females	Mitruka and Rawnsley (1977)
Calcium (mg/dl)				
Serum, 12-hr fast	11.1 ± 0.7	9.8 ± 13.2	30 Males	Maxwell et al. (1985)
	11.0 ± 0.6	9.8 ± 12.4	29 Females	Maxwell et al. (1985)
Serum	9.5 ± 0.9	—	84 Males	Mitruak and Rawnsley (1977)
	10.4 ± 0.9	—	80 Females	Mitruka and Rawnsley (1977)
Chloride (mEq/liter)				
Plasma, nonfasting	100 ± 10	—	32	Bannon and Friedell (1966)
Serum	97 ± 1	—	84 Males	Mitruka and Rawnsley (1977)
	94 ± 1	—	80 Females	Mitruka and Rawnsley (1977)
Magnesium (mg/dl)				
Serum	2.5 ± 0.2	—	84 Males	Mitruka and Rawnsley (1977)
	2.2 ± 0.1	—	80 Females	Mitruka and Rawnsley (1977)
Serum, nonfasted, 223–224 days of age	2.1 ± 0.5	—	9 Males	Thomas et al. (1979)
	1.6 ± 0.4	—	10 Females	Thomas et al. (1979)
Phosphorus (mg/dl)				
Plasma, nonfasted, immature	9.6 ± 1.5	—	Males	Dent (1977)
	8.9 ± 0.9	—	Females	Dent (1977)
Plasma, nonfasted mature	6.3 ± 1.0	—	Males	Dent (1977)
	6.4 ± 0.6	—	Females	Dent (1977)
Serum, 12-hr fast	8.2 ± 1.1	6.2–9.9	31 Males	Maxwell et al. (1985)
	6.3 ± 1.2	3.0–9.4	32 Females	Maxwell et al. (1985)
Serum	5.3 ± 1.0	—	84 Males	Mitruka and Rawnsley (1977)
	6.0 ± 1.1	—	80 Females	Mitruka and Rawnsley (1977)
Lactic acid (mEq/liter)				
Serum, nonfasted				Thomas et al. (1979)
92–93 days of age	8.0 ± 2.3	—	10 Males	
	8.6 ± 2.3	—	10 Females	
140–141 days of age	9.3 ± 1.8	—	9 Males	
	10.2 ± 2.0	—	10 Females	
168–169 days of age	8.1 ± 2.0	—	10 Males	
	8.3 ± 0.8	—	5 Females	

Table XII
Serum Enzymes

Enzyme	Mean ± SD	Range	Number	Reference
Alkaline phosphatase (IU/liter)				
Plasma, nonfasted[a]	218 ± 42	—	Males	Dent (1977)
	369 ± 34	—	Females	Dent (1977)
Serum, 12-hr fast[a]	121 ± 17	99–151	18 Males	Maxwell et al. (1985)
	143 ± 22	86–187	24 Females	Maxwell et al. (1985)
Serum[a]	17.5 ± 6.1	—	84 Males	Mitruka and Rawnsley (1977)
	15.4 ± 4.2	—	80 Females	Mitruka and Rawnsley (1977)
Amylase				
Serum (IU/ml)	7.9 ± 0.9	—	5 Males	Takahashi et al. (1981)
Serum (Somogyi units/dl)	175 ± 21	—	84 Males	Mitruka and Rawnsley (1977)
	196 ± 27	—	80 Females	Mitruka and Rawnsley (1977)
Creatine phosphokinase (CPK) (IU/liter)				
Serum	1.0 ± 0.4	—	84 Males	Mitruka and Rawnsley (1977)
	0.8 ± 0.3	—	80 Females	Mitruka and Rawnsley (1977)
Lactate dehydrogenase (LDH) (IU/liter)				
Plasma, nonfasted	257 ± 64	—	Males	Dent (1977)
	208 ± 55	—	Females	Dent (1977)
Serum, 12-hr fast	211 ± 53	148–412	18 Males	Maxwell et al. (1985)
	217 ± 74	—	15 Females	Maxwell et al. (1985)
Serum	115 ± 20	—	84 Males	Mitruka and Rawnsley (1977)
	110 ± 27	—	80 Females	Mitruka and Rawnsley (1977)
Aspartate aminotransferase (AST, SGOT) (IU/liter)				
Plasma, nonfasted[a]	28 ± 2.0	—	Males	Dent (1977)
	24 ± 2.6	—	Females	Dent (1977)
Serum, 12-hr fast[a]	47 ± 38.3	28–122	18 Males	Maxwell et al. (1985)
	43 ± 22.5	33–92	15 Females	Maxwell et al. (1985)
Serum[a]	124 ± 22	—	84 Males	Mitruka and Rawnsley (1977)
	78 ± 14	—	80 Females	Mitruka and Rawnsley (1977)
Alanine aminotransferase (ALT, SGPT) (IU/liter)				
Plasma, nonfasted[a]	35 ± 9.0	—	Males	Dent (1977)
	32 ± 9.5	—	Females	Dent (1977)
Serum, 12-hr fast[a]	38 ± 26.1	22–128	19 Males	Maxwell et al. (1985)
	49 ± 18.3	28–106	21 Females	Maxwell et al. (1985)
Serum[a]	27 ± 5	—	84 Males	Mitruka and Rawnsley (1977)
	21 ± 4	—	80 Females	Mitruka and Rawnsley (1977)

[a]Different methodologies used; consult reference for details.

XII is unknown, but is probably related to differing methodologies used to determine these reference values.

2. Amylase

Amylase is present in high concentrations in the pancreas, liver, and intestinal mucosa. Isoenzymes of amylase have been described in the hamster and other animals, but resulting patterns following experimental disease have not yielded any effective diagnostic results (Takahashi et al., 1981).

3. Creatine Phosphokinase

Creatine phosphokinase (CPK) is present in striated muscle and nervous tissue. Isoenzymes have been described in other

3. CLINICAL CHEMISTRY AND HEMATOLOGY

animals. CPK catalyzes the phosphorylation of creatine and has become a useful enzyme for investigators studying the hamster myopathies, muscular dystrophies, and cardiac necroses (Bajusz and Homburger, 1966; Eppenberger et al., 1964). Serum CPK values in these conditions may be either increased or decreased from their normal reference values.

4. Lactate Dehydrogenase

Lactate dehydrogenase (LDH) is an enzyme present in all tissues, but is concentrated primarily in muscle, liver, and kidney. Its diagnostic value is sometimes dependent on isoenzyme identification. Increased serum LDH values in hamsters have been observed in the hereditary myopathy of hamsters (Homburger et al., 1966) and after inoculation of hamsters with certain viruses (Eugster et al., 1966). Freezing serum will cause a lowering of LDH activity (Tietz, 1976).

5. Aspartate Aminotransferase

Widely distributed in all body tissues except bone, aspartate aminotransferase (AST, SGOT) has its highest concentration in striated muscle and liver. The activity of the serum enzyme is relatively low but increases rapidly following muscle injury or liver damage. Hamster serum AST values have been increased in neoplastic involvement of the liver and may also be elevated by other changes causing hepatocyte cloudy swelling (Eugster et al., 1966).

6. Alanine Aminotransferase

Alanine aminotransferase (ALT, SGPT) is frequently used as a test for liver damage. Among the laboratory animals, ALT is specific for liver damage in the dog and cat. Specificity for liver damage in the hamster has not been documented. The literature does not contain much information on ALT values from hamsters, nor have changes in ALT levels been reported from hamsters with naturally occurring disease states.

7. Sorbitol Dehydrogenase

Sorbitol dehydrogenase (SDH) activity has been observed in hepatic tissues of a variety of animals, but hamster SDH values have not been reported in the literature. In the rat almost all SDH activity originates from the hepatocytes, and determination of SDH concentrations are becoming more common in rodent toxicological studies (Dooley, 1984).

J. Hormones

Hamster reproductive hormones have been the subject of many publications. For a review of these hormones and their relationships, the reader is referred to the chapters on reproduction.

Table XIII

Thyroid Hormones

Hormone	Means ± SD	Number	Reference
Thyroxin (T_4) (µg/dl)			
Plasma			
3 Months	6.7 ± 0.7	10 Males	Néve et al. (1981)
20 Months	3.6 ± 0.2	9 Males	Néve et al. (1981)
14 hr light	6	8 Females	Vaughn et al. (1982)
10 hr light	3 1/2	8 Females	Vaughn et al. (1982)
Triiodothyronine (T_3) (ng/dl)			
Plasma			
3 Months	62 ± 2	10 Males	Néve et al. (1981)
20 Months	42 ± 3	10 Males	Néve et al. (1981)
14 hr light	65	8 Females	Vaughan et al. (1982)
10 hr light	45	8 Females	Vaughan et al. (1982)
Natural photoperiod	170	40 Males	Vaughan et al. (1984)

1. Thyroid Hormones

Thyroid hormones (Table XIII) play an important role in many facets of growth and metabolism. This role may occur as a primary phenomenon or in synergism with other hormones (catecholamines and glucocorticoids). As with other hormones, many exogenous factors influence the circulating levels of these thyroid hormones.

Chronic exposure to shortened photoperiods has resulted in decreased serum concentration of thyroid-stimulating hormone (TSH), thyroxin (T_4), and triiodothyronine (T_3) (Vaughan et al., 1982). Lower ambient temperatures also lowered the T_4 and T_3 values in hamsters (Vaughan et al., 1984). Basal T_4 and T_3 levels have also been shown to decrease with age in the hamster. Néve and associates (1981) measured T_4 and T_3 levels in 3- and 20-month-old hamsters and reported that T_4 level decreased from 6.75 to 3.59 µg/dl and the T_3 decreased from 62 to 42 ng/dl. These patterns are similar to those in humans and in other aging rodents.

Galton and Galton (1966) stated that pregnant hamsters may metabolize thyroid hormone differently, as decreased protein-bound iodine was seen during pregnancy.

2. Adrenal Hormones

The adrenal cortex of the hamster is similar to other species in that it contains well-delineated glomerulosa, fasciculata, and reticularis zones, and the administration of ACTH produces an increase in adrenal weight (Alpert, 1950).

Cortisol and corticosterone are present in significant amounts in hamster plasma (Ottenweller et al., 1985; Schindler and Knigge, 1959). The pregnant hamster has the capacity to produce unusually large quantities of cortisol, while cortisol in nonpregnant hamsters is quite low in comparison to other species. Brink-Johnsen and associates (1981) reported that plasma cortisol levels increased approximately 100-fold (from 0.3 to 30 µg/dl) as parturition neared in hamsters.

Exposure to cold apparently does not stimulate the pituitary adrenal axis (Wendler and Lyman, 1954). During the 24-hr daily cycle, peak adrenal activity occurs approximately 3 hr before the onset of darkness (Frenkel et al., 1965).

Corticosteroid values of 7.4 ± 1.9 mg/dl have been reported (Cox and Gökcen, 1974).

K. Urinalysis

It is rather difficult to collect good urine specimens (Table XIV) from small laboratory animals, although metabolic cages and microsampling techniques are commonly used. Rodents will frequently urinate when picked up, and this urine can be collected into microcapillary tubes. There is scattered information in the literature about the physical and chemical composition of hamster urine.

Fiszer and associates (1979) reported the average daily urine volume to be about 7 ml. Gerritsen and Blanks (1974), studying 25-gm diabetic Chinese hamsters, measured excretion rates of up to 75 ml urine per day.

Hamster urine is basic, with amorphous calcium carbonate and triple-phosphate crystals occasionally seen (Schuchman, 1980). Urine sodium and potassium concentrations fluctuate daily, but average about 70 mmol/liter and 120 mmol/liter, respectively (Fiszer et al., 1979).

Cox and Gökcen (1974) analyzed hamster urine for lipid content and reported that cholesterol was the main lipid present. The mean excretion rate for cholesterol and total lipid was 11.7 and 10.5 mg/week per animal, respectively. Urine proteins averaged about 9.7 mg/week per animal—about 10 times the protein excretion rate for humans (Tietz, 1976). Electrophoresis of urine pool concentrates yielded three main protein bands (Cox and Gökcen, 1974).

Table XIV
Urine Values in Normal Hamsters

Parameter	Value	Reference
Urine volume	5.1–8.4 ml/24 hr	Fiszer et al. (1979)
Protein	9.7 mg/week	Cox and Gökcen (1974)
pH	Basic	Schuchman (1980)
Potassium	64.9–90.8 mmol/liter	Fiszer et al. (1979)
Sodium	111.8–155.9 mmol/liter	Fiszer et al. (1979)
Sediment	Amorphous carbonate and amorphous phosphate crystals commonly seen	Schuchman (1980)
Cholesterol	11.7 mg/week	Cox and Gökcen (1974)
Total lipids	10.5 mg/week	Cox and Gökcen (1974)

For a routine urinalysis, commercial dipstick kits, specific gravity measuring devices, and centrifuged sediment analysis can be used if enough sample is collected. Multiple collections are usually required.

V. HEMATOLOGY

Desai's chapter (1968) on hamster hematology and microcirculation should be reviewed by readers studying hamster hematology. Since the mouse and the rat have been used to a greater degree than the hamster in hematological research, Bannerman's chapter (1983) in "The Mouse in Biomedical Research" (Vol. 3) is a more current reference on rodent hematology. The section on materials and methods used in hematology and clinical chemistry in Mitruka and Rawnsley's text (1977) is a discussion of the various techniques available and how they can alter test results.

A. General Characteristics

In general, blood collected for routine tests from the heart, tail, or orbital sinus show no significant differences. Fitts and associates (1983), however, noted that the hematocrits of samples taken from jugular sites were consistently higher than those from cardiac collections.

Desai (1968) studied newborn hamster blood parameters and reported the neonate as being relatively anemic with hematocrit values between 34 and 39%. Hypochromic, macrocytic cells were present in smears, with 10–30% of the cells being normoblasts. Total leukocyte counts in the neonate were approximately 2000/µl.

Dent (1977) compared the hamster and the rat and found the values similar, although the mean cell volume (MCV) was slightly higher in the hamster.

Friedell and Bannon (1972) studied hamster anemia and found significantly increased serum iron (310 ± 67 µg/dl) and iron-binding capacity (518 ± 76 µg/dl) values of their control animals when compared to other small animals. More recent studies reported serum iron concentrations of 46.1 µmol/liter, with iron-binding capacities of 104.5 µmol/liter (Rennie et al., 1981).

1. Volume

Hamsters are frequently studied because of their ability to conserve water. Results from these studies yielded most of the reference values in Table XV. Kutscher (1968) found that water deprivation for 72 hr decreased whole-blood volumes to 6.54 ml/100 gm body weight, and plasma volume to 4.18 ml/gm body weight, while the hematocrit remained unaffected.

3. CLINICAL CHEMISTRY AND HEMATOLOGY

Table XV
Blood and Plasma Volumes

Sample[a]	Value ± SD	Number	Reference
Blood volume (ml/100 gm BW)	6.0–9.0		Mitruka and Rawnsley (1977)
0 Hr, water deprived	8.0 ± 0.6	12	Kutscher (1968)
72 Hr, water deprived	6.5 ± 0.7	5	Kutscher (1968)
Red cell volume (ml/kg BW)			
22°C	27.8 ± 0.4	25 Males	Jones et al. (1976)
34°C, Heat acclimated	24.7 ± 12 0.4	24 Males	Jones et al. (1976)
Plasma volume (ml/100 gm BW)	4.6 ± 0.1		Mitruka and Rawnsley (1977)
0 Hr, water deprived	5.1 ± 0.3	12	Kutscher (1968)
72 Hr, water deprived	4.2 ± 0.4	5	Kutscher (1968)
22°C	3.3 ± 0.4	25 Males	Jones et al. (1976)
34°C, Heat acclimated	3.4 ± 1.1	24 Males	Jones et al. (1976)
PCV (%)	52.5 ± 2.3	84 Males	Mitruka and Rawnsley (1977)
	42.9 ± 2.7	96	Dodds et al. (1977)
0 Hr, water deprived	51.9 ± 3.4	12	Kutscher (1968)
72 Hr, water deprived	51.4 ± 4.9	5	Kutscher (1968)

[a]BW, Body weight; PCV, packed-cell volume.

2. Blood pH

The reported venous blood pH for active hamsters is 7.39 ± 0.03 (Volkert and Musacchia, 1970). This value may decrease slightly when the animal is anesthetized or may increase slightly if arterial blood is sampled (O'Brien et al., 1979; Volkert and Musacchia, 1970) or if the animal is hibernating (Kreienbühl et al., 1976). A study of the European hamster by Malan and Arens (1973) showed an increase in blood pH from 7.40 in the nonhibernating state to 7.57 in the hibernating animal.

3. Blood Gases

The hamster's ability to hibernate attracted studies into its metabolism at low levels of oxygen consumption. Hoffman's (1968) and Robinson's (1968) chapters are summaries of the results of these early trials. Arterial and venous blood gases (Table XVI) have been measured in cannulized adult hamsters using blood gas ultramicroanalyzers. During exercise O'Brien and colleagues (1979) measured an increase in the P_aO_2, a decrease in the P_aCO_2, and a decrease in the bicarbonate concentration. Researchers have documented various measurable changes in the blood pH, bicarbonate, and CO_2 levels during hibernation and hypothermia (Kreienbühl et al., 1976; Volkert and Musacchia, 1970).

The blood acid–base status also fluctuated with phenobarbital anesthesia as shown by Holloway and Heath (1984). When compared to the rat, they concluded that hamsters seem to have a greater blood buffering capacity.

B. Peripheral Blood

1. Erythrocytes

Hamster erythrocytes (Table XVII) have a diameter of 5 to 7 μm. Polychromasia is noted more commonly in hamsters than in humans (Desai, 1968). Blood smears from 4- to 8-week-old hamsters will contain slightly microcytic, normochromic red

Table XVI
Blood Gases and pH

Parameter	Value	Number	Reference
P_aO_2 (mm Hg)	71.8 ± 4.9	13 Males	O'Brien et al. (1979)
P_vO_2 (mm Hg)	37.6 ± 4.5[a]	13	Volkert and Musacchia (1970)
	30.4 ± 3.3	13	Volkert and Musacchia (1970)
P_aCO_2 (mm Hg)	41.1 ± 2.4	13 Males	O'Brien et al. (1979)
	59.7 ± 1.7	22	Kreienbühl et al. (1976)
P_vCO_2 (mm Hg)	58.4 ± 4.7[a]	13	Volkert and Musacchia (1970)
	55.5 ± 4.1	13	Volkert and Musacchia (1970)
HCO_3^- (mmol/liter)	29.9 ± 2.9	13 Males	O'Brien et al. (1979)
	31.8[a]	13	Volkert and Musacchia (1970)
	32.6	13	Volkert and Musacchia (1970)
pH (venous)	7.36 ± 0.03[a]	13	Volkert and Musacchia (1970)
	7.39 ± 0.03	13	Volkert and Musacchia (1970)
pH (arterial)	7.48 ± 0.03	13 Males	O'Brien et al. (1979)
	7.3 ± 0.02	22	Kreienbühl et al. (1976)

[a]Under phenobarbital anesthesia.

Table XVII
Erythrocyte Parameters

Parameter[a]	Value	Number	Reference
RBC count ($\times 10^6/\mu l$)	7.5 ± 1.4	84 Males	Mitruka and Rawnsley (1977)
	7.0 ± 1.5	80 Females	Mitruka and Rawnsley (1977)
	6.8 ± 0.3	23	Meyerstein and Cassuto (1970)
Hemoglobin (gm/dl)	16.8 ± 1.2	84 Males	Mitruka and Rawnsley (1977); Desai (1968)
	16.0 ± 0.3	23	Meyerstein and Cassuto (1970)
MCV (fl)	70.0 ± 3.2	84 Males	Mitruka and Rawnsley (1977)
	73.9 ± 3.0	23	Meyerstein and Cassuto (1970)
MCHC (gm/dl)	32.0 ± 2.2	84 Males	Mitruka and Rawnsley (1977); Desai (1968)
	31.6 ± 2.1	23	Meyerstein and Cassuto (1970)
Reticulocytes (%)	2.5 ± 1.2		Desai (1968)
	1.2 ± 0.2	8	Meyerstein and Cassuto (1970)
Sedimentation rate (mm/hr)	1.6	—	Desai (1968)

[a] MCV, mean cell volume; MCHC, mean corpuscular hemoglobin concentration; RBC, red blood cells.

blood cells (RBC) with moderate variation in size and shape. Occasional spherocytes and target cells may also be seen.

The spleen of adult hamsters is not erythropoietically active. Boussios and colleagues (1982) studied the involvement of triiodothyronine (T_3) in the erythroid response of hamsters and concluded that this hormone may be required for this response to occur.

RBC survival time ranges from 50 to 78 days (Brock, 1960; Toffelmire and Boegman, 1980). RBC longevity is increased markedly when the animal hibernates (Desai, 1968). Meyerstein and Cassuto (1970), studying the effects of heat acclimation on hamsters, concluded that heat acclimation has little effect on erythrocyte counts, hemoglobin concentration, or hematocrit values. But cells from these hamsters showed metabolic and fragility patterns common to some hemolytic anemias. They concluded that heat acclimation altered the erythrocytes.

Matsuzawa and Ikarashi (1979) compared the hemolysis fragility of hamster erythrocytes in different solutions with those of other animal erythrocytes. Of the species tested, hamster erythrocytes were grouped midway between the extremely fragile sheep and pig cells and the extremely resistant human and dog cells. Different RBC antigens have not been identified in the hamster, and problems with transfusion reactions have not been described.

2. Hemoglobin

Hemoglobin (Hb) is the protein responsible for transporting oxygen and carbon dioxide in the blood. There are many methods available for determining Hb, and they all vary in accuracy, ease of performance, and reproducibility (Mitruka and Rawnsley, 1977).

Hb levels in hamsters will vary according to age, strain, sex, and health status. Desai (1968) demonstrated a Hb increase with age until maximum values were reached at 8–9 weeks. More recent studies demonstrated little variation in Hb concentration between weeks 1 and 27 (Rennie et al., 1981).

Volkert and Musacchia (1970), studying Hb saturation, concluded that the Hb in arterial blood in hypothermic animals was more fully saturated with oxygen than in the normothermic hamsters.

3. Hematocrit

The hematocrit measures the proportion of RBC to plasma in peripheral blood. The hematocrit reading, which is referred to as the packed-cell volume (PCV) when determined by centrifugal means, is recorded as the volume (percentage) of packed cells per deciliter of blood. The PCV can also be used as an estimate of the RBC count. Although it is not an accurate measurement, by itself the PCV can be used to make judgments concerning the hemoconcentration of blood.

The most significant source of error in using microhematocrit tubes for this measurement is the centrifugation process. Errors up to 8% may result if improper force or incorrect centrifuging times are used (Mitruka and Rawnsley, 1977).

Burns and de Lannoy (1966) reported a range of 30–50% for PCV in hamsters. Kutscher (1968) showed that no relationship exists between the hematocrit and plasma volume when water is withheld. Similar to Hb, the hematocrit increases with age until 8–9 weeks (Desai, 1968).

4. Leukocytes (Table XVIII)

In general the hamster leukogram resembles that of other rodents. The total count varies between 5000 and 10,000/μl, of which about 60 to 75% are lymphocytes. During hibernation these total counts will decrease to about 2500 (Mitruka and Rawnsley, 1977). Neutrophils have a diameter of 10–12 μm, contain annular lobulated nuclei, and have a cytoplasm with dense pinkish granules. These finely granulated acidophilic neutrophils ("heterophils") often resemble eosinophils. They may be either segmented or nonsegmented (Fig. 1).

Monocytes, basophils, and eosinophils resemble those of humans and of rodents.

The predominant leukocyte in the hamster is the lymphocyte. In routinely stained smears lymphocytes are small round cells with dark blue nuclei occupying most of the cell (Fig. 2).

Emminger and associates (1975) found the total leukocyte

3. CLINICAL CHEMISTRY AND HEMATOLOGY

Table XVIII

Leukocyte Values

Parameter	Mean ± SD	Number	Reference
Total (× $10^3/\mu l$)			
	7.61 ± 1.3		Desai (1968)
	7.6 ± 1.3	84 Males	Mitruka and Rawnsley (1977)
	8.6 ± 1.5	80 Females	Mitruka and Rawnsley (1977)
	6.5 ± 2.2	96 Males	Dodds et al. (1977)
	4.9 ± 1.8	96 Females	Dodds et al. (1977)
	5.9 ± 2.1	Males	Dent (1977)
	8.1 ± 1.8	Females	Dent (1977)
Neutrophils			
Segmented (%)	21.9 ± 5.5		Desai (1968)
	22.1 ± 2.5	84 Males	Mitruka and Rawnsley (1977)
	29.0 ± 3.1	80 Females	Mitruka and Rawnsley (1977)
Nonsegmented (%)	8.0 ± 2.5		Desai (1968)
Lymphocytes (%)	73.5 ± 9.4		Desai (1968)
	73.5 ± 9.4	84 Males	Mitruka and Rawnsley (1977)
	67.0 ± 8.5	80 Females	Mitruka and Rawnsley (1977)
Monocytes (%)	2.5 ± 0.8		Desai (1968)
	2.5 ± 0.8	84 Males	Mitruka and Rawnsley (1977)
	2.4 ± 1.0	80 Females	Mitruka and Rawnsley (1977)
Eosinophils (%)	1.1 ± 0.0		Desai (1968)
	0.9 ± 0.3	84 Males	Mitruka and Rawnsley (1977)
	0.7 ± 0.2	80 Females	Mitruka and Rawnsley (1977)
Basophils (%)	1.0 ± 2.0	84 Males	Mitruka and Rawnsley (1977)
	0.5 ± 0.7	80 Females	Mitruka and Rawnsley (1977)

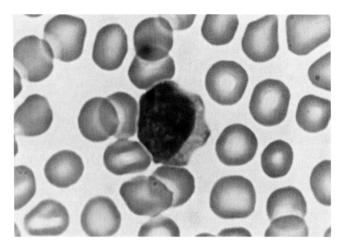

Fig. 2. Hamster lymphocyte surrounded by erythrocytes in peripheral blood.

count in European hamsters higher in the male hamster and higher in older hamsters. The hamsters in their study had 3500 leukocytes/μl at 13 days of age and about 8200 cells/μl at 65 days. Lappenbusch (1972) studied the circadian rhythm and the impact of collecting circulating leukocytes of Chinese hamsters at different times during a 12-hr light–dark cycle (6:00 AM to 6:00 PM). At 1:00 AM leukocytes totaled 6017–6858 cells/μl and at 1:00 PM they had increased to 7753–8838 cells/μl.

Hibernation also drastically reduced the leukocyte and thrombocyte levels in European hamsters (Reznik et al., 1975). Total leukocyte counts were about 1000 cells/μl, with the differentials remaining essentially unchanged.

C. Coagulation

Investigators obtaining hamster blood samples (Table XIX) have noted how quickly coagulation takes place when obtain-

Table XIX

Coagulation Values

Parameter	Mean ± SD	Number	Reference
Bleeding time (sec)	109 ± 19	—	Desai (1968)
Clotting time (sec)	143 ± 50	—	Desai (1968)
Prothrombin time (sec)	10.5 ± 0.2	—	Desai (1968)
	9.9 ± 1.2	Males	Dent (1977)
	9.3 ± 1.8	Females	Dent (1977)
	14.8 ± 1.0	67	Dodds et al. (1977)
Partial thromboplastin time (sec)	24.4 ± 2.7	67	Dodds et al. (1977)
Factor II (%)	100 ± 34	29	Dodds et al. (1977)
Factor VII (%)	113 ± 42	29	Dodds et al. (1977)
Factor VIII (%)	102 ± 10	67	Dodds et al. (1977)
Factor X (%)	105 ± 32	29	Dodds et al. (1977)
Plasminogen (CTA U/ml)	2.63 ± 89	59	Dodds et al. (1977)
Fibrinogen (mg/dl)	252 ± 64	59	Dodds et al. (1977)

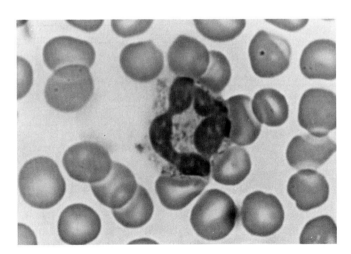

Fig. 1. Hamster neutrophil surrounded by erythrocytes in peripheral blood.

ing these samples. The mean coagulation time is about 142 sec (Desai, 1968).

Dodds and associates (1977), studying coagulation and fibrinolysis in aging hamsters, concluded that female hamsters had shorter one-stage prothrombin times than males, along with longer partial thromboplastin times and higher concentrations of fibrinogen and plasminogen than the males.

D. Bone Marrow

Desai (1968) described adult hamster bone marrow as being richly cellular with a myeloid–erythroid ratio of about 9:1. Hamsters less than 1 week old have predominantly erythroblastic marrow.

The largest cells observed in hamster bone marrow are megakaryocytes. Other cells normally associated with bone marrow are present along with eosinophils, plasma cells, histiocytes, and occasional mast cells.

REFERENCES

Alpert, M. (1950). Observations on the histophysiology of the adrenal gland of the golden hamster. *Endocrinology (Baltimore)* **46**, 166–176.

Bajusz, E., and Homburger, F. (1966). Myopathies. *Science* **152**, 1112–1123.

Bannerman, R. M. (1983). Hematology. *In* "The Mouse in Biomedical Research" (H. L. Foster, J. D. Small, and J. G. Fox, eds.), Vol. 3, pp. 293–312. Academic Press, New York.

Bannon, P. D., and Friedell, G. H. (1966). Values for plasma constituents in normal and tumor bearing golden hamsters. *Lab. Anim. Care* **16**, 417–420.

Bayard, S. P. (1974). Another look at the statistical analysis of changes during storage of serum specimens. *Health Lab. Sci.* **11**, 45–49.

Benjamin, M. M., and McKelvie, D. H. (1978). Clinical biochemistry. *In* "Pathology of Laboratory Animals" (K. Benirschke, F. M. Garner, and T. C. Jones eds.), Vol. 2, pp. 1749–1816. Springer-Verlag, Berlin and New York.

Betts, A., Tanguay, R., and Friedell, G. H. (1964). Effect of necrosis on hemoglobin, serum protein profile and erythroagglutination reaction in golden hamsters. *Proc. Soc. Exp. Biol. Med.* **116**, 66–69.

Boussios, T., McIntyre, W. R., Gordon, A. S., and Bertles, J. F. (1982). Receptors specific for thyroid hormones in nuclei of mammalian erythroid cells: Involvement in erythroid cell proliferation. *Br. J. Haematol.* **51**, 99–106.

Brink-Johnsen, T., Brink-Johnsen, K., and Kilham, L. (1981). Gestational changes in hamster adrenocortical function. *J. Steroid Biochem.* **14**, 835–839.

Brock, M. A. (1960). Production and life span of erythrocytes, during hibernation in the golden hamster. *Am. J. Physiol.* **198**, 1181.

Burns, K. F., and de Lannoy, C. W., Jr. (1966). Compendium of normal blood values of laboratory animals with indication of variations. I. Random-sexed populations of small animals. *Toxicol. Appl. Pharmacol.* **8**, 429–437.

Cincotta, A. H., and Meier, A. H. (1984). Circadian-rhythms of lipogenic and hypoglycemic responses to insulin in the golden-hamster (*Mesocricetus auratus*). *J. Endocrinol.* **103**, 141–146.

Coe, J. E., and Ross, M. J. (1983). Hamster female protein. A divergent acute phase protein in male and female Syrian hamsters. *J. Exp. Med.* **157**, 1421–1433.

Copeland, E. J., and Ginsberg, L. C. (1982). Major plasma glycoproteins of diabetic and nondiabetic Chinese hamsters. *J. Hered.* **73**, 311–313.

Cox, R. A., and Gökcen, M. (1973). A study of golden hamster (*Cricetus cricetus*) alkaline phosphatase isoenzymes. *Comp. Biochem. Physiol. B* **46B**, 99–103.

Cox, R. A., and Gökcen, M. (1974). Circulating lipids in the golden hamster (*Mesocricetus auratus*). *Comp. Biochem. Physiol. B* **49B**, 655–661.

Dent, N. J. (1977). The use of the Syrian hamsters to establish its clinical chemistry and hematology profile. *Clin. Toxicol.* **18**, 321–323.

Denyes, A., and Baumber, J. (1965). Comparison of serum total lipid during cold exposure in hibernating and non-hibernating mammals. *Nature (London)* **206**, 1050–1051.

Desai, R. G. (1968). Hematology and microcirculation. *In* "The Golden Hamster—Its Biology and Use in Medical Research" (R. A. Hoffman, P. F. Robinson, and H. Magalhaes, eds.), pp. 185–191. Iowa State Univ. Press, Ames.

Dodds, W. J., Raymond, S. L., Moynihan, A. C., and McMartin, D. N. (1977). Spontaneous atrial thrombosis in aged Syrian hamsters. II. Hemostasis. *Thromb. Haemostasis* **38**, 457–464.

Dooley, J. F. (1984). Sorbitol dehydrogenase and its use in toxicology testing in lab animals. *Lab Anim.* **13**, 20–21.

Emminger, A., Reznik-Schuller, H., and Mohr, U. (1975). Differences in blood values depending on age in laboratory-bred European hamsters (*Cricetus cricetus* L.). *Lab. Anim.* **9**, 33–42.

Eppenberger, M., Nixon, C. W., Baker, J. R., and Homburger, F. (1964). Serum phosphocreatine kinase in hereditary muscular dystrophy and cardiac necrosis of Syrian golden hamsters. *Proc. Soc. Exp. Biol. Med.* **117**, 465–468.

Eugster, A. K., Albert, P. J., and Kalter, S. S. (1966). Multiple enzyme determinations in sera and livers of tumor bearing hamsters. *Proc. Soc. Exp. Biol. Med.* **123**, 327–331.

Feldman, D. B., McConnell, E. E., and Knapka, J. J. (1982). Growth, kidney disease, and longevity of Syrian hamsters (*Mesocricetus auratus*) fed varying levels of protein. *Lab Anim. Sci.* **32**, 613–618.

Fiszer, M., Stankiewicz, D., and Adamczyk, A. (1979). Normal concentration and excretion values of sodium and potassium ions in urine of Syrian hamsters. *Folia Biol. (Krakow)* **27**, 51–57.

Fitts, D. A., Yang, O. O., Corp, E. S., and Simpson, J. B. (1983). Sodium retention and salt appetite following deoxycorticosterone in hamsters. *Am. J. Physiol.* **244**, R78–R83.

Fox, J. G., Cohen, B. J., and Loew, F. M., eds. (1984). "Laboratory Animal Medicine." Academic Press, New York.

Frenkel, J. G., Cook, K., Grady, H. J., and Pendleton, S. K. (1965). Effects of hormones on adrenocortical secretion of golden hamsters. *Lab. Invest.* **14**, 142–156.

Friedell, G. H., and Bannon, P. D. (1972). Experimental studies of anemia in golden hamsters. *Prog. Exp. Tumor Res.* **16**, 87–97.

Galton, V. A., and Galton, M. (1966). Thyroid hormone metabolism in the pregnant Syrian hamster (*Mesocricetus auratus*). *Acta Endocrinol. (Copenhagen)* **53**, 130–138.

Gerritsen, G. C., and Blanks, M. C. (1974). Characterization of Chinese hamsters by metabolic balance, glucose tolerance and insulin secretion. *Diabetologia* **10**, Suppl., 493–499.

Gleister, C. A., Van Hoosier, G. L., Sheldon, W. G., and Read, W. K. (1971). Amyloidosis and renal paramyloid in a closed hamster colony. *Lab. Anim. Sci.* **21**, 197–202.

Grodsky, G. M., Frankel, B. J., Gerich, J. E., and Gerritsen, G. C. (1974). The diabetic Chinese hamster: In-vitro insulin and glucagon release; the chemical diabetic; and the effect of diet on ketonuria. *Diabetologia* **10**, 521–528.

Handelman, C. T., and Blue, J. (1983). Laboratory data: Read beyond the numbers. *Comp. Contin. Educ. Pract. Vet.* **5**, 687–694.

Hoffman, R. A. (1968). Hibernation and effects of low temperature. *In* "The Golden Hamster—Its Biology and Use in Medical Research" (R. A. Hoffman, P. F. Robinson, and H. Magalhaes, eds.), pp. 25–40. Iowa State Univ. Press, Ames.

Holloway, D. A., and Heath, A. G. (1984). Ventilatory changes in the golden hamster, *Mesocricetus auratus*, compared with the laboratory rat, *Rattus norvegicus*, during hypercapnia and or hypoxia. *Comp. Biochem. Physiol. A* **77A**, 267–273.

Homburger, F., Nixon, C. W., Eppenberger, M., and Baker, J. R. (1966). Hereditary myopathy in the Syrian hamster: Studies on pathogenesis. *Ann. N.Y. Acad. Sci.* **138**, 14–27.

House, E. L., Pansky, B., and Jacobs, M. S. (1961). Age changes in blood of the golden hamster. *Am. J. Physiol.* **200**, 1018–1022.

Jones, A. L., and Armstrong, D. T. (1965). Increased cholesterol biosynthesis following phenobarbital induced hypertrophy of agranular endoplasmic reticulum in liver. *Proc. Soc. Exp. Biol. Med.* **119**, 1136–1139.

Kreienbühl, G., Strittmatter, J., and Ayim, E. (1976). Blood gas analyses of hibernating hamsters and dormice. *Pfluegers Arch.* **366**, 167–172.

Kutscher, C. (1968). Plasma volume change during water-deprivation in gerbils, hamsters, guinea pigs and rats. *Comp. Biochem. Physiol.* **25**, 929–936.

Lappenbusch, W. L. (1972). Effect of circadian rhythm on the radiation response of the Chinese hamster (*Cricetus griseus*). *Radiat. Res.* **50**, 600–610.

Loeb, W., and Quimby, F. (1987). "The Clinical Chemistry of Laboratory Animals." Pergamon Press, Elmsford, New York (in press).

Lyman, C. P., and Leduc, E. H. (1953). Changes in blood sugar and tissue glycogen in the hamster during arousal from hibernation. *J. Cell. Comp. Physiol.* **41**, 471–492.

Malan, A., and Arens, H. (1973). Pulmonary respiration and acid-base state in hibernating marmots and hamsters. *Respir. Physiol.* **17**, 45–61.

Matsuzawa, T., and Ikarashi, Y. (1979). Haemolysis of various mammalian erythrocytes in sodium chloride, glucose and phosphate-buffer solutions. *Lab. Anim.* **13**, 329–331.

Maxwell, K. O., Wish, C., Murphy, J. C., and Fox, J. G. (1985). Serum chemistry reference values in two strains of Syrian hamsters. *Lab. Anim. Sci.* **35**, 67–70.

Meyerstein, N., and Cassuto, Y. (1970). Haematological changes in heat-acclimated golden hamsters. *Br. J. Haematol.* **18**, 417–423.

Mitruka, B. M., and Rawnsley, H. M. (1977). "Clinical Biochemical and Hematological Reference Values in Normal Experimental Animals." Masson, New York.

Néve, P., Authelet, M., and Goldstein, J. (1981). Effect of aging on the morphology and function of the thyroid gland of the cream hamster. *Cell Tissue Res.* **220**, 499–509.

O'Brien, J. J., Jr., Lucey, E. C., and Snider, G. L. (1979). Arterial blood gases in normal hamsters at rest and during exercise. *J. Appl. Physiol.* **46**, 806–810.

Ottenweller, J. E., Tapp, W. N., Burke, J. M., and Natelson, B. H. (1985). Plasma cortisol and corticosterone concentrations in the golden hamster (*Mesocricetus auratus*). *Life Science* **37**, 1551–1557.

Pakes, S. P., Lu, Y. S., and Meunier, P. C. (1984). Factors that complicate animal research. *In* "Laboratory Animal Medicine" (J. G. Fox, B. J. Cohen, and F. M. Loew, eds.), pp. 649–666. Academic Press, New York.

Peterson, R. P., Turbyfill, C. L., Soderwall, A. L., and Yamanaka, J. S. (1961). Electrophoretic serum protein patterns in pregnant hamsters. *Anat. Rec.* **139**, 264 (abstr.).

Rennie, J. S., MacDonald, D. G., and Douglas, T. A. (1981). Haemoglobin, serum iron and transferrin values of adult male Syrian hamsters (*Mesocricetus auratus*). *Lab Anim.* **15**, 35–36.

Reznik, G., Reznik-Schuller, H., Emminger, A., and Mohr, U. (1975). Comparative studies of blood from hibernating and nonhibernating European Hamsters (*Cricetus cricetus* L.). *Lab. Anim. Sci.* **25**, 210–215.

Ringler, D. H., and Dabich, L. (1979). Hematology and clinical biochemistry. *In* "The Laboratory Rat" (H. J. Baker, J. R. Lindsey and S. H. Weisbroth, eds.), Vol. 1, pp. 105–121. Academic Press, New York.

Robinson, P. F. (1968). General aspects of physiology. *In* "The Golden Hamster—Its Biology and Use in Medical Research" (R. A. Hoffman, P. F. Robinson, and H. Magalhaes, eds.), pp. 111–118. Iowa State Univ. Press, Ames.

Rowland, N. (1984). Metabolic fuel homeostasis in golden hamsters: Effects of fasting, refeeding, glucose, and insulin. *Am. J. Physiol.* **247**, R57–R62.

Schindler, W. J., and Knigge, K. M. (1959). Adrenal cortical secretion by the golden hamster. *Endocrinology (Baltimore)* **65**, 739–747.

Schuchman, S. M. (1980). Individual care and treatment of rabbits, mice, rats, guinea pigs, hamsters, and gerbils. *Curr. Vet. Ther.* **7**, 741–767.

Sicart, R., Sable-Amplis, R., and Guiro, A. (1984). Comparative studies of the circulating lipoproteins in hamster (*Mesocricetus auratus*) with a normal or spontaneous high-level of cholesterol in the plasma. *Comp. Biochem. Physiol. A* **78A**, 511–514.

South, F. E., and Jeffay, H. (1958). Alterations in serum proteins of hibernating hamsters. *Proc. Soc. Exp. Biol. Med.* **98**, 885–887.

Statland, B. E., and Winkel, P. (1979). Sources of variation in laboratory measurements. *In* "Clinical Diagnosis and Management by Laboratory Methods" (J. B. Henry, ed.), pp. 3–28. Saunders, Philadelphia, Pennsylvania.

Takahashi, M., Nagase, S., Kokubo, T., and Hayashi, Y. (1981). Changes of amylase during experimental pancreatic carcinogenesis in hamsters. *Gann* **72**, 615–619.

Thomas, R. G., London, J. E., Drake, G. A., Jackson, D. E., Wilson, J. S., and Smith, D. M. (1979). "The Golden Hamster—Quantitative Anatomy with Age." Los Alamos Sci. Lab., University of California (sponsored by the United States Government).

Thompson, M. B., and Kunze, D. J. (1984). Polyethylene glycol-6000 as a cleaning agent for lipemic serum samples from dogs and the effects on 13 serum assays. *Am. J. Vet. Res.* **45**, 2154–2157.

Tietz, N. W. (1976). "Fundamentals of Clinical Chemistry." Saunders, Philadelphia, Pennsylvania.

Toffelmire, E. B., and Boegman, R. J. (1980). Erythrocyte life-span in dystrophic hamsters. *Can. J. Physiol. Pharmacol.* **58**, 1245–1247.

Turner, T. T., and Howards, S. S. (1977). Hyperglycemia in the hamster anesthetized with inactin (5-ethyl-5-(1-methylpropyl)-2-thiobarbiturate). *Lab. Anim. Sci.* **27**, 380–382.

Vaughan, M. K., Powanda, M. C., Richardson, B. A., King, T. S., Johnson, L. Y., and Reiter, R. J. (1982). Chronic exposure to short photoperiod inhibits free thyroxine index and plasma levels of TSH, T_4, triiodothyronine (T_3) and cholesterol in female Syrian hamsters. *Comp. Biochem. Physiol. A* **71A**, 615–618.

Vaughan, M. K., Brainard, G. C., and Reiter, R. J. (1984). The influence of natural short photoperiodic and temperature conditions on plasma thyroid-hormones and cholesterol in male Syrian-hamsters. *Int. J. Biomed.* **28**, 201–210.

Volkert, W. A., and Musacchia, X. J. (1970). Blood gases in hamsters during hypothermia by exposure to He-O_2 mixture and cold. *Am. J. Physiol.* **219**, 919–922.

Wardrop, K. J., and Van Hoosier, G. L., Jr. (1987). The hamster. *In* "The Clinical Chemistry of Laboratory Animals" (W. Loeb and F. Quimby, eds.). Pergamon Press, Elmsford, New York (in press).

Wendler, D. H., and Lyman, C. P. (1954). Body temperature, thyroid and adrenal cortex of hamsters during cold exposure and hibernation, with comparisons to rats. *Endocrinology (Baltimore)* **55**, 300–315.

Williams, D. L., Nun, R. F., and Marks, V. (1978). "Scientific Foundations of Clinical Biochemistry," Vol. 1. Year Book Med. Publ., Chicago, Illinois.

Chapter 4

Care and Management

Melvin W. Balk and Gilbert M. Slater

I. Introduction	61
II. Housing	61
III. Sanitation Procedures	62
IV. Environmental Requirements	62
V. Nutrition and Feeding	63
VI. Practical Breeding	63
VII. Handling, Restraint, Identification, and Sexing	64
A. Handling and Restraint	64
B. Identification	65
C. Sexing	65
References	66

I. INTRODUCTION

The care and management of the laboratory golden Syrian hamster (*Mesocricetus auratus*) has been fairly well defined and described within a variety of publications over the last several decades and is somewhat standard (Slater, 1972; Hoffman *et al.*, 1968). This chapter summarizes known information on this subject especially from the perspective of the authors.

II. HOUSING

The "mortar and brick" aspects of large- and small-scale production and housing of laboratory hamsters has been described in detail by other authors under the broad categories of rodent production and research facilities (Poiley, 1974; National Research Council, 1977). Design criteria of conventional and barrier facilities (Otis and Foster, 1983) exists in the current literature for review. In general, the design of facilities for rodents, though generally for mice and rats, is applicable to the laboratory hamster.

Hamsters may be housed in a variety of cages that are satisfactory for other small rodents. Hamsters have a propensity to escape from many primary enclosures by gnawing on the caging material or by dislodging the cage lid. Therefore, cages designed to house hamsters must be made of smooth, relatively hard material to preclude destruction and escape. Lids must also be designed in such a manner as to have a positive method of closure.

Acceptable hamster caging is generally made of rigid plastic (polycarbonate, polystyrene, polypropylene) and stainless or galvanized steel. Caging materials not acceptable for the hamster are wood, soft plastic, or aluminum.

For routine maintenance housing of hamsters, stainless-steel suspended, open wire mesh-bottomed cages are very practical. Solid-bottom cages with some form of contact bedding, however, provide a more comfortable environment which usually leads to more rapid growth and fewer stress-related deaths. Being a desert animal that instinctively burrows, the hamster, when not feeding or drinking, will generally stay in the darkest part of the cage. These animals also will hoard food and spend a considerable amount of time emptying their feed hopper and putting the food in a selected corner of their cage, which is usually in a different location from where they urinate and defecate.

For breeding colonies, suspended cages are not acceptable. Pregnant females and females with young litters should be placed in solid-bottom cages with a soft bedding material. Bedding materials such as hardwood chips, pine shavings, or ground corn cobs are all acceptable (Poiley, 1950; Festing, 1976). It has been the experience of the authors that the addition of some nesting material, such as facial tissue, will improve litter yields and also serve as an indicator of whether parturition has actually taken place.

Minimal caging size requirements are detailed in the regulations issued pursuant to the Animal Welfare Act (PL89-544 as amended by PL91-579) and the Animal Welfare Act, Code of Federal Regulations (1982), as follows:

1. The interior height of any primary enclosure used to confine hamsters shall be $5\frac{1}{2}$ in. except that in the case of dwarf hamsters, such interior height shall be at least 5 in.

2. A nursing female hamster, together with her litter, shall be housed in a primary enclosure which contains no other hamsters and which provides at least 121 in.^{2} of floor space: Provided, however, that in the case of dwarf hamsters such floor space shall be at least 25 in.^{2}

3. "The minimum amount of floor space per individual hamster and the maximum number of hamsters allowed in a single primary enclosure, except as provided for nursing females shall be:

Age	Minimum space per hamster (in.^{2})		Maximum population per enclosure
	Dwarf	Other	
Weanling to 5 weeks	5.0	10.3	20
5–10 weeks	7.5	12.5	16
≥10 weeks	9.0	15.0	13

Recommendations for caging space are also suggested by the Institute of Laboratory Animal Resources National Research Council Guide for the Care and Use of Laboratory Animals (1985) is as follows:

Weight of hamsters (gm)	Type of housing	Floor area per animal [cm^2(in.^{2})]		Height [cm(in)]	
<60	Cage	64.5	(10.0)	15.2	(6)
60–80	Cage	83.9	(13.0)	15.2	(6)
81–100	Cage	103.2	(16.0)	15.2	(6)
>100	Cage	122.6	(19.0)	15.2	(6)

III. SANITATION PROCEDURES

Sanitation of an animal room and the cages and equipment within the room refers not only to chemical or physical disinfection, but to the housekeeping practices consistent with good husbandry.

In general, weekly cleaning of cages and less frequent but routine cleaning and disinfecting of racks and equipment is standard operating procedure. Details of such sanitation are well described in the literature (Lang, 1983; Otis and Foster, 1983).

IV. ENVIRONMENTAL REQUIREMENTS

Laboratory hamsters are ancestors of original animals captured in Syria over 50 years ago. Similar to most desert animals, they can acclimate to environmental conditions found in the laboratory (Hoffman et al., 1968).

The recommended temperature range for hamsters, both in a maintenance and breeding situation, is between 68° and 75°F (20°–24°C) (Canadian Council on Animal Care, 1984).

A relative humidity range of 45–55% is recommended for hamsters. However, there is no evidence that humidity outside of this range is detrimental to laboratory hamsters (Canadian Council on Animal Care, 1984).

The light cycle which has been successful for optimal reproduction in the hamsters is 14 hr light and 10 hr dark, compared to a standard 12 hr on, 12 hr off for most other laboratory rodents (Slater, 1972). Most hamster matings are accomplished by placing females in estrus with breeding males (and mating) as opposed to the typical monogamous or polygamous family groups set up with mice and rats. The female hamster is quite pugnacious and will frequently not tolerate the male after mating has taken place. When hamsters are mated as a monogamous pair or family group, the males are frequently injured and litter averages reduced due to constant fighting. Hamsters reach peak estrus approximately 1 hr prior to darkness (Slater, 1972). Knowing this, it is relatively simple to set up a semi-reversed lighting schedule which will permit matings to take place toward the end of the normal working day. In the experience of one of the authors (Slater), 24-hr light, either by accident or as a part of experimental design, will radically affect and potentially totally eliminate reproduction.

Hamsters, by nature, will hibernate under appropriate environmental conditions, although normal outbred hamsters are relatively poor hibernators. To induce hibernation in the hamster requires special environmental chambers that provide appropriate temperature (~5°C) and allow the hamsters to remain relatively undisturbed. Under conditions that normally exist in a laboratory, there should be no triggers to induce hibernation.

Under some environmental conditions, especially warm, ambient temperatures, hamsters will go into a deep sleep. In this condition, even normal handling will fail to arouse the animal. Care should be taken in handling hamsters in this condition, as upon awakening they may be pugnacious.

The hamster is an active animal under normal laboratory conditions, especially at night. It has been described that females who are actively reproducing can travel several kilometers per day when measured on an appropriate exercise wheel (Richards, 1966).

Excessive noise, especially with breeding hamsters, is contraindicated. Female hamsters that are nursing young litters may cannibalize their litter if disturbed by loud, unexpected noise (or by rough handling). Females with nursing litters are also particularly aggressive, and care should be taken when disturbing them while they are with their young. The appropriate and safest method of handling an animal in this condition is to allow her to vacate the nest on her own and then pick her up to remove her from the cage.

V. NUTRITION AND FEEDING

The hamster is an animal that has been used at variable levels in the laboratory since its introduction as a research animal. In spite of its significance as a research animal, relatively little recent work has been done to define the specific nutritional requirements of this omnivorous rodent (Cooperman et al., 1943; Hamilton and Hogan, 1944). It is generally accepted that the laboratory diets specifically formulated for rats and mice are adequate for the hamster (National Research Council, 1978).

Most laboratory rodent diets contain approximately 22% protein, which has been shown empirically to be sufficient to support all aspects of the hamster life cycle: growth, reproduction, lactation, and maintenance. From information derived from other laboratory rodents, a protein level of less than 10% is too low, especially for growth and reproductive performance (Harkness and Wagner, 1983; Stoliker, 1981). However, also based on other rodent data, it is most probable that a level of below 22%—probably in the 18–19% range—is sufficient for hamsters (Arrington et al., 1966). Banta and co-workers (1975) described the function of the forestomach of the hamster as a buffer to protein quality changes, making this animal less dependent on dietary protein quality than the laboratory rat.

One of the more common nutritional problems in hamsters is vitamin E deficiency. It has been described by West and Mason (1958) that muscular dystrophy can be produced in the laboratory hamster by feeding a diet deficient in vitamin E beginning at weaning time. Most commercial diets produced today have vitamin E supplementation. However, if special formulas or purified or semipurified diets are used, it is possible to induce this clinical manifestation inadvertently, and it should be considered if muscular weakness is noted in a hamster colony.

Keeler and Young (1979) demonstrated that vitamin E deficiency in diets fed to pregnant hamsters can result in central nervous system hemorrhagic necrosis in the fetal hamster. Since hamsters are useful in teratology studies this condition should be kept in mind.

The hamster, being a hoarding animal by nature, will remove feed pellets from the feeder and relocate them to a location on the cage floor. Although having food in the cage may cause some concern on the part of the animal care staff relative to sanitation, the practice is almost impossible to stop. In addition, hamsters will invariably pile feed in the cage in a corner opposite from the corner used for urination and defecation. There is some question as to whether feeders should be used for hamsters. The Animal Welfare Act states that it is acceptable to place pelleted feed on the floor of a primary enclosure. Additionally, other authors have indicated that hamsters have difficulty eating from feeders (Harkness and Wagner, 1983).

Food is generally offered once weekly in pelleted form, at which time all old feed should be removed if feeders are not used. In the past, pelleted diets have been supplemented with vegetables or fruits. The hamster may prefer the fruits and vegetables and may avoid the nutritionally balanced diet in favor of the supplement. There is no nutritional reason in good colony management for supplemental feeding for hamsters. Additionally, any food which cannot be properly treated to remove potentially pathogenic organisms should not be fed in an animal colony.

The practice of feeding fruits and vegetables originated with the observation that litter averages were increased by providing vegetables to females with young nursing litters (Slater, 1972). It is apparent that the vegetables provided a source of needed additional moisture for young nursing pups which were unable to reach sipper tubes placed at a height convenient for adult animals. Similar reproductive performance can be achieved by placing sipper tubes at a level approximately $\frac{1}{2}$ to $\frac{3}{4}$ in. above the level of bedding in cages containing litters of suckling hamsters between 10 and 20 days of age (Slater, 1972).

Hamsters will eat approximately 5 to 7 gm of food and will drink 10 ml of water daily. Fresh potable water should be available *ad libitum,* supplied by bottles or from an automatic watering system.

VI. PRACTICAL BREEDING

Sexual maturity of the Golden Syrian hamster (*Mesocricetus auratus*) is generally reported to occur at approximately 6 weeks (42 days) of age. Male golden hamsters have spermatozoa present on the glans penis at this age. The penile smear technique to observe for the presence of sperm is a convenient and rapid technique to identify sexual maturity in individual male hamsters (Vandenbergh, 1971). Reproductive activity at a very young age is a hallmark trait of the hamster and, on occasion, it is not uncommon to find that these animals reproduce by 1 month of age. This precocious nature of the hamster leads to operational problems when litters of hamsters are kept together following weaning. Inadvertent pregnancies can and will occur, and this should be recognized by investigators using these animals. In the experience of one of the authors (Slater), maximum reproduction is achieved if female Syrian hamsters are 8–10 weeks of age and male hamsters 10–12 weeks of age prior to the initial mating. This allows the male to be slightly older than the more aggressive female. In addition there is a definite relationship between maternal weight and reproductive efficiency (Robens, 1968).

The female hamsters is a relatively pugnacious animal in the presence of the male and, in general, unless the female is sexually receptive, she will not tolerate the presence of male hamsters. There is some evidence to indicate that olfaction is not necessary for the appearance of sexual receptivity in the female golden hamster as with other rodents (Carter, 1973).

The estrous cycle of the female hamster is quite regular and lasts 4 days. An obvious external indication of this cycle includes the presence of a white, stringy, opaque discharge on the second day of the cycle, followed by a waxy secretion noted on day 3 of the estrous cycle. Determination of the appropriate day for mating of female hamsters can be made by screening groups of females for the stringy, opaque discharge. Since this is indicative of the hamsters having reached peak estrus the day before, one can reliably predict that these animals will achieve peak estrus again on the third day after the appearance of the discharge. In the experience of one of the authors (Slater), large groups of hamsters can be selected in this manner and mated with approximately a 90% take rate on day 3, 7, 11, and so on, postscreening.

There are two lateral pouches lined with cornified epithelial cells in the vagina of the hamster, and this anatomical finding causes confusion if one tries to use the vaginal cytology method for evaluating the phase of the reproductive cycle of the hamster similar to what is used in other mammalian species (Orsini, 1961). Refer to the section on environmental requirements (Section IV) for information on the effects of light on the estrous cycle and hamster reproduction.

At peak estrus, the female hamster tolerates the male's presence and almost immediately upon introduction to the male, lordosis takes place. Shortly after introduction of the female into the cage, copulation will take place and last approximately 30 min. Because of the aggressive nature of the female hamster, most matings are set up as a monogamous system with male and female occupying separate cages. In such a system, 1 male can usually maintain a "harem" of 12 females.

On the sixth day postcoitus, implantation of the fertilized ovum takes place (Hoffman *et al.*, 1968). It is important at this time in the reproductive cycle of the female hamster that minimal handling takes place. Following implantation, embryogenesis proceeds quite rapidly for the next 36 hr, followed by organogenesis, which proceeds for the next 3 days (Robens, 1968; Hoffman *et al.*, 1968).

Teratology studies are often conducted using the hamster because there is rapid differentiation of the embryo during day 8 of gestation (Ferm, 1967). The hamster has the shortest gestation period of any laboratory animal (15.5 days postcoitus), and the gestation period is generally very regular, varying by only 2 or 3 hr under defined ambient conditions.

Gravid females should be transferred to clean cages, with some form of nesting material in addition to bedding (e.g., facial tissue) provided approximately 2 days prior to parturition. Enough food should be made available in the cage to last the female hamster 7–10 days, so that there is minimal interference with the newborn litter. Such interference could result in a high incidence of cannibalism. Although cannibalism is generally high in primiparous females, it does not necessarily disappear in multiparous animals, and relative isolation during the first 7–10 days postparturition is important to minimize litter loss.

Generally, the first litter of a female hamster is smaller in number than subsequent litters. The average litter size of the Syrian hamster is 11, with a range of 4–16 (Slater, 1972). Some developmental characteristics that exist in a newborn hamster include (a) eyes and ears closed at birth; (b) hairless at birth; (c) teeth present at birth; (d) ears first begin to open at 4–5 days; (e) young hamsters begin to eat solid food at around 7–10 days; and (f) eyes begin to open around 14–16 days of age. At approximately 10 days of age, it is important to make sure that water is readily available to the young animals, since they will commence eating solid food at this time.

In most cases, hamsters are weaned at approximately 3 weeks of age, and the female is generally remated at this time. Female hamsters will usually mate on the second or third day postweaning. A postpartum estrus has been described in the literature; however, based on experience (Slater), in practice it does not exist.

In the experience of one of the authors (Slater), the optimal reproductive life of a hamster is 10 months of age, and a significant reduction in reproductive capacity occurs by 1 year of age. During this reproductive lifetime, a female hamster will produce between four and six litters of pups.

VII. HANDLING, RESTRAINT, IDENTIFICATION, AND SEXING

A. Handling and Restraint

Proper handling of hamsters is important and somewhat different from handling rats and mice. As with any animal, it is

4. CARE AND MANAGEMENT

important to avoid startling the hamster and to let him know you are going to handle him.

Protective gloves should not be used when handling hamsters. Gloves inhibit digital dexterity and feeling and generally result in discomfort on the part of the hamster. Additionally, this begins an association between a glove and discomfort, resulting in a naturally aggressive hamster, and an unhappy technician.

There are several acceptable ways to pick up and manipulate hamsters. The easiest and most commonly used for routine inspection and transfer is to grasp the hamster around the head and shoulders as if picking up a baseball. The hamster will often sit up when the cage is opened, which makes this method very natural (Fig. 1).

When it is important that the hamster remain firmly in hand, the following method is usually quite effective. Remove the hamster from his cage as described above to a flat surface such as the cage lid. Place a hand palm down over the hamster with your thumb near the head. Close your hand slowly and you will feel the loose skin begin to bunch under your hand. Continue closing your hand, grasping skin but not body. As you

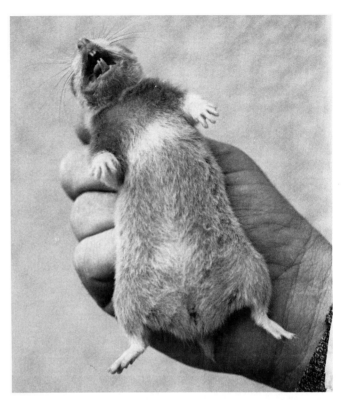

Fig. 2. Hamster restrained by grasping loose skin along entire dorsal area. Ideal for intraperitoneal injection.

pick the hamster up you will find him securely held with the skin taut over chest and abdomen (Fig. 2).

B. Identification

The easiest way to identify hamsters is to house them singly and use an appropriately labeled cage card without any physical marking on the animal itself. Some studies, however, require positive individual identification, and this can be accomplished by ear punching or placing a lightweight, durable ear tag through the pinnae of the animal.

Toe clipping, under appropriate anesthesia, is also a method of identifying an individual animal within a relatively small population.

C. Sexing

Sexing of the hamster, like that of most small laboratory rodents, can be difficult and error-prone to the untrained eye. The scrotum of the male partially occludes the short tail and generally will be filled with the testicles (Fig. 3). The hamster can withdraw his testes into the inguinal canal, and one may only see a wrinkled empty scrotum. The female has an obvious external genital opening, and, since animals generally urinate

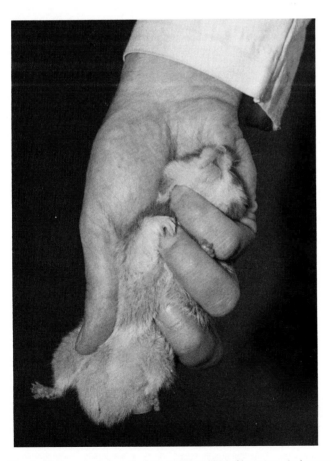

Fig. 1. Most convenient method for rapid handling of hamsters. As long as a hamster feels secure, there is little likelihood of being bitten.

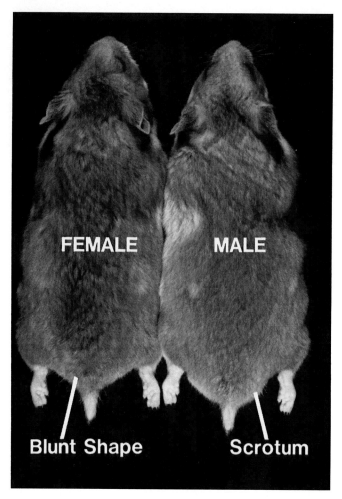

Fig. 3. Dorsal view of female and male hamsters noting anatomical differences.

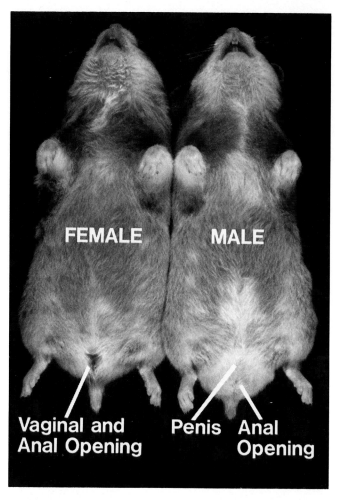

Fig. 4. Ventral view of female and male hamsters noting anatomical differences.

upon handling, this is usually moist (Fig. 4). The distance between the anal and penile openings in the male is much greater than the distance between the anal and vaginal openings in the female.

REFERENCES

Animal Welfare Act, Code of Federal Regulations (1982). Title 9—Animals and Animal Products; Subchapter A—Animal Welfare, Part 3—Standards; Subpart B—Specifications for the Humane Handling, Care, Treatment, and Transportation of Guinea Pigs and Hamsters. U.S. Govt. Printing Office, Washington, D.C.

Arrington, L. R., Platt, J. K., and Shirley, R. L. (1966). Protein requirements of growing hamsters. *Lab. Anim. Care* **16,** 492.

Banta, C. A., Warner, R. C., and Robertson, J. B. (1975). Protein nutrition of the Golden hamster. *J. Nutr.* **105,** 38–45.

Canadian Council on Animal Care (1984). "Guide to the Care and Use of Experimental Animals," Vol. 2, Chapter 15. CCAC.

Carter, C. S. (1973). Olfaction and sexual receptivity in the female Golden hamster. *Physiol. Behav.* **10,** 47–51.

Cooperman, J. M., Waisman, H. A., and Elvehjem, C. A. (1943). Nutrition of the golden hamster. *Proc. Soc. Exp. Biol. Med.* **52,** 50.

Ferm, V. H. (1967). The use of the golden hamster in experimental teratology. *Lab. Anim. Care* **17,** 452.

Festing, M. F. W. (1976). Hamsters. "The UFAW Handbook on the Care and Management of Laboratory Animals," 5th ed. Churchill-Livingstone, Edinburgh and New York.

Hamilton, J. W., and Hogan, A. G. (1944). Nutritional requirements of the Syrian hamster. *J. Nutr.* **27,** 213–219.

Harkness, J. E., and Wagner, J. E. (1983). The Biology and Medicine of Rabbits and Rodents," 2nd ed. Lea & Febiger, Philadelphia, Pennsylvania.

Hoffman, R. A., Robinson, P. F., and Magalhaes, H., eds. (1968). "The Golden Hamster—Its Biology and Use in Medical Research." Iowa State Univ. Press, Ames.

Institute of Laboratory Animal Resources (1985). "Guide for the Care and Use of Laboratory Animals," NIH Publ. No. 85-23. National Research Council, Washington, D.C.

Keeler, R. F., and Young, S. (1979). Role of Vitamin E in the etiology of spontaneous hemorrhagic necrosis of the central nervous system of fetal hamsters. *Teratology* **20,** 127.

Lang, C. M. (1983). Design and management of research facilities for mice. *In* "The Mouse in Biomedical Research" (H. L. Foster, J. D. Small, and J. G. Fox, eds.), Vol. 3, pp. 37–49. Academic Press, New York.

National Research Council (1977). Laboratory animal management. Rodents. *ILAR News* **20,** L1–L15.

National Research Council (1978). "Nutrient Requirements of Laboratory Animals." N.R.C., Natl Acad. Sci., Washington, D.C.

Orsini, M. W. (1961). The external vaginal phenomena characterizing the states of the estrous cycle, pregnancy, pseudopregnancy, lactation, and the anestrous hamster, *Mesocricetus auratus*. *Proc. Anim. Care Panel* **11,** 193–206.

Otis, A. P., and Foster, H. L. (1983). Management and design of breeding facilities. *In* "The Mouse in Biomedical Research" (H. L. Foster, J. D. Small, and J. G. Fox, eds.), Vol. 3, pp. 18–36. Academic Press, New York.

Poiley, S. M. (1950). Breeding and care of the Syrian hamster. *In* "The Care and Breeding of Laboratory Animals" (E. J. Farris, ed.). Wiley, New York.

Poiley, S. M. (1974). Housing requirements—General considerations. *In* "Handbook of Laboratory Animal Science" (E. C. Melby, Jr. and N. H. Altman, eds.), Vol. 1, pp. 21–60. CRC Press, Cleveland, Ohio.

Richards, M. P. M. (1966). Activity measured by running wheels and observation during the oestrous cycle, pregnancy, and pseudopregnancy in the golden hamster. *Anim. Behav.* **14,** 450.

Robens, J. F. (1968). Influence of maternal weight on pregnancy, number of corpora lutea, and implantation sites in the Golden hamster (*Mesocricetus auratus*). *Lab. Anim. Care* **18,** 651–653.

Slater, G. M. (1972). The care and feeding of the Syrian hamster. *Prog. Exp. Tumor Res.* **16,** 42–49.

Stoliker, H. E. (1981). The golden hamster. *Comp. Contin. Educ. Pract. Vet.* **3,** 145.

Vandenbergh, J. G. (1971). The penile smear: An index of sexual maturity in male Golden hamsters. *Biol. Reprod.* **4,** 234–237.

West, W. T., and Mason, K. E. (1958). Histopathology of muscular dystrophy in the Vitamin E deficient hamster. *Am. J. Anat.* **102,** 323.

Chapter 5

Biomethodology

Jerald Silverman

I.	Introduction	70
II.	Clinical Examination	70
	A. General Examination	70
	B. Rectal Temperature	71
III.	Radiographic Techniques	71
	A. Plain Radiographs	71
	B. Intravenous Urography	71
IV.	Injections and Intubations	72
	A. General Safety	72
	B. Subcutaneous Injections	72
	C. Intramuscular Injections	72
	D. Intraperitoneal Injections	72
	E. Intravenous Injections	72
	F. Gastric Intubation	73
	G. Intrarectal Instillations	74
	H. Intratracheal Instillations	74
	I. Intrapulmonary Instillations	74
	J. Central Nervous System	75
	K. Intranasal Injections	75
V.	Collections	75
	A. Blood	75
	B. Saliva	76
	C. Bronchopulmonary Lavage	76
	D. Ejaculate	76
	E. Vaginal Mucus	77
	F. Urine, Feces, and Expired Gases	77
	G. Milk	77
	H. Bile	77
VI.	Anesthesia	77
	A. General Comments	77
	B. Inhalants	78
	C. Injectables	79
	D. Hypothermia	82
VII.	Surgical Procedures	82
	A. Basic Techniques	82
	B. Adrenalectomy	82
	C. Adrenal Demedullation	82

Copyright © 1987 by Academic Press Inc.
All rights of reproduction in any form reserved.

	D. Nephrectomy	83
	E. Renal Papillectomy	83
	F. Cystotomy	83
	G. Experimental Hypertension	83
	H. Thymectomy	83
	I. Thyroparathyroidectomy	84
	J. Hypophysectomy	84
	K. Pinealectomy	84
	L. Ovariectomy	84
	M. Castration	85
	N. Vasectomy	85
	O. Cesarean Derivation	85
	P. Sialectomy	85
	Q. Splenectomy	85
	R. Cholecystectomy and Cholecystotomy	85
	S. Cholecystoduodenostomy	86
	T. Pancreatic Duct Ligation and Pancreatitis	86
	U. Partial Pancreaticocolostomy	86
	V. Superior Cervical Ganglionectomy	86
	W. Neonatal Enucleation	86
	X. Esophageal Constriction	86
VIII.	Physiological Measurements	87
	A. Respiratory Functions	87
	B. Oxygen Consumption	87
	C. Electrocardiogram	88
	D. Arterial Blood Gases	88
	E. Blood Pressure	88
IX.	Perfusions	88
	A. Bilateral Perfusion of Adrenal Glands	88
	B. Liver Perfusion	89
X.	Laparoscopy	89
XI.	Miscellaneous Techniques	89
	A. Esophageal and Pharyngeal Decontamination	89
	B. Cutaneous Application of Chemicals	89
	C. Transplantation into the Anterior Chamber of the Eye	89
XII.	Euthanasia	89
XIII.	Necropsy	90
	References	91

I. INTRODUCTION

Biomethodological procedures using the Syrian golden hamster do not differ significantly from those involving most other laboratory rodents. The reader is referred to a number of such reviews (e.g., Kraus, 1980; Petty, 1982; Cunliffe-Beamer, 1983) for techniques that are not specifically described in this chapter. This chapter will discuss many techniques which have been developed for the hamster, as well as certain techniques used in the author's institute.*

*The inclusion of a literature reference does not imply an endorsement of the technique either by the author or by The Ohio State University.

II. CLINICAL EXAMINATION

A. General Examination

It may be appropriate to ascertain that an apparently dead hamster is truly dead, as the animal may be in a state of estivation or hibernation (Moller, 1968). The hamster should be disturbed or warmed to determine that these conditions do not exist. For a description of routine handling of hamsters, please refer to Chapter 4.

A clinical history of the hamster to be studied is obtained prior to examination. A detailed checklist for the physical examination has been given by Schuchman (1977). From his

5. BIOMETHODOLOGY

checklist and the author's personal experience, it is suggested that particular attention be given to the animal's weight, diet, water consumption, integument, fecal soiling, respiratory system, performance of a fecal flotation, notation of past medication, and examination of its teeth.

The pelage can be examined, and in older animals the signs of *Demodex* infestation may be clearly noticed. If needed, a skin scraping may be performed by having an assistant restrain the animal on the cage or tabletop, while the operator uses standard methodology to perform the scraping. Skin and fur samples may be taken in like manner for other purposes, such as culturing.

After grasping the animal by the dorsum, it may be further examined by one operator, or, if possible, one person restrains the hamster while another proceeds with the examination. Bite wounds and abscesses are not unusual, particularly when the more aggressive females are housed together. The mouth should be examined for overgrown incisors, but the clinician should not forget that on occasion the molars may also become overgrown. A hemostat or other instrument may be used to pry open the mouth gently to examine the back teeth. Johansen (1952) has described the construction of a mechanical apparatus to hold an unanesthetized hamster's mouth open, although caution is advised until experience is gained. If it is necessary to remove food from the cheek pouches of the awake animal, this can frequently be done with a hemostat or by using the cheek pouch everting technique of Haisley (1980). The latter technique should be practiced beforehand, as the beginner may be bitten.

B. Rectal Temperature

By exercising care, a standard thermometer may be used to determine the rectal temperature (normally 37°–38°C in the adult animal) but it is far more convenient to use an electronic thermometer with a semirigid probe. This is particularly true as the lack of a substantial tail frequently creates a problem in attempting to insert the thermometer.

A unique method of recording body temperature has been developed by Pickard *et al.* (1984). Adult hamsters were placed under anesthesia and a temperature-sensitive radiotelemeter, which emitted signals at a rate proportional to the body temperature, was placed in the abdominal cavity. The telemeter was secured to the abdominal muscle with physiological cement. With fresh batteries, the telemeter gave usable signals for 2–3 months. The signals were picked up by wire antennas on the cages and were then transmitted to an AM radio and ultimately to a computer.

In male hamsters, Chaudhry *et al.* (1958) found higher rectal temperatures (~36.8°C) between 1800 and 0600 hr than during daylight periods (0600 to 1800 hr) when it was approximately 35.8°C. In females the recorded temperatures were 37.8°C and 36.8°C, respectively.

Desautels *et al.* (1985) used an electronic thermometer with a semiflexible probe inserted into the rectum and taped to the tail to measure male hamster body temperatures. At room temperature, 6- to 14-day-old animals had a temperature of 35.1°C, 30-day-olds were 37.0°C, 100-day-olds were 37.2°C, and 160-day-olds were 37.4°C.

III. RADIOGRAPHIC TECHNIQUES

A. Plain Radiographs

Satisfactory radiographs may be made with an unsedated animal, but sedation increases the probability of obtaining a diagnostic film. L. A. Bauck (personal communication, 1984) uses approximately 3 mg of ketamine hydrochloride, intramuscularly. Halothane can be given via a small mask in an open system using 0.5–1.5% halothane in 2 liters of oxygen per minute. She notes, however, that one often needs 4% halothane for relatively long periods of time to anesthetize or sedate the hamster initially.

Matthews and Barnhard (1968) note that hamsters can be stretched out on a small rectangular board, which extends an inch or more beyond the extremities, with the animal tied to each corner. A radiograph of the entire body may then be taken in various views. For a whole-body view the beam is directed to the center of the animal. L. A. Bauck (personal communication, 1984) notes that the animal can be taped directly to an intensifying screen with masking tape, limbs fully extended.

Satisfactory radiographs may be obtained using Kodak X-Omatic Fine Intensifying Screen or DuPont Cronex film, with a machine set at 200 mA, 60 kV, and one-fifteenth of a second (L. A. Bauck, personal communication, 1984). General information on laboratory animal radiography may be found in references by Carlson (1965) and Matthews and Barnhard (1968). Table I provides a radiographic technique that can be used for animals the size of hamsters.

B. Intravenous Urography

For intravenous urography, any clinical radiographic intravascular contrast medium can be used. As described by Elkin and Kaplan (1968), direct exposure of the jugular or femoral vein is accomplished, and using a 25- to 27-gauge needle, the medium is injected. Approximately 1 to 2 ml/kg of body weight should prove satisfactory. An appropriate exposure factor is 38 kV at 400 mA at 1/20 of a second. Focal film distance is 36 in., and par speed film is used.

Table I
Radiographic Technique for Small Mammals[a,b]

Sample thickness (cm)	FFD (in.)	kVP	mA	Time (sec)	MaS
Bone					
0.5	36	40	100[c] (Fine focal spot)	1/30	3.3
1		42			
2		44			
3		46			
4		48			
5		50			
6		52			
7		54			
Soft tissue					
1	36	38	100 (Fine focal spot)	1/30	3.3
2		40			
3		42			
4		44			
5		46			
6		48			
7		50			
8		52			
9		54			
Thoracic tissue					
2	36	34	200	1/60	3.3
3		36			
4		38			
5		40			
6		42			
7		44			
8		46			

[a]Modified from Schuchman (1977).
[b]FFD, Focal film distance; MaS, Milliamp-seconds.
[c]If animal is immature, it might be better to use 50 mA.

IV. INJECTIONS AND INTUBATIONS

A. General Safety

With the exception of intravenous injections, most samplings, injections, and intubations are performed as in rats and mice. For laboratory procedures that require the withdrawal of hazardous substances from bottles, an alcohol-moistened cotton pledget can be kept at the point where the needle enters the stopper in order to minimize the inadvertent formation of aerosols. Air bubbles can be dislodged by gently tapping the syringe and slowly expelling the air into absorbent tissue until fluid appears at the needle's end. Needle sizes used will vary with the viscosity of the substance being used, but in general 25- to 27-gauge, $\frac{1}{2}$- to $\frac{5}{8}$-in. (12.5- to 15.6-mm) needles are satisfactory.

B. Subcutaneous Injections

Subcutaneous injections are most frequently given in the inguinal or interscapular areas, although the inguinal area is preferred. It may be necessary to vary the sites if frequent injections are to be given. Usually it is easier to restrain the animal by lifting than by placing it on a solid surface. The needle is advanced through the skin and almost parallel to it. The needle is moved from side to side to ascertain that its location is neither intradermal nor intramuscular, and then the injection is made. For interscapular injections, the dorsal skin is lifted, the needle inserted, and the above procedure followed.

C. Intramuscular Injections

The most common site for intramuscular injection is the muscle mass posterior to the femur, but infrequently the lumbar epaxial muscles are used. The needle is advanced into the muscle, and the injection is made.

D. Intraperitoneal Injections

Intraperitoneal injections are made immediately to the right or left of midline, slightly caudad to the area where the umbilicus would be if it were visible. It is recommended that the animal's head be tilted downward. Eldridge et al. (1982) stated that "to avoid injecting the anesthetic into the abdominal viscera, each animal was held in a supine position with its posterior end slightly elevated. The needle was inserted approximately $\frac{1}{4}$ in. (6–7 mm) through the skin of the posterior abdomen, slightly lateral to the midline." On occasion, the paralumbar fossa (lateral to the lumbar muscles, craniad to the ileum, and caudad to the ribs) has been used for intraperitoneal injections.

E. Intravenous Injections

Due to the lack of a readily accessible tail vein, intravenous injections in hamsters are most frequently made into the cephalic vein, into the jugular vein, or into the lateral vein of the tarsus on the lateral side of the hindlimb, as it courses between the femorotibial and tibiotarsal joints.

1. Lateral Vein of Tarsus (Vena Plantaris Lateralis)

To inject into the lateral vein of the tarsus, the animal is anesthetized or placed in a restraining device. The area over the vein is clipped or shaved, and the vein is dilated by placing pressure on the upper part of the thigh, between the fingers (Grice, 1964). The needle (27- to 30-gauge) is advanced into the vein, almost parallel to it, with the bevel pointing upward, and the injection made. Gentle aspiration on the plunger may allow the visualization of blood in the syringe, but the vein usually collapses before blood enters the syringe.

2. Jugular Vein

Injections or catheterizations into the jugular vein are performed under general anesthesia, using surgical techniques described for other species (e.g., Petty, 1982).

3. Anterior Cephalic Vein

Injections into the anterior cephalic vein have been described by Green *et al.* (1978) and Ransom (1984), and have long been used in the author's laboratory in preference to the lateral tarsal vein, which is more difficult to inject. Using anesthesia, the forelimb is shaved on its anterior surface. Green *et al.* (1978) have used a 26-gauge needle inserted into the vein and taped in place. Ransom (1984) describes a technique for injecting unanesthetized hamsters via the cephalic vein. An assistant holds the animal by the skin of the neck and a hindlimb. A rubber band is placed around the proximal foreleg to dilate the vein, which is swabbed with alcohol. The injection is then made with a 25-gauge needle on a 1.0-ml syringe, and after the injection finger pressure is used to minimize hematomas. The author anesthetizes the hamster and makes the injection using a 27- or 30-gauge needle.

4. Lingual Vein

Lingual vein injections may be made by placing the anesthetized hamster on its back, gently grasping the tongue with gauze, and making the injection using a 1-in. (2.54-cm), 27-gauge needle on a short-barreled syringe. Up to 2.0 ml may be given over a 1-min period (Ferm, 1967).

5. Ophthalmic Venous Sinus

Pinkerton and Webber (1964) describe a method for injecting into the ophthalmic venous sinus, with or without anesthesia. A needle is pushed into the medial canthus of the eye, with the bevel toward the skull. When bone is felt, the needle is slid downward until a bony ledge is felt, and the injection is then made.

F. Gastric Intubation

Gastric intubation (gavage) is easily performed in the adult hamster using an 18-gauge, 4.0- to 4.5-cm ball-tipped feeding needle. With the thumb and index finger around the animal at the axillary region, with the palm dorsal to the animal, the needle is placed into the mouth and gently slid posteriorly along the hard palate and down the esophagus. If resistance is felt, do not advance the needle any further, but begin again. The shaft of the needle is used as a lever to maintain the head in temporary extension. If the animal reacts violently when the needle is in place, let go of the syringe to minimize esophageal damage. As an alternative, the animal may be grasped by the loose dorsal cervical skin.

Rolfe and Iaconis (1983) have used a 23-gauge feeding needle to administer 0.1 ml of cell suspension to hamsters from 1 to 22 days of age. The size of the ball on the needle was not noted. They covered the young animals with fragrant baby powder after treatment to prevent maternal cannibalism. The author has not attempted to use a feeding needle on 1-day-old animals, although using the technique of Smith and Kelleher (1973), as described for rats, he has easily passed a 0.025-in. (0.63-mm O.D.) silicon catheter into the stomach of newborn hamsters. The animal is held as is the neonatal rat in Fig. 1 and, using a small forceps, the tube is slowly and gently passed into the stomach. It should be noted that the use of a silicon catheter is essential.

A 15-in.-long (38 cm) size 8 French infant feeding tube was used by Herrold (1969) to intubate 1-month-old hamsters. The

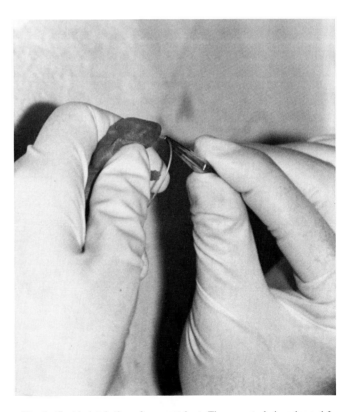

Fig. 1. Gastric intubation of a neonatal rat. The same technique is used for the neonatal hamster.

tip was guided into the forestomach and 0.5 ml of solution was given twice a week for 4 months. David and Harrison (1984) gave hamsters aged 1–30 days 10–100 μl of liquid from the tip of a micropipette. The solution was readily swallowed.

G. Intrarectal Instillations

Due to the relative ease of perforation of the bowel wall, and the large amount of feces found in the rectum, hamsters are lightly anesthetized prior to intrarectal instillation. After administering an anesthetic agent, place the animal in dorsal recumbency, and, using an 18-gauge, 4.0- to 4.5-cm ball-tipped feeding needle attached to a syringe, advance the needle slowly through the anus, usually at the full length of the needle, and slowly make the instillation (Fig. 2). The operator should be careful to hold the skin of the anus somewhat taut, as otherwise this skin can easily be pushed into the rectum with the ball-tipped needle.

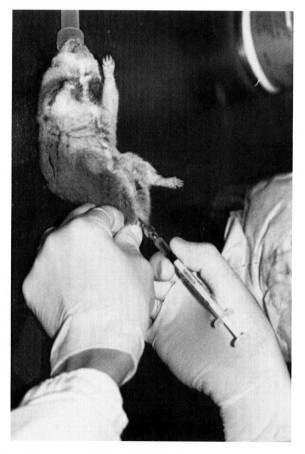

Fig. 2. Intrarectal administration of a liquid in a lightly anesthetized hamster.

H. Intratracheal Instillations

Intratracheal instillations have frequently been performed in the Syrian hamster for cancer research purposes, and a number of different techniques have been described. Hranicka *et al.* (1977) have described a fiberoptic laryngoscope modification that can facilitate tracheal cannulation and intratracheal injections in any small laboratory animal.

Saffiotti *et al.* (1968) placed the anesthetized animal in dorsal recumbency on a slanted board, with its head up, and suspended from its upper incisors. The mouth was kept open by holding the lower incisors down by a wire hook attached to the board. A 0.25-ml tuberculin syringe was fitted with a blunt 19-gauge needle, about 60 mm long, and bent at a 135-degree angle at approximately 45 mm from the tip. A direct-focusing headlight was worn by the operator to give a clear view of the pharynx, and the tongue was gently pulled outward and laterally with a forceps. The needle was inserted between the vocal cords into the tracheal lumen. A light but definite bumping against the tracheal rings ensured proper insertion. The needle was pushed toward the caudal end of the trachea, and the injection made. There was a brief period of apnea after the injection.

A very similar technique is described by Brain and Valberg (1979), who suggest keeping the animal on the slanted board for about 1 min after the instillation. Approximately 0.15 ml of suspension per 100 gm of body weight can be injected. Schreiber *et al.* (1975) anesthetized hamsters with methoxyflurane and used a special apparatus which consisted of a syringe barrel and plunger to which aspiration tubing was mounted. When placed into the trachea (using the slanted-board technique described above), the released fluid flushed a 6-mm length and was then aspirated by vacuum. Using this method, 2 ml of fluid can be delivered into the trachea and aspirated out along with cells. (Schreiber and Nettesheim, 1972).

In order to avoid respiratory depression with standard intratracheal injections, Eldridge *et al.* (1982) recommend waiting until the animal begins to recover from anesthesia before giving the test material. Henry and Port (1978) report that the survival rate of hamsters during 10 weekly intratracheal injections of benzo[*a*]pyrene–ferric oxide was good with methoxyflurane and poor with methohexital anesthesia.

I. Intrapulmonary Instillations

Yabe *et al.* (1963) performed intrapulmonary injections through the chest wall into the peripheral part of the lung of hamsters that were less than 24 hr old. They report that there was usually some leakage from the nose or mouth via the bronchi and trachea. Warren and Gates (1968) implanted a 0.5

× 2 mm cobalt-60 wire into the right lung of anesthetized young adult hamsters by passing it through a trocar placed at the posterior portion of the sixth intercostal space.

J. Central Nervous System

1. Intraspinal Injections

A technique for intraspinal inoculation of mice (Habel and Li, 1951) has also been used in hamsters. A small skin incision is made on the anesthetized animal near the lumbar enlargement of the spinal cord. The animal is held in one hand so that the vertebral column is slightly flexed, and, using a 0.25-in. (6.3 mm), 27-gauge needle with a short bevel on a 0.25-ml tuberculin syringe, the needle point is introduced between the vertebrae at the level of the incision, slightly to the right of midline. As slight pressure is applied, the direction of the needle is approximately 45 degrees toward the head and at a very slight angle toward the midline. As the point where the needle enters the spinal canal, a "giving" sensation is felt. A volume of 0.02 ml can be injected. The injection goes into the central canal and the subarachnoid space. It was reported that best results are obtained with animals near weaning age.

2. Intracerebral Injections

Ferris et al. (1984) describe the appropriate stereotaxic coordinates for microinjections into the adult hamster brain. Stainless-steel guide cannulas (26-gauge) were implanted into the medial preoptic area, above the suprachiasmatic nucleus, the area of the ventromedial and lateral hypothalami, and the lateral ventricle. All microinjections were made by a 33-gauge needle, connected to a 1-μl syringe, and through PE-20 tubing. The needles were easily inserted into the guide cannulas while the animals were restrained without anesthesia.

The lateral ventricle of the brain may also be reached in the anesthetized hamster by trephining a hole in the right parietal bone halfway between the coronal and lambdoidal sutures, and 1.5 mm lateral to the sagittal suture. A 26-gauge needle attached to a 1-ml tuberculin syringe is directed ventrad and slightly laterad through the hole, the meninges, and the cerebral cortex. In most cases the needle will penetrate the right lateral ventricle and the posterodorsal extent of the right inferior horn. Injected material has been found in the contralateral ventricle, in the hypothalamic region of the third ventricle and in the iter (Kent and Liberman, 1949).

K. Intranasal Injections

Intranasal injections are usually performed with the animal lightly anesthetized, by placing the material to be injected at or in the external nares. Taylor (1940) reports being able to administer 0.3–0.4 ml of unfiltered human throat wash to hamsters, and Burr and Nagler (1953) administered 0.3 ml of allantoic fluid intranasally to 7- to 10-week-old hamsters.

V. COLLECTIONS

A. Blood

1. Ophthalmic Venous Sinus

Pansky et al. (1961) used a needle to sample blood from the orbital region. Using a 1-ml heparin-rinsed syringe with a 23-gauge needle, the needle was placed midway along the superior border of the eye, but beneath the upper lid. The point of the needle was directed toward the back of the orbit at a 20- to 40-degree angle, then pushed downward using the medial wall of the orbit as a guide. After 4 mm penetration, the needle is stopped by bone. If it is withdrawn slightly, it will be in the venous sinus. It is now possible to withdraw blood or to make an injection. Pansky et al. (1961) report removing 0.4 ml of blood twice weekly for 5 weeks, 0.1 ml six to eight times within 24 hr, daily for 6 days, then weekly for 7 months. They state that the procedure can be performed with or without anesthesia, but this author recommends use of anesthesia.

A procedure almost identical to that of Pansky et al. (1961) was used by Breckton and Goy (1979), who report being able to obtain approximately 2 ml of blood by the technique, compared to approximately 3 ml noted by Pansky et al. Interestingly, Brecton and Goy state that the depth of penetration required to reach the posterior ocular sinus precludes the use of glass pipettes or plastic cannulas to obtain blood samples.

To obtain orbital blood, the author and others have routinely used a technique similar to that employed with rats. The modification for hamsters was described by Michel (1961), who recommended narcosis but stated that it could be performed without narcosis. This author suggests that it be performed with animals under anesthesia. Michel (1961) placed the hamster on its right side. The index finger and thumb of the left hand grasped the neck and head of the hamster, and the index finger caused compression of the jugular vein, which in turn caused an exophthalmus and swelling of the orbital sinus. The thumb of the left hand pushes the superior eyelid dorsally. As performed in the author's laboratory, a glass capillary pipette is now placed in the lateral canthus and directed medially (but not anteriorly) as the sinus is located in the posterior part of the orbit. Slight pressure is needed to penetrate the tissues surrounding the sinus. Blood is then collected from the free tip of the capillary tube. When the tube is removed, place pressure over the eye.

Michel (1961) was able to collect a maximum of 20 drops of blood. In the author's laboratory 1.5 ml has been collected, and more appears to be possible. Collection from a medial canthus insertion of the tube is also possible, but less blood is obtained. Collection frequency is similar to that described above by Pansky et al. (1961).

2. Lateral Vein of Tarsus (Vena Plantaris Lateralis)

Blood may be sampled from lateral marginal vein of the tarsus. A 25-gauge, ⅝-in. (15.6 mm) needle without a syringe is inserted into the occluded vein. As blood flows into the hub of the needle, it can be collected with a microhematocrit tube, or if skin contamination is not of importance, the vein can be lanced and the blood collected from the skin surface (Schuchman, 1977).

3. Cardiac Puncture

The use of cardiac puncture to obtain blood is discouraged by Pansky et al. (1961). They note a tendency of the heart to roll under the pressure of the needle, as well as high mortality. Turbyfill et al. (1962) refute this statement, noting that they had taken cardiac blood samples from hamsters as often as three times a week for several months. In their method, the adult animal may or may not be anesthetized, but this author prefers to use anesthesia. The animal is placed in dorsal recumbency, and a 23-gauge needle on a 1-ml syringe is introduced immediately to the left of the xiphisternal–sternal junction, at a 30-degree angle. The needle is advanced to the heart, and blood is gently aspirated. For an adult hamster, the removal of 1 ml of blood is practical, and more than 3 ml causes death.

For fetal and young animals, the same authors draw a microhematocrit-sized glass capillary tube to a point, and with a small triangular file they break the tip to the diameter of a 30-gauge needle. The animal is held in one hand between the thumb and index finger, the head elevated, and, using the technique described for an adult animal, the tube is advanced to the heart, with the blood exiting from the tip of the tube.

4. Vena Cava

For collecting larger amounts of blood, Rennie et al. (1982) used an exposed inferior vena cava to obtain up to 5 ml from anesthetized young adult males. Grice (1964) uses the terminal portion of the aorta, as in rats, to obtain large blood quantities. The animal is anesthetized, the abdomen is opened widely, and a needle placed at the iliac bifurcation, pointing craniad. Placing pressure on one of the internal iliac arteries tends to increase the rigidity of the aorta and lead to easier collection. Manning and Giannina (1966) did not use Grice's technique because the male sex organs completely obscure the bifurcation of the iliac artery. The latter authors make a U-shaped incision in the lower abdomen and extend it up both sides to the rib cage. The intestines are placed to the left of the operator, and the aorta is seen at the iliac bifurcation. The aorta is clamped with a hemostat 0.5 cm craniad to the bifurcation, and the hemostat is turned 90 degrees counterclockwise and placed to the operator's right. The entire process takes 2–3 min and 2.75–3.25 ml of blood may be collected from a 50- to 70-gm hamster.

5. Carotid Artery

See discussion of arterial blood gases (Section IX,D).

6. Tail or Nail

Finally, it should be noted that for a routine blood smear an adequate amount of blood can be obtained by clipping the tip of the hamster's short tail (Bauck and Hagan, 1984), or by clipping a nail very short (Burke, 1979).

B. Saliva

Smith et al. (1982) studied saliva volume as related to antibody concentration. Hamsters were anesthetized with ether and injected subcutaneously with 1.0 mg of pilocarpine nitrate per 100 gm of body weight. Fifteen minutes after injection whole saliva was collected by gravity into graduated tubes.

C. Bronchopulmonary Lavage

Mauderly (1977) describes a modified anesthetic system which replaces a breathing mask with a tracheal catheter attachment to allow either free breathing, breathing the anesthetic mixture, or positive-pressure inflation of the lungs by a syringe. The tracheal catheter with its hub can be inserted using the techniques described for intratracheal injection. Lavage is accomplished by hyperventilating the hamster and instilling the calculated volume (~4 ml) of warmed saline, and immediately withdrawing the saline until a slight resistance is felt on the plunger. After clearing residual fluid from the catheter, the animal is ventilated until effective spontaneous breathing is reestablished. The ventilation volume is about 4 ml, and the color of the nose and feet serve as good indicators of the adequacy of ventilation.

D. Ejaculate

Adult male hamsters were induced to ejaculate with an intraperitoneal injection of potassium chloride, given at 400–600 mg/kg body weight. The injection was made into the lower abdomen with no apparent harm. No sperm were found in the ejaculate, possibly because potassium chloride produces only ejaculation of fluid in the seminal vesicle (Rockhold and Hiestand, 1956). Specific references to electroejaculation in hamsters have not been found.

E. Vaginal Mucus

To examine vaginal mucus, a small platinum loop is inserted deeply into the vagina and withdrawn. The mucus collected is mixed with a drop of broth on a slide (Kradolfer, 1954).

F. Urine, Feces, and Expired Gases

Standard glass, metal, or plastic metabolism cages can be used to separate urine from feces, whereas only glass is recommended for the trapping of expired respiratory gases. Adams and Homburger (1983) noted that when hamsters are placed in a circular cage, such as a metabolism cage, which does not give them access to corners, they can become extremely agitated and lose weight. They state that when circular cages are employed, the insertion of a small box will assure normal behavior of the animals. The nesting of hamsters in feeders or other areas of metabolism cages should be avoided when possible. Cages that provide either access for liquid diets or, if necessary, closure of the access to the feeder may be used.

G. Milk

The technique of Haberman (1974) for the collection of milk from mice has been used by the author for collection of hamster milk, with only a minor modification to accommodate the longer nipple of the hamster. A plastic biopipette tip is placed over the nipple (Fig. 3). Stimulants are not routinely used to initiate milk flow. From lactating animals 0.1–0.2 ml total has been collected from two or three nipples on collections spaced a few hours apart. The use of approximately 1 U.S.P. unit of oxytocin, subcutaneously, will, however, increase milk flow if desired. An attempt has not been made to collect milk more than 3 days after birth.

H. Bile

The anesthetized hamster is placed in dorsal recumbency. A midline incision is made from the xiphoid process to the inguinal region. The duodenum is lifted and placed outside of the abdominal cavity. The common bile duct can be seen in the pancreas as it courses from the gallbladder to the duodenum. Ligate the bile duct at its entrance to the duodenum in order to dilate the duct and thereby facilitate placing the collecting tube. Using a 25-gauge needle, puncture the duct a few millimeters from where it enters the duodenum. Carefully insert polyethylene tubing (0.011 in. I.D., 0.024 in. O.D.) with a beveled end, 2.0–2.5 cm into the duct. The length of insertion will depend on the size of the animal. The tubing is secured with two silk ties 1 cm apart. A loop of tubing is left within the abdominal cavity to allow slack during animal movement, while the end is brought through the incision. The muscle incision is closed with 4–0 sutures. If the animal is kept restrained, bile is collected from the tubing with no further procedures needed. If the animal is to be mobile, the catheter is passed subcutaneously to exit between the scapulae. It is then attached to a commercially available tether that combines a jacket for the animal, to which a hollow flexible metal tube and a swivel are attached. The catheter is threaded through the bore of the tube. As hamsters can easily squirm out of the jacket, it may be helpful to remove all but a fringe of it and attach the metal tube to the skin with surgical clips. Approximately 9 ml of bile can be collected from an adult hamster in a 24-hr period.

VI. ANESTHESIA

A. General Comments

A large variety of anesthetics and anesthetic techniques have been used for the Syrian golden hamster. The actual choice of the anesthetic to be used involves not only the criteria normally involved in nonresearch applications, but the special needs of the research community as well, such as the induction of hepatic microsomal enzymes (Valerino et al., 1974) or effects on tumor immunity (Duncan et al., 1977), latency, and body weight (Henry and Port, 1978).

The proper emergency equipment should be available, which in the author's experience is most often needed to stimulate respiration. We routinely keep available an oxygen tank with a

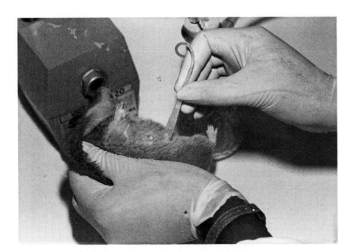

Fig. 3. Collecting milk from an unanesthetized hamster.

small facemask as well as a small hand insufflator which fits over a hamster's face, and a bottle of doxapram. When the latter is needed to stimulate respiration, we usually administer one drop on the mucosa below the tongue.

B. Inhalants

Table II gives a summary of inhalant anesthetic information. Induction times assume that the system has the appropriate anesthetic concentration available to the hamster.

1. Ether

Ether has long been used for rodent anesthesia, including hamsters. The usual technique calls for placing the animal in a jar that contains cotton or other materials that have been wetted with ether. The ether-soaked material should be kept separate from the hamster. Once removed from the jar, the animal may require periodic reexposures to ether placed near its nose in order to complete the procedure being performed. The operator should be aware of the explosive nature of ether, should use ether that is manufactured for anesthetic purposes, and should take note of the expiration date of the product. In addition, the animal must be carefully monitored, as deaths from ether overdosage can occur in a short period of time.

2. Methoxyflurane

Simmons and Smith (1968) described a closed veterinary anesthesia system that is attached to an animal chamber that replaces the facemask. This unit is used with methoxyflurane and has been useful for hamster anesthesia. More recently, a specially built methoxyflurane anesthesia system has been described (Norris and Miles, 1982). Schreiber and Nettesheim (1972) designed a methoxyflurane anesthesia unit having an oxygen flow, in a 35 × 30 × 16 cm chamber, adjusted to 100 ml/min. This met the metabolic requirements of ten 10-week-old male hamsters. For the first 5 min all air was circulated through the vaporizer, after which 90% of the air was allowed to bypass the vaporizer. Hamsters were kept anesthetized under these conditions for more than 2 hr.

Methoxyflurane may also be used in a manner similar to that described for ether and we believe with a greater safety margin. It is more expensive than ether, and the operator must also consider special needs (as with any anesthetic) such as relative ability to induce hepatic microsomal enzymes.

Table II
Inhalant Anesthetics for Hamsters

Drug	Induction concentration	Maintenance concentration	Time to anesthesia (min)	Recovery time (min)	Comments	Reference
Ether	Open drop	—	0.5–2	3–4	Explosive	Henry and Port (1978)
Halothane	4%	1%	—	—	With oxygen	Smith et al. (1973)
	3%	—	—	—	With oxygen	Green et al. (1978)
	2%	0.5%	2–3	3	—	Breckon and Goy (1979)
	3.5%	—	3 min/100 gm body weight	—	Time to loss of pedal reflex; with oxygen	Henderson and Lowrey (1983)
	3.5%	2.0%	4	10–30	8- to 12-week-old animals; with 70:30 (v/v) nitrous oxide–oxygen	Carvell and Stoward (1975)
Methoxyflurane	Room air at 300 ml/min	Room air at 200; then 20 ml/min	5	30	—	Norris and Miles (1982)
	Open drop	—	—	1–2	—	Henry and Port (1978)
	Open drop	—	2–4	15–20	—	Strittmatter (1972)
	Oxygen at 3 liters/min	—	2–3	5–20	For mice; implied to work as well for hamsters	Simmons and Smith (1968)
Carbon dioxide	100%	—	2	Rapid	Caused shallow breathing	Henderson and Lowry (1983)

Strittmatter (1972) placed adult hamsters in a large cylinder which had a perforated raised platform. Then, 1–2 ml of methoxyflurane were sprinkled in. After an initial excitatory phase the animal fell into deep narcosis and was then removed from the closed container. Anesthesia was maintained using a glass tube, 2 cm deep by 10 cm long, bent at a right angle. A cotton plug saturated with methoxyflurane was placed a few centimeters from each end of the tube, and the depth of anesthesia was regulated by adjusting the distance between the nose of the animal and the opening of the tube. The postoperative phase was marked by restlessness and running about. Atropine premedication was not used due to minimal mucosal secretions.

3. Halothane

A variety of vaporizers and anesthetic regimens have been used with halothane. Many are referenced in Table II. Halothane should not be used by the open-drop method, as the depth of anesthesia is difficult to control.

Carvell and Stoward (1975) induced anesthesia in 8- to 12-week-old hamsters in a plastic box using 3.5% halothane vaporized with a 70:30 (v/v) nitrous oxide–oxygen mixture. Anesthesia was maintained with 2% halothane using a small facemask. They report rectal temperature dropped from approximately 37° to 31°C, and respiratory rate fell from 40.3 breaths per minute after 5 min to 21.8 breaths per minute after 25 min. The mean concentration of halothane in the blood was found to be 22.75 mg/dl. Recovery, with full righting reflex and ability to walk and feed, occurred within 10–30 min of the termination of anesthesia. The writer has used a halothane system described by Dudley et al. (1975) for hamsters and other small rodents with satisfactory results.

C. Injectables

Table III lists many injectable anesthetics and anesthetic combinations that have been used with Syrian hamsters.

1. Ketamine

When given alone, ketamine, a cyclohexylamine analog, causes sedation only. When ketamine is combined with diazepam, there is improved muscle relaxation but the combination has no effect on the degree of analgesia (Green et al., 1981). Strittmatter (1972) states that ketamine at 10–30 mg/kg intraperitoneally is adequate for immobilization, but the pain response never completely disappears. The postnarcosis stage is characterized by motor activity. When ketamine was given at 20 mg/kg, no inflammatory lesions were found 1–5 days postinjection.

Schall (1984) suggests the use of ketamine with xylazine or sodium pentobarbital. Dosages are indicated in Table III. In addition, Schall (1984) uses atropine sulfate, 0.5 mg/kg subcutaneously, as preanesthetic. That author notes that the ketamine–xylazine combination is generally good for surgery, but for extremely painful procedures there may be a need for supplemental inhalation anesthesia. Extension of narcosis can be accomplished by administration of one-third of the initial ketamine dose. The postanesthetic effect lasts 3–5 hr.

Curl and Peters (1983) noted muscle hemorrhage and shorter anesthesia times when ketamine–xylazine was given by the intramuscular route as compared to the intraperitoneal route. Subcutaneous injection was not satisfactory. No anesthetized females aborted and no anomalies were noted in any offspring.

Harkness (1980) reports that ketamine–xylazine yields more consistent anesthesia than does sodium pentobarbital. Green et al. (1981) report that the ketamine–xylazine combination can cause significant cardiac arrhythmias.

Ketamine can also be combined with methoxyflurane. Strittmatter (1972) does not use atropine premedication, and gives ketamine at 10–20 mg/kg followed by methoxyflurane. Almost no excitatory phase is seen, and it is easy to control the depth of narcosis. After 20 min of anesthesia, recovery took 10–20 min. This procedure is also satisfactory for hibernating hamsters. When ketamine was combined with promazine and aminopentamide hydrogen sulfate (Ketaset Plus), Mulder et al. (1979) reported some transient hyperexcitability during induction and an increased heart and respiratory rate for the first 5–10 min of anesthesia.

2. Pentobarbital

Sodium pentobarbital (100 mg/kg intraperitoneally) can abruptly lower blood pressure and cardiac and respiratory rates in hamsters, and will also gradually lower the body temperature, sometimes to within a few degrees above room temperature (Berman et al., 1955). Halving the dosage and then giving 1.5-mg supplements was found to reduce side effects which range from depression to hyperexcitability (Berman et al., 1955). Greenblatt et al. (1969) used a dosage of 75 mg/kg intraperitoneally of freshly made pentobarbital and produced approximately 30 min of surgical anesthesia. If more anesthesia was needed, 0.2–0.3 ml of a 15 mg/ml solution was given, but this led to increased postoperative mortality. By sedating hamsters with triflupromazine hydochloride prior to anesthesia, Esparza et al. (1979) found that 96 mg/kg of pentobarbital given 30 min later provided satisfactory anesthesia. Although the triflupromazine was used to decrease the amount of pentobarbital needed for anesthesia, it would appear from Table III that the reduction was minimal.

3. Tiletamine and Zolazepam

Tiletamine is a dissociative anesthetic related to ketamine, and zolazepam is a nonphenothiazine tranquilizer. Silverman

Table III
Injectable Anesthetics for Hamsters[a]

Drug	Concentration	Dosage	Route[b]	Time to anesthesia (min)	Length of anesthesia (min)	Recovery time (min)	Comments	Reference
Pentobarbital		100 mg/kg	IP	—	—	—	—	Berman et al. (1955)
	50 mg/ml	0.1 ml/100 gm	—	—	—	—	>48 hr old	Sherman et al. (1963)
		65 mg/kg	IP	—	—	—	—	Ferm (1967)
	15 mg/ml	0.5 ml/100 gm	IP	—	—	—	—	Greenblatt et al. (1969)
	60 mg/ml	0.1 ml/100 gm	IP	<5	30	<60	—	Willoughby (1973)
		96 mg/kg	IP	—	—	—	With triflupromazine pre-anesthesia	Esparza et al. (1979)
		100 mg/kg	IP	—	—	—	—	Henderson and Lowry (1983)
Methohexital	1.0% Solution	0.4 ml	IP	—	—	—	Young adults	Saffiotti et al. (1968)
	1.0% Solution	0.4–0.8 ml	IP	—	—	—	8–9 Weeks old	Hayes et al. (1977)
		5 mg/100 gm					—	Henry and Port (1978)
	1.0% Solution	0.3–0.42 ml	IP	—	4–10	—	—	Eldridge et al. (1982)
Methohexital Diazepam	7.5 mg/ml 1.25 mg/ml	4.0 ml/kg	IP	2–3	17	60	Good surgical anesthesia	Ferguson (1979)
Secobarbital		45 mg/kg	IP	—	—	—	—	Chen et al. (1945)
Amobarbital		70 mg/kg	IP	5	25	95	—	Chen et al. (1945)
		50 mg/kg	IV	—	—	—	—	Chen et al. (1945)
Chloral hydrate	3.7 gm/100 ml	1 ml/100 gm	IP	—	—	—	—	C. Lee (personal communication, 1984)
Diallylbarbituric acid	0.01 gm/0.1 ml	36 mg/100 gm	—	—	—	—	—	Roosa et al. (1965)
Urethane	0.04 gm/0.1 ml	0.1 ml/100 gm	IP	60	300	—	—	Wyman et al. (1957)
Monoethylurea	0.04 gm/0.1 ml							
Inactin [sodium 5-ethyl-5-(1'-methylpropyl)-2-thiobarbiturate]	100 mg/ml	200 mg/kg	IP	—	30–300	—	—	Turner and Howards (1977)
Ketamine	50 mg/ml	10–30 mg/ml	IP	—	—	—	Immobilization	Strittmatter (1972)
		80 mg/kg	IM	7–9	30	40	Mild sedation	Green et al. (1981)
		50–500 mg/kg	IM	—	—	—	Immobilization	Curl and Peters (1983)
		7.5 mg/animal	—	—	—	—	100-Day-old males	Singhal et al. (1984)
Ketamine Xylazine		80 mg/kg 16 mg/kg	IP	—	—	—	Relaxation and sedation	Green et al. (1981)
		50–200 mg/kg 10 mg/kg	IP	<5	27–71	—	Surgical anesthesia	Curl and Peters (1983)
		87 mg/kg 13 mg/kg	—	—	—	—	—	Dunlap and Grizzle (1984)
		200 mg/kg 5 mg/kg	SC	3–5	30	—	Can intensify with inhalation anesthesia	Schall (1984)

Table III (*Continued*)

Drug	Concentration	Dosage	Route[b]	Time to anesthesia (min)	Length of anesthesia (min)	Recovery time (min)	Comments	Reference
Ketamine		100 mg/kg	SC	—	—	—	Pentobarbital 10 min after ketamine	Schall (1984)
Sodium pentobarbital		50 mg/kg	IP					
Ketamine	100 mg/ml	150 mg/kg	IM	3–4	85	45	Dosage based on ketamine	Mulder et al. (1979)
Promazine hydrochloride	7.5 mg/ml							
Aminopentamide hydrogen sulfate	0.0625 mg/ml							
Tiletamine	Equal parts	~50 mg/kg	IP or IM	—	~13	—	Sedation or restraint	Silverman et al. (1983)
Zolazepam								
Fentanyl citrate	0.2 mg/ml	1 ml/kg	IP	—	—	—	Good analgesia; see text	Green et al. (1975)
Fluanisone	10 mg/ml							
	0.079 mg/ml	0.1 ml	IM	—	—	—	Followed by Mebumal, 10 mg/ml IV	Hellekant and Roberts (1983)
	2.5 mg/ml							
Fentanyl citrate	0.1 mg/ml	1 or 3 ml/kg	IP				Inject diazepam separately	Green (1975)
Fluanisone	10 mg/ml							
Diazepam	5 mg/ml	1 ml/kg	IP	—	150–180	300–360		
Midazolam	1.25 mg/ml	4 ml/kg	IP	5	57	80	Some respiratory distress reversed by naloxane	Flecknell and Mitchell (1984)
Fluanisone	2.5 mg/ml							
Fentanyl citrate	0.079 mg/ml							
Alphaxolone	2 mg/ml	120–160 mg/kg	IP	2	60	120–150	—	Green et al. (1978)
Alphadolone								
		14 mg/kg	IV	<0.5	6–9	20–25	—	Green et al. (1978)
Triflupromazine		80 mg/kg	IP	—	—	—	Preanesthetic	Esparza et al. (1979)

[a]Dash indicates that information is not provided.
[b]IP, intraperitoneal; IM, intramuscular; SC, subcutaneous; IV, intravenous.

et al. (1983), using equal parts of each drug at various dosages, found that dosages above 20 mg/kg body weight produced adequate immobilization but inadequate analgesia. Mild respiratory distress was seen (all animals recovered), and it was suggested that 50 mg/kg might prove satisfactory for immobilization.

4. Methohexital

Methohexital is an ultra-short-acting oxybarbiturate. Eldridge et al. (1982) found that a uniform dosage produced unsatisfactory anesthesia for intratracheal anesthesia in the hamster. For 8-week-old male animals, 0.3 ml of the 1% solution provided adequate anesthesia for intratracheal instillations, while for 11-week-old animals, 0.42 ml was needed. Performing the same procedure, Saffiotti et al. (1968) selected methohexital to avoid the edema and other lung changes associated with ether anesthesia, and to permit rapid resumption of normal respiration.

5. Fluanisone

The butyrophenone tranquilizer fluanisone, when used in combination with fentanyl, produced excellent analgesia but only moderate muscle relaxation and an incomplete loss of the righting reflex (Green, 1975). When diazepam was added as noted in Table III, there was excellent surgical anesthesia, but a prolonged recovery period. Respirations became depressed at both dose levels shown, but all animals retained good color (Green, 1975). Fentanyl and fluanisone can also be mixed with midazolam, a water-soluble benzodiazepine, thus eliminating the need for two separate injections as with diazepam. This neuroleptoanalgesic combination caused a maximum reduction in respiratory rate of 35%, which was reversible by naloxone, 0.1 mg/kg body weight (Flecknell and Mitchell, 1984).

6. Alphaxolone and Alphadolone

The steroid anesthetics alphaxolone and alphadolone, in combination, were evaluated by Green *et al.* (1978). Because of the volumes used, the intramuscular route was inappropriate. Anesthesia can be maintained with this combination for up to 4 hr by giving increments of 4–6 mg/kg at 15-min intervals.

7. Fentanyl and Droperidol

Moller (1968) used fentanyl and droperidol (Innovar-Vet) at approximately 0.01 ml/60 gm body weight, with atropine, for "simple surgery," including cesarean section. Thayer (1972), however, using dosages from 0.046 to 0.694 ml/100 gm body weight (the latter being the LD_{50}), reported unpredictable results with this combination. Thayer (1972) did not find any hyperactivity, as had been reported with other species.

D. Hypothermia

Perhaps the most basic of all anesthesia techniques, hypothermia was described nearly 50 years ago by Pfeiffer (1936) and subsequently adapted for hamsters by other investigators (Ortiz, 1953; Roosa *et al.*, 1965; Rhoades and Fish, 1983). The technique is essentially for very young animals only. Sherman *et al.* (1963) placed hamsters that were up to 48 hr old in plastic ice cube containers and placed the containers in the freezing compartment of a refrigerator for 9–11 minutes. At that time the animals stopped breathing, became slightly pale, and after removal they were satisfactorily thymectomized. Adolph (1951), using thermocouples and immersing animals to the neck for $\frac{1}{2}$–1 hr found that newborns could tolerate temperatures as low as 1°C. until 15 days of age. Previous studies had indicated that adults could tolerate colonic temperatures to 3.8°C.

VII. SURGICAL PROCEDURES

A. Basic Techniques

Clean but not sterile surgical technique is usually applied to procedures performed on hamsters in the laboratory setting. Wound infections in hamsters are not seen as frequently as in dogs and cats, although such problems have been noted. Anesthesia must always be used for painful procedures. When possible, animals should be anesthetized and prepared outside of the surgical area. Hair is clipped from the surgical area; the area is then washed with soap and water, rinsed, and 70% alcohol, "tamed" iodine, or occasionally both are applied to the surgical area. Do not use iodine during certain studies, such as those involving the thyroid gland.

The surgeon should clean his or her hands, using standard surgical scrub technique, and should don sterile gloves. A sterile or clean drape with a small opening may be placed over the surgical site; however, the small size of the hamster sometimes precludes the use of a surgical drape, and it may be necessary to work without one.

Whenever possible, surgical procedures should be performed on a heated operating board and the animals kept warm during recovery and separated from each other until fully awake. During surgery hemostasis can be accomplished by pressure from cotton-tipped swabs, by electrocoagulation using fine-tipped probes, or by absorbable hemostatic material.

Unless referenced otherwise, the surgical techniques described in this section are accomplished on anesthetized animals and are as performed by Dr. C. S. Lee (personal communication, 1984). Muscle sutures are generally not required, and skin incisions can be closed with surgical clips or standard suture materials. Suture removal by the hamster has not proved to be a significant problem.

B. Adrenalectomy

The animal is placed in ventral recumbency, with its tail toward the operator. A 1- to 2-cm incision is made on the dorsal midline, over the area of the kidneys, immediately caudal to the "hump" in the back (the second to fourth lumbar vertebrae). A 4.5-in. (11.4 cm) curved iris forceps is inserted subcutaneously through the incision and pushed through the lateral muscles into the peritoneal cavity. The incision in the muscle is only large enough to observe and remove the adrenals. On the left side the incision places the forceps slightly to the left of the kidney and about two-thirds toward its cranial pole. The adrenal lies under the cranial pole, toward the spine. Clamp the tissue and blood vessels at the base of the adrenal with one forceps and tear the gland away with another. Follow the same procedure for the right gland. It should be noted that due to the smaller size of the female's gland it is more difficult to remove.

It is suggested that hamsters receive 0.1 ml of cortisone acetate (25 mg/ml) just prior to surgery and on the day after surgery. The use of 1% saline solution for drinking water is also recommended. Dunlap and Grizzle (1984) state, however, that the addition of 1% saline is of little benefit, a view shared by Frenkel (1972) and Wyman *et al.* (1953a). This is related to a normal serum sodium level after adrenalectomy. Serum potassium levels are high (19.2–20 mEq/liter compared to the normal 8.5–8.8 mEq/liter). Frenkel (1972) suggests giving a mineralocorticoid postsurgically.

C. Adrenal Demedullation

Exposure of the adrenal gland follows the same procedure described for adrenalectomy. When the adrenal gland is reached, it is grasped with the iris forceps, and the cortex is cut with a Castroviejo ultramicrodissecting scissors. By applying slight pressure on the gland, the medulla is forced out of the cortical incision. Caution must be taken not to disconnect the adrenal gland when cutting into the cortex.

D. Nephrectomy

The hamster is placed in ventral recumbency with its tail toward the surgeon. A paracostal incision is made next to the flank gland. The muscle directly over the kidney is pierced with a 4.5-in. (11.4 cm) iris forceps. The kidney is freed of connective tissue and is gently pulled out, and the iris forceps is carefully placed under the kidney. The adrenal gland can be easily pulled away by tearing the attachments with the forceps and then replacing the gland in the abdominal cavity. A single ligature is placed around the renal blood vessels and the ureter, and tied securely; then the vessels are cut next to the kidney, which is now removed. It should be noted that if bilateral nephrectomy is performed the hamster will survive for 24–48 hr.

E. Renal Papillectomy

Marshall *et al.* (1963), studying the concentrating functions of various parts of the kidney, described a technique for papillectomy. The kidney is exposed by a dorsal midline incision, with dissection of the muscles adjacent and parallel to the spine. The kidney is brought to the incision and the pelvis brought into view by rotation of the kidney. The perirenal fat is cleared away and a longitudinal incision is made over the papilla. About 3 mm of the papilla is excised using a fine-toothed rongeur. The renal pelvis is not sutured and no drains are used. Blood loss is minimal. All survivors (the percentage of survival was not noted) urinated on the first postoperative day. Lower mortality was recorded if one renal papilla was removed 2 weeks after the other.

F. Cystotomy

Cystotomy can be performed by making a 1.5-cm midline abdominal incision, exteriorizing the bladder, and stabilizing it with a 4–0 chromic retraction suture placed at the bladder's apex. Using this technique, Bauck and Hagan (1984) removed uroliths by making a 6-mm incision in the bladder, removing the stones, and closing the bladder with four simple interrupted sutures of 6-0 chromic gut. The abdomen was closed with 4-0 chromic catgut.

G. Experimental Hypertension

Stroia *et al.* (1954) used two methods to induce hypertension in the hamster. In the first method, by using a dorsolateral incision, they placed a cotton thread around both poles of the kidneys and drew it tight. Care must be used not to include the adrenal glands. It should be noted that the hamster kidney is softer than the rat kidney. There was 62% survival over the first 12 days. In their second method they performed a subcutaneous implant of four 15-mg deoxycorticosterone acetate pellets, followed by 1% sodium chloride drinking water. In the unanesthetized animal the mean systolic blood pressure was 108 mm Hg, while the mean systolic pressure in the anesthetized animal was 111 mm Hg. With renal ischemia, pressure rose in most animals to a mean of approximately 150 mm Hg, although the increase was not consistent in all those treated. With cortisone-treated animals, there was a progressive rise in pressure to a maximum of approximately 160 mm Hg 3 weeks after treatment, after which there was a decrease toward the baseline.

H. Thymectomy

Infant hamsters (within 48–72 hours of birth) can be anesthetized by hypothermia as previously described. Using the technique of Van Hoosier *et al.* (1978), the animal is secured to a surgery board with masking tape, and with the use of a dissecting microscope the skin and sternum are opened down to the second or third rib. The width of the incision is extended with a forceps. A cotton-tipped applicator stick is gently rotated about the thymus until both lobes become detached. The sternum and skin are closed by the use of collodion. Chilled animals are rewarmed prior to returning them to their mother.

A similar technique is used by Roosa *et al.* (1965) and Sherman *et al.* (1963) for infant animals. Older animals (to 12 weeks of age) were anesthetized with chloral hydrate or sodium pentobarbital, and in both infants and older hamsters the thymus was gently lifted out with a glass pipette connected to a suction pump. Three to four interrupted sutures were placed in the skin.

Another technique for thymectomy requires placing adult animals in dorsal recumbency with the head firmly stretched with retractors. A 1.5- to 2-cm midline incision is made from the base of the neck posteriorly over the thorax. The connective tissue between the two submaxillary salivary glands is separated. Then, using small, sharp-pointed scissors, the thorax is cut in a V-shaped pattern extending about 5 mm caudad. The V-shaped bone is removed and the thymus is visible. Using a curved iris forceps, the bilobed gland is gently teased out of the thoracic cavity, one lobe at a time to avoid having the gland break apart. Cotton-tipped swabs are used to stop minor bleeding. Sutures are not placed in the sternum. It is to be noted that cutting more than 6 mm deep may cause pneumothorax. Additionally, the incision must be closed rapidly, also to prevent pneumothorax (C. Lee, personal communication, 1984).

Small modifications of the above technique for animals of 3–

12 weeks of age are that of Sherman *et al.* (1963), who removed the gland with forceps and a cotton-tipped applicator (or sometimes with suction), and Yohn *et al.* (1965), who also used forceps and light suction. The latter authors reported 95% survival and a complete thymectomy confirmed in 96% of the hamsters.

I. Thyroparathyroidectomy

Because the parathyroids are deeply imbedded in the thyroids, they cannot be satisfactorily removed without removing the thyroids. The hamster is placed in dorsal recumbency with its head held taut with retractors. A longitudinal midline incision is made along the neck from is base to just below the point of the lower jaw. The fat, connective tissue, salivary glands, and sternohyoid muscles are retracted, exposing the thyroid glands and their isthmus. The isthmus is grasped and broken, separating the thyroids. Each thyroid is gently teased upward toward the head, using caution not to damage the recurrent laryngeal nerve. Bleeding is minimal, no muscle sutures are used, and the incision is routinely closed. Calcium gluconate (1%) is placed in the drinking water to help prevent parathyroid tetany.

J. Hypophysectomy

In this procedure the hamster is placed in dorsal recumbency with its tail toward the operator. The head is firmly held to the operating board with retractors. A 1- to 1.5-cm incision is made on the midline of the neck. After retracting the salivary glands, the sternohyoid muscle is exposed. Using two 4.5-in. (11.4 cm) splinter forceps on the right of the trachea, the forceps are pushed and spread until the cranium is reached. The sternohyoid and omohyoid muscles covering the cranium are scraped away using a dental scaler. A faint blue suture line between the basisphenoid and occipital bones now becomes visible, which is the landmark for trephining. The cranial bone is brittle and easily shatters, so it is very important that the trephines stay perpendicular to the cranium at all times. The bone is gently scored using a small trephine and a second trephine is used to cut and remove the bone. The gland can now be sucked out of the drilled cavity using an 18-gauge needle and 10 lb of vacuum pressure. Usually the gland can be observed while it is being suctioned out. If bleeding occurs, it can be stopped with cotton-tipped swabs. It is important to ascertain that bleeding has stopped before closing the incision.

Thompson (1932) describes a hypophysectomy technique for the rat which was then used by Knigge (1954) for the hamster. In Thompson's method, the soft tissues are cut or retracted until the midline of the base of the skull is seen between the ethmoid bullae. The trachea is retracted and a burr hole is made on the midline posterior to the occipitosphenoid synchondrosis. Once the pituitary is exposed, a small hook is used to tear the capsule, and the gland is then suctioned out. Commenting on the difference between hypophysectomy in the rat and the hamster, Knigge (1954) reported that complete operative removal of the hypophysis is more difficult in the hamster as processes of the pars distalis extend far forward and laterad. Also, in contrast to the rat hypophysis, which is easily "shelled out" of its capsule, the hamster hypophysis tends to remain fixed and usually separates into several pieces upon excision.

Hypophysectomy is considered complete when all of the glandular tissue lying beneath the tectorial membrane is removed, as determined by microscopic examination of the pituitary capsule. In contrast to the rat, hypophysectomized hamsters remain in good physical condition with no apparent pelage change. They are active and aggressive. The testes retract into the peritoneal cavity in 12–15 days, the penis is smaller, and in females the vaginal lips become dry and thin (Knigge, 1954). Finally, it should be noted that because the hypophysis is larger in the female than the male hamster, it is harder to remove.

K. Pinealectomy

The hamster is placed in ventral recumbency with its head held firmly in the surgeon's left hand. A midline incision is made in the scalp skin between the ears and the eyes. The transverse and superior sagittal sinuses can be seen through the bone. Using a dental drill with a contra-angle hand piece and a 5-mm skull trephine, a piece of the cranium is removed. The dura is removed directly over the confluence of the sinuses, using a sharp pair of watchmaker forceps. With the same forceps, the pineal gland is gently removed from the sinus cavity. Caution must be taken not to drill too deeply, thereby damaging the brain. Bleeding is controlled with cotton-tipped swabs and gauze pads. The bone is replaced and dusted with an antibiotic powder and the incision routinely closed.

Hoffman and Reiter (1965) use a technique similar to that just described, but mount the hamster's head in a hand-held stereotaxic holder. They note that the white 0.5-mm gland is not readily visible at the junction of the two sinuses, and it requires practice to grasp its stalk and remove the stalk and the gland in one motion. Nearly 100% survival was reported.

L. Ovariectomy

In this common procedure the hamster is placed on its abdomen with the tail toward the surgeon. A 1- to 1.5-cm incision is made through the skin on the midline, about 1 cm caudad to the place where a hump in the back forms (second to fourth lumbar vertebrae). The peritoneal cavity is entered about 5 mm lateral to the midline by piercing the muscles with a curved iris forceps. The ovaries are pulled out through the muscle by grasping the periovarian fat at the junction of the fallopian tube and the uterus. The ovary with accompanying blood vessels and fat are excised, and the uterine horn is returned to the peritoneal cavity. No ligature is needed when the ovary is removed. This technique was also used by Ravines (1961), who removed part of the uterine horn during ovariectomy.

M. Castration

With the hamster in dorsal recumbency and the tail toward the operator, the skin around the scrotal area is thoroughly cleaned. A small incision (~1 cm) is made at the tip of the scrotum. A 5-mm incision is then made into each scrotal sac at its tip, and the cauda and caput epididymis, testes, vas deferens, and spermatic vessels are pulled out. The incision will expand slightly and then retract. A single ligature is placed around the blood vessels and the vas deferens, which are then cut. The remaining vas deferens and fat are pushed back into the scrotum, which can be closed with one or two 9-mm wound clips. The operator should take care not to close the anus inadvertently when closing the scrotal incision.

N. Vasectomy

The animal is held in dorsal recumbency with its back legs retracted laterally and its tail toward the surgeon. A lower midline abdominal incision is made in the skin. Using curved iris forceps, the abdominal muscle is pierced about 3 mm to the left of the bladder. The vas deferens is lifted out and a 2- to 3-mm section removed. The same procedure is repeated to the right of the bladder. A suture is placed in the abdominal muscle incisions.

In the method of Flickinger (1982), the scrotum is incised on each side and the vas deferens is mobilized with care to avoid damaging the main deferential blood vessels. A 5-mm portion of each vas deferens is removed between 4–0 silk ligatures. The proximal cut end of the vas is ligated a second time to decrease the likelihood of leakage, and the wound closed in layers using 4–0 silk.

O. Cesarean Derivation

A ventral midline incision is made, and the muscle is separated to open the peritoneal cavity. The uterine horns are cut and removed from the body, the horns of the uterus opened, and the pups inside are stimulated by massaging with cotton swabs. (It is implied that the natural mother is sacrificed.) To minimize cannibalism from a foster mother, fresh vegetables are placed in the foster mother's cage. Half of the foster mother's pups are removed and sacrificed, while the pups from the cesarean section are placed with the foster mother's pups for 1 hr; the combined group is then placed with the foster mother (Burke et al., 1970).

The author has performed but one cesarean section (an ovariohysterectomy), where the natural mother was returned to her pups. In this instance, after the abdominal incision was made, the uterus was removed by placing ligatures at the cervix and caudad to the ovaries, and removing that organ. The abdomen and skin were closed in a routine manner, and the mother nursed her pups. A number of attempts to foster-raise baby hamsters on other species have been unsuccessful.

P. Sialectomy

With the hamster on its dorsum and its tail toward the operator, a 1- to 1.5-cm longitudinal midcervical incision is made. The fat and connective tissue are retracted and the salivary glands are exposed. The blood vessels and ducts are clamped with a forceps and a ligature is tied tightly before the glands are cut away. Usually no bleeding occurs and no ligature is needed in the muscle. The skin incision is routinely closed. The submaxillary, parotid, and sublingual glands can be removed in this manner. Gilda and Keyes (1947), performed their sialectomies by making a somewhat longer incision. They also noted that the procedure resulted in a decrease in body weight.

Q. Splenectomy

The hamster is placed in ventral recumbency and a 2-cm paracostal incision is made on the left side. The muscle is pierced with a curved iris forceps and the spleen is gently lifted out with another pair of iris forceps. A ligature is placed around the splenic blood vessels, immediately before they divide, and a cut is made between the ligature and the spleen.

Sherman et al. (1964) placed hamsters in right lateral recumbency, and clamped the blood vessels with a hemostat and excised the spleen. In these 8-week-old animals it was noted that hemostasis was easily accomplished with a hemostat. The peritoneum, abdominal muscles, and fascia were closed with interrupted sutures, and the skin was approximated with continuous sutures.

R. Cholecystectomy and Cholecystotomy

Van der Linden et al. (1959) removed the gallbladder by making a midline incision through which the liver was brought by slight pressure on the lower thorax. The cystic duct was ligated and the gallbladder removed from the fundus downward. Bleeding from the liver was stopped by slight pressure with a tampon. Kowalewski and Todd (1971), using a median epigastric incision of 1.5 cm, exposed the gallbladder and incised the fundus. Cholesterol pellets (3 mm long, 13–15 mg) were placed inside the gallbladder. The gallbladder incision was closed with 6-0 silk, and the muscles and skin were closed with 4-0 chromic catgut.

S. Cholecystoduodenostomy

In this procedure a 2-mm segment of the common bile duct was removed between two silk ligatures placed on the distal portion of the duct just above the opening of the pancreatic duct. A 2- to 3-mm incision was made in the fundus of the gallbladder near the ligatures, and the gallbladder was inserted into the duodenal lumen through a wound made 5–10 mm distal to the opening of the common duct. The hamsters were treated for the next 3 days with a 0.1% solution of oxytetracycline. The postsurgical weight loss was quickly regained (Pour and Donnelly, 1978).

T. Pancreatic Duct Ligation and Pancreatitis

Pour *et al.* (1983a) ligated the duct of the splenic lobe of the pancreas at two points, 3 mm apart. The portion of the duct between the two ligatures was then excised (the artery was left intact) in order to prevent reopening. Oxytetracycline (1%) was given in the drinking water for 3 days postsurgery. The procedure did not affect either survival or body weight gain. At autopsy, the splenic lobe in all hamsters was atrophic and consisted of a tiny tissue string.

In a similar manner pancreatitis can be induced by ligating the common pancreatic duct at a point just before its opening into the duodenum. If it is desired to be able to release this ligature, a suture is placed (not tied) around the distal part of the common duct. After the suture ends are crossed, they are led out through the ventral abdominal wall at the level of the common duct's distal end. Pull both ends of the suture as far as possible (but avoid excess pulling) so that the suture acts as a ligature for the common duct. Optimal time for inducing pancreatitis with minimum mortality is 48 hr, after which time the skin suture can be untied and removed by carefully pulling on one end (Pour *et al.*, 1983b).

U. Partial Pancreaticocolostomy

To aid in their studies of bile reflux, Pour *et al.* (1983c) anastomosed the gastric or splenic lobe of the pancreas to the transverse colon. Either lobe was ligated near the head region of the pancreas with absorbable sutures, leaving the arteries of each lobe intact. A silk suture was carefully threaded in the pancreatic tissue below the ligation, and tied, leaving the ends of the silk thread uncut. The pancreas was dissected into two parts between these ligations, leaving the arteries intact. A hole, smaller than the diameter of the margin of the resected distal pancreatic lobe, was cut on the cranial portion of the transverse colon. The ends of the silk suture were then guided via a suture needle through the hole, into the colonic lumen, and out through the colon wall. The distal, disconnected pancreatic lobe was next inserted into the colon lumen by pulling the free ends of the silk thread. The position of the intracolonic pancreas was secured by tying the silk suture on the outer colon wall. A pancreaticocolonic connection was established after 2 weeks.

V. Superior Cervical Ganglionectomy

The major difficulty in this procedure is the identification of the ganglion. The animal is placed in dorsal recumbency with its tail toward the surgeon, and the head stretched tight with retractors. A 1- to 1.5-cm longitudinal midline incision is made in the neck, and the fat and salivary glands are retracted to expose the sternohyoid muscle, which lies directly ventral to the trachea. With curved iris forceps, work on the left side of the trachea and go through the connective tissue and muscle until the common carotid artery is reached. The ganglion can be found by following the artery toward the branch of the common carotid with the internal carotid artery. It is immediately to the left of this landmark. The ganglion can now be removed, care being taken not to cut any blood vessels in the area.

W. Neonatal Enucleation

To perform this procedure both Rhoades and Fish (1983) and Sengelaub *et al.* (1983) induced mild hypothermia within 24 hr after birth. With the aid of a dissecting microscope, a small incision is made under the line of the prospective eyelid, and the eye, including all pigmented fragments is withdrawn with a fine forceps (Sengelaub *et al.*, 1983).

X. Esophageal Constriction

Dunham and Sheets (1974) studied the narrowing of the esophagus in relation to cancer formation in 10-week-old hamsters. After sodium pentobarbital anesthesia, 3- to 4-mm segments of polyethylene tubing (2.69 mm I.D.) were split lengthwise and cold-sterilized. A 2.5- to 3-cm longitudinal incision was made about 3 mm below the xiphoid process, and the stomach was brought to the skin surface and kept moist. The distal esophagus was freed from surrounding tissues for 8–10 mm above its entry into the forestomach and the split cuff applied to the lower esophagus, using forceps. The 24-hr mortality was reported as less than 10%.

VIII. PHYSIOLOGICAL MEASUREMENTS

A. Respiratory Functions

O'Neil and Raub (1984) have provided an excellent literature review for measuring pulmonary function in small mammals, including hamsters. They discuss lung volume, tests for the distribution of ventilation, tests of pulmonary gas exchange, ventilatory control, and the like. There are particularly helpful comments on equipment, and it is noted that due to a small tidal volume, all tubing, fittings, valves, and connectors should be short with a wide bore whenever rapid responses are important.

Nearly 40 years ago, Guyton (1947b) described various methods for measuring the respiratory volumes of laboratory animals. Using his oscilloscope respirograph method, normal respiratory patterns of hamsters were recorded and from this certain general formulas were developed: The respiratory volume per minute (cm^3) = 2.10 (weight in grams)$^{3/4}$. It averages 60.9 cm^3/min in the hamster. Other formulas (Guyton, 1947a) are as follows:

$$\text{Tidal air (cm}^3\text{)} = 0.0074 \text{ (weight in grams)}$$
$$\frac{\text{Rate of breathing per}}{\text{minute}} = \frac{295}{\text{(weight in grams)}^{1/4}}$$
$$\frac{\text{Inspiration or expiration}}{\text{(cm}^3\text{/sec)}} = 0.15 \text{ (weight in grams)}^{3/4}$$

Guyton states that with his method the calculated deviated from the measured by ±21.5%.

More recently Raub et al. (1983) performed a tracheostomy on hamsters and placed a 15-gauge Luer stub adapter into the tracheal lumen. The animal was then placed in a whole-body plethysomograph for lung function measurements. Briefly, lung volume was obtained by pressure changes in the plethysmograph measured by a differential pressure transducer which was referenced to a chamber of equal volume. Airway pressure was measured by a differential pressure transducer referenced to the plethysmograph providing the pressure difference across the total respiratory system. End-expiratory volume was calculated from recordings of airway and plethysmographic pressure using Boyle's law and was found to be approximately 1.4 ml. The animals were hyperventilated on a small-animal respirator to increase the minute ventilation and to establish a uniform volume history. A pressure–volume curve was obtained during apnea by inflating and deflating the lungs with a syringe pump. Vital capacity was measured between total respiratory system pressure of −15 and +30 cm of water and was 5.7 ml. Other values recorded were residual volume (0.75 ml), total lung capacity (6.5 ml), and respiratory system compliance (0.4 ml/cm water).

Respiratory frequency, tidal volume, and minute volume were measured by Mauderly et al. (1979) using specially constructed nonrebreathing valves. No invasive procedures were used. Animals were trained, beginning at 8 weeks of age, to place their noses in the nonrebreathing valves for 2–3 min. Data were collected at 12–16 weeks of age. Hamsters were readily conditioned by being handled and appeared to respond well to petting and stroking as positive reinforcement. Reported values were respiratory frequency of 76 breaths per minute, tidal volume of 0.66 ml/min, oxygen uptake of 2.3 ml/min, exchange ratio (oxygen uptake/carbon dioxide output) of 0.89, and ventilatory equivalent for oxygen (minute volume/oxygen uptake) of 22. Walker et al. (1985) recorded the ventilation of the unanesthetized, unrestrained hamster by using a sealed plethysmograph and measuring pressure changes in the expansion of inspired gas volumes as they flowed from the ambient environment into the warmer airways of the animal. With 30% oxygen, the respiratory frequency was 58.3/min, the tidal volume was 0.61 ml/100 gm body weight, and the minute ventilation was 35.5 ml/min/100 gm body weight.

A technique used in mice has been described for the remote determination of breathing frequency, using microwave radiation (Gordon and Ali, 1984). The basis of this method is that during inhalation and exhalation an animal's body volume changes and its efficiency in absorbing microwaves also changes. This change can be measured in a waveguide exposure system. Using this system, Gordon and Long (1984) found that mice can tolerate more heat exposure from microwaves than can hamsters.

B. Oxygen Consumption

Harbison and Brain (1983) made an exposure chamber out of Plexiglas to study particle deposition in hamster lungs. Room air was pulled with a vacuum pump through the inlet and outlet ports at flow rates of 1.4–2.7 liters/min. Air was mixed with a fan, and its temperature was monitored. Water and carbon dioxide were removed from the outlet gas, and the oxygen concentration at the outlet was continuously measured. The hamster's ventilation was not monitored. Oxygen consumption was calculated as the difference in oxygen concentration between inlet and outlet air, corrected for carbon dioxide removal and inlet–outlet conditions. It was found the mean oxygen consumption of exercising groups (5.0 ml/min/100 gm body weight) was twice that of resting groups (2.5 ml/min) and four times that of anesthetized groups (1.2 ml/min).

Desautels et al. (1985) measured the rate of oxygen uptake by 6- to 14-day-old males. At standard temperature and pressure (dry), the mean for a 10-gm hamster was approximately 10 ml/hr.

C. Electrocardiogram

A number of researchers (e.g., Wyman et al., 1953b; Jeffrey et al., 1970; Abelmann et al., 1972) have performed electrocardiograms on anesthetized hamsters, using needle electrodes. Lombard (1952), using the three standard leads, found that recording at 9 cm/sec was necessary due to rapid QRST complexes. A hamster was placed in a plastic sling so that it rested on its abdomen with its legs hanging downward through holes in the sling. Voltage was set at 0.5 mV, and voltage and time measurements were made on lead II. Approximate values were a heart rate of 400 beats per minute, an electrical axis of 21 degrees, a QRS duration of 21 msec, R voltage of 0.157 mV, PR interval of 43 msec, P voltage of 0.085 mV, and a visually counted respiration rate of 93 breaths per minute.

D. Arterial Blood Gases

The left carotid artery is exposed (see technique for superior cervical ganglionectomy, Section VII,V). PE-10 polyethylene tubing is inserted so that the tip lies within the aorta. A slight enlargement is made 3 cm from the end of the cannula by having a hot wire surround but not touch it. This is to prevent accidental ejection from the blood vessel. Using heat, a slight bend is made 3–4 mm from the end of the cannula. Once the artery is exposed, a ligature is tied as close to the hamster's head as possible. A small incision is made on the cephalic end of the vessel while it is being stretched by an open forceps. The catheter is inserted. The opening of the cannula must be to the animal's left side. When the bend in the cannula is in the carotid, loosely tie the vessel cephalad to this point. Once the enlargement enters the artery and the distal end of the catheter touches the aorta, the ligatures are tightened. The cannula is heparinized and can be exteriorized behind the ears (Popovic and Popovic, 1960). The incision is closed and a collar is placed around the animal's neck. Blood samples can then be taken from the unanesthetized and unrestrained animal. In male hamsters weighing 100–120 gm, arterial PO_2 was 74.4 torr, arterial PCO_2 was 42.5 torr, pH was 7.42, HCO_3^- was 26.9 mmol/liter, alveolar PO_2 was 99.1 torr, the alveolar–arterial PO_2 difference was 24.8 torr, and the hematocrit was 47.0% (Lucey et al., 1980).

Jeffrey et al. (1970) performed a modified thoracotomy through a left intercostal incision and artificially ventilated tracheotomized hamsters. Right and left ventricular pressures (see following section) were obtained by direct puncture of the exposed ventricular apices (26-gauge needles with an internal diameter of 0.215 mm (Abelmann et al., 1972), and blood was also taken for gas determinations. In these 107-gm animals, the pH was 7.37, the hematocrit was 40.6%, the PCO_2 was 45.0 torr, and PO_2 was 83.0 torr.

E. Blood Pressure

Using the laboratory rat, Kersten et al. (1947) indirectly measured blood pressure by volume changes of the foot. The rat was placed into a restraining apparatus and a modified sphygmomamometer cuff was placed around the lower leg. A photocell–microammeter arrangement measured changes in volume and light intensity. This method was adapted for hamsters in part by Callahan et al. (1959) who measured the return of blood flow in the hindlimb as indicated by a strain-gauge transducer. The cuff is wrapped snugly around the leg above the ankle, and the leg is then placed in a specially constructed plethysmograph chamber. The mean blood pressure in 20 hamsters was 97.8 mm Hg, which compared well when measured against the carotid cannulation technique. Systolic blood pressure was 109.5 mm Hg.

As previously noted, Jeffrey et al. (1970) measured ventricular pressures directly from the ventricular apices. They found a heart rate of 367.2 beats per minute. In the left ventricle the systolic pressure was 124.3 mm Hg, and the end-diastolic pressure was 8.2 mm Hg. In the right ventricle the systolic pressure was 37.1 mm Hg, and the end-diastolic pressure was 4.6 mm Hg. Using the same methodology, Abelmann et al. (1972), using male hamsters 243–370 days old, recorded a stroke volume of 0.047 ml/beat. Other parameters recorded by Abelmann et al. were as follows: hematocrit, 42.2%; indocyanine green dye circulation time from femoral vein injection to arrival at the mesentery, 33.3 sec. The left ventricular stroke work was 5.1 gm-cm/beat, left ventricular minute work was 19.8 gm-m/min, right ventricular stroke work was 1.0 gm-cm/beat, and the right ventricular minute work was 3.90 gm-m/min.

IX. PERFUSIONS

A. Bilateral Perfusion of Adrenal Glands

The general procedure of Knigge (1961) is to cannulate and ligate the abdominal vessels so that blood or other perfusion media go to the adrenals via the abdominal aorta; the adrenal effluent blood is collected from the inferior vena cava. Due to the large amount of abdominal and perirenal fat, the procedure is more difficult in hamsters weighing more than 115 gm.

The animal is placed in dorsal recumbency with the left side to the operator, and after an abdominal incision is made the intestines are reflected to the left. Open loops of silk are placed around the right and left renal arteries and veins, close to the hilus. Two open loops are placed around the aorta, and three around the inferior vena cava, below the kidneys. Some dissection is required to place a loop around the proximal aorta, distal to the superior mesenteric and celiac arteries. A loop is also placed around the proximal inferior vena cava between the caudate and right lobes of the liver. Ligate both right and left testicular vessels. The infusion cannula is placed into the aorta at the most distal aortic loop and secured. The other distal aortic loop is a slipknot which is released. At this time one can place PE-20 polyethylene tubing into the vena cava and secure it with the distal loop to withdraw or give fluids. To infuse fluids, tighten the renal loops and ligatures proximal to the kidneys.

B. Liver Perfusion

The technique described is used in the laboratory of Dr. G. Williams of the American Health Foundation for the isolation of hepatocytes for culture. After sodium pentobarbital anesthesia (50 mg/kg body weight), a ventral midline incision is made from the xiphisternum to the pubic bone and a loose tie is placed around the infrahepatic inferior vena cava. A 21-gauge butterfly needle is inserted in the portal vein and clamped in place. Perfusion with an EDTA solution is then begun using a peristaltic pump, at a rate of 5 ml/min. Immediately following the start of perfusion, ligation of the infrahepatic inferior vena cava is completed and the vein is severed distally to permit the perfusate to drain freely. At this point the blanching of the liver is evident. While this first perfusion is being performed, the right atrium is punctured with a scissor and the perfusate is again allowed to drain freely. The pump speed is increased to 25 ml/min. A second perfusion solution containing collagenase is now used, and the liver is removed for further study.

X. LAPAROSCOPY

The technique described by Dukelow (1978) is for small rodents in general and does not specifically mention the hamster. It is presented nevertheless due to the sparsity of literature on the subject. A 1.7-mm laparoscope (a nonflexible instrument) is preferred. Animals are restrained, after anesthesia, in dorsal recumbency at a 30- to 45-degree angle, with the head down. After surgical preparation of the abdominal area, a small skin incision is made if the scope's trocar does not have a cutting point. Slowly push the trocar slightly under the skin, downward (dorsally) through the abdominal wall. As abdominal insufflation is generally used, this can be performed through the cannula. Carbon dioxide (5%) is most often used, and it is first passed through a water bath to moisten it. The trocar is slowly withdrawn from the cannula and the telescope inserted. If an ancillary instrument is needed, it is placed several centimeters to the right or left of the telescope. At the conclusion, a skin suture or clip can be placed.

XI. MISCELLANEOUS TECHNIQUES

A. Esophageal and Pharyngeal Decontamination

Angulo *et al.* (1978) premixed 0.5 ml of Orabase (Squibb) with neomycin and dihydrostreptomycin and injected it into the cheek pouches, five times a week, by introducing the sterile cone of a syringe into the mouth lateral to the teeth. The Orabase, which adheres to the mucosa, was brought into position with gentle pressure from the outside. It releases the antibiotic over a 2- to 3-hr period.

B. Cutaneous Application of Chemicals

The repeated application of chemicals to the skin on the backs of hamsters ("skin painting") is often performed for carcinogenicity studies. The back of the animal is shaved from the scapular area to the base of the tail, and flank to flank. Using a glass pipette, 0.8 ml of solution is spread out over approximately 8 in.^{2} (51.6 cm^2) of back (Bernfeld and Homburger, 1983). In the author's laboratory, chemicals are usually placed on the clipped skin by means of a microliter pipette, whereas Overman (1985) uses a syringe. It is suggested that the investigator be made aware of the stages of the hamster's hair cycle (Musser and Silverman, 1980), as the stage of the cycle at the time of chemical application may affect carcinogenesis.

C. Transplantation into the Anterior Chamber of the Eye

Using anesthetized adult hamsters, 2% atropine is applied to the cornea. A small incision is made obliquely to the surface of the cornea. The fragment to be transplanted (~1 mm in size) is placed into the anterior chamber through the corneal slit using a modified Pasteur pipette. As the pipette is removed, the slit immediately closes and the transplant is transferred to the lateral angle of the anterior chamber by application of slight pressure to the cornea. This allows vision to be preserved (Nechad and Olson, 1983).

XII. EUTHANASIA

In general, the recommendations of the American Veterinary Medical Association, Panel on Euthanasia (1986) are applicable for the euthanasia of hamsters. However, the writer has been unable to locate hamster-specific references to certain procedures, such as microwave irradiation of the brain.

The most common form of hamster euthanasia used by the writer is carbon dioxide. The hamster is placed in an enclosed chamber and carbon dioxide is let in through an attached hose. Death ensues in approximately 2 min. Although ether may be used for euthanasia, its explosive nature must be considered.

Sodium pentobarbital can be used, intraperitoneally, at approximately 200 mg/kg body weight. Commercially available euthanasia solutions should be used as per manufacturer's directions.

Decapitation has been used by a number of investigators, such as Wexler (1952) and Turner and Howards (1977). The latter found that the mean serum glucose concentration of male hamsters anesthetized with the thiobarbiturate Inactin was 300 mg/dl (blood from heart or carotid artery), whereas in unanesthetized decapitated controls it was 145 mg/dl. Decapitation must be carefully performed to assure that the hamster is properly placed in the guillotine. The use of plastic decapitation cones may prove helpful.

Another acceptable form of euthanasia is cervical dislocation. Henderson and Lowrey (1983) used a commercially available blunt guillotine as a dislocator. As with sharp guillotining, the short neck and loose cervical skin of the hamster must be considered. Rolfe and Iaconis (1983) report euthanizing 72-hr-old hamsters by cervical dislocation after inducing ether anesthesia.

XIII. NECROPSY

The techniques described below are those performed by the author. Reuber (1977) has described a detailed necropsy protocol that is applicable to most all laboratory rodents. Under no circumstances should a necropsy begin until it is ascertained that the animal is dead.

The skin and other external areas are examined for lesions, and if present, they are carefully described and then removed along with adjacent normal tissue. The hair is then wet with 70% alcohol. For animals with skin lesions, a midline incision is made from the mandibular symphysis to the external genitalia, and the skin is reflected and examined. Examine the preputial glands, the mammary glands, salivary glands, and lymph nodes. For animals without overt skin lesions, the operator may choose to make a transverse skin incision across the lumbar area. Then by the use of the fingers or a blunt instrument, the skin is carefully peeled craniad, as if removing a sweater. The subcutaneous tissue, lymph nodes, and glands are examined as above.

Make a midline incision through the linea alba and open the abdominal cavity. Transverse skin flaps may be made to facilitate visualization. Note the amount, color, viscosity, and other characteristics of any abdominal fluid and save samples if required or deemed appropriate. Note if the diaphragm is intact. Puncture the diaphragm and allow the lungs to collapse. Open the thoracic cavity by cutting through the ribs dorsal to the costochondral junction. Save a section of the sternum. Note any fluid accumulations in the same manner as for the abdominal cavity. Examine the positioning of the organs. Remove the salivary glands and associated nodes. The esophagus may be trimmed free if necessary. Remove the tongue, larynx (and thyroid glands), trachea (with esophagus), lung, heart, and thymus in one piece and examine. Examine the lymph nodes of the thoracic area.

Saffiotti *et al.* (1968) describe a technique to remove the lungs at autopsy in order to avoid collapse and to fix them while fully expanded. After anesthesia with a barbiturate, the skin and muscle layers of the anterior neck are dissected. The trachea is exposed and loosely tied off. The animal is bled out by cutting the aorta through an opening in the abdominal wall. The tracheal ligature is now tightened. Open the chest by cutting the diaphragm and ribs on both sides, and remove the sternum. Excise the lungs *en bloc* with the trachea and mediastinal organs by grasping the distal end of the esophagus with forceps and pulling up while dissecting out. Attach a small weight to the tracheal ligature and fix the entire block in 10% neutral buffered formalin.

Next, note the position of the abdominal organs. In the male, remove the testis and epididymis as a unit. Cut the right testis at a right angle and the left testis longitudinally. Split the pubic symphysis and remove the urinary bladder, prostate, seminal vesicles, coagulating gland, and penis in one piece. Inject the urinary bladder (through the urethra) with formalin if desired, and ligate the urinary bladder at the urethra. In the female, remove the ovaries, uteri, cervix, urinary bladder, and vagina in one piece. Inject the urinary bladder as for the male if desired.

Free the stomach using finger pressure or scissors. Separate the gastrointestinal tract from its mesentery to the anus. The stomach and small intestines are separated from the cecum and colon at the ileocolic junction if desired.

Open the stomach along the greater curvature and examine it. Open and examine the remainder of the gastrointestinal tract, including the anus, along the mesenteric line. The mucosal surface is rinsed with formalin and can then be prepared using the Swiss roll technique (Herrold, 1969). Observe the mesenteries and mesenteric area lymph nodes.

Remove the liver along with the gallbladder. Remove the spleen. Remove the pancreas from its attachments to the abdominal organs.

Remove the kidneys, adrenal glands, and, if possible, ureters in one piece. Cut the right kidney at a right angle and the left longitudinally. Observe the cortex, medulla, and pelvis of the kidney. Observe the renal lymph nodes.

Remove the remaining skin from the skull. Open the calvarium if needed, and remove the brain and pituitary gland. Examine the harderian gland if desired. Examine the visible aspects of the nasal cavity and inner ears, and save the skull for decalcification if necessary. To permit further examination of the nasal cavity, remove the skin of the nose and fix the entire nasal cavity *in situ*. After decalcifying the skull, take one or more coronal sections, at least behind the incisors (including maxilloturbinates and nasoturbinates) and, if needed, as in inhalation studies, between the incisors and first molars (to include the maxillary sinus). Also take sections between the first and second molar and through the eyes, to include all the ethmoturbinates, the nasopharyngeal meatus, the eyes, and harderian glands (Ward and Reznik, 1983).

After the necropsy is complete, remaining animal wastes are to be disposed of properly.

REFERENCES

Abelmann, W., Jeffrey, F., and Wagner, F. (1972). Circulatory dynamics in heart failure of Syrian hamsters. *Prog. Exp. Tumor Res.* **16**, 261–273.

Adams, R., and Homburger, F. (1983). Design and logistics of lifetime carcinogenesis bioassay using Syrian hamsters. *Prog. Exp. Tumor Res.* **26**, 202–207.

Adolph, E. F. (1951). Responses to hypothermia in several species of infant mammals. *Am. J. Physiol.* **166**, 75–91.

American Veterinary Medical Association, Panel on Euthanasia (1986). Report of the AVMA Panel on Euthanasia. *J. Am. Vet. Med. Assoc.* **188**, 252–268.

Angulo, A. F., Spaans, J., Zemmouchi, L., and Van der Waaij, D. (1978). Selective decontamination of the digestive tract of Syrian hamsters. *Lab. Anim.* **12**, 157–158.

Bauck, L. A., and Hagan, R. J. (1984). Cystotomy for treatment of urolithiasis in a hamster. *J. Am. Vet. Med. Assoc.* **184**, 99–100.

Berman, H. J., Lutz, B. R., and Fulton, G. P. (1955). Blood pressure of golden hamster as affected by Nembutal sodium and X-irradiation. *Am. J. Physiol.* **183**, 597 (abstr.).

Bernfeld, P., and Homburger, F. (1983). Skin painting studies in Syrian hamsters. *Prog. Exp. Tumor Res.* **16**, 128–153.

Brain, J. D., and Valberg, P. A. (1979). Deportation of aerosol in the respiratory tract. *Am. Rev. Respir. Dis.* **20**, 1325–1373.

Breckon, G., and Goy, P. (1979). Routine chromosome screening in Syrian hamsters (*Mesocricetus auratus*) using orbital sinus blood. *Lab. Anim.* **13**, 301–204.

Burke, J. G., Van Hoosier, G. L., Jr., and Trentin, J. J. (1970). Cesarean derivation and foster nursing of strain LSH inbred hamsters. *Lab. Anim. Care* **20**, 238–246.

Burke, T. J. (1979). Rats, mice, hamsters and gerbils. *Vet. Clin. North Am.* **9**, 473–486.

Burr, M. M., and Nagler, F. P. (1953). Mumps infectivity studies in hamsters. *Proc. Soc. Exp. Biol. Med.* **83**, 714–717.

Callahan, A. B., Degelman, J., and Lutz, B. R. (1959). Systolic blood pressure in the hamster as determined by a strain-gauge transducer. *J. Appl. Physiol.* **14**, 1051–1052.

Carlson, W. D. (1965). Radiography. *In* "Methods of Animal Experimentation" (W. I. Gay, ed.), Vol. 1, pp. 151–166. Academic Press, New York.

Carvell, J. E., and Stoward, P. J. (1975). Halothane anesthesia of normal and dystrophic hamsters. *Lab. Anim.* **9**, 345–352.

Chaudhry, A. P., Halberg, F., Keenan, C. E., Harner, R. N., and Bittner, J. J. (1958). Daily rhythms in rectal temperature and in epithelial mitoses of hamster pinna and pouch. *J. Appl. Physiol.* **12**, 221–224.

Chen, K. K., Powell, C. E., and Maze, N. (1945). The response of the hamster to drugs. *J. Pharmacol. Exp. Ther.* **85**, 348–355.

Cunliffe-Beamer, T. L. (1983). Biomethodology and surgical techniques. *In* "The Mouse in Biomedical Research" (H. L. Foster, J. D. Small, and J. G. Fox, eds.), Vol. 3, pp. 402–437. Academic Press, New York.

Curl, J. L., and Peters, L. L. (1983). Ketamine hydrochloride and xylazine hydrochloride anaesthesia in the golden hamster (*Mesocricetus auratus*). *Lab. Anim.* **17**, 290–293.

David, A. J., and Harrison, J. D. (1984). The absorption of ingested neptunium, plutonium and americum in newborn hamsters. *Int. J. Radiat. Biol.* **46**, 279–286.

Desautels, M., Dulos, R., and Thornhill, J. (1985). Thermoregulatory responses of dystrophic hamsters to changes in ambient temperature. *Can. J. Physiol. Pharmacol.* **63**, 1145–1150.

Dudley, W. R., Soma, L. R., Barnes, C., Smith, T. C., and Marshall, B. E. (1975). An apparatus for anesthetizing small laboratory animals. *Lab. Anim. Sci.* **25**, 481–482.

Dukelow, W. R. (1978). Laparoscopic techniques in mammalian embryology. *In* "Methods in Mammalian Reproduction" (J. C. Daniel, Jr., ed.), pp. 437–460. Academic Press, New York.

Duncan, P. F., Cullen, B. F., and Ray-Keil, L. (1977). Thiopental inhibition of tumor immunity. *Anesthesiology* **46**, 97–101.

Dunham, L. J., and Sheets, R. H. (1974). Effects of esophageal constriction on benzo(a)pyrene carcinogenesis in hamster esophagus and forestomach. *J. Natl. Cancer. Inst. (U.S.)* **53**, 875–881.

Dunlap, N. E., and Grizzle, W. E. (1984). Golden Syrian hamsters: A new experimental model for adrenal compensatory hypertrophy. *Endocrinology (Baltimore)* **114**, 1490–1495.

Eldridge, S. F., McDonald, K. E., Renne, R. A., and Lewis, T. R. (1982). Methohexital anesthesia for intratracheal instillation in the hamster. *Lab Animal* **11**, 50–54.

Elkin, M., and Kaplan, N. (1968). Urologic radiology. *In* "Roentgen Techniques in Laboratory Animals" (B. Felson, ed.), pp. 136–145. Saunders, Philadelphia, Pennsylvania.

Esparza, D. C., Schum, G. M., and Phalen, R. F. (1979). A latex sponge collar for partial body plethysmography using anesthetized rodents. *Lab. Anim. Sci.* **29**, 652–655.

Ferguson, J. W. (1979). Anesthesia in the hamster using a combination of methohexitone and diazepam. *Lab. Anim.* **13**, 305–308.

Ferm, V. H. (1967). The use of the golden hamster in experimental teratology. *Lab. Anim. Care* **17**, 452–462.

Ferris, C. F., Albers, H. E., Wesolowski, S. M., Goldman, B. D., and Luman, S. E. (1984). Vasopressin injected into the hypothalamus triggers a stereotypic behavior in golden hamsters. *Science* **224**, 521–523.

Flecknell, P. A., and Mitchell, M. (1984). Midazolam and fentanyl-fluanisone: Assessment of anesthetic effects in laboratory rodents and rabbits. *Lab. Anim.* **18**, 143–146.

Flickinger, C. J. (1982). The fate of sperm after vasectomy in the hamster. *Anat. Rec.* **202**, 231–239.

Frenkel, J. (1972). Dissecting aneurysms of the aorta and pancreatic islet cell hyperplasia with diabetes in corticosteroid and chlorothiazide-treated hamsters. *Prog. Exp. Tumor Res.* **16**, 300–324.

Gilda, J. E., and Keyes, P. H. (1947). Increased dental caries activity in the Syrian hamster following desalivation. *Proc. Soc. Exp. Biol. Med.* **66**, 28–32.

Gordon, C. J., and Ali, J. S. (1984). Measurement of ventilatory frequency in unrestrained rodents using microwave radiation. *Respir. Physiol.* **56**, 73–79.

Gordon, C. J., and Long, M. D. (1984). Ventilatory frequency of mouse and hamster during microwave induced heat exposure. *Respir. Physiol.* **56**, 81–90.

Green, C. J. (1975). Neuroleptanalgesic drug combinations in the anesthetic management of small laboratory animals. *Lab. Anim.* **9**, 161–178.

Green, C. J., Halsey, M. J., Precious, S., and Wardley-Smith, B. (1978). Alphaxalone–alphadone anesthesia in laboratory animals. *Lab. Anim.* **12**, 85–89.

Green, C. J., Knight, J., Precious, S., and Simpkins, S. (1981). Ketamine alone and combined with diazepam or xylazine in laboratory animals: A 10 year experience. *Lab. Anim.* **15**, 163–170.

Greenblatt, M., Choudari, K. U. R., Sanders, A. G., and Shubik, P. (1969). Mammalian microcirculation in the living animal: Methodologic considerations. *Microvasc. Res.* **1**, 420–432.

Grice, H. C. (1964). Methods for obtaining blood and for intravenous injections in laboratory animals. *Lab. Anim. Care* **14**, 483–493.

Guyton, A. C. (1947a). Measurement of the respiratory volume of laboratory animals. *Am. J. Physiol.* **150**, 70–77.

Guyton, A. C. (1947b). Analysis of respiratory patterns in laboratory animals. *Am. J. Physiol.* **150**, 78–83.

Habel, K., and Li, C. P. (1951). Intraspinal inoculation of mice in experimental poliomyelitis. *Proc. Soc. Exp. Biol. Med.* **76**, 357–361.

Haberman, B. H. (1974). Mechanical milk collection from mice for Bittner virus isolation. *Lab. Anim. Sci.* **24**, 935–937.

Haisley, A. D. (1980). A technique for cheek pouch examination of Syrian hamsters. *Lab. Anim. Sci.* **30**, 107–109.

Harbison, M. C., and Brain, J. D. (1983). Effects of exercise on particle deposition in Syrian golden hamsters. *Am. Rev. Respir. Dis.* **128**, 904–908.

Harkness, J. E. (1980). Anesthetic update. *Synapse* **13**, 10–11.

Hayes, J. A., Christensen, T. G., and Snider, G. L. (1977). The hamster as a model of chronic bronchitis and emphysema in man. *Lab. Anim. Sci.* **27**, 762–770.

Hellenkant, G., and Roberts, T. W. (1983). Study of the effect of gymnemic acid on taste in hamster. *Chem. Senses* **8**, 195–202.

Henderson, R. F., and Lowrey, J. S. (1983). Effect of anesthetic agents on lavage fluid parameters used as indicators of pulmonary injury. *Lab. Anim. Sci.* **33**, 60–62.

Henry, M. C., and Port, C. D. (1978). Effect of anesthetic agent on lung tumor induction in hamsters given Benzo(a)pyrene–Ferric oxide. *JNCI, J. Natl. Cancer. Inst.* **61**, 1221–1227.

Herrold, K. M. (1969). Adenocarcinomas of the intestine induced in Syrian hamsters by N-methyl-N-nitrosourea. *Pathol. Vet.* **6**, 403–412.

Hoffman, R. A., and Reiter, R. J. (1965). Rapid pinealectomy in hamsters and other small rodents. *Anat. Rec.* **153**, 19–22.

Hranicka, L. J., Moy, S. K., and Funahashi, A. (1977). Fiberoptic laryngoscope for small laboratory animals. *Lab. Anim. Sci.* **27**, 1004–1006.

Jeffrey, F. E., Wagner, R., and Abelmann, W. H. (1970). Left and right ventricular pressures in the normal and the cardiomyopathic Syrian hamster. *Proc. Soc. Exp. Biol. Med.* **135**, 940–943.

Johansen, E. (1952). A new technique for oral examination of rodents. *J. Dent. Res.* **31**, 361–365.

Kent, G. C., Jr., and Liberman, M. J. (1949). Induction of psychic estrus in the hamster with progesterone administered via the lateral brain ventricle. *Endocrinology (Baltimore)* **45**, 29–41.

Kersten, H., Brosene, W. G., Jr., Ablondi, F., and Subbarow, Y. (1947). A new method for the indirect measurement of blood pressure in the rat. *J. Lab. Clin. Med.* **32**, 1090–1098.

Knigge, K. M. (1954). The effect of hypophysectomy on the adrenal gland of the hamster (*Mesocricetus auratus*). *Am. J. Anat.* **94**, 225–272.

Knigge, K. M. (1961). A method for *in vivo* cannulation of both adrenal glands of the hamster. *Anat. Rec.* **141**, 145–149.

Kowalewski, K., and Todd, E. F. (1971). Carcinoma of the gallbladder induced in hamsters by insertion of cholesterol pellets and feeding dimethylnitrosamine. *Proc. Soc. Exp. Biol. Med.* **136**, 482–486.

Kradolfer, F. (1954). Experimental vaginal infection of hamster with *Trichomonas foetus*. *Exp. Parasitol.* **1**, 1–8.

Kraus, A. L. (1980). Research methodology. *In* "The Laboratory Rat" (H. J. Baker, J. R. Lindsey, and S. H. Weisbroth, eds.), Vol. 2, pp. 1–43. Academic Press, New York.

Lombard, E. A. (1952). Electrocardiograms of small mammals. *Am. J. Physiol.* **171**, 189–193.

Lucey, E. C., O'Brien, J. J., Pereira, W., Jr., and Snider, G. L. (1980). Arterial blood vas values in emphysematous hamsters. *Am. Rev. Respir. Dis.* **121**, 83–89.

Manning, J. P., and Giannina, T. (1966). A simple method for obtaining blood from hamsters in terminal experiments. *Lab. Anim. Care* **16**, 523–525.

Marshall, S., Miller, T. B., and Farah, A. E. (1963). Effect of renal papillectomy on ability of the hamster to concentrate urine. *Am. J. Physiol.* **204**, 363–368.

Matthews, H. G., and Barnhard, H. J. (1968). Radiographic technique. *In* "Roentgen Techniques in Laboratory Animals" (B. Felson, ed.), pp. 27–80. Saunders, Philadelphia, Pennsylvania.

Mauderly, J. L. (1977). Bronchopulmonary lavage of small laboratory animals. *Lab. Anim. Sci.* **27**, 255–261.

Mauderly, J. L., Tesarek, J. E., and Sifford, L. J. (1979). Respiratory measurements of unsedated small laboratory mammals using nonbreathing valves. *Lab. Anim. Sci.* **29**, 323–329.

Michel, G. (1961). Beitrag zur Blutentnahme beim Syr. goldhamster. *Berl. Muench. Tieraerztl. Wochenschr.* **74**, 418–419.

Moller, A. W. (1968). Diseases and management of the golden hamster (*Mesocricetus auratus*). *Curr. Vet. Ther.* **3**, 418–422.

Mulder, J. B., Johnson, H. B., Jr., McKee, G. S., and Sellers, S. E. (1979). Anesthesia with Ketaset Plus in guinea pigs and hamsters. *VM/SAC, Vet. Med. Small Anim. Clin.* **74**, 1807–1808.

Musser, T. K., and Silverman, J. (1980). The hair cycle of the Syrian golden hamster (*Mesocricetus auratus*). *Lab. Anim. Sci.* **30**, 681–683.

Nechad, M., and Olson, L. (1983). Development of interscapular brown adipose tissue in the hamster. II. Differentiation of transplants in the anterior chamber of the eye: Role of the sympathetic innervation. *Biol. Cell.* **48**, 167–174.

Norris, M. L., and Miles, P. (1982). An improved, portable machine designed to induce and maintain surgical anaesthesia in small laboratory rodents. *Lab. Anim.* **16**, 227–230.

O'Neil, J. J., and Raub, J. A. (1984). Pulmonary function testing in small laboratory mammals. *Environ. Health Perspect.* **56**, 11–22.

Ortiz, E. (1953). The effects of castration on the reproductive system of the golden hamster. *Anat. Rec.* **117**, 65–91.

Overman, D. O. (1985). Absence of embryotoxic effects of formaldehyde after percutaneous exposure in hamsters. *Toxicol. Lett.* **24**, 107–110.

Pansky, B., Jacobs, M., House, E. L., and Tasson, J. P. (1961). The orbital region as a source of blood samples in the golden hamster. *Anat. Rec.* **139**, 409–412.

Petty, C. (1982). "Research Techniques in the Rat." Thomas, Springfield, Illinois.

Pfeiffer, C. A. (1936). Sexual differences of the hypophyses and their determination by the gonads. *Am. J. Anat.* **58**, 195–225.

Pickard, G. E., Kahn, R., and Silver, R. (1984). Splitting of the circadian rhythm of body temperature in the golden hamster. *Physiol. Behav.* **32**, 763–766.

Pinkerton, W., and Webber, M. (1964). A method of injecting small laboratory animals by the ophthalmic plexus route. *Proc. Soc. Exp. Biol. Med.* **116**, 959–961.

Popovic, V., and Popovic, P. (1960). Permanent cannulation of aorta and vena cava in rats and ground squirrels. *J. Appl. Physiol.* **15**, 727–278.

Pour, P. M., and Donnelly, T. (1978). Effect of cholecystoduodenostomy and choledochostomy in pancreatic carcinogenesis. *Cancer Res.* **38,** 2048–2051.

Pour, P. M., Donnelly, T., and Stepan, K. (1983a). Modification of pancreatic carcinogenesis in the hamster model. 6. The effect of ductal ligation and excision. *Am. J. Pathol.* **113,** 365–372.

Pour, P. M., Takahashi, M., Donnelly, T., and Stepan, K. (1983b). Modification of pancreatic carcinogenesis in the hamster model. IX. Effect of pancreatitis., *JNCI, J. Natl. Cancer Inst.* **71,** 607–613.

Pour, P. M., Donnelly, K., and Stepan, K. (1983c). Modification of pancreatic carcinogenesis in the hamster model. 5. Effect of partial pancreatico-colostomy. *Carcinogenesis (London)* **4,** 1327–1331.

Ransom, J. H. (1984). Intravenous injection of unanesthetized hamsters. *Lab. Anim. Sci.* **34,** 200–201.

Raub, J. A., Miller, F. J., Graham, J. A., Gardner, D. E., and O'Neill, J. J. (1983). Pulmonary function in normal and elastase-treated hamsters exposed to a complex mixture of olefin-ozone-sulfur dioxide. *Environ. Res.* **31,** 302–310.

Ravines, H. T. (1961). Effect of castration, hypophysectomy, androgens and estrogens on the spleen and bone marrow of golden hamsters (*Mesocricetus auratus*). *Lab. Invest.* **10,** 341–353.

Rennie, J. S., MacDonald, D. G., and Douglas, T. A. (1982). Experimental iron deficiency in the Syrian hamster. *Lab. Anim.* **16,** 14–16.

Reuber, M. D. (1977). Necropsy of animals for scientific research. *Clin. Toxicol.* **10,** 111–127.

Rhoades, R. W., and Fish, S. E. (1983). Bilateral enucleation alters visual callosal but not corticotectal or corticogeniculate projections in hamster. *Exp. Brain Res.* **51,** 451–462.

Rockhold, W. T., and Hiestand, W. A. (1956). Ejaculatory response induced by potassium chloride in small mammals. *Proc. Soc. Exp. Biol. Med.* **92,** 402–404.

Rolfe, R. D., and Iaconis, J. P. (1983). Intestinal colonization of infant hamsters with *Clostridium difficile*. *Infect. Immun.* **42,** 480–486.

Roosa, R. A., Wilson, D. B., and Defendi, V. (1965). Effect of thymectomy on hamsters. *Proc. Soc. Exp. Biol. Med.* **118,** 584–590.

Saffiotti, U., Cefis, F., and Kolb, L. A. (1968). A method for the experimental induction of bronchogenic carcinoma. *Cancer Res.* **26,** 104–124.

Schall, H. (1984). Chirurgische Behandlungsmöglichkeiten bei Kaninchen, Meerschweinchen und Hamster. *Prakt. Tieraerztl.* **65,** 502–508.

Schreiber, H., and Nettesheim, P. (1972). A new method for pulmonary cytology in rats and hamsters. *Cancer Res.* **32,** 737–745.

Schreiber, H., Schreiber, K., and Martin, D. (1975). Experimental tumor induction in a circumscribed region of the hamster trachea: Correlation of histology and exfoliative cytology. *J. Natl. Cancer. Inst. (U.S.)* **54,** 187–197.

Schuchman, S. M. (1977). Individual care and treatment of rabbits, mice, rats, guinea pigs, hamsters and gerbils. *Curr. Vet. Ther.* **6,** 726–756.

Sengelaub, D. R., Windrem, M. S., and Finlay, B. L. (1983). Increased cell number in the adult hamster retinal ganglion cell layer after early removal of one eye. *Exp. Brain Res.* **52,** 269–276.

Sherman, J. D., Adner, M. M., and Dameshek, W. (1963). Effect of thymectomy on the golden hamster (*Mesocricetus auratus*). I. Wasting disease. *Blood* **22,** 252–271.

Sherman, J. D., Adner, M. M., and Dameshek, W. (1964). Effect of thymectomy on the golden hamster (*Mesocricetus auratus*). II. Studies of immune response in thymectomized and splenectomized non-wasted animals. *Blood* **23,** 375–388.

Silverman, J., Huhndorf, M., Balk, M., and Slater, G. (1983). Evaluation of a combination of tiletamine and zolazepam as an anesthetic for laboratory rodents. *Lab. Anim. Sci.* **33,** 457–460.

Simmons, M. L., and Smith, L. H. (1968). An anesthetic unit for small laboratory animals. *J. Appl. Physiol.* **25,** 324–325.

Singhal, A. K., Cohen, B. L., Finver-Sadowsky, J., McSherry, C. K., and Mosbach, E. H. (1984). Role of hydrophilic bile acids and of sterols on cholelithiasis in the hamster. *J. Lipid Res.* **25,** 564–570.

Smith, C. J., and Kelleher, P. C. (1973). A method for intragastric feeding of neonatal rats. *Lab. Anim. Sci.* **23,** 682–684.

Smith, D. J., Taubman, M. A., Ebersole, J. L., and King, W. (1982). Relationship between frequency of pilocarpine administration and salivary IgA level. *J. Dent. Res.* **61,** 1451–1453.

Smith, D. M., Goddard, K. M., Wilson, R. B., and Newberne, P. M. (1973). An apparatus for anesthetizing small laboratory rodents. *Lab. Anim. Sci.* **23,** 869–871.

Strittmatter, J. (1972). Anesthesie beim Goldhamster mit Ketamine und Methoxyflurane. *Z. Versuchstierkd.* **14,** 129–133.

Stroia, L., Bohr, D. F., and Vocke, L. (1954). Experimental hypertension in the hamster. *Am. J. Physiol.* **179,** 154–158.

Taylor, R. M. (1940). Detection of human influenza virus in throat washings by immunity response in Syrian hamster (*Cricetus auratus*). *Proc. Soc. Exp. Biol. Med.* **43,** 541–542.

Thayer, C. B. (1972). Clinical evaluation of a combination of droperidol and fentanyl as an anesthetic for the rat and hamster. *J. Am. Vet. Med. Assoc.* **161,** 665–668.

Thompson, K. W. (1932). A technique for hypophysectomy of the rat. *Endocrinology (Baltimore)* **16,** 257–263.

Turbyfill, C. L., Peterson, R. P., and Soderwall, A. L. (1962). The cardiac puncture in adult, fetal and young golden hamster. *Turtox News* **40,** 162–163.

Turner, T. T., and Howards, S. S. (1977). Hyperglycemia in the hamster with Inactin [5-ethyl-5-(1-methyl propyl)-2-thiobarbiturate]. *Lab. Anim. Sci.* **27,** 380–382.

Valerino, D. M., Vesell, E. S., Aurori, K. C., and Johnson, A. O. (1974). Effects of various barbiturates on hepatic microsomal enzymes. A comparative study. *Drug Metab. Dispos.* **2,** 448–452.

Van der Linden, W., Christensen, F., and Dam, H. (1959). Cholecystectomy and gallstone formation in the golden hamster. *Acta Chir. Scand.* **118,** 113–116.

Van Hoosier, G. L., Jr., Gist, C., and Trentin, J. J. (1978). Enhancement by thymectomy of tumor formation by oncogenic adenoviruses. *Proc. Soc. Exp. Biol. Med.* **128,** 467–469.

Walker, B., Adams, E., and Voelkel, N. (1985). Ventilatory responses of hamsters and rats to hypoxia and hypercapnia. *J. Appl. Physiol.* **59,** 1955–1960.

Ward, J., and Reznik, G. (1983). Refinements of rodent pathology and the pathologists contribution to evaluation of carcinogenesis bioassays. *Prog. Exp. Tumor Res.* **26,** 266–291.

Warren, S., and Gates, O. (1968). Cancers induced in different species by continuous γ-radiation. *Arch. Environ. Health* **17,** 697–704.

Wexler, B. C. (1952). Adrenal ascorbic acid changes in the gonadectomized golden hamster following a single injection of diethylstilbestrol. *Endocrinology (Baltimore)* **50,** 531–536.

Willoughby, C. R. (1973). Anaesthetising hamsters and small rodents. *Vet. Rec.* **92,** 572–573.

Wyman, L. C., Fulton, G. P., and Shulman, M. H. (1953a). Direct observations on the circulation in the hamster cheek pouch in adrenal insufficiency and experimental hypercorticalism. *Ann. N.Y. Acad. Sci.* **56,** 643–658.

Wyman, L. C., Fulton, G. P., Sudak, F. N., and Patterson, G. N. (1953b). Electrocardiograms and serum electrolyte levels in hamsters with adrenal insufficiency and hypercorticalism. *Proc. Soc. Exp. Biol. Med.* **84,** 280–283.

Wyman, L. C., Drapeau, L. L., Fulton, G. P., and Shulman, M. H. (1957). Pattern of vasomotor activity during histamine intoxication in intact, cortisone-treated and adrenalectomized hamsters. *Angiology* **8,** 170–181.

Yabe, Y., Samper, L., Taylor, G., and Trentin, J. (1963). Cancer induction in hamsters by human type 12 adenovirus. Effect of route of injection. *Proc. Soc. Exp. Biol. Med.* **113,** 221–224.

Yohn, D. S., Funk, C. A., Kalnins, V. I., and Grace, J. T., Jr. (1965). Sex-related resistance in hamsters to adenovirus-12 oncogenesis. Influence of thymectomy at three weeks of age. *J. Natl. Cancer. Inst. (U.S.)* **35,** 617–624.

Chapter 6

Viral Diseases

John C. Parker, James R. Ganaway, and Cynthia S. Gillett

I.	Introduction	95
II.	Viruses	96
	A. Sendai Virus	96
	B. Pneumonia Virus of Mice (PVM)	98
	C. Lymphocytic Choriomenigitis Virus (LCMV)	100
	D. Simian Virus 5 (SV5)	104
	E. Other Viral Agents	105
	References	106

I. INTRODUCTION

Compared to the number of virus infections reported for humans and other animal species, including the laboratory mouse and rat, it is surprising to note the small number of naturally occurring virus infections reported for the hamster. Also, except for rare or unusual cases, these infections are normally subclinical. Apart from the retroviruses (see Chapter 12), and experimental infections such as polyomavirus, there are only four reasonably well-documented virus infections in the hamster (see Table I): Sendai virus, pneumonia virus of mice (PVM), lymphocytic choriomeningitis virus (LCMV), and Simian virus 5 (SV5). Due to the paucity of information, infections with reovirus type 3 (Reo 3), Toolan's H-1 virus, cytomegalovirus, mouse encephalomyelitis virus (GDVII), and a hamster papovavirus (HapV) are described briefly under other viral agents (Section II,E). Of note, there is a definite epizootiological relationship between Sendai, PVM, LCMV, and perhaps Reo 3 infections also harbored by rats, mice, and guinea pigs in that these viruses reciprocally cross-infect all of these commonly used and often adjacently housed laboratory rodent species. A viral etiology has not been associated with the commonly encountered clinical syndrome referred to as "wet-tail" and therefore is not discussed in the chapter.

LCMV presents a potentially serious public health threat if the virus becomes accidentally introduced into a colony of naive animals. LCMV may readily infect humans which come in contact with infected mice or hamsters and has the potential to produce serious disease manifested commonly as an influenzalike fever or more rarely as meningitis or meningoencephalomyelitis. The other virus infections fall into a nuisance category, except that they have the potential to interact with experimental research models and perhaps alter the outcome of the research without the awareness of the investigator.

Table I
Properties of Selected Viruses Infecting Hamsters[a]

Property	Sendai	PVM	LCM	SV5	Reo 3
Classification					
Family	Paramyxoviridae	Paramyxoviridae	Arenaviridae	Paramyxoviridae	Reoviridae
Genus	*Paramyxovirus*	*Pneumovirus*	*Arenavirus*	*Paramyxovirus*	*Reovirus*
Diameter (nm)	150–250	80–200	50–300 (110–130 mean)	158–180	60–90
Morphology	Pleomorphic, roughly spherical, enveloped	Pleomorphic, roughly spherical, filamentous, enveloped	Pleomorphic, round oval or irregularly shaped with 10-nm-long club-shaped surface projection	Pleomorphic, roughly spherical, enveloped	Icosahedral
Nucleic acid type	RNA	RNA	RNA	RNA	RNA
pH Stability	Stable (pH 5.3–9.8)	Acid sensitive (pH ≤3) Alkaline sensitive (pH ≥11)	Acid sensitive (pH ≤5.5)	Acid sensitive	Stable (pH 2.2–8)
Heat stability	Sensitive (48°C, 15 min)	Sensitive (56°C, 30 min)	Sensitive (56°C, 1 hr)	Sensitive (unstable at RT)	Stable (56°C, 2 hr; 60°C, 30 min)
Lipid solvent	Sensitive	Sensitive	Sensitive	Sensitive	Stable
Neuraminidase	Present	Lacks	Lacks	Present	Lacks
Hemolysin	Present	Lacks	Lacks	Present	Lacks
Hemadsorption	Present	Present	Lacks	Present	Lacks
Hemagglutination	Human O, guinea pig, rat, chicken, others (RT, 4°C)	Mouse, rat, hamster (RT, 4°C)	Lacks	Chicken, guinea pig (4°C)	Human O, bovine, geese (RT)
Replication site	Cytoplasm, nucleoli	Cytoplasm	Cytoplasm	Cytoplasm	Cytoplasm
Host range					
In vivo	Mice, rats, guinea pigs, hamsters	Mice, rats, guinea pigs, hamsters	Mice, hamsters, guinea pigs, rats, dogs, monkeys, rabbits, chickens, humans	Nonhuman primates, mice, hamsters, dogs, (HAI antibody observed in numerous species including human)	Mice, rats, guinea pigs, hamsters (some 60 species of animals are infected with revirus 1, 2, or 3)
In vitro	Primary monkey kidney cells, embryonated eggs (serial cell lines with protease treatment)	Primary hamster kidney cells, BHK-21, Vero	Replication without CPE in a wide variety of cell cultures	Primary monkey kidney cells embryonated eggs	Kidney cells of various species (primates, swine, cat, dog, and guinea pig); also FL, human amnion, BS-C-1, KB, and L cells; embryonated eggs (CAM or amniotic route)

[a] Abbreviations: CAM, chorioallantoic membrane; CPE, cytopathic effect; RT, room temperature.

II. VIRUSES

A. Sendai Virus

Sendai virus is a common infection in laboratory rodent colonies and is probably the leading cause of pneumonia in mice (Parker and Richter, 1982). In addition to causing significant morbidity and mortality in mice, Sendai virus induces numerous changes in its host which complicate the use of the animal in research. Research into the pathogenesis and the immunological effects of Sendai virus infection is proceeded at a brisk pace. The reader is referred to the treatise by Parker and Richter (1982) for a more complete discussion of this important virus. Here attention will be focused on the basic biology of Sendai virus and on what is known about Sendai infection in hamsters.

6. VIRAL DISEASES

1. History

Sendai virus was first isolated in Japan from mice inoculated with material from human infant pneumonitis cases (Kuroya et al. 1953; Noda, 1953; Sano et al., 1953). Antibody to the virus was found to be common in laboratory rodents (Fukumi et al., 1954; Matsumoto et al., 1954). Sendai virus was considered a human respiratory pathogen during that time. In 1958, human hemadsorption virus type 2 (HA-2) was isolated and its close relationship and serological cross-reactivity with Sendai virus were discovered (Chanock et al., 1958; Fukumi and Nishikawa, 1961). Subsequently, doubts were raised as to whether Sendai virus was the etiological agent in human newborn pneumonitis cases. Currently, mice, rats, and hamsters are considered the only natural hosts of Sendai. The first reports of Sendai virus isolation from hamsters in Japan (Matsumoto et al., 1954) and in the United States (Soret and Buthala, 1967; Profeta et al., 1969) are among the only such reports.

2. Etiology

a. Characteristics. Sendai virus belongs to the family Paramyxoviridae, genus *Paramyxovirus,* and is classified as the species *parainfluenza* type 1 along with the closely related human virus HA-2 (Matthews, 1982). It is a single-stranded pleomorphic RNA virus measuring 150–250 nm. It possesses a lipoprotein envelope and is thus ether sensitive. Both hemagglutinin and neuraminidase are produced. All isolates of Sendai virus appear antigenically identical, and there is no evidence for the existence of substrains (Parker and Richter, 1982).

b. Propagation. After intranasal inoculation of mice, virus can be recovered from nasal washings, saliva, and lungs for 9–14 days. Virus can occasionally be found in organs such as the kidney, liver, or spleen, but probably replicates only in the respiratory tract (Parker and Reynolds, 1968; Blandford and Heath, 1972). Sendai virus grows to high titer in 8- to 13-day-old chicken embryos and this is the most common method of *in vivo* propagation. Monkey kidney cell cultures readily show cytopathic effect (CPE) and are a sensitive detection method (Parker and Reynolds, 1982).

3. Pathogenesis

a. Clinical Disease. Infection with Sendai virus in laboratory rodents is often clinically silent with low mortality (Parker et al., 1978; Van Hoosier and Ladiges, 1984). Mouse strains that are highly susceptible, such as the 129/J strain or athymic mice (*nu/nu*), may display signs of respiratory distress or, more often, sudden death (Parker et al., 1978). In one hamster colony enzootically infected with Sendai occasional deaths in suckling hamsters were the only clinical sign (Profeta et al., 1969). Fatal pneumonia due to Sendai has been reported in Chinese hamsters (*Cricetulus griseus*) (Chun and Chu, 1956). The authors are aware of no other reports of clinical disease due to natural Sendai virus infection in hamsters.

b. Pathology. Complete consolidation of the lung lobes was seen in newborn hamsters dying after intranasal inoculation with hamster lung suspensions. Sendai virus was isolated from the consolidated lungs as well as from normal nonmanipulated hamster litters within the infected colony (Profeta et al., 1969). In mice, grossly reddened lung lobes representing extensive pulmonary consolidation with sharp demarcations between affected and unaffected tissue are seen (Ward, 1974; Zurcher et al., 1977).

Microscopically, acute Sendai infection in mice results in interstitial pneumonitis. Secondary bacterial infection of the lung often complicates the pathological picture (Jakab, 1981). The histopathology of experimental nonlethal Sendai infection in hamsters agrees with that of infection in mice: patchy pulmonary consolidation, bronchial edema and epithelial desquamation with basement membrane destruction, and inflammatory cell infiltrates (Blandford and Charlton, 1977). Concurrent cyclophosphamide administration and Sendai inoculation in hamsters resulted in clinically apparent disease, increased viral replication, abolishment of humoral antibody response, lack of typical microscopic lesions, and deaths. Additionally, bronchial basement membranes remained undisturbed and renal immune complex deposition seen in Sendai infected hamsters was not seen with cyclophosphamide treatment. It was concluded that humoral immunity was necessary for viral elimination but also mediated the resultant inflammatory lesions.

4. Epizootiology

a. Host Range. As previously discussed, Sendai virus was initially considered a human respiratory virus, but later studies indicated that its only natural hosts are laboratory mice, rats, and hamsters. Virus isolation has not been documented in guinea pigs, although many colonies do have antibody titers to a parainfluenza virus serotype (Van Hoosier and Robinette, 1976). Antibody against Sendai virus has not been found in gerbils (Parker et al., 1978).

b. Prevalence. Antibody to Sendai virus is commonly found in hamster colonies. The percentage of seropositive animals in infected colonies ranges from 2 to 4% (Reed et al., 1974; Van Hoosier et al., 1970) to 64% (Parker and Richter, 1982). The percentage of tested colonies in the United States with antibody titers has been reported to be 83% (Parker et al., 1978) and 94% (18 colonies tested) (Parker and Richter, 1982). In Japan two of three hamster colonies tested had antibody titers (Suzuki et al., 1982).

c. Natural History and Transmission. The epizootiology of natural Sendai infection in hamsters is largely unknown. In one hamster colony in which Sendai virus was isolated from newborns, 80% of the adult animals had high complement fixation (CF) and hemagglutination inhibition (HAI) titers (Profeta *et al.*, 1969). Mice and guinea pigs in the facility also had a high prevalence of Sendai antibody. Virus could not be isolated from seropositive hamsters. New hamsters arriving at the facility did not have titers but seroconverted within 5 weeks. The infection was clinically inapparent except for occasional deaths in neonates (Profeta *et al.*, 1969). This enzootic type of infection was most likely perpetuated by ongoing breeding and the introduction of susceptible animals as is the case in mouse enzootics (Fujiwara *et al.*, 1976; Goto and Shimizu, 1978).

5. Diagnosis

Serological diagnosis of Sendai infection in hamsters is readily accomplished through the use of either HAI (Reed *et al.*, 1974) or CF tests. The CF test is considered more sensitive than HAI, although it is technically more difficult and time-consuming (Parker *et al.*, 1978). CF titers remain elevated after recovery from Sendai infection unlike CF titers to PVM (pneumonia virus of mice). An enzyme-linked immunosorbent assay (ELISA) is available for mice and rats which has high sensitivity and reliability (Ertl *et al.*, 1979; Parker *et al.*, 1979).

Viral isolation is best done either by inoculating embryonated chicken eggs (Fukumi *et al.*, 1962) or monkey kidney cell cultures (Parker and Reynolds, 1968). Suspect animals are sampled by taking mouth swabs, nasopharyngeal swabs or washings, or lung tissue. Virus can be isolated between 7 and 14 days postinfection (van der Veen *et al.*, 1970; Parker *et al.*, 1978). Viral identification is made by the observation of CPE and the use of HA and HAI, CF, or immunofluorescent antibody techniques (IFAT) with inoculated cell cultures. Hemagglutination and specific inhibition is used for specific identification with allantoic fluids from inoculated chicken embryos.

6. Control and Prevention

The authors are unaware of a published report which describes the elimination of Sendai virus from an infected hamster colony. The best approach is to prevent the introduction of Sendai virus into the colony. This is accomplished by the purchase of Sendai virus-free hamsters and by quarantining and testing them prior to introduction into the colony. Additionally, if the facility houses conventional rodents suspected of harboring Sendai infections, the Sendai-free colony must be maintained under strict isolation or barrier conditions. Routine colony surveillance should be established. Cesarean derivation of mice, while expensive, has been used to eliminate successfully many horizontally transmitted viruses from the colony (Van Hoosier *et al.*, 1966, 1970). A formalin-killed vaccine has been developed which is effective in suppressing Sendai virus infection (Fukumi and Takeuchi, 1975; Eaton *et al.*, 1982). Testing, culling, vaccination, and cessation of breeding has been used successfully to eliminate Sendai virus from a mouse colony (Eaton *et al.*, 1982). In practice, many hamster colonies used for research have a high prevalence of antibody titers to Sendai (Parker *et al.*, 1978).

7. Significance

The significance of Sendai infection in hamsters is unknown. Although there is a paucity of reports of clinically apparent disease in hamsters, inapparent effects of the virus could have serious consequences. In mice, a large number of physiological parameters remain altered long after recovery from acute infection with Sendai (Kay, 1978; Kay *et al.*, 1979). Sendai has also been shown to suppress T-cell mitogenesis in rats (Garlinghouse and Van Hoosier, 1978) and alter carcinogenesis studies in the mouse (Peck *et al.*, 1983). Numerous other effects of Sendai have been reported: the reader is referred to Pakes *et al.* (1984). Therefore, it is possible that Sendai infection in hamsters could alter the animals' responses in various investigative efforts.

B. Pneumonia Virus of Mice (PVM)

1. History

Pneumonia virus of mice (PVM) was first isolated from mice by Horsfall and Hahn (1939, 1940) during experiments designed to isolate human influenza viruses from patients with respiratory disease. Case material was inoculated intranasally into mice, and the lungs were serially passaged. Lung tissue from both inoculated and uninoculated passage mice showed patchy pulmonary consolidation, and a filterable agent (PVM) was recovered which could reproduce the lesions. PVM was also recovered from apparently healthy mice by serial lung passage. PVM was subsequently found to be a common infection in mouse colonies (Horsfall and Hahn, 1940).

PVM was isolated from the lungs of hamsters which died after intranasal inoculation of mouse-passage influenca virus (Pearson and Eaton, 1940). It is probable that the virus was present in the mouse lung suspensions, although the possibility that the hamsters were naturally infected prior to the inoculation cannot be excluded.

2. Etiology

PVM is classified in the family Paramyxoviridae, genus *Pneumovirus* (Matthews, 1982). The human and bovine respiratory syncytial viruses are the only other members of this genus. Lack of neuraminidase is one feature which distinguishes the pneumoviruses from parainfluenza viruses, whereas hemagglutinin production is a common characteristic (Matthews, 1982). PVM is antigenically distinct and there is no evidence for the existence of substrains. PVM is found most often in filamentous forms but also as spherical forms, 100 nm or 80–200 nm in diameter, respectively. The virus is a single-stranded RNA virus, has a lipoprotein envelope, and is ether sensitive. The virus is unstable at room temperature and thus intimate contact is probably required for transmission. A more detailed description of PVM can be found in the treatise by Parker and Richter (1982).

PVM has long been thought to grow *in vivo* only in the lung (Horsfall and Hahn, 1940; Parker and Richter, 1982). However, a recent report describes viral isolation from the nasal turbinates of experimentally infected weanling mice (Smith *et al.*, 1984). PVM can be grown in hamster kidney cell cultures (HKCC, BHK-21), Vero cells, and hamster embryo cultures (Tennant and Ward, 1962; Harter and Choppin, 1967; Reed *et al.*, 1975). CPE is seen as cellular degeneration and small intracytoplasmic inclusions. CPE is not seen in Swiss mouse embryo or kidney cultures, or rat embryo cultures (Reed *et al.*, 1975). Primary hamster kidney cell culture (HKCC) is not recommended for PVM isolation because of low viral titers and slow CPE. PVM does not replicate in embryonated chicken or duck eggs (Volkert and Horsfall, 1947).

3. Pathogenesis

a. Clinical Disease. There are no documented reports of natural clinical disease or mortality attributed to PVM in mice or hamsters. There is mention of one case of PVM associated with a disease outbreak in hamsters, but clinical signs were not described (Carthew *et al.*, 1978; Carthew and Sparrow, 1980). Descriptions of disease caused by experimental manipulations can be found in articles describing the original isolations from mice (Horsfall and Hahn, 1939, 1940) and hamsters inoculated intranasally with mouse-passaged lung suspensions of human influenza virus (Pearson and Eaton, 1940). The hamsters were seen to be sneezing, dyspneic, and weak. Death occurred 6–15 days post inoculation.

b. Pathology. Patchy plum-colored consolidation of 50–75% of the lung occurred during experimental PVM infection in hamsters (Pearson and Eaton, 1940). Thickening of alveolar walls with a predominantly mononuclear cell infiltrate and atelectasis were seen (Pearson and Eaton, 1940). The overall picture is one of interstitial pneumonia, as it is in mice (Parker and Richter, 1982). The presence of swollen hyperemic nasal turbinates and mild erosive rhinitis described in experimentally inoculated hamsters and mice (Pearson and Eaton, 1940: Smith *et al.*, 1984) suggests that PVM may not be strictly pneumotropic.

4. Epizootiology

a. Host Range. Laboratory mice, rats, and hamsters are considered the natural hosts of PVM (Horsfall and Hahn, 1939, 1940; Pearson and Eaton, 1940; Eaton and van Herick, 1944; Parker and Richter, 1982). Guinea pigs may have significant antibody titers, but the virus has not been isolated (Van Hoosier and Robinette, 1976). The only reported isolation from hamsters was in animals inoculated with lung tissue suspensions from mice (Pearson and Eaton, 1940). The possibility that the mice used were infected with PVM should be kept in mind.

b. Prevalence. PVM antibody is frequently found in laboratory mice, hamsters, and rats. Prevalence in hamster colonies has been reported to be 86% (Parker *et al.*, 1967), 29% (Van Hoosier *et al.*, 1970), 10% (Reed *et al.*, 1974), and 45% (Parker and Richter, 1982). A serological survey of young Chinese hamsters (*Cricetus griseus*) indicated a low prevalence of PVM antibody (Schiff *et al.*, 1973). In a survey of rodent colonies which spanned several years, the prevalence of PVM infection (determined by HAI) was higher in hamster (45%) and rat (62%) colonies than in mouse colonies (20%) (Parker and Richter, 1982).

c. Natural History and Transmission. As mentioned previously, there have been no documented reports of natural clinical disease due to PVM. Current thinking on the mode of PVM transmission in mice is that it is not a highly contagious virus, and this results in a series of self-limiting focal infections within an infected colony. The higher antibody titers and incidence of positive hamster colonies suggest a difference in the infection pattern and/or physiological response to infection in the hamster.

Horsfall stated that PVM exists in a latent state within the lung because blind passage of lung tissue in mice resulted in isolation of virus from clinically normal animals (Horsfall and Hahn, 1939, 1940). However, more extensive studies have shown that latent infections do not occur in mice (Tennant *et al.*, 1966). Immunoperoxidase studies of experimental PVM in ex-germ-free mice have shown the virus to be present in the lung between days 2 and 7 (Carthew and Sparrow, 1980). The epizootiology of PVM in hamsters is unknown, although the success of cesarean derivation in eliminating antibody titers in subsequent generations suggests that vertical transmission does not occur (Van Hoosier *et al.*, 1970). Acute, self-limiting enzootics as occur in mice probably do not occur in hamsters.

5. Diagnosis

PVM possesses strong hemagglutinin activity similar to other paramyxoviruses. Until recently, the HAI test was the most commonly used serological test (Parker and Richter, 1982) and is still the primary test for hamsters. The ELISA has supplanted HAI as the preferred serological test for the diagnosis of PVM in mice and rats (Small, 1984). It is more sensitive than HAI (Descoteaux et al., 1980). Seroconversion takes place 9–10 days after experimental infection in mice, and antibody persists several months or more.

Infected cell cultures absorb mouse erythrocytes readily (Tennant and Ward, 1962), but virus from ground lung suspensions does not cause hemagglutination without treatment to release the virus from cell association (Ginsberg and Horsfall, 1951; Curnen and Horsfall, 1947). Viral isolation is best accomplished by inoculation of BHK-21, HKCC, Vero cells, or hamster embryo culture. BHK-21 is more sensitive to PVM than HKCC (Reed et al., 1975). Diagnosis of PVM in culture is made by observation of CPE, hemadsorption of mouse erythrocytes, IFAT, or hemagglutination (Parker and Richter, 1982). A plaque assay has also been developed for use with BHK-21 cells (Shimonaski and Came, 1970).

Mouse inoculation may be used to isolate the virus but is tedious and expensive. Virus-free mice, and presumably hamsters, may be used in the mouse antibody production (MAP) test for PVM diagnosis (Parker and Richter, 1982).

6. Control and Prevention

Cesarean derivation and foster rearing is an effective procedure for eliminating infection in a hamster colony (Van Hoosier et al., 1970). Development of a vaccine has not been reported. Several reports indicate that infection in hamsters results in immunity (Pearson and Eaton, 1940; Tennant et al., 1966).

7. Significance

Morbidity and mortality due to PVM in hamsters are unknown. The significance of PVM infection as a research variable has not been elucidated. However, investigators using hamsters, especially in areas such as toxicology or inhalation studies, should be forewarned about potential complications. Many murine virus infections are known to interfere with research results.

C. Lymphocytic Choriomeningitis Virus (LCMV)

1. History

The origin of early isolates of lymphocytic choriomeningitis virus (LCMV) (Armstrong and Lillie, 1934; Traub, 1935; Rivers and Scott, 1935) remains unknown because of the common practice in the past of attempting viral isolations and passaging of these isolates in mice which were not defined microbiologically. Similarity of these viral isolates resulted in the adoption of the present name (Armstrong and Dickens, 1935).

LCMV infection of mice has been studied extensively because it is an excellent model for investigating such phenomena as persistent viral infections, virus-specific immunological tolerance, and pathological immune reactions in virus diseases. Excellent reviews of this large body of knowledge have been published (Hotchin, 1971; Pfau et al., 1974; Casals, 1975; Buchmeier et al., 1980; Lehmann-Grube, 1982, 1984).

Intense interest in LCMV infection of hamsters originated from a report by Lewis et al. (1965), who were studying the complement-fixing antibody response to specific tumor antigens and found that the reactivity was due to LCMV, an inapparent contaminant of the tumor. The virus had previously been isolated from a number of primary and transplanted tumors of mice (Stewart and Hass, 1956) and guinea pigs (Jungeblut and Hodza, 1963) but not from hamsters, either normal or tumor-bearing. Laboratory personnel contracted LCMV infection from working with the infected hamsters. During the ensuing decade, numerous LCMV infections in humans were traced to infected pet and laboratory hamsters (see Section IIC,8). These events, when coupled with the general lack of information regarding the pathogenesis of LCMV infection of hamsters, prompted the experimental studies of Parker et al. (1976), which provide the majority of our knowledge concerning the pathogenesis of LCMV in hamsters.

2. Etiology

LCMV is the type species of the genus *Arenavirus*, which belongs to the family Arenaviridae (Fenner, 1976). In addition to LCMV, this genus includes *Lassa* virus and members of the Tacaribe complex. By CF and indirect IFAT, these viruses were found to be serologically related (Rowe et al., 1970). They also have biological and morphological similarities (Dalton et al., 1968; Murphy, 1977; Lehmann-Grube, 1982). Virus particles are round, oval, or pleomorphic; they have a mean diameter of 110–130 nm with individual variations from 50 to 300 nm. Surface projections are 10 nm long and club shaped. The projections cover a unit membrane envelope derived from the plasma membrane of the host cell. The interior of the particle contains a variable number of 20- to 25-nm electron-dense granules which morphologically and physiologically are indistinguishable from ribosomes. The virus particle matures via budding from the plasma membrane of infected cells. Intracytoplasmic inclusions are prominent in infected cells and are made up of ribosome masses in a moderately electron-dense matrix. The single-stranded segmented RNA of LCMV has a molecular weight of approximately 3.5×10^6. Infectivity is adversely affected by lipid solvents, detergents, acid treatment (pH 5.5), heat, and radiation (UV and γ) (Buckley and Casals, 1970; Carter et al., 1973; Pfau et al., 1974). Defective interfering particles, a by-product of infectious virus replication both *in vitro* and *in vivo*, are more resistant to UV light than infectious virus particles. They have the ability to prevent CPE, and they can reduce the yield of infectious virus (Lehmann-Grube, 1982). The WE, E-350, W, and CA 1371 strains of LCMV are most widely used experimentally. However, numerous strains exist which may have only minor antigenic differences but which may differ widely with respect to their pathogenicity for adult mice following extracerebral inoculation, the induction of immune response in mice, the distribution and replication in tissues of acutely infected adult mice, ability to kill guinea pigs, and replication in cell cultures (Dutko and Oldstone, 1983). The variability among various strains of virus and the susceptibility of various strains of animals emphasize the necessity of accurately defining experimental conditions.

Historically, the wild house mouse (*Mus musculus*) was considered to be the only significant reservoir of LCMV (Maurer, 1958). This was modified during the past couple of decades when numerous cases of LCMV infection of humans could be traced to their association with asymptomatically infected pet and laboratory hamsters. The host range also includes monkeys, dogs, guinea pigs, rats, rabbits, and chickens (Parker *et al.*, 1976; Skinner and Knight, 1979).

3. Pathogenesis

The pathogenic mechanism of murine LCMV infection has been reviewed (Hotchin, 1971; Cole and Nathanson, 1974; Rawls *et al.*, 1981; Lehmann-Grube, 1982). Although this knowledge is derived from studies in mice, experimental studies indicate that similar mechanisms are operative in LCMV infection of hamsters (Smadel and Wall, 1942; Parker *et al.*, 1976; Skinner *et al.*, 1976; Thacker *et al.*, 1982). Noninbred Syrian golden hamsters inoculated neonatally by the subcutaneous or intraperitoneal route with the Fortner strain of LCMV appear to respond earlier and produce higher levels of CF antibody than do mice (Parker *et al.*, 1976). The pathogenesis of LCMV infection in hamsters resulting from natural transmission of the virus in a colony setting has not been studied.

In its natural host (mice), LCMV is basically a harmless agent capable of inducing a homograft response which results in sickness and death. The absence of a significant immune response in the newborn host explains the immunologically tolerant state in which there is persistent infection (high-titer viremia and viruria). Acute LCM disease and its associated pathological effects are prevented by immunosuppression (e.g., X irradiation, cortisone, thymectomy, chemical immunosuppressants, and antilymphocyte sera). Thus the host may respond immunologically in a positive manner (active immunity with virus suppression) or in a negative manner (immunological tolerance or paralysis). Factors which affect the outcome of LCMV infection leading to immunological disease (positive response) are age, subcutaneous sensitization with LCMV, small virus dose, and neurotropism of virus strain. Factors which affect the outcome of LCMV infection leading to immunological tolerance (negative response) are infancy, neonatal thymectomy, X irradiation, cortisone, amethopterin, antilymphocyte serum, high virus doese, and viscerotropism of virus strain.

Virus titer and fluorescent antibody titer have been useful in classifying the responses to LCMV infection (Benson and Hotchin, 1969). In the neonatal persistent state, cellular immunity is absent so the virus titer is high; fluorescent antibody level is low. In the high-dose immune paralysis state (a laboratory-induced phenomenon), cellular immune response is slight so virus titer is high but it declines to zero in a matter of months; fluorescent antibody levels are high. In acute adult disease, the cellular immune response is very active so virus is eliminated; fluorescent antibody levels are high.

4. Epizootiology

The virus shed in the urine constitutes the major source of infection for other animals including humans. The virus titer in the urine is high (10^5 mouse ID_{50} per 0.03 ml) and persists for long periods of time (6 months to a year in a few experimentally infected hamsters) (Parker *et al.*, 1976; Skinner *et al.*, 1976). The origin of infection in a colony might be other infected hamsters or mice (Centers for Disease Control, 1974), or a variety of contaminated cell culture materials, animal tissues, or transplantable tumors (Lewis *et al.*, 1965; Wiktor *et al.*, 1966; Armstrong *et al.*, 1969; Skinner and Knight, 1969; Hotchin, 1971; Hotchin *et al.*, 1974; Gregg, 1975).

Congenital infections occur in hamsters similar to those observed in mice (Parker *et al.*, 1976); persistent tolerant infections were observed through three generations.

The prevalence of LCMV infection in commercial and laboratory colonies of hamsters in the United States is unknown, although frequent infections have been reported in hamster colonies in Germany (Ackermann, 1973, 1977; Forster and Wachenforfer, 1973). Of 12 colonies in the United States which supplied pet hamsters to a major distributor also in the United States, only one was found to be infected with LCMV (Centers for Disease Control, 1974).

An outbreak of LCMV infection in laboratory personnel (Hinman *et al.*, 1975) was associated with proximity to LCMV-infected Swiss golden hamsters; both aerosols and actual hamster contact were considered important in human acquisition of the disease. Although LCMV in high titer is shed in the urine of persistently infected tolerant hamsters (and mice), the ease with which the virus spreads and infects other susceptible animals is not clearly established. Bowen *et al.* (1975) noted a lack of spread of LCMV infection to mice, guinea pigs, gerbils, cats, monkeys, and dogs maintained in a vivarium where there was ample opportunity for fomite and contact spread from infected hamsters; caretakers move from room to room, different species were housed in the same room, and animals were moved to and from procedure rooms. Thacker et al. (1982) observed cagemate contact spread but not to hamsters in separate cages covered with filter tops. Skinner and Knight (1973) studied a persistently infected mouse colony and concluded that there was no evidence to suggest that infection was readily airborne, nor that ingestion was a likely route of infection except in newborn mice. Mice of all ages were readily infected with saliva or urine contaminating sacrificed skin. Fifty percent of docile young mice contracted the infection within a week with close contact whereas 70% contracted infection within 30 min if the mice were involved in fighting with persistently infected tolerant mice.

5. Clinical Signs and Pathology

The common finding in natural LCMV-infected hamsters is chronic subclinical infections with prolonged viruria (Lewis et al., 1965; Baum et al., 1966; Ackermann, 1973; Forster and Wachenforfer, 1973; Hirsch et al., 1974; Biggar et al., 1975a,b). Similar infections were observed in experimentally infected hamsters (Smadel and Wall, 1942; Petrovic and Timm, 1968; Parker et al., 1976; Skinner et al., 1976; Thacker et al., 1982). All experimentally inoculated hamsters did not respond in like manner, however.

Two distinct types of infection were noted when neonatal hamsters were inoculated subcutaneously with the Fortner strain of LCMV (Parker et al., 1976). Approximately half of the hamsters remained healthy, cleared their initial viremia and viruria, and developed only moderate and transient lymphocytic infiltration of the visceral organs. The other half began to show signs of illness beginning from the seventh week. The illness was progressive and gave the appearance of late LCM or wasting disease similar to that seen in mice (Hotchin, 1962a). The hamsters developed a persistent viremia and viruria, and, at necropsy, lymphocytic inflammation, widespread vasculitis, and progressive glomerulonephritis was observed. The marked lymphocytic infiltrates occurred throughout the liver, lung, pancreas, spleen, kidney, meninges, and brain. The aggregates of lymphocytes were most marked in the periportal areas of the liver. Subacute and chronic interstitial pancreatitis was accompanied by acinar atrophy and interstitial fibrosis. Focal collections of lymphocytic cells in the meninges and perivascular mononuclear infiltrates in the cortex of the brain were noted. The inflammatory infiltrates became prominent in the heart, skeletal muscle, urinary bladder, and uterus. The most striking finding between 26 and 42 weeks of infection was a widespread panarteritis (Fig. 1) that involved the kidney, uterus, heart, pancreas, and sometimes the gonads, urinary bladder, intestine, and skeletal muscle. After 42 weeks of infection the most significant finding was a progressive

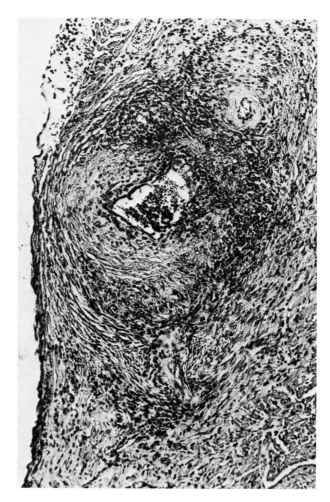

Fig. 1. Marked panarteritis of the heart of a clinically ill hamster 6 months after LCM virus infection. This animal had a widespread visceral vascular disease as well as glomerulonephritis. There is thickening, hyalinization, and fibrosis of the vessel walls and perivascular chronic inflammation. Hematoxylin–eosin stain. Magnification: ×~100. From Parker et al. (1976), with permission of publisher.

chronic glomerulonephropathy (Fig. 2) with lobular simplification of the tufts, concentric adhesions, basement membrane thickening, and glomerular sclerosis and obliteration. γ-Globulin and LCMV-specific antigen were demonstrated by immunofluorescence along the glomerular basement membranes and the walls of involved arterioles. In contrast to the Fortner strain of LCMV, the CA 1371 strain produced an acute leghal infection in newborn Syrian hamsters. Intraabdominal lymphoma was associated with an epizootic of LCMV infection of Syrian hamsters treated with 7,12-dimethylbenz[a]anthracene (Garman et al., 1977). Lymphoreticular infiltrates were observed in the liver especially but also in the kidney of LCMV-infected hamsters, but LCMV isolation was not dependently related to the occurrence of lymphomas.

As is true of most infections, the response of hamsters to LCMV will vary with such factors as age and strain of the

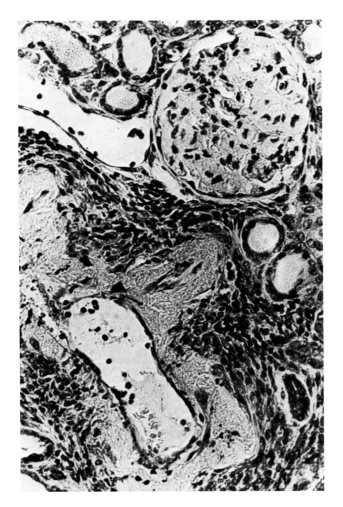

Fig. 2. Chronic glomerulonephropathy of a clinically ill hamster 10 months after LCM virus infection. There is marked glomerular hyalinization, basement membrane thickening, and arteriolar thickening, hyalinization, and chronic perivascular inflammation. Hematoxylin–eosin stain. Magnification: ×~250. From Parker *et al.* (1976), with permission of publisher.

animal, strain of virus, route of inoculation, and dose of the virus.

6. Diagnosis

Since the pathogenesis of LCMV infection in mice and hamsters appears to have much in common and natural LCMV infection of hamsters has not been adequately studied, there is a tendency to extrapolate from the vast knowledge concerning the diagnosis of LCMV infections of mice to the diagnosis of LCMV infections of hamsters. Such a premise could lead to erroneous procedures and conclusions, however, on the basis of the experimental studies of Parker *et al.* (1976), which suggest that the serological response to LCMV infection in hamsters may differ from that which occurs in mice. Hamsters inoculated neonatally with the Fortner strain of LCMV produced CF antibody which coexisted with persistent viremia and viruria. Anticomplementary activity was frequently encountered, however, which could be caused by *in vivo* virus–antibody complexing. Thus the frequent occurrence of anticomplementary activity during serological screening of a hamster colony could be suggestive of LCMV infection. Increasing the sample size to be tested would provide a broader basis for judgment. In addition to confirming questionable CF results (e.g., anticomplementary activity) by other serological procedures (e.g., immunofluorescence), virus isolation can be attempted in a properly equipped laboratory.

An aspect of major importance in the detection of LCMV infection in the mouse colony is the possibility of encountering tolerant persistently infected mice. Serological survey (complement fixation or immunofluorescence) for detection of LCMV-specific circulating antibody in small samples of animals from such a colony may not be rewarding in that very low to nonexistent titers might be encountered (Benson and Hotchin, 1969). In mice, this is not due to B-cell tolerance; antibody is present but normally undetected owing to its binding to virus (persistent viremia) with the formation of virus–antibody immune complexes (Oldstone and Dixon, 1969; Volkert *et al.*, 1975; Buchmeier and Oldstone, 1978; Buchmeier *et al.*, 1980). In theory, the only way to determine whether a mouse colony is free of LCMV is to test all its members for viremia (Lehmann-Grube, 1982), which is not practical. Lehmann-Grube (1982) offers two alternative methods of detecting LCMV infection in mice: either (1) challenge adult animals intracerebrally with LCMV of known potency (in which case the inexperienced mouse usually dies; survival indicates either persistent infection or previous experience resulting in immunity); or (2) inoculate mice (from stock known to be free of LCMV infection) intracerebrally with pooled blood from mice to be tested. If they were viremic, the inoculated mice usually will die. Death, however, cannot be used as an end point (Lehmann-Grube, 1982); the survivors should be challenged with an LCMV strain known to produce 100% mortality in nonimmune animals. The footpad inoculation of mice is a sensitive method for detection of LCMV (Hotchin, 1962b); the reaction can be neutralized with known antibody (Hotchin *et al.*, 1969). This T cell-mediated (Cole *et al.*, 1972) delayed-type hypersensitivity reaction results in a marked swelling of the foot pad 6-7 days following inoculation.

LCMV multiplies but without CPE in a variety of cell cultures including BHK-21 (commonly used), KB, HeLa, Detroit 6, human intestine, human liver, human kidney, FL, primary human amnion, and L cells (Benson and Hotchin, 1960; van der Zeijst *et al.*, 1983). Oldstone and Dixon (1968) suggested that assay of LCMV by immunofluorescent tissue culture technique is more sensitive, rapid, and economical than the mouse titration assay. Hamsters which acquire LCMV infection as adults develop humoral antibody which can be detected by virus neutralization, CF, or indirect IFAT (Lewis and

Clayton, 1969; Lewis *et al.*, 1975; Parker *et al.*, 1976; Thacker *et al.*, 1982). Indirect IFAT is preferred (Lewis and Clayton, 1969; Lewis *et al.*, 1975; Thacker *et al.*, 1982), as the antibody appears early, reaches high titer, and persists for long periods after infection.

7. Control and Prevention

The documented transmission of LCMV infection to numerous people who have contacted or worked with inapparent chronically LCMV-infected pet or laboratory hamsters (see Section IIC,8) emphasizes the necessity of prevention rather than control. The prevention of LCMV infection of hamsters is a straightforward matter of testing animals, animal tissues, and animal-derived or -passaged materials (e.g., blood, organ suspensions, cell cultures, virus pools, transplantable tumors, and tissue extracts) for the presence of LCMV in an appropriately equipped laboratory. It is difficult to overemphasize the need to test any suspect material which may have originated from or have been passaged in animals (especially mice or hamsters) that were not defined microbiologically; LCMV is not an uncommon contaminant of such materials (see Section II,C,8).

Feral mice should be excluded from hamster colonies. Once the hamster colony has been tested and is known to be free of viremic animals (persistent tolerant infections), serum samples from the colony should be tested periodically for detection of anti-LCMV antibody.

8. Significance

Unless one is intentionally studying LCMV in an adequately equipped laboratory, measures should be taken to exclude this virus from all experimental studies, because the virus can complicate and interfere with the interpretation of experimental results (Hotchin, 1971) and cause one of the more important zoonotic diseases (Lewis *et al.*, 1965). Furthermore, it is necessary to perform tests purposely and routinely to detect the presence of LCMV as a contaminant in studies with hamsters, because the infection is likely to be clinically silent in the animal and CPE is not a feature of LCMV infection of cell cultures, even though the virus replicates in a wide variety of cell cultures (Benson and Hotchin, 1960; van der Zeijst *et al.*, 1983). The necessity of such testing procedures cannot be overemphasized, because LCMV-infected hamsters may appear healthy clinically, while persistently shedding large quantities of infectious virus in the urine.

An estimated 4000 human cases of LCMV infection resulted from contact with LCMV-infected pet hamsters supplied by a single commercial breeder (Biggar *et al.*, 1975a,b; Thacker *et al.*, 1982). Epidemics of LCMV infection have occurred in laboratory personnel who unknowingly handled LCMV-contaminated materials such as transplantable tumors, cell lines, animal tissues, or infected animals (Lewis *et al.*, 1965; Baum *et al.*, 1966; Armstrong *et al.*, 1969; Hotchin *et al.*, 1974; Bowen *et al.*, 1975; Hinman *et al.*, 1975; Vanzee *et al.*, 1975; Biggar *et al.*, 1976, 1977). Human cases have not been reported in numerous other instances of LCMV being detected as a contaminant during studies of canine distemper, mycoplasmas, rabies virus, murine poliovirus, lymphosarcoma, Ehrlich's carcinoma, *Toxoplasma gondii*, and cell cultures (see review by Hotchin, 1971). LCMV infection in humans may be debilitating; fatalities are rare (Baum *et al.*, 1966; Bowen *et al.*, 1975; Deibel *et al.*, 1975; Maetz *et al.*, 1976; Murphy, 1977). The infection may rarely follow a subclinical course or may appear in one of three clinical forms: (1) grippe or influenzalike illness, (2) meningitis, or (3) meningoencephalomyelitis. In extraneural infections there is fever, malaise, muscular pain, and, in some cases, coryza and bronchitis. The most common syndrome is meningitis, which may be preceded by the grippe and is marked by stiff neck, fever, headache, malaise, and muscular pain; symptoms may be mild or severe (Deibel *et al.*, 1975).

D. Simian Virus 5 (SV5)

1. History

Simian virus 5 (SV5) was isolated by Hull and co-workers in 1956 after observing CPE in "normal" uninoculated rhesus (*Macaca mulatta*) and cynomolgus (*Macaca fascicularis*) kidney cell cultures. Subsequent to the development of hemadsorption techniques for cell culture systems, it was found that SV5 was prevalent in apparently normal monkey kidney cell cultures. High antibody titers to SV5 were found in monkeys, hamsters, guinea pigs, and humans (Hsiung and Swack, 1972).

2. Etiology

Simian virus 5 belongs to the family Paramyxoviridae, genus *Paramyxovirus*, and is variously classified as parainfluenza type 2 (Kingsbury *et al.*, 1978) or parainfluenza type 5 (Matthews, 1982). Using monoclonal antibody directed against viral nucleoprotein, SV5 has been found to cross-react with parainfluenza 2 viruses but not with the viruses of parainfluenza 1 (Sendai) or parainfluenza 3 (shipping fever) (Goswami and Russell, 1983).

SV5 is a single-stranded RNA virus measuring 158–180 nm (Hsiung, 1972). It has a lipoprotein envelope with club-shaped projections which consist of hemagglutinins, is relatively unstable at temperatures over 37°C, and is ether sensitive.

While many mammalian cell cultures can support replication of SV5, rhesus and African green (*Cercopithecus aethiops*) monkey kidney cell cultures are the most sensitive and commonly used. Human, baboon, calf, dog, hamster, and guinea pig kidney cell cultures have also been successfully infected (Hsiung, 1972). SV5-Infected cell cultures rarely exhibit CPE, but cells do detach from glass earlier than noninfected cell cultures (Hsiung, 1972). Syncytial cell formation in culture, a type of CPE characteristic of paramyxoviruses, may occasionally be seen in SV5-infected cultures after 7–14 days. Holmes and Choppin (1966) reported an inverse relationship between syncytia and virus yield in baby hamster kidney cell cultures. The virus can also be grown in 7- to 8-day embryonated chicken eggs. Adaptation to the amniotic sac prior to allantoic cavity propagation increases virus yield (Hsiung, 1972).

3. Pathogenesis

In the hamster, naturally occurring SV5 infection associated with clinical disease is unknown. There is one report of SV5 associated with encephalitis in hamsters, but is not documented (Carthew *et al.*, 1978). Experimentally inoculated baboons exhibit a mild respiratory disease (Hsiung and Swack, 1972). SV5 has been shown to be a factor in the canine respiratory disease complex known as kennel cough (Appel and Bemis, 1977) and has produced respiratory disease with lesions in experimentally inoculated beagles (Wagener *et al.*, 1983). Hamsters exposed *in utero* via amniotic sac inoculation to another parainfluenza 2 virus, acute laryngotracheobronchitis virus, may develop hydrocephalus, inclusion bodies, and pneumonitis (Margolis and Kilham, 1977).

4. Epizootiology

a. Host Range. Nonhuman primates are considered the natural hosts for SV5. The authors are not aware of any published reports of SV5 isolation from hamsters. However, hamsters do have a high prevalence of antibody to SV5 (Hsiung and Swack, 1972; Parker *et al.*, 1967). Virus can be recovered from the blood, liver, kidney, brain, and urine of hamsters inoculated intranasally with the closely related parainfluenza 2 virus, DA (Chang and Hsiung, 1965). Mice are more resistant to experimental infection than hamsters (Hsiung, 1972). Dogs also can be naturally infected with the virus and exhibit disease.

b. Prevalence. A high prevalence of HAI titers to SV5 has been found in guinea pigs, hamsters, monkeys, dogs, sheep, goats, cows, and humans (Hsiung and Swack, 1972). Of 850 hamsters from 14 colonies tested in 1967, 43% had positive SV5 titers (Parker *et al.*, 1967). SV5 antibody titers were found in 21% of an inbred strain of hamsters but not in a random-bred line (Van Hoosier *et al.*, 1970). Cesarean derivation of this strain eliminated SV5 titers in subsequent generations.

c. Natural History and Transmission. Wild or newly captured nonhuman primates have little or no antibody to SV5, but titers rise significantly during quarantine and then fall. This pattern is similar to that of measles virus titers in newly acquired monkeys and suggests that monkeys may acquire SV5 through human contact (Hsiung and Swack, 1972). Monkey tissue cultures often carry latent infections of SV5 (Hsiung and Swack, 1972). Interspecies transmission between dogs, cats, and humans has also been postulated (Appel and Bemis, 1977). High SV5 titers in hamsters and guinea pigs have been associated with resistance to experimental infection (Hsiung and Swack, 1972), and tissues from monkeys with high SV5 titers are reportedly free of virus (Hsiung, 1972). Thus humoral immunity appears to play a role in host resistance and viral elimination.

5. Diagnosis

Antibody levels are most commonly assayed by the HAI method (Parker *et al.*, 1965; Small, 1984). The presence of antibody in guinea pig sera or on erythrocytes can interfere with CF and HAI tests and produce erroneous results (Hsiung, 1972). SV5 in tissue cultures is revealed by the capacity of cells in culture to absorb guinea pig erythrocytes. Isolation of virus can best be done by inoculating SV5 virus-free cell cultures with urine or blood or by initiating cell cultures from kidney tissues of suspect animals.

6. Control and Prevention

Little is known about the control or prevention of SV5 infection in hamsters. Cesarean derivation and foster rearing have been reported to eliminate SV5 antibody titers in progeny (Van Hoosier *et al.*, 1970).

7. Significance

While antibody to SV5 is common in hamsters, an association between SV5 and naturally occurring disease has not been documented, and the virus has not been isolated from the hamster. The serological reactivity could be caused by a serologically related virus rather than SV5. The possibility of disease or zoonosis must be considered but is probably remote, and no confirmed reports of either have been published.

E. Other Viral Agents

Brief comment only on several viral infections of hamsters is warranted due to the paucity of information.

Reovirus type 3 (Reo 3) infection is suggested by serological surveys of hamster colonies using HAI (Parker *et al.*, 1967; Suzuki *et al.*, 1982; Van Hoosier *et al.*, 1970; Reed *et al.*, 1974). Prevalence within infected colonies was low (2–5%). Nonspecific inhibitors are frequently encountered (Stanley, 1977). There are no reports of virus isolation or disease due to

Reo 3 in hamsters. Reovirus murine models were reviewed by Stanley (1974). Reoviruses are ubiquitous in nature and are known to infect at least 60 different species of mammals, marsupials, birds, and reptiles, in which infections are usually inapparent. A notable exception is Reo 3 infection in suckling mice (reviewed by Kraft, 1982). The three mammalian serotypes (1, 2, and 3) cross-react by CF and IFAT but are distinguishable by neutralization and HAI (Stanley, 1977; Rosen, 1979). HAI is commonly used for serological survey and typing of isolates. Experimental teratological studies (Kilham and Margolis, 1974, 1976) show that Reo 3 infection during the first 4 days of gestation causes fetal resorption. The virus was recovered from the uterus with peak titers reached during the second half of gestation.

Serological evidence of Toolan's H-1 virus infection was associated with epizootic disease in a breeding colony of Syrian hamsters (Gibson et al., 1983). Clinical disease, characterized by high mortality and malformed or missing incisors, was seen in suckling and weanling hamsters only. Necrosis and inflammation of incisor dental pulp, monocytic cell infiltration of dental lamina, and increased osteoclastic activity in the alveolus were seen histologically.

Morphological evidence of cytomegalovirus (CMV) infection in hamsters has been reported (Kuttner and Wang, 1934) and reviewed (Lussier, 1975). Typical acidophilic intranuclear inclusion bodies were found in the submaxillary salivary gland of over 90% of an unspecified number of full-grown Chinese hamsters. Typical inclusions were demonstrated in the brain of young hamsters inoculated intracerebrally and in the submaxillary gland of hamsters inoculated subcutaneously, intraperitoneally, or intraglandularly (submaxillary) with submaxillary gland suspension prepared from reputed infected adult hamsters.

Serological evidence of infection with mouse encephalomyelitis virus (GDVII) has been reported (Parker et al., 1967; Reed et al., 1974). There are no reports of virus isolation or spontaneous disease in hamsters.

A hamster papovavirus (HapV) has been recovered from multiple skin epitheliomas of Syrian hamsters (Graffi et al., 1967, 1968a,b, 1969, 1970). The transplantable tumors occurred spontaneously in 5–10% of the hamsters from 3 months to more than 1 year of age in a separate colony and were characterized by numerous coalescing nodules in the cutis and subcutis. The nodules formed massive layers up to 5 mm thick and weighed 30 gm per animal. The epithelial, partly keratinized neoplasms originated from hair follicle cells, and some were rich in melanin. Unequivocal evidence of cell-free virus transmission has not been demonstrated. The HapV may have a leukemogenic effect indirectly by activation of a latent leukemia virus in the hamster strain studied (Graffi et al., 1969; Scherneck et al., 1979). The HapV particles are 35–40 nm in size (Graffi et al., 1970) and have a molecular weight of about 3.1×10^6 (Bottger et al., 1971). HapV differs from other papovaviruses (Shope papilloma, polyoma, and SV40) in tumor induction in appropriate animal species, and antisera against other papovaviruses did not inactivate HapV (Graffi et al., 1970). However, the amino acid sequences of the viral DNA show an average of 50% homology with polyoma-coded polypeptides (Delmas et al., 1985). The cloned HapV DNA can immortalize rodent cells and transform established rodent cells; the HapV DNA is stably integrated in the host genome (Delmas et al., 1985).

REFERENCES

Ackermann, R. (1973). Epidemiologic aspects of lymphocytic choriomeningitis in man. In "Lymphocytic Choriomeningitis Virus and Other Arenaviruses" (F. Lehmann-Grube, ed.), pp.233–237. Springer-Verlag, Berlin and New York.

Ackermann, R. (1977). [Risk to humans through contact with golden hamsters carrying lymphocytic choriomeningitis virus.] Dtsch. Med. Wochenschr. **39,** 1367–1370.

Appel, M., and Bemis, D. (1977). Canine respiratory disease complex. Curr. Vet. Ther. **6,** 7287–7292.

Armstrong, C., and Dickens, P. F. (1935). Benign lymphocytic choriomeningitis (acute aseptic meningitis). A new disease entity. Public Health Rep. **50,** 831–842.

Armstrong, C., and Lillie, R. D. (1934). Experimental lymphocytic choriomeningitis of monkeys and mice produced by a virus encountered in studies of the 1933 St. Louis encephalitis epidemic. Public Health Rep. **49,** 1019–1027.

Armstrong, D., Fortner, J. G., Rowe, W. P., and Parker, J. C. (1969). Meningitis due to lymphocytic choriomeningitis virus endemic in a hamster colony. JAMA, J. Am. Med. Assoc. **209,** 265–266.

Baum, S. G., Lewis, A. M., Rowe, W. P., and Huebner, R. J. (1966). Epidemic nonmeningitic lymphocytic-choriomeningitis-virus infection. An outbreak in a population of laboratory personnel. N. Engl. J. Med. **274,** 934–936.

Benson, L. M., and Hotchin, J. E. (1960). Cytopathogenicity and plaque formation with lymphocytic choriomeningitis virus. Proc. Soc. Ex. Biol. Med. **103,** 623–625.

Benson, L. M., and Hotchin, J. E. (1969). Antibody formation in persistent tolerant infection with lymphocytic choriomeningitis virus. Nature (London) **222,** 1045–1047.

Biggar, R. J., Woodall, J. P., Walter, P. D., and Haughie, G. E. (1975a). Lymphocytic choriomeningitis outbreak associated with pet hamsters. Fifty-seven cases from New York State. JAMA, J. Am. Med. Assoc. **232,** 494–500.

Biggar, R. J., Douglas, R. G., and Hotchin, J. (1975b). Lymphocytic choriomeningitis associated with hamsters. Lancet **1,** 856–857.

Biggar, R. J., Deibel, R., and Woodall, J. P. (1976). Implication, monitoring, and control of accidental transmission of lymphocytic choriomeningitis virus within hamster tumor cell lines. Cancer Res. **36,** 551–553.

Biggar, R. J., Schmidt, T. J., and Woodall, J. P. (1977). Lymphocytic choriomeningitis in laboratory personnel exposed to hamsters inadvertently infected with LCM virus. J. Am. Vet. Med. Assoc. **171,** 829–832.

Blandford, G., and Charlton, D. (1977). Studies of pulmonary and renal immunopathology after nonlethal primary Sendai virus infection in normal and cyclophosphamide-treated hamsters. Am. Rev. Respir. Dis. **115,** 305–314.

Blandford, G., and Heath, R. B. (1972). Studies on the immune response and

6. VIRAL DISEASES

pathogenesis of Sendai virus infection of mice. I. The fate of viral antigens. *Immunology* **22,** 637–649.

Bottger, M., Bierwolf, D., Wunderlich, V., and Graffi, A. (1971). New calibration correlations for molecular weights of circular DNA: The molecular weight of the DNA of an oncogenic papovavirus of the Syrian hamster. *Biochim. Biophys. Acta* **232,** 21–31.

Bowen, G. S., Calisher, C. H., Winkler, W. G., Kraus, A. L., Fowler, E. H., Garman, R. H., Fraser, D. W., and Hinman, A. R. (1975). Laboratory studies of a lymphocytic choriomeningitis virus outbreak in man and laboratory animals. *Am. J. Epidemiol.* **102,** 233–240.

Buckley, S. M., and Casals, J. (1970). Lassa fever, a new virus disease of man from West Africa. III. Isolation and characterization of the virus. *Am. J. Trop. Med. Hyg.* **19,** 680–691.

Buchmeier, M. J., and Oldstone, M. B. A. (1978). Virus-induced immune complex disease: Identification of specific viral antigens and antibodies deposited in complexes during chronic lymphocytic choriomeningitis virus infection. *J. Immunol.* **120,** 1297–1304.

Buchmeier, M. J., Welsh, R. M., Dutko, F. J., and Oldstone, M. B. A. (1980). The virology and immunobiology of lymphocytic choriomeningitis virus infection. *Adv. Immunol.* **30,** 275–331.

Carter, M. F., Murphy, F. A., Brunschwig, J. P., Nooman, C., and Rawls, W. E. (1973). Effects of actinomycin D and ultraviolet and ionizing radiation on Pichinde virus. *J. Virol.* **12,** 33–38.

Carthew, P., and Sparrow, S. (1980). A comparison in germ-free mice of the pathogenesis of Sendai virus and mouse pneumonia virus infection. *J. Pathol.* **130,** 153–158.

Carthew, P., Sparrow, S., and Verstraete, A. P. (1978). Incidence of natural virus infections of laboratory animals 1976–1977. *Lab. Anim.* **12,** 245–246.

Casals, J. (1975). Arenaviruses. *Yale J. Biol. Med.* **48,** 115–140.

Center for Disease Control (1974). Follow-up on hamster-associated LCM infection—United States. *Morbid. Mortal. Wekly. Rep.* **23,** 131–132.

Chang, P. W., and Hsiung, G. D. (1965). Experimental infection of parainfluenza virus type 5 in mice, hamsters, and monkeys. *J. Immunol.* **95,** 591–601.

Chanock, R. M., Parrott, R. H., Cook, K., Andrews, B. E., Bell, J. A., Reichelderfer, T. E., Kapikian, A. Z., Mastrota, F. M., and Huebner, R. J. (1958). Newly recognized myxoviruses from children with respiratory disease. *N. Engl. J. Med.* **258,** 207–213.

Chun and Chu (1956). Fatal pneumonitis in Chinese hamsters. *Acta Microbiol. Sin.* **4,** 47.

Cole, G. A., and Nathanson, N. (1974). Lymphocytic choriomeningitis pathogenesis. *Prog. Med. Virol.* **18,** 94–110.

Cole, G. A., Nathanson, N., and Prendergast, R. A. (1972). Requirement for O-bearing cells in lymphocytic choriomeningitis virus-induced central nervous system disease. *Nature (London)* **238,** 335–337.

Curnen, E. C., and Horsfall, F. L. (1947). Properties of pneumonia virus of mice (PVM) in relation to its state. *J. Exp. Med.* **85,** 39–53.

Dalton, A. J., Rowe, W. P., Smith, G. H., Wilsnack, R. E., and Pugh, W. E. (1968). Morphological and cytochemical studies on lymphocytic choriomeningitis virus. *J. Virol.* **2,** 1465–1478.

Deibel, R., Woodall, J. P., Decher, W. J., and Schryver, G. D. (1975). Lymphocytic choriomeningitis virus in man. Serologic evidence of association with pet hamsters. *J. Am. Vet. Med. Assoc.* **232,** 501–504.

Delmas, V., Bastien, C., Scherneck, S., and Feunteun, J. (1985). A new member of the polyomavirus family: The hamster papovavirus. Complete nucleotide sequence and transformation properties. *EMBO J.* **4,** 1279–1286.

Descoteaux, J.-P., Payment, P., and Trudel, M. (1980). Comparison of the hemagglutination inhibition procedures and an enzyme-linked immunosorbent assay for detection of specific antibodies to pneumonia virus of mice in experimentally infected laboratory rats. *J. Clin. Microbiol.* **11,** 162–166.

Dutko, F. J., and Oldstone, M. B. A. (1983). Genomic and biological variation among commonly used lymphocytic choriomeningitis virus strains. *J. Gen. Virol.* **64,** 1689–1698.

Eaton, G. J., Lerro, A., Custer, P. R., and Crane, A. R. (1982). Eradication of Sendai pneumonitis from a conventional colony. *Lab. Anim. Sci.* **32,** 384–386.

Eaton, M. D., and van Herick, W. (1944). Demonstration in Cotton rats and rabbits of a latent virus related to pneumonia virus of mice. *Proc. Soc. Exp. Biol. Med.* **57,** 89–92.

Ertl, H. C. J., Gerlich, W., and Koszinowski, U. H. (1979). Detection of antibodies to Sendai virus by enzyme-linked immunosorbent assay (ELISA). *J. Immunol. Methods* **28,** 163–176.

Fenner, F. (1976). Classification and nomenclature of viruses. Second Report of the International Committee on Taxonomy of Viruses. *Intervirology* **7,** 1–115.

Forster, U., and Wachenforfer, G. (1973). Inapparent infection of Syrian hamsters with the virus of lymphocytic choriomeningitis. In "Lymphocytic Choriomeningitis Virus and Other Arenaviruses" (F. Lehmann-Grube, ed.), pp. 113–120. Springer-Verlag. Berlin and New York.

Fujiwara, K., Takenaka, S., and Shumiya, S. (1976). Carrier state of antibody and viruses in a mouse breeding colony persistently infected with Sendai and mouse hepatitis viruses. *Lab. Anim. Sci.* **26,** 153–159.

Fukumi, H., and Nishikawa, F. (1961). Comparative studies of Sendai and HA-2 viruses. *Jpn. J. Med. Sci. Biol.* **14,** 109–120.

Fukumi, H., and Takeuchi, Y. (1975). Vaccination against parainfluenza 1 virus (*typus muris*) infection in order to eradicate this virus in colonies of laboratory animals. *Dev. Biol. Stand.* **28,** 477–481.

Fukumi, H., Nishikawa, F., and Kitayama, T. (1954). A pneumotropic virus from mice causing hemagglutination. *Jpn. J. Med. Sci. Biol.* **7,** 345–363.

Fukumi, H., Mizutani, H., Takeuchi, Y., Tajima, Y., Imaizumi, K., Tanaka, T., and Kaneko, J. I. (1962). Studies on Sendai virus infection in laboratory mice. *Jpn. J. Med. Sci. Biol.* **15,** 153–163.

Garlinghouse, L. E., Jr., and Van Hoosier, G. L., Jr. (1978). Studies on adjuvant-induced arthritis, tumor transplantability and serologic response to bovine serum albumin in Sendai infected rats. *Am. J. Vet. Res.* **39,** 297–300.

Garman, R. H., Bowen, G. S., Fowler, E. H., Kraus, A. L., Newman, A. I., Rifkin, B. R., Andrews, E. J., and Winkler, W. G. (1977). Lymphoma associated with an epizootic of lymphocytic choriomeningitis in Syrian hamsters (*Mesocricetus auratus*). *Am. J. Vet. Res.* **38,** 497–502.

Gibson, S. V., Rottinghaus, A. A., and Wagner, J. E. (1983). Mortality in weanling hamsters associated with tooth loss. *Lab. Anim. Sci.* **33,** Abstr. 66, 497.

Ginsberg, H. S., and Horsfall, F. L. (1951). Characteristics of the multiplication cycle of pneumonia virus of mice (PVM). *J. Exp. Med.* **93,** 151–160.

Goswami, K. K., and Russell, W. C. (1983). Monoclonal antibodies against human paramyxovirus type 3 and against SV5 virus. *J. Gen. Virol.* **64,** 204–215.

Goto, H., and Shimizu, K. (1978). The role of maternal antibody in contact infection of mice with Sendai virus. *Exp. Anim.* **27,** 423–426.

Graffi, A., Schramm, T., Bender, E., Bierwolf, D., and Graffi, I. (1967). Über einen neuen virushaltigen Hauttumor beim Goldhamster. *Arch. Geschwulstforsch.* **30,** 277–283.

Graffi, A., Schramm, T., Bender, E., Graffi, I., Horn, K. H., and Bierwolf, D. (1968a). Cell-free transmissible leukosis in Syrian hamsters, probably of viral etiology. *Br. J. Cancer* **22,** 577–581.

Graffi, A., Schramm, T., Graffi, I., Bierwolf, D., and Bender, E. (1968b). Virus-associated skin-tumors of Syrian hamsters: Preliminary note. *J. Natl. Cancer Inst. (U.S.)* **40,** 867–873.

Graffi, A., Bender, E., Schramm, T., Kuhn, W., and Schneiders, F. (1969). Induction of transmissible lymphomas in Syrian hamsters by application of DNA from viral hamster papovavirus-induced tumors and by cell-free filtrates from human tumors. *Proc. Natl. Acad. Sci. U.S.A.* **64,** 1172–1180.

Graffi, A., Bender, E., Schramm, T., Graffi, I., and Bierwolf, D. (1970). Studies on the hamster papilloma and the hamster virus lymphoma. *Bibl. Haematol. (Basel)* **36,** 293–303.

Gregg, M.B. (1975). Recent outbreaks of lymphocytic choriomeningitis in the United States of America. *Bull. W.H.O.* **52,** 549–553.

Harter, D. H., and Choppin, P. W. (1967). Studies on pneumonia virus of mice (PVM) in cell culture. I. Replication in baby hamster kidney cells and properties of the virus. *J. Exp. Med.* **126,** 251–266.

Hinman, A. R., Frazer, D. W., Douglas, R. G., Bowen, G. S., Kraus, A. L., Winkler, W. G., and Rhodes, W. W. (1975). Outbreak of lymphocytic choriomeningitis virus infections in medical center personnel. *Am. J. Epidemiol.* **101,** 103–110.

Hirsch, M. S., Moellering, R. C., Pope, H. G., and Poskanzer, D. C. (1974). Lymphocytic-choriomeningitis-virus infection traced to a pet hamster. *N. Engl. J. Med.* **291,** 610–612.

Holmes, K. V., and Choppin, P. W. (1966). On the role of the response of the cell membrane in determining virus virulence. Contrasting effects of the parainfluenza virus SV5 in two cell types. *J. Exp. Med.* **124,** 501–520.

Horsfall, F. L., and Hahn, R. G. (1939). A penumonia virus of Swiss mice. *Proc. Soc. Exp. Biol. Med.* **40,** 684–686.

Horsfall, F. L., and Hahn, R. G. (1940). A latent virus in normal mice capable of producing pneumonia in its natural host. *J. Exp. Med.* **71,** 391–408.

Hotchin, J. (1962a). The biology of lymphocytic choriomeningitis infection: Virus-induced immune disease. Basic mechanisms in animal virus biology. *Cold Spring Harbor Symp. Quant. Biol.* **27,** 479–499.

Hotchin, J. (1962b). The foot pad reaction of mice to lymphocytic choriomeningitis virus. *Virology* **17,** 214–215.

Hotchin, J. (1971). The contamination of laboratory animals with lymphocytic choriomeningitis virus. *Am. J. Pathol.* **64,** 747–769.

Hotchin, J., Benson, L., and Sikora, E. (1969). Detection of neutralizing antibody to lymphocytic choriomeningitis virus in mice. *J. Immunol.* **102,** 1128–1135.

Hotchin, J., Sikora, E., Kinch, W., Hinman, A., and Woodall, J. (1974). Lymphocytic choriomeningitis in a hamster colony causes infection of hospital personnel. *Science* **185,** 1173–1174.

Hsiung, G. D. (1972). Parainfluenza 5 virus. Infection of man and animal. *Prog. Med. Virol.* **14,** 241–274.

Hsiung, G. D., and Swack, N. S. (1972). Myxovirus and pseudomyxo- virus group. In "Pathology of Simian Primates" (T. W. Fiennes, ed.), Part II, pp. 537–561. Karger, Basel.

Hull, R. N., Minner, J. R., and Smith, J. W. (1956). New viral agents recovered from tissue cultures of monkey kidney cells. *Am. J. Hyg.* **63,** 204–215.

Jakab, G. J. (1981). Interaction between Sendai virus and bacterial pathogens in the murine lung: A review. *Lab. Anim. Sci.* **31,** 172–177.

Jungeblut, C. W., and Hodza, H. (1963). Interference between lymphocytic choriomeningitis virus and the leukemia-transmitting agent of leukemia L_2C in guinea pigs. *Arch. Gesamte Virusforsch.* **12,** 552–560.

Kay, M. M. B. (1978). Long term subclinical effects of parainfluenza (Sendai) infection on immune cells of aging mice. *Proc. Soc. Exp. Biol. Med.* **158,** 326–331.

Kay, M. M. B., Mendoza, J., Hausman, S., and Dorsey, B. (1979). Age-related changes in the immune system of mice of eight medium and long-lived strains and hybrids. II. Short- and long-term effects on natural infection with parainfluenza type 1 virus (Sendai). *Mech. Ageing Dev.* **11,** 347–362.

Kilham, L., and Margolis, G. (1974). Congenital infections due to reovirus type 3 in hamsters. *Teratology* **9,** 51–64.

Kilham, L., and Margolis, G. (1976). Comparative susceptibilities of pregnant, and lactating hamsters to infection with H-1 and reoviruses. *Teratology* **13,** 179–1984.

Kingsbury, D. W., Bratt, M. A., Choppin, P. W., Hanson, R. P., Hosaka, Y., ter Meulen, V., Norrby, E., Plowright, W., Rott, R., and Wunner, W. H. (1978). Paramyxoviridae. *Intervirology* **10,** 137–152.

Kraft, L. M. (1982). Viral diseases of the digestive system. In "The Mouse in Biomedical Research" (H. L. Foster, J. D. Small, and J.G. Fox eds.), vol. 2, pp. 159–191. Academic Press, New York.

Kuroya, M., Ishida, N., and Shiratori, T. (1953). Newborn virus pneumonitis (type Sendai). II. Report: The isolation of a new virus possessing hemagglutinin activity. *Yokohama Med. Bull.* **4,** 217–233.

Kuttner, A. G., and Wang, S. (1934). The problem of the significance of the inclusion bodies in the salivary glands of infants, and the occurrence of inclusion bodies in the submaxillary glands of hamsters, white mice, and wild rats (Peiping). *J. Exp. Med.* **60,** 773–791.

Lehmann-Grube, F. (1982). Lymphocytic choriomeningitis. In "The Mouse in Biomedical Research" (H. L. Foster, J. D. Small, and J. G. Fox eds.), Vol. 2, pp. 231–266. Academic Press, New York.

Lehmann-Grube, F. (1984). Portraits of viruses: Arenaviruses. *Intervirology* **22,** 121–145.

Lewis, A. M., Rowe, W. P., Turner, H. C., and Huebner, R. J. (1965). Lymphocytic-choriomeningitis virus in hamster tumor: Spread to hamsters and humans. *Science* **150,** 363–364.

Lewis, V. J., and Clayton, D. (1969). Detection of lymphocytic choriomeningitis virus antibody in murine sera by immunofluorescence. *Appl. Microbiol.* **18,** 289–290.

Lewis, V. J., Walter, P. D., Thacker, W. L., and Winkler, W. G. (1975). Comparison of three tests for the serological diagnosis of lymphocytic choriomeningitis virus infection. *J. Clin. Microbiol.* **2,** 193–197.

Lussier, G. (1975). Murine cytomegalovirus (MCMV). *Adv. Vet. Sci. Comp. Med.* **19,** 223–247.

Maetz, H. M., Sellers, C. A., Bailey, W. C., and Hardy, G. E. (1976). Lymphocytic choriomeningitis from pet hamster exposure: A local public health experience. *Am. J. Public Health* **66,** 1082–1085.

Margolis, G., and Kilham, L. (1977). Induction of congenital hydrocephalus in hamsters with parainfluenza type 2 virus. *Exp. Mol. Pathol.* **27,** 235–248.

Matsumoto, T., Nagata, I., Kariya, Y., and Ohashi, K. (1954). Studies on a strain of pneumotropic virus of hamster. *Nagoya J. Med. Sci.* **17,** 93–97.

Matthews, R. E. F. (1982). Classification and nomenclature of viruses. *Intervirology* **17,** 104–105.

Maurer, F. D. (1958). Lymphocytic choriomeningitis. *J. Natl. Cancer Inst. (U.S.)* **20,** 867–870.

Murphy, F. A. (1977). Arenaviruses: Diagnosis of lymphocytic choriomeningitis, Lassa, and other arenaviral infections. In "Comparative Diagnosis of Viral Diseases" (E. Kurstak and C. Kurstak eds.), Vol. 1, pp. 759–791. Academic Press, New York.

Noda, K. (1953). Newborn virus pneumonitis (type Sendai). III. Report: Pathological studies on the 9 autopsy cases and the mice inoculated with the newfound virus. *Yokohama Med. Bull.* **4,** 281–287.

Oldstone, M. B. A., and Dixon, F. J. (1968). Direct immunofluorescence tissue culture assay for lymphocytic choriomeningitis virus. *J. Immunol.* **100,** 1135–1138.

Oldstone, M. B. A., and Dixon, F. J. (1969). Pathogenesis of chronic disease associated with persistent lymphocytic choriomeningitis viral infection. I. Relationship of antibody production to disease in neonatally infected mice. *J. Exp. Med.* **129,** 483–505.

Pakes, S. P., Lu, Y.-S., and Meunier, P. C. (1984). Factors that complicate animal research. In "Laboratory Animal Medicine" (J. G. Fox, B. J. Cohen, and F. M. Loew, eds.), pp. 649–665. Academic Press, New York.

Parker, J. C., and Reynolds, R. K. (1968). Natural history of Sendai virus infection in mice. *Am. J. Epidemiol.* **88,** 112–125.

Parker, J. C., and Richter, C. B. (1982). Viral diseases of the respiratory system. In "The Biology of the Laboratory Mouse" (H. L. Foster, J. D.

Small, and J. G. Fox, eds.), Vol. 2, pp. 107–155. Academic Press, New York.

Parker, J. C., Tennant, R. W., and Ward, T. G. (1965). Virus studies with germfree mice. I. Preparation of serologic diagnostic reagents and survey of germfree and monocontaminated mice for indigenous murine viruses. *J. Natl. Cancer Inst. (U.S.)* **34**, 371–380.

Parker, J. C., Hercules, J. I., and von Kaenel, E. (1967). The prevalence of some indigenous viruses of rat and hamster breeder colonies. *Bacteriol Proc.* Abstr. No. 172, p. 163.

Parker, J. C., Igel, H. J., Reynolds, R. K., Lewis, A. M., and Rowe, W. P. (1976). Lymphocytic choriomeningitis virus infection in fetal, newborn, and young adult Syrian hamsters (*Mesocricetus auratus*). *Infect. Immun.* **13**, 967–981.

Parker, J. C., Whiteman, M. D., and Richter, C. B. (1978). Susceptibility of inbred and outbred mouse strains to Sendai virus and prevalence of infection in laboratory rodents. *Infect. Immun.* **19**, 123–130.

Parker, J. C., O'Beirne, A. J., and Collins, M. J., Jr. (1979). Sensitivity of the enzyme-linked immunosorbent assay, complement fixation, and hemagglutination-inhibition serologic tests for detection of Sendai virus antibody in laboratory mice. *J. Clin. Microbiol.* **9**, 444–447.

Pearson, H. E., and Eaton, M. D. (1940). A virus pneumonia of Syrian hamsters. *Proc. Soc. Exp. Biol. Med.* **45**, 677–679.

Peck, R. M., Eaton, G. J., Peck, E. B., and Litwin, S. (1983). Influence of Sendai virus on carcinogenesis in strain A mice. *Lab. Anim. Sci.* **33**, 154–156.

Petrovic, M., and Trimm, H. (1968). [Latent infection of the Syrian hamster (*Mesocricetus auratus*) with lymphocytic choriomeningitis virus (LCM).] *Zentralbl. Bakteriol., Parasitenkd., Infektionskr. Hyg., Abt. 1: Orig.* **207**, 435–442.

Pfau, C. J., Bergold, G. H., Casals, J., Johnson, K. M., Murphy, F. A., Pederson, I. R., Rawls, W. E., Rowe, W. P., Webb, P. A., and Weissenbacher, M. C. (1974). Arenaviruses. *Intervirology* **4**, 207–213.

Profeta, M. L., Lief, F. S., and Plotkin, S. A. (1969). Enzootic Sendai infection in laboratory hamsters. *Am. J. Epidemiol.* **89**, 316–324.

Rawls, W. E., Chan, M. A., and Gee, S. R. (1981). Mechanisms of persistence in arenavirus infections: A brief review. *Can. J. Microbiol.* **27**, 568–574.

Reed, J. M., Schiff, L. J., Shefner, A. M., and Henry, M. C. (1974). Antibody levels to murine viruses in Syrian hamsters. *Lab. Anim. Sci.* **24**, 33–38.

Reed, J. M., Schiff, L. J., Shefner, A. M., and Poiley, S. M. (1975). Murine virus susceptibility of cell cultures of mouse, rat, hamster, monkey, and human origin. *Lab. Anim. Sci.* **25**, 420–424.

Rivers, T. M., and Scott, T. F. M. (1935). Meningitis in man caused by a filterable virus. *Science* **81**, 439–440.

Rosen, L. (1979).Reovirus. *In* "Diagnostic Procedures for Viral, Rickettsial and Chlamydial Infections" (E. H. Lennette and N. J. Schmidt, eds.), 5th ed., pp. 577–584. Am. Public Health Assoc., Washington, D.C.

Rowe, W. P., Pugh, W. E., Webb, P. A., and Peters, C. J. (1970). Serological relationship of the Tacaribe complex of viruses to lymphocytic choriomeningitis virus. *J. Virol.* **5**, 289–292.

Sano, T., Niitsu, I., Nakagawa, I., and Ando, T. (1953). Newborn virus pneumonitis (type Sendai). I. Report: Clinical observation of a new virus pneumonitis of the newborn. *Yokohama Med. Bull.* **4**, 199–216.

Scherneck, S., Bottger, M., and Feunteun, J. (1979). Studies on the DNA of an oncogenic papovavirus of the Syrian hamster. *Virology* **96**, 100–107.

Schiff, L. J., Shefner, A. M., Barbera, P. W., and Poiley, S. M. (1973). Microbial flora and viral contact status of Chinese hamsters (*Cricetus griseus*). *Lab. Anim. Sci.* **23**, 899–902.

Shimonaski, G., and Came, P. E. (1970). Plaque assay for pneumonia virus of mice. *Appl. Microbiol.* **20**, 775–777.

Skinner, H. H., and Knight, E. H. (1969). Studies on murine lymphocytic choriomeningitis within a partially infected colony. *Lab. Anim.* **3**, 175–184.

Skinner, H. H., and Knight, E. H. (1973). Natural routes for postnatal transmission of murine lymphocytic choriomeningitis. *Lab. Anim.* **7**, 171–184.

Skinner, H. H., and Knight, E. H. (1979). The potential role of Syrian hamsters and other small animals as reservoirs of lymphocytic choriomeningitis virus. *J. Small Anim. Pract.* **20**, 145–161.

Skinner, H. H., Knight, E. H., and Buchley, L. S. (1976). The hamster as a secondary reservoir host of lymphocytic choriomeningitis virus. *J. Hyg.* **76**, 299–306.

Smadel, J. E., and Wall, M. J. (1942). Lymphocytic choriomeningitis in the Syrian hamster. *J. Exp. Med.* **75**, 581–591.

Small, J. D. (1984). Rodent and lagomorph health surveillance—quality assurance. *In* "Laboratory Animal Medicine" (J. G. Fox, B. J. Cohen, and F. M. Loew, eds.), pp. 709–723. Academic Press, New York.

Smith, A. L., Carrano, V. A., and Brownstein, D. G. (1984). Response of weanling random-bred mice to infection with pneumonia virus of mice (PVM). *Lab. Anim. Sci.* **34**, 35–37.

Soret, M. G., and Buthala, D. A. (1967). Enzootic Sendai virus infection in hamsters. *Fed. Proc., Fed. Am. Soc. Exp. Biol.* Abstr. **175**, 163.

Stanley, N. F. (1974).The reovirus murine models. *Prog. Med. Virol.* **18**, 257–272.

Stanley, N. F. (1977).Diagnosis of reovirus infections: Comparative aspects. *In* "Comparative Diagnosis of Viral Diseases" (E. Kurstak and C. Kurstak eds.), Vol. 1, pp. 385–421. Academic Press, New York.

Stewart, S. E., and Hass, V. H. (1956). Lymphocytic choriomeningitis virus in mouse neoplasms. *J. Natl. Cancer Inst. (U.S.)* **17**, 233–245.

Suzuki, E., Matsubara, J., Saito, M., Muto, T., Nakagawa, M., and Imaizumi, K. (1982). Serological survey of laboratory rodents for infection with Sendai virus, reovirus type 3 and adenovirus. *Jpn. J. Med. Sci. Biol.* **35**, 249–254.

Tennant, R. W., and Ward, T. G. (1962). Pneumonia virus of mice (PVM) in cell culture. *Proc. Soc. Exp. Biol. Med.* **111**, 395–398.

Tennant, R. W., Parker, J. C., and Ward, T. G. (1966). Respiratory virus infections of mice. *Natl. Cancer Inst. Monogr.* **20**, 93–104.

Thacker, W. L., Lewis, V. J., Shaddock, J. H., and Winkler, W. G. (1982). Infection of Syrian hamsters with lymphocytic choriomeningitis: Comparison of detection methods. *Am. J. Vet. Res.* **43**, 1500–1502.

Traub, E. (1935). A filterable virus recovered from white mice. *Science* **81**, 298–299.

van der Veen, J., Poort, Y., and Birchfield, D. J. (1970). Experimental transmission of Sendai virus infection in mice. *Arch. Gesamte Virusforsch.* **31**, 237–246.

van der Zeijst, A. A., Noyes, B. E., Mirault, M. E., Parker, B., Osterhaus, A. K., Swyryd, E. A., Bleumink, N., Horzinek, M. C., and Stark, G. R. (1983). Persistent infection of some standard cell lines by lymphocytic choriomeningitis virus; transmission of infection by an intracellular agent. *J. Virol.* **48**, 249–261.

Van Hoosier, G. L., Jr., and Ladiges, W. C. (1984). Biology and diseases of hamsters. *In* "Laboratory Animal Medicine" (J. G. Fox, B. J. Cohen, and F. M. Loew, eds.), pp. 124–147. Academic Press, New York.

Van Hoosier, G. L., Jr., and Robinette, L. R. (1976). Viral and chlamydial diseases. *In* "The Biology of the Guinea Pig" (J. E. Wagner and P. J. Manning, eds.), pp. 137–150. Academic Press, New York.

Van Hoosier, G. L., Jr., Trentin, J. J., Shields, J., Stephens, K., Stenback, W. A., and Parker, J. C. (1966). Effect of caesarean-derivation, gnotobiotic foster nursing and barrier maintenance of an inbred colony on enzootic virus status. *Lab. Anim. Care* **16**, 119–128.

Van Hoosier, G. L., Jr., Stenback, W. A., Parker, J. C., Burke, J. G., and Trentin, J. J. (1970). The effects of cesarean derivation and foster nursing procedures on enzootic viruses of the LSH strain of inbred hamsters. *Lab. Anim. Care* **20**(2), Part 1, 232–237.

Vanzee, B. E., Douglas, R. G., Betts, R. F., Bauman, A. W., Fraser, D. W., and Hinman, A. R. (1975). Lymphocytic choriomeningitis in university hospital personnel. *Am. J. Med.* **58,** 803–809.

Volkert, M., and Horsfall, F. L. (1947). Studies on a lung tissue component which combines with pneumonia virus of mice (PVM). *J. Exp. Med.* **86,** 393–407.

Volkert, M., Bro-Jorgensen, K., Marker, O., Rubin, B., and Trier, L. (1975). The activity of T and B lymphocytes in immunity and tolerance to the lymphocytic choriomeningitis virus in mice. *Immunology* **29,** 455–464.

Wagener, J. S., Minnich, L., Sobonya, R., Taussig, C. G. R., and Fulginiti, V. (1983). Parainfluenza Type II infection in dogs. *Am. Rev. Respir. Dis.* **127,** 771–775.

Ward, J. M. (1974). Persistent parainfluenza (Sendai) virus infection of athymic nude mice. *Lab. Invest.* **34,** 336.

Wiktor, T. J., Kaplan, M. M., and Koprowski, H. (1966). Rabies and lymphocytic choriomeningitis virus (LCMV). Infection of tissue culture; enhancing effect of LCMV. *Ann. Med. Exp. Fenn.* **44,** 290–296.

Zurcher, C., Burek, J. D., van Nunen, M. C. J., and Mcihuizen, S. P. (1977). A naturally occurring epizootic caused by Sendai virus in breeding and aging rodent colonies. I. Infection in the mouse. *Lab. Anim. Sci.* **27,** 955–962.

Chapter 7

Bacterial and Mycotic Diseases

Craig S. Frisk

I. Introduction	112
II. Proliferative Ileitis, Hamster Enteritis, Atypical Ileal Hyperplasia	112
A. Introduction	112
B. History	112
C. Etiology	113
D. Pathogenesis	114
E. Epizootiology	115
F. Clinical Signs	115
G. Pathology	116
H. Diagnosis	120
I. Treatment and Control	120
J. Significance	120
III. Cecal Mucosal Hyperplasia	121
A. History, Etiology, and Epizootiology	121
B. Clinical Signs and Pathology	121
C. Diagnosis and Control	121
IV. Tyzzer's Disease	121
A. History	121
B. Etiology	122
C. Pathogenesis	122
D. Epizootiology	123
E. Clinical Signs	123
F. Pathology	123
G. Diagnosis	124
H. Control and Prevention	124
I. Significance	125
V. Salmonellosis	125
A. History, Etiology, and Epizootiology	125
B. Clinical Signs and Pathology	125
C. Pathogenesis	126
D. Diagnosis	126
E. Control and Prevention	126
F. Significance	127
VI. Campylobacteriosis	127
A. History, Etiology, and Epizootiology	127
B. Clinical Signs and Pathogenesis	127
C. Treatment	127
D. Significance	128

Copyright © 1987 by Academic Press Inc.
All rights of reproduction in any form reserved.

VII.	Tularemia	128
	A. Etiology and Epizootiology	128
	B. Clinical Signs and Pathology	128
	C. Diagnosis	128
	D. Control, Prevention, and Significance	128
VIII.	Pseudotuberculosis	129
IX.	Streptococcal Infections	129
X.	*Pasteurella* Infections	129
XI.	Bacterial Respiratory Infections	129
XII.	Actinomycosis	130
XIII.	Cutaneous Bacterial Abscesses	130
XIV.	Postirradiation Infections	131
XV.	Miscellaneous Bacterial Infections	131
XVI.	Mycotic Infections	131
	References	131

I. INTRODUCTION

Because of increasing evidence, proliferative ileitis is assumed to be caused by bacteria. With the inclusion of this disease, bacteria are the most significant cause of disease, mortality, and economic loss in hamsters. Mycotic infections, on the other hand, are extremely rare, being the least significant category of infectious disease of hamsters. This chapter will discuss the various diseases and syndromes that have been attributed to bacterial and mycotic agents.

II. PROLIFERATIVE ILEITIS, HAMSTER ENTERITIS, ATYPICAL ILEAL HYPERPLASIA

A. Introduction

Proliferative ileitis is the most commonly recognized disease of golden hamsters and usually results in high morbidity and mortality. The disease is characterized clinically by diarrhea in weanling hamsters and pathologically by enteritis, most often the lesions being proliferative or hyperplastic. Proliferative ileitis has also been referred to as regional enteritis, terminal ileitis, enzootic intestinal adenocarcinoma, atypical ileal hyperplasia, hamster enteritis, and "wet-tail." All of the above synonyms apparently describe the same disease syndrome, which will be referred to as proliferative ileitis in this section. The term "wet-tail" has been used to denote this disease; however, this appearance can result from any disease which may cause diarrhea, and therefore, the term's use can result in confusion. The literature on proliferative ileitis has been reviewed by Frisk and Wagner (1977a) and Jacoby and Johnson (1981).

B. History

The first major outbreak of proliferative ileitis recorded in the literature was in 1958 (Friedman, 1965). Since that time, numerous articles have proposed bacteria, viruses, parasites, and diet as the suspected cause; however, Koch's postulates have never been fulfilled with any agent or combination of agents.

1. Bacterial Origin

Varela (1953) reported that proliferative ileitis was reproduced by inoculation of *Proteus morganii,* which was recovered from affected hamsters; however, this result has not been repeated. Chesterman (1972) isolated *Proteus mirabilis* from 90% of affected hamsters in an outbreak, but attempts to reproduce disease with this isolate were not reported. *Proteus* sp. and *Pseudomonas* sp. were isolated from necrotic intestinal lesions by Boothe and Cheville (1967), who thought these organisms were secondary invaders.

Inoculation of hamsters with many different bacteria have been carried out in attempts to reproduce the disease. *Escherichia coli, Clostridium* sp., and a gram-negative rod resembling *Pasteurella* sp. were isolated by Friedman (1965) from diseased hamsters and inoculated into weanling hamsters without success. *Clostridium perfringens* type D and *E. coli* were studied in pure culture separately and in combination by Goldman *et al.* (1972). Hamsters were inoculated either orally or by intraluminal ileal injections via laparotomy. Some of the inoculated hamsters died. However, ileal lesions were not observed, the deaths being attributed to surgical technique. Jacoby and Johnson (1981) attempted to reproduce lesions of proliferative ileitis with several bacterial isolates. *Bacteroides fragilis, Fusobacterium varium, Fusobacterium* sp., *Clostridium sordelli, E. coli,* and a pool of these organisms were inoculated into hamsters orally and intracecally without the production of disease.

7. BACTERIAL AND MYCOTIC DISEASES

Escherichia coli was isolated from the intestines of hamsters with enteritis by Sheffield and Beveridge (1962). They postulated that *E. coli* was the cause of the fatal diarrhea and compared the condition to infantile diarrhea of humans; scouring in piglets, chickens, and calves; and mucoid enteritis of rabbits. Thomlinson (1975) isolated *E. coli* O 117:K? from a group of hamsters with lesions of acute enteritis. This serotype was identified in 8 of 10 affected and 1 of 6 unaffected hamsters. The organism was cultured from stomach, intestine, and spleen. This O serotype, 117, has also been reported as an intestinal pathogen of calves. Thomlinson (1975) felt that stress was a part of the pathogenesis of *E. coli* enteritis in hamsters and compared it to colibacillosis of calves and pigs. Although both Sheffield and Beveridge (1962) and Thomlinson (1975) isolated *E. coli* from hamsters with a fatal form of enteritis, neither described proliferation associated with the intestinal lesions. It is possible that these reports describe a different disease, acute enteropathogenic *E. coli* enteritis, rather than proliferative ileitis.

2. Viral Origin

Jonas *et al.* (1965) isolated two viruses from intestinal lesions of hamsters with proliferative ileitis. These viruses produced cytopathic changes in embryonic hamster fibroblast cultures and were lethal to suckling hamsters. The viruses were shown to contain DNA and were not neutralized with antisera to known murine viruses. These viruses did not produce intestinal lesions when inoculated by Jonas *et al.* (1965) or by Tomita and Jonas (1968). Their role in the pathogenesis of proliferative ileitis could not be established.

Eosinophilic intranuclear inclusion bodies were observed in intestinal epithelial cells by Lussier and Pavilanis (1969). These inclusions, which resembled type A inclusions produced by herpesviruses, suggested that a virus was involved in the pathogenesis of the disease. Kilham's rat virus, however, was suspected as being latent in the colony, and viral isolation attempts were not made.

Viral particles have not been found in proliferative lesions examined by electron microscopy (Wagner *et al.*, 1973; Frisk and Wagner, 1977b; Johnson and Jacoby, 1978).

3. Parasitic Origin

Several investigators have found high numbers of protozoan parasites in the intestine of hamsters with proliferative ileitis (Sheffield and Beveridge, 1962; Jackson and Wagner, 1970; Boothe and Cheville, 1967). *Hymenolepis nana* has also been observed in hamsters with lesions of proliferative ileitis (Sheffield and Beveridge, 1962; Friedman, 1965; Boothe and Cheville, 1967). The significance of parasite infections in the pathogenesis of proliferative ileitis has not been shown. It appears that their presence is unrelated, because parasitized hamsters are routinely observed without lesions of proliferative ileitis. Also, Amend *et al.* (1976) and Johnson and Jacoby (1978) were experimentally able to reproduce proliferative ileitis with intestinal homogenates which were parasite free.

4. Nutritional Origin

Goldman *et al.* (1972) suggested that initial intestinal hyperplasia could be caused by a deficiency of pantothenic acid. Outbreaks of proliferative ileitis, however, have occurred in colonies fed commercial diets recommended for hamsters (Boothe and Cheville, 1967; Jackson and Wagner, 1970). Also, Jacoby *et al.* (1975), Amend *et al.* (1976), and Frisk and Wagner (1977b) have experimentally reproduced proliferative ileitis in hamsters fed complete commercial diets free choice.

C. Etiology

The search for the etiological agent of proliferative ileitis has met with only partial success. Conventional methods of bacteriology and virology have resulted in several promising isolates; however, the causative organism or the method by which an identified organism causes the disease continues to elude investigators. The etiology can be studied by experimentally reproducing the disease by orally inoculating weanling hamsters with ileal lesions ground in saline (Jacoby *et al.*, 1975; Amend *et al.*, 1976; Frisk and Wagner, 1977b; Jacoby, 1978; Johnson and Jacoby, 1978).

Jacoby *et al.* (1975) showed that the responsible organism could be removed by filtration. Filtration of the ground ileal suspension through 0.45- and 0.22-μm filters caused a reduction or loss of infectivity. Also, infectivity was lost if homogenates were heated to 56°C for 30 min or treated with chloroform.

Jacoby and Johnson (1981) used several different established and primary cell culture lines to attempt isolation of the causative organism. They were able to isolate an organism, which by immunofluorescence was shown to be present within proliferative epithelial cells. Isolation was best using a chicken–hamster (CER) cell line. Organisms were viable up to 3 weeks, but at 2 weeks they began to degenerate. Isolated organisms were inoculated into hamsters, but were unable to produce disease. The taxonomic classification of this particular isolate was not determined.

Proliferation, which is typical of this disease, has been shown to be associated with an intracytoplasmic organism (Frisk and Wagner, 1977b; Johnson and Jacoby, 1978). Jacoby *et al.* (1975) showed by indirect immunofluorescence that experimentally infected animals produced antibodies to this intracytoplasmic organism. Antibody to this intracellular organism can be detected by 10 days after infection (Jacoby, 1978). Attempts at protecting hamsters by passive immunization with hyperimmune serum have failed (Jacoby and Johnson, 1981). The role of the antibody produced to this intracytoplasmic organism and the animal's immunological response to the disease remain obscure.

1. *Campylobacter*

Recent studies have investigated the role of *Campylobacter* sp. in the etiology of proliferative ileitis. *Campylobacter jejuni* has been isolated from hamsters with lesions of proliferative ileitis (Lentsch *et al.*, 1982; La Regina and Lonigro, 1982). Organisms which morphologically resemble *Campylobacter* sp. have been observed by electron microscopy within intestinal epithelial cells (Frisk and Wagner, 1977b). Although the presence of *Campylobacter jejuni* has been demonstrated, isolated organisms have not been able to produce proliferative lesions in hamsters inoculated with pure cultures (La Regina and Lonigro, 1982: Lentsch *et al.*, 1982).

The frequency of isolation and identification of *Campylobacter* in affected tissue sections strongly suggests that it has a direct role in the pathogenesis of the disease syndrome. It remains possible, however, that there may be many serotypes of *Campylobacter jejuni*, some pathogenic and some commensal. Also, the causative organism may be a *Campylobacter* different from the ones thus far studied. For instance, *C. hyointestinalis* was isolated by Chang *et al.* (1984) from a hamster with diarrhea. This organism was antigenically identical with those isolated from swine with proliferative enteritis. The exact role of *Campylobacter*, therefore, in the etiology of proliferative ileitis is still speculative at this point in time. For a more complete discussion of *Campylobacter* infections of hamsters, see Section VI.

2. *Escherichia coli*

Escherichia coli isolated from hamsters with proliferative ileitis has been studied extensively. Jackson and Wagner (1970) isolated a hemolytic *E. coli* from hamsters with proliferative ileitis, but they also reported similar cultures from nonaffected hamsters. From a separate outbreak, Wagner *et al.* (1973) isolated from proliferated ileal lesions *E. coli* which did not ferment lactose. Serologically, this organism cross-reacted with antisera to *E. coli* O 138 and *Shigella boydii* 11 and 12. Pure cultures inoculated orally into weanlings did not produce lesions.

Jacoby *et al.* (1975) isolated a slow lactose-fermenting *E. coli* from hamsters with proliferative lesions. Pure cultures of *E. coli*, as well as *Streptococcus faecalis* and *Clostridium* sp., were orally inoculated into three groups of weanling hamsters. Hamsters developed proliferative lesions only in the group inoculated with *E. coli*. The authors, however, felt that this was of doubtful significance, because mixed cultures containing the same *E. coli*, as well as attempts at repeating this result with identical *E. coli* isolates, were unsuccessful.

Although pure cultures of *E. coli* have been unable to produce proliferative changes within the ileum, they have produced acute enteritis. Boothe and Cheville (1967) isolated an *E. coli* from hamsters with typical lesions of proliferative ileitis. Weanling hamsters were orally inoculated with pure cultures of the isolated *E. coli*, and diarrhea and generalized enteritis occurred. Amend *et al.* (1976) isolated a slow lactose-fermenting *E. coli* from hamsters with proliferative ileitis. This organism was not typeable with available antisera. Oral inoculations of this *E. coli* into weanling hamsters resulted in lesions of acute enteritis.

Frisk and Wagner (1977b) showed by isolation, electron microscopy, and indirect fluorescent antibody techniques that a non-lactose-fermenting *E. coli* was present within ileal epithelial cells in experimentally infected hamsters within the first 9 days after inoculation with proliferated ilea (Frisk and Wagner, 1977b). Lesions of acute enteritis were present concurrent with these organisms. Frisk *et al.* (1981) inoculated pure cultures of this *E. coli*, which was similar biochemically to the organisms isolated by Wagner *et al.* (1973) into weanling hamsters. The organism was able to produce clinical signs of diarrhea and lesions of acute enteritis. By electron microscopy and indirect fluorescent antibody techniques, *E. coli* were observed within intestinal epithelial cells. They were able to divide within cells and produced a disease which was comparable to shigellosis. The enteritis produced, however, never progressed to proliferation.

Frisk *et al.* (1978) investigated the enteropathogenicity of *E. coli* isolates from hamsters with proliferative ileitis. In an intestinal loop technique developed in weanling hamsters, three *E. coli* isolates from hamsters with proliferative ileitis were shown to be enteropathogenic. An *E. coli* isolate that was isolated from a healthy hamster was found to be nonenteropathogenic.

The possibility of a coincidental infection of an enteropathogenic *E. coli*, which can cause enteritis but is unrelated to the pathogenesis of proliferative ileitis, must be considered when reviewing these *E. coli* reports. The finding of an enteropathogenic *E. coli* is not consistent: Johnson and Jacoby (1978) did not observe such an organism. Also, inhibition of intestinal flora of donors infected with proliferated ilea by neomycin, which should eliminate *E. coli*, did not reduce the incidence of experimental lesions (Jacoby and Johnson, 1981).

D. Pathogenesis

Because proliferative ileitis has not been reproduced by a single organism or any defined combination of organisms, one can only speculate on the exact pathogenesis of the disease. It appears, however, that the organism consistently observed, presumably *Campylobacter* sp., has a direct and causative role in the disease process. Its definite identification, mechanism of invasion into the cell, and physiological interaction with the cell to cause hyperplastic lesions remain to be discovered.

Escherichia coli appears to have a secondary role, if any, in the disease syndrome. The organism has been shown to be enteropathogenic and can cause acute enteritis, but its contribution to the production of ileal proliferation is doubtful.

Because of the presence of both an intracellular organism, presumed to be *Campylobacter* sp., and an enteropathogenic *E. coli*, it is possible that a synergism between organisms may be responsible for the entire spectrum of lesions observed with proliferative ileitis. This was proposed by Frisk and Wagner (1977b), because they were able to demonstrate both organisms within the same diseased animals and, at times, within the same cell. They proposed that an acute enteritis, in their investigation caused by an enteropathogenic *E. coli*, was necessary to alter the integrity of ileal epithelial cells to allow infection by *Campylobacter* sp. In this proposed pathogenesis, enteropathogenic *E. coli* would not necessarily be the only organism or cause of the initial alterations of ileal epithelium needed for *Campylobacter* sp. to become established within cells. This supposition is further supported by the observation that from some natural outbreaks of proliferative ileitis, ileal cultures have not revealed enteropathogenic *E. coli*, but other intestinal pathogens such as *Salmonella* sp. (R. E. Flatt, personal communication, 1978). Also, Andrews (1975) measured several intestinal enzymes of hamsters affected with proliferative ileitis, and found that those enzymes which increased were associated with cytolysis rather than proliferation. Increased cellular destruction early in the disease was proposed as an important aspect of the disease, the proliferation being an exaggerated regeneration of mucosa.

E. Epizootiology

Proliferative ileitis is a contagious disease. The disease is most likely transmitted by the fecal–oral route. Cannibalism of affected hamsters by cagemates may also contribute to morbidity (Jacoby and Johnson, 1981).

Proliferative ileitis is characterized by variable morbidity (20–60%) and high mortality (≤90%) (Friedman, 1965; Jonas *et al.*, 1965; Boothe and Cheville, 1967; Goldman *et al.*, 1972). In colony outbreaks, proliferative ileitis is first manifested as an epizootic disease (Friedman, 1965; Jackson and Wagner, 1970). Usually only a few hamsters in an outbreak recover, and, in some instances, no survivors have been reported (Jonas *et al.*, 1965). Boothe and Cheville (1967) and Jacoby *et al.* (1975) reported that some animals initially recovered, but later succumbed to ileal obstruction associated with chronic scarring. Jackson and Wagner (1970) noted that months after an initial outbreak of proliferative ileitis occurred, the disease became enzootic, with only sporadic cases being recorded.

Hamsters affected with proliferative ileitis are usually between 3 and 8 weeks of age. Jonas *et al.* (1965) and Trum and Routledge (1967) reported that hamsters of all ages were affected, including adults, but animals over 12 weeks of age were not affected in an outbreak reported by Boothe and Cheville (1967). In experimental transmission studies, hamsters were susceptible to infection by oral inoculation of ileal homogenates up to 9 weeks of age (Frisk, 1976). Jacoby and Johnson (1981) showed that susceptibility to experimentally induced disease is age related. Animals were less susceptible beginning at 6 weeks of age and resistant at 10 weeks.

Chesterman (1972) reported that males were affected with greater frequency than females, but most reports have not indicated the percentages of males and females affected. Litters from primiparous females seem to be more susceptible than subsequent litters (Friedman, 1965; Frenkel, 1972). Increased severity and development of the disease have been associated with overcrowding, excess noise, transport, surgery, limited diets, purified diets (Decker and Henderson, 1959), transplantation of neoplasms (Lussier and Pavilanis, 1969), and experimental visceral leishmaniasis (Frenkel, 1972).

F. Clinical Signs

Early clinical signs of proliferative ileitis include lethargy, anorexia, irritability, ruffled hair coat, and rapid weight loss. As the disease progresses, there is the appearance of a fetid, watery diarrhea which causes moist matted fur on the perineum, tail, and ventral abdomen. Signs associated with the diarrhea include dehydration, inactivity, and a hunched back, which has been proposed to be indicative of abdominal pain (Friedman, 1965). A severe drop in body temperature, abdominal distention, and convulsions can occur just prior to death. Prolapse of the rectum or more commonly, intussusceptions with extension through the anus can be observed. Fresh blood around the anal region usually indicates an internal intussusception; the prognosis in such instances is grave. Hamsters usually die 24–48 hr after clinical signs of proliferative ileitis are noted, and deaths may occur in animals without premonitory signs.

Jacoby *et al.* (1975) closely observed hamsters after experimental transmission of proliferative ileitis and divided clinical signs into acute, subacute, and chronic. Acute signs occurred in 10% of hamsters 7–10 days after inoculation, the primary sign being hemorrhagic diarrhea. Subacute signs of retarded growth and diarrhea appeared 21–30 days after transmission.

The chronic disease did not produce clinical signs, these animals showing normal growth rates.

G. Pathology

1. Gross Lesions

Gross lesions of proliferative ileitis vary from acute enteritis to chronic inflammation with fibrosis. Lesions of acute enteritis include hyperemia of the ileum, which contains a fetid, yellow-gray fluid, blood, and/or gas (Sheffield and Beveridge, 1962; Boothe and Cheville, 1967). Sheffield and Beveridge (1962) reported that the cecum was sometimes ulcerated.

Segmental proliferative ileitis is generally observed in outbreaks of the disease. Typically, the posterior 4–8 cm of small intestine is progressively thickened, which abruptly terminates at the ileocecal junction (Fig. 1). Occasionally, the proximal colon or cecum can be involved. Affected intestines are 2–3 mm thick, turgid, edematous, and friable. White patches of necrotic mucosa are commonly observed. Ileal serosa appears hyperemic, rough, or granular. Subserosal abscesses are occasionally present, which can rupture, causing focal or generalized peritonitis (Friedman, 1965; Boothe and Cheville, 1967). There may be focal fibrinous adhesions between affected intestinal segments and adjacent intestines, mesentery, or peritoneum. Intussusceptions of the ileum are observed, some of which extend into the cecum, colon, and at times through the anus. Intussusceptions of the colon and prolapses of the rectum have also been described.

Chronic ileal lesions are characterized by a circumferential fibrotic scar, which at times can cause partial or complete obstruction.

Peyer's patches and mesenteric lymph nodes of hamsters with proliferative ileitis are typically enlarged two to three times, hyperemic, and edematous. Focal hepatic necrosis and hepatic fatty change are sometimes observed.

2. Microscopic Lesions

Microscopic lesions have been classified into the categories of acute enteritis, proliferative ileitis, proliferative ileitis with epithelium in muscle layers, and chronic ileitis (Frisk and Wagner, 1977a). Lesions of acute enteritis range from mild catarrhal enteritis to severe diffuse hemorrhagic necrosis of ileal mucosa. Proliferation of absorptive epithelial cells is not observed in these cases.

Proliferative lesions, which are characteristic of the disease, typically show ileal villi that are wider and longer than normal and may eventually fuse (Fig. 2). There is usually a distinct demarcation between the affected ileum and normal-appearing cecum. Boothe and Cheville (1967) reported that proliferation of epithelial cells starts at the extrusion zone at the tips of villi. In studies involving the experimental production of the lesions, epithelial cell proliferation is first observed in crypt epithelium (Frisk and Wagner, 1977b; Jacoby, 1978). "Focal proliferation" (Jackson and Wagner, 1970) is characterized by individual crypts, which show proliferative changes among normal-appearing crypts (Fig. 3).

Hyperplasia first occurs with the movement of crypt epithelium onto the villus within 10 days after experimental infection (Jacoby, 1978). These hyperplastic epithelial cells are enlarged two to four times and contain large hyperchromatic nuclei with prominent nucleoli (Jonas *et al.*, 1965). Their cytoplasm appears more basophilic than unaffected ileal epithelial cells. Immature absorptive epithelium is markedly thickened and pseudostratified beginning near the villar base and ascending to eventually replace normal villar epithelium (Jacoby, 1978). Increased numbers of mitoses are observed within crypts, and mitotic figures are also observed on affected villi (Fig. 4). The number of epithelial cells from the crypt to the villar tip is increased two to three times, 3–4 weeks after infection (Jacoby, 1978).

Intracytoplasmic inclusions within epithelial cells have been

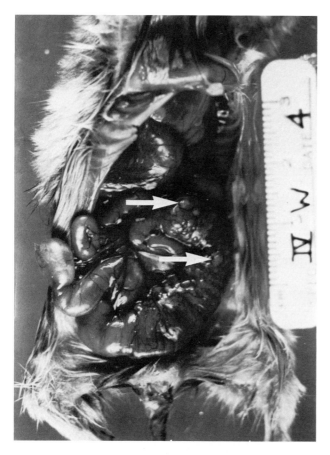

Fig. 1. Abdominal viscera of a hamster with proliferative ileitis showing marked enlargement of the ileum. Subserosal nodules are prominent on the surface of the ileum (arrows).

7. BACTERIAL AND MYCOTIC DISEASES

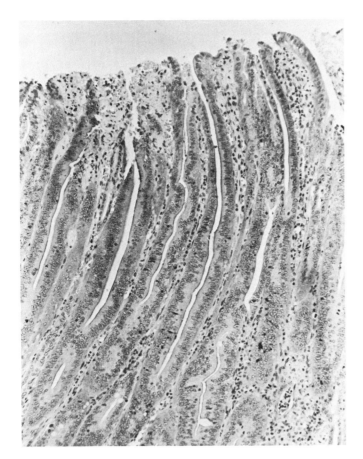

Fig. 2. Ileum from a hamster with proliferative ileitis showing hyperplastic epithelial cells covering elongated villi. Hematoxylin–eosin stain. Magnification: ×130.

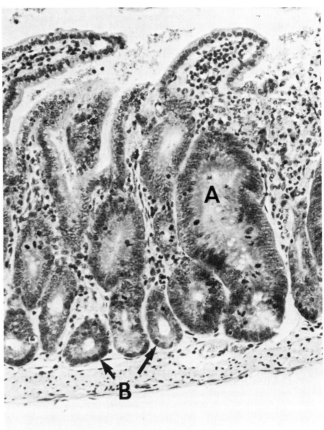

Fig. 3. Hamsters with early lesions of proliferative ileitis may have proliferated crypts (A) adjacent to apparently normal crypts (B). Hematoxylin–eosin stain. Magnification: ×190.

described (Jonas *et al.*, 1965; Frisk and Wagner, 1977b; Jacoby, 1978); however, the finding is not consistent. Lussier and Pavilanis (1969) described intranuclear inclusions within affected epithelial cells; however, they suspected that the animals were coincidentally infected with a virus unrelated to the causative agent of proliferative ileitis.

Necrosis of villi with inflammation of the lamina propria is common. Lesions can sometimes extend into muscle layers with accompanying inflammatory changes. This inflammation becomes granulomatous as the lesions progress. Subserosal abscesses may develop, which may communicate with the lumen via crypts. Hyperplastic epithelial cells may extend downward into inflamed tissue layers and produce pseudodiverticula, especially apparent when the animal survives for several weeks after the initial infection (Fig. 5). This lesion led Jonas *et al.* (1965) to classify the disease as neoplastic rather than inflammatory, describing this change as "carcinoma *in situ*." Jacoby and Johnson (1981) critically interpreted this "intrusive behavior," and felt it was due to mechanical pressures from proliferation and was not invasion of a neoplastic epithelium.

Chronic lesions are characterized by marked thickening and hypertrophy of muscle layers, which is believed to be secondary to partial luminal obstruction (Boothe and Cheville, 1967). Scattered granulomas and epithelial-lined diverticula are present in the submucosa and subserosa. The ileal epithelium contains large numbers of goblet cells.

Mesenteric lymph nodes may show reticuloendothelial hyperplasia, edema, granulomatous inflammation, and focal necrosis. Epithelial cells have not been observed within lymph nodes. Focal lesions of hepatitis are occasionally observed.

3. Transmission Electron Microscopy

Intracytoplasmic bacteria, morphologically identical to *Campylobacter* sp., have been observed within hyperplastic ileal epithelial cells in natural and experimental cases of proliferative ileitis (Wagner *et al.*, 1973; Frisk and Wagner, 1977b; Johnson and Jacoby, 1978). Cells containing organisms have been observed as early as 5 days after experimental inoculation of hamsters (Frisk and Wagner, 1977b; Johnson and Jacoby, 1978). These intracellular organisms are slightly

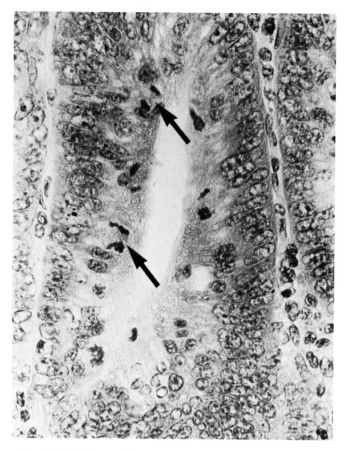

Fig. 4. This ileal crypt, from a hamster with proliferative ileitis, contains several cells in mitosis (arrows). Hematoxylin–eosin stain. Magnification: ×510.

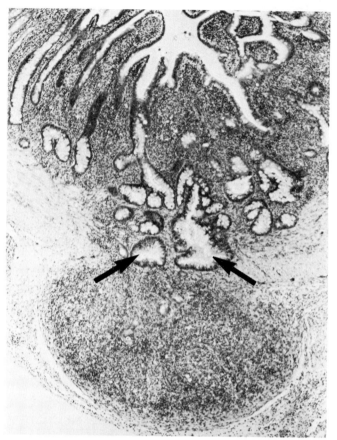

Fig. 5. A hamster with chronic lesions of proliferative ileitis shows a subserosal nodule of inflammation and crypts (arrows) that have extended into ileal muscle layers. Hematoxylin–eosin stain. Magnification: ×50.

curved, rod-shaped, measure $0.3–0.4 \times 1–2$ μm, and divide by binary fission (Fig. 6). Organisms are most prominent in the apical portion of hyperplastic cells and are more numerous in epithelial cells that are further out on the villus (Johnson and Jacoby, 1978). Organisms are usually free within the cytoplasm, although occasionally they are observed bounded by lysosome membranes (Frisk and Wagner, 1977b). Clusters of the organisms observed by electron microscopy within the cytoplasm of hyperplastic epithelial cells could account for the intracytoplasmic inclusion bodies observed by light microscopy (Fig. 7). Degenerative or toxic changes of the invaded cell are not observed. The organisms, in fact, appear to have a stimulating effect on infected cells, being reported primarily within hyperplastic cells and within cells in apparently normal stages of mitosis (Frisk and Wagner, 1977b; Johnson and Jacoby, 1978). Organisms are not observed in unaffected portions of intestine or in the intestines from normal hamsters (Jacoby and Johnson, 1981).

Hyperplasia of ileal epithelial cells is observed by 10 days after experimental inoculation (Johnson and Jacoby, 1978). Hyperplastic cells are elongated and abut the basement membrane. Microvilli of affected cells are much shorter, less regularly arranged, and less densely packed than normal cells (Kim and Jourden, 1977). Most studies have found that hyperplastic cells always contain the intracytoplasmic organisms described previously (Frisk and Wagner, 1977b; Johnson and Jacoby, 1978). Kim and Jourden (1977) did not observe organisms within affected epithelial cells; however, only four animals were examined, and the disease described was atypical, occurring in hamsters approximately 1 year of age, and intestinal lesions were complicated by amyloid deposits. There may be focal necrosis of crypt epithelial cells and penetration of the muscularis muscosa, as evidenced by bacteria between myofibers. Macrophages and leukocytes infiltrate crypts, lamina propria, submucosa, and adjacent muscle layers.

In chronic cases, goblet cells appear which do not contain intracytoplasmic bacteria. Absorptive epithelial cells contain increased numbers of lysosomes and large membrane-bound vacuoles containing degenerative organisms (Frisk and Wagner, 1977b).

In addition to the previously described organism, large intracytoplasmic organisms were observed within epithelial cells by

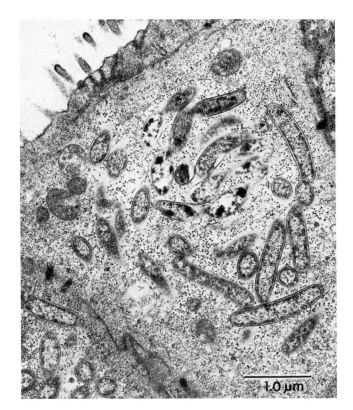

Fig. 6. Intracytoplasmic organisms are prominent in the apical portion of a hyperplastic ileal epithelial cell. With permission from Frisk and Wagner (1977b), and the *American Journal of Veterinary Research.*

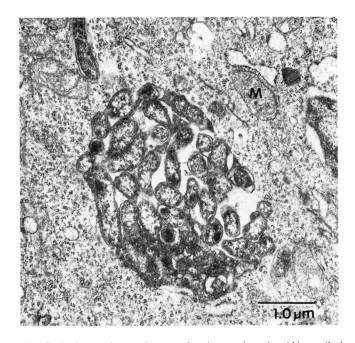

Fig. 7. A cluster of many intracytoplasmic organisms is within an ileal epithelial cell from a hamster with proliferative ileitis. Mitochondrion (M). With permission from Frisk and Wagner (1977b), and the *American Journal of Veterinary Research.*

Frisk and Wagner (1977b). These organisms were larger (0.5–1 × 2.2 μm) and by indirect fluorescent antibody techniques were identified as *E. coli*. They were present only in hamsters with experimentally produced disease, being observed alone in hamsters 1–9 days after inoculation and in combination with the other previously described organism 6–14 days after inoculation (Fig. 8). Detection of *E. coli* was associated with microscopic lesions of acute enteritis. Their presence in combination with the previously described organism, sometimes even being within the same cell (Frisk and Wagner, 1977b), may suggest a synergism between the two organisms (see Section II,D). Johnson and Jacoby (1978) did not observe a second organism by electron microscopy in their experimental transmission studies. In naturally occurring cases of proliferative ileitis, *E. coli* was not observed by electon microscopy (Wagner *et al.,*

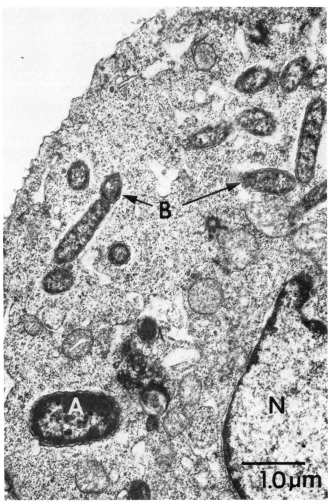

Fig. 8. Two different organisms are present within the same hyperplastic epithelial cell from a hamster with early lesions of proliferative ileitis. The larger organism (A) is *Escherichia coli*, and the smaller organisms (B) are presumed to be *Campylobacter* sp. Nucleus (N). With permission from Frisk and Wagner (1977b), and the *American Journal of Veterinary Research.*

1973; Frisk and Wagner, 1977b). Wagner *et al.* (1973), however, did isolate an *E. coli* organism, which was serologically related to *Shigella boydii,* from tissues examined by electron microscopy.

4. Scanning Electron Microscopy

By scanning electron microscopy, Johnson and Jacoby (1978) noted that villi from hamsters with proliferative ileitis were elongated, broader, and flatter than normal. The crypt–villus junction was elevated above the basal lamina. In fractured blocks examined with the scanning electron microscope, crypts were observed to be four to five times normal length with only rudimentary villi at the surface. Villi sometimes appeared leaflife and were fused into plateaus, with the surface obscured by bacteria and cellular debris. Microvilli were sparse on villi covered by hyperplastic cells.

H. Diagnosis

Proliferative ileitis should always be suspected with the observance of clinical signs of diarrhea in young hamsters. Diarrhea may occur with other diseases such as salmonellosis, hymenolepiasis, Tyzzer's disease, or cecal musocal hyperplasia. Their exclusion is primarily based on gross and microscopic lesions. The proliferative changes involving the ileum, so typically observed, are pathognomonic for the disease. Palpable, ropelike intraabdominal masses can be detected 2 weeks postexposure, prior to the observation of clinical signs at 3 weeks (Jacoby *et al.*, 1975). Culture of *Campylobacter jejuni* can support the diagnosis; however, without Koch's postulates being fulfilled, its presence cannot be relegated as causative.

I. Treatment and Control

Antibiotics administered orally have been the most commonly recommended treatment for proliferative ileitis but are only moderately successful. Tetracycline hydrochloride (400 mg/liter drinking water) was found to be most effective in reducing the numbers of hamsters contracting proliferative ileitis after experimental inoculation (La Regina *et al.*, 1980). Dimetridazole (500 mg/liter drinking water) was much less effective than tetracycline, and neomycin (125 mg/liter drinking water or 10 mg per hamster) was considered ineffective when compared to untreated controls. Jacoby and Johnson (1981) reported that metronidazole (2 mg/ml drinking water) was nearly as effective as tetracycline and neomycin was ineffective even at a dose rate of 2000 mg/liter of drinking water. Sheffield and Beveridge (1962) reported that oral administration of neomycin (10 mg per hamster) resulted in 78% survival as compared to 41% survival of untreated hamsters in a natural outbreak. Frenkel (1972) reported that neomycin (5 gm/liter) and sulfadiazine (1.2 gm/liter) administered in the water are more effective than neomycin alone. Williams (1976) recommended pediatric chloramphenicol palmitate orally at 6–9 mg per hamster twice a day. The routine administration of a soluble oxytetracycline (0.025%) in the drinking water to hamsters from the fourth to the sixth week of life has been proposed as a control measure (Sebesteny, 1979).

Strict quarantine of affected hamsters and good sanitation practices are useful in controlling proliferative ileitis. Amend *et al.* (1976) stated that filter bonnets were effective in preventing cage to cage transmission of the disease. Feeding alfalfa has been recommended as a control measure (Brush *et al.*, 1957; Ershoff, 1956). Jacoby and Johnson (1981) reported that different commercial diets could affect the number of hamsters developing lesions after experimental transmission. Colony depopulation, thorough facility fumigation, and repopulation with hamsters from a source without a history of the disease is probably the most effective way to eliminate proliferative ileitis completely.

J. Significance

Proliferative ileitis continues to be the most significant disease that occurs in hamsters. The disease is still quite prevalent even with increased sanitation and surveillance procedures. High morbidity ($\leq 60\%$) and mortality (can be 100%) in colony outbreaks ranks proliferative ileitis as the most feared disease of commercial hamster breeders. It has complicated research projects by causing clinical signs and death in hamsters on experimentation.

Proliferative ileitis is not regarded as a disease which is transmissible to humans. However, with the emerging significance of *Campylobacter* sp. in the pathogenesis of the disease, this attitude may not continue. Human campylobacteriosis has not been thus far associated with hamsters affected with proliferative ileitis. The hamster has been proposed as a model for studying human *Campylobacter* infections (see Section VI).

Proliferative ileitis has been proposed as an animal model for regional enteritis or Crohn's disease of humans. Hyperplasia of ileal epithelial cells, which is so characteristic of proliferative ileitis, does not occur in Crohn's disease (Crohn *et al.*, 1932; Wheelock, 1974; Mayberry *et al.*, 1980). Also, in humans with Crohn's disease, the absorptive epithelial cells do not contain organisms (Cook and Turnbull, 1975), whereas, they do in hamsters with proliferative ileitis. The epizootiology of the diseases are also different. Although lesions of the two diseases may occur in the same region of intestine and are segmental, the pathogenesis of the two diseases are most likely quite different.

The experimental production of proliferative ileitis of hamsters could be exploited as a model for studying the regulatory mechanism of mucosal proliferation (Jacoby and Johnson,

1981). Crypt hyperplasia occurs in humans as a response to mucosal injury, such as with cytolytic agents, radiation injury, and tropical and celiac sprue. The mechanism by which this hyperplastic lesion is produced has been proposed to be through a feedback-regulatory system, and this particular hamster lesion may give investigators valuable insights into this system (Jacoby and Johnson, 1981).

Proliferative ileitis of hamsters has similarities to porcine proliferative enteritis and proliferative colitis of ferrets. The clinical and epizootiological aspects of porcine proliferative enteritis are very similar to those of hamster proliferative ileitis. The gross, microscopic, and electron-microscopic changes observed in swine with proliferative enteritis are identical to those in hamsters with proliferative ileitis (Rowland and Lawson, 1975; Lomax and Glock, 1982). In the ferret, proliferative lesions are found in the colon, but they are similar in character to those found in hamster proliferative ileitis (Fox et al., 1982a). The definitive etiological agent and complete pathogenesis of proliferative enteritis of swine and proliferative colitis of ferrets have not been elucidated, but *Campylobacter* has been incriminated in both diseases. The hamster, swine, and ferret diseases most likely have very similar pathogenesis, and thus data and information learned about the disease in one species may be helpful in understanding the disease in the others.

III. CECAL MUCOSAL HYPERPLASIA

A. History, Etiology, and Epizootiology

Cecal mucosal hyperplasia has been reported once. It occurred in a colony of 800 breeding adult hamsters (Barthold et al., 1978). The disease affected suckling (8–14 days of age) and weanling hamsters with a mortality rate of approximately 80%. The disease was transmitted by close contact between hamsters, its incubation period being approximately 4 days.

The epizootiology of the disease was characteristic of an infectious organism; however, none was identified. Aerobic cultures of affected ceca revealed presumed nonpathogenic organisms from the families Enterobacteriaceae and Pasteurellaceae. Bacteria or viruses could not be demonstrated in cecal lesions by transmission electron microscopy. Hamsters experimentally inoculated orally with ground cecal lesions and by direct cecal injection did not develop the disease.

B. Clinical Signs and Pathology

The most prominent clinical sign of cecal mucosal hyperplasia is diarrhea. Fur at the base of the tail is matted and covered with liquid feces. Other signs include runting and dehydration, with affected animals progressing to a moribund state.

Affected ceca had thickened, opaque walls and hyperemic serosal vessels. Ceca were contracted containing a small amount of ingesta. Histologically, there was cecal mucosal hyperplasia. Crypts were two to three times normal height, mitoses were numerous, and cellular differentiation was diminished. Mucosal erosion with acute inflammation was also occasionally observed within the cecum and lower portions of the small intestine. Mild nonsuppurative portal hepatitis was often observed.

C. Diagnosis and Control

The differential diagnosis should include any disease which causes diarrhea. Proliferative ileitis occurs in the same age group of hamsters, and proliferation in the large intestine has been associated with typical ileal lesions (Jonas et al., 1965; Jackson and Wagner, 1970; Frisk and Wagner, 1977a). Intracytoplasmic organisms, however, could not be demonstrated in lesions of cecal mucosal hyperplasia, nor were proliferative ileal lesions observed (Barthold et al., 1978). Also, hyperimmune sera to the intracytoplasmic organism of proliferative ileitis was reacted with hyperplastic cecal lesions and was negative. It appears, therefore, that cecal mucosal hyperplasia is distinct from proliferative ileitis; however, this statement is speculative because the etiologies and pathogeneses of neither disease have been elucidated. Other diagnoses which can be differentiated by culture and lesions include salmonellosis, Tyzzer's disease, and hymenolepiasis. This single outbreak of cecal mucosal hyperplasia was eradicated by colony depopulation.

IV. TYZZER'S DISEASE

A. History

In 1917, Tyzzer reported on a fatal epizootic diarrheal disease of Japanese waltzing mice. The disease was characterized by the demonstration of organisms (*Bacillus piliformis*) within hepatocytes and intestinal epithelial cells, and pathologically by focal hepatic necrosis. Although more descriptive terminology has been used to identify the disease, such as epizootic hepatitis (Rights et al., 1974), transmissible enterocolitis (Nakayama et al., 1975), and typhlohepatitis (Nakayama et al., 1976), the eponym "Tyzzer's disease" is most accepted.

Prior to the report of spontaneous outbreaks of Tyzzer's disease in golden hamsters, it was shown that hamsters could be infected experimentally. Rights et al. (1947) injected hamsters intracerebrally with brain tissue from mice that were infected with Tyzzer's organism. Hamster cerebral lesions consisted of liquefaction necrosis accompanied by acute inflammation, these areas containing many organisms. Takagaki et al. (1966) produced focal hepatic necrosis in hamsters by injecting them intravenously, intraperitoneally, and subcutaneously with liver homogenates from mice with Tyzzer's disease. When hamsters were infected orally with these same homogenates, they developed enteritis. Craigie (1966) experimentally infected hamsters intraperitoneally using yolk sac infected with Tyzzer's organisms.

In England, White and Waldron (1969) reported unconfirmed cases of Tyzzer's disease in hamsters during an outbreak of the disease in a colony of gerbils. Since that time, naturally occurring cases have been sporadic, being limited to two outbreaks in Japan (Nakayama et al., 1975; Takasaki et al., 1974) and two in the United States (Zook et al., 1977; Timmons and Dickey, 1981).

B. Etiology

Bacillus piliformis is the causative agent of Tyzzer's disease. The taxonomic position of this organism remains undetermined. The organism is a gram-negative, motile, pleomorphic bacillus measuring 0.5 × 8–10 μm, with some organisms being up to 40 μm in length. Bacillus piliformis can produce spores characterized by monilial subterminal swellings. The vegetative form of the organism is very unstable. Spores most likely play an important role in establishing infection by the oral route. The spores are responsible for residual infectivity of contaminated materials up to 1 year at room temperature (Craigie, 1966).

Bacillus piliformis has not been grown on cell-free artificial media. The organism can be grown in the yolk sac of 6-day-old embryonated chicken eggs. Yolk sac smears from inoculated eggs are examined for organisms with Giemsa stain 5–9 days postinoculation (Zook et al., 1977). The antigenicity of B. piliformis isolated from hamsters has been compared to isolates from other species by complement fixation, immunofluorescence, and double-diffusion techniques (Fujiwara et al., 1974). Hamster organisms and the organism isolated from a kitten during a hamster outbreak showed strong cross-reaction. There was weak or no cross-reaction with Tyzzer's organisms isolated from rats or mice. Nakayama et al. (1975) showed common antigens by complement fixation when organisms from hamsters with enteric lesions were compared with organisms from hamsters with hepatic lesions. It appears that the antigenic specificity of B. piliformis is dependent on the host species from which it is derived.

In tissue sections, organisms are intracytoplasmic, at times beaded, and are most often arranged in parallel bundles or starburst arrays. Spores are not usually observed in tissue. Tyzzer's organism has a predilection for multiplication in certain cells such as the hepatocyte and epithelial cell of the intestine. Bacillus piliformis are poorly visible in tissues stained with hematoxylin–eosin and Gram's stain. The organisms are most easily visible, as light-brown to black rods, with the silver stains of Warthin–Starry or Levaditi (Ganaway et al., 1971). Other effective stains are Giemsa, toluidine blue, Gomori methenamine silver, and periodic acid–Schiff.

Bacillus piliformis is able to infect a very wide host range. In addition to hamsters, Tyzzer's disease has been reported in the mouse, rat, rabbit, gerbil, cat, rhesus monkey, horse, guinea pig, dog, coyote, muskrat, and lesser panda (Wallach and Boever, 1983).

C. Pathogenesis

Tyzzer's disease in hamsters has been studied by inoculating B. piliformis contained within liver homogenates, cecal contents, or embryonated chicken eggs. Infections in hamsters and mice have been induced by many routes including intravenous, intraperitoneal, subcutaneous, intrasplenic, and oral. The inoculation of B. piliformis orally is probably the method of studying Tyzzer's infection which gives results which are the closest to the naturally occurring disease.

Nakayama et al. (1976) orally inoculated hamster liver homogenates into 6- to 8-week-old hamsters to study the pathogenesis of the disease. Hamsters were necropsied daily to determine pathological lesions and for the detection and location of organisms. One day after inoculation, mild inflammation was noted in the cecum with a low percentage of epithelial cells containing organisms. By day 2 inflammation had spread to the entire small and large intestine, with organisms being detected in the cecum and colon. Lesions progressed to more severe inflammation, involving the submucosa by day 3. Organisms could be detected from the jejunum to the rectum. Liver lesions could be observed as mild foci of necrosis with the infiltration of a few organisms. Hamsters necropsied on days 4 and 5 showed severe necrotic changes throughout the intestinal tract with moderate to large numbers of organisms. Liver necrosis was judged to be moderate. On days 6–8 intestinal lesions were severe, including necrosis and desquamation of epithelial cells. There were also areas of severe hepatic necrosis with moderate numbers of organisms present. Organisms had disappeared from liver lesions by day 10. Colon and cecum showed shortening of villi and inflammatory infiltrate. Other organs showing lesions at this time included pancreatitis, pericarditis, and gastric ulceration.

The organ in which Tyzzer's bacillus first grows is the cecum, its physiology possibly being favorable to the proliferation of B. piliformis (Nakayama et al., 1976). Factors possibly involved include its anaerobic environment, presence of other microflora, and the movement of contents as compared with other areas of the alimentary tract. Dissemination of

7. BACTERIAL AND MYCOTIC DISEASES

the organism is rapid throughout the remainder of the intestinal tract, affecting all regions from duodenum to rectum by the third day postinoculation (Nakayama et al., 1976).

The route of *B. piliformis* to the liver has been investigated (Nakayama et al., 1976). It has been proposed that the organism may infect the liver from the duodenum via the bile duct. This can be supported by observations of organisms in the epithelium of the bile duct and gallbladder of mice (Tyzzer, 1917). This evidence was only circumstantial, however, and the cecal ligation experiments in hamsters by Nakayama et al. (1976) were important in establishing the route via portal circulation and lacteals. Hamsters had a portion of their cecum ligated which was then inoculated with *B. piliformis*. Organisms could be observed within cecal epithelial cells, lamina propria, and muscle layers prior to being detected in hepatic lesions which developed 3–4 days postinoculation. No organisms could be detected in other parts of the intestine, including the proximal side of the ligated cecum, showing that it was portal circulation and not ascending bile duct infection which was the organism's route to the liver.

D. Epizootiology

Although there have been widespread reports of Tyzzer's disease in mice from many countries, there have been only scattered reports of the disease in hamsters. Because the organism can spread to hamsters from other species such as mice, rabbits, or gerbils, there is a possibility of Tyzzer's disease occurring in hamsters whenever they are housed near susceptible species. It is possible that the prevalence of Tyzzer's disease in hamsters may be higher because some cases go undiagnosed. Diagnosis may not be confirmed in some cases because the typical lesions of focal hepatic necrosis are not present in all cases, *Bacillus piliformis* will not grow on artificial media, and special stains are needed to observe the organism histologically.

The transmission of Tyzzer's organism from animal to animal is most probably by the oral route. Zook et al. (1977) noted during an outbreak that if a hamster in a cage became affected usually its cagemates became ill as well. Nakayama et al. (1975) exposed hamsters to soiled bedding 3 months after the hamsters housed on it died of Tyzzer's disease, and after a few days exposure, all hamsters died with enteric lesions. This result was in contrast to that of Zook et al. (1977), who reported no disease development on exposing hamsters and mice to contaminated bedding.

E. Clinical Signs

Tyzzer's disease outbreaks can be characterized by the sudden unexpected loss of animals with few premonitory signs. Hamsters may die within 24 hr to a few weeks after onset. Diarrhea is the most characteristic sign, although it is not always observed. The affected hamster has pale, yellow, watery feces which result in wet hairs on the tail and ventral abdomen. Other clinical signs include a ruffled hair coat, lethargy, anorexia, and dehydration. The diseased hamster may appear hunched and can progress to a moribund condition.

F. Pathology

Lesions of Tyzzer's disease can vary in the hamster. In some outbreaks the predominant lesions involve the liver (Takasaki et al., 1974); in others enterocolitis occurs without liver lesions (Nakayama et al., 1975), and in yet others heart lesions occur (Zook et al., 1977). Typical liver lesions consist of organ enlargement with a few to many white-yellow or white-gray plaques 0.5–2 mm in diameter. Gross lesions involving the intestinal tract most commonly occur in the lower ileum, cecum, and colon. These regions show edema, hyperemia to hemorrhage, and dilatation with semiliquid, foamy, yellow contents. Focal necrotic gray plaques, 1–3 mm in diameter, can be observed on the mucosal surface. Multiple white nodules, 2–5 mm in diameter, may be observed in the heart (Zook et al., 1977). These lesions bulge from the surface of the heart and, in some cases, extend the full thickness of the ventricle. Other less notable gross lesions include markedly enlarged and congested mesenteric lymph nodes and enlarged spleens with demonstrable follicles.

Histologically, liver lesions consist of foci of necrosis of variable size distributed from the center of the lobule to the median zone (Fig. 9). There are pyknosis and karyorrhexis of hepatocytes, and infiltration of affected areas by lymphocytes and neutrophils. In some cases, a zone of neutrophils and large mononuclear cells surround foci of necrosis. Portal areas usually do not show necrosis but are infiltrated by lymphocytes and plasma cells. Organisms are observed primarily in living hepatocytes immediately adjacent to necrotic foci with fewer numbers in necrotic hepatocytes or the center of necrotic foci (Fig. 10). The appearance of Tyzzer's bacillus in histopathological sections is discussed in Section IV,B.

Intestinal lesions can generally be classified as catarrhal inflammation with focal mucosal erosions. The mucosa of the ileum, cecum, and colon shows degenerative and necrotic changes. There is edema, hyperemia, and inflammatory cell infiltration of the lamina propria. The most severe mucosal erosions include edema and inflammation of the submucosa and muscle layers. Organisms can be observed within the cytoplasm of mucosal epithelial cells in the intermediate zone of villi and less predominantly in crypts. Organisms are occasionally present within smooth muscle cells and rarely in areas of necrosis.

Myocardial lesions described by Zook et al. (1977) consist of ill-defined granulomas. There are focal areas of necrosis

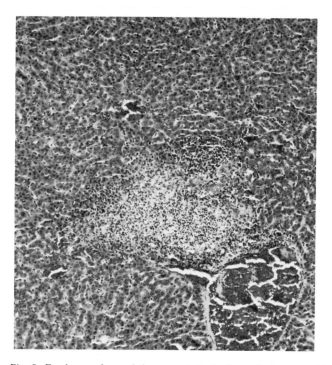

Fig. 9. Focal area of coagulative necrosis in the liver of a hamster with Tyzzer's disease. Hematoxylin–eosin stain. Magnification: ×64. Courtesy of Dr. K. S. Waggie, National Institutes of Health.

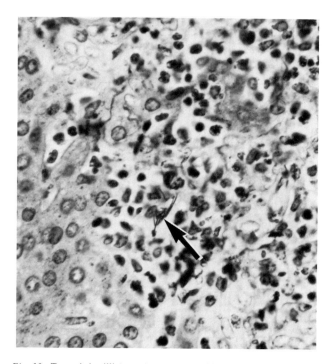

Fig. 10. Tyzzer's bacilli (arrow) are present within an area of hepatic necrosis. Warthin–Starry silver Stain. Magnification: ×640. Courtesy of Dr. K. S. Waggie, National Institutes of Health.

with accompanying inflammation characterized by infiltration by large macrophages, neutrophils, lymphocytes, plasma cells, and fibroblasts. Organisms can be observed within these lesions.

Histologically, mesenteric lymph nodes show sinus catarrh and hyperplasia without the demonstration of organisms. Nakayama *et al.* (1975) reported kidney lesions in 2 of 13 hamsters consisting of tubule dilatation, degeneration of epithelial cells, and tubular casts. The association of these renal lesions with the hepatic and intestinal lesions was not discussed, nor was there mention of organisms being present.

G. Diagnosis

The diagnosis of Tyzzer's disease is dependent on the demonstration of typical bacilli within the cytoplasm of hepatocytes, intestinal epithelial cells, or myocardial cells adjacent to regions of necrosis. This is most commonly done with special stains of tissue sections. Wet impressions of liver stained with Giemsa may reveal light-blue, rod-shaped bacteria within hepatocytes. *Bacillus piliformis* also can be grown in the yolk sac of embryonated chicken eggs. The lack of growth of other organisms from affected tissues, feces, or inoculated eggs is important in ruling out other pathogenic bacteria.

A definitive diagnosis of Tyzzer's disease can also be substantiated by inoculation of experimental animals. Affected tissue, such as liver or cecum, can be homogenated with phosphate-buffered saline for inoculation. Mice treated with 2.5 mg cortisone have been injected intravenously with liver homogenates from affected hamsters (Nakayama *et al.*, 1975). Upon spontaneous death at 4 days or necropsy at 10 days postinoculation, such animals showed focal liver lesions with bacteria within hepatocytes. A cecal homogenate from hamsters with suspected Tyzzer's disease was inoculated intraperitoneally into hamsters that subsequently died in 5 days from a necrotizing enterocolitis (Zook *et al.*, 1977).

The differential diagnosis for Tyzzer's disease in hamsters should include any disease which can cause diarrhea. Proliferative ileitis would be the most important disease to differentiate. Tyzzer's disease would lack the typical proliferative ileal lesions of the former but have *B. piliformis* within intestinal and hepatic lesions. Hymenolepiasis and salmonellosis can be easily differentiated by the demonstration or isolation of either of these organisms. Timmons and Dickey (1981) reported that the initial isolation of *Pseudomonas* confused the tentative diagnosis of a Tyzzer's disease outbreak in hamsters.

H. Control and Prevention

The most important factors in the control and prevention of Tyzzer's disease involve sanitation and isolation. In the first reported outbreak, Tyzzer (1917) was only able to save some

7. BACTERIAL AND MYCOTIC DISEASES

strains of mice by restocking the colony into sterilized caging, sterilizing the food, and isolating the colony from ill or exposed mice. These practices of providing adequate sanitation with strict isolation still apply. Spores of *B. piliformis* are likely the most important factor in maintaining an outbreak. Cleaning the environment to eliminate spores and preventing cage to cage spread with filter covers or individual isolation units are helpful in controlling spread of the disease. Other important aspects of control involve adequate vermin control, reducing overcrowding, and preventing contamination of food and bedding.

Treatment of hamsters with Tyzzer's disease was not described in any of the reported outbreaks. The recommended treatment in other species is tetracycline. Craigie (1966) showed that penicillin had a beneficial effect in controlling experimental and naturally occurring Tyzzer's infection of mice. Ganaway *et al.* (1971) reported that streptomycin, erythromycin, penicillin, and chlortetracycline have a "partial" effect, and sulfonamide and choramphenicol have no effect on infection of embryonated eggs with *B. piliformis* isolated from rabbits.

Experimentally, a killed formalin (1%) vaccine was developed with Tyzzer's organisms from hamster origin (Fujiwara *et al.*, 1974). When this vaccine was used in mice, it was protective against the hamster organism but did not provide protection against challenge with organisms of mouse origin. In experiments using mouse antisera to hamster organisms, mice were protected from challenge of hamster organisms but not to mouse organisms (Fujiwara *et al.*, 1974). Mouse titers to hamster organisms were completely protective at 1:16, 1:8, and 1:4 (Fujiwara *et al.*, 1974). Experimentally, hamsters that had survived oral inoculation were resistant to challenge with Tyzzer's organisms, demonstrating that recovered animals have some immunological protection (Nakayama *et al.*, 1975).

I. Significance

Although there have been sporadic outbreaks of Tyzzer's disease in hamsters, the disease can be quite significant because it has a high morbidity and high mortality. If an outbreak occurs in any species in an animal care unit, other susceptible species including the hamster are at risk of infection. This has been a very real problem, with outbreaks in hamsters being associated with diseased gerbils in one outbreak (White and Waldron, 1969) and a kitten concurrently infected in another (Takasaki *et al.*, 1974). As mentioned (Section IV, B), *B. piliformis* may infect many species of animals, but it is not considered to be transmissible to humans.

Although Nakayama *et al.* (1975) and Zook *et al.* (1977) proposed that Tyzzer's organism could be at least partly responsible for the pathogenesis of "wet-tail" disease of hamsters, the two appear to be separable by distinctive lesions and morphology of organisms present. As mentioned, however, Tyzzer's disease should always be part of the differential diagnosis of hamsters developing diarrheal signs.

Investigators who carry out experimental protocols that involve tumor transplantation or immunosuppression in hamsters, especially with cortisone or irradiation, should be watchful of Tyzzer's disease outbreaks. Tyzzer's disease in hamsters can be exaggerated in its effects by cortisone or irradiation treatment. Cortisone administered subcutaneously at 2.5 mg per hamster enhanced infection (Nakayama *et al.*, 1976). Organisms were more abundant in affected tissues, and lesions involving the alimentary tract and liver were accentuated. Hamsters treated with cortisone died 4–7 days after inoculation as compared with 7–13 days if not given cortisone (Nakayama *et al.*, 1976). Takagaki *et al.* (1966) noted that no hamster died after inoculation with Tyzzer's organisms unless they were treated with cortisone. Cortisone's effect can probably be attributed to its immunosuppressive effects, especially its reduction in the activity of the animal's phagocytic system (Nakayama *et al.*, 1976). Hamsters exposed to whole-body irradiation (500 R) at the time of inoculation with Tyzzer's organisms developed more severe lesions (Takagaki *et al.*, 1966). In contrast to these findings, Timmons and Dickey (1981) reported that cecal lesions were more pronounced in hamsters nonimmunosuppressed as compared to those given 2.5 mg of cortisone acetate twice per week or exposed to irradiation (1000 R, cheek pouch).

V. SALMONELLOSIS

A. History, Etiology, and Epizootiology

Salmonellosis was a significant disease of hamsters, as in other rodents, prior to the improvement of sanitation practices, and the commercial production of diets and bedding. In recent years, the disease in hamsters is rarely diagnosed (Renshaw *et al.*, 1975). In India as recently as 1970, *Salmonella* organisms were isolated from 46% of hamsters that were moribund or found dead (Ray and Mallick, 1970).

Both *S. enteritidis* and *S. typhimurium* have been isolated from diseased hamsters. Organisms can be cultured from heart blood, liver, spleen, lungs, and feces. Incubation period is usually 3–6 days. Transmission is by the oral route, most commonly by contaminated food or bedding. A carrier state exists wherein apparently normal hamsters shed *Salmonella* organisms in the feces (Soave, 1963).

B. Clinical Signs and Pathology

Salmonellosis in hamsters can produce signs and lesions which are very similar to salmonellosis of other rodents

(Habermann and Williams, 1958; Soave, 1963). Innes *et al.* (1956), however, reported that hamsters were more susceptible to *Salmonella* infection than mice and described a fulminating disease.

Clinical signs include a reduction in normal activity, rough hair coat, anorexia, weight loss, and increased respiratory rate. Feces may appear as normal pellets or may be lighter in color and soft. Hamsters showed intussusception of the colon with extension through the anus after experimental inoculation of *Salmonella* (Pollock, 1975). With chronic salmonellosis, there may be no observable signs or animals may show progressive cachexia. Hamsters that survive weeks may show distended abdomens from enlargement of the liver and spleen.

Gross necropsy reveals small white foci in the liver, patchy hemorrhagic lungs, and reddened lymph nodes. Enteritis may be observed, although it may not be apparent in all cases (Innes *et al.*, 1956). Mediastinal lymph nodes may show congestion, hemorrhage, erythrophagocytosis, and foci of necrosis. Pericapsular regions may be edematous and infiltrated by inflammatory cells.

C. Pathogenesis

The pathogenesis of *Salmonella* infection of hamsters was studied by Innes *et al.* (1956). They found that the hamster was highly susceptible to infection, with death occurring in all inoculated hamsters by the second day after intraperitoneal injection and by the ninth day after oral infection. After intraperitoneal injection with *S. enteritidis*, hamsters developed peritonitis, perihepatitis, and perisplenitis. Upon histological examination, embolic glomerular lesions, purulent pericarditis, and necrotizing placentitis were observed. Acute enteritis with necrosis of Peyer's patches developed after oral inoculation.

Innes *et al.* (1956) proposed two possibilities for the pathogenesis of the phlebothrombosis which was so prominent in the lungs of hamsters. They theorized that a bacterial septicemia developed and bacteria agglutinated on venular walls to form the nidus of a thrombus (Fig. 11). A second proposal involved toxin damage to the venous intima resulting in thrombus formation at the site of damage.

D. Diagnosis

Salmonellosis should be suspected in hamsters whenever there are unexpected deaths along with necrotic lesions in the liver. Because diarrhea is observed as an inconsistent sign, it should not be relied on when including salmonellosis as part of the differential diagnosis. Other diseases which have to be differentiated include Tyzzer's disease, tularemia, pseudotuberculosis, pasteurellosis, and streptococcosis. A definitive

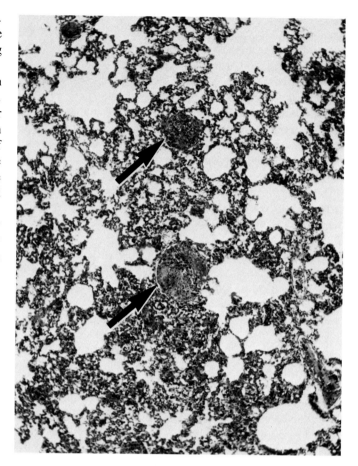

Fig. 11. Thrombi (arrows) occluding pulmonary vessels in a hamster with salmonellosis. Hematoxylin–eosin stain. Magnification: ×100. Courtesy of Dr. R. M. Bunte, Armed Forces Institute of Pathology.

diagnosis of *Salmonella* infection can be easily made because the organism can be readily cultured from many parenchymal organs.

E. Control and Prevention

The best way to manage a *Salmonella* outbreak in hamsters is through preventive measures rather than treatment. Antibiotics and drug treatments have no lasting benefit (Short and Woodnott, 1969). Eradication of the disease from a colony, once it is established, is best done by depopulation, sanitizing all equipment and the environment, and restocking with unaffected hamsters. To prevent the introduction of *Salmonella* into an existing colony, incoming hamsters should be quarantined and their feces cultured. Species separation should also be practiced because other species such as mice, rats, and guinea pigs may be carriers and could serve as a source for infection.

Good sanitation practices and the use of commercially prepared diets and bedding will reduce the chances of introducing salmonellosis. Bacteriological screening of food, bedding, and water can be useful in detection of contaminants. In screening food samples for *Salmonella* contamination, Habermann (1959) found only 1 in 375 non-heat-treated samples positive. With the routine heat treatment of all commercially prepared pelleted rodent diets, this rate of occurrence would be expected to be lower. Stress, such as overcrowding or feeding an inadequate diet, can change unrecognized salmonellosis in a colony to an epizootic with high mortality (Soave, 1963).

F. Significance

Salmonellosis is transmissible to many species including humans. In India, salmonellosis in laboratory animals was described as a "serious public health hazard" (Ray and Mallick, 1970). The disease can be latent or produce high morbidity and high mortality in hamster colonies. Innes *et al.* (1956) stated that the disease "flared up with startling rapidity."

The hamster has been proposed as an animal model to study the association of salmonellosis and schistosomiasis in humans (Mikhail *et al.*, 1981, 1982). *Schistosoma mansoni* infections could enhance and prolong the growth of *Salmonella paratyphi* A in hamsters. It was shown that the worm itself carries *Salmonella*, and, therefore, can facilitate the establishment and growth of *Salmonella*.

VI. CAMPYLOBACTERIOSIS

A. History, Etiology, and Epizootiology

Campylobacter jejuni has been recognized as a major cause of diarrhea in humans and has been isolated from a wide range of domestic and wild animals (Fox, 1982). There appears to be a high incidence of infection in hamsters from the United States, up to 100% of hamsters tested in some colonies (Table I).

Campylobacter jejuni is a motile, slender, curved, gram-negative rod which requires a microaerophilic environment (5–10% carbon dioxide). Organisms produce pinpoint to spreading colonies after 24–48 hr incubation at 42°C. The organism can be best isolated with the use of selective media or after filtration of infected material through a 0.65-μm filter. Isolates are oxidase and catalase positive, sensitive to nalidixic acid, grow in 1% glycine, produce hydrogen sulfide on acetate disks, and fail to grow at 25°C or in 3% sodium chloride.

Route of transmission from animal to animal is by the fecal–oral route. Food and water contamination may also be a factor in the spread of the organism (Fox, 1982).

B. Clinical Signs and Pathogenesis

Most hamsters infected with *Campylobacter jejuni* are asymptomatic. Signs of diarrhea, however, have been observed in a small percentage of animals (Fox *et al.*, 1983). The pathogenesis of *Campylobacter*-caused diarrhea in any species is unknown (Fox, 1982). Clinical signs of infected humans, including bloody diarrhea, suggest that the organism is invasive. The mechanisms involved in infection, which causes some individuals to be asymptomatic shedders of organisms while others show clinical signs, are unknown (Fox, 1982).

Attempts at production of disease in hamsters by oral inoculation of *Campylobacter jejuni* have been reported. Suckling hamsters (4 days old) were resistant to colonization of a strain isolated from humans with diarrhea (Field *et al.*, 1981). Humphrey *et al.* (1985) produced enteritis and cecitis in adult hamsters inoculated directly into the cecum with a *Campylobacter* organism isolated from a human. Lesions were also produced in a group inoculated orally after intestinal purgation with magnesium sulfate and pretreatment with cimetidine hydrochloride and sodium bicarbonate. Intestinal lesions consisted of focal edema to diffuse hyperemia with inflammation and mucosal cell hyperplasia. *Campylobacter*-like bacteria were observed within intestinal epithelial cells by electron microscopy. In comparing the experimental disease produced by Humphrey *et al.* (1985) with proliferative ileitis (see Section II), hyperplasia of the former was not as marked. Although the *Campylobacter* organisms were similar morphologically, they were not as numerous within the cell in the disease produced by Humphrey *et al.* (1985) as they are in proliferative ileitis.

Other experiments inoculating organisms from human infections into weanling hamsters have been successful in establishing intestinal colonization, detectable 16–22 weeks postinoculation (Fox *et al.*, 1982b). Enteritis in these culture-positive animals, however, was not a significant finding. Inoculation of *Campylobacter jejuni* isolated from hamsters with proliferative ileitis have also been attempted (Lentsch *et al.*, 1982; La Regina and Lonigro, 1982). Colonization of the intestinal tract was established in all inoculated hamsters, mild enteritis being noted in a few (Lentsch *et al.*, 1982; La Regina and Lonigro, 1982). The possible role of *Campylobacter* in the pathogenesis of proliferative ileitis of hamsters is discussed elsewhere (Section II).

C. Treatment

Orally administered erythromycin is the drug of choice for treatment of humans with *Campylobacter* diarrhea. The organism *in vitro* is sensitive to erythromycin, aminoglycosides, tetracycline, chloromycetin, furazolidone, and clindamycin. Resistance has been demonstrated for penicillin, polymyxin B, cephalosporins, trimethoprim, and vancomycin.

Table I
Incidence of *Campylobacter jejuni* in Hamsters

Country and source	Incidence (%)		Comments	Reference
England, laboratory	0/10		No signs	Fernie and Park, (1977)
United States, commercial	18/27	(66)	No signs	Fox *et al.* (1981)
United States, commercial	38/44	(86)	No signs	
United States, commercial	12/12	(100)	No signs	
United States, commercial	0/11		No signs	
United States, commercial	93/126	(74)	Majority asymptomatic	Ackerman *et al.* (1982)
United States, commercial	74/75	(99)	Proliferative ileitis	Lentsch *et al.* (1982)
United States, commercial	53/57	(93)	Proliferative ileitis	
United States, commercial	0/40		No signs	
United States, commercial	0/10		Weanling—history of proliferative ileitis	La Regina *et al.* (1980)
United States, commercial	9/10	(90)	Adult—history of proliferative ileitis	
United States, commercial	0/10		Weanling	
United States, commercial	0/10		Adult	
United States, commercial	0/10		Weanling	
United States, commercial	0/10		Adult	

D. Significance

Because of the relatively high incidence of *Campylobacter* carriers, hamsters should be considered as a potential source for human infections. Although biochemical reactions are identical between human and hamster isolates, different serotypes and biotypes of the organism may be an important factor in the spread of disease from one species to another (Fox, 1982). The hamster has been proposed as an animal model for studying the pathogenesis of humans with campylobacteriosis (Humphrey *et al.*, 1985). The experimental disease results in diarrhea and intestinal lesions which are similar to the findings of humans with the disease.

VII. TULAREMIA

A. Etiology and Epizootiology

There has been only one reported outbreak of tularemia in hamsters (Perman and Bergeland, 1967). This highly fatal disease occurred in a closed hamster breeding colony over a 1-month period.

The particular *Fancisella tularensis* strain isolated in this outbreak was shown to be moderately virulent by standard virulence titration tests. Bacterial growth was barely visible on 5% sheep or bovine blood agar after 48 hr, and small colonies were visible after 96 hr incubation. Colony growth was enhanced by subculture on Seller's medium with 50% glucose and glucose cysteine blood agar.

Tularemia can affect many species of animals, including all wild rodents and lagomorphs (Wallach and Boever, 1983). The organism is readily transmissible among susceptible animals. In this outbreak, hamsters 4–6 weeks of age in different areas of the room were affected; however, breeding animals were unaffected.

B. Clinical Signs and Pathology

The tularemia outbreak described was characterized by sudden onset. In the disease's course of less than 48 hr, hamsters were found dead or moribund. The only signs noted on the day prior to death were "huddled" behavior and roughened hair coats.

Pathological lesions were uniform between hamsters. Grossly, lungs were mottled by subpleural petechial and ecchymotic hemorrhages. Livers were enlarged and pale, spleens were enlarged with miliary white foci, and lymph nodes were enlarged and mottled. Peyer's patches within the intestine were prominent and chalky white.

Histopathologically, the most consistent lesion was necrosis of lymphoreticular tissue. Blood vessels within organs, phagocytes, and necrotic lymphoreticular tissue contained large numbers of small gram-negative bacteria.

C. Diagnosis

Tularemia should be considered along with other diagnoses such as salmonellosis, Tyzzer's disease, and proliferative ileitis, in any disease outbreak which causes mortality. A presumptive diagnosis can be made from pathological lesions and a definitive diagnosis from bacteriological cultures.

Diagnosis of tularemia in hamsters can be assisted by impression smears which reveal small, gram-negative coccobacilli both within macrophages and free in cellular debris. Bacterial isolates can be inoculated into laboratory animals (mice, guinea pigs, and hamsters) orally or intraperitoneally. Death occurs in these species 2–5 days postinoculation.

D. Control, Prevention, and Significance

The source of infection in the tularemia outbreak reported was not definitely identified. Hamsters in this colony were given a diet supplemented with carrots, apples, and lettuce. When this supplement was eliminated, no additional cases of tularemia had been reported for over 2 years. Another possible source of introduction of tularemia into the colony could have been contamination of bedding by wild animals.

A tularemia outbreak, although very uncommon, would be very significant because the organism is infectious to humans. Also, the mortality rate of all affected hamsters in the only reported outbreak was 100%.

VIII. PSEUDOTUBERCULOSIS

Infections of *Yersinia pseudotuberculosis* in golden hamsters are very rare and are not recognized in well-managed colonies. The chronic form of the disease produces lesions which resemble tubercular lesions (Handler, 1965). The organism primarily affects the digestive system. Infections progress slower in hamsters than in other species (Bjotvedt and Tufts, 1963). Clinical signs include intermittent diarrhea and cachexia prior to death. Small necrotic nodules can occur in the intestines, mesenteric lymph nodes, liver, spleen, and lungs, from which the organism can be cultured. Introduction of the organism into a colony is usually attributed to contamination of food or water by the feces of wild rodents or birds (Gleiser *et al.*, 1974). Sources and husbandry practices concerning food and water should, therefore, be critically reviewed. Control of pseudotuberculosis is best accomplished by elimination of diseased animals, sanitation of all equipment, and repopulation with unaffected hamsters (Bjotvedt and Tufts, 1963).

IX. STREPTOCOCCAL INFECTIONS

A β-hemolytic *Streptococcus* was isolated from two female hamsters with acute bacterial mastitis (Frisk *et al.*, 1976). The infection occurred 7–10 days after parturition. Litters were noted to be runted until 1 week of age, when they were cannibalized by the female. Mastitis in one of these females involved the posterior four mammary glands. Affected glandular tissue was hemorrhagic and firm, and contained suppurative exudate. Myriads of polymorphonuclear inflammatory cells had infiltrated glandular parenchyma. The frequency of occurrence of streptococcal mastitis in this colony was very low. Only two cases were observed in a colony of 350 breeding females over a 3-year period.

Streptococci can also cause pneumonia in hamsters. Renshaw *et al.* (1975), in a survey of naturally occurring diseases of hamsters, reported that diplococci (*Streptococcus pneumoniae*) were the cause of pneumonia in hamsters from 1 of 14 laboratories. Diplococci and *Streptococcus* sp. have also been reported to cause pneumonia and cervical lymphadenitis in hamsters (Wescott, 1969).

Streptococcal pneumonias are relatively uncommon. When they occur, they are frequently associated with stress. The route of infection appears to be through aerosols. Clinical signs include depression, anorexia, nasal and ocular discharge, dehydration, and weight loss. The course of the disease is short (3 days). Diagnosis is made by clinical signs, lesions, and culture. Treatment can be attempted with antibiotics that are specifically effective against the organism.

Streptococcus agalactiae has been reported to cause acute pleuropneumonia and septicemia in hamsters (Kummeneje *et al.*, 1975). *Streptococcus agalactiae* was isolated from the lungs, liver, and spleen of moribund hamsters. The organism was β hemolytic and cAMP positive, thus being classified in Lancefield's group B. It was suspected that the organism was transmitted to the hamsters from a human carrier.

X. *PASTEURELLA* INFECTIONS

Pasteurella pneumotropica can occasionally cause pneumonia in hamsters. The organism can cause acute or chronic respiratory infections or be present in carrier hamsters that do not show signs (Gleiser, 1974; Harkness and Wagner, 1977). Stress may be associated with the precipitation of clinical disease.

Pasteurella pneumotropica is a small, gram-negative, nonmotile bipolar rod. The organism is lactose, urease, indole, and oxidase positive (Brennan *et al.*, 1969). The organism is very pneumotropic, especially for mice (Jawetz, 1948). Hamsters experimentally inoculated with *P. pneumotropica* by intranasal, intravenous, intraperitoneal, and subcutaneous routes could not be infected, nor were lesions produced (Jawetz, 1948, 1950). Pneumonia, cutaneous abscessation (Section XIII), conjunctivitis, and otitis interna, however, have been reported in natural infections (Harkness and Wagner, 1977). Clinical signs associated with pneumonia may include labored breathing, nasal exudate, and weight loss. Pneumonic lesions are characterized by red foci of consolidation.

Systemic antibiotic therapy may be instituted to treat individual hamsters. Chloromycetin and ampicillin are usually found to be effective. The organism is usually resistant to tetracycline and streptomycin. In a colony, the most effective control measure is elimination of carrier animals. New animals should be isolated and cultured before introduction into an existing population.

XI. BACTERIAL RESPIRATORY INFECTIONS

Bacterial respiratory infections of hamsters do occur, but they have not been extensively studied experimentally. Pneumonia of hamsters was recognized by 43% (6 of 14) of laboratories surveyed in the United States and was listed as the second most common clinical condition (Renshaw *et al.*, 1975). In a survey of clinical conditions of hamsters in Germany, respiratory infections were noted in 8% (10 of 121) of cases (Lindt, 1958). In both of these surveys, viruses as well as bacteria could account for some of the respiratory infections reported. Certainly, bacterial respiratory infections occur with much less frequency in golden hamsters than in mice, rats, or guinea pigs. Because of this reduced incidence, the hamster has been proposed as an animal model for respiratory pathology studies (Kleinerman, 1972).

Bacterial respiratory infections, including pneumonia, conjunctivitis, otitis media, otitis interna, and encephalitis, can be caused by several organisms, including *Pasteurella pneumotropica* (Harkness and Wagner, 1977; Gleiser, 1974; Silverman and Chavannes, 1983; Brennan *et al.*, 1965), *Pasteurella* sp. (Castleman, 1983), *Streptococcus pneumoniae* (Renshaw *et al.*, 1975; Wescott, 1969), *Streptococcus agalactiae* (Kummeneje *et al.*, 1975), *Streptococcus* sp. (Wescott, 1969), and *Salmonella* sp. (Innes *et al.*, 1956). Hamsters may also carry organisms in their respiratory tract which are believed to be potentially pathogenic without showing clinical signs. Such organisms include *Pasteurella pneumotropica, Pasteurella multocida* (Orcutt, 1980), *Streptococcus pneumoniae, Staphylococcus aureus* (Castleman, 1983), *Klebsiella pneumoniae* (Castleman, 1983), *Bordetella* sp. (Castleman, 1983), and *Salmonella* sp.

Mycoplasma pulmonis has been only rarely isolated from hamsters. In a study that exposed hamsters to aerosolized graphite, *M. pulmonis* was isolated from the respiratory system from one of the test groups (Battigelli *et al.*, 1971). The significance of the organism's presence in this particular group was not established, nor were lesions attributed to it. *Mycoplasma pulmonis* certainly does not appear to be pathogenic for hamsters. Also, the number of hamsters which carry the organism is minuscule.

Several other outbreaks of apparent bacterial pneumonias have been described. Kleinerman (1972) noted that some hamsters succumbed to an acute bacterial pneumonitis characterized by focal areas of pneumonia to consolidation of an entire lobe of lung. Chronic pneumonia was rarely seen. Hagen *et al.* (1965) reported that many hamsters showed evidence of pneumonitis, although they appeared clinically in good health. Upon arrival from the supplier, 50% had small plaques scattered throughout lung parenchyma. Lesions were classified histologically as patchy areas of interstitial bronchopneumonia, areas of atelectasis, and areas of compensatory emphysema. Cultures from affected lungs were not reported; therefore, the exact etiology of these lesions is unknown.

Additional discussion of respiratory infections, including specific organisms and the disease which they produce, are found in the sections on *Pasteurella* infections (Section X), streptococcal infections (Section IX), and salmonellosis (Section V).

XII. ACTINOMYCOSIS

A tumorlike growth in the submaxillary salivary gland of a hamster was attributed to *Actinomyces bovis* infection (Gorlin and Chaudhry, 1959). The salivary gland duct in this animal had been ligated 3 days previously. Grossly, the enlargement was 1.5 cm in diameter and revealed several foci of liquefaction necrosis on cross section. Microscopically, the nodule contained suppurative exudate with demonstrable sulfur granules that appeared as raylike fungi. The organism was identified as *A. bovis* on the basis of its gram-positive staining with the Brown–Brenn stain.

The origin of the infection in this case was unknown. Experimentally, 75% of hamsters inoculated intraperitoneally with *A. bovis* from human cases developed clinical infection (Hazen *et al.*, 1952). Experimental infection by this route resulted in abscess formation within the abdominal cavity and skin lesions in some animals. The occurrence of *A. bovis* infection in hamsters appears to be extremely rare. Infection with this organism should be considered whenever an abscess is found.

XIII. CUTANEOUS BACTERIAL ABSCESSES

Skin abscesses can occur in hamsters at any location, but are most prevalent near the head, sometimes involving the cheek pouch (Burke, 1979). The most commonly reported organism isolated from skin abscesses is *Staphylococcus aureus*. Other organisms associated with cutaneous abscesses are *Streptococcus* sp. (Wescott, 1969), *Actinomyces bovis* (Gorlin and Chaudhry, 1959), and *Pasteurella pneumotropica* (Harkness and Wagner, 1977; McKenna *et al.* 1970) (Sections X, XII, XIV).

Abscesses can contain a variable amount of purulent exudate. They can be associated with superficial injuries caused by poor caging or fighting (Gleiser, 1974; Slater and Musser, 1984). The oral route of infection may predominate when lesions involve the submandibular and cervical lymph nodes (Wescott, 1969). The most predominant clinical sign is a swelling which may rupture and cause matting of the fur. Cervical lymphadenitis takes weeks to develop and may become severe enough to cause anorexia and death (Wescott, 1969).

Exudate from abscesses may be surgically drained and the cavity flushed with iodine solution, hydrogen peroxide, or tincture of merthiolate. Systemic antibiotic therapy may also be employed. Caging should be examined for sharp edges or protruding wires, and affected hamsters should be isolated from others to prevent fighting.

XIV. POSTIRRADIATION INFECTIONS

Bacterial infections may occur in hamsters after whole-body irradiation. Smith *et al.* (1955) irradiated groups of hamsters with a range of 760–10,000 R. These animals had 67–90% positive heart blood cultures, up to 28 days following exposure. *Pasteurella, Streptococcus,* or both were the species of organisms most frequently encountered. Infection with *Pseudomonas* caused the shortest survival time postexposure, 6 days as compared to 15 days for uninfected hamsters after exposure to 720–760 R. Most hamsters showed evidence of bacterial invasion on the fifth day following irradiation. The site of initial invasion is the nasopharynx, with rapid progression to salivary glands.

Hamsters showed rapid deterioration 4–5 days after being exposed to between 700 and 2000 R (Mayo *et al.*, 1962). Clinical signs in these animals were diarrhea, weight loss, dehydration, and edema of the face. Bacterial infections were possibly the cause of frequently observed suppurative blepharitis and orbital abscesses. Necrotic lesions were also noted in the oral cavity.

McKenna *et al.* (1970) reported isolation of *Pasturella pneumotropica* from abscesses of hamsters that received 500–5000 R. These abscesses appeared 2–3 weeks following exposure in only 15 of 582 (2.6%) of irradiated animals. The most remarkable feature of these abscesses was the consistency of their location, being midway between the eye and ear.

There is a possibility of bacterial infections altering results of experiments utilizing hamsters in whole-body irradiation studies. The significance of these infections may depend on the animal's bacterial microflora, especially that of the upper respiratory tract. It would, therefore, be in the investigator's best interest to screen the nasopharynx of a percentage of hamsters for *Pasturella pneumotropica, Pseudomonas* sp., and β-hemolytic *Streptococcus* prior to radiation exposure.

XV. MISCELLANEOUS BACTERIAL INFECTIONS

Several diseases have been mentioned that have a bacteria associated with them; however, specific information on the disease conditions is sparse. An "unidentified coryneform organism" has been associated with vaginitis (Sebesteny, 1979), and kidney lesions may be due to *Proteus vulgaris, Proteus mirabilis,* and *E. coli* (Sebesteny, 1979). Among other previously mentioned bacteria, *Pseudomonas aeruginosa* has been listed as a cause of fatal generalized septicemia (Sebesteny, 1979). *Staphylococcus aureus* may cause ulcerative dermatitis (Gleiser *et al.*, 1974) as well as cutaneous abscesses (Section XIII).

XVI. MYCOTIC INFECTIONS

Although the golden hamster has been used as an animal model for fungal infection (see Chapter 14), spontaneously occurring fungal diseases are extremely rare.

Ringworm caused by *Trichophyton mentagrophytes* has occurred in the Djungarian hamster (see Chapter 21), but cases in golden hamsters are not well documented. Sebesteny (1979) includes *Trichophyton* and *Microsporum* as causes of skin lesions in hamsters. He states that infections with these organisms may cause dry scaly lesions, encrustations with broken hair, or no clinical signs. Diagnosis can be made by culture, examination of hair shafts cleared with 10% potassium hydroxide, or in the case of *Microsporum,* the examination of affected areas with a Wood's lamp. Topical fungicides and oral grieseofulvin may be used to treat ringworm. Sebesteny (1979) reported that treatment with antifungal agents was unreliable, but he did not describe the incidence of infection or present individual cases. Ringworm infections are transmissible to humans. A case of *Trichophyton mentagrophytes* infection occurred on the arm of an employee working with hamsters in Roumania (Alteras, 1966). Although hamsters were reported as the source of this human infection, cultures from suspected carrier hamsters were not recorded.

Cryptococcal infection occurred concurrently with tuberculosis in a group of hamsters that received cortisone and a tumor transplant from a human with tuberculosis (Chesterman, 1972). Multiple nodules containing mycobacteria and fungi "morphologically indistinguishable" from *Cryptococcus neoformans* were observed. These nodules occurred in the peritoneum, spleen, mesentery, and lungs 4 months after the tumor transplant.

REFERENCES

Ackerman, J. I., Newcomer, C. E, and Fox, J. G. (1982). Intestinal carriage of *Campylobacter fetus* subsp. *jejuni* in laboratory animals. *Lab. Anim. Sci.* **32,** 442.

Alteras, I. (1966). Human dermatophyte infections from laboratory animals. *Sabouraudia* **4,** 143–145.

Amend, N. K., Loeffler, D. G., Ward, B. C., and Van Hoosier, G. L. (1976). Transmission of enteritis in the Syrian hamster. *Lab. Anim. Sci.* **26,** 566–572.

Andrews, E. J. (1975). Alterations of selected intestinal enzymes in hamsters with hamster enteritis syndrome. *Am. J. Vet. Res.* **36,** 889–891.

Barthold, S. W., Jacoby, R. O., and Pucak, G. J. (1978). An outbreak of cecal mucosal hyperplasia in hamsters. *Lab. Anim. Sci.* **28,** 723–727.

Battigelli, M. C., Frasen, D. A., and Cole, H. (1971). Microflora of the respiratory surface of rodents exposed to "inert" particulates. *Arch. Intern. Med.* **127,** 1103–1104.

Bjotvedt, G., and Tufts, J. (1963). Laboratory animal disease and its control. In "Manual for Laboratory Animal Care," p. 21. Ralston Purina Co., St. Louis, Missouri.

Boothe, A. D., and Cheville, N. F. (1967). The pathology of proliferative ileitis of the golden Syrian hamster. *Pathol. Vet.* **4**, 31–44.

Brennan, P. C., Fritz, T. E., and Flynn, R. J. (1965). *Pasteurella pneumotropica*: Cultural and biochemical characteristics, and its association with disease in laboratory animals. *Lab. Anim. Care* **15**, 307–312.

Brennan, P. C., Fritz, T. E., and Flynn, R. J. (1969). Murine pneumonia: A review of the etiologic agents. *Lab. Anim. Care* **19**, 360–371.

Brush, M. K., McCoy, J. R., Rosenthal, H. L., Stauber, L. A., and Alison, J. B. (1957). The addition of non-ionic surface-active agents of the polyoxyethylene type to the diet of the hamster, the mouse and the dog. *J. Nutr.* **62**, 601–619.

Burke, T. J. (1979). Rats, mice, hamsters and gerbils. *Vet. Clin. North Am.* **9**, 473–486.

Castleman, W. L. (1983). Spontaneous pulmonary infections in animals affecting studies on comparative lung morphology and function. *Am. Rev. Respir. Dis.* **128**, 583–587.

Chang, K., Kurtz, H. J., Ward, G. E., and Gebhart, C. J. (1984). *Campylobacter* microagglutination tests of swine with proliferative enteritis. *Am. J. Vet. Res.* **45**, 1373–1378.

Chesterman, F. C. (1972). Background pathology in a colony of golden hamsters. *Prog. Exp. Tumor Res.* **16**, 50–68.

Cook, M. G., and Turnbull, G. J. (1975). A hypothesis for the pathogenesis of Crohn's disease based on an ultrastructural study. *Virchows Arch. A: Pathol. Anat. Histol.* **365**, 327–336.

Craigie, J. (1966). *Bacillus piliformis* (Tyzzer) and Tyzzer's disease of the laboratory mouse. II. Mouse pathogenicity of *B. piliformis* grown in embryonated eggs. *Proc. R. Soc. London, Ser. B* **165**, 61–77.

Crohn, B. B., Ginzburg, L., and Oppenheimer, G. D. (1932). Regional ileitis. *JAMA, J. Am. Med. Assoc.* **99**, 1323–1329.

Decker, R. H., and Henderson, L. M. (1959). Hydroxyanthranilic acid as a source of niacin in the diets of the chick, guinea pig, and hamster. *J. Nutr.* **68**, 17–24.

Ershoff, B. A. (1956). Beneficial effects of alfalfa, aureomycin and cornstarch on the growth and the survival of hamsters fed highly purified ration. *J. Nutr.* **59**, 579–585.

Fernie, D. S., and Park, R. W. A. (1977). The isolation and nature of campylobacters (microaerophilic vibrios) from laboratory and wild rodents. *J. Med. Microbiol.* **10**, 325–329.

Field, L. H., Underwood, J. L., Pope, L. M., and Berry L. J. (1981). Intestinal colonization of neonatal animals by *Campylobacter fetus* subsp. *jejuni. Infect. Immun.* **33**, 884–892.

Fox, J. G. (1982). Campylobacteriosis—a "new" disease in laboratory animals. *Lab. Anim. Sci.* **32**, 625–637.

Fox, J. G., Zanotti, S., and Jordan, H. V. (1981). The hamster as a reservoir of *Campylobacter fetus* subsp. *jejuni. J. Infect. Dis.* **143**, 856.

Fox, J. G., Murphy, J. C., Ackerman, J. I., Prostak, K. S., Gallagher, C. A., and Rambow, V. J. (1982a). Proliferative colitis in ferrets. *Am. J. Vet. Res.* **43**, 858–864.

Fox, J. G., Zanotti, S., and Jordan, H. V. (1982b). Implantation and distribution of *Campylobacter fetus* subsp. *jejuni* in hamsters. *Lab. Anim. Sci.* **32**, 442.

Fox, J. G., Hering, A. M., Ackerman, J. I., and Taylor, N. S. (1983). The pet hamster as a potential reservoir of human campylobacteriosis. *J. Infect. Dis.* **147**, 784.

Frenkel, J. K. (1972). Infection and immunity in hamsters. *Prog. Exp. Tumor Res.* **16**, 326–367.

Friedman, M. H. (1965). "Wet-tail disease" of hamsters. *Lab. Anim. Dig.* **1**, 18–19.

Frisk, C. S. (1976). The etiology and pathogenesis of hamsters enteritis. Ph.D. Thesis, University of Missouri, Columbia.

Frisk, C. S., and Wagner, J. E. (1977a). Hamster enteritis: A review. *Lab. Anim.* **11**, 79–85.

Frisk, C. S., and Wagner, J. E. (1977b). Experimental hamster enteritis: An electron microscopic study. *Am. J. Vet. Res.* **38**, 1861–1868.

Frisk, C. S., Wagner, J. E., and Owens, D. R. (1976). Streptococcal mastitis in golden hamsters. *Lab. Anim. Sci.* **26**, 97.

Frisk, C. S., Wagner, J. E., and Ownes, D. R. (1978). Enteropathogenicity of *Escherichia coli* isolated from hamsters (*Mesocricetus auratus*) with hamster enteritis. *Infect. Immun.* **20**, 319–320.

Frisk, C. S., Wagner, J. E., and Owens, D. R. (1981). Hamster (*Mesocricetus auratus*) enteritis caused by epithelial cell-invasive *Escherichia coli. Infect. Immun.* **31**, 1232–1238.

Fujiwara, K., Takasaki, Y., Kubokawa, K., Takenaka, S., Kubo, M., and Sato, K. (1974). Pathogenic and antigenic properties of the Tyzzer's organisms from feline and hamster cases. *Jpn. J. Exp. Med.* **44**, 365–372.

Ganaway, J. R., Allen, A. M., and Moore, T. D. (1971). Tyzzer's disease. *Am. J. Pathol.* **64**, 717–732.

Gleiser, C. A. (1974). Diseases of laboratory animals—bacterial. *In* "Handbook of Laboratory Animal Science" (E. C. Melby, Jr. and N. H. Altman, eds.), Vol. 2, pp. 271–285. CRC Press, Cleveland, Ohio.

Gleiser, C. A., Andrews, E. J., Pick, J. R., Small, J. D., Weisbroth, S. H., Wescott, R. B., and Wilsnack, R. E. (1974). "A Guide to Infectious Diseases of Guinea Pigs, Gerbils, Hamsters and Rabbits." Natl. Acad. Sci., Washington, D.C.

Goldman, P. M., Andrews, E. J., and Lang, C. M. (1972). A preliminary evaluation of *Clostridium* sp. in the etiology of hamster enteritis. *Lab. Anim. Sci.* **22**, 721–724.

Gorlin, R. J., and Chaudhry, A. P. (1959). A note on the occurrence of *Actinomyces bovis* in the hamster. *J. Dent. Res.* **38**, 842.

Habermann, R. T. (1959). Spontaneous diseases and their control in laboratory animals. *Public Health Rep.* **74**, 165–169.

Habermann, R. T., and Williams, F. P. (1958). Salmonellosis in laboratory animals. *J. Natl. Cancer Inst. (U.S.)* **20**, 933–948.

Hagen, C. A., Shefner, A. M., and Ehrlich, R. (1965). Intestinal microflora of normal hamsters. *Lab. Anim. Care* **15**, 185–193.

Handler, A. H. (1965). Spontaneous lesions of the hamster. *In* "The Pathology of Laboratory Animals" (W. E. Ribelin and J. R. McCoy, eds.), pp. 210–240. Thomas, Springfield, Illinois.

Harkness, J. E., and Wagner, J. E. (1977). "The Biology and Medicine of Rabbits and Rodents." Lea & Febiger, Philadelphia, Pennsylvania.

Hazen, E. L., Little, G. N., and Resnick, B. S. (1952). The hamster as a vehicle for the demonstration of pathogenicity of *Actinomyces bovis. J. Lab. Clin. Med.* **40**, 914–918.

Humphrey, C. D., Montag, D. M., and Pittman, F. E. (1985). Experimental infection of hamsters with *Campylobacter fetus* subsp. *jejuni. J. Infect. Dis.* **151**, 485–493.

Innes, J. R. M., Wilson, C., and Ross, M. A. (1956). Epizootic *Salmonella enteritidis* infection causing septic pulmonary phlebothrombosis. *J. Infect. Dis.* **98**, 133–141.

Jackson, S. J., and Wagner, J. E. (1970). Proliferative ileitis in Syrian hamsters (*Mesocricetus auratus*). *Lab. Anim. Dig.* **6**, 12–15.

Jacoby, R. O. (1978). Transmissible ileal hyperplasia of hamsters I. Histogenesis and immunocytochemistry. *Am. J. Pathol.* **91**, 433–450.

Jacoby, R. O., and Johnson, E. A. (1981). Transmissible ileal hyperplasia. *Adv. Exp. Med. Biol.* **134**, 267–289.

Jacoby, R. O., Osbaldiston, G. W., and Jonas, A. M. (1975). Experimental transmission of atypical ileal hyperplasia of hamsters. *Lab. Anim. Sci.* **25**, 465–473.

Jawetz, E. (1948). A latent pneumotropic *Pasteurella* of laboratory animals. *Proc. Soc. Exp. Biol. Med.* **68**, 46–48.

Jawetz, E. (1950). A pneumotropic *Pasteurella* of laboratory animals. I. Bacteriological and serological characteristics of the organism. *J. Infect. Dis.* **86**, 172–183.

Johnson, E. A., and Jacoby, R. O. (1978). Transmissible ileal hyperplasia of hamsters. II. Ultrastructure. *Am. J. Pathol.* **91**, 451–468.

Jonas, A. M., Tomita, Y., and Wyand, D. S. (1965). Enzootic intestinal adenocarcinoma in hamsters. *J. Am. Vet. Med. Assoc.* **147**, 1102–1108.

7. BACTERIAL AND MYCOTIC DISEASES

Kim, J. C. S., and Jourden, M. (1977). Ultrastructure of proliferative ileitis in hamsters (*Mesocricetus auratus*). *Lab. Anim.* **11,** 171–174.

Kleinerman, J. (1972). Some aspects of pulmonary pathology in the Syrian hamster. *Prog. Exp. Tumor Res.* **16,** 287–299.

Kummeneje, K., Nesbakken, T., and Mikkelsen, T. (1975). *Streptococcus agalactiae* infection in a hamster. *Acta Vet. Scand.* **16,** 554–556.

La Regina, M., and Lonigro, J. (1982). Isolation of *Campylobacter fetus* subsq. *jejuni* from hamsters with proliferative ileitis. *Lab. Anim. Sci.* **32,** 660–662.

La Regina, M., Fales, W. H., and Wagner, J. E. (1980). Effects of antibiotic treatment on the occurrence of experimentally induced proliferative ileitis of hamsters. *Lab. Anim. Sci.* **30,** 38–41.

Lentsch, R. H., McLaughlin, R. M., Wagner, J. E., and Day, T. J. (1982). *Campylobacter fetus* subsp. *jejuni* isolated from Syrian hamsters with proliferative ileitis. *Lab. Anim. Sci.* **32,** 511–514.

Lindt, V. S. (1958). Über Krankheiten des syrischen Goldhamsters (*Mesocricetus auratus*). *Schweiz, Arch. Tierheilkd.* **100,** 86–97.

Lomax, L. G., and Glock, R. D. (1982). Naturally occurring porcine proliferative enteritis: Pathologic and bacteriologic findings. *Am. J. Vet. Res.* **43,** 1608–1614.

Lussier, G., and Pavilanis, V. (1969). Presence of intranuclear inclusion bodies in proliferative ileitis of the hamster (*Mesocricetus auratus*), a preliminary report. *Lab. Anim. Care* **19,** 387–390.

McKenna, J. M., South, F. E., and Musacchia, X. J. (1970). *Pasteurella* infection in irradiated hamsters. *Lab. Anim. Care* **20,** 443–446.

Mayberry, J. F., Rhodes, J. and Heatley, R. V. (1980). Infections which cause ileocolic disease in animals: Are they relevant to Crohn's disease? *Gastroenterology* **78,** 1080–1084.

Mayo, J. Carranza, F. A., Epper, C. E., and Cabrini, R. L. (1962). The effect of total body irradiation on the oral tissues of the Syrian hamster. *Oral Surg., Oral Med. Oral. Pathol.* **15,** 739–745.

Mikhail, I. A., Higashi, G. I., Mansour, N. S., Edman, D. C., and Elwan, S. H. (1981). *Salmonella paratyphi* A in hamsters concurrently injected with *Schistosoma mansoni*. *Am. J. Trop. Med. Hyg.* **30,** 385–393.

Mikhail, I. A., Higashi, G. I., Edman, D. C., and Elwan, S. H. (1982). Interaction of *Salmonella paratyphi* A and *Schistosoma mansoni* in hamsters. *Am. J. Trop. Med. Hyg.* **31,** 328–334.

Nakayama, M., Saegusa, J., Itoh, K., Kiuchi, Y., Tamura, T., Ueda, K., and Fujiwara, K. (1975). Transmissible enterocolitis in hamsters caused by Tyzzer's organism. *Jpn. J. Exp. Med.* **45,** 33–41.

Nakayama, M., Machii, K., Goto, Y., and Fujiwara, K. (1976). Typhlohepatitis in hamsters infected perorally with the Tyzzer's organism. *Jpn. J. Exp. Med.* **46,** 309–324.

Orcutt, R. P. (1980). Bacterial diseases: Agents, pathology, diagnosis and effects on research. *Lab Anim.* **9,** 28–43.

Perman, V., and Bergeland, M. E. (1967). A tularemia enzootic in a closed hamster breeding colony. *Lab. Anim. Care* **17,** 563–568.

Pollock, W. B. (1975). Prolapse of invaginated colon through the anus in golden hamsters (*Mesocricetus auratus*). *Lab. Anim. Sci.* **25,** 334–336.

Ray, J. P., and Mallick, B. B. (1970). Public health significance of *Salmonella* infections in laboratory animals. *Indian Vet. J.* **47,** 1033–1037.

Renshaw, H. W., Van Hoosier, G. L., and Amend, N. K. (1975). A survey of naturally occurring diseases of the Syrian hamster. *Lab. Anim.* **9,** 179–191.

Rights, F. L., Jackson, E. B., and Smadel, J. E. (1947). Observations on Tyzzer's disease in mice. *Am. J. Pathol.* **23,** 627–636.

Rowland, A. C., and Lawson, G. H. K. (1975). Intestinal adenomatosis in the pig: A possible relationship with a haemorrhagic enteropathy. *Res. Vet. Sci.* **18,** 263–268.

Sebesteny, A. (1979). Syrian hamsters. *In* "Handbook of Diseases of Laboratory Animals" (J. M. Hime and P. N. O'Donoghue, eds.), pp. 111–113. Heinemann Veterinary Books, London.

Sheffield, F. W., and Beveridge, E. (1962). Prophylaxis of "wet-tail" in hamsters. *Nature (London)* **196,** 294–295.

Short, D. J., and Woodnott, D. P. (1969). "The I. A. T. Manual of Laboratory Animal Practice and Techniques." Thomas, Springfield, Illinois.

Silverman, J., and Chavannes, J. (1983). What's your diagnosis? Balding. *Lab Anim.* **12,** 14–16.

Slater, J. G., and Musser, T. K. (1984). The hamster. *In* "Manual for Assistant Laboratory Animal Technicians" (W. B. Sapanski and J. E. Harkness, eds.), pp. 184–197. Am. Assoc. Lab. Anim. Sci., Joliet, Illinois.

Smith, W. W., Marston, R. Q., Gonshery, L., Alderman, I. M., and Ruth, H. J. (1955). X-Irradiation in hamsters, and effects of streptomycin and marrow–spleen homogenate treatment. *Am. J. Physiol.* **183,** 98–110.

Soave, O. A. (1963). Diagnosis and control of common diseases of hamsters, rabbits, and monkeys. *J. Am. Vet. Med. Assoc.* **142,** 285–290.

Takagaki, Y., Ito, M., Naiki, M., Fujiwara, K., Okugi, M., Maejima, K., and Tajima, Y. (1966). Experimental Tyzzer's disease in different species of laboratory animals. *Jpn. J. Exp. Med.* **36,** 519–534.

Takasaki, Y., Oghiso, Y., Sato, K., and Fujiwara, K. (1974). Tyzzer's disease in hamsters. *Jpn. J. Exp. Med.* **44,** 267–270.

Thomlinson, J. R. (1975). "Wet-tail" in the Syrian hamster: A form of colibacillosis. *Vet. Rec.* **96,** 42.

Timmons, E. H., and Dickey, K. M. (1981). Tyzzer's disease in Syrian hamsters. *Ann. Sess. Am. Assoc. Lab. Anim. Sci.* **32,** 16.

Tomita, Y., and Jonas, A. M. (1968). Two viral agents isolated from hamsters with a form of regional enteritis: A preliminary report. *Am. J. Vet. Res.* **29,** 445–453.

Trum, B. F., and Routledge, J. K. (1967). Common disease problems in laboratory animals. *J. Am. Vet. Med. Assoc.* **151,** 1886–1896.

Tyzzer, E. E. (1917). A fatal disease of the Japanese waltzing mouse caused by a spore-bearing bacillus (*Bacillus piliformis*, N. SP.). *J. Med. Res.* **37,** 307–338.

Varela, G. (1953). Aislamiento de *Proteus morganii* en hamsters (*Mesocricetus auratus auratus*) con diarrhea. *Medicina (Mexico City)* **33,** 479–480.

Wagner, J. E., Owens, D. R., and Troutt, H. F. (1973). Proliferative ileitis of hamsters: Electron microscopy of bacteria in cells. *Am. J. Vet. Res.* **34,** 249–252.

Wallach, J. D., and Boever, W. J. (1983). "Diseases of Exotic Animals: Medical and Surgical Management." Saunders, Philadelphia, Pennsylvania.

Wescott, R. B. (1969). Diseases of hamsters. *In* "An Outline of Diseases of Laboratory Animals," pp. 61–66. University of Missouri, Columbia.

Wheelock, F. C. (1974). Surgical management of regional ileitis, ulcerative colitis, and granulomatous colitis. *Surg. Clin. North Am.* **54,** 675–688.

White, D. J., and Waldron, M. M. (1969). Naturally-occurring Tyzzer's disease in the gerbil. *Vet. Rec.* **85,** 111–114.

Williams, C. S. F. (1976). Hamster. *In* "Practical Guide to Laboratory Animals," pp. 26–37. Mosby, St. Louis, Missouri.

Zook, B. C., Huang, K., and Rhorer, R. G. (1977). Tyzzer's disease in Syrian hamsters. *J. Am. Vet. Med. Assoc.* **171,** 833–836.

Chapter 8

Parasitic Diseases

Joseph E. Wagner

    I. Introduction 135
   II. Parenteral Parasites 136
       A. Protozoa 136
       B. Nematodes 136
       C. Cestodes 136
 III. Parasites of the Alimentary System 137
       A. Protozoa 137
       B. Nematodes 143
       C. Cestodes 146
 IV. Parasites of the Integument 149
       A. Mites 149
       B. Hexapods 153
       References 153

I. INTRODUCTION

The goal of this chapter is to present an up-to-date review of naturally or spontaneously occurring parasitisms of the laboratory golden Syrian hamster (*Mesocricetus auratus*). Descriptions of naturally occurring parasitisms of hamsters are disproportionally deficient and superficial when one considers the great importance of this rodent in contemporary human health-related research. While hamsters are uniquely susceptible to a wide variety of induced parasitisms, discussion of these experimental infections was not within the purview of this author. (Refer to Chapter 14 for such discussions.)

Saxe (1954a) reported success in establishing protozoa-free breeding colonies of laboratory rats and golden hamsters through treatment of 27- and 28-day-old animals with carbarsone for 5 days (day 1, 1000 mg/kg body weight; days 2, 3, and 4, 300 mg/kg; and day 5, 600 mg/kg). This treatment eliminated all protozoa except *Giardia muris*. Treatment with Atabrine (Winthrop Laboratories, Aurora, Ontario) for 7 days with daily transfer to a sterile cage eliminated *Giardia*. Atabrine dosage on days 1, 2, and 7 was 1.5 mg per hamster and 1.0 mg per hamster on days 3, 4, 5, and 6. Producers desiring pathogen and protozoa-free hamsters may want to consider Saxe's protocol, or variations of it (i.e., use of dimetridazole or ipronidazole) in rederivation attempts, since dozens of attempts at rederivation based on caesarean section have failed. Through judicious use of antibiotics to eliminate bacteria, stop-breeding techniques to eliminate viral infections such as pneumonia virus of mice (PVM) and Sendai viruses, and protozoan eradication coupled with isolator–barrier production methods, it should be possible to derive protozoan- and pathogen-free hamsters.

II. PARENTERAL PARASITES

A. Protozoa

Little is known about apparently infrequent infections of hamsters with *Encephalitozoon cuniculi* (order Microsporidia), a sporozoan. The significance of this agent in its "classical" host, the rabbit, has been reviewed in another monograph of the ACLAM series (Pakes, 1974).

Chalupsky *et al.* (1979) used an indirect fluorescent antibody test to survey sera of 110 hamsters from five Czechoslovakian colonies for antibody to *E. cuniculi*. From 14 to 80% of the animals in four colonies were positive. One colony was negative. Other reports of apparent encephalitozoonosis in hamsters are by Hartmann (1967) and Kinzel and Meiser (1968). Meiser *et al.* (1971) described an *Encephalitozoon* infection in a transplantable plasmacytoma of hamsters. The infection was successfully transmitted to rats, mice, and guinea pigs. Parasites were seen only in tumor cells, not in parenchymatous organs.

There are few if any differences between *Encephalitozoon* organisms from different mammalian species (Shadduck and Pakes, 1971). Rabbit choroid plexus cells support replication of microsporidian isolates of hamster origin (Shadduck, 1969). If a single species of *E. cuniculi* infects different animal species, then laboratory hamsters could acquire infections from other laboratory animal species, especially rabbits, which are commonly infected. In the rabbit, spores of the parasite are excreted in the urine, and transmission is by contact with contaminated material (Frenkel, 1972). Vertical transmission has also been demonstrated.

Although the classification of some of the following agents as protozoa is in doubt, they are listed here because they were originally reported as protozoan infections of hamsters. Baker *et al.* (1971) list *Grahamella cricetuli* as infecting the hamster. Weinman and Kreier (1977), Ristic and Lewis (1977), and Gothe and Kreier (1977), in reviews of *Bartonella, Grahamella, Aegyptianella, Eperythrozoon,* and *Haemobartonella,* listed four species of *Grahamella* in hamsters, none of which were golden Syrian hamsters; *Mesocricetus auratus*. *Grahamella* parasites and hosts listed were *G. criceti domestici* in *Cricetus domesticus, G. cricetuli* in *Cricetulus griseus, G. dudtschenkoi* in *Cricetulus* sp., and *G. ninsehohl-yakimov* in *Cricetus phoca*. Griesemer (1958) reported that bartonellosis is rare in hamsters, with the organisms resembling *Haemobartonella muris*. Ristic and Lewis (1977) list *Babesia cricetuli* as occurring in the Daurain hamster, *Cricetulus furunculus*.

Pneumocytis carinii is a ubiquitous agent which can cause pneumonitis in humans and animals. *Pneumocystis carinii* infection is usually subclinical but can cause significant morbidity and mortality in humans with chronic conditions that result in debilitation and immunosuppression. The taxonomic status of *P. carinii* is undecided; either fungus or sporozoa (protozoan). *Pneumocystis* organisms cause a foamy filling of alveoli of the lungs of humans, dogs, cats, mice, rats, guinea pigs, rabbits, pigs, goats, sheep, monkeys, and horses. *Pneumocystis carinii* infections are classically provoked through administration of cortisone (Frenkel *et al.*, 1966). The golden hamster, however, tends either not to be subclinically infected or is remarkably nonsusceptible to *P. carinii* even when corticosteroids are given (Yoshida *et al.*, 1981; Frenkel, 1976). The latter observation has potential research application in studies of mechanisms of resistance to *Pneumocystis* infection. This observation may also have application in animal colony management; that is, the hamster probably does not need to be considered as a significant source of this infection for rats, immune-deficient mice, and other research animal species.

B. Nematodes

Chesterman and Buckley (1965) reported *Trichosomoides* sp., possibly *T. nasalis,* in parasagittal sections through the nose and adjacent tissues of 5-week-old hamsters in England. Some hamsters had a peculiar pug-faced deformity believed associated with inoculation of sarcoma material as newborns. Ova with polar plugs were seen in the feces of 37 of 185 stock hamsters. The ova were infective for rats when given orally (Chesterman, 1972). Later, Redha and Horning (1980) reported finding *T. nasalis* in the nasal cavities of hamsters in Switzerland.

Trichosamoides nasalis was originally described and named by Biocca and Aurizi in 1961 in the nasal cavities of *Epimys (Rattus) norvegicus* in Rome. Females of Trichosomoidinae carry the small male within the reproductive tract. Males possess no demonstrable copulatory organ. The type species, *T. crassicauda,* is parasitic in the urinary tract of rats. No intermediate host is required. Ova are infective when passed in the urine. Experimentally, pulmonary granulomatous reactions result from larval migration (von Fricsay, 1956; Zubaidy and Majeed, 1981). Presumably, *T. nasalis* of hamsters has a migratory phase in its life cycle which may produce lesions in the lungs or other organs. Although there have been no reports of this parasite in the United States, this author (Wagner) has been consulted about an instance wherein "doubly operculated, capillarialike" eggs were found in the feces of hamsters from a commercial source. Retrospectively, the question arises: Could this have been a *T. nasalis* infection?

C. Cestodes

Chesterman (1972) saw fragments of *Taenia crassicolis* larvae in hamster livers. Tripathi and Ray (1976) noted *Cysticercus fasciolaris,* the larval stage of the cat tapeworm *Taenia taeniaformis,* in the livers of golden hamsters in India, where *C. fasciolaris* is common in rodents. The animals had ascites and died. In 1953, Wantland tried unsuccessfully to infect hamsters with ova of *T. taeniformis* (Wantland, 1955).

III. PARASITES OF THE ALIMENTARY SYSTEM

A. Protozoa

One tends to find large numbers of protozoon forms, particularly flagellates, in the lumen of the hamster intestine, whether the animal is healthy or diseased. The etiological role of intestinal protozoa of hamsters, as pathogens causing enteritis and other more subtle adverse effects, is poorly defined. Van Hoosier and Ladiges (1984) described fecal smears from hamsters as a "'gold mine' for a protozoologist, as one can observe a large number and variety of organisms."

Saxe (1954a), in a remarkable piece of work, developed breeding colonies of hamsters and rats, free of intestinal protozoa, and did a series of transfaunation studies to determine host specificity of enteric protozoa. Interestingly, the faunas of rats and hamsters were freely interchangeable and morphologically identical. Many intestinal protozoan species found in rats and hamsters also occurred in the mouse, and could be successfully transfaunated. Attempts to transfer fauna of the guinea pig to rats and hamsters failed, which suggested that the protozoan fauna of the guinea pig is physiologically isolated from that of other laboratory rodents. Saxe suggests that the intestinal protozoan fauna of hamsters, as we know it, is not an indigenous fauna but one acquired from other rodents since domestication. Protozoan species found in hamsters from eight sources (Saxe, 1954b) and successfully transfaunated between rats and hamsters by Saxe (1954a) included *Giardia lamblia*, *G. muris*, *Hexamitas muris*, *Octomitus pulcher*, *Trichomonas muris*, *T. minuta*, *T. hominis*, *T. microti*, *T. wenyoni*, *Hexamastix muris*, *Chilomastix* sp. (small), *C. bettencourti*, *Entamoeba muris*, *Enteromonas hominis*, and *Monocercomonoides* sp.

1. Flagellates

a. Spironucleus (Hexamita) muris

History. Generally, no clinical signs are seen in "healthy" adult animals carrying infections. Rats and hamsters appear more resistant to *S. muris* infections than are mice, especially nude mice and mice of select inbred strains.

Etiology. The genus name of *Spironucleus* (syn. *Hexamita muris*, *Octomitus muris*, *Syndoyomita muris*) has been widely adopted in accord with Brugerolle's suggestions (1975). Much of the older literature carries the *Hexamita* genus name. *Spironucleus muris* is a non-host-specific diplomonadida flagellate with a distinctive ecological niche in the crypts of the anterior small intestine or the pyloric portion of the stomach of conventional animals of various species of the family Muridae, especially rats, mice, hamsters, and other rodents. *Spironucleus muris* is commonly encountered in the hamster, according to Kunstyr and Friedhoff (1980), Sebesteny (1979a), Saxe (1954a,b), Wagner *et al.* (1974), Stone and Manwell (1966), and Chesterman (1972).

Pathogenesis. *Spironucleus* sp. feeds on intestinal bacteria, may damage microvilli, and occasionally penetrates the lamina propria. Wagner *et al.* (1974) reported finding *Spironucleus* sp. in Syrian hamsters from four of four commercial sources in the United States. The finding was considered incidental, that is, not producing disease or death. An interesting observation by the Wagner group (1974) was the observation of motile *Spironucleus* sp. organisms in fresh blood smears. The animals had acute enteritis and proliferative ileitis. *Spironucleus* sp. apparently entered the bloodstream through mucosa damaged by the agents causing the acute enteritis.

The *S. muris* trophozoite is elongated, carrot-shaped, and bilaterally symmetrical, measuring 7–9 μm × 2–3 μm. It has four pair of flagella arising from a pair of blepharoplasts between the nuclei. Three pair of flagella appear as anterior flagella, while a fourth pair emerges from the rear of the body and appears as taillike posterior flagella (Fig. 1). Multiplication is by longitudinal fission. Infection is acquired by ingestion of resistant cysts that have entered the environment from fecal contamination.

Clinical Signs and Pathology. There are numerous reports documenting the substantial pathogenicity and research complications associated with *S. muris* for mice (Hsu, 1982), but few similar reports concern hamsters. Stressed and recently weaned mice appear to be the most susceptible. In enzootically

Fig. 1. Scanning electron micrograph of *Spironucleus (Hexamita) muris*. Mouse origin. Magnification: ×4000.

infected breeding colonies of susceptible strains of mice, recently weaned young may be listless and depressed, have a roughened hair coat, lose weight or fail to gain weight, have a hunched posture, have a distended abdomen, and have a low to high mortality. Presumably, affected hamsters could show similar clinical signs. At necropsy the anterior small intestine may contain an increased amount of watery, catarrhal fluid. Histopathologically, on the basis of the extensive, entangled mass of large numbers of organisms one sees covering villi in the intestinal lumen (Fig. 2), one would expect to see significant intestinal lesions microscopically. The amount of host response to heavy infections is surprisingly minimal. Hyperplasia of the absorptive epithelium and an associated increase in mitotic activity seem to accompany *Spironucleus* infections.

Significance. While little has been done to document adverse effects of *Spironucleus* sp. infections on research using Syrian hamsters, a substantial body of knowledge is accumulating about adverse effects in mice in a variety of research settings. In mice, *Spironucleus* sp. infection may increase macrophage activity, interfere with the immune response of the host (Ruitenberg and Kruyt, 1975), and increase sensitivity to irradiation (Myers, 1973).

Diagnosis. Usually, hamsters have more than one protozoan infection, and the search for intestinal protozoa should not end after one species has been identified; that is, other areas of the intestine should be searched for additional protozoa. When checking for *Spironucleus* sp. in enzootically infected colonies, it is important to use weanling animals 3–5 weeks old because older hamsters develop immunity and may have far fewer parasites. Diagnosis can be established by finding trophozoites in wet mounts (a loop full of intestinal contents mixed in two drops of saline with a coverslip added) of the anterior small intestine. Use of a phase-contrast microscope aids detection of protozoan forms. Also, the characteristic rapid, direct, straight-line movement of the trophozoites helps to distinguish them from other less mobile forms such as *Giardia* sp. or *Trichomonas* sp. that have slower rolling or tumbling movements.

To examine fecal material for cysts, a loop full of intestinal content or a fresh fecal pellet is collected from the cage or rectum and "rubbed around" in a few drops of saline on a glass slide, then coverslipped (Giemsa stain is optional) and examined by phase-contrast microscopy for the 4- by 7-μm banded cysts (Kunstyr, 1977).

Histologically, the various protozoa described in this chapter can be differentially distinguished from their location in the intestine and various morphological features. *Spironucleus* tends to be found deep in the crypts of Lieberkuhn in the anterior small intestine and pylorus of the stomach. The crypts may appear packed and distended with protozoa. In highly susceptible young, stressed, or immunodeficient animals, numerous trophozoites can be seen between villi and suspended in the abundant mucus contents of the intestinal lumen (Fig. 3). *Spironucleus* tends to be smaller than *Giardia* and does not have the sucking disk and undulating membrane. Special stains such as periodic acid–Schiff (PAS) and methenamine silver aid in visualization of the organisms in section.

Treatment, Prevention, and Control. In the case of mice, use of 0.04–0.1% dimetridazole in the drinking water for 10 to 14 consecutive days may give control of clinical disease but cannot be relied on to eliminate parasites from infected animals (Sebesteny, 1969, 1979b; Flatt *et al.*, 1978). Kunstyr *et al.* (1977) reported unsuccessful attempts to cure spironucleosis with four chemotherapeutic agents. Sebesteny (1979a) and Kunstyr *et al.* (1977) documented transmission of various in-

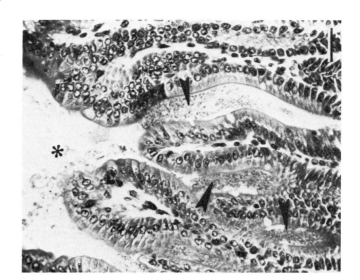

Fig. 2. Tip of intestinal villus (arrowhead) covered partially with *Spironucleus muris*. Bar = 13 μm. From Eisenbrandt and Russell (1979).

Fig. 3. Light microscopy of *Spironucleus muris* trophozoites in lumen (asterisk) and between villi (arrowheads) of small intestine. Bar = 30 μm. From Einsenbrandt and Russell (1979).

testinal flagellates between rats, mice, and hamsters. Interspecies transference of *Spironucleus* among rodents is common and must be controlled as part of an effective disease prevention program in a research animal facility. Cysts of *S. muris* are sensitive to many disinfectants, if concentration and exposure are sufficient, and to high temperature (Kunstyr and Ammerpohl, 1978). Of particular interest was the observation that 45°C for 30 min destroyed cysts' infectivity. It is this author's view that decontamination of animal facilities by superheating (pasteurization) may be more efficacious and practical than use of disinfectants, some of which are easily neutralized by organic material. Also, it is difficult to disinfect chemically everything in an animal facility, especially air ducts.

b. Giardia

History. Historically, *Giardia* have been speciated according to the host of origin. This scheme is being challenged, and more and more work refutes the strict host specificity of *Giardia* sp. Filice (1952) named the *Giardia* sp. he found in hamsters *G. muris* race *mesocricetus;* that is, though somewhat larger, it resembled *G. muris* of mice. Grant and Woo (1978b) also found hamsters infected with *Giardia* sp. whose trophozoites were similar to, but larger than, those of *Giardia* from mice. They proposed *G. muris* for the agent found in the mouse and *G. mesocricetus* for the hamster agent. Grant and Woo (1978b) found that *G. microti* from meadow voles infected hamsters and *G. mesocricetus* from hamsters infected laboratory rats. *Giardia muris* from wild rats infected hamsters, and cysts from hamsters infected laboratory rats and mice. On the basis of cyst morphology and the species from which they were obtained, Roberts-Thomson *et al.* (1976) decided that the *Giardia* they found in a hamster were *G. muris*. The taxonomic issue is currently unsettled. The commercial availability of specified disease-free laboratory rodents and their high level of susceptibility to *Giardia* (and other intestinal protozoa) should be of value in providing answers to questions about host specificity. Reports of *Giardia* in hamsters are common (Wenrich, 1946; Castellino and DeCarneri, 1964; Kunstyr and Friedhoff, 1980; Chesterman, 1972; Stone and Manwell, 1966; Sebesteny, 1979a; Hagen *et al.*, 1965).

Etiology. *Giardia* sp. are found worldwide and affect many classes of vertebrates. Among mammals infected are most laboratory animal species, including rats, mice, hamsters, chinchillas, rabbits (*G. duodenalis*), subhuman primates, and humans (*G. lamblia*). *Giardia muris* is a pear-shaped to ellipsoidal, bilaterally symmetrical organism that resides in the lumen of the anterior small intestine of laboratory rodents, especially mice, hamsters, rats, and wild rodents. Trophozoites are about 7–13 μm × 5–10 μm. The large sucking disk is found in the broadly rounded anterior end. The posterior end is drawn out or pointed (Fig. 4).

Pathogenesis and Epizootiology. Transmission is by inges-

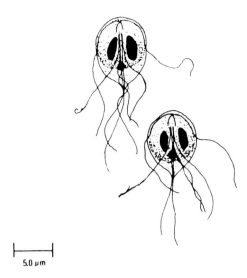

Fig. 4. Camera lucida drawings of trophozoites of *Giardia mesocricetus* that were wet-fixed in 100% ethanol and 10% buffered formalin prior to staining with Giemsa stain. From Grant and Woo (1978a).

tion of resistant cysts from the environment. Reproduction is by binary fission in the anterior small intestine. All hamsters necropsied by Castellino and DeCarneri (1964) in Italy carried *Giardia* sp. that was easily transmissible to young rats previously freed of parasites by treatment with Atabrine.

Clinical Signs and Pathology. *Giardia* infections of hamsters appear not to cause enteritis of sufficient magnitude to result in diarrhea or other clinical signs. But the association of large numbers of *Giardia* sp. in the anterior small intestine of weanling hamsters with abundant catharral fluid suggests that these agents are potentially pathogenic. Additional work is needed to determine whether *Giardia* sp. of hamsters are opportunistic pathogens as they are of mice (Hsu, 1982).

Diagnosis. Diagnosis can be achieved by recognizing characteristic pear-shaped trophozoites that have a rolling and tumbling movement in phosphate-buffered saline (PBS) wet-mount preparations from the anterior small intestine examined microscopically (×100). The parasite is easily recognized in profile on histological examination of hematoxylin–eosin stained sections of the small intestine. It is frequently seen adhered or capped onto villar columnar epithelial cells of the small intestine by means of a concave sucking disk. Also, thick-walled ellipsoidal "cysts" (15 × 17 μm) with four nuclei can be found in the large intestine or in Lugol's iodine-stained fecal smears. Cysts can be concentrated using zinc sulfate solution in a flotation technique (Lennette *et al.*, 1974). If feces are diarrheic, trophozoites may be seen.

Control and Prevention. Grant and Woo (1978b) established *Giardia*-free colonies of hamsters by administering an aqueous solution of quinacrine hydrochloride (Atabrine) to lactating females by stomach tube (0.75 mg quinacrine base per 10 gm body weight) once every other day until weaning, starting when young were 3 days old. Unlike mice and young rats, pregnant hamsters would not accept water containing 0.5% metronidazole. Cage cleaning, autoclaving, filter tops, and a high level of sanitation were also used in the successful eradication of *Giardia* sp. from hamsters.

Preventing the introduction of infected animals through quarantine and diagnostic evaluations and a sound management/sanitation program are paramount to controlling infections such as giardiasis. Addition of 0.1% of dimetridozole to the drinking water for 14 days can be used as a treatment (Sebesteny, 1969). Chlorocrine, quinacrine, and amodiaquin can also be used in treatment. Dry temperatures above 50°C or 2.5% phenol can be used to kill cysts in the environment (Hsu, 1982).

Significance. Evidence of human-to-animal transmission is easier to obtain than animal-to-human data; that is, feeding *Giardia* cysts from a human to eight animal species produced infections in all eight species (Fox *et al.*, 1984). Evidence of animal to human transmission of *Giardia* is increasing, however. *Giardia* cysts from beavers in drinking water have been implicated in several outbreaks of human and canine giardiasis (Dykes *et al.*, 1980). In humans, giardiasis can be a severe, chronic disease characterized by intermittent diarrhea, steatorrhea, malaise, anorexia, and weight loss. The stool is likely to be mucus-laden, light-colored, and soft but not watery.

While rodents can be experimentally infected with *Giardia* pathogenic for humans, *G. lamblia*, they do not appear to harbor these organisms naturally. Therefore, laboratory hamsters are not likely to be a source of infection for humans; however, the literature is not entirely clear on this point. If animal-to-human infection is suspected, the species of *Giardia* infecting animal and human should be established.

c. *Tritrichomonas* sp.

History and Taxonomy. Trichomonas, subfamily Trichomonadinae, with a typical pelta or shieldlike structure ahead of the blepharoplast, and *Tritrichomonas,* subfamily Tritrichomonadinae, without a pelta, are of the family Trichomonadidae (trichomonads). *Trichomonas* has four anterior flagella and no trailing posterior flagella. *Tritrichomonas* sp. has three anterior flagella and a posterior flagellum. The parabasal body of Tritrichomonadinae is sausage-shaped.

The nomenclatural, taxonomic, and host relations of the trichomonads are not clear. Two relatively nonpathogenic species of *Tritrichomonas* that occur commonly in the hamster and in many other wild and laboratory rodents are *T. muris* Grassi, 1879 (syn. *T. criceti*), and *T. minuta* Wenrich, 1924. They are egg- or pear-shaped but somewhat ellipsoidal, with a slightly drawn-out pointed posterior. *Tritrichomonas muris* measures 16–26 μm × 10–14 μm. *Tritrichomonas minuta* is 4–9 μm × 2–5 μm. Both have three equal anterior flagella and a prominent recurrent posterior flagellum incorporated into the margin of a well-developed undulating membrane extending the entire length of the body. They have a single nucleus in the anterior end. The flagella arise from the blepharoplast anterior to the nucleus. Figure 5 depicts several trichomonads.

Wantland in 1956 further described the large *T. muris*-like flagellates from the hamster cecum and colon and proposed the name *Trichomonas cricetus.* They measured 12–25 μm × 5–10 μm. Spherical cystlike forms found in the feces measured 7–13 μm in diameter. Excessive numbers were found in feces of hamsters fed a sunflower seed diet.

Ray and Sen Gupta (1958) described a trichomonad from the cecae of hamsters and proposed a new species, *Trichomonas criceti.* In an extension of Ray and Sen Gupta's work, Chakraborty *et al.* (1961) described the fine structure of the large *T. muris*-like organism of hamster origin and referred to it as *T. criceti.* The organism they described shared features with the *T. muris*-type flagellate described by others (Honigberg *et al.,* 1971) as reviewed by Daniel *et al.* (1971). The parasite described by Ray and Sen Gupta (1958) and Chakraborty *et al.* (1961) appears to have been *Trichomonas cricetus* as described

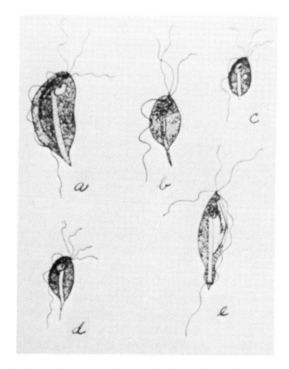

Fig. 5. Photomicrographs of camera lucida drawings of iron–hematoxylin stained trichomonad specimens. Magnification: ×1500. (A) *Tritrichomonas muris* trophozoite; note thick, pear shape and heavy, full-body length costa. (B) *Trichomonas wenyoni* trophozoite; note free posterior portion of undulating membrane and long posterior projection of axostyle. (C) *Tritrichomonas minuta* trophozoite; note small size, short posterior projection of axostyle, and chromatic ring at point of emergence of axostyle from body. (D) *Trichomonas microti* trophozoite; note presence of four anterior flagella and long filamentous end of axostyle. (E) *Trichomonas cricetus* trophozoite; note long, slender body, posterior bulbar expansion, and chromatic ring at point of emergence of axostyle. Magnification: ×1500. From Wantland (1956).

by Wantland (1956). Levine (1961) emended "*cricetus*" to "*criceti*" and placed the species in the genus *Tritrichomonas* Kofoid, as *Tritrichomonas criceti* Wantland.

Honigberg, in Daniel *et al.* (1971), reviewed the above-described literature on *T. cricetus* and *T. criceti* and Saxe's transfaunation experiments (1954a), which showed the lack of *T. muris* host specificity among rodents, including hamsters. He concluded that all of these organisms belong to *Tritrichomonas muris* Grassi, 1879. In 1973, Levine continued to consider *T. muris* and *T. criceti* synonymous. Trichomonads are commonly reported from hamsters (Wenrich, 1946; Chesterman, 1972; Stone and Manwell, 1966; Sebesteny, 1979a; Hagen *et al.*, 1965; Brugerolle, 1981).

Pathogenesis. None of the cecal trichomonads have ever been shown pathogenic (Levine, 1973). They flourish in a fluid or semifluid medium. Therefore, enteric diseases create conditions favorable to trichomonad growth, which may lead some people to assume that trichomonads may have caused enteric disease.

Mattern and Daniel (1980) found hamsters were not infected with *T. muris* at 3 days of age. By 7 days nearly all had infected ceca concomitant with cecal enlargement and the appearance of adult-type feces. Wantland (1956) could not find *T. criceti* in hamsters under 6 days of age. He found marked hyperemia of the intestinal wall, with pinpoint entorrhagia in heavily infected hamsters. Erythrocytes were demonstrated in trophozoites. (Caution must be used in interpreting older reports of enteritis in hamsters, since bacteria such as *Escherichia coli* and *Campylobacter*-associated proliferative ileitis may cause enteritis in weanling hamsters.)

On the basis of electron-microscopic examinations of preparations of spherical quiescent bodies resembling cysts in the feces of trichomonad-infected hamsters, Schnur (1975) concluded these bodies were not true cysts because they lack structures equivalent to a cyst wall. Based on cytochemical, morphological, and electron-microscopic study of *T. muris*, Seliukaite (1977), on the other hand, described pseudocysts or precysts that formed in the large intestine. Later cysts form in the environment, where they survive for "a very long time."

Epizootiology. Examination of wet smears from 412 hamsters from eight U.S. sources for protozoan forms by Wantland (1955) revealed the following protozoa, their location, and incidence: Trichomonad flagellates, cecum, 99%; *Entamoeba muris*, cecum, 33%; *Chilomastix bettencourti*, cecum, 9%; and *Giardia muris*, small intestine, 12%. Most of the trichomonads were identified as *Tritrichomonas muris*, with smaller numbers of *Trichomonas wenyoni, Tritrichomonas minuta,* a form resembling *Trichomonas microti,* and an unidentified fifth trichomonad unlike any previously reported.

Transmission is by ingestion of trophozoites or pseudocysts that reach the environment as a result of fecal contamination. Pseudocysts exencyst several hours after ingestion and once again function as trophozoites, dividing by binary fission.

Trichomonads have a simple life cycle. They reproduce by longitudinal binary fission. There are no sexual stages and there are no true cysts, but there is a pseudocyst that has some of the features of a cyst. The immobile pseudocysts represent a resident form of trophozoite that has internalized its anterior and recurrent flagella and undulating membranes.

Diagnosis (including Differential). The appearance of *Tritrichomonas* sp. in the feces, cecum, or fecal pellets in the colon will vary greatly depending on whether one is looking at a trophozoite, a pseudocyst, or a partially "encysted" or "exencysted" active trophozoite; that is, many morphological markers are missing in pseudocysts. The flagella, undulating and marginal lamellae are internalized and not apparent. Pseudocysts appear as oval to round bodies in fecal suspensions, usually in large numbers.

It is important to note that while these organisms occur primarily in the cecum, one can also observe small numbers almost anywhere in the digestive tract, including cheek pouches, lower esophagus, stomach, and duodenum (Saxe and Batipps, 1950). By hemocytometer count, Saxe and Batipps (1950) found over 1 million trophozoites of *T. muris* per 100 mg of fresh hamster cecal contents.

Control and Prevention. Feeding a high-protein diet (42% ground beef, 45% beef liver, 11% lard, and 2% $CaCO_3$) and carbarsone was effective in removing flagellates from hamsters (Wantland and Johansen, 1954). Wilhelm and Pope (1974) also found hamsters on a high-protein diet were trichomonad-free after 13 days. These hamsters also had high mortality, weight loss, and thinning of hair. Angulo *et al.* (1978) failed to eliminate intestinal flagellates, trichomonads, *Spironucleus muris,* and *Giardia muris* by treatment with dimetridazol for 8 weeks.

Significance. Trichomonads can be found in the cecum and colon of almost all birds and mammals. The trichomonads of rodents are generally considered nonpathogenic commensal flagellate inhabitants of the cecum, colon, and lower small intestine of hamsters, rats, mice, and other rodents. They appear to be rather rodent-specific and apparently are not infectious to humans, that is, do not have public health consequences.

d. Other Flagellates

Trichomonas wenyoni (Wenrich, 1946) (syn. *T. parva*), a nonpathogenic trichomonad, has a pelta and three anterior flagella and a posterior free-trailing flagellum that extends beyond the body. It measures 4–15.4 μm × 2.5–5.5 μm (Wenrich and Nie, 1949) and is found in the hamster and other rodents.

Tetratrichomonas microti Honigberg, 1963, was originally described and named *Trichomonas microti* in 1950 by Wenrich and Saxe after they found it in several hamsters, *Microtus pennsylvanicus, Peromyscus leucopus,* and a wild Norway rat. It was also transmitted to rats, guinea pigs, and hamsters.

Trichomonas microti was 4–9 μm long and had four anterior free flagella. The margin of the undulating membrane contained the posterior, trailing flagellum. Honigberg in 1963 proposed the genus name *Tetratrichomonas,* which was accepted by Levine in 1973. It may be fairly common in laboratory and wild rodents and possibly other species such as the dog, cat, and ground squirrel (Simitch *et al.,* 1954), which could serve as a source of infection for laboratory hamsters.

Pentatrichomonas hominis Davaine, 1980, has been found in humans, primates, dogs, cats, rats, mice, hamsters, and oxen (Levine, 1973). Saxe (1954a) transmitted it from hamsters to rats, and vice versa. These nonpathogenic trichomonads occur in the cecum and colon and have five anterior flagella and a pelta (Levine, 1973). Four flagella are grouped and point anteriorly, while a fifth is separate and directed posteriorly and a sixth runs along the undulating membrane and extends beyond it as a free-trailing flagellum. The undulating membrane runs the full length of the body. They are piriform and measure 8–20 μm × 3–14 μm (Flynn, 1973).

Chilomastix bettencourti da Foneeca, 1915, occurs in the cecum of the golden hamster, wild Norway rat, and domestic mouse (Levine, 1973). It has been reported from U.S. hamsters by Saxe (1954b) and Wantland (1955). The piriform trophozoites have three anterior flagella and a short fourth flagellum that undulates in a large cytosomal groove.

Parasites of the genus *Hexamastix* Alexeieff, 1912, are piriform with six flagella, of which one trails. They have a pelta, conspicuous axostyle, and a prominent parabasal body (Levine, 1973). They are about 5–12 μm long by 4–10 μm. They are nonpathogenic cecal inhabitants and reproduce by binary fission. They occur in the cecum of rats, hamsters, and other rodents (Kirby and Honigberg, 1949; Wenrich, 1946).

Caviomonas mobilis is a nonpathogenic flagellate reported to occur in the cecum of guinea pigs and hamsters (Saxe, 1954b). This rudimentary naked-appearing trichomonad has but a single anterior flagellum. It is 2–7 μm long and 2–3 μm wide (Flynn, 1973).

Parasites of the genus *Monocercomonoides* Travis, 1932, have two pair of anterior flagella, a pelta, and a filamentous axostyle. Saxe (1954b) reported *Monocercomonoides* sp. in rats and hamsters. It can be transmitted from the hamster to the rat. The incidence of this parasite is unknown.

Parasites of the genus *Octomitus* von Prowazek, 1904, have a symmetrical piriform body with two nuclei near the anterior end. They have six anterior and two posterior flagella. This genus is distinguished from *Spironucleus (Hexamita)* by the structure of its axostyles (Gabel, 1954). In *Octomitus,* two axostyles originate at the anterior end, fuse posteriorly, and emerge as a single central rod from the middle of the posterior end. *Octomitus pulcher* Becker, 1926, Gabel, 1954 (syn. *O. intestinalis* and *H. pulcher*) is found in the cecum of hamsters, rats, mice, and ground squirrels (Levine, 1973). It is 6–10 μm long and 3–7 μm wide. The incidence of this parasite in contemporary hamster stocks is unknown.

2. Ciliates

Sheffield and Beveridge (1962) found *Balantidium coli* in hamsters with "wet-tail." As a result of studies in which about 600 cysts of *B. coli* of swine origin were inoculated orally into hamsters, it was concluded that "*B. coli* is not a typical parasite for *M. auratus* since a state of balance is formed in which two components may exist for longer periods of time" (Humiczewska and Skotarczak, 1984).

3. Sporozoa

It appears that there have been no published reports of the coccidian genus *Cryptosporidium* in the golden hamster as a spontaneous occurrence. Tzipori (1983) reviewed the literature on cryptosporidiosis in animals and humans. No reference was made to natural or experimental infections of hamsters. Schloemer (1982) successfully transmitted *Cryptosporidium* sp. from calves to hamsters. This author (Wagner) in 1985 talked to two individuals, with reports in preparation, that document the occurrence of *Cryptosporidium* sp. in the intestines of weanling hamsters (J. L. Bryant, personal communication, 1985; A. J. Davis, personal communication, 1985). In neither case was there an obvious association with clinical disease. Davis has confirmed his observation through electron-microscopic examination of affected hamster intestines.

Assuming hamster cryptosporidia resemble those seen in other species, the life cycle probably involves a small sporocyst 3–6 μm in diameter that appears in the feces. Use of Sheather's sugar solution for flotation will enhance recovery of parasites from the feces. Giemsa or dichromate stains will help in visualizing the agent. Sporozoites, 3–7 μm in diameter, are likely to be found attached to the striated border of enterocytes in hematoxylin–eosin stained sections of the intestine.

4. Amoebae

Entamoeba muris Grassi, 1879, is generally considered a common nonpathogenic inhabitant of the lumen of the cecum of a variety of rodents, including the hamster. It feeds on food, bacteria, other protozoa, and desquamated cells. *Entamoeba muris* is apparently readily transferred among mice, rats, and hamsters and was first reported in hamsters by Neal (1947). Of 85 hamsters examined, 50 (59%) were infected. Neal (1947) and Saxe (1954a) both succeeded in infecting rats with amebae from hamsters and vice versa. In the course of work on *E. histolytica* in rats, Neal (1947) found rats became infected with *E. muris* carried by hamsters in the same room.

Fulton and Joyner (1948) found 53% of 58 hamsters harbored *E. muris.* Orkney voles, cotton rats, and white rats in the same animal facility in England were also infected with *E. muris.* Mudrow-Reichenow (1956) found *E. muris* in 7% of the hamsters he examined in Germany, while Wantland (1955) found 33% of 412 hamsters from several U.S. sources infected.

Trophozoites of *E. muris* are 8–30 μm long. The spherical nucleus is from 3 to 9 μm in diameter and may be surrounded by refractile granules. Trophozoites in fresh specimens may be active, with pseudopodia being extended and withdrawn without directive locomotion. Mature cysts have eight nuclei with strongly eccentric compact karyosomes. When ingested by a new host, eight trophozoites are produced. Immature cysts contain one to eight nuclei, and tend to contain more and larger vacuoles than mature cysts. Cysts vary greatly in size from 9 to 22 μm in diameter.

B. Nematodes

1. *Syphacia* sp. Seurat, 1916 (Pinworms)

While pinworms of the family Oxyuridae, order Ascarida, are important and common helminth parasites of laboratory rats (*Syphacia muris*), mice (*Syphacia obvelata*), and rabbits (*Passalurus ambiguus*), they are less prevalent in the laboratory hamster. This may be in part because the hamster is not reinfected with disease flora and fauna from his wild or feral conspecific counterparts as are laboratory rats and mice.

Syphacia obvelata (formerly *Oxyuris obvelata* Rudolphi, 1802), the common mouse pinworm, also affects hamsters (Taffs, 1976a; Wantland, 1955; Watson, 1946). Chesterman (1972) reported seeing *S. obvelata* in the large bowel of hamsters, but it did not produce disease. Kirschenblatt (1949) reported *S. obvelata* common in laboratory hamsters, as did Stone and Manwell (1966) and deRoever-Bonnet and Rijpstra (1961). The latter authors also reported finding immature forms of *S. obvelata* in the brain of adult hamsters. Kellogg and Wagner (1982) demonstrated that *S. obvelata* could be transmitted from infected mice to rats, hamsters, and gerbils. Infection in rats, hamsters, and gerbils was determined by the appearance of *S. obvelata* eggs on the perianum of these animals. Fahmy *et al.* (1967) found *S. obvelata* among laboratory golden hamsters in Egypt. Watson (1946) found 52% of his hamsters in London carried *S. ovelata.* Ogden (1971) reported that *S. obvelata* is found in bank voles and field voles in England. Thus, these species could possibly serve as a source of infection for laboratory rodents, including hamsters.

Ross *et al.* (1980) demonstrated that *Syphacia muris,* the rat pinworm, could be transmitted from infected rats to uninfected Syrian hamsters, mice, and Mongolian gerbils. Further, they demonstrated that infections could be transmitted from the latter three species to uninfected naive rats.

Epizootiology. Extrapolating from knowledge about *Syphacia* infections in mice (*S. obvelata*) (Chan, 1952) and rats (*S. muris*) (Stahl, 1961), we can speculate that *Syphacia* in the hamster is probably of low pathogenicity, and with a direct life cycle. Within hours after ingestion of an infective egg, larvae hatch in the small intestine and rapidly move or are carried to the cecum, where they develop to the adult stage in 4 or 5 days. Then, copulation occurs and the male nematodes die and disappear. After several more days the fertilized gravid females migrate through the colon to the perianal area, where large numbers of eggs are deposited and the females die. The eggs are infective within a day. Eggs collected from female worms in the cecum are not infective. The minimal prepatent period, though not well established, is probably around 8 or 9 days. Treatment regimes must take the life cycle into account, since few drugs are ovicidal. Therefore, treatments must be repeated to eliminate infections developing after a previous treatment; colonies should usually be treated at weekly intervals. Treatments should be accompanied by efforts to eliminate worms from animals (perianum) and from the environment by transfer to clean cages.

Diagnosis. Best approaches to antemortem and postmortem diagnosis of pinworm infections of hamsters must be extrapolated from the literature on similar infections of the mouse and rat. Antemortem diagnosis of *S. muris* and *S. obvelata* is best accomplished by examinations of low-power magnifications of perianal cellophane tape impression smears for eggs with species-specific morphology. Eggs can also be found on fecal flotations. Diagnosis can be confirmed by morphological examination of nematodes, preferably adults, from the cecal and colonic contents. Prevalence varies with age and sex. Use of 4- to 5-week-old animals maximizes chances of finding eggs.

Control. The effects of a large number of anthelmintics on rodent pinworms are known as a result of efficacy studies performed to find drugs effective against *Enterobius vermicularis,* the human pinworm. Procedures for control of pinworms in the hamster can probably be adapted from those used for rats and mice. Wagner (1970) reported elimination of mouse pinworms, *S. obvelata,* by treatment with dichlorvos in the feed at the rate of 500 mg/kg feed for 24 hr. Taffs (1976b) and Comley (1980) reported that 0.3% (w/w) of thiabendazole (TBZ) in the diet of mice for 3 days resulted in 100% expulsion of *Syphacia obvelata* and *Aspicularis tetraptera,* with *S. obvelata* eliminated earlier than *A. tetraptera.* An oral dose of 100 mg/kg was without effect. While many drugs work on the neuromuscular system of the parasite and cause paralysis or killing and expulsion of worms within a few hours, TBZ requires up to 3 days of continuous treatment for accumulation of adequate drug in the parasite. MacArthur and Wood (1978) reported that continuous medication through addition of 0.1%

TBZ to the diet resulted in elimination of mouse oxyurids after 24 days. They cautioned that success of this compound in totally eradicating worms from a colony was dependent on continuous medication until eggs in the environment are no longer infectious. Similarly, Owen and Turton (1979) found 0.1% TBZ in the diet continuously for 3 months eliminated *S. obvelata*. Unay and Davis (1980) reported eradication of *S. obvelata* in hamsters using two 7-day treatments with 10 mg of piperazine citrate per milliliter of drinking water separated by 5 days. TBZ is a satisfactory drug to incorporate into diets which will be autoclaved, because TBZ is relatively heat stable, while dichlorvos would be destroyed.

2. *Syphacia mesocriceti*

Syphacia mesocriceti was originally described by Quentin in 1971. He examined female worms collected in 1953 by Dr. Rausch, Arctic Health Research, Anchorage, Alaska. No males were available for description. Through use of scanning electron microscopy, Dick *et al.* in 1973 more completely described *S. mesocriceti* Quentin, 1971, after males and additional females became available from a commercial animal breeding establishment in Ontario, Canada. In 1981, Hasegawa described *S. mesocriceti* collected from cecae of golden hamsters bought from a dealer in Niigata City, Japan, in October 1979. The described host range of *S. mesocriceti* currently is limited to *Mesocricetus auratus*. The numbers in parentheses in the following descriptions are those of Hasegawa (1981), but the descriptions are based primarily on the scanning electron microscopy studies of *S. mesocriceti* by Saeki *et al.* (1982).

Figure 6 depicts various views of *S. mesocriceti*. The small, thick, colorless worms of cecal origin have a simple cuticle with annulations or transverse striations. The excretory pore posterior to the esophageal bulb is prominent. The shape of the cephalic extremities is characteristic: *S. mesocriceti* has reduced lips surrounding a distinctly triangular stoma, whereas *S. muris* and *S. obvelata* have distinct lips.

Males are 1.2–1.5 mm (1.39–1.88 mm) long and up to 80–99 μm (130–167 μm) wide with no lateral alae. *Syphacia obvelata* males are 1.1–1.5 mm by 120–140 μm, and *S. muris* is 1.2–1.3 mm by about 100 μm wide. Three mamelons have prominent annulations, with up to four rows of spines on each annule. The spicule and gubernaculum are prominent, and the tail is relatively slender.

Female worms are 5–6.9 mm (6.9–7.5 mm) long and up to 165 μm wide (232–376 μm). *Syphacia obvelata* is 3.4–5.8 mm × 240–400 μm, and *S. muris* is 2.5–4.0 mm long. The body is tapered to both extremities. Lateral alae are prominent. The uterus of mature females may be full of large embryonated operculate eggs 113–118 μm long × 31–46 μm wide. *Syphacia obvelata* eggs are flat on one side and measure 118–153

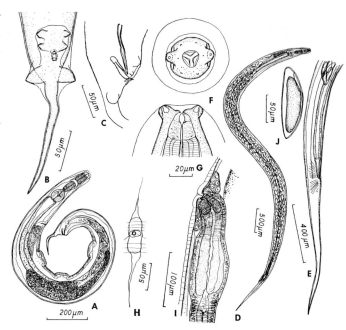

Fig. 6. Syphacia mesocriceti Quentin, 1971. (A) Male, lateral view. (B,C) Posterior part of male, ventral (B) and lateral (C) views. (D) Female, lateral view. (E) Posterior part of female, lateral view. (F,G) Anterior extremity of female, apical (F) and ventral (G) views. (H) Excretory pore, ventral view. (I) Vulva and vagina, lateral view. From Hasegawa (1981).

μm × 33–55 μm, while *S. muris* eggs are 72–82 μm × 25–36 μm.

Eggs of *S. obvelata* and *S. mesocriceti* are of about the same size. The eggs of *S. obvelata* are asymmetrically elliptical and measure 118–153 μm (129–137 μm) × 33–55 μm (33–39 μm). The eggs of *S. muris* are smaller and less elongated, measuring 72–82 μm × 25–36 μm.

There are but a few reports of *S. mesocriceti*. Therefore, topics such as pathogenicity, host range, public health significance, and research complications are incompletely described. One can assume that much of the information published about other pinworm species in the genus *Syphacia* applies to *S. mesocriceti*. *Syphacia obvelata* and *S. muris*, of the mouse and rat, respectively, have been particularly well described and studied (Wescott, 1982; Hsu, 1979). Table I is a comparison of these *Syphacia* species and another nematode pinworm, *Aspicularis tetraptera*.

3. Other Nematodes

Hamsters may be resistant to *Aspicularis tetraptera*. Stone and Manwell (1966) were unable to infect hamsters using *A. tetraptera* eggs. Similarly, in their survey of mice and hamsters from pet shops, department stores, and other places, they failed to find *A. tetraptera* in hamsters. However, 62% of mice from these sources were infected. At the time of this writing it

Table I
Comparison of *Syphacia mesocriceti, Syphacia obvelata, Syphacia muris,* and *Aspicularis tetraptera*[a]

Characteristic	*Syphacia mesocriceti*	*Syphacia obvelata*	*Syphacia muris*	*Aspicularis tetraptera*
Host	Hamster (???)	Mouse (rat, gerbil, hamster)	Rat (mouse, gerbil, hamster)	Mouse (Rat???, wild rodents)
Egg				
Location	In cecum or colon	In colon and on perianal skin	In colon and on perianal skin	In feces only (not found on perianal region); eggs appear intermittently
Shape	Large operculate	Asymmetrical, flat on one side, reniform, or crescent-shaped	Oval, slightly flattened on one side	Symmetrically ellipsoidal, thin shell
Length (μm)	113–118	118–153	72–82	89–93
Width (μm)	31–46	33–55	25–36	36–42
Content	?	Unembryonated when laid but soon embryonate	Larval stage	Morula stage
Infectivity	?	5–20 Hr after laid	5–20 Hr after laid	6 Days after laid
Worm				
Appearance	Pointed, thick, colorless	Pointed, white, opaque, glistening	Pointed, white, opaque, glistening	Pointed, white, opaque
Location	Cecum	Cecum (colon)	Cecum (colon)	Adults in proximal colon and cecum
Adult female				
Length (mm)	5.0–6.9	3.4–5.8	2.8–4.0	2.6–4.7
Width (μm)	149–165	110–270	?	190–250
Adult male				
Length (mm)	1.21–1.51	1.1–1.5	1.2–1.3	2.0–2.6
Width (μm)	80–90	120–140	100	150–170
Cervical alae	Lateral alae prominent in female; none in male	Present, small	?	Present—broad, end abruptly behind level of esophageal bulb
Mouth	Simple reduced lips around a distinct triangular stoma	Three distinct lips	Three lips	?
Buccal capsule	Conspicuous cephalic inflations	Absent	?	?
Esophagus	Prominent nerve ring, prominent excretory pore posterior to esophageal bulb	Club-shaped esophagus subspherical globular bulb	?	Oval bulb, club-shaped esophagus
Vulva	At 850 μm from front end (anus at 625 μm from tip of tail)	In anterior one-sixth of body	In anterior one-fourth of body	In anterior one-third of body
Spicula	Present—prominent	Single, prominent, slender, long	Present	Absent
Gubernaculum	Present	Present	Present	Absent
Ventral male mamelons	Three on ventral surface with prominent annulation, center mamelon near middle of body	Three on ventral surface, center mamelon near middle of body	Present on ventral surface, anterior mamelon near middle of body	Absent
Tail	Anal papillae	Long and sharply pointed, extremely curved ventrally in male, length equal to width of body, narrows abruptly behind cloaca	Thin and twice as long as body width	Conical—broad caudal alae
Life cycle	Direct	Direct	Direct	Direct
Infection route	Ingestion	Ingestion of infective eggs	Ingestion of infective eggs	Ingestion of embryonated eggs
Prepatent period	?	9–15 Days	≥7 Days	23 Days

(continued)

Table I (Continued)

Characteristic	Syphacia mesocriceti	Syphacia obvelata	Syphacia muris	Aspicularis tetraptera
Pathological effects	?	Minimal—usually inapparent; reduced weight gain and growth, impaction, rectal prolapse; sticky stools associated with mucoid enteritis	Minimal—usually inapparent; reduced weight gain and growth, impaction, rectal prolapse; sticky stools associated with mucoid enteritis	Minimal pathogenicity, larvae in crypts of Lieberkuhn; may occur in very large numbers in nude mice
Diagnosis	Observe worms in cecum or colon at necropsy; flotation technique to examine feces for eggs	Worms in cecum or colon at necropsy; demonstrate eggs on perianal area with cellophane tape; flotation technique to examine feces for eggs	Worms in cecum or colon at necropsy; demonstrate eggs on perianal area with cellophane tape; flotation technique to examine feces for eggs	Worms in cecum or colon at necropsy; cellophane tape method of no value
Control	? Should respond to anthelmintics like other Syphacia	Repeated anthelmintic treatment	Repeated anthelmintic treatment	Repeated anthelmintic treatment; frequent cage cleaning is effective (6 days for infectivity) and there are no perianal eggs
Public health considerations	?	May have a low level of infectivity for humans	May have a low level of infectivity for humans	Not known to infect humans

[a]From Dick et al. (1973); Flynn (1973); Quentin (1971); and Taffs (1976a).

is this reviewer's impression that there are no substantiated reports of the spontaneous occurrence of *A. tetraptera* Schultz, 1924, Nitzsch, 1981 (syn. *Ascaris tetraptera*) in hamsters. The parasite does, however, occur in rats and other rodents throughout the world (Yamaguti, 1961). Documentation of its occurrence in hamsters will not come as a surprise.

Heligmosomum juvenum is a trichostrongylid nematode parasite of the class Secernentrasida, order Strongylorida, family Trichostrongylidae, subfamily Heligmosominae, which also contains the genus *Nematospiroides* (Flynn, 1973; Kirschenblatt, 1949; Yamaguti, 1961) that affects the small intestine of hamsters in western Asia. Infection is probably acquired through ingestion of infectious larvae. It appears not to occur in contemporary laboratory hamsters. Its public health importance, pathogenicity, and incidence in nature are unknown.

Mastophorus muricola and *M. muris,* spirurids, and *Gongylonema neoplasticum,* a thelaziid, are found in a variety of monkeys, rabbits, hares, voles, and rodents, including hamsters, from many areas of the world. They occur in the stomach, and infection is acquired through ingestion of intermediate hosts (cockroaches, mealworms, and fleas). *Gongylonema neoplasticum* also affects the esophagus and may cause gastric ulcers. They are common in nature but have not been reported as a spontaneously occurring disease of laboratory rodents. The public health importance and pathological effects of these parasites are unknown (Flynn, 1973).

Mauer et al. (1953) reported finding a single hamster from Maryland that was heavily infected with *Protospirura muris.* The cardiac end of the esophagus and the stomach were greatly distended by the worms. The hamster was believed to have acquired the infection by eating infected insects.

C. Cestodes

Tapeworms of the family Cataenotaenidae are all found in rodents, but they are not strictly species-specific. They are common in wild rats and mice (Owen, 1976).

1. *Hymenolepis* spp.

History. *Hymenolepis* sp. in small numbers are relatively nonpathogenic, and infections may go unnoticed clinically. More severely infected animals may have catarrhal enteritis, retarded growth, and emaciation. Intestinal occlusion and impaction with ensuing death may occur in hamsters. Stunkard (1945) may have been the first to report *H. nana* as a naturally occurring parasite of hamsters. He found several laboratory hamsters, purchased from a commercial supply company, infected on their arrival in his New York laboratory. He warned that laboratory workers and animal attendants should be aware of possible contaminative infections.

Read (1951) may have been the first to report *H. diminuta* in laboratory hamsters. He reported finding *H. diminuta* in hamsters from a commercial source in the Los Angeles area. The parasites were easily established in rats. Likewise, hamsters were easily infected with *H. diminuta* parasites that had been maintained in rats for 14 years. Hughes *et al.* (1973) reported rats dually infected with *H. nana* and *H. diminuta*. Presumably, hamsters could also be dually infected. The potential for dual infection emphasizes the need for speciating these parasites in the course of diagnostic evaluations. Though not reported to occur naturally in laboratory hamsters, several other species of *Hymenolepis* readily infect hamsters. Their presence in hamsters could easily be overlooked by people making presumptive diagnoses not based on thorough examinations and morphological measurements that provide sufficient information to allow one to speciate infecting tapeworms accurately. *Hymenolepis microstoma* (Litchford, 1963), *H. citelli*, and *H. peromysci* (Stallard and Arai, 1978) are in this class. All resemble *H. nana* and *H. diminuta* and have an indirect life cycle that could use the confused flour beetle, *Tribolium confusum*, as an intermediate host.

Etiology, Characteristics of Agent, and Host Range. *Hymenolepis nana*, a cosmopolitan cestode of rodents, occurs frequently in laboratory hamsters (Soave, 1963). Although *H. nana* organisms of rodents and humans appear morphologically identical, the issue of their sameness and the potential for cross-infections has been debated. Grassi (1887) infected a child with eggs derived from a rodent. Others have also reported infecting children with *H. nana* eggs from rats (Kiribayashi, 1933; Ogura, 1936). Conversely, rodents have been infected with *H. nana* eggs obtained from human sources (Saeki, 1920; Uchimura, 1922; Woodland, 1924; Tsuchiya and Rohlfing, 1932). Epidemiological studies also provide evidence that rodents are the source of many human infections with *H. nana;* that is, homes of children with *H. nana* are often infested with rodents. It is now generally accepted that *H. nana* passes readily between hosts of a wide variety of species.

Hymenolepis nana adults usually develop in the lower small intestine (Fig. 7) and are slender white worms up to 1 mm wide and of variable length, 7–100 mm long. The muscular inverted scolex has four suckers, and the rostellum is armed with a single row or crown of 20–27 hooks. Mature proglothids are wider than long, and trapezoidal.

Hymenolepis diminuta is generally found in the upper small intestine and is a much larger worm than *H. nana*. It measures up to 60 cm in length and 3–4 mm in breadth. The scolex of *H. diminuta* is unarmed; that is, rostellar hooks are absent. *Hymenolepis diminuta,* like *H. nana,* occurs in the intestine of wild and feral rodents and uncommonly in the monkey and human (references cited in Flynn, 1973). *Hymenolepis diminuta* has been reported in laboratory mouse, rat, and hamster colonies (Handler, 1965; Read, 1951, Sasa *et al.,* 1962; Stone and Manwell, 1966). Because of *H. diminuta*'s mandatory indirect life cycle, human infections are infrequent, usually resulting from ingesting dried fruits or cereals contaminated with grain insects that became infected from rodent feces.

Larsh (1944, 1946) reported that 2.2 times as many *H. nana* worms developed in the gut of Syrian hamsters (7.4%) as developed in white mice (3.4%) following infection with a standardized egg dose. Compared to hamsters and mice, rats, deer mice, and guinea pigs showed a striking resistance, with much less than 1% of eggs developing into worms. Further, the length of 11-day worms in 1- and 1.5-month-old animals averaged 17.5 mm for hamsters and 7.5 mm for mice. The shortest prepatent period in hamsters was 9 days, whereas it was 14 days in mice. The patent period ranged from 19 to 32 days in hamsters and from 27 to 62 days in mice.

Epizootiology and Pathogenesis. Hymenolepid tapeworms are found in the small intestine of a variety of birds and mammals. All have an indirect life cycle except for *H. nana*, which can also be transmitted directly. Insects of the family Tenebrionidae, grain or flour beetles (*Tribolium confusum*), moths, earwigs, dung beetles, and fleas serve as intermediate hosts. Success in transmission depends on the presence or absence of appropriate intermediate hosts and environmental conditions. When infection originates from ingestion of eggs voided in the feces of a definitive host (direct life cycle), after hatching in the small intestine, the hexacanth larvae enter the villi of the upper small intestine and form cysticercoides. This form is highly immunogenic, rendering the animal resistant to subsequent challenging doses of eggs. In 4 or 5 days the young worms leave the villi and attach their scolices further down the intestine and develop into mature worms 10–11 days later (14- to 16-day life cycle in the mouse).

The indirect life cycle, on the other hand, involves cysticercoid development in the intermediate host (Tenebrionidae), which becomes infected by eating rodent feces or foods contaminated with rodent feces. The definitive host acquires infection through consumption of the insect in which the cysticercoid has developed, thereby bypassing the stage of larval development in the villi of the definitive host. Thus, infections acquired through ingestion of cysticercoids in fleas or flour

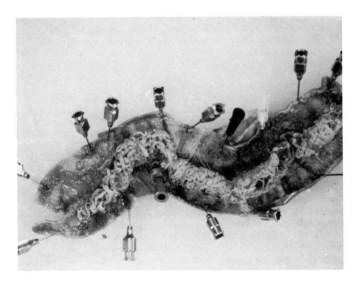

Fig. 7. Gross view of *Hymenolepis nana* infestation in hamster.

beetles are minimally immunogenic. The rate of larval development in the insect varies with the environmental temperature. For example, at 30°C it takes 8 days for larval stages to develop in the confused flour beetle. After ingestion of infective insects by the definitive host it takes 16–19 days to develop an adult worm (Flynn, 1973). Eggs of both *H. nana* and *H. diminuta* are expelled in the feces. Adult worms in the intestine may produce eggs throughout the life of the host.

Autoreinfection with *H. nana* may follow initial infections acquired either through the direct or indirect life cycle routes. Autoreinfections following ingestion of infected grain or flour beetles or fleas are thought to result in very heavy *H. nana* infections (Heyneman, 1961, 1963).

Ronald and Wagner (1975) reported that 27 of 250 (10.9%) retired breeder female hamsters from a commercial producer of hamsters were infected (average 20 scolices per animal). Sano *et al.* (1975) reported spontaneously occurring hymenolepiasis in hamsters in Japan. Some individuals had as many as 290 worms, and numerous cysticercoids were seen in villi. Of 40 hamsters examined, 16 (40%) were infected. Heavily infected hamsters had weight loss, and some died. Microscopically, chronic inflammation and destruction of intestinal mucosa were seen. Verster and Brooker (1970) reported *H. nana* in hamsters at the Veterinary Research Institute in Onderstepoort, South Africa.

Possible sources of *H. nana* infection for hamsters in a research animal facility might be insect hosts of the intermediate stage or wild or feral rodents, humans, other hamsters, and laboratory rodents. It is uncommon in the rhesus monkey, squirrel monkey, and chimpanzee (references cited in Flynn, 1973).

Clinical Signs and Pathology. Most infected hamsters have low-level infections and fail to show clinical signs. Heavy infections have caused intestinal occlusion, impaction, and death (Soave, 1963). Catarrhal enteritis (Habermann and Williams, 1958), chronic enteritis, and abscesses of mesenteric lymph nodes (Simmons *et al.*, 1967) may occur. There may be hyperplasia of Peyer's patches (Handler, 1965). Chesterman (1972) observed an enzootic of *Hymenolepis* sp. infection in a large colony of hamsters. Stages in the life cycle of the parasites were seen in the duodenum, small intestine, pancreatic ducts, extra- and intrahepatic bile ducts, and mesenteric nodes, and it was suggested that they may be a factor in liver cirrhosis observed in the colony.

Diagnosis. Diagnosis is established by identification of ova in the feces or by finding adult worms in the intestines at necropsy. Histopathological observation of cysticercoids in the lamina propria of intestinal villi or of the scolex with characteristic suckers also aids diagnosis.

The oval egg of *H. nana* measures 37–47 μm × 50–55 μm in diameter. They may contain an oncosphere (16–25 μm × 24–30 μm) enclosed in an inner envelope with two polar thickenings. Four to eight filaments attach to each of these thickenings. The oncosphere contains three pair of hooklets.

The spherical ova of *H. diminuta* are about twice as large as those of *H. nana*, measuring 50–79 μm × 72–86 μm. Like *H. nana*, they have an inner membrane around an oncosphere or embryo and two polar thickenings but no filaments. The oncosphere measures 24–30 μm × 16–23 μm. The eggs are slightly yellowish with a transparent shell. The hooklets in the oncosphere are arranged in a fan-shaped pattern. Figure 8 shows both *H. nana* and *H. diminuta* eggs.

Control and Prevention. Chute (1961) reported that many 50- to 80-gm hamsters purchased from a commercial supplier were heavily infected with *H. diminuta* and had high mortality in a dark, cold room (5°C, 90–97% relative humidity). By inducing hibernation in *Citellus citellus* (European ground squirrel), Chute (1961) reported elimination of *H. nana* after 10 days.

Control is difficult because infection (with *H. nana* in particular) spreads in so many ways. A high level of sanitation and insect control are a necessary part of the control program. Continuing abatement programs to control feral and wild rodents will reduce chances for introduction of *H. nana* infections. *Hymenolepis nana* infections may be introduced through fecal contamination of food, cages, hands, water, and bedding by wild rodents or arthropod vectors (intermediate hosts). Once a colony is free of *Hymenolepis* infections, new animals should be introduced only after they have been proven free of infection. This can be based on testing before animals are

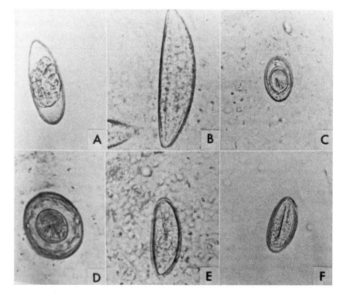

Fig. 8. Eggs of helminths. (A) *Aspicularis tetraptera;* (B) *Syphacia obvelata;* (C) *Hymenolepis nana;* (D) *Hymenolepis diminuta;* (E) *Syphacia muris;* (F) *Enterobius vermicularis.* Magnification: ×336. From Stone and Manwell (1966).

shipped to the colony or after the animals arrive at the destination facility but while they are held in quarantine away from the known pathogen-free colony. Ronald and Wagner (1975) reported successful treatment of hamsters infected by *H. nana* using 0.33% active niclosamide (Yomesan, Farbenfabriken, Bayer GmbH, Leverkusen, Germany) fed *ad libitum* in ground feed to hamsters for a period of 7 days. Taffs (1975, 1976b) reported 0.3% TBZ in feed is a safe and effective anthelmintic for treatment of *H. nana* and *Syphacia obvelata*. Praziquantel may be used as an alternative to TBZ. Treatment with TBZ for 7 days eliminated pinworms, but 14 days were required to eliminate *H. nana*. Taffs emphasized that unless animals are kept in an egg- and worm-free environment after treatment, they are likely to become reinfected.

Niclosamide fed at the rate of 1 mg/gm of pelleted feed (~100 mg/kg body weight) for 2 nonconsecutive weeks separated by a week of no treatment eliminated *H. nana* and *H. diminuta* infections from rats (Hughes *et al.*, 1973). The authors emphasized that because of the long prepatent period of *H. nana* (15–30 days), the second treatment was essential to eliminate immature forms not killed in the initial treatment.

Significance. Possibly the major threat from *H. nana* infections is in the potential for transmission to humans. Stunkard (1945) and Watson (1946) stressed the risk of *H. nana* infections being acquired by people handling hamsters. Stoll, in 1947, estimated that 100,000 persons in North America and 20 million worldwide were infected with *H. nana*. Lightly parasitized individuals may have no symptoms, but in heavier infections headaches, anorexia, convulsions, nervousness, enteritis and diarrhea, pruritus, and abdominal distress may be seen. *Hymenolepis diminuta,* on the other hand, is a rare parasite of humans because its eggs are infective only to arthropods, which must be ingested by the definitive host. Hamsters and other rodents from pet stores, department stores, and commercial producers are commonly infected with *H. nana* and *Syphacia obvelata* (Stone and Manwell, 1966). Animal technicians, because of intensity of exposure, and children, because of unhygienic habits, immunological incompetence, and a more suitable intestinal environment, are particularly susceptible to hamster-to-human transmission of *H. nana* (Faust and Russell, 1964). Children also have more exposure to rodents as pets and in classrooms than adults. The public is generally unaware of the threat to human health that *H. nana* infections of pet rodents constitute. Likewise, many investigators responsible for the use of these species, particularly hamsters with *H. nana* infections, are unaware of the potential severity of the human health hazard they pose. Additionally, worms produce unexpected variables in experiments. Further, the wormy rodent as a source of infection is a threat to the research of all others using animals of the same or other susceptible species. To preclude the above-described adverse effects on research careers and human health, it behooves personnel with administrative and operational responsibilities continually to assure the disease-free status of animals under their aegis.

Laboratory personnel and animal technicians should be made aware of the risks involved in working with infected hamsters. Further, they should be instructed in personal hygiene. Waste from infected animals should be treated in accord with its biohazard potential to personnel and handled in such a manner as to prevent spread of *H. nana* eggs to other animals in the facility. *Hymenolepis diminuta* should not present a threat to human health unless infected insects are consumed.

2. Other Cestodes

A case of *Cataenotaenia pusilla* infection in a pet hamster colony was reported by Owen in 1976. The intermediate hosts are grain mites. Mites are infectious about 15 days after ingestion of ova. The larvae attach to the wall of the small intestine of the rodent host, where a fairly large tapeworm develops, 30–160 mm long. Joyeaux and Baer (1945) have given a detailed account of this tapeworm in the laboratory mouse. It is unlikely that this tapeworm would be found in laboratory hamster colonies. Because this parasite is dependent on an intermediate host, good hygienic measures will preclude colony infections.

Mikhail and Fahmy (1968) reported finding a single specimen of *Mathevotaeni* sp. in a single laboratory hamster in Egypt. More material and life history information is necessary before this can be accepted as a new parasite species–host combination.

IV. PARASITES OF THE INTEGUMENT

A. Mites

1. *Demodex* spp.

History. *Demodex criceti* was discovered and provisionally described in the course of B. Nutting's thesis research at Cornell University in 1950. It was established as a distinct species in 1958 (Nutting and Rauch, 1958). In 1958 they noted, but did not describe, distinct elongate specimens of *Demodex* in their hamsters; that is, the animals had dual infections with an elongate and a squat (*D. criceti*) form. In 1961 Nutting described this new thin, elongate species as *D. aurati*.

Etiology (Including Characteristics of the Agent and Host Range). The Syrian hamster, *Mesocricetus auratus,* is the host for both *D. criceti* and *D. aurati*. It appears that no other animals serve as hosts to these two species of *Demodex* mites. These *Demodex* mites may represent one of the few parasites or other fauna of the hamster to accompany it on its journey from the wild through many research and production laboratories worldwide.

Demodex criceti (Nutting and Rauch, 1958) is a relatively small member of the genus. The body is broadly rounded in all stages. Total body length of adult males averages 87 μm and females 103 μm. Body width is about 32 μm for males and 34 μm for females. Ova (38 × 20 μm) are thin-shelled and transparent. Larval forms (56 × 29 μm) are very small with six rudimentary "legs" or plates that are incapable of moving the animal. Mouthparts are well developed. Nymphal forms have four pair of plate "legs" or "holdfasts" and measure about 84 × 30 μm. There may be substantial size variations in larval and nymph forms because several molts occur, and there may be differences in size between males and females.

Unlike small, rounded *D. criceti*, *D. aurati* is very thin and elongate, with a sharp "tail" (terminal opisthosomae). It somewhat resembles *D. folliculorum*. Adult males average 183 μm long and females 192 μm. They are 21–23 μm wide. Ova are 65 × 17 μm, spindle-shaped, thin-shelled, and transparent. The end of the egg, which will become the posterior of the adult, is pointed. Figure 9 depicts both *D. criceti* and *D. aurati*.

Pathogenesis. All stages of the life cycle of *D. criceti* and *D. aurati* occur in the host epidermis (Nutting and Rauch, 1958). *Demodex criceti* mites are found in epidermal pits that are "cut in the epidermis" and are rarely larger than the body of the mite. The pits extend to the stratum germinativum but rarely to the dermis. Larvae, nymphs, and adults have their piercing mouthparts directed toward the dermis. *Demodex criceti* mites are not found in normal hair follicles. Nutting and Rauch (1958) suggested that the mites feed on epidermal cell contents, not sebaceous secretions.

Adult mites of *D. aurati* and ova, on the other hand, are found in the pilosebaceous system (up to five male and female mites per hair follicle) at or above the openings of the sebaceous canal. Larvae and nymphs are found below this opening. Finding adult mites with mouthparts directed into sebaceous glands and the destruction of adjacent sebaceous cells indicates these cells are eaten by mites. Unembryonated eggs of *D. aurati* are laid singly in a hair follicle near the level of a sebaceous gland outlet, with the blunt "head end" of the egg pointed toward the sebaceous canal. Adult mites and ova are usually located more superficially in the philosebaceous follicle, while larvae and nymphs are found deep in the hair follicle and occasionally penetrate the follicular epithelium. Usually only adult mites are removed when hairs are plucked.

Outbreaks of overt disease caused by demodicosis in Syrian hamsters may require stressing or resistance-lowering factors (Owen and Young, 1973). *Demodex aurati* far outnumbers *D. criceti* in aged hamsters, especially on the dorsum (Nutting, 1961). Nutting and Rauch (1963) reported the parasite load with *D. aurati* to be much heavier in male than female hamsters—180 versus 50 mites per standardized area of skin. Mite

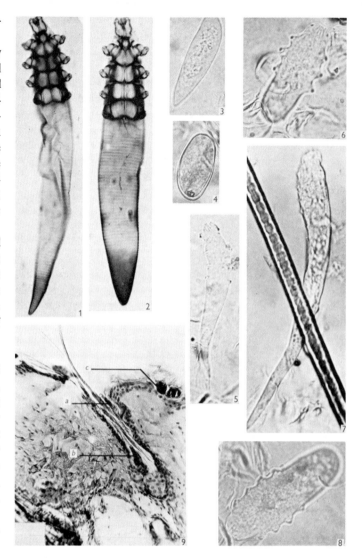

Fig. 9. (1) Male of *Demodex aurati*. This is an unusually long specimen. (2) Female of *D. aurati*. Note vulva situated behind epimeral plates. (3) Ovum of *D. aurati*. (4) Ovum of *Demodex criceti*. (5) Larva of *D. aurati*. (6) Larva of *D. criceti*. (7) Nymph of *D. aurati*. (8) Nymph of *D. criceti*. (9) Section of skin of *Mesocricetus auratus* showing (a) adult of *D. aurati*, (b) larva of *D. aurati*, and (c) larva of *D. criceti*. Magnification: ×231. From Nutting (1961). Reprinted courtesy of Cambridge University Press.

load did not decrease following gonadectomy of males (Nutting, 1964).

Epizootiology (Including Prevalence and Transmission). Larvae and nymphs develop in the epidermal pits. Direct host-to-host transfer is from mother to young at the time of suckling. Mites have been obtained from young at 5 days of age (Nutting and Rauch, 1958). All hamsters had *D. aurati* by 40 days of age (Nutting and Rauch, 1961), and by 5 days in another report (Nutting, 1964).

Owen and Young (1973) examined 65 adult and recently weaned hamsters, mostly females, from seven colonies in

Great Britain, for *D. criceti* and *D. aurati*. Between 66 and 100% of animals in all seven colonies were infected with *D. criceti* and/or *D. aurati*. *Demodex aurati* was found in 60% and *D. criceti* in 33% of hamsters. Twenty-five percent had both species of *Demodex*. All seven colonies had both species of mites. Surprisingly, this was the first report of *Demodex* sp. in hamsters from Great Britain. The delay in reporting this parasite in hamsters no doubt relates to the low pathogenicity of these parasites.

In a major survey of naturally occurring diseases of the Syrian hamster in 24 U.S. institutions (11 diagnostic resource centers, 7 research laboratories, and 6 commercial breeders), demodicosis was not specifically listed as a "disease known to occur in hamsters." "External parasitism" and "dermatitis" were listed once each. This information indicates that demodicosis is not generally recognized as a major disease of hamsters in the United States (Renshaw *et al.*, 1975). Heath and Bishop (1984) reported the occurrence of *D. aurati* and *D. criceti* from laboratory hamsters in New Zealand. Retnasabapathy and Lourdusamy (1974) reported finding them in hamsters in Malaysia.

Clinical Signs and Pathology. In their original description of *D. criceti* in 1958, Nutting and Rauch noted that even in severe cases no evidence of tissue reaction or altered pigmentation was found and that gross symptoms were wanting. With *D. aurati*, Nutting (1961) stated that there was no gross evidence of mite activity in heavily infested skin. Experimental biotin deficiency of hamsters caused a marked reduction in numbers of *D. criceti* and *D. aurati* (Nutting and Rauch, 1961). Nutting and Rauch (1963) found that despite large numbers of *D. aurati* mites, there was no apparent depilatory effect. Flatt and Kerber (1968) described an outbreak of clinical skin disease in a colony of aged golden hamsters being used in a lymphosarcoma transmission study. Hair loss occurred over the back and hindquarters, which had underlying dry, scaly skin. Scabs 2–4 mm in size were scattered throughout the area. Pruritus was not apparent. Large numbers of both *D. aurati* and *D. criceti* mites were observed. Flatt and Kerber felt that the *Demodex* mites and debility related to aging were responsible for the clinical skin disease. Welshman (1973) similarly suggested that age and intercurrent infection may be predisposing factors in clinical demodectic mange of hamsters.

Estes *et al.* (1971) described clinical skin disease due to *D. aurati* in 50% of 348 control and experimental golden Syrian male hamsters over 30 weeks of age (many were approaching 2 years of age) on long-term exposure to carcinogenic chemicals. Some hamsters became almost hairless, but no pruritus was observed. The alopecia was most common over the rump, and in many cases extended over the entire back. The denuded areas were dry and scaly with multiple small scab formations. Microscopically, many hair follicles were dilated and filled with debris and mites. There was little evidence of inflammatory cells around the distorted follicles. Heath and Bishop (1984) reported hair loss, scabs, and scurfiness, particularly on the neck and abdomen, of laboratory hamsters. *Demodex aurati* was more abundant than *D. criceti* on these hamsters.

Diagnosis (including Differential). Brockis (1973) discussed difficulties involved in establishing a diagnosis of demodicosis in pet hamsters and suggested that the disease is underreported in pet hamsters. According to Nutting (1961), the presence or absence of *Demodex* sp. can be easily discovered by plucking hair from areas of alopecia and examining for parasites. Mites can also be observed in skin scrapings in warm 5–10% NaOH or KOH. It is important to rule out possible bacterial infections, bite wounds and injuries, other parasites, and dermatophytes by direct examination, skin scrapings, histopathological examination, and culture.

Upon application of heat to excised skin, both *D. criceti* and *D. aurati* back out and leave their intradermal habitats (Nutting and Rauch, 1963), an observation that may be used to advantage in attempting to collect specimens for diagnostic evaluations. The displaced mites may ascend hairs, presumably to facilitate transference to another host. Adult mites move from hair follicles by 16 hr postmortem without heat, which could also be a process to increase chances of transference to other hosts.

In light infestations it may be difficult to find mites on microscopic examinations of skin sections. In such cases adult mites are found more reliably by digestion (boil in a 10% solution of KOH for 10 min and examine the deposit) of skin samples (Owen and Young, 1973). Males should be tested instead of females, since they tend to have heavier infestations (Nutting and Rauch, 1963). Chances of finding mites may be further enhanced by searching aged animals with other factors predisposing them to heavier infestations such as stress or poor nutrition (Flatt and Kerber, 1968). Nutting (1961) found that maximal *Demodex* mite populations were found on the dorsum of hamsters, while reduced numbers were found in skin of the ventrum, genitalia, mammae, muzzle, ear, eyelids, perianal area, and axillae. Professional judgment is needed to interpret diagnostic findings. The presence of *Demodex* sp. in lesions does not necessarily mean that the agent is the cause of the lesion; rather, its presence could be incidental.

Control and Prevention. Our knowledge about the natural history of *Demodex* sp. in hamsters is very limited, as is our knowledge about the prevalence of infections in commercial breeding colonies of hamsters. However, several basic husbandry practices will help in control and prevention of *Demodex* infections. These include minimizing stress, maintaining high levels of sanitation, avoiding introduction of overtly infected animals, isolating obviously infected animals, and treating affected animals. A long-range program to identify infected colonies and to work toward derivation of *Demodex*-free colonies would appear to be a noble goal for commercial producers of golden Syrian hamsters.

There are no specifically recommended treatment regimes

for hamster *Demodex* infections. Nutting (1950), after maintaining *D. aurati* in machine oil for 4 days and, finding them alive, noted they are "very hardy organisms." Presumably, however, treatments recommended for *Demodex* infestation of other animal species, such as the topical application of Amitraz (Mitaban Liquid Concentrate, Upjohn, Kalamazoo, Michigan) in dogs may be effective. A trial treatment to check toxicity and efficacy should be carried out.

Significance. *Demodex* species appear to be highly host-specific. Therefore, transfer between hamsters and other laboratory animal species and humans is unlikely. Nutting (1976) remarked that little basic biology of demodicids has been explored in a scientific manner. The bulk of over 1200 published accounts (in 1976) were of a superficial nature concerned with simple observations or case reports. This author foresees *Demodex* sp. assuming greater importance as an etiological agent of hamster disease as the use of hamsters for life span and other chronic studies increases. A promising route to explore in attempting to derive colonies free of *Demodex* sp. is an attempt to break the life cycle at the point where dam-to-young transfer of mites occurs (Nutting, 1964).

2. *Notoedres notoedres*

Although *Notoedres* infections are probably not seen in contemporary commercial and research colonies of hamsters, such infections were apparently fairly commonplace in the 1940s in England. Hindle (1947) stated that "Notoedric mange is the most common infestation" in the hamster in England.

The *Notoedres* mite is a small burrowing mite. The females of this mite are 0.3–0.4 mm long, while males are 0.1–0.2 mm long. Eggs are laid by females in the burrows. Larvae hatch in 3 or 4 days. Larvae molt to nymphs that burrow into the skin, molt several times, and become adults—the entire cycle taking 6–10 days. Infection is by direct contact, and the threat to humans is minimal.

In 1943, Fulton reported successful treatment of *Notoedres* mange in hamsters. He indicated that they had very heavy infestations with much scabbing and that the disease had a great tendency to spread. The naturally acquired infections involved the ears, snout, feet, and perianal region. Some animals had scabious masses the size of a marble around the anus. There were numerous mites, nymphs, larvae, and eggs in sections of infected tissues. Fulton (1943) reported a complete cure was effected in all 24 animals treated with one of two sulfur drugs in coconut oil.

3. *Notoedres cati*

An outbreak of scab infection caused by *Notoedres cati* has been reported in a group of golden hamsters (Baies *et al.*, 1968). Skin lesions were primarily on the ears, nasal and genital areas, tail and limb extremities. Males were more severely infected than females. Lesions of females were frequently limited to the ears. Animals died 2–3 months after infection, with chronic inflammatory modifications of the epidermis.

4. Nasal Mites

Speleorodens (*Paraspeleognathopsis*) *clethrionomys* were observed in epizootic proportions in the nasal cavities of nearly all animals in three breeding colonies of Syrian hamsters in Sweden (Bornstein and Iwarsson, 1980). Affected animals did not show clinical signs. The mites, about 150 × 350 μm, were easily seen with a stereomicroscope on the mucosa of the posterior nasal cavities. This species of mite was originally reported from field voles in Holland and seems not to have been otherwise reported in laboratory rodents.

5. *Ornithonyssus*

Ornithonyssus (*Liponyssus* or *Bdellonyssus*) *bacoti,* the tropical rat mite or red mite, is a free-living, avid bloodsucking mite. This mite is widespread in tropical and subtropical regions. It has an unusually wide host range, infecting a wide variety of wild rodents and small mammals as well as laboratory animals. Female mites move on and off the host for a day or so until fully fed, at which time they move to the nesting material and other surroundings to lay eggs during the next 3 days. Within a day or so a nonfeeding six-legged larva hatches and then molts in a couple of days to an eight-legged feeding nymph that must get a blood meal from a host over the next 3 weeks or die. After several more molts, adults emerge and move on and off the host to feed and oviposit for weeks (Chesterman, 1972). The minimum egg-to-egg life cycle of this very prolific mite is about 14 days. Since this parasite, for the most part, lives off of the host, it is readily controlled and can be eradicated by provision of clean cages with clean bedding twice per week. Transfer of nesting material from cage to cage may perpetuate infestation. The mite will readily attack humans when necessary to obtain a blood meal. Keefe *et al.* (1964) reported dermatitis and skin rash in five animal technicians, presumably due to the tropical rat mite found in the mice and hamsters under their care.

It may be possible to preclude infestations by putting the legs of cage racks in liquid traps containing insecticides or odorless kerosene. Keefe *et al.* (1964) recommended spraying bedding with 5 ml of a 1% suspension of malathion at 2- to 4-week intervals to control red mites in mice and hamsters. The infestation was believed to have been introduced via wild rats in the animal facility in Thailand.

Scott (1958) reported heavy infestations of hamsters, 50 mites per animal, with *O. bacoti.* Olson and Dahms (1946) reported *Liponyssus bacoti* infestations so heavy as to cause exsanguination of mice and hamsters. They found 28,000 mites on a single hamster, which represented only a small fraction of the mites present in the wood shavings of the cage. Rats and guinea pigs were also affected, but hamsters appeared particularily susceptible.

Ornithonyssus sylviarum, the northern fowl mite, resembles *O. bacoti* and is a common bloodsucking mite of chickens that occurs in temperate regions. Humans and laboratory rodents may become incidental hosts (Flynn, 1973).

Brockis (1973) reported a mite that resembled *Cheletophanes rugosa* in pet hamsters.

B. Hexapods

Wantland and Lichtenstein (1954) and Wantland (1955) reported the infestation of the rectal area of a laboratory hamsters with larvae of *Musca domestica.*

ACKNOWLEDGMENTS

The editorial assistance of John Ballenot and Karen Boillot and the word processing capabilities of Sylvia Bradfield are gratefully acknowledged. Also acknowledged is the valuable assistance of Trenton Boyd, librarian, University of Missouri College of Veterinary Medicine, for his cooperation in acquiring reprints of many of the more difficult-to-find papers cited in this chapter. This work was supported in part by NIH grant DHHS RR00471.

REFERENCES

Angulo, A. F., Spaans, J., Zemmouchi, L., and Van der Waaij, D. (1978). Selective decontamination of the digestive tract of Syrian hamsters. *Lab. Anim.* **12,** 157–158.

Baies, A., Suteu, I., and Klemm, W. (1968). *Notoedres* scabies of the golden hamster. *Z. Versuchstierkd.* **10,** 251–257.

Baker, H. J., Cassell, G. H., and Lindsey, J. R. (1971). Research complications due to *Haemobartonella* and *Eperythrozoon* infections in experimental animals. *Am. J. Pathol.* **64,** 625–652.

Biocca, E., and Aurizi, A. (1961). On a new parasitic nematode *Trichosomoides nasalis* n. sp., from the nasal cavities of *Epimys norregieus;* and considerations on the family Trichosomoididae Yorke and Maplestone, 1926. *J. Helminth.* **35** (Suppl.), 5–8.

Bornstein, S., and Iwarsson, K. (1980). Nasal mites in a colony of Syrian hamsters. *Lab. Anim.* **14,** 31–33.

Brockis, D. C. (1973). Demodectic mites in the Syrian hamster. *Vet. Rec.* **92,** 349–350.

Brugerolle, G. (1975). Contribution à l'étude cytologique et phyletique des diplozoaires. *Protistologica* **11,** 111–118.

Brugerolle, G. (1981). The ultrastructure of the undulating membrane as a feature differentiating the 2 species of tritrichomonas *T. muris* and *T. minuta* parasites of rodents. *Protistologica* **17,** 431–438.

Castellino, S., and DeCarneri, I. (1964). Frequenza delle giardiasi in vari roditori de laboratorie e attechimento delle giardie dell hamster nel ratta. *Parassitologia (Rome)* **6,** 55.

Chakraborty, J., DasGupta, N. N., and Ray, H. N. (1961). An electron microscope study of *Trichomonas criceti. Cytologia* **26,** 320–326.

Chalupsky, J., Vaura, J., and Bedrnik, P. (1979). Encephalitozoonosis in laboratory animals: A serological survey. *Folia Parasitol. (Prague)* **26,** 1–8.

Chan, K. F. (1952). Life cycle studies on the nematode *Syphacia obvelata. Am. J. Hyg.* **56,** 14–21.

Chesterman, F. C. (1972). Background pathology in a colony of golden hamsters. *Prog. Exp. Tumor Res.* **16,** 50–68.

Chesterman, F. C., and Buckley, H. C. (1965). *Trichosomoides* sp. from the nasal cavities of a hamster. *Trans. R. Soc. Trop. Med. Hyg.* **59,** 8.

Chute, R. M. (1961). Infections of *Trichinella spiralis* in hibernating hamsters. *J. Parasitol.* **47,** 25–29.

Comley, J. C. W. (1980). The expulsion of *Aspiculuris tetraptera* and *Syphacia* spp. from mice after anthelmintic treatment. *Int. J. Parasitol.* **10,** 205–211.

Daniel, W. A., Mattern, C. F. T., and Honigberg, B. M. (1971). Fine structure of the mastigant system in *Tritrichomonas muris* (Grassi). *J. Protozool.* **18,** 575–586.

deRoever-Bonnet, H., and Rijpstra, A. C. (1961). *Syphacia obvelata* in the brain of a golden hamster. *Trop. Geogr. Med.* **13,** 167–170.

Dick, T. A., Quentin, J. C., and Freeman, R. S. (1973). Description of *Syphacia mesocriceti,* parasite of the golden hamster. *J. Parasitol.* **59,** 256–259.

Dykes, A. C., Juranek, D. D., Lorenz, R. A., Sinclair, S., Jakuboski, W., and Davies, R. (1980). Municipal water-borne giardiasis: An epidemiologic investigation. *Ann. Intern. Med.* **92,** 165–170.

Eisenbrandt, D. L., and Russell, R. J. (1979). Scanning electron microscopy of *Spironucleus* (Hexamita) *muris* infection in mice. *Scanning Electron Microsc.* **3,** 23–27.

Estes, P. C., Richter, C. B., and Franklin, J. A. (1971). Demodectic mange in the golden hamster. *Lab. Anim. Sci.* **21,** 825–828.

Fahmy, M. A., Mikhail, J. W., and McConnell, E. (1967). Some nematode parasites collected from small mammals in the U.A.R. *J. Vet. Sci. U.A.R.* **4,** 153–164.

Faust, E. C., and Russell, P. F. (1964). "Craig and Faust's Clinical Parasitology," 7th ed. Lea & Febiger, Philadelphia, Pennsylvania.

Filice, F. P. (1952). Studies on the cytology and life history of a *Giardia* from the laboratory rat. *Univ. Calif., Berkeley, Publ. Zool.* **57,** 53–143.

Flatt, R. E., and Kerber, W. T. (1968). Demodectic mite infestation in golden hamsters. *Lab. Anim. Dig.* **4,** 6–7.

Flatt, R. E., Halvonsen, J. A., and Kemp, R. L. (1978). Hexamitiasis in a laboratory mouse colony. *Lab. Anim. Sci.* **28,** 62–65.

Flynn, R. J. (1973). "Parasites of Laboratory Animals." Iowa State Univ. Press, Ames.

Fox, J. G., Newcomer, C. E., and Rozmiarek, H. (1984). Selected zoonoses and other health hazards. *In* "Laboratory Animal Medicine" (J. G. Fox, B. J. Cohen, and F. M. Loew, eds.), pp. 613–648. Academic Press, New York.

Frenkel, J. K. (1972). Infection and immunity in hamsters. *Prog. Exp. Tumor Res.* **16,** 326–367.

Frenkel, J. K. (1976). *Pneumocystis jironeci* n. sp. from man: Morphology, physiology and immunology in relation to pathology. *Natl. Cancer Inst. Monogr.* **43,** 13–30.

Frenkel, J. K., Good, J. F., and Shultz, J. A. (1966). Latent *Pneumocystis* infection of rats, relapse and chemotherapy. *Lab. Invest.* **15,** 1559–1577.

Fulton, J. D. (1943). The treatment of *Notoedres* infections in golden hamsters (*Cricetus auratus*) with dimethyldiphenylene disulphide and tetraethylthiuram monosulphide. *Vet. Rec.* **55,** 219.

Fulton, J. D., and Joyner, L. P. (1948). Natural amoebic infections in laboratory rodents. *Nature (London)* **161,** 66–68.

Gabel, J. R. (1954). Morphology and taxonomy of intestinal protozoa of American woodchuck. *J. Morphol.* **94(3),** 473–549.

Gothe, R., and Kreier, J. P. (1977). *Aegyptianella, Eperythrozoon,* and *Haemobartonella. In* "Parasitic Protozoa" (J. P. Kreier, ed.), Vol. 4, pp. 251–294. Academic Press, New York.

Grant, D. R., and Woo, P. T. K. (1978a). Comparative studies of *Giardia* spp. in small mammals in southern Ontario. I. Prevalence and identity of the parasites with a taxonomic discussion of the genus. *Can. J. Zool.* **56,** 1348–1359.

Grant, D. R., and Woo, P. T. (1978b). Comparative studies of *Giardia* spp. in small mammals in southern Ontario. II. Host specificity and infectivity of stored cysts. *Can. J. Zool.* **56,** 1360–1366.

Grassi, B. (1887). Entwicklungscyclus der *Taenia nana*. *Zentrabl. Bakteriol., Parasitenkd. Infektionskr.* **2,** 305–312.

Griesemer, R. A. (1958). Bartonellosis. *J. Natl. Cancer Inst. (U.S.)* **20,** 949–955.

Habermann, R. T., and Williams, F. P. (1958). The identification and control of helminths in laboratory animals. *J. Natl. Cancer Inst. (U.S.)* **29,** 979–1009.

Hagen, C. A., Shefner, A. M., and Ehrlich, R. (1965). Intestinal microflora of normal hamsters. *Lab. Anim. Care* **15,** 185–193.

Handler, A. H. (1965). Spontaneous lesions of the hamster. In "The Pathology of Laboratory Animals" (W. E. Ribelin and J. R. McCoy, eds.), pp. 210–240. Thomas, Springfield, Illinois.

Hartmann, P. (1967). Activation of toxoplasmosis infections with tumor cell free extracts. *Proc. Am. Assoc. Cancer Res.* **8,** 26.

Hasegawa, H. (1981). Two nematode species of the genus *Syphacia* (Oxyuridae) collected from rodents in Niigata perfecture, Japan. *Jpn. J. Parasitol.* **30,** 325–329.

Heath, A. C., and Bishop, D. M. (1984). Treatment of mange in guinea pigs, hamsters, and hedgehogs. *N. Z. Vet. J.* **32,** 120.

Heyneman, D. (1961). Studies on helminth immunity. III. Experimental verification of autoinfection for *cysticercoids* of *Hymenolepis nana* in the white mouse. *J. Infect. Dis.* **109,** 10–18.

Heyneman, D. (1963). Host–parasite resistance patterns—some implications from experimental studies with helminths. *Ann. N.Y. Acad. Sci.* **113,** 114–129.

Hindle, E. (1947). The golden hamster. In "The U.F.A.W. Handbook on the Care and Management of Laboratory Animals" (A. N. Worden, ed.), pp. 196–202. Williams & Wilkins, Baltimore, Maryland.

Honigberg, B. M. (1963). Evolutionary and systematic relationships in the flagellate order Trichomonadida Kirby. *J. Protozool.* **10,** 20–63.

Honigberg, B. M., Mattern, C. F. T., and Daniel, W. A. (1971). Fine structure of the mastigont system of *Tritrichomonas foetus*. *J. Protozool.* **18,** 183–198.

Hsu, C. K. (1979). Parasitic diseases. In "The Laboratory Rat" (H. J. Baker, J. R. Lindsey, and S. H. Weisbroth, eds.), vol. 1, pp. 307–331. Academic Press, New York.

Hsu, C. K. (1982). Protozoa. In "The Mouse in Biomedical Research" (H. L. Foster, J. D. Small, and J. G. Fox, eds.), Vol. 2, pp. 359–372. Academic Press, New York.

Hughes, H. C., Barthel, C. H., and Lang, C. M. (1973). Niclosamide as a treatment for *Hymenolepis nana* and *Hymenolepis diminuta* in rats. *Lab. Anim. Sci.* **23,** 72–73.

Humiczewska, M., and Skotorczak, B. (1984). The parasitizing effect of '*Balantidium coli* (Malmsten) on the liver metabolism of *Mesocricetus auratus* (Waterhouse). *Folia Biol. (Krakou)* **32,** 189–194.

Joyeaux, C., and Baer, J. G. (1945). Morphologie, évolution et position systématique de *Catenotaenia pusilla* (Goeze, 1782), parasite de rongeurs. *Rev. Suisse Zool.* **52,** 13–51.

Keefe, T. J., Scanlon, J. E., and Wetherald, L. D. (1964). *Ornithonyssus bacoti* (Hirst) infestation in mouse and hamster colonies. *Lab. Anim. Care* **14,** 366–369.

Kellogg, H. S., and Wagner, J. E. (1982). Experimental transmission of *Syphacia obvelata* among mice, rats, hamsters and gerbils. *Lab. Anim. Sci.* **32,** 500–501.

Kinzel, V., and Meiser, J. (1968). Mikrosporidien-infektion (*Nosema cuniculi*) an einern transplantabeln Plasmocytom-Demonstration. *Verh. Dtsch. Ges. Pathol.* **52,** 453–455.

Kirby, H., and Honigberg, B. (1949). Flagellates of the caecum of ground squirrels. *Univ. Calif. Publ. Zool.* **53,** 315–350.

Kiribayashi, S. (1933). Studies on the growth of *Hymenolepis nana*, with special reference to the possibility of differentiation of *H. nana* var. *fraterna*. *Taiwan Igakkai Zasshi* **32,** 1175–1190.

Kirschenblatt, I. D. (1949). On the helminth fauna of *Mesocricetus auratus brandti* Nehr. *Uch. Zap. Leningr. Gos. Univ. im. A. A. Zhdanova, Ser. Biol. Nauk* **19,** 110–127.

Kunstyr, I. (1977). Infectious form of *Spironucleus* (*Hexamita*) *muris:* Banded cysts. *Lab. Anim.* **11,** 185–188.

Kunstyr, I., and Ammerpohl, E. (1978). Resistance of faecal cysts of *Spironucleus muris* to some physical factors and chemical substances. *Lab. Anim.* **12,** 95–97.

Kunsytr, I., and Friedhoff, K. T. (1980). "Parasitic and mycotic Infections of Laboratory Animals, 7th ICLAS Symp. Fischer, Stuttgart.

Kunstyr, I., Ammerpohl, E., and Meyer, B. (1977). Experimental spironucleosis (hexamitiasis) in the nude mouse as a model for immunologic and pharmacologic studies. *Lab. Anim. Sci.* **27,** 782–788.

Larsh, J. E., Jr. (1944). Comparative studies on a mouse strain of *Hymenolepis nana* var. *fraterna* in different species and varieties of mice. *J. Parasitol.* **30,** 21–25.

Larsh, J. E., Jr. (1946). A comparative study of *Hymenolepis* in white mice and golden hamsters. *J. Parasitol.* **32,** 477–479.

Lennette, E. H., Spaulding, E. H., and Truant, J. P. (1974). "Manual of Clinical Microbiology," 2nd ed., Am. Soc. Microbiol., Washington, DC.

Levine, N. D. (1961). "Protozoan Parasites of Domestic Animals and of Man." Burgess, Minneapolis, Minnesota.

Levine, N. D. (1973). "Protozoan parasites of Domestic Animals and of Man." Burgess, Minneapolis, Minnesota.

Litchford, R. G. (1963). Observations on *Hymenolepis microstoma* in three laboratory hosts: *Mesocricetus auratus, Mus musculus* and *Rattus norvegicus*. *J. Parasitol.* **49,** 403–410.

MacArthur, J. A., and Wood, M. (1978). Control of oxyurids in mice using thiabendazole. *Lab. Anim.* **12,** 141–143.

Mattern, C. F. T., and Daniel, W. A. (1980). *Tritrichomonas muris* in the hamster: Pseudocysts and the infection of newborn. *J. Protozool.* **27,** 435–439.

Mauer, S. I., Stanber, L. A., and Grun, J. (1953). The Syrian hamster, *Cricetus auratus,* host of *Protospirura muris* (Note). *J. Parasitol.* **39,** 227–228.

Meiser, J., Kinzel, V., and Jirovec, O. (1971). Nosematosis as an accompanying infection of plasmacytoma ascites in Syrian golden hamsters. *Pathol. Microbiol.* **37,** 249–260.

Mikhail, J. W., and Fahmy, M. A. M. (1968). Two new records of the genus *Mathevotaenia* (cestodes) with description of a new species and a review of the genus. *Zool. Anz.* **180,** 335–344.

Moller, T. (1968). A survey of toxoplasmosis and encephalitozoonosis in laboratory animals. *Z. Versuchstierkd.* **10,** 27–38.

Mudrow-Reichenow, V. (1956). Spontanes vorkommen von amoben und ciliaten bei laboratoriumstiesen. *Tropenmed. Parasitol.* **7,** 198–216.

Myers, D. D. (1973). Sensitivity to X-irradiation of mice infected with *Hexamita muris*. *Abstr. Pap., 24th Annu. Meet. Am. Assoc. Lab. Anim. Sci.*

Neal, R. A. (1947). *Entamoeba* sp. from the Syrian hamster (*Cricetus auratus*). *Nature (London)* **159,** 502.

Nutting, W. B. (1950). Studies of the genus *Demodex,* Owen (Acari, Demodicodea, Demodicidae). Ph.D. Thesis, Cornell University, Ithaca, New York.

Nutting, W. B. (1961). *Demodex aurati* sp. nov. and *D. criceti*, ectoparasites of the golden hamster. *Parasitol.* **51,** 515–522.

Nutting, W.B. (1964). Demodicidae—status and prognostics. *Acarologia* **6,** 441–454.

Nutting, W. B. (1965). Host–parasite relations: Demodicidae. *Acarologia* **7,** 301–317.

Nutting, W. B. (1976). Hair follicle mites (*Demodex spp.*) of medical and veterinary concern. *Cornell Vet.* **66,** 214–231.

Nutting, W. B., and Rauch, H. (1958). *Demodex criceti* n. sp. (Acarina: Demodicidae) with notes on its biology. *J. Parasitol.* **44,** 328–333.

Nutting, W. B., and Rauch, H. (1961). The effect of biotin deficiency in *Mesocricetus auratus* on parasites of the genus *Demodex. J. Parasitol.* **47,** 319–322.

Nutting, W. B., and Rauch, H. (1963). Distribution of *Demodex aurati* in the host (*Mesocricetus auratus*) skin complex. *J. Parasitol.* **49,** 323–329.

Ogden, C. G. (1971). Observations on the systematics of nematodes belonging to the genus *Syphacia* Seurat, 1916. *Bull. Br. Mus. (Nat. Hist.), Zool.* **20,** 253–280.

Ogura, K. (1936). Studies on the *Hymenolepis nana* in Korea. *J. Chosen Med. Assoc.* **26,** 649–668.

Olson, T. A., and Dahms, R. G. (1946). Observations on the tropical rat mite, *Liponyssus bacoti*, as an ecto-parasite of laboratory animals and suggestions for its control. *J. Parasitol.* **32,** 56–60.

Owen, D. (1976). Cestodes in laboratory mice: Isolation of *Cataemotaenia pusilla. Lab. Anim.* **10,** 59–64.

Owen, D., and Turton, J. A. (1979). Eradication of the pinworm *Syphacia obvelata* from an animal unit by anthelmintic therapy. *Lab. Anim.* **13,** 115–118.

Owen, D., and Young, C. (1973). The occurrence of *Demodex aurati* and *Demodex criceti* in the Syrian hamster in the United Kingdom. *Vet. Rec.* **92,** 282–284.

Pakes, S. P. (1974). Protozoal diseases. *In* "The Biology of the Laboratory Rabbit" (S. H. Weisbroth, R. E. Flatt, and A. L. Kraus, eds.), pp. 263–286. Academic Press, New York.

Quentin, J. C. (1971). Morphologie comparée des structures céphaliques et genitales des oxyures du genre Syphacia. *Am. Parasitol.* **46,** 15–60.

Ray, H. N., and Sen Gupta, P. C. (1958). On *Trichomonas criceti* n. sp. found in the caecum of the golden hamster. *Bull. Calcutta Sch. Troop. Med.* **6,** 16.

Read, C. P. (1951). *Hymenolepis diminuta* in the Syrian hamster. (Note) *J. Parasitol.* **37,** 324.

Redha, von F., and Horning, B. (1980). Haarwarmbefall (*Trichosomoides nasalis*) der Nasenhohlen eines Goldhamsters. *Schweiz. Arch. Tierheilkd.* **122,** 357–358.

Renshaw, H. W., Van Hoosier, G. L., and Amend, N. K. (1975). A survey of naturally occurring diseases of the Syrian hamster. *Lab. Anim.* **9,** 179–191.

Retnasabapathy, A., and Lourdusamy, D. (1974). *Demodex aurati* and *Demodex criceti* in the golden hamster (*Mesocricetus auratus*). *Southeast Asian J. Trop. Med. Public Health.* **5,** 460.

Ristic, M., and Lewis, G. E. (1977). Babesia in man and other mammals. *In:* "Parasitic Protozoa" (J. P. Kreier, ed.), Vol. 4, pp. 53–76. Academic Press, New York.

Roberts-Thompson, I. C., Stevens, D. P., Mahmond, A. F., and Warren, K. (1976). Giardiasis in the mouse: An animal model. *Gastroenterology* **71,** 57–61.

Ronald, N. C., and Wagner, J. E. (1975). Treatment of *Hymenolepis nana* in hamsters with Yomesan. *Lab. Anim. Sci.* **25,** 219–220.

Ross, C. R., Wagner, J. E., Wightman, S.R., and Dill, S. E. (1980). Experimental transmission of *Syphacia muris* among rats, mice, hamstersand gerbils. *Lab. Anim. Sci.* **30,** 35–37.

Ruitenberg, E. J., and Kruyt, B. C. (1975). Effect of intestinal flagellates on immune response of mice. *Parasitology* **71,** R30.

Saeki, H., Imai, S., Hiyama, M., Fujita, J., and Ishii, T. (1982). An oxyurid nematode of the genus *Syphacia* from the laboratory golden hamster. *Jpn. J. Vet Sci.* **44,** 115–124.

Saeki, Y. (1920). Experimental studies on the development of *Hymenolepis nana. Jikwa Zasshi* **238,** 203–244.

Sano, M., Majumder, S., Watarai, S., and Odaka, S. (1975). Spontaneous infection of *Hymenolepis nana* in hamsters. *Exp. Anim. (Jpn.)* **24,** 41–44.

Sasa, M., Tanka, H., Fukui, M., and Takata, A. (1962). Internal parasites of laboratory animals. *In* "The Problems of Laboratory Animal Disease" (R. J. L. Harris, ed.), pp. 195–214. Academic Press, New York.

Saxe, L. H. (1954a). Transfaunation studies on the host specificity of the enteric protozoa of rodents. *J. Protozool.* **1,** 220–230.

Saxe, L. H. (1954b). The enteric protozoa of laboratory golden hamsters. Abstract 26, *J. Parasitol.* **40,** Suppl., 20.

Saxe, L. H., and Batipps, F. W. (1950). The distribution of *Trichomonas muris* in the gut of the golden hamster. *Proc. Am. Soc. Protozool.* **1,** 6–7.

Schloemer, L. (1982). Die Übertagung von *Cryptosporidium* sp. des Kalbes auf Mause. Hamster and Meerschweinchen sowie Schweine, Schafe und Ziegen. Inaug. Diss., p. 44. Ludwig-Maximilians Univ., Munchen.

Schnur, L. F. (1975). Observations on the transfer stages of trichomonads from Syrian hamsters. *J. Protozool.* **22,** 55A.

Scott, H. G. (1958). Control of mites on hamsters. *J. Econ. Entomol.* **51,** 412–413.

Sebesteny, A. (1969). Pathogenicity of intestinal flagellates in mice. *Lab. Anim.* **3,** 71–77.

Sebesteny, A. (1979a). Transmission of *Spironucleus* and *Giardia* spp. and some nonpathogenic intestinal protozoa from infested hamsters to mice. *Lab. Anim.* **13,** 189–191.

Sebesteny, A. (1979b). Syrian hamsters. *In* "Handbook of Diseases of Laboratory Animals" (J. M. Hime and P. N. O'Donoghue, eds.), pp. 111–136. Heinemann Veterinary Books, London.

Seliukaite, Z. (1977). Trichomonas of rodents: Host–parasite relations (in mouse, rat, hamster, field vole and suslik). *Acta Parasitol. Litu.* **15,** 25–34.

Shadduck, J. A. (1969). *Nosema cuniculi in vitro* isolation. *Science* **166,** 1516–517.

Shadduck, J. A., and Pakes, S. P. (1971). Encephalitozoonosis (nosematosis) and toxoplasmosis. *Am. J. Pathol.* **64,** 657–674.

Sheffield, F. W., and Beveridge, E. (1962). Prophylaxis of "wet-tail" in hamsters. *Nature (London)* **196,** 294–295.

Simitch, T., Petrovitch, Z., and Lepech, T. (1954). Contribution à la connaissance de la biologie des Trichomonas. II. Différenciation de *T. microti* Wenrich et Saxe, 1950 et de *T. intestinalis* Leuckart 1879, pars leurs caractères biologiques. *Ann. Parisitol.* **29,** 199–205.

Simmons, M. L., Richter, C. B., Franklin, J. A., and Tennant, R. W. (1967). Prevention of infectious diseases in experimental mice. *Proc. Soc. Exp. Biol. Med.* **126,** 830–837.

Soave, O. A. (1963). Diagnosis and control of common diseases of hamsters, rabbits and monkeys. *J. Am. Vet. Med. Assoc.* **142,** 285–290.

Stahl, W. (1961). *Syphacia muris*, the rat pinworm. *Science* **133,** 576–577.

Stallard, H. E., and Arai, H. P. (1978). The growth and development of *Hymenolepis peromysci* Tinkle, 1972 (Cestoda: Cyclophyllidea). *Can. J. Zool.* **56,** 90–93.

Stoll, N. R. (1947). This wormy world. *J. Parasitol.* **33,** 1–18.

Stone, W. B., and Manwell, R. D. (1966). Potential helminth infections in humans from pet or laboratory mice and hamsters. *Public Health Rep.* **81,** 647–653.

Stunkard, H. W. (1945). The Syrian hamster, *Cricetus auratus*, host of *Hymenolepis nana. J. Parasitol.* **31,** 151.

Taffs, L. F. (1976a). Pinworm infections in laboratory rodents: A review. *Lab. Anim.* **10,** 1–13.

Taffs, L. F. (1976b). Further studies on the efficacy of thiabendazole given in the diet of mice infected with *H. nana, S. obvelata* and *A. tetraptera. Vet. Rec.* **99,** 143–144.

Taffs, L. F. (1975). Continuous feed medication with thiabendazole for the removal of *Hymenolepis nana, Syphacia obvelata*, and *Aspicularis tetraptera* in naturally infected mice. *J. Helminthol.* **49,** 173–177.

Tripathi, K. P., and Ray, D. K. (1976). A note on the occurrence of *Cysticercus fasciolaris* in golden hamsters. *Indian Vet. J.* **53,** 329–330.

Tsuchiya, H., and Rohlfing, E. H. (1932). *Hymenolepis nana. Am. J. Dis. Child.* **43,** 865–872.

Tzipori, S. (1983). Cryptosporidiosis in animals and humans. *Microbiol. Rev.* **47,** 84–96.

Uchimura, R. (1922). On the development of *Hymenolepis nana* and *Hymenolepis murina*. *Jikwa Zasshi* **240,** 268.

Unay, E. S., and Davis, B. J. (1980). Treatment of *Syphacia obvelata* in the Syrian hamster with piperazine citrate. *Am. J. Vet. Res.* **41,** 1899–1900.

Van Hoosier, G. L., and Ladiges, W. C. (1984). Biology and diseases of hamsters. *In* "Laboratory Animal Medicine" (J. G. Fox, B. J. Cohen, and F. M. Loew, eds.), pp. 123–147. Academic Press, New York.

Verster, A. J. M., and Brooker, D. (1970). Helminth parasites of small laboratory animals at the Veterinary Research Institute, Onderstepoort. *J. S. Afr. Vet. Med. Assoc.* **41,** 183–184.

von Fricsay, M. (1956). Lungenveränderungen bei Laboratoriumstratten infolge Infektion mit *Trichosomoides crassicauda*. *Schweiz. Z. Allg. Pathol. Bakteriol.* **19,** 351–355.

Wagner, J. E. (1970). Control of mouse pinworms, *Syphacia obvelata* utilizing dichlorvos. *Lab. Anim. Care* **20,** 39–44.

Wagner, J. E., Doyle, R. E., Ronald, N. C., Garrison, R. G., and Schmitz, J. A. (1974). Hexamitiasis in laboratory mice, hamsters and rats. *Lab. Anim. Sci.* **24,** 349–354.

Wantland, W. W. (1955). Parasitic fauna of the golden hamster. *J. Dent. Res.* **34,** 631–649.

Wantland, W. W. (1956). Trichomonads in the golden hamster. *Trans. Ill. State Acad Sci.* **48,** 197–201.

Wantland, W. W., and Johansen, E. (1954). Effect of carbasone, chiniofon and high protein diet on trichomonads in the intestine of the golden hamster. *J. Parasitol.* **40,** 479–480.

Wantland, W. W., and Lichtenstein, E. P. (1954). Intestinal myiasis in the Syrian hamster. *J. Parasitol.* **40,** 365.

Watson, J. M. (1946). Helminths infective to man in the Syrian hamster. *Br. Med. J.* **2,** 578.

Weinman, D., and Kreier, J. P. (1977). *Bartonella* and *Grahamella*. *In* "Parasitic Protozoa" (J. P. Kreier, ed.), Vol. 4, pp. 198–234. Academic Press, New York.

Welshman, M. D. (1973). Demodectic mange in hamsters. *Vet. Rec.* **92,** 684.

Wenrich, D. H. (1924). Trichomonad flagellates in the cecum of rats and mice. *Anat. Rec.* **29,** 118.

Wenrich, D. H. (1946). Culture experiments on intestinal flagellates. I. Trichomonad and other flagellates obtained from man and certain rodents. *J. Parasitol.* **32,** 40–53.

Wenrich, D. H., and Nie, D. (1949). The morphology of *Trichomonas wenyoni*. *J. Morphol.* **85,** 518–531.

Wenrich, D. H., and Saxe, L. H. (1950). *Trichomonas microti*, n. sp. (Protozoa, Mastigophora). *J. Parasitol.* **36,** 261–269.

Wescott, R. B. (1982). Helminths. *In* "The Mouse in Biomedical Research" (H. L. Foster, J. D. Small, and J. G. Fox, eds.), Vol. 2, pp. 373–384. Academic Press, New York.

Wilhelm, W. E., and Pope, D. C. (1974). *Tritrichomonas muris* from the golden hamster. *J. Protozool.* **21,** 434–435.

Woodland, W. N. F. (1924). *Hymenolepis nana* and *Hymenolepis fraterna*. *Lancet,* Vol. 1, May 3, 922–923.

Yamaguti, S. (1961). The nematodes of vertebrates. *In* "Systema Helminthum" (S. Yamaguti, ed.), Vol. 3, Parts 1 and 2. Wiley (Interscience), New York.

Yoshida, Y., Yamada, M., Shiota, T., Ikai, T., Takeuchi, S., and Ogino, K. (1981). Provocation experiment: *Pneumocystis carinii* in several kinds of animals. *Zentralbl. Bakteriol., Parasitenkd., Infektionskr. Hyg., Abt. 1: Orig., Reine A* **250,** 206–212.

Zubaidy, A. J., and Majeed, S. K. (1981). Pathology of the nematode *Trichosomoides crassicauda* in the urinary bladder of laboratory rats. *Lab. Anim.* **15,** 381–384.

# Chapter 9

Neoplastic Diseases

John D. Strandberg

I. Introduction 157
II. Respiratory System 158
III. Gastrointestinal System 159
IV. Urinary System 161
V. Musculoskeletal System 161
VI. Tumors of Soft Connective Tissues 161
VII. Integumentary System 161
VIII. Circulatory System 163
IX. Hematopoietic System 163
X. Endocrine System 164
XI. Reproductive System 166
XII. Nervous System 167
    References 167

I. INTRODUCTION

The purpose of this chapter is to provide an illustrated overview of the spontaneous tumors of Syrian hamsters. It is based on previous summaries by Van Hoosier and Trentin (1979), Squire *et al.* (1978), Kirkman and Algard (1968), and Handler and Chesterman (1968), and on the description of tumors of the hamster edited by Turusov (1982). This last document is a detailed survey of both spontaneous and induced tumors by a number of scientists who use hamsters in experimental oncology. Following brief introductory comments on overall tumor incidence, neoplastic lesions are described by major tissue of origin, and selective illustrations are presented.

In terms of total numbers of laboratory rodents used in biomedical research, Syrian hamsters fall a distant third behind rats and mice. They are still important research subjects, however, and in special cases such as in the use of their cheek pouches for tumor induction and transplantation and in other forms of experimental carcinogenesis, they contribute significantly to scientific advancement. While there are several strains of inbred hamsters in various laboratories, most hamsters used in biomedical research are outbred animals.

Descriptions of spontaneous tumors in hamsters are scattered through the literature and vary in detail. Van Hoosier and Trentin (1979) summarized most of the reports of hamster tumors appearing prior to 1976 and presented them in tabular form. They divided the tumors into major categories of benign and malignant lesions and further subdivided them by site of origin and histological morphology. This schema was modified from one published by Stewart *et al.* (1959). The review emphasized the point that hamsters have a relatively low tumor incidence (3.7% malignancies) when compared to laboratory rats and mice.

A comprehensive review of spontaneous and induced neoplasms in hamsters was published by the International Agency for Research on Cancer (Turusov, 1982). This publication includes excellent illustrations of both gross and histological features of lesions classified by site or origin. It highlights the point that most of the tumors observed in hamsters have been experimentally induced; spontaneous tumor levels are low.

Other reports of hamster neoplasia include the early work of Chesterman (1972), who described the background pathology of the hamster colony at the Mill Hill laboratory; the animals were derived from stock brought to England from Syria in 1931. Primary tumors encountered were presented in tabular form and included a limited variety of lesions similar to those noted by other subsequent investigators. No illustrations were provided. In a summary of spontaneous lesions occurring in control animals, Mohr (1970) reported malignant tumors including a hemangiosarcoma of the spleen, a "rhabdomyosarcoma of the skin," fibrosarcoma, cholangiocarcinoma, and myeloid leukemia. Most of these occurred as individual cases and were not further described. In a beautifully illustrated description of the lesions of the endocrine glands and sex organs, a low tumor incidence was also noted by Russfield (1966). Squire and colleagues (1978) included hamster lesions in their description of tumors of laboratory animals, but emphasis was placed on lesions of other species. A recent review of tumors and other lesions was presented by Schmidt and co-workers (1983), who studied 750 control male and 91 female Syrian hamsters of an unspecified strain. They presented an illustrated listing of the tumors and other lesions encountered in these animals.

The low tumor incidence in the above reports contrasted sharply with a series of reports issued by Parviz Pour and associates (1976a–d, 1979; Pour and Birt, 1979) and further described below. In a survey of different hamster colonies, they found considerably higher tumor incidences. It is important to note that these studies were the result of intensive pathological examination of step sections of several organs. These methods had not been applied to hamster materials previously, nor are they routinely used by most laboratories. Significant differences in overall tumor incidence, occurrence of malignant tumors, and sites of primary tumors occurred between two colonies. The tumor incidence in this life-span study was found to be 32% for animals from the Eppley colony and 41% for those from the Hannover colony. Many animals had more than one tumor; most tumors were not seen grossly. In fact, from 75 to 83% of the neoplasms noted were only found on microscopic examination. The majority of these small tumors were in the upper and lower respiratory tracts and the endocrine, digestive, and genital systems. The importance of examination of serial sections of organs in detecting such neoplasms was clearly demonstrated. Other points raised by Pour and collaborators included the fact that tumor incidence was higher in females than in males, even though the males had a greater life span.

Life expectancy varies considerably from one hamster colony to another and ranges from 1 to 3 years. As with neoplasia in many species, increased numbers of tumors are seen in older individuals, and neoplasms are uncommon in animals under 1 year of age.

Life-span studies of cream, white, and albino hamsters revealed strain differences in predominant tumor types and incidences (Pour *et al.*, 1979). Of note is the fact that melanomas were found only in pigmented lines. It was also found that more tumors developed in those strains with longer life spans. Certain tumors were restricted to one strain; that is, squamous cell carcinomas of the oral cavity, tumors of the submandibular gland, nephroblastomas and chondromatosis of the external ear were found only in the white strain.

Homburger (1983) noted the infrequent use of hamsters for bioassays and emphasized the point that there is not necessarily a correlation between incidence and type of spontaneous tumors and those induced by carcinogens. He also made reference to the use of step sections to evaluate tumor incidence more effectively. However, it was felt that the lower tumor incidence levels summarized by Van Hoosier and Trentin (1979) better relate to the more usual monitoring practices. Homburger pointed out the low spontaneous cancer incidence in hamsters relative to rats and noted that they still respond to carcinogens with a high incidence of tumors.

It is known that in many systems nutritional status affects tumor incidence and type. In a dietary study several groups of hamsters were fed fat at low, medium, and high levels at early and late stages of life (Birt and Pour, 1983). Diets in most groups were altered at 8 weeks and continued at the second level for the animals' lives. High-fat diets after 8 weeks caused an increased incidence of thyroid adenomas, vaginal papillomas, and adrenocortical adenomas, but the first two were in accordance with increased life span. Salivary gland adenocarcinomas were found only in animals on high-fat diets. Hyperplasia of various cellular elements was also documented, but striking changes in incidence of common neoplasms were not observed.

Primary tumors of several organ sites have been noted to occur in a single animal. This phenomenon was stressed by Pour and co-workers (1976b), who found incidences of 40 and 32% of multiple neoplasms in the two colonies which they reported. In many cases there was no obvious relationship between tumors, and benign and malignant lesions occurred together. In other cases, tumors of various endocrine glands were found in animals which also possessed neoplasms of the liver and other portions of the digestive tract.

II. RESPIRATORY SYSTEM

Despite their frequent and productive use in the investigation of carcinogens acting on the respiratory system, hamsters, like rats and guinea pigs, have relatively few primary tumors of this system. Even in the intensive studies of Pour and associates (1976b, 1979), such lesions were relatively uncommon and found only in up to 3% of the animals. Many of these lesions were tumors of the nasal cavity and upper trachea; these included benign polyps covered with stratified squamous or cuboidal epithelium overlying a delicate fibrous stroma. Among malignancies reported are clear-cell carcinomas of the larynx; these lesions are of unknown cellular origin and have been seen in hamsters in several colonies. They are not usually observed grossly. Rare malignancies of the nasal cavity included carcinoma of the nasoturbinates; these demonstrated local invasion and were typified by formation of rosettes and by mucus production.

Spontaneous malignant epithelial tumors of the lower bronchial tree and alveoli are extremely uncommon. Those which have been described include polyps of the trachea covered with mucus-producing tracheal epithelium, bronchogenic adenomas with mucus production, and rare bronchial carcinomas.

In contrast to the low incidence of spontaneous tumors are many reports of experimental pulmonary carcinogenesis. In these, intratracheal instillation of polynuclear hydrocarbons results in benign and malignant squamous cell lesions of the tracheobronchial lumen and bronchiolar–alveolar tumors which are usually benign (Safiotti, 1970). The squamous lesions are accompanied by anaplastic carcinomas and adenocarcinomas of the bronchi.

The lung is, of course, a common site for tumor metastasis. Many pulmonary tumors have arisen elsewhere; they include carcinomas of the adrenal cortex, melanomas of the skin, lymphomas, and sarcomas from a variety of sites.

III. GASTROINTESTINAL SYSTEM

Considerable variability in incidence of tumors of the digestive tract is encountered in the literature. This is encountered in reports from a range of authors as well as in publications emanating from different colonies studied by the same investigator. Van Hoosier and Trentin (1979) reported numerous polyps of the intestine and squamous papillomas of the stomach (Fig. 1) among benign lesions and noted that adenocarcinomas of the bowel were the most common malignancy reported. The cecum and colon are the usual sites for these lesions. These findings were mirrored in the Hannover colony surveyed by Pour *et al.* (1976b), with a 23% incidence of gastrointestinal neoplasms; this contrasted with the observation by the same authors of a 7% incidence in the Eppley colony. These figures included tumors of the liver (hemangioendotheliomas, cholangiomas, cholangiocarcinomas) and gallbladder. Other reported tumors of the digestive tract include squamous cell papillomas of the oral cavity as well as a squam-

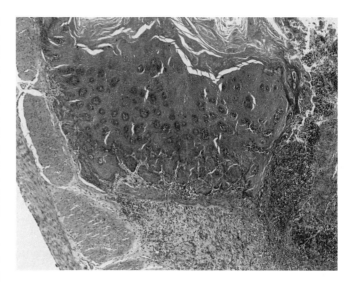

Fig. 1. A benign gastric papilloma is ulcerated and shows proliferation of rete pegs and accumulation of surface keratin. Hematoxylin–eosin stain. Magnification: ×42.

ous cell carcinoma and several odontomas of the molar teeth (Pour *et al.*, 1979).

Lesions of the stomach are usually benign and are, with few exceptions, gastric squamous papillomas. These lesions often develop near the glandular portion of the stomach and appear grossly as roughened, irregular protrusions from the squamous surface. On microscopic examination, they are seen to be composed of papillary outgrowths of parakeratotic squamous epithelial cells covered with thickened layers of keratinized cells. Bacteria may be numerous in these dead surface cells. The basilar portion of the epithelium is thickened, with proliferation of epithelial cords, but usually overlies an intact muscularis mucosae. There is relatively orderly maturation of the squamous epithelium above the basal layer. Focal ulcerations with associated inflammation are not uncommon (Fig. 1). Squamous carcinomas of the stomach and other gastric malignancies occur very rarely as spontaneous events.

Epithelial tumors of the bowel occur as both adenomas and adenocarcinomas and are reported in varying incidence from one study to another. The intestinal lesions are usually well-differentiated tumors with glandular or cystic glandular patterns and mucus production; most arise in the colon, but they can be found in the small bowel as well (Fig. 2).

Adenomas are usually papillary and are seen as cauliflowerlike growths attached to the mucosal surface by a thin fibrovascular stalk. The epithelium is thrown up into folds and is hypercellular with enlarged epithelial cells with increased basophilia. Mucus production may or may not be present. Adenomas show limitation of cellular proliferation to the area above the basement membrane. Malignant lesions have more downward growth of epithelium into the stroma and out into the colonic epithelium surrounding the polyp. First metastasis

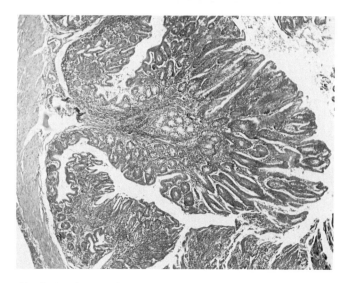

Fig. 2. An adenoma of the small intestine shows the usual polypoid pattern with thickened, hyperchromatic epithelium and a stalk connecting the mass to the bowel wall. Hematoxylin–eosin stain. Magnification: ×22.5.

is usually to regional mesenteric lymph nodes. Intestinal neoplasms have been responsible for significant morbidity and mortality in hamsters (Pour *et al.*, 1976b; Fortner, 1961). A relatively high incidence of intestinal tumors is usually only seen in those studies in which intensive microscopic scrutiny is employed. Primary benign lesions may be small, and the bowel undergoes rapid autolysis.

In the past there has been confusion over the nature of an epizootic disease of young hamsters characterized by extreme glandular proliferation of the epithelium of the ileum. This condition, known as proliferative ileitis and a variety of other names, has been observed by several investigators and may be in part responsible for some, but by no means all, of the historical reports of intestinal adenocarcinoma. It is discussed in Chapter 7 of this volume.

Connective tissue tumors of the digestive system are reported in low frequency; most of those seen have been malignancies of one of the native connective tissue elements of the tract. Thus small numbers of leiomyosarcomas, angiosarcomas, liposarcomas, and fibrosarcomas have been reported; they arise most commonly in the cheek pouches and intestines. Tumors composed of "granular cells" presumably of neural origin have been described in the bowel wall in white hamsters by Pour and co-workers (1973).

The incidence of tumors of the liver and biliary tree is moderate, about 1% in most series, and is thus similar to that of their counterparts in rats and mice. In the study of Pour *et al.* (1979) of three inbred strains, liver tumors included hepatocellular adenomas, cholangiomas, and hemangioendotheliomas. Tumors of hepatocytes occur as single or multiple nodules which vary in color depending on the amount of fat within the tumor cells. They may closely resemble the surrounding hepatic parenchyma or may stand out as gray, tan, or yellow nodules compressing the normal adjacent tissue. There is a range of histological differentiation; well-differentiated tumors are composed of hepatocytes arranged in trabecular patterns and often forming thickened cords and plates of cells. More poorly differentiated tumors can be composed of disorganized masses of cells with open, vesicular nuclei, bizarre and often numerous mitotic figures, and pleomorphic, basophilic cytoplasm. Differentiation of benign from malignant hepatocellular tumors on histological grounds may be difficult. Indeed, many reports do not specify whether a lesion termed a hepatoma is felt to be benign or malignant. Intravascular invasion may be noted in the hepatic sections, such neoplasms are clearly malignant. In other cases tumor size, differentiation, presence or absence of encapsulation, and mitotic activity are all used to arrive at a benign or malignant diagnosis. Collagen is not a prominent component of any of these lesions, although necrosis may be extensive in large malignancies.

Cholangiomas are commonly encountered hepatic neoplasms, and smaller numbers of cholangiocarcinomas have also been found. Sixteen percent of the animals with intestinal neoplasms described by Pour *et al.* (1976b) also had simultaneous tumors of the bile duct. Biliary tumors were also associated with cystic dilatation of the bile ducts. Tumors arising from the intrahepatic bile ducts are most commonly observed as firm white nodules, occasionally cystic, within the substance of the liver. When malignant, they are often multiple. The histological appearance is typified by the presence of fibrous stroma supporting tubules and ducts lined by epithelial cells of variable differentiation. These ducts may be cystic; their epithelium is hyperchromatic and they have a relatively high nuclear to cytoplasmic ratio. When malignant, the connective tissue stroma often increases in relative amount, and neoplastic epithelial cells form greatly distorted tubular formations which grow outward into the surrounding hepatocytes.

Cirrhosis was found by Chesterman (1972) in his colony in 33% of the females and in 20% of the males; 6% of the cirrhotic livers were the site of tumor development. By far the majority of these tumors were cholangiocarcinomas, but hepatomas were also observed.

A survey of changes occurring spontaneously in aging outbred hamsters revealed proliferation of ducts and ductules, and formation of new islets with islet cell hyperplasia, but neoplasms were not described (Takahashi and Pour, 1978). Russfield (1966) noted pancreatic exocrine neoplasms in her review of the endocrine tumors of 1000 hamsters. Pour *et al.* (1976b) also reported two pancreatic adenomas. Exocrine neoplasms of the pancreas are unusual lesions however, and when they occur are found as nodules of relatively well-differentiated pancreatic acinar tissue with variable amounts of supporting fibrous stroma.

IV. URINARY SYSTEM

The kidney is an uncommon site for spontaneous neoplasia. Pour and co-workers (1976c) were able to discover only one renal tubular adenoma in two colonies of hamsters despite an extensive search using step sections. In another study, animals of the albino (*AH*) strain developed multiple tubular adenomas, and two nephroblastomas were also found (Pour *et al.*, 1979). The latter tumors were large and caused renal vein thrombosis. The largest group of primary renal tumors was noted by Kirkman and Algard (1968), who reported renal tubular adenocarcinomas and nephroblastomas in their group of animals. Benign tubular tumors are small, circumscribed nodules of relatively well-differentiated tubular cells located within the cortex. Carcinomas tend to retain their tubular differentiation but expand radially and are accompanied by moderate to large amounts of stromal fibroblastic response. Nephroblastomas are microscopically similar to those found in other species and are highly cellular masses with large nuclei and formations of disorganized tubules and masses of cells resembling glomeruli. Necrosis of these lesions may be extensive.

Similarly, tumors of the urinary bladder are also extremely uncommon. Van Hoosier and Trentin (1979) were able to find a single report of a leiomysarcoma occurring in the bladder wall, and a transitional-cell papilloma has also been described (Turusov, 1982).

V. MUSCULOSKELETAL SYSTEM

Tumors of the musculoskeletal system are encountered only rarely. Osteosarcomas are the most frequently reported neoplasms of this system and can be found in bones anywhere in the body (Turusov, 1982). They are nodular masses composed of polygonal cells embedded in a hyaline matrix. In the survey of the Eppley and Hannover colonies, Pour and colleagues (1976a) reported a vertebral chondroma in one male animal; this animal also had chondromatosis of the vertebral cartilage which resulted in kyphosis with compression of the spinal cord. Neoplasms of synovial joints, bursae, and tendon sheaths seem to be extremely uncommon; of interest is the description of two giant-cell tumors of tendon sheath origin cited by Van Hoosier and Trentin (1979) and others.

VI. TUMORS OF SOFT CONNECTIVE TISSUES

Connective tissue tumors are not often reported; when they are, they tend to be histologically malignant with evidence of

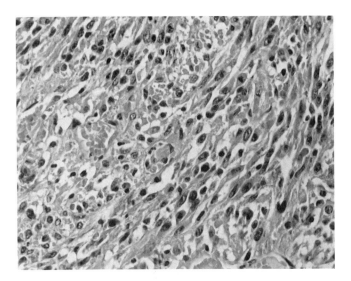

Fig. 3. Cells of a fibrosarcoma have elongated nuclei and variable amounts of eosinophilic cytoplasm. They are arranged in irregular bundles; little collagen is seen in this illustration. Hematoxylin–eosin stain. Magnification: ×30.

local tissue invasion or hematogenous spread (Turusov, 1982). Such lesions can be classified on the basis of their histological appearance into a variety of tissue types including fibrosarcomas, hemangiosarcomas, neurofibrosarcomas, and liposarcomas; undifferentiated sarcomas are also reported (Figs. 3–5). These all tend to be rather expansive lesions at their site of origin and to exhibit metastasis late in their courses. The histological features are essentially those used in differentiating connective tissue neoplasms in all other species. A few benign soft-tissue neoplasms, including retroperitoneal hemangioendothelioma and lipoma, have also been seen.

VII. INTEGUMENTARY SYSTEM

Tumors of the skin are infrequently encountered as spontaneous lesions in aging hamsters, but a variety of neoplasms of the squamous epithelium and skin adnexal structures have been cited (Pour *et al.*, 1976a; Streilein *et al.*, 1981; Van Hoosier and Trentin, 1979). Of the several lesions described, melanomas appear to be the most common (Fortner and Allen, 1958; Squire *et al.*, 1978). There is considerable difference in incidence of melanotic tumors reported from different laboratories as well as in different strains of hamsters. Small skin nodules composed of nests of pigmented cells located just at the basement membrane of the epithelium are termed nevi. Obviously, malignant melanomas are also encountered; these are variably pigmented, and most exhibit junctional activity with malignant cells in small clusters invading the overlying

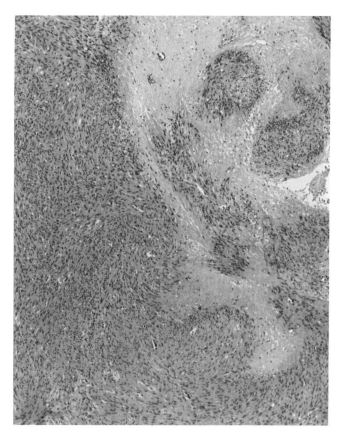

Fig. 4. A malignant schwannoma at low magnification exhibits palisading of cells around irregular areas of necrosis. Hematoxylin–eosin stain. Magnification: ×62.

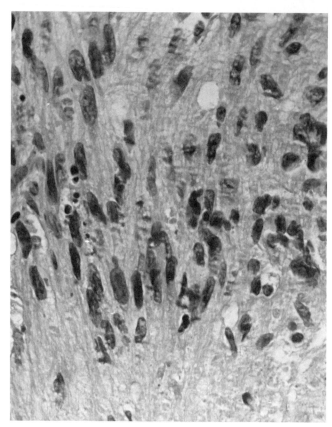

Fig. 5. Higher magnification of malignant Schwann cells reveals elongated nuclei and poorly demarcated pale cytoplasm. An area of necrosis is at the bottom of the figure. Hematoxylin–eosin stain. Magnification: ×618.

squamous epithelium (Fig. 6). Malignant melanomas can be widely metastatic, and nests of neoplastic cells may be found in regional lymph nodes and sites throughout the body.

A wide range of tumors of the skin adnexal structures has also been cited, although their appearance is very sporadic. These neoplasms resemble their counterparts in other species and include benign tumors of the hair follicles and sebaceous glands, keratoacanthomas, squamous papillomas and carcinomas, and basal cell tumors. One of the latter was reported to exhibit metastatic behavior (Kirkman and Algard, 1968). Tumors of the Harderian gland were found in both male and female hamsters (Pour *et al.*, 1976d); they were all benign tumors with papillary glandular patterns composed of large cuboidal or columnar cells. "Warts" were noted with a viroid disease described by Coggin and co-workers (1981, 1983). These papillary lesions appeared as disseminated foci of hyperkeratotic squamous epithelium thrown up in folds overlying a delicate fibrovascular stroma.

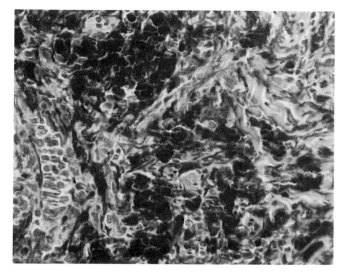

Fig. 6. A cutaneous melanoma has densely pigmented cells infiltrating the dermal collagen. Granules obscure most cellular detail. Hematoxylin–eosin stain. Magnification: ×270.

VIII. CIRCULATORY SYSTEM

Hemangioendotheliomas are noted above in the discussion of tumors of soft tissues. Tumors of vascular origin tend to be the most commonly observed soft-tissue neoplasms. While they may be encountered in any body site, they are usually found as single or multiple lesions of the spleen, liver, and retroperitoneum. Benign vascular tumors take the form of cavernous hemangiomas made up of distended, disorganized vascular channels filled with erythrocytes. They may contain organizing thrombi. The endothelial cells are flat, and the channels are separated by delicate fibrous stroma. There is considerably more variability in morphology of the malignant vascular lesions. These include poorly circumscribed nodules of tissue similar to that just described but with more cellular vascular walls lined by plump, pleomorphic endothelial cells. More undifferentiated tumors may be composed of sheets and bundles of spindle cells with only small areas showing clear vascular differentiation.

IX. HEMATOPOIETIC SYSTEM

In contrast to the systems just described, the hematopoietic system, particularly the lymphoid segment, is the most frequent site of development of spontaneous neoplasms. In one report 10 of 256 females and 14 of 270 males had malignant lymphomas of a spectrum of cell types, and tumors seemed to originate in the mesenteric lymph nodes (McMartin, 1979). These neoplasms have several different patterns of differentiation and can be found in several body sites. The most common situation is that in which there is enlargement of one, or more commonly, several peripheral lymph nodes. Other organs often involved are the bowel (Figs. 7 and 8), liver, kidney, and spleen, although tumor masses can be found anywhere (Fig. 9). Grossly, the tumor infiltrates cause diffuse involvement of the affected organ, or they can produce white nodular masses. Tumor cell differentiation varies from case to case and from report to report (Pour *et al.* 1976d; Van Hoosier and Trentin, 1979; Squire *et al.*, 1978). Tumor impression smears stained with Wright's stain may be used to differentiate tumor cell types better. The most common pattern recognized is that of lymphosarcoma, in which there are sheets of more or less well-differentiated lymphocytes replacing normal structures. There may be interspersed macrophages with cytoplasmic basophilic bodies reflecting phagocytosis of cell debris. The amount of cytoplasm may be variable, but is usually not great in sectioned material. Other cases may show histiocytic differentiation, in which the tumor cells are larger and have irregular, vesicular nuclei surrounded by pale, eosinophilic cytoplasm.

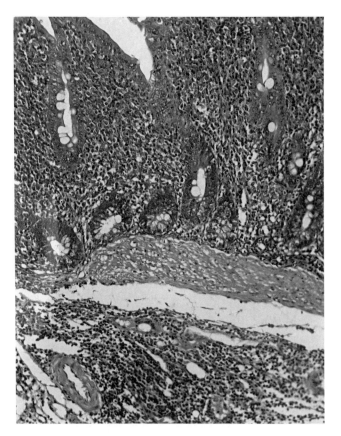

Fig. 7. A focus of differentiated lymphoma in the small intestine involves both the mucosa and submucosa. Neoplastic cells separate glands and blunt villi. Hematoxylin–eosin stain. Magnification: ×138.

Plasma cell differentiation is also reported; in these cases the cells have moderate amounts of basophilic cytoplasm, and prominent Golgi zones may be seen. The patterns of differentiation have varied from report to report and from hamster colony to hamster colony. In the Eppley colony, most of these tumors were of the histiocytic and plasmacytic type; in the Hannover colony, lymphocytic tumors were also found.

The widespread occurrence of the lymphoid tumors and their appearance in epizootic outbreaks have given cause for suspicion that they may have an underlying viral etiology. This was borne out by several reports on horizontally transmitted malignant lymphomas (Coggin *et al.*, 1981, 1983; Manci *et al.*, 1984; Streilein, 1984; Mashiba *et al.*, 1983; Gershon *et al.*, 1967). No retrovirus could be identified, but C-type particles were observed by electron microscopy in both affected and control animals. A novel agent with viroid characteristics was proposed as the cause. Viroids are small infectious structures composed of nucleic acid (DNA or RNA) but free of capsids or envelopes. Cell-free transmission was accomplished; the agent caused lymphoma in five recurrent epizootics, with a 50–90% incidence in young animals. Other clinical syndromes associ-

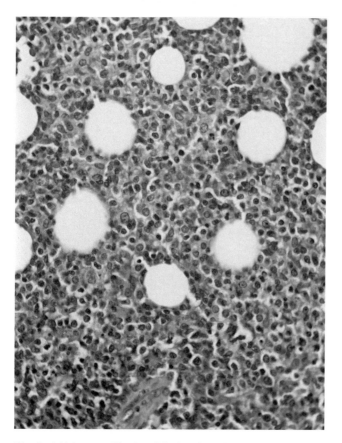

Fig. 8. A higher magnification of the lymphoma in Fig. 7 demonstrates the pleomorphic nature of the malignant lymphoid cells surrounding entrapped lipocytes. Hematoxylin–eosin stain. Magnification: ×428.

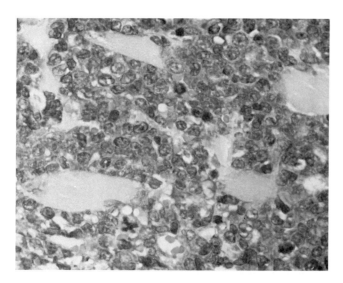

Fig. 9. Scattered skeletal muscle fibers are separated by an infiltrate of neoplastic lymphocytes. Hematoxylin–eosin stain. Magnification: ×506.

ated with the agent were severe enteritis, pyelonephritis, warts, poor breeding efficiency, and intussusceptions. The disease was an important clinical problem, and in one case was responsible for the destruction of a colony of 6000 animals. In a more complete clinical and pathological description (Coggin *et al.*, 1983), it was reported that disease develops in newborn animals; resistance is acquired with age. Tumors occur as spontaneous disease and have involvement of a variety of body sites. Intestinal or intraperitoneal location (46–54%) is most common, followed by liver (33–24%). Other viscera are affected, as is the eye and inguinal area. In the intestine there were two patterns: infiltration from the peritoneum and formation within Peyer's patches. Lymph nodes are affected in many cases, and their architecture is obliterated by the presence of neoplastic lymphoid cells. Hyperplasia of lymphoid tissue is also common. There is some variability in the histological appearance of tumor cells; some were monotonous populations of immature lymphocytes, while others are composed of more pleomorphic cells. Tumors with plasma cellular differentiation were also reported. Tumors appeared over a prolonged time course (to over a year) following exposure to the infectious agent as neonates.

Nonlymphoid neoplasms are considerably less frequent; an eosinophilic leukemia has been described in an animal which was part of a bioassay of benzo[*a*]pyrene–hematite. The animal had a large chest mass which involved much of the lungs, pericardium, and mediastinal lymph nodes (Port and Richter, 1977). Myeloid tumors are not frequently seen.

X. ENDOCRINE SYSTEM

Adrenal Cortex. Like other authors, Russfield (1966) noted that tumors of the adrenal cortex are the most characteristic spontaneous neoplasma of hamsters. There was a range from hyperplasia through adenoma to adenocarcinoma. They are most commonly seen in males, and the rates vary from one colony to another. In the Stanford colony, 667 of 4575 animals (14.5%) had proliferative lesions of the adrenal cortex; these ranged from adenomatous nodules to lesions with histological characteristics of adenocarcinoma (Kirkman and Algard, 1968). One of the latter had evidence of androgen secretion. In this study, these were clearly shown to be lesions of older animals; only 1.7% occurred in the first year of life, while incidences in years 2 and 3 were 35.5% and 40%, respectively. Many of these adrenocortical proliferations occur in the zona glomerulosa (Fig. 10). A similar high incidence of adrenal tumors was reported in an inbred line (Homburger and Russfield, 1970). Adrenocortical tumors were relatively common and were of two major types, one originating in the zona glomerulosa and the other more closely associated with the zona fasciculata.

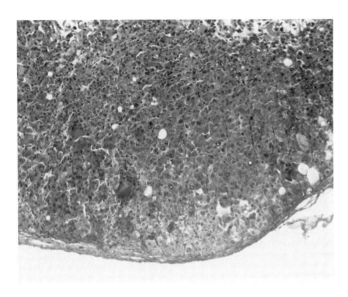

Fig. 10. This adreno cortical adenoma in the outer cortex is well differentiated, compresses the surrounding cortex, and elevates the capsule. Hematoxylin–eosin stain. Magnification: ×90.

A second type of adrenal adenoma was noted by Pour and co-workers (1976c); it was more common, found in the inner portion of the cortex, and composed of cells which resemble those of the zona fasciculata (Fig. 11). These benign lesions have glandular formations made up of cells with abundant cytoplasm and occasional intranuclear inclusions. In addition, these investigators noted a proliferation of the subcapsular cells of the adrenal. These were not felt to be neoplastic and were found in 17 to 31% of the older animals. Carcinomas of the adrenal cortex often had glandular and trabecular patterns.

There was evidence of invasion of local vascular structures and other tissues. Some had metastasized to regional lymph nodes and one to the lungs.

Adrenal Medulla. Benign tumors of the adrenal medulla are only reported occasionally. They are composed of polygonal cells with amphiphilic to slightly basophilic cytoplasm, and they are usually located within the medulla. Occasionally they may be found in the cortex. Since proliferative lesions of the cortex are common, it is not surprising to find a pheochromocytoma adjacent to a cortical neoplasm.

Thyroid. Tumors of the thyroid are relatively uncommon in most series; most are noted to be of the follicular type and to have a benign histological appearance (McMartin, 1979). A 1.5% incidence of well-differentiated follicular carcinoma was found; there was an increased incidence of these lesions on an iodine-deficient diet (Russfield, 1966). In contrast, thyroid lesions were found to compose over 30% of the endocrine tumors of three inbred hamster strains (Pour *et al.*, 1979). Both adenomas and carcinomas were found, and solid and papillary patterns were observed histologically. The reason for this high incidence in these studies is probably due to the fact that most of the lesions were not observed grossly and were only detected by intensive histological examination. Follicular and papillary adenomas (Fig. 12) and solid carcinomas of the thyroid were found in 20 animals by Pour and associates (1976c); only 4 were observed grossly. They included both follicular and Huerthle cell types. Metastases were only found in regional lymph nodes. Solid tumors often possessed large intranuclear inclusion bodies.

Parathyroid. Adenomas range in incidence from 1.7 to 7.7%

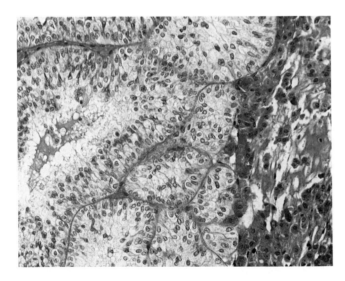

Fig. 11. A very florid tumor of the inner adrenal cortex is made up of cords of cells with extensive pale cytoplasm. Hematoxylin–eosin stain. Magnification: ×255.

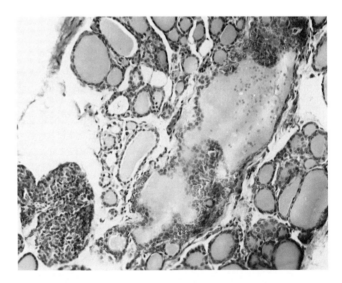

Fig. 12. This papillary cystadenoma of the thyroid is a small lesion and was not grossly apparent. It consists of irregular follicular formations filled with colloid and lined by epithelium of variable thickness. An adjoining solid nodule of adenomatous cells is also present. Hematoxylin–eosin stain. Magnification: ×169.

with evidence of strain variation; carcinomas appear to be very rare despite extensive searches on the part of several investigators (Pour, 1983). Other summaries have noted that adenomas of the parathyroid are of moderate incidence (Van Hoosier and Trentin, 1979). When found, they are well-differentiated enlargements of the glands and are sometimes noted to be bilateral, raising the question of hyperparathyroidism secondary to chronic renal disease.

Pancreatic Islets. Most reports indicate the presence of low numbers of benign and malignant tumors of the pancreatic islet cells. They tend to be relatively well differentiated and are seen as enlarged nodules of histologically normal islet cells compressing the surrounding pancreatic tissue. Islet cell lesions were found as adenomas in very high incidence in males of the Eppley colony and much less frequently in other groups of animals (Pour *et al.*, 1976c). A tumor was isolated from a male hamster treated with a subcutaneously implanted testosterone pellet. The primary tumor showed the presence of histochemically demonstrable insulin which was retained through transplants over a 12-year period (Grillo *et al.*, 1967). The original tumor weighed over 2 gm, but no metastases were found. Fluorescent-antibody staining revealed the presence of insulin in tumor cells. After this initial report, the tumor was passed serially, with shortening of latent period and development of multiple visceral metastases. Grillo also reviewed Fortner's original study of islet cell tumors, in which he found a 60% incidence of benign and malignant lesions in 620 animals.

Tumors of the pancreatic islets were described in both of the colonies surveyed by Pour and collaborators (1976c). The lesions were relatively infrequent but were seen as proliferations of one or more islets within the pancreas. Hyperplasia was differentiated from neoplastic lesions on the basis of size. Lesions which had only one cell type and which were larger than 1 mm were classified as adenomas. Tumor cells were of a relatively uniform population with round nuclei and polyhedral or columnar cells; they resembled either β or α cells of the islets. In the reports of inbred animals, islet cell tumors were among the most common endocrine neoplasms. Most were found as microscopic lesions in male animals, and the majority of the lesions were adenomas. Carcinomas are also found; the usual lesions have mixed trabecular–papillary patterns and demonstrate invasion of local structures. Occasionally these tumors metastasize to the liver (Fig. 13) and regional lymph nodes. Islet cell hyperplasia was also common in older animals.

Pituitary. In their review, Kirkman and Algard (1968) noted the low incidence of spontaneous tumors of the pituitary as well as the presence of estrogen-induced adenomas and adenocarcinomas of the intermediate lobe. Tumors of the intermediate lobe are sometimes functional, secreting gonadotropins. Transplants of these induced tumors were noted to survive in the recipient animals but did not grow well. In

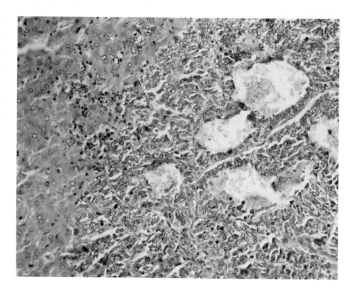

Fig. 13. A focus of metastatic islet cell tumor in the liver retains its endocrine pattern, with cells occurring in small clusters. There are scattered clefts and spaces also lined by neoplastic cells. The mass is not encapsulated. Hematoxylin–eosin stain. Magnification: ×223

contrast to other rodents, pituitary neoplasms were found in only two females in one large survey; one was an otherwise unspecified adenoma, and the other was a small carcinoma with invasion of the sella turcica (Pour *et al.*, 1976c). However, hyperplasia of the pars intermedia was noted with about 12% of females and fewer male animals.

XI. REPRODUCTIVE SYSTEM

The female reproductive tract has occasioned somewhat variable reports on its importance as a site of tumor occurrence. In the two colonies of hamsters reviewed by Pour and co-workers (1976c), the female genital tract tumor incidences were 3.5 and 11%. In conjunction with this marked variation in incidence, there was also a considerable difference in the types of tumors which were observed. In one colony, the majority of tumors were found in the ovaries, whereas in the other the uterus and vagina were by far the most common sites of origin.

Tumors of the ovary are usually unilateral and arise from granulosa or theca cells. Both cell types may be found in a single lesion. Thecomas were found in up to 2% of the females in one life-span study (Russfield, 1966). Granulosa cell tumors are composed of cells arranged in rosettes, glandular formations, ribbons, and cords. Metastasis is uncommon.

Tumors of the uterus fall into two major groups: those of stromal origin and those arising from the endometrial epithelium. Endometrial stromal polyps vary considerably in incidence from one report to another and are often noted microscopically. These polyps consist of papillary proliferations of uterine epithelium and stroma in the upper and middle portions of the uterus; both elements tend to be well differentiated, and the stroma may be edematous. The adjoining uterine epithelium may be either atrophic or hyperplastic, and the stroma often contains granular leukocytes. Leiomyomas of the myometrium have been found in the lower portion of the uterus extending into the lumen of the organ or into the vagina itself. These smooth muscle tumors are nodules of well-differentiated smooth muscle cells which have outward expansile growth and compress surrounding structures. Spontaneous leiomyosarcomas have also been noted. These are less well differentiated and are not encapsulated, extending into the other pelvic organs.

Epithelial neoplasms are occasionally noted in the upper portions of the tubular reproductive tract; these include cystadenomas of the fallopian tubes with focal proliferation of the tubal epithelium, forming cysts containing eosinophilic secretion. Endometrial adenocarcinomas are also reported by several investigators (Pour *et al.*, 1976c; Van Hoosier and Trentin, 1979; Squire *et al.*, 1978). These usually occur in animals over 100 weeks of age. These lesions have been reported to compose up to 5–10% of the spontaneous tumors in a group of hamsters. This is in contrast to a report of a very high incidence of uterine adenocarcinoma (25%) in female Chinese hamsters cited by Squire *et al.* (1978). Uterine adenocarcinomas are composed of irregular ductules lined by atypical endometrial lining cells; the supporting stroma is varying in amount, and some of these masses may be very scirrhous.

Van Hoosier and Trentin (1979) encountered three cervical carcinomas in their survey of the literature, and squamous papillomas of the vagina have also been observed.

In most studies, tumors of the mammary gland are infrequent, unlike the situation in laboratory rats and mice. When found, they are usually poorly differentiated adenocarcinomas with a scirrhous pattern. They may metastasize to the lungs and other tissues, and several of these tumors have been capable of being transplanted.

Tumors of the male reproductive tract were exceedingly rare, with the exception of one report of a high incidence of adenoma of the epididymis in one group of animals (Pour *et al.*, 1976c).

Testicular tumors are also infrequent. Kirkman, as cited by Squire *et al.* (1978), reported seminomas and an adenoma of the rete testis.

The secondary sex glands of the male hamster are not often the site of neoplasia, despite the report of benign hypertrophy of the prostate in several inbred strains (Homburger, 1970). Russfield (1966) reported two adenocarcinomas of the prostate in 94 animals, and a cystadenocarcinoma of Cowper's gland was illustrated. Cowper's (bulbourethral) gland adenomas were found in white and albino hamsters as well, and all had a papillary cystic pattern. In the summary of Squire *et al.* (1978), an adenocarcinoma of the epididymis and two prostatic adenocarcinomas were mentioned, along with seven cystadenocarcinomas of the bulbourethral glands. Pour *et al.* (1976c) reported a single adenoma of the epididymis. In all these cases the histological characteristics of the lesions were like those encountered in similar tumors in other rodent species.

XII. NERVOUS SYSTEM

Tumors of the central nervous system are extremely uncommon spontaneous events, even in those studies in which the animals have been scrutinized closely. Peripheral nervous lesions have been noted, although they, too, are not frequently observed. The latter include neurofibromas, which appear as collections of spindle cells arranged along peripheral nerves.

REFERENCES

Birt, D. F., and Pour, P. (1983). Influence of dietary fat on spontaneous lesions of Syrian golden hamsters. *JNCI, J. Natl. Cancer Inst.* **71,** 401–406.

Chesterman, F. C. (1972). Background pathology in a colony of golden hamsters. *Prog. Exp. Tumor Res.* **16,** 50–68.

Coggin, J. H., Jr., Oakes, J. E., Huebner, R. J., and Gilden, R. (1981). Unusual filterable oncogenic agent isolated from horizontally transmitted Syrian hamster lymphomas. *Nature (London)* **290,** 336.

Coggin, J. H., Jr., Bellomy, B. B., Thomas, K. V., and Pollock, W. J. (1983). B-Cell and T-cell lymphomas and other associated diseases induced by an infectious DNA viroid-like agent in hamsters (*Mesocricetus auratus*). *Am. J. Pathol.* **110,** 254–266.

Fortner, J. G. (1961). The influence of castration on spontaneous tumorigenesis in the Syrian (golden) hamster. *Cancer Res.* **21,** 1491–1498.

Fortner, J. G., and Allen, A. C. (1958). Hitherto unreported malignant melanomas in the Syrian hamster: An experimental counterpart of the human malignant melanomas. *Cancer Res.* **18,** 98–104.

Gershon, R. K., Carter, R. L., and Kondo, K. (1967). On concomitant immunity in tumour-bearing hamsters. *Nature (London)* **213,** 674–676.

Grillo, T. A. I., Whitty, A. J., Kirkman, H., Foa, P. P., and Kobernick, S. D. (1967). Biological properties of a transplantable islet-cell tumor of the golden hamster. I. Histology and histochemistry. *Diabetes* **16,** 409–414.

Handler, A., and Chesterman, A. (1968). "Spontaneous Disease in the Syrian Golden Hamster." Iowa State Univ. Press, Ames.

Homburger, F. (1970). Pathology of the Syrian hamster. *Prog. Exp. Tumor Res.* **16.**

Homburger, F. (1983). Background data on tumor incidence in control animals (Syrian hamsters). *Prog. Exp. Tumor Res.* **26,** 259–265.

Homburger F., and Russfield, A. B. (1970). An inbred line of Syrian hamsters with frequent spontaneous adrenal tumors. *Cancer Res.* **30,** 305–308.

Kirkman, H., and Algard, F. T. (1968). Spontaneous and nonviral-induced neoplasms. In "The Golden Hamster—Its Biology and Use in Medical Research" (R. A. Hoffman, P. F. Robinson, and H. Magalhaes, eds.), pp. 227–240. Iowa State Univ. Press, Ames.

McMartin, D. N. (1979). Morphologic lesions in aging Syrian hamsters. *J. Gerontol.* **34,** 502–511.

Manci, E. A., Heath, L. S., and Coggin, J. H., Jr. (1984). Lymphoma-associated ulcerative bowel disease in the hamster (*Mesocricetus auratus*) induced by an unusual agent. *Am. J. Pathol.* **116,** 1–8.

Mashiba, H., Matsunaga, K., Hata, K., Hosoi, M., and Nomoto, K. (1983). The role of macrophages in preventing metastasis of a homotransplantable hamster lymphoma. *Gann* **74,** 548–553.

Mohr, U. (1970). Effects of diethylnitrosamine in the respiratory system of Syrian golden hamsters. *In* "Morphology of Experimental Respiratory Carcinogenesis" (P. Nettesheim, M. G. Hanna, Jr., and J. W. Deatherage, Jr., eds.), pp. 255–265. Natl. Tech. Inf. Serv., U.S. Dept. of Commerce, Washington, D.C.

Port, C. D., and Richter, W. R. (1977). Eosinophilic leukemia in a Syrian hamster. *Vet. Pathol.* **14,** 283–286.

Pour, P. (1983). Adenoma, carcinoma, parathyroid, hamster. *In* "Endocrine System" (T. C. Jones, V. Mohr, and R. D. Hunt, eds.), pp. 275–281. Springer-Verlag, Berlin and New York.

Pour, P., and Birt, D. (1979). Spontaneous diseases of Syrian hamsters—their implications in toxicological research: Fact, thoughts and suggestions. *Prog. Exp. Tumor Res.* **24,** 145–156.

Pour, P., Althoff, J., and Cardesa, A. (1973). Granular cells in tumors and in nontumorous tissue. *Arch. Pathol.* **95,** 135–138.

Pour, P., Kmoch, N., Greiser, F., Mohr, U., Althoff, J., and Cardesa, A. (1976a). Spontaneous tumors and common diseases in two colonies of Syrian hamsters. I. Incidence and sites. *J. Natl. Cancer Inst. (U.S.)* **56,** 931–935.

Pour, P., Mohr, U., Cardesa, A., Althoff, J., and Kmoch, N. (1976b). Spontaneous tumors and common diseases in two colonies of Syrian hamsters. II. Respiratory tract and digestive system. *J. Natl. Cancer Inst. (U.S.)* **56,** 937–948.

Pour, P., Mohr, U., Althoff, J., Cardesa, A., and Kmoch, N. (1976c). Spontaneous tumors and common diseases in two colonies of Syrian hamsters. III. Urogenital system and endocrine glands. *J. Natl. Cancer Inst. (U.S.)* **56,** 949–961.

Pour, P., Mohr, U., Althoff, J., Cardesa, A., and Kmoch, N. (1976d). Spontaneous tumors and common diseases in two colonies of Syrian hamsters. IV. Vascular and lymphatic systems and lesions of other sites. *J. Natl. Cancer Inst. (U.S.)* **56,** 963–974.

Pour, P., Althoff, J., Salmasi, S., and Stepan, K. (1979). Spontaneous tumors and common diseases in three types of hamsters. *JNCI, J. Natl. Cancer Inst.* **63,** 797–811.

Russfield, A. (1966). Tumors of the endocrine gland and secondary sex organs. *Public Health Serv.* **1332,** 26.

Safiotti, U. (1970). Morphology of respiratory tumors induced in Syrian golden hamsters. *In* "Morphology of Experimental Respiratory Carcinogenesis" (P. Nettesheim, M. G. Hanna, Jr., and J. W. Deatherage, Jr., eds.), pp. 245–254. Natl. Tech. Inf. Serv., U.S. Dept. of Commerce, Washington, D.C.

Schmidt, R. E., Eason, R. L., Hubbard G. B., Young, J. T., and Eisenbrandt, D. L. (1983). "Pathology of Aging Syrian Hamsters." CRC Press, Boca Raton, Florida.

Squire, R. A., Goodman, D. G., Valerio, M. G., Fredrickson, T. N., Strandberg, J. D., Levitt, M. H., Lingemen, C. H., Harshbarger, J. C., and Dawe, C. J. (1978). Tumors. *In* "Pathology of Laboratory Animals" (C. K. Benirschke, F. M. Garner, and T. C. Jones, eds.), Vol. 2, pp. 1051–1262. Springer-Verlag, Berlin and New York.

Stewart. H. L., Snell, K. C., Dunham, L. J., and Schylen, S. M. (1959). "Transplantable and Transmissible Tumors of Animals," p. 378. Armed Forces Inst. Pathol. Washington, D.C.

Streilein, J. W. (1984). Ileal hyperplasia, wet tail and infectious lymphoma: The black plague of Syrian hamsters colonies. *Hamster Inf. Serv.* **6,** 2–3.

Streilein, J. W., Duncan, R. W., Hart, D. A., Stein-Streilein, J. S., and Billingham, R. E. (1981). "Hamster Immunity and Infectious and Oncologic Diseases." Plenum, New York.

Takahashi, M., and Pour, P. (1978). Spontaneous alterations in the pancreas of the aging Syrian golden hamster. *J. Natl. Cancer Inst. (U.S.)* **60,** 355–364.

Turusov, V. S. (1982). "Pathology of Tumours in Laboratory Animals," Vol. 3. Int. Agency Res. Cancer, Lyon, France.

Van Hoosier, G. L., Jr., and Trentin, J. J. (1979). Naturally occurring tumors of the Syrian hamster. *Prog. Exp. Tumor Res.* **23,** 1–12.

Chapter 10

Noninfectious Diseases

Gene B. Hubbard and Robert E. Schmidt

I.	Introduction	169
II.	Diseases Associated with Aging	169
	A. Amyloidosis	169
	B. Systemic Pathology	171
III.	Nutritional Disease	175
	A. Spontaneously Occurring Disease	175
	B. Experimentally Induced Disease	175
IV.	Polycystic Disease	175
V.	Traumatic Disease	175
VI.	Prenatal Mortality	175
	References	177

I. INTRODUCTION

With a few exceptions, noninfectious diseases of hamsters have received little attention until recent years, although many noninfectious conditions are of considerable importance in hamster colony management and toxicological studies. In this chapter we discuss noninfectious diseases of hamsters in the order of our perception of their relative importance.

II. DISEASES ASSOCIATED WITH AGING

A. Amyloidosis

Amyloidosis is the noninfectious disease of primary importance in the Syrian hamster. Deposition of amyloid is a common occurrence in aging hamsters (Franks and Chesterman, 1957) and is considered to be due to a defect of the immune system with initial deposition of fragments of immune globulins in subendothelial locations in vessels (Glenner et al., 1972).

The incidence of amyloidosis varies by hamster colony, with one colony having five times the incidence of another in one study (Pour et al., 1976a).

Amyloid may be diagnosed in histological sections: green birefringence when stained with Congo red, metachromasia when stained with crystal or methyl violet, and secondary fluorescence when stained with thioflavin S and thioflavin T (Jacob, 1971).

As seen with the electron microscope, amyloid is composed of nonbranching fibrils approximately 100 Å in diameter, which are composed of two or more twisted subfilaments about 40 Å diameter and separated by a 20-Å gap, which form a double helix of 100-Å periodicity (Jones and Hunt, 1983).

Schmidt et al. (1983) documented a marked increase in mortality in 15- to 20-month-old male hamsters when the incidence of amyloidosis of the liver and kidney was the highest.

Amyloidosis is associated with early mortality. Schmidt *et al.* (1983) reported significant kidney lesions in hamsters at 1 year of age, and amyloidosis was experimentally induced in adult hamsters by Gruys *et al.* (1979) at 8 weeks.

There appears to be a sex difference in the occurrence of hamster amyloidosis. There is a higher incidence, increased severity, and earlier onset of amyloidosis of the liver, kidney, stomach, adrenal, thyroid, and spleen of female hamsters (Schmidt *et al.*, 1983).

The hepatic amyloidosis was detected as early as 5 months. Deposits were seen in vascular walls and portal areas first. As severity increased, filling of portal areas and adjacent parenchyma occurred, with accentuation of the lobular architecture, which could often be seen grossly (Fig. 1).

The renal amyloidosis reported by Schmidt *et al.* (1983) was presented primarily as glomerular depositions (Fig. 2); but interstitial and combined amyloidosis was often seen. Grossly, the kidneys were often pale tan, enlarged and/or misshapen. Histologically, the amyloid deposition occurred in basement membranes. Ultrastructurally, the amyloid deposited on basement membranes and in the glomerular mesangium.

Redman *et al.* (1979) reported glomerular amyloidosis as the main cause of death in Syrian hamsters. Gleiser *et al.* (1970) reported amyloidosis of the liver, kidneys, and adrenal gland, and found that 88% of the amyloidosis occurred in hamsters over 18 months of age. This amyloidosis was accompanied by a rise in serum globulins and was first detected morphologically in the kidney.

Adrenal gland degeneration was considered to be an end stage of amyloidosis and associated changes by Schmidt *et al.* (1983). Amyloid appeared to deposit initially in the zona fasciculata and zona reticularis, with eventual collection through-

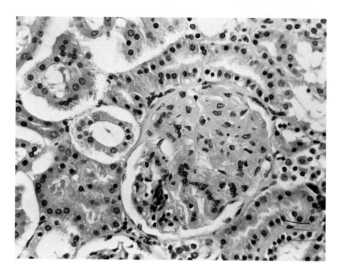

Fig. 2. Glomerular amyloidosis of the kidney. Hematoxylin–eosin stain. Magnification: ×245.

out the cortex (Fig. 3). This deposition was seen as early as 5 months and became more frequent after 15 months.

Besides amyloid deposition in the kidney, liver, and adrenal, it has been documented in salivary glands, stomach, small intestine, thyroid gland, parathyroid gland, spleen, meninges, trachea, lung, gallbladder, and pancreas (Schmidt *et al.*, 1983; Herrold and Dunham, 1963; Chesterman, 1972; Pour *et al.*, 1976a; Frenkel, 1958; Fortner, 1957; McMartin, 1979).

According to Schmidt *et al.* (1983), amyloidosis of the small intestine was rare (Fig. 4), but was typical of amyloid deposition in stomach and other intestinal sites.

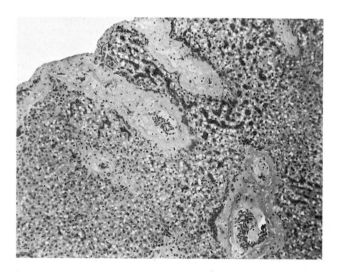

Fig. 1. Typical hepatic amyloid deposition in the portal areas and adjacent parenchyma. Hematoxylin–eosin stain. Magnification: ×123.

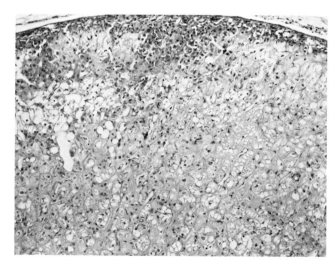

Fig. 3. Adrenocortical amyloidosis with associated degenerative change. Hematoxylin–eosin stain. Magnification: ×105.

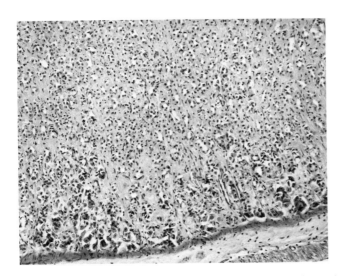

Fig. 4. Intestinal amyloidosis with almost complete obliteration of normal architecture. Hematoxylin–eosin stain. Magnification: ×133.

Experimental induction of amyloidosis in hamsters leads to lesions and distribution similar to that seen in natural disease. Gruys *et al.* (1979) induced secondary amyloidosis in adult hamsters by daily subcutaneous injections of 1 ml of 50% casein Hammerstan in 0.3 M sodium bicarbonate, pH 7.5, five times per week. The casein solution also contained bacteria. Amyloidosis was induced in all hamsters in 8 weeks. Amyloid appeared first in the liver and spleen, and then in the kidneys. In the hepatic lobule, both mononuclear phagocytes and hepatocytes had a topographical relationship to the first deposits of amyloid. Amyloid fibrils were seen in macrophages and between microvilli of hepatocytes.

B. Systemic Pathology

1. Genitourinary System

Hamster nephrosis is considered by some (Newberne, 1978) to be histologically identical to that observed in aged rats, and may be directly related to the level of dietary protein. Slausen *et al.* (1978) reported spontaneous renal lesions that resembled arteriolar nephrosclerosis in humans and differed from other spontaneous or viral-induced renal diseases in other rodent species. Fibrinoid necrosis of intrarenal arterioles, glomerular amyloidosis, and uremia were characteristics of the end-stage kidneys. Marked alterations of the glomerular basement membranes in apparently healthy hamsters are described in detail by Van Marck *et al.* (1978). IgG deposits or amyloid in the glomeruli were not found. Klei *et al.* (1974) reported similar changes in his hamsters. An atrophic glomerular lesion occurred spontaneously in Syrian hamsters, and appeared to be age related, and increased in severity with stilbestrol treatment (Maquire *et al.*, 1974).

Chesterman (1972) reported gross and microscopic changes in hamster kidneys. Grossly, the kidneys were pale tan with irregular surfaces. The microscopic appearance was variable but included glomerular fibrosis, atrophy, and amyloidosis; interstitial fibrosis; inflammatory cells; and tubular epithelial atrophy. Pour *et al.* (1976a) reported renal amyloidosis, nephrocalcinosis, acute to chronic pyelonephritis, and cortical cysts.

Schmidt *et al.* (1983) consider amyloidosis as the most prevalent renal lesion, but also describe a progressive degenerative disease with tubular atrophy leading to an end-stage nephrosis (Fig. 5). The most prominent changes included interstitial fibrosis, tubular atrophy, tubular dilatation, tubular collapse, and occasional shrunken glomerular tufts. Filling of tubules with proteinaceous fluid and glomeruli with increased mesangium and thickened basement membranes were demonstrated. No vascular lesions were seen. Additionally, nephrocalcinosis, renal infarcts, pyelonephritis, and pyelitis were reported. The nephrocalcinosis may be related to dietary mineral imbalance (Schmidt *et al.*, 1983). Pour *et al.* (1976a) reported the incidence of nephrocalcinosis in females was almost twice that seen in males.

Cystitis has been reported by Pour *et al.* (1976d) and Schmidt *et al.* (1983). Bauck and Hagen (1984) described a case of urolithiasis in a 1-year-old teddy bear hamster. Pour *et al.* (1976c) saw occasional epithelial hypertrophy in renal pelves and urinary bladders.

Lesions of the male reproductive system are well described

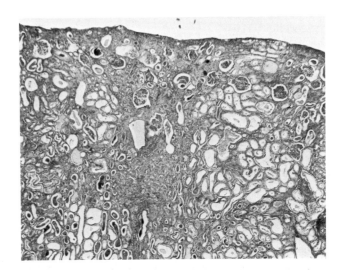

Fig. 5. Typical end-stage nephrosis of the hamster. Note the interstitial fibrosis, dilated, often protein-filled tubules, irregular capsular margin, and occasional atrophied glomeruli. Hematoxylin–eosin stain. Magnification: ×280.

by Schmidt et al. (1983), with most changes related to aging. The predominant lesions of the testicles and epididymides were reduced numbers of sperm, atrophy, degeneration, and lipofuscinosis. Adenitis and mineralization were the most common changes in the prostate. Pour et al. (1976c) reported epididymitis, orchitis, spermatoceles, and spermatic granulomas. Chesterman (1972) reported epididymitis as a rare condition. Handler and Chesterman (1968) reported atrophic testicles and epididymitis. Prostatic adenitis, atrophy, and mineralization were reported by Schmidt et al. (1983).

Pour et al. (1976c) found 7% of the females with ovarian cysts and squamous metaplasia of the cervical epithelium. Hydrometra in aged females was considered to be common by Chesterman (1972). The most complete data on the female reproductive tract is furnished by Schmidt et al. (1983). They report the most frequent diagnosis as fibrinoid degeneration of blood vessels which was frequently accompanied by periarteritis. These lesions were found in ovarian and uterine arteries, and arteries of the broad ligament. Uterine and cervical inflammation, and squamous metaplasia of the cervix were documented.

Spontaneous material mortality of approximately 1 in 300 pregnancies occurred at a commercial colony of hamsters, with approximately 25% of the deaths occurring after the fourteenth day of gestation. Nearly half of the hamsters dying at this time had renal lesions characteristic of the generalized Schwartzman reaction. Histologically, there were renal glomerular fibrin thrombosis and renal cortical necrosis. This disease resembled human eclampsia (Galton and Slater, 1965). Richter et al. (1984) reported a similar eclampsia syndrome in late-term pregnant hamsters. Abdominal pregnancies have been reported in hamsters (Peters, 1982).

2. Endocrine System

There is minimal information concerning noninfectious diseases of the endocrine glands. The best single source is Schmidt et al. (1983), which covers the adrenals, parathyroids, thyroid, and pituitary glands. Amyloidosis is the most severe problem in the hamster adrenal gland, followed in importance by cortical hyperplasia, degeneration, congestion, pigmentation, cysts, and mineralization. Handler and Chesterman (1968) mention severe damage or complete destruction of the adrenocortical zonae reticularis and fasciculata by lipid deposition, amyloid, congestion, and/or hemorrhage.

Nodular hyperplasia of the parathyroids was the main lesion recorded by Schmidt et al. (1983). Amyloid was seen in one parathyroid.

Amyloidosis preceded the development of thyroid follicular cysts and mineralization (Schmidt et al., 1983). Hamsters fed diets low in iodine developed goiter (Anderson and Capen, 1978).

Pituitary lesions found in aging hamsters were as follows in order of decreasing frequency: cysts, mineralization, and degeneration (Schmidt et al., 1983). Aplasia and hypoplasia of the pituitary occur in hamsters whose mothers were fed high-vitamin A diets (Anderson and Capen, 1978).

3. Liver, Gallbladder, and Pancreas

Aside from amyloidosis, the most frequent hepatic lesions encountered by Schmidt et al. (1983) were hemosiderosis, biliary cysts (Fig. 6), cholangitis, congestion, and bile duct hyperplasia (Fig. 7). Other less frequently observed lesions included fatty change, fibrosis, necrosis, hematopoiesis, vasculitis, telangiectasis, and nonspecific degeneration. Hamsters with bile duct hyperplasia were first detected at 8 months of age. Biliary cysts were not seen until the hamsters were 18 months of age. Lymphocytes and plasma cells were generally seen with hepatic cholangitis (Schmidt et al., 1983).

Pour et al. (1976b) reported cystic bile ducts as common in both sexes. Handler and Chesterman (1968) found vasculitis, amyloid deposition, bile duct proliferation, cyst formation, fatty change, and necrosis in hamsters. Chesterman and Pomerance (1965) documented cirrhotic livers characterized by bile duct proliferation and fibrosis. Multilocular cysts were reported by Chesterman (1972) in hamsters over 1 year of age. Cholangitis was reported in two colonies by Pour et al. (1976b). Cholangiofibrosis varying from minimal inflammatory cell infiltrates to compression of bile ducts to biliary hyperplasia was documented by Hamilton et al. (1983).

Lesions of the gall bladder were rarely seen by Schmidt et al.

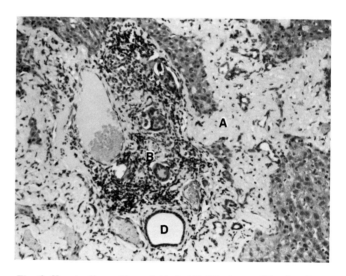

Fig. 6. Hamster liver with amyloidosis (A), bile duct proliferation (B), and dilation (D). Note loss of hepatocytes (H). Hematoxylin–eosin stain. Magnification: ×172.

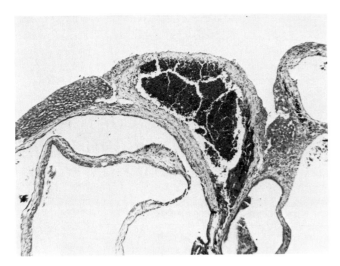

Fig. 7. Hamster liver parenchyma is replaced by marked distention of bile ducts. Hematoxylin–eosin stain. Magnification: ×57.

(1983), who reported only cholecystitis and edema. Chesterman (1972) described hard, darkly pigmented gallstones in hamsters.

Pancreatic lesions are not rare, and as reported by Schmidt *et al.* (1983), include interstitial amyloid, cystic dilatation of pancreatic ducts, exocrine degeneration, and pancreatitis. Handler and Chesterman (1968) reported fat deposition in pancreatic acinar cells. Pour *et al.* (1976d) recorded local multiplication and distention of pancreatic ducts without cell proliferation in male and female hamsters. Hypertrophied eosinophilic pancreatic islet cells similar to those seen in the Chinese hamster with spontaneous diabetes were documented by Murgatroyd and Tucker (1981) in the Syrian hamster. There was no clinical indication of the hyperglycemic condition. Takahashi and Pour (1978) have thoroughly discussed spontaneous alteration in the pancreas of aging Syrian hamsters. The primary lesions were eosinophilic alteration, degeneration, and atrophy of pancreatic acinar cells, ductular proliferation, ductular dilation, adiposis, and fibrosis.

4. Cardiovascular System

Thrombosis of the heart chambers is a common cardiovascular disease of the Syrian hamster (Pour *et al.*, 1976c; Handler and Chesterman, 1968; Schmidt *et al.*, 1983; Sichuk *et al.*, 1964). Atrial thrombosis was the most frequent cardiovascular diagnosis made by Schmidt *et al.* (1983). It was often visible grossly, was first seen at 13 months of age, and was most severe at 27 months of age. Myxomatous and fibrotic valvular changes were part of the early histological changes. According to Pour *et al.* (1976b), cardiac thromboses developed more frequently in the atria, particularly the left atria, and the incidence of this varies between colonies.

Dodds *et al.* (1977) studied spontaneous atrial thrombosis in aged Syrian hamsters and found the process is a consumption coagulopathy. They reported a 73% incidence for both sexes of spontaneous atrial thrombosis in a hamster colony. However, the females develop the condition earlier than males. McMartin (1977) saw bilateral ventricular hypertrophy in thrombosed hearts and myxomatous valvular thickening. He theorized the thrombi occurred from local blood stasis, secondary to heart failure. Sichuk *et al.* (1964) found atrial thrombi in the left heart only, and he observed clinical signs including hyperpnea, tachycardia, and cyanosis. Gonadectomy, as well as estrogen and testosterone therapy, variably influenced the incidence and age of onset. Handler and Chesterman (1968) reported thrombi in peripheral vessels and auricles, hyaline necrosis of arterioles, endocardial alterations, myocardial necrosis, and fibrosis.

The most common noninfectious lesion of the cardiovascular system seen by Pour *et al.* (1976c) was calcifying vasculopathy. As also reported by Schmidt *et al.* (1983), this mineralization was generally confined to the elastic layers, but sometimes extended to the medias and intimas. The most severely affected vessels were the aorta, and the cardiac, renal, and gastric arteries.

Arteriolar nephrosclerosis resembling the condition seen in humans was reported by Slausen *et al.* (1978). The renal changes were progressive, and related to fibrinoid necrosis of intrarenal arterioles. Fibrinoid necrosis of small arteries was also found in the uterus, ovaries, testes, mesenteries, and coronary arteries, and occurred earlier and were more severe in females. He also found cardiac thromboses of the left atrium and left atrioventricular valves. Other lesions of the cardiovascular system reported by Schmidt *et al.* (1983) were cardiomyopathy, myocarditis, endocarditis, epicarditis, and mineralization. Mineralization was most frequently seen in the wall of the thoracic aorta.

5. Alimentary System

Noninfectious lesions of the oral cavity are not common in the Syrian hamster. Schmidt *et al.* (1983) found only occasional cases of periodontitis. Pour *et al.* (1976b) saw dental abscesses with occasional extension to the meninges. Chesterman (1972) reported malopposition of incisors, as well as broken, crooked, and missing teeth. These dental abnormalities were more common in pug-faced hamsters. Cheek pouch inflammation, malocclusion, and odontogenic abscesses were reported in hamsters in a survey conducted by Renshaw *et al.* (1975).

Focal and diffuse esophagitis characterized by inflammatory cell infiltrates of lymphocytes or macrophages have been seen in hamsters. Sialoadenitis, amyloidosis, and megalocytosis were seen in the salivary gland (Schmidt et al., 1983).

Gastritis, mineralization, amyloidosis, squamous hyperplasia, edema, ulcers, vascular mineralization, adenomyosis, and bloat have been reported in the stomachs of Syrian hamsters by various authors (Schmidt et al., 1983; Renshaw et al., 1975; Pour et al., 1976d; Handler and Chesterman, 1965). Nelson (1975) described fatal tricholithiasis in a male long-haired hamster. The hairball measured 1 cm in diameter.

Lesions seen in the hamster small intestine include amyloid deposition, enteritis, intussusception, and ulcers (Schmidt et al., 1983; Pour et al., 1976d; Handler and Chesterman, 1968). Colitis, typhlitis, intussusception, focal squamous metaplasia of the rectum, prolapses of invaginated colon, and rectal prolapses have been reported (Schmidt et al., 1983; Pour et al., 1976b; Renshaw et al., 1975; Pollock, 1975). Cecal mucosal hyperplasia is a rare condition with no determined etiology. Barthold et al. (1978) described an outbreak with diarrhea, runting, and high mortality in young hamsters in a breeding colony.

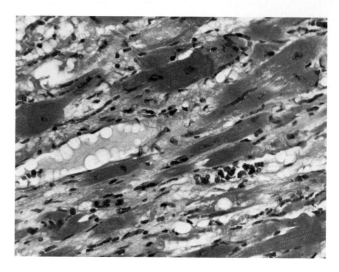

Fig. 8. Skeletal muscle degeneration with fragmentation, sarcolemmal nuclear rowing, edema, and necrosis. Hematoxylin–eosin stain. Magnification: ×330.

6. Respiratory System

Noninfectious diseases of the respiratory tract of the hamster are not of major significance. A comprehensive review by Schmidt et al. (1983) lists many morphological diagnoses, but low percentages for the lesions. Listed for the upper respiratory tract were tracheal amyloidosis, tracheal mucosal gland adenitis, ossifications of tracheal rings, tracheal gland degeneration, tracheitis, rhinitis, bronchitis, and a nasal cavity parasitism. Lung lesions included congestion, adenomatosis, edema, histiocytosis, mineralization, hemorrhage, amyloidosis, emphysema, atelectasis, heterotopic bone, hemosiderosis, and an infarct.

Pour et al. (1976d) made reference to hyperplastic, metaplastic, and dysplastic changes in respiratory epithelium of hamsters in two colonies. Fluid accumulation in the pleura of the hamster lungs was seen by Handler and Chesterman (1968). Rhinitis of unknown etiology was observed in control hamsters in one study (Herrold and Dunham, 1963).

7. Musculoskeletal System

Skeletal muscle degeneration (Fig. 8) and mineralization was variable in location and severity, and occurred as early as 8 months of age in the hamsters studied by Schmidt et al. (1983). Typical morphological changes included rowing of sarcolemmal nuclei and fragmentation of fibers. Edema, swelling, necrosis, and mineralization often accompanied the more severe lesions. The degenerative lesions were morphologically similar to those reported in inbred hamsters with muscular dystrophy. Myositis was usually associated with severe skin inflammation. Atrophy of skeletal muscle in 14 hamsters was reported by McMartin (1979).

Silberberg and Silverger (1941) reported degenerative joint diseases in Syrian hamsters. They found lesions of varying severity in 15 males and 20 females. Osteoarthritis was considered to be a common finding by Alspauch and Van Hoosier (1973). The femorotibial articulation was the joint most commonly involved, although more than one joint was often affected. Sokoloff and Habermann (1958) reported idiopathic bone necrosis in hamsters.

8. Central Nervous System

Schmidt et al. (1983) listed their morphological diagnoses in the central nervous system of aging Syrian hamsters to be mineralization, encephalitis, encephalomalacia, meningitis, keratitis, amyloidosis, heterotopic bone, and melanosis. Other authors have recorded hydrocephalus, Buscaino bodies, skull abscesses, amyloidosis, vacuolization in both white and gray matter, axonal dystrophy, and infarcts (Renshaw et al., 1975; Pour et al., 1976a; McMartin, 1979).

Hemorrhagic necrosis of the central nervous system of fetal hamsters has been studied by Margolis and Kilham (1976) and Young and Keeler (1978). The disease occurred in litters of random-bred, timed pregnant hamsters from three commercial colonies. The typical morphological changes included edema, malacia, and hemorrhage which were restricted to neural tissue, retina, and the internal ear. Marked neuroepithelial proliferation with rosette formation was considered to reflect a healing process. However, no inflammatory cell response was seen. Keeler and Young (1979) indicated that a vitamin E deficiency of the pregnant mothers may be involved in the disease.

9. Hematopoietic and Lymphatic Organs

There is minimal information concerning noninfectious disease of the hematopoietic and lymphatic organs, with most data recorded by Schmidt *et al.* (1983). The only bone marrow lesions were amyloid hyperplasia and necrosis. Lymph node changes included plasmacytosis, lymphoid hyperplasia, lymphadenitis, hemorrhage, edema, and mineralization. Splenic amyloidosis, hemosiderosis, congestion, lymphoid hyperplasia, reticular hyperplasia, and splenitis were reported. Thymic hemorrhage and necrosis were rarely seen.

McMartin (1979) reported severe thymic atrophy in aged Syrian hamsters.

10. Pleura, Peritoneum, and Mesentery

Noninfectious lesions of the pleura, peritoneum, and mesentery are rare. Schmidt *et al.* (1983) reported a low percentage of peritonitis and pleuritis in control hamsters. Fortner (1957) found fluid accumulation in the pleural and peritoneal cavities.

11. Skin

Noninfectious skin disorders listed by Schmidt *et al.* (1983) include edema, dermatitis, hyperkeratosis, and adnexal atrophy. The only other reference to skin lesions in hamsters was a case of dermatitis reported by Renshaw *et al.* (1975).

III. NUTRITIONAL DISEASE

A. Spontaneously Occurring Disease

Spontaneous nutritional disease is not well described in the Syrian hamster. Keeler and Young (1979) discuss the possibility of a vitamin E deficiency in pregnant hamsters that have litters with spontaneous hemorrhagic necrosis of the central nervous system of fetal hamsters. Inoculation of the mothers with vitamin E during the first half of the gestation decreased the spectrum of effects and the severity of the disease. Basic research on hamster nutrition is meager, and the requirements of the hamster, rat, and mouse are probably essentially the same (Granados, 1968; Newberne, 1978).

B. Experimentally Induced Disease

Most information on nutritional and metabolic disease in hamsters is available in "The Golden Hamster, Its Biology and Use in Medical Research" (Hoffman *et al.*, 1968) and in "Pathology of Laboratory Animals" (Benirschke *et al.*, 1978). Table I summarizes experimental dietary work and effects in the Syrian hamster.

IV. POLYCYSTIC DISEASE

Gleiser *et al.* (1970) described polycystic disease in a colony of golden hamsters. True cysts were found in the liver, epididymis, ovary, and adrenal. Cystic dilations were seen in the seminal vesicles, renal pelvis, endometrium, pancreas, and esophagus. The cysts often occurred at multiple sites; most were found in hamsters 13–27 months of age. The authors noted the similarity of this condition to congenital cystic disease of humans.

V. TRAUMATIC DISEASE

Hamsters have a tendency to fight among themselves, and cannibalism can be a problem. Prevention may be related to improved husbandry, including clean bedding, fresh greens, reduced grain feeding, and noise reduction (Ponsford and Whyley, 1954; Haley, 1951; Pour *et al.*, 1976a; Magalhaes, 1968). A survey conducted by Renshaw *et al.* (1975) indicated that 97.5% of total preweaning mortality was due to cannibalism. Shenefelt (1978) states that malformed hamsters are often eaten by the mother. Magalhaes (1968) recommends that animals that destroy all or part of litters not be used as breeding stock.

VI. PRENATAL MORTALITY

Prenatal mortality can have a significant impact on hamster production and research. Purdy and Hillemann (1950) reported overall prenatal mortality varying from 0 to 77%. The average percentage mortality for each day of gestation ranged from 6 to 35%, with the higher averages in days 12–16. The most frequently occurring litter size of viable embryos was 10. Resorptions of embryos occurred on days 9–16. No mummies were seen.

Yamamoto and Ingalls (1972) reported polyploidy and aneuploidy in embryos as manifestations of pregnancy wastage when the time between ovulation and fertilization is increased.

Shenefelt (1978) reports postimplantation intrauterine deaths of 70%. Preimplantation death rates are difficult to determine, or separate, from loss due to lack of fertilization. He reports

Table I
Experimental Nutritional Disease

Experimental nutrient	Effect	Reference
Semipurified diet with 60% casein	Decreased growth	Newberne (1978)
High protein	Nephrosis	Feldman et al. (1982)
High-sucrose diet	Diarrhea	Newberne (1978)
Apple supplement	Increased body weight	Knapka and Judge (1974)
Essential fatty acid deficiency	Hair loss; increased ear wax production; tight skin	Christensen and Dam (1953)
High fat intake (butterfat)	Convulsions	Swank and Engel (1958)
Crude fat (7.3 and 9.2%)	Increased mortality	Knapka and Judge (1974)
Cholesterol- and fat-free diets	Various effects on gallstone formation	Dam and Christensen (1952)
5% Polyoxyethylene monostearate preparation	Reduced weight gain; diarrhea	Schweigert et al. (1950)
Methionine	Fatty change in liver	Feldman et al. (1982)
Phenylalanine excess	Plasma phenylalanine levels comparable to those found in phenylketonuric children; increased plasma tyrosine levels	Horowitz and Waisman (1966)
Vitamin A deficiency	Weight loss; thin, coarse hair; xerophthalmia; gut hemorrhages; keratinizing squamous metaplasia; stomach ulcers	Hirschi (1950); Salley et al. (1959); Harada et al. (1982); Salley and Bryson (1957)
Vitamin D excess	Decreased body weight; anorexia; nephrocalcinosis	Fahmy et al. (1961)
Vitamin D and phosphorus deficiency	Rickets	Jones (1945)
Vitamin E deficiency	Muscle degeneration and weakness—may be related to etiology of spontaneous hemorrhagic necrosis of central nervous system of fetal hamsters; testicular atrophy; DNA synthesis and turnover increased in muscle and bone	Schweigert et al. (1950); Gerber et al. (1962); Keeler and Young (1979); Mason and Mauer (1975); Gerber et al. (1962)
Vitamin K deficiency	Depressed growth rate; hemorrhages in muscle and subcutaneous tissue	Hamilton and Hogan (1944)
Thiamin (vitamin B_1) deficiency	Polyneuritis	Routh and Houchin (1942)
Riboflavin (vitamin B_2) deficiency	Diarrhea; dermatitis, alopecia; reduced growth; stupor	Niggeman (1946)
Pyridoxine (vitamin B_6) deficiency	Weight loss; poor growth; anorexia; hair loss; achromotrichia; increased urinary excretion of xanthuremic acid	Schwartzman and Strauss (1949)
Vitamin B_{12} deficiency	Increased excretion of urinary methylmalonic and formiminoglutamic acids; increased levels of glutathione in the blood serum and liver	Cohen et al. (1967); Scheid et al. (1950)
Niacin (nicotinic acid) deficiency	Alopecia; rough coat; death	Routh and Houchin (1942)
Folic acid deficiency	Anemia; urinary excretion of aminoimidazole carboxamide and forminoglutamic acid; reduction in blood and liver folates, more severe in females	Newberne (1978); Cohen et al. (1971)
Pantothenic acid deficiency	Weight loss; porphyrin secretion; death	Newberne (1978)
Choline deficiency	Poor growth; fatty change in the liver	Granados (1951); Handler and Bernheim (1949)
Iron deficiency	Anemia	Rennie et al. (1982)
Iodine deficiency	Thyroid hyperplasia; thyroiditis	Follis (1964)
Iodine excess	Accumulation of colloid	Newberne (1978)
Copper deficiency	Hair depigmentation	Newberne (1978)

the start of reproductive senescence in multiparous hamsters at approximately 1 year of age, with the reduction largely due to prenatal death. The cause of intrauterine deaths may be due to endocrine abnormalities in older hamsters.

ACKNOWLEDGMENTS

The research reported in this chapter was accomplished by personnel of the Veterinary Sciences Division, USAF School of Aerospace Medicine, Aerospace Medical Division (AFSC), United States Air Force, Brooks Air Force

Base, Texas 78235-5301. The animals involved in this study were procured, maintained, and used in accordance with the Animal Welfare Act and the "Guide for the Care and Use of Laboratory Animals" prepared by the Institute of Laboratory Animal Resources—National Research Council.

REFERENCES

Alspauch, M. A., and Van Hoosier, G. L., Jr. (1973). Naturally-occurring and experimentally-induced arthritides in rodents: A review of the literature. *Lab. Anim. Sci.* **23**(1), 724–742.

Anderson, M. P., and Capen, C. C. (1978). The endocrine system. *In* "Pathology of Laboratory Animals" Vol. 1, (C. K. Benirschke, F. M. Garner, and T. C. Jones, eds.), pp. 423–499. Springer-Verlag, Berlin and New York.

Barthold, S. W., Jacoby, R. D., and Pucak, G. J. (1978). An outbreak of cecal mucosal hyperplasia in hamsters. *Lab. Anim. Sci.* **28**(6), 723–727.

Bauck, L. A., and Hagen, R. J. (1984). Cystotomy for treatment of cholelithiasis in a hamster. *J. Am. Vet. Med. Assoc.* **184**(1), 99–100.

Benirschke, C. K., Garner, F. M., and Jones, T. C., eds. (1978). "Pathology of Laboratory Animals," Vol. 2. Springer-Verlag, Berlin and New York.

Chesterman, F. C. (1972). Background pathology in a colony of golden hamsters. *Prog. Exp. Tumor Res.* **16**, 50–68.

Chesterman, F. C., and Pomerance, A. (1965). Cirrhosis and liver tumors in a closed colony of golden hamsters. *Br. J. Cancer* **19**, 802–811.

Christensen, F., and Dam, H. (1953). A new symptom of fat deficiency in hamsters: Profuse secretion of cerumen. *Acta Physiol. Scand.* **27**, 204–205.

Cohen, N. L., Reyes, P. S., Typpo, J. T., and Briggs, G. M. (1967). Vitamin B_{12} deficiency in the golden hamster. *J. Nutr.* **91**(4), 482.

Cohen, N. L., Reyes, P. S., and Briggs, G. M. (1971). Folic acid deficiency in the golden hamster. *Lab. Anim. Sci.* **21**, 350–355.

Dam, H., and Christensen, F. (1952). Alimentary production of gallstones in hamsters. *Acta Pathol. Microbiol. Scand.* **30**, 236–242.

Dodds, W. J., Raymond, S. L., Moynihan, A. C., and McMartin, D. W. (1977). Spontaneous atrial thrombosis in aged Syrian hamsters. II. Hemostasis. *Thromb. Haemostasis* **38**, 457–464.

Fahmy, H., Rogers, W. E., Mitchel, D. F., and Brewer, H. E. (1961). Effects of hypervitaminosis D on the periodontium of the hamster. *J. Dent. Res.* **40**, 870–877.

Feldman, D. B., McConnel, E. E., and Knapka, J. J. (1982). Growth, kidney disease and longevity of Syrian hamsters (*Mesocricetus auratus*) fed varying levels of protein. *Lab. Anim. Sci.* **32**(6), 613–618.

Follis, F. H. J. (1964). Further observations on thyroiditis and colloid accumulation in hyperplastic thyroid glands of hamsters receiving excess iodine. *Lab. Invest.* **13**, 1590–1599.

Fortner, J. G. (1957). Spontaneous tumors, including gastrointestinal neoplasms and malignant melanomas, in the Syrian hamster. *Cancer (Philadelphia)* **10**, 1153–1156.

Franks, L. M., and Chesterman, F. C. (1957). Adrenal degeneration and tumour formation in the golden hamster following treatment with stilbesterol and methylcholanthrene. *Br. J. Cancer* **11**, 105–111.

Frenkel, J. K. (1958). Hormonal effects on a sarcoid-like response with Schaumann bodies and amyloid in golden hamsters infected with photochromogenic mycobacteria. *Am. J. Pathol.* **34**(3), 586–587.

Galton, M., and Slater, S. M. (1965). Naturally occurring fatal disease of the pregnant golden hamster. *Proc. Soc. Exp. Biol. Med.* **120**(3), 873–876.

Gerber, G. B., Aldrich, W. G., Koszalka, T. R., and Gerber, G. (1962). Biochemical and autoradiographic studies on DNA metabolism in vitamin E deficient hamsters. *J. Nutr.* **78**, 307–308.

Gleiser, C. A., Van Hoosier, G. L., Jr., and Sheldon, W. G. (1970). A polycystic disease of hamsters in a closed colony. *Lab. Anim. Care* **20**(5), 923–929.

Glenner, G.G., Ein, D., and Terry, W. D. (1972). The immunoglobulin origin of amyloid. *Am. J. Med.* **52**, 141–147.

Granados, H. (1951). Nutritional studies on the growth and reproduction of the golden hamster. *Acta Physiol. Scand.* **24**, Suppl. 87, 1–113.

Granados, H. (1968). *In* "The Golden Hamster—It's Biology and Use in Medical Research" (R. A. Hoffman, P. F. Robinson, and H. Magalhaes, eds.), pp. 157–170. Iowa State Univ. Press, Ames.

Gruys, E., Timmermans, H. J. F., and Van Ederen, A. M. (1979). Desposition of amyloid in the liver of hamsters: An enzyme-histochemical and electron-microscopical study. *Lab. Anim.* **13**, 1–9.

Haley, J. (1951). Cannibalism in hamsters. *Am. Small Stock Farmer* **35**, 10–11.

Hamilton, J. M., and Hogan, A. G. (1944). Nutritional requirements of the Syrian hamster. *J. Nutr.* **27**, 213–224.

Hamilton, J. M., Reynolds, T., and Sahija, P. B. (1983). Cholangiofibrosis in the Syrian golden hamster. *Vet. Rec.* **112**, 359–360.

Handler, A. H., and Chesterman, F. C. (1965). Spontaneous lesions of the hamster. *In* "The Pathology of Laboratory Animals" (W. I. Ribelin and J. R. McCoy, eds.), pp. 210–240. Thomas, Springfield, Illinois.

Handler, A. H., and Chesterman, F. C. (1968). Spontaneous diseases. *In* "The Golden Hamster—It's Biology and Use in Medical Research" (R. A. Hoffman, P. F. Robinson, and H. Magalhaes, eds.), p. 213. Iowa State Univ. Press, Ames.

Handler, P., and Bernheim, F. (1949). Choline deficiency in the hamster. *Proc. Soc. Exp. Biol. Med.* **72**, 569–571.

Harada, T., Yamashiro, S., Mende, P. D., and Basnur, P. K. (1982). Stomach ulcers in vitamin A deficient Syrian golden hamsters. *Jpn. J. Vet. Sci.* **44**, 267–274.

Herrold, K. M., and Dunham, L. J. (1963). Induction of tumors in the Syrian hamster with diethylnitrosamine (*N*-nitrosodiethylamine). *Cancer Res.* **23**(Pt. 2), 773–777.

Hirschi, R. (1950). Post-extraction healing in vitamin A deficient hamsters. *J. Oral Surg.* **8**, 3–11.

Hoffman, R. A., Robinson, P. F., and Magalhase, H., eds. (1968). "The Golden Hamster—Its Biology and Use in Medical Research." Iowa State Univ. Press, Ames.

Horowitz, I., and Waisman, H. A. (1966). Some biochemical changes in the hamsters fed excess phenylalanine diets. *Proc. Soc. Exp. Biol. Med.* **122**, 750–754.

Jacob, W. (1971). Spontaneous amyloidosis of mammals. *Vet. Pathol.* **8**, 292–306.

Jones, C. J., and Hunt, R. D. (1983). "Veterinary Pathology," pp. 43–47. Lea & Febiger, Philadelphia. Pennsylvania.

Jones, J. H. (1945). Experimental rickets in the hamster. *J. Nutr.* **30**, 143–146.

Keeler, R. F., and Young, S. (1979). Role of vitamin E in the etiology of spontaneous hemorrhagic necrosis of the central nervous system of fetal hamsters. *Teratology* **20**, 127–132.

Klei, T. R., Crowell, W. A., and Thompson, P. E. (1974). Ultrastructural glomerular changes associated with filariasis. *Am. J. Trop. Med. Hyg.* **23**, 609–618.

Knapka, J. J., and Judge, F. J. (1974). The effects of various levels of dietary fat and apple supplementation on growth of golden hamsters (*Mesocricetus auratus*). *Lab. Anim. Sci.* **24**, 318–325.

McMartin, D. N. (1977). Spontaneous atrial thrombosis in aged Sryian hamsters. I. Incidence and pathology. *Thromb. Haemostasis* **38**, 447–456.

McMartin, D. N. (1979). Morphologic lesions in aging Syrian hamsters. *J. Gerontol.* **34**(4), 502–511.

Magalhaes, H. (1968). The golden hamster as a laboratory animal. *Lab. Anim. Bur. Congr. Meet.*, pp. 39–44.

Maguire, S., Hamilton, J. M., and Fulker, M. J. (1974). A spontaneous glomerular lesion in the Syrian hamster (*Mesocricetus auratus*). *Br. J. Exp. Pathol.* **55,** 562–569.

Margolis, G., and Kilham, L. (1976). Hemorrhagic necrosis of the central nervous system. A spontaneous disease of fetal hamsters. *Vet. Pathol.* **13,** 250–263.

Mason, K. E., and Mauer, S. I. (1975). Reversible testis injury in the vitamin E-deficient hamster. *J. Nutr.* **105**(4), 484–490.

Murgatroyd, L. B., and Tucker, M. J. (1981). Pancreatic islet cell hypertrophy in the Syrian hamster. *J. Comp. Pathol.* **91,** 455–459.

Nelson, W. B. (1975). Fatal hairball in a long-haired hamster. *VM/SAC, Vet. Med. Small Anim. Clin.* **70**(Pt. 2), 1193.

Newberne, P. M. (1978). Nutritional and metabolic diseases. *In* "Pathology of Laboratory Animals" (C. K. Benirschke, F. M. Garner, and T. C. Jones, eds.), pp. 2065–2154. Springer-Verlag, Berlin and New York.

Niggeman, B. M. (1946). Histological and cytological effects of thiamin-, riboflavin- and pyridoxine-free diets upon the gastrointestinal tract, spinal cord, and sciatic nerve of the golden hamster. *Univ. Colo. Stud.* **27,** 95.

Peters, L. J. (1982). Abdominal pregnancy in a golden hamster (*Mesocricetus auratus*). *Lab. Anim. Sci.* **32**(4), 392–393.

Pollock, W. B. (1975). Prolapse of invaginated colon through the anus in golden hamsters (*Mesocricetus auratus*). *Lab. Anim. Sci.* **25**(3), 334–336.

Ponsford, P. S., and Whyley, G. A. (1954). The golden hamster (*Mesocricetus auratus*). *J. Med. Lab. Technol.* **12,** 142–145.

Pour, P., Kmoch, N., Greiser, E., Mohr, U., Althoff, J., and Cardesa, A. (1976a). Spontaneous tumors and common diseases in two colonies of Syrian hamsters. I. Incidence and sites. *J. Natl. Cancer Inst. (U.S.)* **56**(5), 931–935.

Pour, P., Mohr, U., Cardesa, A., Althoff, J., and Kmoch, N. (1976b). Spontaneous tumors and common diseases in two colonies of Syrian hamsters. II. Respiratory tract and digestive system. *J. Natl. Cancer Inst. (U.S.)* **56**(5), 937–948.

Pour, P., Mohr, V., Althoff, J., Cardesa, A., and Kmoch, N. (1976c). Spontaneous tumors and common diseases in two colonies of Syrian hamsters. III. Urogenital and endocrine glands. *J. Natl. Cancer Inst. (U.S.)* **56**(5), 949–961.

Pour, P., Mohr, U., Althoff, J., Cardesa, A., and Kmoch, N. (1976d). Spontaneous tumors and common diseases in two colonies of Syrian hamsters. IV. Vascular and lymphatic systems and lesions of other sites. *J. Natl. Cancer Inst. (U.S.)* **56**(5), 963–974.

Purdy, D. M., and Hillemann, H. H. (1950). Prenatal mortality in the golden hamster (*Mesocricetus auratus*). *Anat. Rec.* **106,** 577–583.

Redman, H. C., Hobbs, C. H., and Rebar, A. H. (1979). Survival distribution of Syrian hamsters (*Mesocricetus auratus*) (Sch:Syr) used during 1972–1977. *Prog. Exp. Tumor Res.* **24,** 108–117.

Rennie, J. S., McDonald, D. G., and Douglas, T. A. (1982). Experimental iron deficiency in the Syrian hamster. *Lab. Anim.* **16,** 14–16.

Renshaw, H. W., Van Hoosier, G. L., Jr., and Amend, N. K. (1975). A survey of naturally occurring diseases of the Syrian hamster. *Lab. Anim.* **9**(3), 179–191.

Richter, A. G., Lansen, N. C., and Lage, A. L. (1984). Pregnancy toxemia (eclampsia) in Syrian golden hamsters. *J. Am. Vet. Med. Assoc.* **185**(11), 1357–1358.

Routh, J. I., and Houchin, O. B. (1942). Some nutritional requirements of the hamster. *Fed. Proc., Fed. Am. Soc. Exp. Biol.* **1,** 191–192.

Salley, J. J., and Bryson, W. F. (1957). Vitamin A deficiency in the hamster. *J. Dent. Res.* **36,** 935–944.

Salley, J. J., Bryson, W. F., and Eshleman, J. R. (1959). The effect of chronic vitamin A deficiency on dental caries in the Syrian hamster. *J. Dent. Res.* **38,** 1038–1043.

Scheid, H. E., McBride, B. H., and Schwergert, B. S. (1950). The vitamin B_{12} requirement of the Syrian hamster. *Proc. Soc. Exp. Biol. Med.* **75,** 236–239.

Schmidt, R. E., Eason, R. L., Hubbard, G. B., Young, J. T., and Eisenbrandt, D. L. (1983). "Pathology of Aging Syrian Hamsters." CRC Press, Boca Raton, Florida.

Shwartzman, G., and Strauss, L. (1949). Vitamin B_6 deficiency in the Syrian hamster. *J. Nutr.* **38,** 131–154.

Schweigert, B. S., McBride, B. H., and Carlson, J. A. (1950). Effect of feeding polyoxyethylene monostearates on the growth rate and gross pathology of weanling hamsters. *Proc. Soc. Exp. Biol. Med.* **73,** 427–432.

Shenefelt, R. E. (1978). Developmental abnormalities. *In* "Pathology of Laboratory Animals" (C. K. Benirschke *et al.*, eds.), Vol. 2, pp. 1866–1869. Springer-Verlag, Berlin and New York.

Sichuk, G., Bettigole, R. E., Der, B. K., and Fortner, J. B. (1964). Influence of sex hormone on thrombosis of the left atrium in Syrian (golden) hamsters. *Am. J. Physiol.* **208**(3), 465–470.

Silberberg, M., and Silverger, R. (1941). Age changes of bones and joints in various strains of mice. *Am. J. Anat.* **68,** 69–96.

Slausen, D. O., Hobbs, C. H., and Crain, C. (1978). Arteriolar nephrosclerosis in the Syrian hamster. *Vet. Pathol.* **15,** 1–11.

Sokoloff L., and Habermann, R. T. (1958). Idiopathic necrosis of bone in small laboratory animals. *Arch. Pathol.* **65,** 323–330.

Swank, R. L., and Engel, R. (1958). Production of convulsions in hamsters by high butter fat intake. *Trans. Am. Neurol. Assoc.* **83,** 33–36.

Takahashi, M., and Pour, P. (1978). Spontaneous alterations in the pancreas of the aging Syrian golden hamsters. *J. Natl. Cancer Inst. (U.S.)* **60**(2), 355–364.

Van Marck, E. A. E., Jacob, W., Deelder, A. M., and Gigase, P. L. J. (1978). Spontaneous glomerular basement membrane changes in the golden Syrian hamster (*Mesocricetus auratus*): A light and electron microscope study. *Lab. Anim.* **12,** 207–211.

Yamamoto, M., and Ingalls, T. H. (1972). Delayed fertilization and chromosome anomalies in the hamster embryo. *Science* **176,** 518–521.

Young, S., and Keeler, R. F. (1978). Hemorrhagic necrosis of the central nervous system of fetal hamsters: Litter incidence and age-related pathological changes. *Teratology* **17**(3), 293–302.

Chapter 11

Drugs Used in Hamsters with a Review of Antibiotic-Associated Colitis

J. David Small

  I. Introduction 179
 II. Drug Therapy in Hamsters 180
III. Antibiotic-Associated Enterocolitis 180
     A. Introduction 180
     B. Etiology 180
     C. Pathogenesis 187
     D. Epizootiology 187
     E. Signs and Lesions 193
     F. Diagnosis 193
     G. Hamster Models of Human Antibiotic-Associated Enterocolitis ... 196
     References 196

I. INTRODUCTION

Data on the therapeutic use of drugs in hamsters kept in the laboratory or as pets are few. Most references to therapeutic use of drugs have involved problems of the digestive system under the terms "wet-tail" (Cosar and Kolsky, 1955; Sheffield and Beveridge, 1962) and transmissible ileal hyperplasia (La Regina *et al.*, 1980; Jacoby and Johnson, 1981). A few references to treatment of hamsters for parasitic infections and infestations are also available (Ronald and Wagner, 1975; Sebesteny, 1979; Heath and Bishop, 1984). Most substances listed in Tables I–V were studied for their toxicity or their effect on a specific parasite, or they were used to induce a physiological response. Thus much of the data are not directly applicable for therapeutic purposes in the hamster. It is recommended that the references be consulted prior to using listed drugs for the first time for therapeutic purposes. Often, in the studies cited, the hamsters were not maintained long enough to observe toxic effects, nor were the necessary observations made to rule out toxicity. The dosages and comments for the substances listed were mostly gleaned from published reports, which included original observations in hamsters. Recommendations from sources not containing original animal data have been identified. Reference to treatment of specific diseases and conditions elsewhere in this text or in current editions of other texts (e.g., Holmes, 1984; Harkness and Wagner, 1983; Russell *et al.*, 1981; Sebesteny, 1979; Williams, 1976) is suggested.

Data from a series of reports on the chemotherapeutic activity of nitro compounds and anthraquinones by Winkelmann and Raether (1978, 1979) and Winkelmann *et al.* (1975, 1977a,b, 1978) were not included in Tables I–V because of the large number of experimental compounds tested. However, these reports serve as a useful source of information concerning the use of these compounds in hamsters infected with *Entamoeba histolytica* as well as animals other than the hamster infected with a variety of agents and with neoplasms. Also, results of studies with laetrile (DL-amygdalin) (Willhite, 1982) and the antioxidants, quercetin and butylated hydroxytoluene (Harada *et al.*, 1984) are not included in Tables I–V. An extensive number of steroids plus other substances were studied for their effect on adrenocortical secretion (Frenkel *et al.*, 1965). This work has been reviewed previously (Frenkel, 1972b).

II. DRUG THERAPY IN HAMSTERS

The determination of clinically useful drug dosages for laboratory rodents is often made empirically, usually as some factor of the human dose or published dose for another animal species. The dose is usually in milligrams per kilogram body weight. Such comparisons may not provide adequate amounts of drug. Using the 10% lethal dose (LD_{10}) for three species and the maximum tolerated dose (MTD) for the dog, monkey, and human, Freireich *et al.* (1966) developed formulas to compare the toxicity of several antineoplastic agents for the animals (dog, monkey, mouse, rat, and hamster) and humans. On a milligrams per kilogram basis, the MTD in humans is about one-ninth the LD_{10} in hamsters, one-twelfth the LD_{10} in mice, one-seventh the LD_{10} in rats, one-third the MTD in rhesus monkeys, and one-half the MTD in dogs: however, when compared on the basis of surface area (mg/m^2), the MTD and the LD_{10} are about the same, regardless of species. Applying Freireich's formulas to antimicrobials, Bartlett *et al.* (1978a) calculated that doses of antibiotics for 70- to 90-gm hamsters should be six times those of adult humans on a milligrams per kilogram basis. Failure to adjust the dosage of a drug to the species and its weight or body area may result in too little drug for a therapeutic effect or a toxic overdose (Killby and Silverman, 1967).

III. ANTIBIOTIC-ASSOCIATED ENTEROCOLITIS

A. Introduction

Many antimicrobial agents, even when given in single minute doses, are highly toxic for hamsters as well as for guinea pigs. Hamre *et al.* (1943) were the first to report mortality in guinea pigs given penicillin. The first report of antibiotic toxicity in hamsters was by Kaipainen and Faine (1954), who observed mortality in hamsters given erythromycin. Tetracycline and chlortetracycline were reported toxic for hamsters by Cosar and Kolsky (1955). They were able to reduce mortality from 70–80% to 15–20% with concomitant administration of sulfaguanidine. Schneierson and Perlman (1956) reported penicillin to be toxic for hamsters. Several reports of mortality in hamsters and guinea pigs given antimicrobial agents appeared during the 1950s and 1960s, but the cause of death was not determined. Following the report of lincomycin's toxicity in the hamster (Small, 1968a), the hamster's exquisite sensitivity to several antimicrobial agents and the resulting enterocolitis led to its use in the study of antimicrobial-associated colitis (AAC), as found in humans. Several synonyms have been used for this condition in humans: pseudomembranous enterocolitis, antibiotic-associated colitis, antibiotic-associated pseudomembranous enterocolitis, antibiotic-associated enterocolitis, and antimicrobial-induced enterocolitis. The potential value of treating hamsters with antimicrobial agents must be weighed against the risk of inducing AAC.

B. Etiology

The early reports of antimicrobial toxicity in guinea pigs and hamsters have been reviewed (Small, 1968a). The reasons for the hamster's sensitivity to many antimicrobial agents is uncertain; however, alterations in the microbial flora of the intestinal tract are suspected as the cause of AAC. Hagen *et al.* (1965) established that the predominant bacterial genera of the normal hamster's intestinal content (presumably cecum) are *Lactobacillus* and *Bacteroides*. Relatively few coliforms were isolated. Early studies of changes in the microbial flora following administration of antimicrobials to hamsters focused on the marked increase in coliforms (Small, 1968a). While the specific cause of antimicrobial toxicity is now recognized to be primarily due to toxins from *Clostridium difficile*, the means by which this bacterium and its toxins are suppressed in the normal hamster remains unclear. A component of the normal intestinal flora has been postulated as providing protection, but the organism(s) responsible have not been identified. Wilson *et al.* (1981) regularly killed (95%) hamsters with a single 200 mg/kg oral dose of vancomycin; however, if similarly treated hamsters were given homogenates of cecal content from normal hamsters, 80% survive. This protective effect was eliminated by prior incubation of the homogenate with clindamycin, but not gentamicin or vancomycin. Heated or filtered homogenates did not provide protection. Based on these observations plus the relative resistance of gram-negative anaerobes as well as some clostridia to gentamicin and vancomycin and their sensitivity to clindamycin, it was postulated that these organisms are responsible for suppressing *C. difficile*. Rolfe *et al.* (1981), in an *in vitro* study of 401 bacterial isolates, showed inhibition of *C. difficile* by species of *Bifidobacterium*,

Table I
Antimicrobial Agents

Drug	Dose and route[a]	Comment	References
Amphotericin B	1 mg per hamster sid SC, 5 d/wk for 3 wk	Effective against experimental infection with *Histoplasma capsulatum*; untreated hamsters died with disseminated infection after 3–6 wk	J. K. Frenkel, personal communication
Ampicillin	10 mg/kg 1× PO; 100 mg/kg 1× PO	5/15 (10 mg) and 15/15 (100 mg) died within 4–7 d	Fekety *et al.* (1979a)
	0.5–10 mg per hamster per d PO until death (1–24 d)	All died (1–24 d), but longevity increased wtih increasing doses	Rolfe and Finegold (1983)
	100 mg/kg 1× SC	10/12 Died within 21 d	Small (1968b)
	5 mg tid PO for 5 d	9/10 Died (8–12 d)	Bartlett *et al.* (1978b)
Bacitracin	3 mg/ml drinking water for 14 d starting 2 d before giving clindamycin IM for 3 d	Blocked clindamycin-induced mortality (0/10); no posttreatment observation period	Bartlett *et al.* (1977b)
Carbenicillen	100 mg 1× PO	9/10 Died (6–8 d)	Fekety *et al.* (1979a)
	100, 200, 400, 800 mg/kg 1× SC	4/11, 3/6, 4/6, 5/6 Died within 21 d	Small (1968b)
Cefamandole	1 mg 1× SC	0/10 Died	Ebright *et al.* (1981)
	10 mg 1× SC	2/10 Died (3 d)[b]	Ebright *et al.* (1981)
	100 mg 1× SC	10/10 Died (5 d)[b]	Ebright *et al.* (1981)
	10 mg 1× PO	10/10 Died (5 d)[b]	Ebright *et al.* (1981)
Cefachlor	10 mg 1× PO; 10 mg (125 mg/kg) sid PO for 11 d	10/10 Died (3 d)[b] and (5 d)[b]	Ebright *et al.* (1981)
Cefazolin	1 mg 1× SC; 10 mg (125 mg/kg) sid PO for 11 d	0/10 Died; survivors (PO) died within 4 d of discontinuing drug	Ebright *et al.* (1981)
	10 mg 1× SC	5/10 Died (2 d)[b]	Ebright *et al.* (1981)
	10 mg 1× PO	10/10 Died (4 d)[b]	Ebright *et al.* (1981)
	100 mg 1× SC	10/10 Died (5 d)[b]	Ebright *et al.* (1981)
Cefoxitin	1 mg 1× SC	8/10 Died (3 d)[b]	Ebright *et al.* (1981)
	10 or 100 mg, 1× SC	10/10 Died 2 d and 5 d)[b]	Ebright *et al.* (1981)
	10 mg 1× PO	10/10 Died (4 d)[b]	Ebright *et al.* (1981)
	10 mg (125 mg/kg sid PO for 11 d	0/10 Died, 5 necropsied, remaining 5 died within 4 d of discontinuing drug	Ebright *et al.* (1981)
	10 mg tid IM 5 d	10/10 Died	Bartlett *et al.* (1978b)
Cephalexin	10 mg (125 mg/kg) sid PO for 11 d	9/10 Died (5 d)[b]	Ebright *et al.* (1981)
	10 mg 1× PO	10/10 Died (3 d)[b]	Ebright *et al.* (1981)
	5 mg tic PO for 5 d	9/10 Died	Bartlett *et al.* (1978b)
	50 mg/kg sid PO for 34 consecutive days	0/8 Died; weight same as controls	J. D. Small (unpublished data)
	100 mg/kg 1× PO	5/8 Died within 4 d; weight of survivors equaled controls after 34 d	J. D. Small (unpublished data)
Cephaloridine	1 mg 1× SC; 10 mg (125 mg/kg) sid PO for 11 d	1/10 Died; survivors (PO) died within 4 d of discontinuing drug	Ebright *et al.* (1981)
	10 mg 1× SC	0/10 Died	Ebright *et al.* (1981)
	100 mg 1× SC	9/10 Died (7 d)[b]	Ebright *et al.* (1981)
	10 mg 1× PO	5/10 Died (4 d)[b]	Ebright *et al.* (1981)
Cephalothin	10 mg (125 mg/kg) sid PO for 11 d	1/10 Died during treatment, 5 survivors necropsied, remaining 4 died within 4 d of discontinuing drug	Ebright *et al.* (1981)
	10 mg 1× PO	10/10 Died (3 d)[b]	Ebright *et al.* (1981)
	1 or 10 mg 1× SC	0/10 Died	Ebright *et al.* (1981)
	100 mg, 1× SC	10/10 Died (4 d)[b]	Ebright *et al.* (1981)
	20 mg tid IM 5 d	8/10 Died (4–7 d)	Bartlett *et al.* (1978b)

(continued)

Table I (Continued)

Drug	Dose and route[a]	Comment	References
Cephradine	5 mg tid PO 5 d	9/10 Died	Bartlett et al. (1978b)
	100 mg/kg 1× SC	1/12 Died (3 d), observed 21 d	Small (1968b)
	1 mg and 10 mg, 1× SC	0/10 Died	Ebright et al. (1981)
	100 mg, 1× SC, 10 mg, 1× PO	10/10 Died (3 d)[b] and (3d)[b]	Ebright et al. (1981)
	10 mg (125 mg/kg) sid PO 11 d	6/10 Died (5 d)[b]	Ebright et al. (1981)
Chloramphenicol	<1000 mg/kg 1× SC (succinate) ≥300 mg/kg 1× PO (palmitate)	"Did not reliably induce typhlitis"; doses <300 mg/kg did not produce disease in most hamsters	Fekety et al. (1979a)
	10 mg tid PO for 53 d	2/10 Died within 19-d observation period	Bartlett et al. (1978b)
	128 mg/kg sid SC for 2 d	0/5 Died in 10 d, uninfected controls	Kemp (1965)
	100 mg/kg 1× SC, or sid SC for 10 d (succinate)	0/24 Died	Small (1968b)
	100 mg/kg sid PO for 12 d, 21 d, and 43× in 49 d (palmitate)	1/12 (12 d), 0/9 (21 d), 0/6 (49 d) Died	Small (1968b)
	6–9 mg (0.2–0.3 ml) per hamster bid PO (palmitate)	Suggested for "wet-tail"; no data presented	Williams (1976)
	30 mg per adult hamster sid SC for 5 d	Observed 15 d; no undesirable clinical signs or postmortem changes in five animals	Sebesteny (1979)
	0.01% Active soluble preparation in only drinking water	Observed 15 d; no undesirable clinical signs or postmortem changes in five animals. Note: No measurement of water intake reported. This is a very bitter compound, and drug intake needs to be assured [Author, JDS]	Sebesteny (1979)
Chlortetracycline	50 mg/kg sid PO for 3 d	69% Mortality; simultaneous administration of sulfaguanidine (250 mg/kg) PO reduced mortality to 20%	Cosar and Kolsky (1955)
	4 or 8 mg/kg sid SC for 2 d	4 mg, 1/10 died; 8 mg, 0/5 died	Kemp (1965)
	16 or 32 mg/kg sid PO for 2 d	16 mg, 35/89 died; 32 mg, 8/57 died; all received Leptospira pomona 3 d before drug; observed for 10 d after last dose	Kemp (1965)
	20 mg/kg bid SC for 14 d	Remained well	Heilman (1948)
	0.03125–0.5 mg/70–100 gm hamster bid SC for 10 d Approximate dose = 1.4 mg/kg	1/30 Leptospira icterohaemorragica-infected hamsters died (lowest dose); all others survived; 15/15 untreated controls died by d 11	Heilman (1948)
Clindamycin	1 mg 1× IP	15/15 Died within 7 d	Bartlett et al. (1978a)
	3 mg tid PO for 5 d	10/10 Died within 19-d observation period	Bartlett et al. (1978b)
	1 mg or 200 mg/kg 1× SC	12/15 Died within 7 d	Lusk et al. (1978)
	5, 10, 25, 50, 75, or 100 mg/kg 1× SC	15/15 Died per dose group within 7 d	Lusk et al. (1978)
	10 mg/kg or 430 mg/kg 1× PO	10/10 Died within 3–8 d	Lusk et al. (1978)
	0.01, 0.1, 1.0, 10, or 40 mg per hamster sid dermal	1, 10, 40 mg; all died within 3–7 d; 0.1 mg, 4/7 died; 0.01 mg, no deaths; 5 mg tetracycline sid blocked death in hamsters given 10 mg/d only as long as tetracycline was given. Note: Dermal administration such that oral ingestion not possible; skin shaved.	Feingold et al. (1979)

Table I (*Continued*)

Drug	Dose and route[a]	Comment	References
Colistin	128 mg/kg sid SC for 2 d	0/5 Died in 10 d, uninfected controls	Kemp (1965)
Demethylchlortetracycline-HCl (demeclocycline)	8 mg/kg sid SC for 2 d; 16 mg/kg sid PO for 2 d	0/10 and 1/10 died in 10 d, respectively	Kemp (1965)
Erythromycin	16 mg/kg sid SC for 2d; 256 mg/kg sid PO for 2 d	1/10 and 0/5 died in 10 d, repectively; all received *L. pomona* 3 d before drug; observed for 10 d after last dose	Kemp (1965)
	30–200 mg/kg "one or more doses IP"	". . . died after approximately 13 d . . ."	Kaipainen and Faine (1954)
	LD_{50} 3.5 (2.4–5.1) mg/kg 1× PO; 5.8 (3.9–8.7) mg/kg 1× SC	2-Wk observation period	De Salva *et al.* (1969)
Furazolidone		See Table III, Antiparasitic Agents	
Gentamicin	1 mg/sid PO	20/20 Died in 4–14 d with typhlitis	Fekety *et al.* (1979a)
	1 mg/tid PO for 5 d	10/10 Died	Bartlett *et al.* (1978b)
	5 mg/kg sid IM	Dilute 1:10 before dosing (no comment on concentration of stock; assumed 50 mg/ml; no data)	Russell *et al.* (1981)
Ketoconazole		See Table III, Antiparasitic Agents	
Lincomycin	5–400 mg/kg 1× SC	10–400 mg/kg, 92/102 Died within 19 d; 5 mg/kg, 1/23 died (9 d)	Small (1968a)
Methacycline	2.5 or 5.0 mg/ml in drinking water *ad libitum* for 60 d	On a cariogenic diet (CD); no mortality reported; drug + CD produced most caries	Simard-Savoie and Dupuis (1968)
Methicillin	100, 200, 400, 800 mg/kg 1× SC	3/12, 3/6, 5/6, 4/6 Died within 21 d	Small (1968b)
Metronidazole		See Table III Antiparasitic Agents	
Minocycline	500 mg/kg sid PO for 10 d	5/5 Died; infected with *Babesia microti*, which did not cause mortality	Miller *et al.* (1978)
Nafcillin	100 mg 1× PO	10/10 Died (1–9 d)	Fekety *et al.* (1979a)
	100 mg 1× SC	11/12 Died within 9 d; observed 21 d	Small (1968b)
Neomycin	285 mg/kg 1× PO	Death within 5 d	Fekety *et al.* (1979a)
	2 mg/ml drinking water for 2 wk	No report of mortality, did not prevent transmissible ileal hyperplasia (TIH)	Jacoby and Johnson (1981)
	0.125 mg/ml drinking water for 10 d	No report of mortality, did not prevent TIH	La Regina *et al.* (1980)
	0.25 mg/ml drinking water for 5 wk commercial diet	No diarrhea (0/8)	Michelich *et al.* (1981)
	0.25 mg/ml drinking water for 5 wk purified diet	Diarrhea, *Clostridium difficile* isolated, and positive for toxin	Michelich *et al.* (1981)
	10 mg/40 gm hamster sid PO in cheese pellet; initial dose in 0.25 ml water by gavage	50 mg per hamster tolerated; 10-mg dose used successfully for treatment of "wet-tail"	Sheffield and Beveridge (1962)
	64 mg/kg sid SC for 2 d	0/10 Died in 10 d; all received *L. pomona* 3 d before drug	Kemp (1965)
	10 gm/gal drinking water for 5 d, then 5 gm/gal for 5 d	Suggested for treatment of "wet-tail"; no data	Russell *et al.* (1981)
Oleandomycin	64 mg/kg sid SC for 2 d; 256 mg/kg sid PO for 2 d	0/10 Died (SC and PO) in 10 d; all received *L. pomona* 3 d before drug; observed for 10 d after last dose	Kemp (1965)
Oxacillin	100 mg/kg 1× SC	4/6 Died within 3 d; observed 21 d	Small (1968b)
Oxytetracycline-HCl	16 mg/kg sid SC for 2 d; 64 mg/kg sid PO for 2 d	0/21 Died (SC and PO) in 10 d; all received *L. pomona* 3 d before drug	Kemp (1965)

(*continued*)

Table I (*Continued*)

Drug	Dose and route[a]	Comment	References
Paromomycin sulfate, Humatin	0.025% Active drug in drinking water age 4–6 wk	"Incidence of 'wet-tail' kept at a satisfactory low level"	Sebesteny (1979)
	See Table III, Antiparasitic Agents		
Penicillin G	100 mg 1× PO; 600 mg 1× SC	Died within 5 d	Fekety et al. (1979a)
	1–25 × 10³U, 1× SC or IP	8/40 (SC) and 11/40 (IP) died within 7 d	Schneierson and Perlman (1956)
	50–100 × 10³U, 1× SC or IP	14/20 (SC) and 19/20 (IP) died within 12 d; observed 21 d	
	100 mg/kg 1× SC	6/6 Died within 13 d	Small (1968b)
Ribavirin	25 mg/kg IP sid or bid for two to six total doses over d 10–15 of gestation	Cerebellar hypoplasia, tooth germ defects, epidermal defects, retinal defects	Margolis and Kilham (1978)
	0.1–0.25 mg sid for 5 d, neonates		
	2.5 or 5.0 mg/kg 1× IP on d 8 of gestation	2.5 mg/kg, 47/177 (27%) abnormal; 5.0 mg/kg, 65/118 (55%) abnormal	Willhite and Ferm (1978)
Streptomycin	LD₅₀ 400 mg/kg 1× PO; >500 mg/kg 1× SC	2 Wk observation	De Salva et al. (1969)
	40–1280 mg/kg PO (diet)	Progressing to highest dose over 12 wk; no mortality (0/20), normal weight gain	De Salva et al. (1969)
	16 mg/kg sid SC for 2 d	0/10 Died in 10 d; all received *L. pomona* 3 d before drug	Kemp (1965)
	50, 75, 150, 300, 400, 1000 mg/kg diet for 162 d	5 Hamsters per dose; 33/35 survived; cariogenic diet high in glucose	Jordan and van Houte (1972)
Streptomycin (dihydrostreptomycin)	2 gm/liter drinking water for 4 wk 27 mg/gm Orabase in cheek pouches sid 5×/wk for 4 wk	Orabase also contained 27 mg/gm neomycin; drinking water contained 800 mg/liter dimetridazole; eliminated aerobic gram-negative bacteria from throat by d 5 and from feces during wk 2; enterococci increased; protozoa not eliminated; No gram-negative aerobes 7 wk after stopping drugs; enterococci returned to pretreatment levels in 2 wk; housed in laminar flow bench	Angulo et al. (1978)
Sulfadiazine, sodium	0.6–1.2 gm/liter drinking water for 30 d	Blocked lethality of *Besonitia jellisoni*	Frenkel and Lunde (1966)
Sulfadiazine (S) + pyrimethamine (P)	(S) 600 mg + (P) 50 mg/kg for 10 d	No anti-*Babesia microti* activity; no adverse effects reported	Miller et al. (1978)
Sulfaguanidine	250 mg/kg PO 250 mg/kg sid PO for 3 d	Reduced mortality of tetracycline-induced cecitis when given together	Cosar and Kolsky (1955)
Sulfamethoxazole (S) + trimethoprim (T)	200 mg/kg (167 mg S + 33 mg T) 1× PO	6/20 Died with typhlitis in one experiment; inconsistent results	Fekety et al. (1979a)
	5 ml/kg (200 mg S + 40 mg T) 5×/wk for 6 wk	0/4 Died in 6 wk, 0/3 died in 14 wk; no observable effect on intestinal protozoa	J. D. Small (unpublished data)
Sulfasalazine	100 and 300 mg/kg bid for 4 d	Blocked lethal effect of clindamycin-induced colitis; observed 10 d after last dose	Will et al. (1985)
Tetracycline-HCl	LD₅₀ > 400 mg/kg 1× PO; >400 mg/kg SC	2 Wk observation	De Salva et al. (1969)
	100 mg/kg 1× PO; sid for 7 d PO; sid for ≥10 d PO	1×, Most died within 3–4 d; 7 d, most died 2–3 d after discontinuance; ≥10 d, began to die during treatment	Fekety et al. (1979a)
	5 mg tid PO for 5 d	0/10 Died, weighed 70–90 gm (males)	Barlett et al. (1978b)

Table I (*Continued*)

Drug	Dose and route[a]	Comment	References
	1 mg/kg or 10 mg/kg 1× PO	0/10 Died, weighed 70–90 gm (males)	Toshniwal *et al.* (1979)
	100 mg/kg 1× PO	5/7 Died, 8 d, 60–100 gm (males)	Toshniwal *et al.* (1979)
	10 mg/kg sid PO for 7 or 14 d	0/10 Died, 8 d, 60–100 gm (males)	Toshniwal *et al.* (1979)
	100 mg/kg sid PO for 3, 7, or 14 d	8/9 (5 d); 6/9 (10 d); 9/10 (11 d) died	Toshniwal *et al.* (1979)
	50 mg/kg sid PO for 34 consecutive days	0/8 Died; weight gain same as controls	J. D. Small (unpublished data)
	100 mg/kg 1× PO	2/8 Died between 15 and 12 d; weight of survivors same as controls	J. D. Small (unpublished data)
	400 mg/liter water for 10 d beginning 3 d postinoculation	No report of mortality, reduced incidence of transmissible ileal hyperplasia	La Regina (1980)
	50 mg/kg sid PO for 3 d	82% Mortality; simultaneous administration of sulfaguanidine (250 mg/kg) PO reduced mortality to 20%	Cosar and Kolsky (1955)
Ticarcillin	10 mg 1× PO	10/10 Died (5–6 d)	Fekety *et al.* (1979a)
Thienamycin	25 mg/kg 1× SC	30/40 Died	Hawkins *et al.* (1984)
Tylosin	4 mg/kg sid SC for 2 d; 256 mg/kg sid PO for 2 d	1/42 and 0/5 died in 10 d, respectively; all received *L. pomona* 3 d before drug; observed for 10 d after last dose	Kemp (1965)
	2–8 mg/kg bid IM or 100 mg/kg PO (2.5 gm/gal drinking water for 21 d)	May be beneficial for pneumonia due to *Streptococcus pneumoniae*, *Streptococcus* sp., and *Pasteurella*; no data	Russell *et al.* (1981)
	10 mg sid SC for 5 d	Observed 15 d; no clinical signs or postmortem lesions observed in five hamsters	Sebesteny (1979)
Vancomycin	50–70 mg/kg sid PO for 7 d	5/10 Died 5 d after stopping vancomycin	Fekety *et al.* (1979a)
	3 mg per hamster (60–80 gm) sid PO for 7 d	17/17 Died, with a MST of 12.9 d, after stopping vancomycin	Browne *et al.* (1977)
	3 mg per hamster (60–80 gm) sid SC for 7 d	2/10 Given 100 mg/kg clindamycin 1× SC died, with a MST of 13.5 d; clindamycin (100 mg/kg 1× SC) (MST, 3.4 d and 3.2 d, respectively)	Browne *et al.* (1977)
	50–70 mg/kg) sid PO for 7 d	17/17 Died, with a MST of 6.9 d, after stopping vancomycin; study done in "conventional room"	Fekety *et al.* (1980)
	200 µg/ml drinking water for 10 or 20 d	15/20 Died in "conventional room," with a MST of 5.5 d, after stopping vancomycin; 1/20 died (d 14) in "clean room"	Toshniwal *et al.* (1981)
	3 mg/ml drinking water for 14 d 10 mg per hamster bid IM for 14 d	Blocked clindamycin-induced mortality 0/20 and 0/10, respectively; no posttreatment observation period	Bartlett *et al.* (1977b)
	0.30 mg/ml drinking water for 5 wk + purified diet; also treated 4× in cheek pouch with dimethylbenzanthracene	9/9 Developed diarrhea 3 wk following discontinuance of drug	Michelich *et al.* (1981)

[a]Abbreviations used in Table I–V: bid, twice a day; CD_{50}, curative dose 50%; d, day; ED_{50}, effective dose 50%; F, female; IM, intramuscular; IP, intraperitoneal; LD, lethal dose; LD_{50}, lethal dose 50%; LD_{10}, lethal dose 10%; M, male; MST, mean survival time; NS, not significant; PO, orally; q, every; qid, four times a day; SC, subcutaneous; SD, survivor dose; sid, once a day; TI, therapeutic index; tid, three times a day; U.M.R.C., units, Medical Research Council; 1× one time.

[b]Mean day of death.

Table II

Antimetabolic Agents

Drug	Dose and route[a]	Comment	References
Actinomycin D	$LD_{10} = 0.044$ mg/kd sid IP for 7 d	Observation period 14 d	Freireich et al. (1966)
BCNU	$LD_{10} = 8.3$ mg/kg sid IP for 7 d	Observation period 14 d	Freireich et al. (1966)
Cyclosphosphamide, cytoxan	$LD_{10} = 56$ mg/kg sid IP for 7 d	Observation period 14 d	Freireich et al. (1966)
	10, 25, or 50 mg/kg 3×/wk IP for 2–4 wk	Chinese hamster; lymphocytes reduced 41, 44 and 74%; PMNs, 0, 28, and 85%, respectively ($N = 3$); 50 mg/kg dose decreased food consumption; normal and diabetic animals used	Gishizky et al. (1984)
	50, 25, 25 mg/kg loading dose wk 1 (4.5 wk old), then 15 mg/kg 3×/wk IP from 5 to 22 wk of age	Depressed lymphocytes 50%, did not alter granulocytes, food consumption, or body weight; did not alter onset of hyperglycemia or glucosuria	Gishizky et al. (1984)
	2.5 mg per hamster (~25 mg/kg) sid for 3 d posttumor inoculation	Limited tumor development (TBD 932 lymphocarcinoma) and death; all survived with no signs of tumor up to 50 d posttumor inoculation; activity blocked by interferon (IFN-αA). 50 μg per hamster not effective	Lee et al. (1984)
5-FUDR	$LD_{10} = 39$ mg/kg sid IP for 7 d	Observation period 14 d	Freireich et al. (1966)
5-Fluorouracil	$LD_{10} = 12$ mg/kg sid IP for 7 d	Observation period 14 d	Freireich et al. (1966)
	100 mg sid IP or PO for 4 d	5/5 Died in each group by 4 d with "wet-tail"; ascites in 6 hamsters, hemorrhagic cecitis in 2, mucositis of eyes and rectum in 2	Silva et al. (1984)
	10 mg sid IP, PO, SC for 4 d	10/10 Died in each group; 3/15 positive for *Clostridium difficile* (2) or its cytotoxin (1); death attributed to inhibition of mucosal replication, not *Cl. difficile* overgrowth	Silva et al. (1984)
Melphalan	50 μg per hamster (~0.5 mg/kg) sid for 3d posttumor inoculation	Limited tumor development (TBD 932 lymphocarcinoma) and death; drug activity not blocked by interferon (IFN-α)	Lee et al. (1984)
6-Mercaptopurine	$LD_{10} = 56$ mg/kg sid IP for 7 d	Observation period 14 d	Freireich et al. (1966)
Methotrexate (Amethopterin) (MTX)	$LD_{10} = 18$ mg/kg sid IP for 7 d	Observation period 14 d	Freireich et al. (1966)
	0.7 mg bid PO for 10 d (MTX alone)	10/10 Died with cecitis by 10 d	Silva et al. (1984)
	0.7 mg bid PO for 10 d MTX + 200 mg vancomycin per ml water	3/10 Died with cecitis by 10 d; 9/10 died within 3 d	Silva et al. (1984)
	0.7 mg bid PO for 10 d MTX + 1 mg sid SC folinic acid	0/10 Died	Silva et al. (1984)
Methyl GAC	$LD_{10} = 29$ mg/kg sid IP for 7 d	Observation period 14 d	Freireich et al. (1966)
Mitomycin C	$LD_{10} = 1.2$ mg/kg sid IP for 7 d	Observation period 14 d	Freireich et al. (1966)
Nitrogen mustard (NH_2)	$LD_{10} = 0.90$ mg/kg sid IP for 7 d	Observation period 14 d	Freireich et al. (1966)
Thio-TEPA	$LD_{10} = 7.3$ mg/kg sid IP for 7 d	Observation period 14 d	Freireich et al. (1966)
Vinblastine	$LD_{10} = 0.38$ mg/kg sid IP for 7 d	Observation period 14 d	Freireich et al. (1966)
Vincristine	$LD_{10} = 0.90$ mg/kg sid IP for 7 d	Observation period 14 d	Freireich et al. (1966)

[a]See Table I, footnote a, for abbreviations.

Lactobacillus, *Streptococcus*, *Pseudomonas*, and *Bacteroides*. Eight isolates of *C. difficile* inhibited species of three genera of anaerobic bacteria (*Bacteroides*, *Peptococcus*, and *Peptostreptococcus*). Several other isolates belonging to these genera were not inhibited. Toothaker and Elmer (1984) reported the prevention of clindamycin-induced mortality in hamsters by the yeast, *Saccharomyces boulardii*. Bohnhoff et al. (1964) and Meynell (1963) reported lowering resistance of mice to *Salmonella* sp. by prior treatment with streptomycin. Reduced levels of acetic, butyric, and propionic acids were observed along with a rise in pH and Eh (oxidation–reduction potential) in the cecal content. Further, they observed the disappearance of *Bacteroides* sp., a gram-negative anaerobe which produces acetic and butyric acid. Using an *in vitro* system, Hentges

(1967) demonstrated the inhibition of *Shigella flexneri* by formic and acetic acids produced by *Klebsiella* sp.

C. Pathogenesis

Studies on the pathogenesis of AAC received a boost with the demonstration of a toxin in feces of humans with AAC (Larson *et al.*, 1977). Rifkin *et al.* (1977) identified a toxin in stool filtrates of humans with AAC which was neutralized by *Clostridium sordellii* antitoxin. Bartlett *et al.* (1977a) isolated a toxin-producing strain of *Clostridium* (similar to *C. difficile*) from clindamycin-treated hamsters which produced the lesions of AAC. AAC was succesfully reproduced following intracecal inoculation of (1) cecal content from a hamster given clindamycin followed by five serial passages in hamsters not given clindamycin; (2) 0.45-μm and 0.02-μm filtrates of cecal content from a hamster with AAC; (3) cecal contents from normal hamsters incubated with clindamycin prior to dialysis; (4) a pure culture of a strain of *Clostridium* designated BV 17 HF 1-9 with characteristics of *C. difficile* and isolated from the cecum of a hamster with AAC; and (5) the cell-free supernatant of the bacterium. In contrast, AAC was not seen in hamsters given filtrates of (1) cecal contents prepared with a PM-10 membrane filter (Amicon Corp.); (2) cecal content from hamsters with AAC following incubation with gas gangrene polyvalent antitoxin against five species of *Clostridium;* and (3) cultures of other bacteria isolated from cecal content, including five strains of *Clostridium*, two strains of anaerobic gram-negative bacilli, and one strain of gram-positive coccus. Chang *et al.* (1978) and Bartlett *et al.* (1978c) reported comparable toxic activity in stool specimens from patients with AAC and in cecal content from hamsters given clindamycin. *Clostridium difficile* was isolated from both patients and hamsters. Rifkin *et al.* (1978a) demonstrated the presence of heat-labile toxin(s) in sterile filtrates of cecal contents from hamsters with AAC induced with clindamycin. The toxin(s) produced morphological changes in Y-1 adrenal cell cultures; the activity was neutralized by *in vitro* incubation with polyvalent clostridial antitoxin; and administration of sterile filtrates of cecal content by both oral and intraperitoneal routes produced deaths.

Rifkin *et al.* (1978b) identified antitoxin to *C. sordellii* as the component of the polyvalent gas gangrene antitoxin responsible for neutralizing toxins in sterile filtrates of cecal content from hamsters with AAC. Antitoxins to *Clostridium perfringens, Clostridium septicum, Clostridium histolyticum,* and *Clostridium oedematiens* were ineffective in blocking mortality. Antitoxin to *C. sordellii* blocked death induced by filtrates derived from both *C. sordellii* and *C. difficile*. Culture filtrates from three reference strains of *C. sordellii* proved lethal, whereas filtrates of five isolates of *C. sordellii* from hamsters with AAC did not. Filtrates of all *C. difficile* isolates (seven) proved lethal for both hamsters and cell cultures. Only the strain of *C. sordellii* (Lederle S-4) used in producing the polyvalent gas gangrene antitoxin was cytotoxic. While *C. sordellii* antitoxin neutralized toxins produced by both *C. sordellii* and *C. difficile*, the lesions following the intraperitoneal injection of culture filtrates of *C. difficile* more clearly mirrored those observed in hamsters given clindamycin subcutaneously. *Clostridium sordellii* antitoxin passively immunizes hamsters against AAC (Allo *et al.*, 1979).

Toxoids prepared from both *C. difficile* and *C. sordellii* toxins induce varying degrees of resistance to challenge with clindamycin; however, *C. difficile* toxoid is more effective (Fernie *et al.*, 1983). Injection of *C. difficile* culture filtrates or stool filtrates from clindamycin hamsters into the hamster cecum produces lesions similar to those seen in clindamycin-treated animals (Abrams *et al.*, 1980). *Clostridium difficile* toxin consists of two components, toxin A and toxin B (Taylor and Bartlett, 1980; Taylor *et al.*, 1980, 1981). Further, toxins A and B are antigenically distinct and produce different disease patterns. Toxin A causes enlarged, moderately hemorrhagic ceca, while toxin B causes hemorrhagic ceca of normal size (Libby *et al.*, 1982). Lowe *et al.* (1980) isolated *C. difficile* from guinea pigs with cecitis and demonstrated the presence of *C. difficile* toxin following administration of penicillin.

D. Epizootiology

Discrepancies exist among reports of antimicrobial use in the hamster, with some indicating marked toxicity while others mention no problems. These variations in response most probably result from differences in the gut bacterial flora of the hamsters studied, especially the presence or absence of *C. difficile* in the hamster's cecum or its environment. Fekety *et al.* (1980) isolated *C. difficile* from only 3 of 78 normal hamsters. *Clostridium difficile* was routinely isolated from hamsters given clindamycin, and they postulated that hamsters acquired the organism from infected hamsters or from the environment. Of 73 cultures from their hamster rooms, 14 yielded *C. difficile*. It was isolated from the hands of 6 of 12 people handling the hamsters, cage equipment, food pellets in the cage, and from one of four air samples. *Clostridium difficile* was not isolated from food in newly opened bags or from 50 samples of drinking water. Later work from the same laboratory reported the isolation of *C. difficile* from cecal content of only 5 of 92 normal hamsters (Toshniwal *et al.*, 1981).

Compared to a conventional animal room, the use of a clean room with sterilization of supplies and equipment plus garbing in sterile gowns, gloves, and masks did not significantly reduce mortality in hamsters given clindamycin alone. Differences in mortality were observed between the two rooms in hamsters given vancomycin alone ($p<0.01$) or vancomycin before ($p<0.01$) or after clindamycin ($p<0.05$). Of those which died, time to death in the clindamycin alone groups was

Table III

Antiparasitic Agents

Drug	Dose and route[a]	Comment[a]	References
Aminitrozole	CD_{50} = 25 mg/kg sid PO for 4 d	Active against *Entamoeba histolytica* but more toxic than metronidazole; (TI = CD_{50}/LD_{50} = 1/32	Benazet *et al.* (1970)
1,4-Bis[1-methyl-5-nitroimidazoyl-(2-methyleneimino)] piperazine (HOE 316)[b]	2.1 or 5.7 mg/kg 4× PO in 3 d	ED_{50} and ED_{95}, respectively, against extraintestinal *E. histolytica*; no comments on toxicity	Raether *et al.* (1977)
Chloroquine	100 mg/kg sid PO for 5 d, 12 d (three and five hamsters, respectively)	No anti-*Babesia microti* activity	Miller *et al.* (1978)
	50 mg/kg sid IM for 11 d 16.1 (10.9–22.2) mg/kg sid PO for 5 d 13.9 (11.3–16.5) mg/kg sid SC for 5 d	*Plasmodium berghei* (K173) 28-d test; "cured" (i.e., no detectable parasitemia after 28 d)	Raether and Fink (1979)
Cythioate (10 mg/ml)	1 drop PO q3d for 3× or 9×	10 Hamsters per group; 3×, 8/10 still had hair loss and scabs, no mites demonstrated; 9×, free of scurf and no sign of hair loss, deep scrapings revealed two mites	Heath and Bishop (1984); A.C.G. Heath (personal communication)
4,4'-Diazoaminobenzamidine, Ganaseg, Berenil	6 mg/kg sid IM for 11 d	Anti-*B. microti* activity which ceased when treatment stopped; four animals	Miller *et al.* (1978)
Dimetridazole, Emtryl	500 mg/liter drinking water for 10 d	No report of mortality; ~50% reduction in incidence of transmissible ileal hyperplasia	La Regina *et al.* (1980)
	0.1% Solution active drug in drinking water for 10 d	Observed 15 d; did not result in ill effects visible clinically	Sebesteny (1979)
Entamide furoate	200 mg/kg sid PO for 5 d	Hamsters infected with *Schistosoma mansoni*; portal shift of 50–75% of worms + a 60–70% reduction of the first-stage ova 3–7 d after end of treatment	Girgis and McConnell (1966)
Flexinidazole (HOE 239)[b]	12.6 or 17.0 mg/kg 4× PO in 3 d	ED_{50} and ED_{95}, respectively, against extraintestinal *E. histolytica*; no comments on toxicity; 380 hamsters used	Raether and Seidenath (1983)
Floxacrine (HOE 991)[b]	LD_5 >12,000 mg/kg 5× PO >10,000 mg/kg 1× PO 5000 mg/kg 5× SC LD_{50} >5000 mg/kg 5× SC CD_{50} = 4.3 (2.1–5.2) mg/kg sid PO for 5 d CD_{50} = 1.3 (1.1–1.6) mg/kg sid SC for 5 d	Slight to moderate local inflammatory reaction of subcutis and musculature after multiple SC doses; well tolerated by both routes; CD_{50} vs *P. berghei* K173	Raether and Fink (1979)
Furazolidone	100 mg/kg sid PO for 5 d	Hamsters infected with *Shistosoma mansoni*; portal shift of 50–75% of worms + a 60–70% reduction of the first-stage ova 3–7 d after end of treatment	Girgis and McConnell (1966)
	30 mg/kg sid or bid PO for 2 d	Eradicated experimental *Trichomonas foetus* vaginal infection	Natt *et al.* (1960)
	30 mg/kg sid PO for 34 consecutive days	0/8 Died; final weight same as controls	J. D. Small (unpublished data)

Table III (Continued)

Drug	Dose and route[a]	Comment[a]	References
γ-BHC, Lindane	1× "Treatment" with 0.1% concentration of adult females and weanlings	Stopped hair loss; adult females and weanlings treated 3× with 1 drop cythionate q3d without change in physical appearance	Heath and Bishop (1984); A.C.G. Heath (personal communication)
Hydroxychloroquine	100 and 300 mg/kg bid for 4 d	Blocked lethal effect of clindamycin-induced colitis; observed 10 d after last dose	Will et al. (1985)
Ipronidazole-HCl, Ipropran (81.33% Drug)	125–250 mg/liter drinking water for 30–38 d	No observable effect on cecal protozoa; drank 6 ml/d; at 2 gm/liter refused to drink	J. D. Small (unpublished data)
	2 mg sid PO for 2–23× (5×/wk)	No observable effect on cecal protozoa	J. D. Small (unpublished date)
	100 mg/kg sid PO (gavage) (5×/wk) + 225 mg ipronidazole + 2 gm/liter neomycin sulfate in drinking water for 11–32 d; drug weights based on active ingredients	Severe depression of appetite with weight loss; no cecal protozoa detected after 11, 18, and 32 d; 1/1 positive on d 24; neomycin (2 gm/liter) alone, normal weight and positive for protozoa	J. D. Small (unpublished data)
Ketoconazole	200 mg/kg 5× PO or SC 100 mg/kg 5× SC	PO; No effect on *Leishmania donovani* in liver; SC, marked reduction of parasites in liver; 100 mg no effect	Raether and Seidenath (1984)
	100 or 150 mg/kg 4× PO	150 but not 100 mg/kg partially effective against *E. histolytica*; followed 7 d after infection	Raether and Seidenath (1984)
Metronidazole	167–233 mg/kg (10 mg per hamster) sid PO for 11 d	4/5 Died 2 d after discontinuing drug	Fekety et al. (1979b)
	167–233 mg/kg (10 mg per hamster) 1× SC	10/10 died within 4 d	Fekety et al. (1979a)
	200 mg/kg sid PO for 5 d	No effect on *B. microti* or hamsters ($N = 3$)	Miller et al. (1978)
	40 mg/kg 4× PO in 3 d	Total reduction of hepatic necrosis in *E. histolytica*-infected hamsters; followed 7 d	Raether and Seidenath (1984)
	23.9 or 43.8 mg/kg 4× PO in 3 d	ED_{50} and ED_{95}, respectively, against extraintestinal *E. histolytica*; no comments on toxicity; 430 hamsters used	Raether and Seidenath (1983)
	7.5 mg per hamster (70–90 gm) tid PO for 5 d	0/10 Died; observed 14 d following last dose	Bartlett et al. (1978b)
	2 mg/ml drinking water for 14 d	No report of mortality; reduced incidence of transmissible ileal hyperplasia	Jacoby and Johnson (1981)
	$CD_{50} = 15$ mg/kg sid PO for 4 d	Active against *E. histolytica*; TI = 1/200	Benazet et al. (1970)
N-Methylglucamine antimoniate	75, 150, or 300 mg/kg 1× SC (5 hamsters per dose) 12.5, 25, 50 mg/kg 6× SC (5 hamsters per dose)	Slight effect on *L. donovani* at highest dose only; examined after 28 d; maximum tolerated SC dose 6× > 1000 mg/kg	Raether et al. (1978)
Niclosamide, Yomesan	0.33% Active drug in ground feed fed *ad libitum* for 7 d	0/250 Treated and 27/250 control retired females positive for *Hymenolepis nana*; no toxicity observed in 6- to 8-wk-old hamsters fed 3–10× the therapeutic dose	Ronald and Wagner (1975)
	100 mg/kg PO repeated after 7 d	*H. nana*, "fairly effective"; no data	Patton (1979)

(continued)

Table III (*Continued*)

Drug	Dose and route[a]	Comment[a]	References
	1.3 gm powder (75% active ingredient) suspended in 100 ml of water; 1 ml/100 gm bid PO q3wk, 3×	No ill effects observed; "appeared to reduce but not fully eliminate *H. nana*"	Sebesteny (1979)
Niridazole	CD_{50} = 100 mg/kg sid PO for 4 d	Active against *E. histolytica*; TI = 1/4	Benazet *et al.* (1970)
Nitrimidazine	71.4 or 152 mg/kg 4× PO in 3 d	ED_{50} and ED_{95}, respectively, against extraintestinal *E. histolytica*; no comments on toxicity; 40 hamsters used	Raether and Seidenath (1983)
Ornidazole	17.1 or 26.0 mg/kg 4× PO in 3 d	ED_{50} and ED_{95}, respectively, against extraintestinal *E. histolytica*; no comments on toxicity; 30 hamsters used	Raether and Seidenath (1983)
p-(4-Amindophenoxy) benzaldehyde-*p*-amidinophenyl-hydrazone·2HCl (HOE 668)[b]	75, 150, or 300 mg/kg 1× SC (5 hamsters per dose) 12.5, 25, or 50 mg/kg sid SC for 6 d (10 hamsters per dose)	Infected with *L. donovani* at time of first dose; parasite burden examined on d 28; 300 mg/kg 1× or 50 mg/kg 6× was superior to all antileishmanial compounds examined; toxic effects on liver and kidney; proposed as a laboratory reference drug; maximum tolerated SC dose: 800 mg/kg (1×) or 75 mg/kg (6×)	Raether *et al.* (1978)
Paromomycin sulfate, Humatin	440 mg/kg bid PO for 4 d; 13.2–300 mg/kg/d (2–6 divided doses) SC; dose expressed as paromomycin base	PO, 19/19 survived infection with *E. histolytica*; SC, 53/53 survived infection; reduction of hepatic lesions dose dependent; killed 7 d following infection. *Note:* Available today only for oral use; nephrotoxic by parenteral routes.	Thompson *et al.* (1959)
Pentamidine isethionate	24 mg/kg sid IM for 11 d 50 mg/kg sid IM for 17 d	5/5 Died (50 mg/kg) during or shortly after treatment ended; both doses showed anti-*B. microti* activity which ceased when treatment stopped; reaction at injection site	Miller *et al.* (1978)
	150 or 300 mg/kg 1× SC (5 hamsters per dose) 12.5, 25, 50 mg/kg sid SC for 6 d (5 hamsters per dose)	300 mg/kg 1× toxic; reduction but not elimination of *L. donovani*; examined after 28 d	Raether *et al.* (1978)
	14.4 mg/kg sid SC for 5 d	"Total reduction of *L. donovani*"	Raether and Seidenath (1984)
Piperazine citrate	10 mg/ml drinking water for 7 d on, 5 d off, 7 d on (1×)	Eradicated *Syphacia obvelata*; no toxicity observed when given same dose for 4 wk	Unay and Davis (1980)
Quinine	125 mg/kg sid SC for 4 d 250 mg/kg sid PO for 7 d	No anti-*B. microti* activity alone; ulceration at injection sites; enhanced activity of clindamycin against parasite; combined with vancomycin orally	Rowin *et al.* (1982)
Ronnel	45 ml of the 33⅓% ronnel suspension added to 250 ml propylene glycol. Apply to ⅓ of the body daily until skin scrapings are negative	Excellent results reported against dermatitis associated with *Dermodex aurati* and *D. crieti*. No cases of organophosphate toxicity observed	Burke (1979)

Table III (Continued)

Drug	Dose and route[a]	Comment[a]	References
Sodium stibogluconate	75, 150, or 300 mg/kg 1× SC (5 hamsters per dose) 12.5, 25, 50 mg/kg 6× SC (5 hamsters per dose)	Reduction but not elimination of *L. donovani* when examined after 28 d; maximum tolerated SC dose, 400 mg/kg (6×)	Raether *et al.* (1978)
Thiabendazole	0.3% In diet for 7–10 d	Eliminated all *Syphacia obvelata* from 110 hamsters	Sebesteny (1979)
	0.3% In diet for 7–14 d	Reduced level of *H. nana* when fed in diet for 7–14 d	Sebesteny (1979)
Tinidazole	25.9 or 48.0 mg/kg 4× PO	ED_{50} and ED_{95}, respectively, against extraintestinal *E. histolytica*; no comments on toxicity; 160 hamsters used	Raether and Seidenath (1983)
Trichlorfon, Masoten	2.1 gm/liter drinking water (80% active drug)	Eradicated *Syphacia obvelata*; no toxicity observed; treatment, 2 wk on, 1 wk off, repeat 1–4×; Thorough cleanup of facility and equipment usually required	Small *et al.* (1985)
	20 mg/kg 1×/wk IP for 90 wk from 8 wk of age	No statistically significant differences in tumor incidences; 22 M and 23 F	Teichmann and Schmidt (1978)

[a]See Table I, footnote a, for abbreviations.
[b]Experimental compound, not available commercially.

the same; however, in the other groups, time to death was more than twice as long in the clean room than in the conventional room. The case is made for acquisition of *C. difficile* from the environment rather than persistence in the hamster. Hawkins *et al.* (1984) found *C. difficile* in 12 of 90 hamsters cultured immediately on arrival from the supplier. Using a bacteriophage and bacteriocin typing system, they characterized *C. difficile* isolates from hamsters and their environments. The same bacteriocin type of *C. difficile* was isolated from hamsters on arrival, hamsters given antibiotics, and from the environment. Hamsters culture negative for *C. difficile*, treated with thienamycin, and housed in cages which previously held hamsters with AAC, developed AAC with the same bacteriocin type. *Clostridium difficile* in the hospital environment and its association with cases of AAC in humans has been studied. It appears that infections with *C. difficile* can be of nosocomial origin and transmitted among patients (Wüst *et al.*, 1982; Kim *et al.*, 1981; Fekety *et al.*, 1981). Larson *et al.* (1978) were able to colonize the ceca of hamsters and induce death with 2–25 colony-forming units (CFU) of *C. difficile* if the hamsters had been pretreated with vancomycin. Without vancomycin, administration of 2000 CFU of *C. difficile* did not result in colonization or mortality. *C. difficile* is sensitive to vancomycin, and it is the treatment of choice for AAC in humans. Deaths in hamsters follow the cessation of administration of vancomycin, as the bowel becomes repopulated with *C. difficile* from the environment (Browne *et al.*, 1977). In subsequent work, Larson *et al.* (1980) failed to isolate *C. difficile* from feces of 36 normal hamsters. Also, *C. difficile* was isolated from only 1 of 10 hamsters inoculated orally and none of 10 hamsters 30 days following intracecal inoculation of *C. difficile*. No cecal or colonic lesions were noted. Pretreatment of 171 hamsters with a single subcutaneous 5-mg dose or 0.5 mg/ml drinking water of clindamycin for 7 days, followed by placement in various areas of the animal facility, research laboratories, and hospital, gave widely different levels of mortality. Generally, areas where AAC had been previously studied resulted in a high level of mortality. Other areas, including a hospital ward where a patient had been treated for AAC, resulted in no deaths (0/8) after 10 days association. The infectious and epizootic nature of AAC was observed when four cages with two hamsters each inside an isolator progressively developed AAC. The progression from cage to cage followed the direction of the air flowing through the isolator (Larson *et al.*, 1980).

These data support the contention that AAC is an infectious disease acquired from the environment after the host is rendered susceptible by treatment with antimicrobial agents. Additional evidence for AAC being an infectious disease comes from the report of enterocolitis with the isolation of *C. difficile* and the demonstration of toxin in untreated hamsters kept in the same room with hamsters which had received tetracycline 5 months earlier (Rehg and Lu, 1982). The recent recognition of methotrexate as an inciter of colitis associated with *C. difficile* and its toxins suggests that drugs other than conventional antimicrobial agents can alter the gut flora (Silva *et al.*, 1984; Cudmore *et al.*, 1980). A similar lesion of the mucosa was produced by 5-fluorouracil, but *C. difficile* was not isolated. In

Table IV
Corticosteroids

Drug	Dose and route[a]	Comment	References
Cortisone acetate	2 mg/sid SC 6×/wk to death; 0.25, 0.5, 1 mg doses also examined	2 mg, death in 35 d, 22 d, and 11 d in normal, *Toxoplasma*-infected, and *Besnoitia*-infected hamsters, respectively; dose-related mortality; deaths beyond 30 d associated with several bacteria and aortic aneurysms	Frenkel (1960)
	≤2 mg sid SC 6×/wk to death	Infected with *Besnoitia jellisoni*; potency ratio: dose producing fatal relapse on d 12 (mean) used in comparison with hydrocortisone (1.3 mg); potency ratio = 0.7	Frenkel (1960)
	0, 0.5, 1.0 mg sid SC 4 hamsters per dose	Mean day of death: 374 (0 mg), 495 (0.5 mg), 121 (1.0 mg)	Frenkel and Havenhill (1963)
	2.5, 5, 10 mg 2×/wk SC, or ~25, ~50, and ~100 mg/kg	Mean day of death: 36 (2.5 mg), 14 (5 mg), 17 (10 mg)	Frenkel and Havenhill (1963)
	50 mg/kg sid IM for 10 d	M, decreased body and thymus ($p < 0.01$), carcass ($p < 0.001$), spleen ($p < 0.02$), adrenal, gonads ($p < 0.40$) weights; F, decreased body and carcass ($p < 0.05$), thymus ($p < 0.001$), spleen ($p < 0.30$) weights; all had multiple acute gastric ulcers, many bleeding; near-normal weight gain with up to 3.5 mg/kg sid; 7 mg/kg sid gave 50% weight loss and death within 2 months. *Note:* This reference also discussed mice and rats	Frenkel and Havenhill (1963)
Dexamethasone	0.2 mg or 0.02 mg/kg SC q48 hr 4× (20–25 3-month-old F hamsters per group)	*Babesia microti* (I) and uninfected control (C) used; 0.2-mg but not 0.02-mg dose increased parasitemia; both doses in all groups had neutrophilia; on d 9 only 0.2 mg I had neutrophilia; lymphocytopenia in both I groups, only 0.2 mg C group, and only after first dose; spleen weight reduced in both C groups	Eckblad *et al.* (1984)
	20 mg/kg sid SC for 4 d, then 1×/wk for 5 wk, then 1×/wk q2wk to end of experiment	Reduced growth rate (drug vs control); *Mycobacterium paratuberculosis*-infected hamsters had higher counts of bacteria in tissues; dexamethasone not associated with clinical signs	Larsen and Miller (1978)
	≤0.5 mg sid SC 6×/wk to death	See cortisone acetate; potency ratio = 7.0	Frenkel (1960)
	24 μg sid IM or by Alzet pump to males	Suppressed ACTH secretion	Dunlap and Grizzle (1984)
9-Fluoroprednisolone	≤0.5 mg sid SC 6×/wk to death	See cortisone acetate; potency ratio = 7.0	Frenkel (1960)
Hydrocortisone	≤2 mg sid SC 6×/wk to death	See cortisone acetate; potency ratio = 1.0; more effective than cortisone in causing death whether infected or not	Frenkel (1960)
	0.25, 0.5, 1.0, and 2 mg sid SC 6×/wk to death		Frenkel (1960)

Table IV (Continued)

Drug	Dose and route[a]	Comment	References
	50 mg/kg sid SC for 4 d	Increased maltase, sucrase, alkaline phosphatase, and leucine aminopeptidase, but not lactase, activity in mucosa of small intestines; γ-glutamyltranspeptidase activity increases slightly in females only; increased absorption of proline and glycine but not lysine	Andres et al. (1984)
2-Methyl-9-d-fluorohydrocortisone acetate	≤0.5 mg sid SC 6×/wk to death	See cortisone acetate; potency ratio = 2.4	Frenkel (1960)
6-Methylprednisolone	0.15–20 mg/kg sid SC unstated multiple doses	0/15 Died; 1/5 at 20 mg/kg developed diarrhea; no effect on clindamycin-induced colitis	Fekety et al. (1979b)
	0.6 mg tid IM for 5 d	No effect on clindamycin-induced colitis	Bartlett et al. (1978a)
	≤0.5 mg sid SC 6×/wk to death	See cortisone acetate; potency ratio = 3.7	Frenkel (1960)
Prednisolone	≤2 mg sid SC 6×/wk to death	See cortisone acetate; potency ratio = 1.2	Frenkel (1960)
Prednisolone acetate	≤1 mg sid SC 6×/wk to death	See cortisone acetate; potency ratio = 1.3	Frenkel (1960)
Triamcinolone	≤1 mg sid SC 6×/wk to death	See cortisone acetate; potency ratio = 3.8	Frenkel (1960)
Triamcinolone acetonide	≤0.5 mg sid SC 6×/wk to death	See cortisone acetate; potency ratio = 14.2	Frenkel (1960)

[a]See Table I, footnote a, for abbreviations.

addition, diet may play a role. Neomycin sulfate given in the water induced AAC in hamsters fed a purified diet but not a ground commercial diet. Toxin-producing *C. difficile* were isolated (Michelich et al., 1981).

With some drugs the route of administration is important, as it may determine whether or not the cecal bacteria are exposed to the drug. Some drugs administered by injection are excreted via the kidneys without reaching significant levels in the cecum and colon. Some orally administered drugs are almost completely absorbed before reaching the lower bowel and are excreted via the kidneys without attaining high levels in the cecum. The concentration of chloramphenicol in hamster feces 2–24 hr following an oral dose of chloramphenicol palmitate was 12 μg/gm. Subcutaneous doses of chloramphenicol succinate <1000 mg/kg often did not result in detectable levels in the feces (Fekety et al., 1979a).

E. Signs and Lesions

Regardless of the antimicrobial agent given, if AAC is induced, the clinical signs and lesions are similar. Anorexia, ruffled fur, dehydration, and diarrhea are almost always present within a few days, and most animals die within 4–10 days. The fur around the anus is soiled, and the nonspecific term "wet-tail" is applicable. However, "wet-tail" has been applied to the clinical sign diarrhea regardless of cause. Because of this, AAC, an induced condition, must be differentiated from spontaneous enteritides not associated with antimicrobial agents. In AAC the characteristic gross lesion in the hamster is a hemorrhagic ileocolitis. Microscopically, severe congestion of the villous and submucosal capillaries and venules are seen in the small intestine (most intense in the distal ileum), cecum, and colon. Early there is an accumulation of acute inflammatory cells in the lamina propria, with an outpouring of fluid and many neutrophils into the lumen. The mucosal surface remains mostly intact, although the tips of the villi are frequently distended with cellular exudate, or edema fluid and focal erosion of the mucosal surface occurs. There is increased secretory activity in the large bowel (Price et al., 1979; Lusk et al., 1978; Small, 1968a).

F. Diagnosis

Diagnosis of AAC in hamsters is usually based on a history of exposure to antimicrobial agents, animals treated with these drugs, or exposure to an environment which previously housed

Table V
Miscellaneous Agents

Drug	Dose and route[a]	Comment	References
Brodifacoum	0.005% of diet for 1–21 d	100% Mortality (5 M, 5 F) with 3 d of feeding; 1/5 M consumed drug for 21 d before death; 1 and 2 d feeding caused high mortality; LD (mg/kg), M (3 d), 8.7 (7.7–10.6); F (3 d), 6.1 (5.3–6.7)	Bradfield and Gill (1984)
Caffeine	200 mg/liter drinking water for 60 d, M Chinese hamsters	30 Treated, 30 controls; sex ratio of pups from untreated F, 61.4% F vs 49.2% F ($p < 0.025$)	Weathersbee et al. (1975)
Calciferol	0.1% of diet for 1 or 2 d	100% Mortality (10 M, 10 F); MST, 5–5.8 d, range 3–13 d (1 d), 4–7 (2 d); LD (mg/kg), M (1 d), 87 (60–115); F (1 d), 64 (48–73); M (2 d), 108 (75–147); F (2 d), 128 (81–186)	Bradfield and Gill (1984)
Calcitonin (salmon)	3.5 U.M.R.C./kg sid IP 6×/wk for 6 wk beginning 45 d after induction of periodontitis	Increased bone formation 400% ($p < 0.03$), reduced bone resorption (NS), and reversal phase decreased 75% ($p < 0.02$)	Saffar and LasFargues (1984)
Castor oil	0.5 ml sid PO for 7 d	10/10 Survived	Silva et al. (1984)
	1.0 ml sid PO for 5 d	4/5 Survived, 1 death with "wet-tail"	Silva et al. (1984)
	0.5 ml sid PO for 6 d + clindamycin (100 mg/kg 1× SC)	8/15 Survived (1/10 clindamycin only); cecitis in dying animals only	Silva et al. (1984)
Cholestyramine	4 mg in 1 ml water tid PO for 5 d	10/10 Died; treatment begun 24 hr after 1 mg clindamycin IP; survival time prolonged; 4/10 died within 5 d treatment period; clindamycin-treated controls, 10/10 died	Taylor and Bartlett (1980)
	5 mg tid PO + 10 mg/kg 1× SC clindamycin	20/20 Died; MST, 3.7 ± 1.0 d (2–6 d); clindamycin-treated controls, 20/20 died; MST, 2.6 ± 0.5 d (2–3 d)	Fekety et al. (1979b)
Chlorothiazide (Diuril)	5 mg per hamster daily PO 2.5 mg/kg 1× SC	Slight increase in blood glucose in normal and drug-pretreated hamsters; if pretreated (5 d) with cortisol (2.5 mg 2×/wk) alone or with chlorthiazide, administration of 5 mg chlorthiazide gave rise to hyperglycemia; see publication for SC use	Frenkel (1972a) Frenkel (1972a)
Colestipol	100 mg tid PO for 5 d	No effect on clindamycin-induced colitis	Bartlett et al. (1978a)
Difenacoum	0.005% of diet for 21 d	1/5 M, 4/6 F died; MST, M (17 d); F (12.3 d), range 8–20 d; LD (mg/kg), M (41), F (17.5)	Bradfield and Gill (1984)
Epinephrine	0.05 mg or 0.2 mg SC	Used to study effects of chlorthiazide and corticosteroids on blood glucose levels	Frenkel (1972a)
Indomethacin	2 mg/kg sid PO 6×/wk for 6 wk beginning 45 d after induction of periodontitis	Increased bone formation 270% ($p < 0.05$), reduced bone resorption (NS), and reversal phase decreased 33% (NS)	Saffar and LasFargues (1984)
Insulin, protamine zinc or regular	2 units 1× SC	Used in a glucose–epinephrine–insulin tolerance study	Frenkel (1972a)

Table V (*Continued*)

Drug	Dose and route[a]	Comment	References
Lomotil (0.5 mg diphenoxylate-HCl + 0.005 mg atropine/ml)	5 mg lomotil per milliliter drinking water (estimated intake 0.3 mg/kg/d)	Did not block effects of clindamycin. Controls—no effect noted	Fekety *et al.* (1979b)
Metyrapone	260 mg/kg IP 1 hr before cannulation of adrenal vein	Abolished secretion of cortisol and corticosterone. *Note:* See reference for long list of substances tested.	Frenkel *et al.* (1965)
Milk of magnesia	0.5 ml sid PO for 5 d	Diarrhea within 3 hr, but normal weight gain and formed stools by d 4; did not alter response to clindamycin (100 mg/kg 1× SC)	Silva *et al.* (1984)
Retinol	5×10^3, 10×10^3, 15×10^3, 20×10^3, 30×10^3 IU 1× PO on d 8 of gestation	Production of the Arnold–Chiari malformation types I and II in 0, 6.6, 57, 56, and 85%, respectively, of all fetuses examined	Willhite (1984)
Streptozotocin	50 mg/kg (1×?) route? 9-wk-old M	Induced diabetes mellitus	Muratsu and Morroka (1984)
Tolbutamide	2.5 mg or 5.0 mg per hamsters sid PO	Reduced blood glucose following loading dose of glucose + epinephrine	Frenkel (1972a)
Vegetable oil (soybean, Krogers)	1.0 ml sid PO for 7 d	4/5 Survived, 1 death with "wet-tail"	Silva *et al.* (1984)
	1.0 ml sid PO for 9 d beginning 3 d before clindamycin (100 mg/kg 1× SC)	6/10 Survived	Silva *et al.* (1984)
Verapamil	0.25 mg bid SC wk 1; 0.36 mg bid SC wk 2; 0.5 mg bid SC wk 3 and 4	Prevented severe myocardial lesions found in untreated 2-month-old B1053:58 female myopathic hamsters	Kuo *et al.* (1984)
Warfarin	0.025% or 0.25% of diet for 28 d	No mortality (0/8 and 0/4 M); mean total dose, 474 mg/kg and 4544 mg/kg	Bradfield and Gill (1984)
	0.5% of diet for 56 d	3/4 (M) Died; MST, 27.3 d (26–29 d); LD (mg/kg), 5891 (5784–6081); SD (mg/kg), 15037.	Bradfield and Gill (1984)
Zinc phosphide	3, 4, 5.0% Of diet for 1 day	3%, 1/2 M Died; LD (mg/kg); 231; SD (mg/kg), 104 4%, 6/7 M and 3/3 F died; LD (mg/kg), M, 482 (183–674), F, 113 (69–198); SD (mg/kg), M, 70 5%, 6/6 M and 4/4 F died; LD (mg/kg), M, 192–542; F, 254–385	Bradfield and Gill (1984)

[a] See Table I, footnote *a*, for abbreviations.

animals treated with antimicrobial agents. The history coupled with clinical signs and gross necropsy findings are usually sufficient to support a diagnosis. Isolation of *C. difficile* from cecal content or feces is not sufficient evidence for a diagnosis of AAC, as a small percentage of hamsters will be asymptomatic carriers. However, if *C. difficile* is isolated in large numbers, a diagnosis of AAC is probably justified. Selective media for isolation of *C. difficile* have been described (George *et al.*, 1979; Bartlett, 1981; Buggy *et al.*, 1983). Cytotoxic activity (toxin B) of stool or cecal content filtrates has been measured in several cell cultures (Chang *et al.*, 1979). Enzyme-linked immunosorbent assays (ELISA) have been developed for antibodies to *C. difficile* (Yolken *et al.*, 1981) and its A and B toxins (Viscidi *et al.*, 1983; Laughon *et al.*, 1984). Diagnosis of AAC has been recently reviewed (Fekety, 1984; Bartlett, 1981).

G. Hamster Models of Human Antibiotic-Associated Enterocolitis

The hamster as a model for AAC has been the subject of several recent reviews and a symposium (Rehg, 1985; Rifkin and Fekety, 1985; Bartlett and Taylor, 1982; Schlessinger, 1979). While the hamster is useful for the study of AAC as found in humans, two differences in the disease need to be recognized. Hamsters rarely develop pseudomembranes as do humans, and in humans the major lesions are found in the distal colon. In hamsters the distal ileum, cecum, and proximal colon are the target tissues.

ACKNOWLEDGMENTS

The assistance of several investigators in furnishing data and reprints is acknowledged. It is a pleasure to acknowledge the cooperation of Ms. Jean H. Gordner in typing the manuscript.

REFERENCES

Abrams, G. D., Allo, M., Rifkin, G. D., Fekety, R., and Silva, J., Jr. (1980). Mucosal damage mediated by clostridial toxin in experimental clindamycin-associated colitis. *Gut* **21**, 493–499.

Allo, M., Silva, J., Jr., Fekety, R., Rifkin, G. D., and Waskin, H. (1979). Prevention of clindamycin-induced colitis in hamsters by *Colistridium sordelli* antitoxin. *Gastroenterology* **76**, 351–355.

Andres, M. D., Rebolledo, E., Taboada, M. C., and Fernandez Otero, M. P. (1984). The effect of cortisol on stimulation of enzymatic activity and absorption of amino acids in the small intestine of adult hamsters. *Comp. Biochem. Physiol. A* **79A**, 525–528.

Angulo, A. F., Spaans, J., Zemmouchi, L., and van der Waaij, D., (1978). Selective decontamination of the digestive tract of Syrian hamsters. *Lab. Anim.* **12**, 157–158.

Bartlett, J. G. (1981). Laboratory diagnosis of antibiotic-associated colitis. *Lab. Med.* **12**, 347–351.

Bartlett, J. G., and Taylor, N. S. (1982). Antibiotic-associated colitis. *Med. Microbiol.* **1**, 1–48.

Bartlett, J. G., Onderdonk, A. B., Cisneros, R. L., and Kasper, D. L. (1977a). Clindamycin-associated colitis due to a toxin-producing species of *Clostridium* in hamsters. *J. Infect. Dis.* **136**, 701–705.

Bartlett, J. G., Onderdonk, A. B., and Cisneros, R. L. (1977b). Clindamycin-associated colitis in hamsters: Protection with vancomycin. *Gastroenterology* **73**, 772–776.

Bartlett, J. G., Chang, T.-W., and Onderdonk, A. B. (1978a). Comparison of five regimens for treatment of experimental clindamycin-associated colitis. *J. Infect. Dis.* **138**(1), 81–86.

Bartlett, J. G., Chang, T.-W., Moon, N., and Onderdonk, A. B. (1978b). Antibiotic-induced lethal enterocolitis in hamsters: Studies with eleven agents and evidence to support the pathogenic role of toxin-producing clostridia. *Am. J. Vet. Res.* **39**, 1525–1530.

Bartlett, J. G., Chang, T. W., Gurwith, M., Gorbach, S. L., and Onderdonk, A. B. (1978c). Antibiotic-associated pseudomembranous colitis due to toxin-producing clostridia. *N. Engl. J. Med.* **298**, 531–534.

Benazet, F., Lacroix, L., Godard, C., Guillaume, L., and Leroy, J.-P. (1970). Laboratory studies of the chemotherapeutic activity and toxicity of some nitroheterocycles. *Scand. J. Infect. Dis.* **2**, 139–143.

Bohnhoff, M., Miller, C. P., and Martin, W. R. (1964). Resistance of the mouse's intestinal tract to experimental *Salmonella* infection. I. Factors which interfere with the initiation of infection by oral inoculation. *J. Exp. Med.* **120**, 805–816.

Bradfield, A. A. G., and Gill, J. E. (1984). Laboratory trials of five rodenticides for the control of *Mesocricetus auratus* Waterhouse. *J. Hyg.*, **93**, 389–394.

Browne, R. A., Fekety, R., Jr., Silva, J., Jr., Boyd, D. I., Work, C. O., and Abrams, G. D. (1977). The protective effect of vancomycin on clindamycin-induced colitis in hamsters. *Johns Hopkins Med. J.* **141**, 183–192.

Buggy, B. P., Wilson, K. H., and Fekety, R. (1983). Comparison of methods for recovery of *Clostridium difficile* from an environmental surface. *J. Clin. Microbiol.* **18**, 348–352.

Burke, T. J. (1979). Rats, mice, hamsters, and gerbils. *Vet. Clin. North Am.: Small Anim. Pract.* **9**, 473–486.

Chang, T. W., Bartlett, J. G., Gorbach, S. L., and Onderdonk, A. B. (1978). Clindamycin-induced enterocolitis in hamsters as a model of pseudomembranous colitis in patients. *Infect. Immun.* **20**, 526–529.

Chang, T. W., Lauerman, M., and Bartlett, J. G. (1979). Cytotoxicity assay in antibiotic-associated colitis. *J. Infect. Dis.* **140**, 765–770.

Cosar, C., and Kolsky, M. (1955). Etude chez le hamster des diarrhées cholériformes consécutives à l'administration d'antibiotiques par voie buccale. Essais de traitement par la sulfaquanidine. *C. R. Seances Soc. Biol. Ses. Fil.* **149**, 1163–1167.

Cudmore, M., Silva, J., Jr., and Fekety, R. (1980). Clostridial enterocolitis produced by antineoplastic agents in hamsters and humans. *In* "Current Chemotherapy and Infectious Disease" (J. D. Nelson and C. Grassi, eds.), pp. 1460–1461. Am. Soc. Microbiol., Washington, D.C.

De Salva, S. J., Evans, R. A., and Marcussen, H. W. (1969). Lethal effects of antibiotics in hamsters. *Toxicol. Appl. Pharmacol.* **14**, 510–514.

Dunlap, N. E., and Grizzle, W. E. (1984). Golden Syrian hamsters: A new experimental model for adrenal compensatory hypertrophy. *Endocrinology (Baltimore)* **114**, 1490–1495.

Ebright, J. R., Fekety, R., Silva, J., and Wilson, K. H. (1981). Evaluation of eight cephalosporins in hamster colitis model. *Antimicrob. Agents Chemother.* **19**, 980–986.

Eckblad, W. P., Siller, D., Woodard, L. F., and Kuttler, K. L. (1984). Effect of dexamethasone on babesiasis in hamsters. *Am. J. Vet. Res.* **45**, 1880–1882.

Feingold, D. S., Chen, W. C., Chou, D.-L., and Chang, T.-W. (1979). Induction of colitis in hamsters by topical application of antibiotics. *Arch. Dermatol.* **115**, 580–581.

Fekety, R. (1984). Antibiotic-associated enterocolitis. *In* "Infectious Diarrheal Diseases—Current Concepts and Laboratory Procedures" (P. D. Ellner, ed.), pp. 77–92. Dekker, New York.

Fekety, R., Silva, J., Toshniwal, R., Allo, M., Armstrong, J., Browne, R., Ebright, J., and Rifkin, G. (1979a). Antibiotic-associated colitis: Effects of antibiotics on *Clostridium difficile* and the disease in hamsters. *Rev. Infect. Dis.* **1**, 386–396.

Fekety, R., Silva, J., Browne, R. A., Rifkin, G. D., and Ebright, J. R. (1979b). Clindamycin-induced colitis. *Am. J. Clin. Nutr.* **32**, 244–250.

Fekety, R., Kim, K.-H., Batts, D. H., Browne, R. A., Cudmore, M. A., Silva, J., Jr., Toshniwal, R., and Wilson, K. H. (1980). Studies on the epidemiology of antibiotic-associated *Clostridium difficile* colitis. *Am. J. Clin. Nutr.* **33**, 2527–2532.

Fekety, R., Kim, K.-H., Brown, D., Batts, D. H., Cudmore, M., and Silva, J., Jr. (1981). Epidemiology of antibiotic-associated colitis—isolation of *Clostridium difficile* from the hospital environment. *Am. J. Med.* **70**, 906–908.

Fernie, D. S., Thomson, R. O., Batty, I., and Walker, P. D. (1983). Active and passive immunization to protect against antibiotic-associated caecitis in hamsters. *Dev. Biol. Stand.* **53**, 325–332.

Freireich, E. J., Gehan, E. A., Rall, D. P., Schmidt, L. H., and Skipper, H. E. (1966). Quantitative comparison of toxicity of anticancer agents in mouse, rat, hamster, dog, monkey, and man. *Can. Chemother. Rep.* **50**, 219–244.

Frenkel, J. K. (1960). Evaluation of infection-enhancing activity of modified corticoids. *Proc. Soc. Exp. Biol. Med.* **103**, 552–555.

Frenkel, J. K. (1972a). Dissecting aneurysms of the aorta and pancreatic islet cell hyperplasia with diabetes in corticosteroid- and chlorothiazide-treated hamsters. *Prog. Exp. Tumor Res.* **16**, 300–324.

Frenkel, J. K. (1972b). Infection and immunity in hamsters. *Prog. Exp. Tumor Res.* **16**, 326–367.

Frenkel, J. K., and Havenhill, M. A., II (1963). The corticoid sensitivity of golden hamsters, rats, and mice. *Lab. Invest.* **12**, 1204–1220.

Frenkel, J. K., and Lunde, M. N. (1966). Effects of corticosteroids on antibody and immunity in *Besonita* infection of hamsters. *J. Infect. Dis.* **116**, 414–424.

Frenkel, J. K., Cook, K., Grady, H. J., and Pendleton, S. K. (1965). Effects of hormones on adrenocortical secretion of golden hamsters. *Lab. Invest.* **14**, 142–156.

George, W. L., Sutter, V. L., Citron, D., and Finegold, S. M. (1979). Selective and differential medium for isolation of *Clostridium difficile*. *J. Clin. Microbiol.* **9**, 214–219.

Girgis, N. I., and McConnell, E. (1966). Anti bilharzial activity of two non antimonial drugs. *J. Egypt. Public Health Assoc.* **41**, 243–253.

Gishizky, M. L., Frankel, B. J., and Grodsky, G. M. (1984). Cyclophosphamide treatment of prediabetic Chinese hamsters. *Acta Physiol. Scand.* **121**, 81–84.

Hagen, C. A., Shefner, A. M., and Ehrlich, R. (1965). Intestinal microflora of normal hamsters. *Lab. Anim. Care* **15**, 185–193.

Hamre, D. M., Rake, G., McKee, C. M., and MacPhillamy, H. B. (1943). The toxicity of penicillin as prepared for clinical use. *Am. J. Med. Sci.* **206**, 642–652.

Harada, T., Maita, K., Odanaka, Y., and Shirasu, Y. (1984). Effects of quercetin and butylated hydroxytoulene on cigarette smoke inhalation toxicity in Syrian golden hamsters. *Jpn. J. Vet. Sci.* **46**, 527–532.

Harkness, J. E., and Wagner, J. E. (1983). "The Biology and Medicine of Rabbits and Rodents," 2nd ed. Lea & Febiger, Philadelphia, Pennsylvania.

Hawkins, C. C., Buggy, B. P., Fekety, R., and Schaberg, D. R. (1984). Epidemiology of colitis induced by *Clostridium difficile* in hamsters: Application of a bacteriophage and bacteriocin typing system. *J. Infect. Dis.* **149**, 775–780.

Heath, A. C. G., and Bishop, D. M. (1984). Treatment of mange in guinea pigs, hamsters and hedgehogs. *N. Z. Vet. J.* **32**, 120.

Heilman, F. R. (1948). Aureomycin in the treatment of experimental relapsing fever and leptospirosis ichterohaemorrhagica (Weil's Disease). *Proc. Staff Meet. Mayo Clin.* **25**, 569–573.

Hentges, D. J. (1967). Inhibition of *Shigella flexneri* by the normal intestinal flora. I. Mechanisms of inhibition by *Klebsiella*. *J. Bacteriol.* **93**, 1369–1373.

Holmes, D. D. (1984). "Clinical Laboratory Animal Medicine." pp. 27–33, 94–95. Iowa State Univ. Press, Ames.

Jacoby, R. O., and Johnson, E. A. (1981). Transmissible ileal hyperplasia. *Adv. Exp. Med. Biol.* **134**, 267–289.

Jordan, H. V., and van Houte, J. (1972). The hamster as an experimental model for odontopathic infections. *Prog. Exp. Tumor Res.* **16**, 539–556.

Kaipainen, W. J., and Faine, S. (1954). Toxicity of erythromycin. *Nature (London)* **174**, 969–970.

Kemp, G. (1965). Therapy of experimental leptospirosis. *In* "Antimicrobial Agents and Chemotherapy—1964" (J. C. Sylvester, ed.), pp. 746–751. Am. Soc. Microbiol. Washington, D.C.

Killby, V. A. A., and Silverman, P. H. (1967). Toxicity of antibiotics in laboratory rodents. *Science* **156**, 264.

Kim, K.-H., Fekety, R., Batts, D. H., Brown, D., Cudmore, M., Silva, J., Jr., and Waters, D. (1981). Isolation of *Clostridium difficile* from the environment and contacts of patients with antibiotic-associated colitis. *J. Infect. Dis.* **143**, 42–50.

Kuo, T. H., Ho, K. L., and Wiener, J. (1984). The role of alkaline protease in the development of cardiac lesions in myopathic hamsters: Effect of verapamil treatment. *Biochem. Med.* **32**, 207–215.

La Regina, M., Fales, W. H., and Wagner, J. E. (1980). Effects of antibiotic treatment on the occurrence of experimentally induced proliferative ileitis of hamsters. *Lab. Anim. Sci.* **30**, 38–41.

Larsen, A. B., and Miller, J. M. (1978). Effect of dexamethasone on *Mycobacterium paratuberculosis* infection in hamsters. *Am. J. Vet. Res.* **39**, 1866–1867.

Larson, H. E., Parry, J. V., Price, A. B., Davies, D. R., Dolby, J., and Tyrrell, D. A. J. (1977). Undescribed toxin in pseudomembranous colitis. *Lancet* **1**, 1246–1248.

Larson, H. E., Price, A. B., Honour, P., and Borriello, S. P. (1978). *Clostridium difficile* and the aetiology of pseudomembranous colitis. *Lancet* **1**, 1063–1066.

Larson, H. E., Price, A. B., and Borriello, S. P. (1980). Epidemiology of experimental enterocecitis due to *Clostridium difficile*. *J. Infect. Dis.* **142**, 408–413.

Laughon, B. E., Viscidi, R. P., Gdovin, S. L., Yolken, R. H., and Bartlett, J. G. (1984). Enzyme immunoassays for detection of *Clostridium difficile* toxins A and B in fecal specimens. *J. Infect. Dis.* **149**, 781–788.

Lee, S. H., Chiu, H., Renton, K. W., and Stebbing, N. (1984). Modulation by human interferon of antitumor effects of cyclophosphamide against a lymphosarcoma in hamsters. *Biochem. Pharmacol.* **33**, 3439–3444.

Libby, J. M., Jortner, B. S., and Wilkins, T. D. (1982). Effects of the two toxins of *Clostridium difficile* in antibiotic-associated cecitis in hamsters. *Infect. Immun.* **36**, 822–829.

Lowe, B. R., Fox, J. G., and Bartlett, J. G. (1980). *Clostridium difficile*-associated cecitis in guinea pigs exposed to penicillin. *Am. J. Vet. Res.* **41**, 1277–1279.

Lusk, R. H., Fekety, R., Silva, J., Browne, R. A., Ringler, D. H., and Abrams, G. D. (1978). Clindamycin-induced enterocolitis in hamsters. *J. Infect. Dis.* **137**, 464–475.

Margolis, G., and Kilham, L. (1978). Cerebellar, epidermal, and dental defects induced by ribavirin in perinatal hamsters and rats. *Exp. Mol. Pathol.* **29**, 44–54.

Meynell, G. G. (1963). Antibacterial mechanisms of the mouse gut. II. The role of E_h and volatile fatty acids in the normal gut. *Br. J. Exp. Pathol.* **44**, 209–219.

Michelich, V. J., Nunez-Montiel, O., Schuster, G. S., Thompson, F., and Dowell, V. R., Jr. (1981). Diet as a coadjuvant for development of antibiotic-associated diarrhea in hamsters. *Lab. Anim. Sci.* **31**, 259–262.

Miller, L. H., Neva, F. A., and Gill, F. (1978). Failure of chloroquine in human babesiosis (*Babesia microti*)—Case report and chemotherapeutic trials in hamsters. *Ann. Intern. Med.* **88**, 200–202.

Muratsu, K., and Morioka, T. (1984). Elevated salivary lactoferrin levels in streptozotocin-induced diabetic hamster. *J. Dent. Res.* **63**, Abstr. No. 55, 560.

Natt, M. P., Gustafson, R. R., Schultes, L., and Moynihan (1960). Comparative antitrichomonal activity of furazolidone [3-(5-nitrofurfurylideneamino)-2-oxazolidone] and aminothiazol (2-acetamido-5-nitrothiazol) *in vitro* against experimental trichomonal infections in mice and hamsters. *J. Parasitol.* **46**, Suppl., 7–8.

Patton, N. M. (1979). What every practitioner should know about rabbits and rodents. *Calif. Vet.* **33,** 25–33.

Price, A. B., Larson, H. E., and Crow, J. (1979). Morphology of experimental antibiotic-associated enterocolitis in the hamster: A model for human pseudomembranous colitis and antibiotic-associated diarrhoea. *Gut* **20,** 467–475.

Raether, W., and Fink, E. (1979). Antimalarial activity of floxacrine (HOE 991). I. Studies on blood schizontocidal action of floxacrine against *Plasmodium berghei, P. vinckei* and *P. cynomolgi. Ann. Trop. Med. Parasitol.* **73,** 505–526.

Raether, W., and Seidenath, H. (1983). The activity of fexinidazole (HOE 239) against experimental infections with *Trypanosoma cruzi,* trichomonads and *Entamoeba histolytica. Ann. Trop. Med. Parasitol.* **77,** 13–26.

Raether, W., and Seidenath, H. (1984). Ketoconazole and other potent antimycotic azoles exhibit pronounced activity against *Trypanosoma cruzi, Plasmodium berghei* and *Entamoeba histolytica in vivo. Z.Parasitenkd.* **70,** 135–128.

Raether, W., Seidenath, H., and Windelmann, E. (1977). 1,4-bis-1-methyl-5-nitroimidazolyl-(2-methyleneimino)-piperazine, a new drug with marked activity against *Entamoeba histolytica. Arzneim.-Forsch.* **27,** 968–969.

Raether, W., Seidenath, H., and Loewe, H. (1978). Action of *p*-(4-amidinophenoxy)-benzaldehyde-*p*-amidino-phenylhydrazone dihydrochloride on *Leishmania donovani* infections in the golden hamster. *Ann. Trop. Med. Parasitol.* **72,** 543–547.

Rehg, J. E. (1985). Clostridial enteropathies, hamster. *In* "Monographs on Pathology of Laboratory Animals—Digestive System" (T. C. Jones, U. Mohr, and R. D. Hunt, eds.), pp. 340–346. Springer-Verlag, Berlin and New York.

Rehg, J. E., and Lu, Y.-S. (1982). *Clostridium difficile* typhlitis in hamsters not associated with antibiotic therapy. *J. Am. Vet. Med. Assoc.* **181,** 1422–1423.

Rifkin, G. D., and Fekety, F. R., Jr. (1985). Antibiotic-induced enterocolitis in animals. *In* "Animal Models of Intestinal Disease" (C. J. Pfeiffer, ed.), pp. 123–133. CRC Press, Boca Raton, Florida.

Rifkin, G. D., Fekety, F. R., Silva, J., Jr., and Sack, R. B. (1977). Antibiotic-induced colitis: Implication of a toxin neutralized by *Clostridium sordellii* antitoxin. *Lancet* **2,** 1103–1106.

Rifkin, G. D., Silva, J., Jr., and Fekety, R. (1978a). Gastrointestinal and systemic toxicity of fecal extracts from hamsters with clindamycin-induced colitis. *Gastroenterology* **74,** 52–57.

Rifkin, G. D., Fekety, R., and Silva, J., (1978b). Neutralization by *Clostridium sordelli* antitoxin of toxins implicated in clindamycin-induced cecitis in hamsters. *Gastroenterology* **74,** 422–424.

Rolfe, R. D., and Finegold, S. M. (1983). Intestinal β-lactamase activity in ampicillin-induced, *Clostridium difficile*-associated ileocecitis. *J. Infect. Dis.* **147,** 227–235.

Rolfe, R. D., Helebian, S., and Finegold, S. M. (1981). Bacterial interference between *Clostridium difficile* and normal fecal flora. *J. Infect. Dis.* **143,** 470–475.

Ronald, N. C., and Wagner, J. E. (1975). Treatment of *Hymenolepis nana* in hamsters with Yomesan® (niclosamide). *Lab. Anim. Sci.* **25,** 219–220.

Rowin, K. S., Tanowitz, H. B., and Wittner, M. (1982). Therapy of experimental babesiosis. *Ann. Intern. Med.* **97,** 556–558.

Russell, R. J., Johnson, D. K., and Stunkard, J. A. (1981). "A Guide to Diagnosis, Treatment and Husbandry of Pet Rabbits and Rodents," pp. 39–41. Vet. Med. Publ. Co., Edwardsville, Kansas.

Saffar, J. L., and Lasfargues, J. J. (1984). A histometric study of the effect of indomethacin and calcitonin on bone remodelling in hamster periodontitis. *Arch. Oral Biol.* **29,** 555–558.

Schlessinger, D. (1979). Animal models of antibiotic-induced colitis. *In* "Microbiology—1979" (D. Schlessinger, ed.), pp. 256–279. Am. Soc. Microbiol., Washington, D.C.

Schneierson, S. S., and Perlman, E. (1956). Toxicity of penicillin for the Syrian hamster. *Proc. Soc. Exp. Biol. Med.* **91,** 229–230.

Sebesteny, A. (1979). Syrian hamsters. *In* "Handbook of Diseases of Laboratory Animals" (J. M. Hime and P. N. O'Donoghue, eds.), pp. 111–136. Heinemann Veterinary Books, London.

Sheffield, F. W., and Beveridge, E. (1962). Prophylaxis of "wet-tail" in hamsters. *Nature (London)* **196,** 294–295.

Silva, J., Fekety, R., Werk, C., Ebright, J., Cudmore, M., Batts, D., Syrjamaki, C., and Lukens, J. (1984). Inciting and etiologic agents of colitis. *Rev. Infect. Dis.* **6,** Suppl. 1, S214–S221.

Simard-Savoie, S., and Dupuis, R. (1968). Effet de la methacycline sur la carie dentaire expérimentale chez le hamster doré syrien. *Rev. Can. Biol.* **27,** 121–26.

Small, J. D. (1968a). Fatal enterocolitis in hamsters given lincomycin hydrochloride. *Lab. Anim. Care* **18,** 411–420.

Small, J. D. (1968b). Toxicity of antibiotics for the hamster. *19th Annu. Meet., Am. Assoc. Lab. Anim. Sci.,* Abstr. No. 94.

Small, J. D., Brodie, K. R., Fogelson, L. D., and Tucker, A. N. (1985). The effects of trichlorfon on the immune system of the mouse. *Lab. Anim. Sci.* **35,** 545.

Taylor, N. S., and Bartlett, J. G. (1980). Binding of *Clostridium difficile* cytotoxin and vancomycin by anion-exchange resins. *J. Infect. Dis.* **141,** 92–97.

Taylor, N. S., Thorne, G. M., and Bartlett, J. G. (1980). Separation of an enterotoxin from the cytotoxin of *Clostridium difficile. Clin. Res.* **28,** 285.

Taylor, N. S., Thorne, G. M., and Bartlett, J. G. (1981). Comparison of two toxins produced by *Clostridium difficile. Infect. Immun.* **34,** 1036–1043.

Teichmann, B., and Schmidt, A. (1978). Test of *O,O,*methyl(-1-hydroxy-2,2,2-trichlorethyl)phosphonate (Trichlorfon) for carcinogenic activity in Syrian golden hamsters (*Mesocricetus auratus* Waterhouse) by intraperitoneal administration. *Arch. Geschwulstforsch.* **48,** 718–721.

Thompson, P. E., Bayles, A., Herbst, S. F., Olszewski, B., and Meisenhelder, J. E. (1959). Antiamebic and antitrichomonal studies on the antibiotic paromomycin (Humatin) *in vitro* and in experimental animals. *Antibiot. Chemother. (Washington, D.C.)* **9,** 618–626.

Toothaker, R. D., and Elmer, G. W. (1984). Prevention of clindamycin-induced mortality in hamsters by *Saccharomyces boulardii. Antimicrob. Agents Chemother.* **26,** 552–556.

Toshniwal, R., Fekety, R., and Silva, J., Jr. (1979). Etiology of tetracycline-associated pseudomembranous colitis in hamsters. *Antimicrob. Agents Chemother.* **16,** 167–170.

Toshniwal, R., Silva, J., Jr., Fekety, R., and Kim, K.-H. (1981). Studies on the epidemiology of colitis due to *Clostridium difficile* in hamsters. *J. Infect. Dis.* **143,** 51–54.

Unay, E. S., and Davis, B. J. (1980). Treatment of *Syphacia obvelata* in the Syrian hamster (*Mesocricetus auratus*) with piperazine citrate. *Am. J. Vet. Res.* **41,** 1899–1900.

Viscidi, R. P., Yolken, R. H., Laughon, B. E., and Bartlett, J. G. (1983). Enzyme immunoassay for detection of antibody to toxins A and B of *Clostridium difficile. J. Clin. Microbiol.* **18,** 242–247.

Weathersbee, P. S., Ax, R. L., and Lodge, J. R. (1975). Caffeine-mediated changes of sex ratio in Chinese hamsters, *Cricetulus griseus. J. Reprod. Fertil.* **43,** 141–143.

Will, P. C., Femia, P., Witt, C., Lin, A., and Gaginella, T. S. (1985). Sulfasalazine and hydroxychloroquine protect against clindamycin-induced colitis in hamsters. *Pharmacologist* **27,** Abstr. No. 65, 128.

Willhite, C. G. (1982). Congenital malformations induced by laetrile. *Science* **215,** 1513–1515.

Willhite, C. C. (1984). Dose–response relationships of retinol in production of the Arnold-Chiari malformation. *Toxicol. Lett.* **20,** 257–262.

Willhite, C. C., and Ferm, V. H. (1978). Potentiation of ribavirin-induced teratogenesis by natural purines. *Exp. Mol. Pathol.* **28,** 196–201.

Williams, C. S. F. (1976). "Practical Guide to Laboratory Animals," pp. 26–34. Mosby, St. Louis, Missouri.

Wilson, K. H., Silva, J., and Fekety, F. R. (1981). Suppression of *Clostridium difficile* by normal hamster cecal flora and prevention of antibiotic-associated cecitis. *Infect. Immun.* **34,** 626–628.

Winkelmann, E., and Raether, W. (1978). Chemotherapeutically active nitro compounds. 4. 5-Nitroimidazoles. Part III. *Arzneim.-Forsch.* **28,** 739–749.

Winkelmann, E., and Raether, W. (1979). Chemotherapeutically active anthraquinones. I. Aminoanthraquinones. *Arzneim.-Forsch.* **29,** 1504–1509.

Winkelmann, E., Raether, W., Dittmar, W., Duwell, D., Gericke, D., Hohorst, W., Rolly, H., and Schrinner, E. (1975). Chemotherapeutisch wirksame Nitroverbindungen. 1. Mitteilung; Nitroaniline. *Arzneim.-Forsch.* **25,** 681–708.

Winkelman, E., Raether, W., Hartung, H., and Wagner, W.-H. (1977a). Chemotherapeutisch wirksame Nitroverbindungen. 3. Mitteilung: Nitropyridine, Nitroimidazopryidine und verwandte Verbindungen. *Arzneim.-Frosch.* **27,** 82–89.

Winkelmann, E., Raether, W., Gebert, U., and Sinharay, A. (1977b). Chemotherapeutically active nitro compounds. 4. 5-Nitroimidazoles. Part I. *Arzneim.-Forsch.* **27,** 2251–2263.

Winkelmann, E., Raether, W., and Gebert, U. (1978). Chemotherapeutically active nitro-derivatives. 4. 5-Nitroimidazoles. Part IV. *Arzneim.-Forsch.* **28,** 1682–1684.

Wüst, J., Sullivan, N. M., Hardegger, U., and Wilkins, T. D. (1982). Investigation of an outbreak of antibiotic-associated colitis by various typing methods. *J. Clin. Microbiol.* **16,** 1096–1101.

Yolken, R. H., Whitcomb, L. S., Marien, G., Bartlett, J. D., Libby, J., Ehrich, M., and Wilkins, T. (1981). Enzyme immunoassay for the detection of *C. difficile* antigen. *J. Infect. Dis.* **144,** 378.

Chapter 12

Experimental Biology: Use in Oncological Research

John J. Trentin

I.	Introduction	201
II.	The Cheek Pouch as a Privileged Tumor Transplant Site	202
III.	Use of Conditioned Hamsters for Growth and Transplantation of Human Tumors	202
IV.	Transmissible (Epizootic) Leukemia of Hamsters	202
V.	Nonrandom Chromosome Abnormalities in Transformed Syrian Hamster Cell Lines	203
VI.	Role of Adenovirus Type 2-Induced Early Tumor Antigens in Natural Cell-Mediated Defense against Such Tumors in Hamsters	204
	References	205
	Appendix 1. Spontaneous Tumor Development: Benign Tumors	207
	Appendix 2. Spontaneous Tumor Development: Malignant Tumors	208
	Appendix 3. Induced Tumors: Viral Oncogenesis	210
	Appendix 4. Induced Tumors: Survival and Tumor Incidence after Irradiation	214

I. INTRODUCTION

The great utility of the Syrian golden hamster, *Mesocricetus auratus*, in biomedical research has long been firmly established (Hoffman *et al.*, 1968). Over the years it has been used primarily in one, then another, field of research, as trends changed and as its advantages for one or another field of research were recognized (Magalhaes, 1977). Although now used in a number of diverse fields, its predominant use as of 1976 was in cancer research, as documented in Table I. This is due in part to (a) its relatively low incidence of spontaneous tumors, especially as compared to the rat and mouse (Appendixes 1 and 2; Chapter 9; Homburger *et al.*, 1978), (b) its relatively low number of indigenous viruses (McCormick and Trentin, 1979), (c) its very high susceptibility to viral oncogenesis by viruses of other species, including human viruses (Trentin *et al.*, 1962; Appendix 3), (d) the fact that whereas random-bred hamsters are relatively resistant to chemical carcinogens, a number of inbred strains are now available, some of which are resistant, but others of which are highly susceptible to one or another chemical carcinogen (Homburger *et al.*, 1978, 1979; Chapters 15 and 16), (e) its susceptibility to estrogen-induced tumors, including renal adenocarcinomas in

Table I

Classification of Publications Utilizing Hamster, from 1963 to 1976, into a Single Subject Area[a,b]

Subject	Number of publications	Percentage of total publications	Change since previous tabulation in 1963 (%)
Cancer	1510	30.0	+25.2
Reproduction	699	13.9	+8.3
Pathology	492	9.8	+5.2
Physiology	460	9.2	+4.8
Virology	436	8.7	−8.2
Parasitology	258	5.1	−6.8
Biochemistry	245	4.8	+4.8
Endocrinology	247	4.0	+0.6
Anatomy	156	3.1	−0.4
Development	153	3.0	−1.2
Dental disease	138	2.8	−11.3
Pharmacology	120	2.4	+0.4
Behavior	113	2.3	−0.9
Genetics	72	1.4	−0.4
Microbiology	65	1.3	−10.0
Nutrition	67	1.3	−3.6
Animal care	36	0.7	−0.4
Ecology	5	0.04	−2.2

[a] After Magalhaes (1977). (Copyright by United Business Publications, Inc.)
[b] Although immunology is not listed as a subject heading, this area had become quite active by 1976, as witness the symposium sponsored by the Transplantation Society in 1977 and organized by Streilein (1978). Undoubtedly much published work of this kind has been included under the heading of cancer or, as in the recollection of H. Magalhaes (personal communication), that of pathology. Any future classification would undoubtedly merit a separate category of immunology.

male hamsters (Kirkman and Bacon, 1949; Kirkman, 1957; Letourneau et al., 1975), and (f) its cheek pouch, providing an evertable and privileged site for tumor transplantation, including human tumors (Handler and Shepro, 1968). These attributes have made the hamster especially useful in studies of both viral oncogenesis and chemical carcinogenesis (Trentin and Homburger, 1979). In contrast, and not generally appreciated, is the relatively high resistance of the hamster to radiation carcinogenesis (Appendix 4). The unique susceptibility or resistance of hamsters to one or another form of oncogenic transformation undoubtedly involves genetic and immunogenetic factors, although the exact mechanisms are not yet fully understood. For current status of the rapidly developing area of hamster genetics and immunogenetics, see Chapter 13.

The tumor-associated viruses of the hamster, most of them retroviruses (type R, intracytoplasmic type A, type C), and their relationship to neoplasia, have been reviewed by McCormick and Trentin (1979).

II. THE CHEEK POUCH AS A PRIVILEGED TUMOR TRANSPLANT SITE

The hamster cheek pouch is lined by a thin, well-vascularized membrane that is easily everted, and has provided a useful technology for a variety of studies (Handler and Shepro, 1968). Important among these is its use as a transplantation site for a variety of tumors and normal tissues. Lutz et al. (1951) first reported the use of the hamster cheek pouch for the transplantation of a methylcholanthrene-induced hamster sarcoma. Later it was unexpectedly found that the hamster cheek pouch offered an immunologically privileged site for the prolonged survival of transplanted normal foreign tissues. The mechanisms of this privilege have been variously reported to be related to the unique character of its connective tissue and/or its lack of lymphatic drainage into regional lymph nodes (Shepro et al., 1960, 1963; Witte et al., 1965; Billingham et al., 1964; Chapters 13 and 14).

III. USE OF CONDITIONED HAMSTERS FOR GROWTH AND TRANSPLANTATION OF HUMAN TUMORS

Toolan was first to report the growth and transplantation of human tumors in X-irradiated rats and mice (Toolan, 1951), and later in cortisone-treated rats and hamsters (Toolan, 1953, 1954). The use of cortisone gave more uniform results than irradiation, and was much more widely available than the X-ray machine (H. W. Toolan, personal communication). Some of these tumors are still being used, in particular HEP 3 and HS 1. HEP 3 has proved of special interest, because it grows so vigorously, eroding blood vessels as it spreads. Another, HEP 2, has become a widely used tissue culture cell line.

In more recent years, the genetically athymic nude (nu/nu) mouse has gained favor as a host for transplantation of human tumors.

IV. TRANSMISSIBLE (EPIZOOTIC) LEUKEMIA OF HAMSTERS

In 1975, Ambrose and Coggin reported an unusual epizootic of lymphomas with an apparent infectious cause in Syrian golden hamsters held in one of their animal facilities. More than 50% of 1-day to 3-week-old hamsters placed in that facility developed lymphoma within 4–30 weeks! On two occasions, the disease reappeared in the same facility after destruction of all hamsters and disinfection. On the second trial, after all hamsters were killed, exhaustive cleaning and disinfection were performed, including UV and virucidal gases, and the facility left vacant for more than 7 weeks. New animal personnel were employed. New hamsters introduced into the facility developed lymphomas in high frequency after 4 months, whereas control animals held at another facility were lymphoma-free after more than 2 years.

Later the infectious lymphoma epizootic spread to a new containment facility established at another site. The disease was highly active among younger hamsters. Newborn hamsters introduced (room exposure) developed tumors as early as 31 days, with 100% incidence possible in LSH hamsters and 60–70% among LVG hamsters (Coggin et al., 1983).

Studies over several years established that (a) an infectious agent is involved; (b) conventionally prepared cell-free extracts of the primary lymphomas, injected into newborn hamsters, not only did not induce lymphomas, but actually conferred protection against horizontal infection and lymphoma induction by room exposure; (c) however, treatment of the extract with protamine sulfate during the initial tissue blending yielded cell-free extracts that induced 50–80% lymphomas within 5–9 weeks of injection into neonates; (d) electron-microscopic examination of these oncogenic filtrates revealed no virus particles, although thin sections of the tumor from which they were extracted revealed picodna-like particles budding from the nuclear membrane; (e) the oncogenic capacity of the original protamine sulfate preparations could be completely inactivated by exposure to DNase, but was resistant to RNase or protease, and to UV. It was environmentally stable and not reproducibly susceptible to antibody neutralization (Coggin et al., 1978); (f) most of the intraperitoneal lymphomas had the properties of B cells, whereas primary lymphomas that arose in the thymus had properties of T cells, that is, the agent produced both B- and T-cell lymphomas (Coggin et al., 1983); (g) several other specific pathological conditions prevailed in the diseased colonies over a 6-year period. These included a nonbacterial enteritis associated with intussusception of both the small and large intestine (Manci et al., 1984), pyelonephritis in a majority of the animals autopsied in the lymphomatous colony, cholangitic disease of the liver, in addition to cirrhosis, benign hemangiomas, and primary hepatic lymphomas. Based on their studies of five epizootics of this lymphoma, the authors concluded that the B-cell and T-cell lymphomas and associated diseases were induced by an infectious DNA viroid-like agent (Coggin et al., 1983). The causative agent is characterized as a nonencapsulated, DNase-sensitive, low molecular weight, disease-causing, self-replicating, naturally infectious nucleic acid (Coggin et al., 1981).

Hamsters with this horizontally transmitted lymphoma, or others housed with them, occasionally developed epitheliomas containing an unclassified papovavirus. Studies of this virus confirmed that it was the causative agent of the epitheliomas, but indicated that it is not the agent responsible for the lymphomas (J. H. Coggin, Jr., personal communication). Similar outbreaks of horizontally transmitted lymphoma in hamsters have occurred at Yale and at Southwestern Medical School in Dallas.

This interesting series of investigations reveals a phenomenon (epizootic transmission of leukemia) of unique importance to the study of vir(oid)ology, infectious diseases, cancer, and as yet unrevealed factors contributing to their interactions! For further discussion, see Chapter 9.

V. NONRANDOM CHROMOSOME ABNORMALITIES IN TRANSFORMED SYRIAN HAMSTER CELL LINES

The advent of chromosome banding techniques in recent years has permitted the detailed recognition of individual chromosomes and chromosome segments, thereby facilitating the detection of translocations or deletions of even minute chromosome segments, not to mention monosomy or trisomy of entire chromosomes (Hsu, 1983). This resulted in the recognition that many malignant tumors of humans, mice, hamsters, and other species have an associated specific, nonrandom chromosome anomaly, belatedly providing strong support to the concept of Boveri (1912) that abnormal distribution of chromatin is responsible for some malignancies. The first such consistent chromosome anomaly associated with a particular malignancy was the Philadelphia chromosome (Ph[1]) of human chronic myelogenous leukemia. This is an abnormally small chromosome detected before the advent of banding techniques (Nowell and Hungerford, 1960). Subsequent analysis by banding methods showed it to result from an unequal translocation between chromosome 22q and usually 9q, although occasionally a chromosome other than 9q (Pathak, 1983; Trujillo, 1983). A number of other translocations of a minute portion of a specific segment of one particular chromosome to another specific portion of a different particular chromosome have been described in human and mouse (Pathak, 1983; Klein and Lenoir, 1982). In human retinoblastomas, an interstitial deletion in chromosome 13 appears to be the specific anomaly (Wilson et al., 1973). In human meningiomas, a partial or complete monosomy of chromosome 22 has been discovered (Zankl and Zang, 1972). In T-cell leukemias of the mouse, both spontaneous and induced, trisomy 15 is the dominant alteration.

Several investigators have examined the chromosomal characteristics of chemically and virally transformed Syrian hamster embryo cells. Specific chromosome involvement has been demonstrated with the expression and suppression of malignancy in transformed Syrian hamster cells (Yamamoto et al., 1973; Bloch-Shtacher and Sachs, 1976; Benedict et al., 1975). Trisomy of an acrocentric chromosome has been demonstrated in two clones of Syrian hamster cells transformed by herpes simplex type 2 virus (Kessous et al., 1979). In a study of 18 tissue culture cell lines of Syrian hamster cells transformed either in vivo or in vitro by either DNA or RNA viruses, the most consistent chromosome anomaly found was monosomy of C15, which occurred in 15 of the 18 cell lines (Pathak et al., 1981; Table II). Monosomy C15 was also reported by DiPaolo and his associates in Syrian hamster cells transformed either by virus or by chemical carcinogens (DiPaolo et al., 1973; DiPaolo and Popescu, 1976; Popescu et al., 1974). The missing C15 may either be lost or broken up in the process of translocation.

In recent years, it has been discovered that certain viral nucleic acid sequences capable of inducing transformation when integrated into the genome of the infected cell, may also be present in the normal cell genome in smaller amounts or slightly different form. These sequences, known as oncogenes, can be activated by a variety of mechanisms, such as virus infection, DNA transfection, or translocation from one chromosome location to another location proximal to an immunoglobulin gene (Klein, 1981; Klein and Lenoir, 1982). A summary of the known human oncogenes located on chromosomes involved in rearrangements in various neoplasms is presented in Table III.

Table II

Characteristics of the Transformed Syrian Hamster Cell Lines and Tumor-Derived Cultures[a]

Identification no.	Passage no.	Transforming agent	Modal chromosome no.	Monosomy C15
A	22	SA7	44	+
B	22	SA7	42	+
I-A	22	SA7	42	+
III-A	22	SA7	43	+
TU-26	5	SA7	40	+
HEL1	32	HSV-2	45	+
HEL2	5	HSV-2	46	+
HEL4	10	HSV-2	45	+
J19L	25	SV40	53	−
Ha4A28	75	SV40	44	+
IV	42	CELO	42	+
IX	42	CELO	44	+
X	21	CELO	56	−
VI	8	Spontaneous	43	−
VII	NA	MSV	41	+
XI	NA	MSV	78	−
T-71-30	5	AD7	50	+
AD-7	NA	AD7	48	+
AD-7 (TU)	NA	AD7	46	+

[a]From Pathak et al. (1981), Nonrandom chromosome abnormalities in transformed Syrian hamster cell lines. In "Genes, Chromosomes and Neoplasia" (Arrighi, F. E., Rao, P. N. and Stubblefield, E., eds.), pp. 405–418. Raven Press, New York. NA, Information not available; +, monosomy C15 present; and −, monosomy C15 not present. SA7, simian adenovirus, type 7; HSV-2, herpes simplex virus, type 2; SV40, simian virus, type 40 (a papovavirus); CELO, chicken embryo lethal orphan virus; MSV, murine sarcoma virus; and AD7, human adenovirus, type 7.

Table III

Summary of Human Oncogenes' Location on Chromosomes and Their Involvement in Specific Neoplasms[a]

Oncogenes	Chromosomes	Type of Tumor
Sis	22	CGL, meningioma, Burkitt's lymphoma
Src (Sarc)	20	Myeloproliferative disorders
Ras[k]	12	CLL
RAS[H]	11	Wilms' tumor, breast tumors
Mos, Myc	8	AML, Burkitt's lymphoma
Myb	6	ALL, melanoma
Fes	15 t (15;17)	Acute promyelocytic leukemia
Abl	9	CGL
Erb	?	Unknown
?	13	Retinoblastoma
?	2	Cervical carcinoma
Lym	1	Neuroblastoma, breast tumor, melanoma
?	7	Hematological disorders
?	3	Renal cell carcinoma, small-cell carcinoma of the lung
?	17 i (17q)	CML
?	21	Primary thrombocythemia, AML

[a]From Pathak (1983). (Courtesy of the Medical Arts Publishing Foundation, Houston.) ?, Oncogenes are not yet mapped on these chromosomes. CGL, Chronic granulocytic leukemia; CLL, chronic lymphocytic leukemia; AML, acute myeloid leukemia; ALL, acute lymphocytic leukemia; and CML, chronic myeloid leukemia.

VI. ROLE OF ADENOVIRUS TYPE 2-INDUCED EARLY TUMOR ANTIGENS IN NATURAL CELL-MEDIATED DEFENSE AGAINST SUCH TUMORS IN HAMSTERS

As indicated in Appendix 3, newborn hamsters have been a valuable experimental animal for induction of tumors by viruses of other species, such as polyomavirus of the mouse, SV40 virus of monkeys, and several of the more than 30 serotypes of human adenoviruses. The tumors induced by these DNA viruses have transplantation antigens that elicit T cell-mediated transplantation immunity. In addition, such tumors have virus-induced and virus-specific "early" neoantigens that are not viral structural antigens. These early antigens, which appear within hours of either lytic or nonlytic and transforming infection, are detected as distinctive intranuclear flecks by immunofluorescence microscopy with serum of tumor-bearing hamsters. Because they were not involved in the T cell-mediated transplantation immunity, they were at first of only diagnostic significance. Serum of hamsters bearing tumors induced by human adenovirus type 12 would detect the adeno 12 early antigens in tumors induced by adeno 12 in either hamsters or mice and presumably in other species, making them useful in screening human cancers for possible viral etiology (Van Hoosier and Trentin, 1968; McCormick et al., 1968). Although the adenovirus that induces the tumor does not replicate complete virus in the hamster cell, portions of the viral genome, including the genes for early antigens, do integrate into the hamster genome. It has been shown, by mutant viruses and endonuclease-cleaved fragments of the viral DNA, that the early-antigen genes can transform normal cells to neoplastic cells (Tooze, 1980).

Syrian hamsters can discriminate between the oncogenic potential of different DNA tumor viruses (Lewis and Cook, 1982, 1985). Paradoxically, some of the adenoviruses, such as adenovirus type 2, that are quite capable of transforming hamster cells *in vitro* are nononcogenic *in vivo* in newborn hamsters. Some lines of hamster embryo cells transformed *in vitro* by adenovirus type 2 will grow as tumors in nude mice but not in hamsters of any age. Other such cell lines will grow in nude mice and newborn but not adult hamsters. Hamster cells transformed by adenovirus type 12 *in vitro* will grow in syngeneic hamsters of any age but not in adult allogeneic hamsters. Hamster cells transformed *in vitro* by SV40 or polyoma viruses will grow in either syngeneic or allogeneic hamsters of any age (Lewis and Cook, 1985). Yet all of these transformed cell lines are highly immunogenic in terms of being able to elicit high levels of virus-specific, T-lymphocyte-mediated, transplantation immunity (Allison, 1980; Tevethia, 1980; Lewis and Cook, 1984). However, histopathological studies of the rejection of transplanted adeno 2-transformed hamster cells suggested that lymphocytes and macrophages appearing at the transplant site in less than 3 days (i.e., before T-cell immunity is induced) were responsible for rejection (Cook et al., 1979). This same early mononuclear cell infiltrate was seen in both rejected adeno 2 and nonrejected adeno 12-transformed cell transplants, suggesting differences in susceptibility to early non-T lymphocyte and/or macrophage cell-mediated responses. Indeed, when DNA virus-transformed hamster cell lines of different tumorigenic phenotypes were exposed *in vitro* to hamster natural killer (NK) lymphocytes and activated hamster macrophages, there was a direct correlation between resistance to lysis by NK cells and ability to form tumors in immunocompetent animals (Cook et al., 1980, 1982; Lewis and Cook, 1985). Adeno 12- and SV40-transformed cells were resistant, while adeno 2-transformed cells were susceptible to lysis by NK cells. Adeno 12-transformed hamster cell lines, while resistant to hamster NK cells, were highly susceptible to lysis by Bacillus Calmette-Guérin (BCG)-activated macrophages, whereas SV40-transformed hamster cell lines were resistant to lysis by activated macrophages (Cook et al., 1980). Rat embryo cells transformed *in vitro* by either adeno 2 or adeno 12 were also found to have a direct correlation between resistance to lysis by rat NK cells, and *in vivo* tumorigenic growth potential in rats (Raska and Gallimore, 1982; Sheil et al., 1984).

Of considerable interest has been the finding that expression of adenovirus type 2 early gene-encoded polypeptides (so-called T antigens, T for "tumor" antigens, *not* for targets of T cell-mediated transplantation immunity) on transformed hamster cells correlates with and may be responsible for susceptibility to lysis by hamster NK cells and macrophages (Cook et al., 1983; Cook and Lewis, 1984). Those somatic cell hybrids formed between hamster cells transformed by adeno 2 and by SV40, which expressed adeno 2 early T antigens showed much greater susceptibility to lysis by either hamster NK cells or activated macrophages, as compared to those hybrids that did not express adeno 2 early antigen, regardless of whether SV40 early antigen was expressed or not. This indicates that the adeno 2 early antigens, in addition to having significance for the transformed state, have significance for susceptibility to host early cell-mediated surveillance mechanisms against transformed cells. In addition, this provides an important clue to the much sought-after targets of susceptibility of tumor cells to recognition and lysis by NK cells and macrophages.

It is perplexing and disappointing that not all viral-induced early antigens confer high susceptibility to early detection and lysis by NK cells and macrophages. We must remember, however, that what we see as spontaneous or induced tumors represent the failures of the host immune surveillance mechanisms of all kinds, which probably operate successfully more often than not, as indicated by the greatly increased susceptibility to both spontaneous and induced tumors in a variety of immunological deficiency states of both humans and animals (Szentivanyi and Friedman, 1986).

REFERENCES

Allison, A. C. (1980). Immune responses to polyoma virus and polyoma-induced tumors. *In* "Viral Oncology" (G. Klein, ed.), pp. 481–488. Raven Press, New York.

Ambrose, K. R., and Coggin, J. H., Jr. (1975). An epizootic in hamsters of lymphomas of undetermined origin and mode of transmission. *J. Natl Cancer Inst. (U.S.)* **54**, 877–880.

Benedict, W., Rucker, N., Mark, C., and Kouri, R. (1975). Correlation between balance of specific chromosomes and expression of malignancy in hamster cells. *J. Natl. Cancer Inst. (U.S.)* **54**, 157–162.

Billingham, R. E., and Silvers, W. K. (1964). Studies on homografts of foetal and infant skin and further observations of the anomalous properties of pouch skin grafts in hamsters. *Proc. R. Soc. London, Ser. B* **161**, 168–190.

Bloch-Shtacher, N., and Sachs, L. (1976). Chromosome balance and the control of malignancy and cell transformation. *J. Cell. Physiol.* **87,** 89–100.

Boveri, T. (1912). Bietrag zum Studium des Chromatins in den Epithelzellen der Carcinome. *Beitr. Pathol. Anat. Allg. Pathol.* **14,** 249.

Coggin, J. H., Jr., Thomas, K. V., and Huebner, R. J. (1978). Horizontally transmitted lymphomas of Syrian hamsters. *Fed. Proc., Fed. Am. Soc. Exp. Biol.* **37,** 2086–2088.

Coggin, J. H., Jr., Oakes, J. E., Huebner, R. J., and Gilden, R. (1981). Unusual filterable oncogenic agent isolated from horizontally transmitted Syrian hamster lymphomas. *Nature (London)* **290,** 336–338.

Coggin, J. H., Jr., Bellomy, B. B., Thomas, K. V., and Pollock, W. J. (1983). B-Cell and T-cell lymphomas and other associated diseases induced by an infectious DNA viroid-like agent in hamsters *Mesocricetus auratus*. *Am. J. Pathol.* **110,** 254–266.

Cook, J. L., and Lewis, A. M., Jr. (1984). Differential NK cell and macrophage killing of hamster cells infected with nononcogenic or oncogenic adenovirus. *Science* **224,** 612–615.

Cook, J. L., Kirkpatrick, C. H., Robson, A. S., and Lewis, A. M., Jr. (1979). Rejection of adenovirus 2-transformed cell tumors and immune responsiveness in Syrian hamsters. *Cancer Res.* **39,** 4949–4955.

Cook, J. L., Hibbs, J. B., Jr., and Lewis, A. M., Jr. (1980). Resistance of simian virus 40-transformed hamster cells to the cytolytic effect of activated macrophages: A possible factor in species-specific viral oncogenicity. *Proc. Natl. Acad. Sci. U.S.A.* **77,** 6773–6777.

Cook, J. L., Hibbs, J. B., Jr., and Lewis, A. M., Jr. (1982). DNA virus-transformed hamster cell–host effector cell interactions: Level of resistance to cytolysis correlated with tumorigenicity. *Int. J. Cancer* **30,** 795–803.

Cook, J. L., Hauser, J., Patch, C. T., Lewis, A. M., Jr., and Levine, A. S. (1983). Adenovirus 2 early gene expression promotes susceptibility to effector cell lysis of hybrids formed between hamster cells transformed by adenovirus 2 and simian virus 40. *Proc. Natl. Acad. Sci. U.S.A.* **80,** 5995–5999.

DiPaolo, J. A., and Popescu, N. C. (1976). Relationship of chromosome changes to neoplastic cell transformation. *Am. J. Pathol.* **85,** 709–738.

DiPaolo, J. A., Popescu, N. C., and Nelson, R. L. (1973). Chromosomal banding patterns and *in vitro* transformation of Syrian hamster cells. *Cancer Res.* **33,** 3250–3258.

Handler, N. H., and Shepro, D. (1968). Cheek pouch technology: Uses and applications. *In* "The Golden Hamster—Its Biology and Use in Medical Research" (R. A. Hoffman, P. F. Robinson, and H. Magalhaes, eds.), pp. 195–201. Iowa State Univ. Press, Ames.

Hoffman, R. A., Robinson, P. F., and Magalhaes, H., eds. (1968). "The Golden Hamster—Its Biology and Use in Medical Research." Iowa State Univ. Press, Ames.

Homburger, F., Adams, R. A., Soto, E., and Van Dongen, C. G. (1978). Chemical carcinogenesis in Syrian hamsters. *Fed. Proc., Fed. Am. Soc. Exp. Biol.* **37,** 2090.

Homburger, F., Adams, R. A., and Soto, E. (1979). The special suitability of inbred Syrian hamsters for carcinogenesis testing. *Arch. Toxicol., Suppl.* **2,** 445–450.

Hsu, T. C. (1983). The rise and present status of human cytogenetics: An introduction. *Cancer Bull. Univ. Tex. M. D. Anderson Hosp. & Tumor Inst. Houston, Tex.* **35,** 101–105.

Kessous, A., Bibor-Hardy, V., Suh, M., and Simard, R. (1979). Analysis of chromosomes, nucleic acids, and polypeptides in hamster cells transformed by herpes simplex virus type 2. *Cancer Res.* **39,** 3225–3234.

Kirkman, H. (1957). Steroid tumorigenesis. *Cancer (Philadelphia)* **10,** 757–764.

Kirkman, H., and Bacon, R. L. (1949). Renal adenomas and carcinomas in diethylstilbestrol-treated male golden hamster. *Anat. Rec.* **103:**475–476 (abstr.).

Klein, G. (1981). The role of gene dosage and genetic transpositions in carcinogenesis. *Nature (London)* **294,** 313–318.

Klein, G., and Lenoir, G. (1982). Translocations involving Ig-locus-carrying chromosomes: A model for genetic transposition in carcinogenesis. *Adv. Cancer Res.* **37,** 381–387.

Letourneau, R. J., Li, J. J., Roser, S., and Viller, C. A. (1975). Functional specialization in estrogen-induced renal adenocarcinomas of the golden hamster. *Cancer Res.* **35,** 6–10.

Lewis, A. M., Jr., and Cook, J. L. (1982). Spectrum of tumorigenic phenotypes among adenovirus 2-, adenovirus, 12-, and simian virus 40-transformed Syrian hamster cells defined by host cellular immune–tumor cell interactions. *Cancer Res.* **42,** 939–944.

Lewis, A. M., Jr., and Cook, J. L. (1984). The interface between adenovirus transformed cells and cellular immune responses in the challenged host. *Curr. Top. Microbiol. Immunol.* **110,** 1–22.

Lewis, A. M., Jr., and Cook, J. L. (1985). A new role for DNA virus early proteins in viral carcinogenesis. *Science* **227,** 15–20.

Lutz, B. R., Fulton, G. P., Patt, D. I., Handler, A. H., and Stevens, D. F. (1951). The cheek pouch of the hamster as a site for the transplantation of methylcholanthrene-induced sarcoma. *Cancer Res.* **11,** 64–66.

McCormick, K. J., and Trentin, J. J. (1979). Tumor-associated viruses of the Syrian hamster: Relation to neoplasia. *Prog. Exp. Tumor Res.* **23,** 13–55.

McCormick, K. J., Van Hoosier, G. L., Jr., and Trentin, J. J. (1968). Attempts to find human adenovirus type 12 tumor antigens in human tumors. *J. Natl. Cancer Inst. (U.S.)* **40,** 255–261.

Magalhaes, H. (1977). From taxonomy to toxicology: 138 years of hamster research. *Lab Anim.* **6,** 21–24.

Manci, E. A., Heath, L. S., Leinbach, S. S., and Coggin, J. H., Jr. (1984). Lymphoma-associated ulcerative bowel disease in the hamster (*Mesocricetus auratus*) induced by an unusual agent. *Am. J. Pathol.* **116,** 1–8.

Nowell, P. C., and Hungerford, D. A. (1960). A minute chromosome in human chronic granulocytic leukemia. *Science* **132,** 1497.

Pathak, S. (1983). Chromosome constitution of human solid tumors. *Cancer Bull. Univ. Tex. M. D. Anderson Hosp. & Tumor Inst. Houston, Tex.* **35,** 126–131.

Pathak, S., Hsu, T. C., Trentin, J. J., Butel, J. S., and Panigraphy, B. (1981). Non-random chromosome abnormalities in transformed Syrian hamster cell lines. *In* "Genes, Chromosomes, and Neoplasia" (F. Arrighi, P. Rao, and E. Stubblefield, eds.), pp. 405–418. Raven Press, New York.

Popescu, N. C., Olinici, C. D., Casto, B. C., and DiPaolo, J. A. (1974). Random chromosome changes following SA7 transformation of Syrian hamster cells. *Int. J. Cancer* **14,** 461–472.

Raska, K., Jr., and Gallimore, P. H. (1982). An inverse relation of the oncogenic potential of adenovirus-transformed cells and their sensitivity to killing by syngeneic natural killer cells. *Virology* **123,** 8–18.

Sheil, J. M., Gallimore, P. H., Zimmer, S. G., and Sopori, M. L. (1984). Susceptibility of adenovirus 2-transformed rat cell lines to natural killer (NK) cells: Direct correlation between NK resistance and *in vivo* tumorigenesis. *J. Immunol.* **132,** 1578–1582.

Shepro, D., Eidelhoch, L. P., and Patt, D. I. (1960). Lymph node responses to malignant homo- and heterografts in the hamster. *Anat. Rec.* **136,** 393–405.

Shepro, D., Kula, N., and Halkett, J. (1963). The role of the cheek pouch in effecting transplantation immunity in the hamster. *J. Exp. Med.* **117,** 749–754.

Streilein, J. W. (1978). Symposium: Hamster immune responses: Experimental models linking immunogenetics, oncogenesis and viral immunity. *Fed. Proc., Fed. Am. Soc. Exp. Biol.* **37,** 2023–2109.

Szentivanyi, A., and Friedman, H., eds. (1986). "Viruses, Immunity and Immunodeficiency." Plenum Press, New York (in press).

Tevethia, S. S. (1980). Immunology of simian virus 40. *In* "Viral Oncology" (G. Klein, ed.), pp. 581–601. Raven Press, New York.

Toolan, H. W. (1951). Successful subcutaneous growth and transplantation of human tumors in X-irradiated laboratory animals. *Proc. Soc. Exp. Biol. Med.* **77,** 572–578.

Toolan, H. W. (1953). Growth of human tumors in cortisone-treated laboratory animals: The possibility of obtaining permanently transplantable human tumors. *Cancer Res.* **13,** 389–394.

Toolan, H. W. (1954). Transplantable human neoplasms in cortisone-treated laboratory animals: H.S. #1; H. Ep. #1; H. Ep. #2; H. Ep. #3; and H. Emb. Rh. #1. *Cancer Res.* **14,** 660–666.

Tooze, J., ed. (1980). "DNA Tumor Viruses." Cold Spring Harbor Lab., Cold Spring Harbor, New York.

Trentin, J. J., and Homburger, F., eds. (1979). "Progress in Experimental Tumor Research," Vol. 23. Karger, Basel.

Trentin, J. J., Yabe, Y., and Taylor, G. (1962). The quest for human cancer viruses. A new approach to an old problem reveals cancer induction in hamsters by human adenovirus. *Science* **137,** 835–841.

Trujillo, J. M. (1983). Chromosomal alterations in hematologic neoplastic diseases. *Cancer Bull. Univ. Tex. M. D. Anderson Hosp. & Tumor Inst. Houston, Tex.* **35,** 119–126.

Van Hoosier, G. L., Jr., and Trentin, J. J. (1968). Reactivity of human sera with adenovirus type-12 hamster tumor antigens in the complement-fixation test. *J. Natl. Cancer Inst. (U.S.)* **40,** 249–253.

Wilson, M. G., Towner, J. W., and Fujimoto, A. (1973). Retinoblastoma and D-chromosome deletions. *Am. J. Hum. Genet.* **25,** 57–61.

Witte, S., Goldenberg, D. M., and Schricker, T. (1965). Mangel an lymphgefassen als Ursache der immunologischen Privilegierung der hamster Backentasche. *Klin. Wochenschr.* **43**(21), 1182.

Yamamoto, T., Rabinowitz, Z., and Sachs, L. (1973). Identification of the chromosomes that control malignancy. *Nature (London), New Biol.* **243,** 247–250.

Zankl, H., and Zang, K. D. (1972). Cytological and cytogenetic studies on brain tumors. IV. Identification of the missing G chromosome in human meningiomas as No. 22 by fluorescence technique. *Humangenetik* **14,** 167–169.

APPENDIX 1. SPONTANEOUS TUMOR DEVELOPMENT: BENIGN TUMORS

The classification of tumors in the appendixes is according to Stewart *et al.* (ref. 22). Included are all tumors reported in the literature as "malignant," (see Appendix 2) or those reported by histological type and site. Excluded are tumors of unspecified site, or metastatic tumors with the primary site un-

Table A.I

Benign Tumors[a]

Tissue and group	Tumor type	Site	Relative frequency	Reference
Epithelial: I	Polyp	Stomach	<10	7
		Intestine	>100	5–7,13,20,23
		Uterus	<10	3,7
	Adenoma	Harderian gland	<10	3
		Nasal cavity	<10	3
		Bile duct	<10	3
		Pancreas	<10	7,13,26
		Ovary	<10	10
		Thyroid	30–100	13,15,27
		Parathyroid	10–30	7,13
		Adrenal cortex	>100	2,3,5,7,13,15,16,24,26,27
	Adenomatosis	Lung	<10	13,20,27
	Papilloma	Skin	<10	11
		Pharynx	<10	3
		Palate	<10	3
		Stomach	30–100	3,7,15,17,20,26,27
	Keratoacanthoma	Skin	<10	13
		Vagina	<10	7,9,25
Connective: III	Hemangioma	Skin	<10	13
		Subcutis	<10	13
		Liver	10–30	3,7,13,17,23,26
		Spleen	30–100	3,6,7,13,23
	Lymphangioendothelioma	Mesentery	<10	13
	Cholangioma	Bile duct	10–30	13,20,24,26,27
	Thecoma	Ovary	10–30	5,7,13
	Fribroma	Face	<10	16
		Uterus	<10	22
	Leiomyoma	Prostate	<10	12
		Uterus	<10	3,12
Melanin-forming: IV	Cellular blue nevus	Skin	10–30	12,19
Neural: V	Pheochromocytoma	Adrenal	<10	27
	Ganglioneuroma	Adrenal	<10	12
	Schwannoma	Facial nerve	<10	25

[a]Relative frequency gives the total number of tumors reported for the specified type and site.

known. The material in this appendix is from G. L. Van Hoosier, Jr. and J. J. Trentin, Biological handbooks. III. In "Inbred and Genetically Defined Strains of Laboratory Animals" (P. L. Altman and D. D. Katz, eds.), Part 2, pp. 480–484. Fed. Am. Soc. Exp. Biol., Bethesda, Maryland, 1979. With the exception of the lymphomas reported in reference 1, most of the spontaneous tumors were observed after the animals were 1 year old. References 1, 4, 8, 9, 14, and 21 suggest the possibility of viral etiology of hamster lymphomas, skin papillomas, and melanomas. For detailed discussions of tumor development, consult references 18 and 28.

Table A.I gives a listing of benign tumors.

References to Appendix 1

1. Ambrose, K. R., and Coggin, J. H., Jr. (1975). *J. Natl. Cancer Inst. (U.S.)* **54**(4), 877–879.
2. Cox, C. B., et al. (1972). *Appl. Microbiol.* **23**, 675–678.
3. Dontenwill, W., et al. (1973). *Z. Krebsforsch.* **80**, 127–158.
4. Epstein, W. L., et al. (1968). *Nature (London)* **219**, 979–980.
5. Fortner, J. G. (1957). *Cancer* **10**, 1153–1156.
6. Fortner, J. G. (1958). *Arch. Surg. (Chicago)* **77**, 627–633.
7. Fortner, J. G. (1961). *Cancer Res.* **21**, 1491–1498.
8. Graffi, A., et al. (1968). *Br. J. Cancer* **22**, 577–581.
9. Graffi, A., et al. (1968). *J. Natl. Cancer Inst., (U.S.)* **40**, 867–868.
10. Heubner, R. J., et al. (1965). *Proc. Natl. Acad. Sci. U.S.A.* **54**, 381–388.
11. Horn, K. H., and Siewert, R. (1968). *Acta Biol. Med. Ger.* **20**, 103–110.
12. Kirkman, H. (1950). *Anat. Rec.* **106**, 227.
13. Kirkman, H., and Algard, F. T. (1968). In "The Golden Hamster" (R. A. Hoffman et al., eds.), pp. 227–240. Iowa State Univ. Press, Ames.
14. Kistler, G. S. (1975). Universität Zürich, Anatomisches Institut, Switzerland (unpublished).
15. Lee, K. Y., et al. (1963). *Proc. Soc. Exp. Biol. Med.* **114**, 579–582.
16. Muto, M. (1974). *Exp. Anim.* **23**, 100–102.
17. Porta, G. D. (1961). *Cancer Res.* **21**, 575–579.
18. Pour, P., et al. (1976). *J. Natl. Cancer Inst. (U.S.)* **56**, 931–974.
19. Sherman, J. D., et al. (1963). *Cancer Res.* **23**, 1689–1693.
20. Shubik, P., et al. (1962). *Henry Ford Hosp. Int. Symp.* **12**, 285–297.
21. Stenback, W. A., et al. (1966). *Proc. Soc. Exp. Biol. Med.* **122**, 1219–1223.
22. Stewart, H. L., et al. (1959). *U.S. Armed Forces Inst. Pathol. Fasc.* **40**, 378.
23. Stewart, S. E., and Irwin, M. (1960). *Proc. Natl. Cancer Conf.* **4**, 539–557.
24. Tomatis, L., et al. (1961). *Cancer Res.* **21**, 1513–1517.
25. Toolan, H. W. (1967). *Nature (London)* **214**, 1036.
26. Toth, B. (1967). *Cancer Res.* **27**, 1430–1442.
27. Toth, B., et al. (1961). *Cancer Res.* **21**, 1537–1541.
28. Van Hoosier, G. L., Jr. (1979). *Prog. Exp. Tumor Res.* **23**, 1–12.

APPENDIX 2. SPONTANEOUS TUMOR DEVELOPMENT: MALIGNANT TUMORS

Of the laboratory rodents, it appears that the hamster, in common with the guinea pig (ref. 3), develops fewer spontaneous malignant tumors than the mouse (ref. 36) or the rat (refs. 21, 45). Where the total number of hamsters observed was reported, the cumulative incidence of spontaneous malignant neoplasms approximated 3.7% (435 tumors in 11,792 animals). In Table A.II the total number of tumors per site is given only when it differs from the sum of the tumors of each type at the specified site; when the difference occurs, it does so because some tumors were reported only as malignant, with histological type unspecified. The material in this appendix is from G. L. Van Hoosier, Jr. and J. J. Trentin, Biological handbooks. III. In "Inbred and Genetically Defined Strains of Laboratory Animals" (P. L. Altman and D. D. Katz, eds.), Part 2, pp. 480–484. Fed. Am. Soc. Exp. Biol., Bethesda, Maryland, 1979.

Table A.II

Malignant Tumors

Tissue and group	Site	Tumor Type	No.	Reference
Epithelial: IA	Salivary gland	Undifferentiated carcinoma	1	38,39,43
		Total no. of tumors	3	
	Stomach	Carcinoma	3	9,15
		Adenocarcinoma	3	
	Intestine	Adenocarcinoma	75	10,13–15,32,37,46
	Liver and intrahepatic bile duct	Adenocarcinoma	1	9,13,15,30,32,40,44,52
		Hepatocarcinoma	6	
		Cholangiocarcinoma	11	
	Extrahepatic bile duct	Adenocarcinoma	1	15
	Pancreas	Carcinoma	3	2,13,15
		Adenocarcinoma	2	
	Kidney	Carcinoma	1	10,13,15,22,33,52,54
		Adenocarcinoma	5	
		Nephroblastoma	2	
		Renal cell carcinoma	1	
	Epididymis	Adenocarcinoma	1	16
	Prostate	Adenocarcinoma	2	15
	Cowper's gland	Cystadenocarcinoma	7	15

(continued)

Table A.II (*Continued*)

Tissue and group	Site	Tumor Type	No.	Reference
Epithelial: IA (*continued*)	Ovary	Granulosa cell	9	10,33,41,52
		Total no. of tumors	11	
	Uterus	Carcinoma	2	9,10,15,33,47,52
		Adenocarcinoma	10	
		Adenoacanthoma	1	
	Cervix	Carcinoma	3	
	Mammary gland	Adenocarcinoma	4	10,11,18
	Thyroid	Spindle cell carcinoma	6	10,15,17,33,52
		Follicular cell carcinoma	5	
		Total no. of tumors	18	
	Parathyroid	Carcinoma	2	33
	Adrenal	Carcinoma	26	9,10,13,15,23,31,51,52
		Adenocarcinoma	4	
		Pheochromocytoma	1	
Epithelial: IB	Lower respiratory tract	Carcinoma	2	10,13,33,47
		Bronchogenic adenocarcinoma	1	
		Total no. of tumors	8	
Epithelial: IC	Skin	Squamous-cell carcinoma	5	8,10,28,33
		Basal-cell carcinoma	3	
	Esophagus	Squamous-cell carcinoma	1	6,33
Lymphoid[a]: IIA	Skin	Lymphosarcoma	3	55
	Small intestine	Lymphosarcoma	79	1
	Liver	Lymphosarcoma	20	1
	Spleen	Lymphosarcoma	12	1
	Lymph node	Lymphosarcoma	113	1,9,10,12,25,29,30,37,47,51,54
	Kidney	Lymphosarcoma	18	1
Lymphoid: IIC	Lymph node	Reticulum cell sarcoma	121	5,10,13,15,23,27–29,39,47,48,50–55
Lymphoid: IID	Extramedullary	Plasma cell tumor	13	9,10,12,15,20,28,39
Connective: IIIA	Subcutis	Fibrosarcoma	1	7,33,34,54
		Leiomyosarcoma	1	
		Osteosarcoma	1	
		Undifferentiated sarcoma	1	
	Cheek pouch	Myxofibrosarcoma	2	19,20,50,54
		Myxoma	1	
		Myxoid liposarcoma	1	
		Hemangiopericytoma	1	
	Intestine	Leiomyosarcoma	4	13–15,23
		Angiosarcoma	1	
		Liposarcoma	1	
	Liver	Hemangioendothelioma	2	9
	Liver and spleen	Angiosarcoma	1	15
	Spleen	Hemangioendothelioma	3	9,31
	Urinary bladder	Leiomyosarcoma	1	51
	Testis	Fibrosarcoma	1	55
	Uterus	Leiomyosarcoma	4	15,33,52
	Internal chest wall	Hemangioendothelioma	1	55
	Rib	Osteosarcoma	1	18
	Humerus	Chondrosarcoma	1	49
	Femur	Osteosarcoma	2	10,55
	Tendon sheath	Giant-cell tumor	2	42
Connective: IIIB	Subcutis	Sarcoma	4	10,35,53
	Pleura	Sarcoma	2	10,50
	Peritoneum	Endothelioma	1	10,33
		Sarcoma	1	
Melanin-forming: IV	Skin	—	30[b]	4,13,15,16,24,26,33,51,52

[a]Approximately 75% of lymphosarcomas reported resulted from an epizootic of horizontally transmitted lymphomas in one hamster colony (ref. 1).

[b]Incidence was 0.4 (0.1–2.76)%; occurrence by sex—male/female = 10:1.

References to Appendix 2

1. Ambrose, K. R., and Coggin, J. H., Jr. (1975). *J. Natl. Cancer Inst. (U.S.)* **54**(4), 877–879.
2. Ashbel, R. (1945). *Nature (London)* **155**, 607.
3. Blumenthal, H. T., and Rogers, J. B. (1965). *In* "The Pathology of Laboratory Animals" (W. E. Ribelin and J. R. McCoy, eds.), pp. 183–209. Thomas, Springfield, Illinois.
4. Bomirski, A., *et al.* (1962). *Acta Unio Int. Cancrum* **18**, 178–180.
5. Brindley, D. C., and Banfield, W. G. (1961). *J. Natl. Cancer Inst. (U.S).* **26**, 949–954.
6. Chesterman, F. C. (1963). *Proc. Int. Cancer Congr., 8th, 1962,* p. 486.
7. Cox, C. B., *et al.* (1972). *Appl. Microbiol.* **23**, 675–678.
8. Crabb, E. D., and Kelsall, M. A. (1952). *Cancer Res.* **12**, 256.
9. Dontenwill, W., *et al.* (1973). *Z. Krebsforsch.* **80**, 127–158.
10. Dunham, L. J., and Herrold, K. M. (1962). *J. Natl. Cancer Inst. (U.S.)* **29**, 1047–1067.
11. Eddy, B. E., *et al.* (1958). *J. Natl. Cancer Inst. (U.S.)* **20**, 747–762.
12. Finkel, M. P., *et al.* (1968). *Proc. Natl. Acad. Sci. U.S.A.* **60**, 1223–1230.
13. Fortner, J. G. (1957). *Cancer (Philadelphia)* **10**, 1153–1156.
14. Fortner, J. G. (1958). *Arch. Surg. (Chicago)* **77**, 627 633.
15. Fortner, J. G. (1961). *Cancer Res.* **21**, 1491–1498.
16. Fortner, J. G., and Allen, A. C. (1958). *Cancer Res.* **18**, 98–104.
17. Fortner, J. G. (1960). *Endocrinology (Baltimore)* **66**, 364–376.
18. Fortner, J. G., *et al.* (1961). *Cancer Res.* **21**, 199–229.
19. Friedell, G. H., *et al.* (1960). *Transplant. Bull.* **7**, 97–100.
20. Garcia, H., *et al.* (1961). *J. Natl. Cancer Inst. (U.S.)* **27**, 1323–1333.
21. Gilbert, C., and Gillman, J. (1958). *S. Afr. J. Med. Sci.* **23**, 257–272.
22. Girardi, A. J., *et al.* (1962). *Proc. Soc. Exp. Biol. Med.* **111**, 84–93.
23. Girardi, A. J., *et al.* (1964). *Proc. Soc. Exp. Biol. Med.* **115**, 1141–1150.
24. Greene, H. S. N. (1958). *Cancer Res.* **18**, 422–425.
25. Greene, H. S. N., and Harvey, E. K. (1960). *Cancer Res.* **20**, 1094–1100.
26. Gye, W. E., and Foulds, L. (1939). *Am. J. Cancer* **35**, 108.
27. Haemmerli, G., *et al.* (1966). *Int. . Cancer* **1**, 599–612.
28. Handler, A. H. (1965). *In* "The Pathology of Laboratory Animals" (W. E. Ribelin and J. R. McCoy, eds.), pp. 210–240. Thomas, Springfield, Illinois.
29. Handler, A. H., *et al.* (1960). *Acta Unio Int. Cancrum* **16**, 1175–1177.
30. Horn, K. H., and Siewert, R. (1968). *Acta Biol. Med. Ger.* **20**, 103–110.
31. Kesterson, J. W., and Carlton, W. W. (1970). *Lab. Anim. Care* **20**, 220–225.
32. Kirkman, H. (1950). *Anat. Rec.* **106**, 227.
33. Kirkman, H., and Algard, F. T. (1968). *In* "The Golden Hamster" (R. A. Hoffman *et al.,* eds.), pp. 227–240. Iowa State Univ. Press, Ames.
34. Klein, M. (1961). *J. Natl. Cancer Inst. (U.S.)* **26**, 1381–1390.
35. Lee, K. Y., *et al.* (1963). *Proc. Soc. Exp. Biol. Med.* **114**, 579–582.
36. Murphy, E. D. (1966). *In* "Biology of the Laboratory Mouse" (E. L. Green, ed.), 2nd ed., pp. 521–562. McGraw-Hill, New York.
37. Muto, M. (1974). *Exp. Anim.* **23**, 100–102.
38. Patterson, W. B. (1963). *Acta Unio Int. Cancrum* **19**, 640–643.
39. Porta, G. D. (1961). *Cancer Res.* **21**, 575–579.
40. Porta, G. D., *et al.* (1959). *J. Natl. Cancer Inst. (U.S.)* **22**, 463–471.
41. Rolle, G. K., and Charipper, H. A. (1949). *Anat. Rec.* **105**, 281–297.
42. Ruffolo, P. R., and Kirkman, H. (1965). *Br. J. Cancer* **19**, 573–580.
43. Sherman, J. D., *et al.* (1963). *Cancer Res.* **23**, 1689–1693.
44. Shubik, P., *et al.* (1962). *Henry Ford Hosp. Int. Symp.* **12**, 285–297.
45. Snell, K. C. (1965). *In* "The Pathology of Laboratory Animals" (W. E. Ribelin and J. R. McCoy, eds.), pp. 241–302. Thomas Springfield, Illinois.
46. Stewart, H. L., *et al.* (1959). *U.S. Armed Forces Inst. Pathol. Fasc.* **40**, 378.
47. Stewart, S. E., and Irwin, M. (1960). *Proc., Natl. Cancer Conf.* **4**, 539–557.
48. Strauli, V. P. (1962). *Pathol. Microbiol.* **25**, 301–305.
49. Taylor, D. O. N. (1968). *Cancer Res.* **28**, 2051–2055.
50. Tomatis, L., *et al.* (1961). *Cancer Res.* **21**, 1513–1517.
51. Toolan, H. W. (1967). *Nature (London)* **214**, 1036.
52. Toth, B. (1967). *Cancer Res.* **27**, 1430–1442.
53. Toth, B., *et al.* (1961). *Cancer Res.* **21**, 1537–1541.
54. Van Hoosier, G. L., Jr., *et al.* (1971). *Defining Lab. Anim., Symp. 4th, 1969,* pp. 450–473.
55. Yabe, Y., *et al.* (1972). *Gann* **63**, 329–336.

APPENDIX 3. INDUCED TUMORS: VIRAL ONCOGENESIS

Newborn hamsters are probably the most susceptible animals to the oncogenic viruses of other species. They are therefore the most sensitive *in vivo* test system for the oncogenic viruses of other mammalian species. Newborn hamsters injected with certain human viruses, such as reovirus types 1 and 2 and herpes simplex viruses, may undergo acute morbidity and mortality (ref. 21). A transplanted hamster sarcoma became a carrier of lymphocytic choriomeningitis virus of mice and transmitted the infection to other hamsters and humans (ref. 11). In addition, hamsters have relatively few indigenous viruses (ref. 17), and a low incidence of spontaneous tumors (see Appendixes 1 and 2 and consult ref. 25), making them ideal assay animals for oncogenic viruses. Although most hamster colonies harbor certain viruses "borrowed" from other species—such as pneumonia virus of mice, simian virus type 5, and parainfluenza 1 (Sendai virus)—specific pathogen-free (SPF) hamsters uncontaminated by these viruses are available, both random-bred and inbred (ref. 24). The material in this appendix is from J. J. Trentin, Biological handbooks. III. *In* "Inbred and Genetically Defined Strains of Laboratory Animals" (P. L. Altman and D. D. Katz, eds.), Part 2, pp. 484–488. Fed. Am. Soc. Exp. Biol., Bethesda, Maryland, 1979.

In Table A.III the following notations are used. Virus titer or dose: $TCID_{50}$, tissue culture infective dose, 50%; $MTCID_{100}$, minimum tissue culture infective dose, 100%; GMK indicates assayed in GMK cell cultures; HeLa indicates assayed in HeLa cell cultures; HEK indicates assayed in HEK cell cultures; $CELD_{50}$, chicken egg lethal dose, 50%; PFU, plaque-forming units. Route: ip, intraperitoneal; iv, intravenous; sc, subcutaneous. Hamster: Tumor incidence of the concurrent controls is not given, since in most cases it was zero; in some cases, where controls were kept beyond a year, they showed the low incidence of spontaneous latearising tumors typical of untreated hamsters (see Appendixes 1 and 2 and consult ref. 25).

Table A.III

Viral Oncogenesis

Virus name, (natural host), and modifying factors	Virus titer or dose (route)	Hamster Age at injection[a]	Hamster Tumorous/ surviving no. (%)	Tumor Type and site	Tumor Latent period[a]	Reference
Papovaviruses						
JC virus (human)[b]	10^6 TCID$_{50}$ (intracranial, sc)	<1	52/63 (83)	Multiple malignant brain tumors: glioblastomas (most), medulloblastomas, unclassified primitive tumors, papillary ependymomas (2 only); no sc tumors, no metastases	4–6 mo	26
SV40 virus (monkey)						
Effect of virus dose	32 TCID$_{50}$(sc)	Newborn	0/25 (0)	—	—	5
	3200 TCID$_{50}$ (sc)	Newborn	5/28 (18)	Malignant sc sarcomas	344 (262–416)	5
	320,000 TCID$_{50}$ (sc)	Newborn	52/54 (96)	Malignant sc sarcomas	201 (115–403)	5
Effect of route of injection	10^7 TCID$_{50}$ (sc)	Newborn	17/22 (77)	Malignant sc sarcomas	—	5
	(ip)	Newborn	0/25 (0)	—	—	
	(intrapulmonary)	Newborn	2/18 (11)	Malignant sc sarcomas	—	
	10^7 TCID$_{50}$[c] (intracerebral)	<1	4/9 (44)	Ependymomas	(100–124)	9
	10^8 TCID$_{50}$ (iv)	(17–18)	90/95 (95)	Reticulum cell sarcomas, 78% Osteogenic sarcomas, 72% sc Sarcomas, 6% Lymphosarcomas, 3% Malignant adrenal tumors, 3%	(3–6 mo)	2
	$10^{8.5}$TCID$_{50}$ (iv)	(21–22)	125/143 (87)	Reticulum cell sarcomas, 72% Osteogenic sarcomas, 55% sc Sarcomas, 4% Lymphosarcomas, 4% Lymphocytic leukemia, 1%	(4–6 mo)	2
Effect of hamster age	0.2 ml of $10^{-5.75}$ titer (GMK) (sc)	Newborn	52/54 (96)	Malignant sc sarcomas	201 (115–403)	5
	0.2 ml of $10^{-5.5}$ titer (GMK) (sc)	7	24/40 (60)	Malignant sc sarcomas	314 (178–487)	5
	0.2 ml of 10^{-6} titer (GMK) (sc)	30	6/26 (23)	Malignant sc sarcomas	360 (188–494)	5
		90	0/35 (0)	—	—	5
Polyomavirus (mouse)	0.2 ml unfiltered tissue culture fluid (sc)	<1	34/40 (85)	Sarcoma of kidney, sc tissue, heart, stomach, or intestines; hemangiomas of liver, ovaries, and lung[d]	(4–28 wk)	3
Adenoviruses						
Human adenoviruses[e]						
Types 1, 3, 7, 8 (?), 11, 14, 16, 21, 24 (?)	—	<1	Weakly oncogenic	Most were undifferentiated sarcomas at site of injection; some were lymphosarcomas at site of injection or remote site	—	4, 7, 10, 20, 22
Types 12, 18, 31	—	<1	Strongly oncogenic	Undifferentiated sarcomas at site of injection	—	6, 15, 19

(continued)

Table A.III (Continued)

Virus name, (natural host), and modifying factors	Virus titer or dose (route)	Hamster Age at injection[a]	Tumorous/ surviving no. (%)	Tumor Type and site	Latent period[a]	Reference
Type 12						
Effect of virus dose	0.5 MTCID$_{100}$ (HeLa) (intrapulmonary)	<1	1/5 (20)	Undifferentiated sarcomas at site of injection	73	27
	50 MTCID$_{100}$ (HeLa) (intrapulmonary)	<1	16/19 (84)	Undifferentiated sarcomas at site of injection	(33–91)	27
	500 MTCID$_{100}$ (HeLa) (intrapulmonary)	<1	26/27 (96)	Undifferentiated sarcomas at site of injection	59 (35–157)	27
Effect of route of injection	50 MTCID$_{100}$ (HeLa) (ip)	<1	5/6 (83)	Undifferentiated sarcomas at site of injection and/or in liver	—	28
	500 MTCID$_{100}$ (HeLa) (sc)	<1	3/3 (100)	Undifferentiated sarcomas at site of injection and/or in liver	(39–98)	28
	(intrapleural)	<1	6/6 (100)	Undifferentiated sarcomas at site of injection and/or in liver	(29–45)	
	(intrapulmonary)	<1	4/5 (80)	Undifferentiated sarcomas at site of injection and/or in liver	(42–45)	
	(iv)	<1	2/3 (67)	Undifferentiated sarcomas at site of injection and/or in liver	(50, 70)	
	1000 MTCID$_{100}$ (HeLa) (iv)	<1	7/11 (64)	Undifferentiated sarcomas at site of injection and/or in liver	(36–103)	28
Effect of hamster age	500 MTCID$_{100}$ (HeLa) (intrapulmonary)	<1	26/27 (96)	Undifferentiated sarcomas at site of injection	59 (35–157)	27
		5	5/7 (71)	Undifferentiated sarcomas at site of injection	57 (41–78)	
		8	2/9 (22)	Undifferentiated sarcomas at site of injection	86 (70, 102)	
	1000 MTCID$_{100}$ (HeLa) (intrapulmonary)	14	1/8 (12)	Undifferentiated sarcomas at site of injection	56	27
		(21–77)	0/20 (0)	—	—	
Effect of sex: ♂	10^6 TCID$_{50}$ (HEK) (sc)	<1	19/17 (27)	—	≤120	29
♀	10^6 TCID$_{50}$ (HEK) (sc)	<1	47/82 (57)	—	≤120	29
Effect of hamster strain						
Noninbred	10^6 TCID$_{50}$ (HEK) (sc or ip)	<1	26/29 (90)	—	61	23
Inbred: LSH	10^6 TCID$_{50}$ (HEK) (sc or ip)	<1	15/31 (48)	—	51	23
Simian adenoviruses[f]						
SA7 (Cercopithecus monkey)	0.05 ml undiluted virus (sc)[g]	<1	46/47 (98)	—	(28–48)	8
Effect of hamster strain						
Noninbred	10$^{7.5}$ TCID$_{50}$ (sc or ip)	<1	21/23 (91)	—	43	23
Inbred: LSH	10$^{7.5}$ TCID$_{50}$ (sc or ip)	<1	32/34 (94)	—	49	23 8
(Rhesus or cynomolgus monkey)						
SV20	0.05 ml undiluted virus (sc)[g]	<1	54/141 (38)	—	(40–211)	

(continued)

Table A.III (Continued)

Virus name, (natural host), and modifying factors	Virus titer or dose (route)	Hamster Age at injection[a]	Tumorous/ surviving no. (%)	Tumor Type and site	Latent period[a]	Reference
SV33	0.05 ml undiluted virus (sc)[g]	<1	2/25 (8)	—	(82, 229)	
SV34	0.05 ml undiluted virus (sc)[g]	<1	3/48 (6)	—	(110–330)	
SV37	0.05 ml undiluted virus (sc)[g]	<1	1/20 (5)	—	231	
SV38	0.05 ml undiluted virus[g]	<1	3/14 (21)	—	(154–280)	
Bovine adenovirus Types 1, 2, 4–10	0.2 ml virus[g] (sc)	Newborn	0/234 (0)	—	>1 yr[h]	13
Type 3	($10^{4.0}$–$10^{5.2}$) (sc, ip, intrathoracic, intracerebral)	<1	22/45 (49)	Firm or cystic undifferentiated sarcomas	(24–67)	1
	2.3×10^5 PFU (sc)	Newborn	17/19 (90)	—	48	14
Chicken adenovirus: CELO virus	10^6 $CELD_{50}$[i] (sc)	<1	23/69 (33)	Well-differentiated fibrosarcomas	(88–195)	16
	($10^{6.6}$–10^8) $TCID_{50}$ or PFU (sc or ip or intracerebral)	<1[j]	130/277 (47)	Most were fibrosarcomas at site of injection; also 3 brain tumors (glioma, papilloma, and ependymoma), and 12 liver tumors (5 hepatocellular carcinomas, 5 hepatic adenocarcinomas, 2 hepatic sarcomas)	—	18

[a]Days, unless otherwise specified. Under "Latent period," values in parentheses are ranges.
[b]Isolated from brain of a case of progressive multiple leukoencephalopathy.
[c]Stated as "0.02 ml of $18^{8.7}$ $TCID_{50}$/ml" in reference 9.
[d]Range of tumor types more limited than in mouse.
[e]Of the 31 human adenovirus serotypes, all those not listed were nononcogenic in newborn hamsters, but some transform rat embryo cells or the NIL-2 hamster cell line in vitro (refs. 12, 20).
[f]Twelve additional serotypes tested were nononcogenic (ref. 8).
[g]Titer unspecified.
[h]No tumors after 1 yr of observation.
[i]Stated as "0.1 ml of 10^7 $CELD_{50}$/ml" in reference 16.
[j]Both noninbred and inbred LSH.

References to Appendix 3

1. Darbyshire, J. H. (1966). *Nature (London)* **211**, 102.
2. Diamondopoulos, G. T. (1973). *J. Natl. Cancer Inst. (U.S.)* **50**, 1347–1365.
3. Eddy, B. E., et al. (1958). *J. Natl. Cancer Inst. (U.S.)* **20**, 747–756.
4. Fujinaga, K., and Green, M. (1967). *J. Virol.* **1**, 576–582.
5. Girardi, A. J., et al. (1963). *Proc. Soc. Exp. Biol. Med.* **112**, 662–667.
6. Huebner, R. J., et al. (1962). *Proc. Natl. Acad. Sci. U.S.A.* **48**, 2051–2058.
7. Huebner, R. J., et al. (1965). *Proc. Natl. Acad. Sci. U.S.A.* **54**, 381–388.
8. Hull, R. N., et al. (1965). *Science* **150**, 1044–1046.
9. Kirchstein, R. L., and Gerber, P. (1962). *Nature (London)* **195**, 299–300.
10. Larson, V. M., et al. (1965). *Proc. Soc. Exp. Biol. Med.* **118**, 15–24.
11. Lewis, A. M., et al. (1965). *Science* **150**, 363–364.
12. McAllister, R. M., et al. (1969). *J. Natl. Cancer Inst. (U.S.)* **43**, 917–923.
13. Mohanty, S. B. (1971). *Am. J. Vet. Res.* **32**, 1899–1905.
14. Panigrahy, B., et al. (1975). *J. Natl. Cancer Inst. (U.S.)* **54**, 449–451.
15. Pereira, M. A., et al. (1965). *Lancet* **1**, 21–23.
16. Sarma, P. S., et al. (1965). *Science* **149**, 1108.
17. Stenback, W. A., et al. (1970). *Bibl. Haematol. (Basel)* **36**, 559–565.
18. Stenback, W. A., et al. (1973). *J. Natl. Cancer Inst. (U.S.)* **50**, 963–970.
19. Trentin, J. J., et al. (1962). *Science* **137**, 835–841.
20. Trentin, J. J., et al. (1968). *Proc. Soc. Exp. Biol. Med.* **127**, 683–689.
21. Trentin, J. J., et al. (1969). *Proc. Soc. Exp. Biol. Med.* **132**, 912–915.
22. Van Hoosier, G. L., Jr., et al. (1968). *Proc. Soc. Exp. Biol. Med.* **128**, 467–469.
23. Van Hoosier, G. L., Jr., et al. (1970). *Proc. Soc. Exp. Biol. Med.* **134**, 427–429.
24. Van Hoosier, G. L., Jr., et al. (1970). *Lab. Anim. Care* **20**, 232–237.

25. Van Hoosier, G. L., Jr., *et al.* (1971). *Defining Lab. Anim., Symp., 4th, 1971*, pp. 450–473.
26. Walker, D. L., *et al.* (1973). *Science* **181**, 674–676.
27. Yabe, Y., *et al.* (1962). *Proc. Soc. Exp. Biol. Med.* **111**, 343–344.
28. Yabe, Y., *et al.* (1963). *Proc. Soc. Exp. Biol. Med.* **113**, 221–224.
29. Yohn, D. S., *et al.* (1965). *J. Natl. Cancer Inst. (U.S.)* **35**, 617–624.

APPENDIX 4. INDUCED TUMORS: SURVIVAL AND TUMOR INCIDENCE AFTER IRRADIATION

The hamster is relatively resistant to radiation-induced tumors. Table A.IV lists some data pertinent to survival and tumor incidence of irradiated hamsters. The material in this appendix is from J. J. Trentin, Biological handbooks. III. *In* "Inbred and Genetically Defined Strains of Laboratory Animals" (P. L. Altman and D. D. Katz, eds.), Part 2, pp. 484–488. Fed. Am. Soc. Exp. Biol., Bethesda, Maryland, 1979.

Reference to Appendix 4

Stenback, W. A., *et al.* (1979). *Prog. Exp. Tumor Res.* **23**, 89–99.

Table A.IV

Survival and Tumor Incidence after Irradiation

X Radiation[a]			Hamster no.[b]		Days survival from first irradiation[c]	Tumors		
Dose rate (R/wk)	Duration (wk)	Total (R)	Total at beginning	Surviving last irradiation		%	Total no.	Type and site
20	10	200	10	10	(241–442)	0	0	—
	20	400	10	10	(206–423)	0	0	—
	30	600	20	20	(284–485)	5	1	Undifferentiated neoplasm of spleen and liver
40	10	400	10	10	(122–449)	10	1	Adenocarcinoma of liver
	20	800	20	19	(216–440)	0	0	—
	30	1600	20	19	(233–492)	0	0	—
80	10	800	10	10	(108–340)	0	0	—
	20	1200	15	15	(161–423)	13	1	Reticulum cell sarcoma of spleen and liver
							1	Melanoma of neck
	30	2400	25	24	(267–333)	13	1	Undifferentiated neoplasm of spleen
							1	Squamous cell carcinoma of lung
							1	Undiagnosed transplantable neoplasm of spleen
150	10	1500	15	15	(172–429)	13	2	Reticulum cell sarcoma of spleen
	20	3000	45	33	(149–269)	12	1	Small-cell sarcoma of spleen
							1	Small-cell sarcoma of spleen and liver
							2	Undifferentiated neoplasm of spleen and liver

[a]250 kVp X irradiation, whole body, once a week. R, roentgen.
[b]Equal numbers of males and females were used.
[c]Values in parentheses are ranges.

Chapter 13

Experimental Biology: Use in Immunobiology

J. Wayne Streilein

I.	Introduction	215
	A. Origin and Genetics of Hamsters as They Relate to Immune Responsiveness	216
II.	Ontogeny of Immunity	216
	A. Thymus and T Cells	217
III.	Immunochemistry	217
IV.	Immunogenetics	218
V.	Lymphoreticular Cells	219
	A. T Lymphocytes	219
	B. B Lymphocytes	220
	C. Natural Killer Cells	220
	D. Macrophages and Dendritic Cells	221
VI.	Recent Advances in Hamster Immunobiology	221
	A. Contact Hypersensitivity	221
	B. Pulmonary Immunity	221
	C. Tumor Immunity	222
	References	223

I. INTRODUCTION

The Syrian hamster has been used as an experimental animal for more than 50 years. It was early appreciated that the hamster would prove to be important in the study of *Leishmania* infestations, suggesting that proper study of the immune system of hamsters might be interesting. During this period of time, a great deal of experimental work has examined the immunobiology of this interesting species, and as a consequence a number of unique and important contributions have been made to our general understanding of immunobiology. These contributions have largely sprung from unique properties of the hamster. During the same span of time, orders of magnitude more research have been conducted on the immune systems of mice, rats, rabbits, chickens, and primates, including humans. This chapter will highlight those unique contributions made from the study of Syrian hamsters to the general field of immunobiology. An attempt will be made to indicate those features which have yet to be explored fully and may afford additional contributions to our understanding.

A. Origin and Genetics of Hamsters as They Relate to Immune Responsiveness

Elsewhere in this text, a detailed description of the initial capture, husbandry, and development of laboratory hamster strains is described. It is important for this chapter merely to highlight those aspects of the breeding and husbandry of hamsters which relates to their unique immunobiological properties. It has been proposed that an important feature of the Syrian hamster is the unusual nature of its major histocompatibility complex (MHC) and the impact this genetic region has on the immune responses of hamsters (Streilein and Duncan, 1983). Thus, it is pertinent to point out that to the present all experiments conducted on Syrian hamsters have utilized progeny from three independent captures of wild hamsters in Syria. The vast majority of these experiments have been conducted on animals descended from three littermates caught in 1930 (Billingham, 1978). In fact, all inbred hamster strains available in the United States, Japan, and Europe derive from this very restricted gene pool. In 1970, 11 wild hamsters were caught in Syria (Murphy, 1971). Experiments conducted by Murphy and colleagues and Streilein and colleagues have utilized the progeny from this second genetic sampling. Finally, in 1978, Duncan captured two additional wild hamsters in Syria, the genes of one of which were preserved in progeny that were used in subsequent immunogenetic testing. Thus, the hamster gene pool has been sampled on three separate occasions covering 50 years, and the sites of capture are known to span hundreds of kilometers in the Syrian desert. Since the current environmental range of the hamster is thought to be restricted to the Syrian desert, it is likely that a significant and presumably adequate proportion of the hamster gene pool has been subjected to experimental scrutiny.

The study of immunobiology was transformed into the 1940s and 1950s by the introduction of genetically defined inbred strains of various laboratory animals, especially rodents. During that period of time, closed breeding colonies of hamsters were located in the United States, England, Germany, France, and Japan. Billingham and colleagues (1960) set up hamsters in inbreeding protocols, the results of which were the production of three isogenic lines. Virtually all animals used as inbred hamsters in the United States today are derived from the pioneering work of these investigators. Inbred lines have also been developed in France, chiefly under the influence of Dr. Christiane de Vaux Saint Cyr; several inbred strains have been produced and are currently maintained in Japan. The pedigrees of all of these inbred strains can be traced to the 1930 capture of wild hamsters. A number of new inbred lines were also developed from the 1970 catch of hamsters, and these lines proved to be critical to recent studies describing hamster immunogenetics. However, this colony of inbred hamsters was decimated by the disease known as infectious hamster lymphoma (Coggin et al., 1978, 1983), and the stocks are no longer available for study. At the present time, therefore, the inbred lines originally derived from the 1930 catch, which are now located in United States, France, England, and Japan, are the only genetic stocks available for study.

Within the past decade two international symposia have been held in Dallas, Texas pertaining to the immunobiology of the Syrian hamster (Streilein, 1978, 1981). Numerous presentations at these symposia summarized the current status with respect to various components of the hamster immune response. Accordingly, it is beyond the scope of this chapter to recapitulate those summary data, and the reader is referred to the bibliography, where these specific references are listed. The mission of this chapter, therefore, is 2-fold: to summarize briefly the hamster immune system as it is currently understood, and to emphasize important observations made concerning hamster immunobiology since 1980.

II. ONTOGENY OF IMMUNITY

The extremely short gestation period of the Syrian hamster—approximately 16 days postcoitus—equips the newborn Syrian hamster with a very immature immune apparatus (Solomon, 1978). While postnatal immaturity is shared by other rodents and mammals, it is extreme in the hamster. The first evidence of an epithelial anlage forming the rudimentary thymic gland is observed on day 9 of gestation (Adner et al., 1965; Linna, 1968; Solomon et al., 1972). By the day of birth (day 16) the thymus gland is homogeneously infiltrated with lymphocytes, and at that time tiny structures barely identifiable as lymph nodes and spleen are observed in the periphery. In the spleen, virtually all of the cellular elements appear to be hematopoietic rather than lymphocytic. Differentiation of the thymus into a cortex and medulla is delayed until 3 days after birth. The typical adult mammalian morphology is not achieved until day 25, fully 9 days after birth. Thus, the newborn hamster is forced to survive in an alien environment for at least 9 days without morphological evidence of a completely formed immune system.

The delay in anatomical maturation of the immunological organs of the hamster is matched by a comparable delay in maturation of immune responsiveness. Mitogenically active thymocytes are first detectable immediately after birth (Solomon, 1978). Cells able to respond in a mixed lymphocyte reaction are not observed within the thymus until 2 days later. Cells capable of responding to phytohemagglutinin (PHA) appear for the first time in the spleen on day 19 after birth. Immunization with a variety of antigens, most notably sheep erythrocytes (SRBC), indicates that the first evidence of antibody-forming cells can be detected in the hamster spleen on the sixth postnatal day. Earliest evidence that antibody is secreted into the serum is not observed until 8 days later (Young et al., 1963). However, the ability to reject skin grafts emerges on the eighth postnatal day, which coincides with the end of the "tolerance-responsive" interval (Billingham and Silvers, 1964).

A. Thymus and T Cells

A number of publications in the 1960s documented several unusual features of the development of the hamster's thymus and T-cell system: (1) autoradiographic evidence indicated that few lymphocytes migrate from the thymus to the spleen in the first 3 weeks after birth (Linna, 1968), and (2) thymectomy as late as 4 weeks after birth still produces a wasting syndrome similar to that produced in mice by thymectomy within the first 2 days of birth (Adner *et al.*, 1965).

Within the last decade serious questions have been raised about the capacity of hamsters to generate either cytotoxic T cells or suppressor T cells in response to immunological challenge (Streilein and Duncan, 1983). These observations instigated a renewed study of the development of the hamster's thymus and T cells on the part of Dr. Pamela Witte, at the University of Texas Health Science Center at Dallas (Witte *et al.*, 1980). Witte undertook a detailed electron-microscopic and light-microscopic study of the hamster thymus during development. Making allowances for differences in rate of fetal development, and comparing hamsters with mice, she observed that both species displayed lymphocyte infiltration within the reticular epithelial component of the thymus at comparable developmental times. However, the differentiation of thymus gland into a cortex and medulla, which occurs prior to birth in the mouse, is delayed in the hamster, considerably beyond the time at which it takes place in the mouse. Moreover, a full adult morphology of the hamster thymus is not acquired until day 25 postfertilization, as compared to day 20 in the mouse.

An intriguing component of Witte's work was the observation that at no time during the development of the hamster thymus could C-type virus particles be identified within either the epithelial cells or the proliferating thymocytes. Since the expression of C-type particles appears to be characteristic (Jordan, 1974; Reisenfeld and Alm, 1977) of rodent (mice and rats) and human thymuses during ontogeny, it is surprising that this feature should be missing from the hamster thymus. It has been proposed that retroviruses may play a role in embryological development (Jaenisch and Berns, 1977). Thus, development of the hamster thymus may be a fruitful area for research into the possible relationships between expression of retroviral sequences and ontogeny.

The effects of neonatal thymectomy on subsequent immune capability of hamsters are profound. Johnson and colleagues (1975) have shown that subacute sclerosing panencephalitis (SSPE), a disease that hamsters typically develop following infection with a hamster-adapted measles virus, is prevented in animals that have been neonatally thymectomized. In thymec-

tomized animals, complete virus particles are formed and they die within 8–17 days from the ravages of central nervous system viral infection. Similarly, Cook *et al.* (1979) have shown that acquisition of resistance to tumors after injection of adenovirus-transformed cell lines, which occurs during the first 21 days of life in normal hamsters, is prevented if thymectomy is carried out shortly after birth. In the Johnson studies, it was proposed that in the absence of a thymic influence, there was no production of antibody which causes alteration of the virus into an SSPE-inducing form. The studies of Cook and Lewis suggest that the acquisition of resistance to virus-induced tumors is thymus dependent and perhaps even T-cell mediated.

III. IMMUNOCHEMISTRY

The immunoglobulins produced by Syrian hamsters appear largely to be representative of those that have been described in other mammals. In the late 1960s and during the 1970s several laboratories participated in the description of hamster serum immunoglobulins (Coe, 1978). IgM, and IgG_1 and IgG_2 isotypes were readily identified. IgA was also detected, first from gastrointestinal secretions and later from the serum. The first documentation of hamster IgE was reported in 1980 (Sullivan *et al.*, 1980). Thus, all of the major forms of secreted immunoglobulins seem to be produced by the Syrian hamster.

The existence and identification of a hamster IgD equivalent is still in question. A candidate molecule has been described on the surface of certain hamster B lymphocytes (Robles *et al.*, 1980). Recently, as a portion of her doctoral dissertation, Kathy McGuire has used mouse immunoglobulin heavy-chain complex-specific genomic and cDNA probes to characterize corresponding loci in the hamster genome (McGuire, 1985). Her studies reveal that the heavy-chain gene complex of this species is very similar to that of the mouse, confirming the presence of J_H-like sequences as well as mu switch recombination sequences. The mu (μ) heavy-chain constant-region gene was cloned and sequenced. It was found to share extensive homology with the corresponding mouse gene. However, attempts to identify a δ-chain gene failed.

Descriptions of our current knowledge concerning hamster complement components and acute-phase reactants are included in the bibliography (Barta *et al.*, 1980). The most interesting recent observation in this general area has been in the description of the so-called female protein found in the serum of Syrian hamsters (Coe, 1980). This sex-limited serum protein has been sequenced, its gene identified, and studies are currently under way to understand its expression. It is probable that this protein is in some way related to the amyloid substance found in chronic inflammatory states (Coe and Ross, 1985).

The identification, description, isolation, and molecular study of factors released by activated lymphocytes is a rapidly moving field in immunobiology. Studies of this type are also in progress in the hamster, although much remains to be done. Lymphotoxin, a lymphokine that nonspecifically kills target cells, has been studied in relation to its antitumor properties in the hamster (Evans and DiPaolo, 1981; DiPaolo et al., 1984). Hamster interferon has also been partially purified and characterized (Bollin and Sulkowski, 1981).

IV. IMMUNOGENETICS

In the strict sense in which the term is employed in this section of this chapter, immunogenetics refers to those genes which encode cell surface molecules that function as histocompatibility antigens. By definition, at least two allelic forms of the gene must exist within a breeding population in order for it to be detected. Since the discovery of histocompatibility genes by Gorer (1937), much has been learned about these genes and their products in many species. Predominant attention has been directed to the unique chromosomal segment designated as the major histocompatibility complex (MHC). Numerous and extensive reviews of the MHC in mouse and human and other species have been written (Gotze, 1977; Klein et al., 1981). It is against this background of information concerning the MHC that Syrian hamsters have proved to be of more than passing interest.

With the production of inbred lines of Syrian hamsters in the late 1950s and early 1960s, it became possible to study the nature of histocompatibility systems in this species. Several distinctive and unusual features were immediately apparent. First, the number of genetic loci encoding for transplantation antigens revealed in skin grafting was very few, the number being placed somewhere between two and three (Billingham and Hildemann, 1958; Billingham et al., 1960). In addition, the putative hamster major histocompatibility locus appeared to possess only two alleles. Most surprisingly, immunization failed to demonstrate the production of alloantibodies in this species (Palm et al., 1967). Two alternative possibilities were raised to account for this spectrum of findings. On the one hand, it was proposed that hamsters had an anomalous MHC which was pauci polymorphic, and serologically silent (Duncan and Streilein, 1977, 1978a,b). On the other hand, it was suggested that the apparent lack of extensive polymorphism reflected an inadequate sampling of the wild hamster gene pool, since all animals examined to that point derived from the three littermates caught in Aleppo, Syria in 1930.

To address this important question, new inbred strains of hamster were produced from independent captures of wild animals in Syria in 1970 and 1978. By a series of cross-alloimmunizations among these strains, it became possible to define the Syrian hamster MHC in terms of diverse functional properties, (in vivo, in vitro), serologically, immunochemically, and most recently at the molecular genetic level (Streilein and Duncan, 1979; Duncan and Streilein, 1981; Phillips et al., 1981). The MHC in the hamster has been designated *Hm-1* (Duncan and Streilein, 1977). As in other mammalian MHCs, genes within this region encode two different types of cell surface molecules. One set of molecules, the hamster homologs of class II MHC molecules in other species, are encoded by four distinct loci (Phillips et al., 1981; Streilein et al., 1981; Sung et al., 1982). Two loci encode α chains and the other two encode β chains. One pair of α/β-chain loci contain genes which are polymorphic. In fact, at the serological, immunochemical, and molecular genetic levels, extensive polymorphism has been described for this paired locus. The products appear to be the equivalents of the murine I-A set of molecules. At least six allelic forms have been described among the hamsters that have been studied to date (Streilein et al., 1981), and the alloreactivities initiated by these alleles account for essentially all of the observed MHC-dependent alloreactivity in this species: acute skin graft rejection, graft versus host reactivity, mixed lymphocyte reactivity, alloantibody production. The other set of class II loci, also composed of genes encoding α and β chains, are monomorphic among all the hamsters that have been studied so far; they appear to be the hamster equivalents of the I-E molecules and genes of the mouse.

Hamster class I genes and molecules do not appear to be polymorphic in terms of the ability to induce alloreactivities including production of alloantisera (Duncan and Streilein, 1977). In fact, unusual experimental strategies had to be employed to document that hamster class I molecules (and genes) actually exist. Using xenogeneic anti-hamster antisera, as well as cross-reacting anti-human and anti-mouse alloantisera, it has been documented that hamster cell surfaces express at least two and perhaps three different class I MHC molecules (Phillips et al., 1969; Darden and Streilein, 1985). These molecules have differential expression on somatic tissues, and they are *monomorphic* among all hamsters so far tested. The lack of polymorphism at class I loci in the Syrian hamsters is unprecedented among mammals that have been tested to date (Streilein et al., 1984). It cannot be ascribed to a sampling error or a genetic bottleneck, as has been proposed by some, since no comparable monomorphism exists at class II hamster loci. Possible reasons for the absence of polymorphism for class I genes in the hamster have been discussed elsewhere (Streilein et al., 1984; Reimann and Miller, 1983; Ohno and Wallace, 1983).

Using both mouse and hamster genomic probes, it has been possible to analyze the hamster class I gene family through Southern hybridization, and by gene cloning and sequence analysis. It has been revealed that the hamster possesses a similar number of class I gene loci to that found in mice, and these genes share extensive sequence homology with the murine genes (Atherton et al., 1984; McGuire, 1985; McGuire et al., 1986). Hamster class I genes have a similar exon–intron structure to those of mouse and human. However, unlike mouse and human, the hamster demonstrates only limited restriction endonuclease polymorphism in its class I genes. This finding corroborates the serological observation that this species lacks functionally detectable polymorphism of its class I products.

The unusual aspects of the hamster MHC have been the source of controversy and a stimulus to reconsider our ideas concerning the nature and biological meaning of the MHC. Much more remains to be done in this arena. At the present, it has not been shown formally that the hamster class I and class II encoding genes are located within a linked chromosomal segment, although there is no reason to believe that they are not. The absolute number, arrangement, and extent of polymorphism at the DNA level has not been described for class II molecules in this species. Moreover, no experiments have been reported concerning other MHC-related genes, such as those encoding complement components.

Very little is known concerning non-MHC minor histocompatibility systems in the hamsters. That they exist is not questioned, but the number of such loci is controversial. While early studies suggested that only a few histocompatibility loci segregate in the Syrian hamster, subsequent data using small exchanged skin allografts suggest that the number may be considerably higher when this more sensitive assay system is employed (Silvers et al., 1975). A minor histocompatibility locus governing mixed lymphocyte responses, independent of those derived from MHC disparities, has also been described in the hamster (Streilein et al., 1980a), but little is known beyond that. Nothing is known concerning possible polymorphism of hamster immunoglobulin molecules, or other cell surface determinants expressed on lymphocyte subsets.

V. LYMPHORETICULAR CELLS

Some of the most important advances in modern immunology have resulted from our abilities to distinguish successfully among lymphoreticular cells—a variety of functionally heterogeneous subpopulations. Distinctive cell surface molecules have been discovered which correlate in many instances with specific functional properties. This realization has made unequivocal description and identification of lymphoreticular subpopulations possible.

It has long been recognized that lymphoreticular cells, irrespective of their functional diversity, ultimately derive from a common hematopoietic precursor cell. With regard to immunology, two separate lineages are recognized: one in which the cells possess cell surface recognition structures specific for nominal antigens, and the other representing cells without antigenic specificity but with specialized effector function. The first lineage comprises T lymphocytes and B lymphocytes, while the second includes macrophages, natural killer (NK) cells, antigen-presenting cells, and dendritic cells.

A. T Lymphocytes

The existence of a distinct T-lymphocyte lineage in Syrian hamsters was appreciated in the early 1960s when it was observed that neonatal thymectomy produced a runting syndrome similar to that obtained in thymectomized mice and rats (Adner et al., 1965). Mature hamsters possess the capability of rejecting skin allografts (Billingham and Hildemann, 1958), of mounting contact and delayed-hypersensitivity (DTH) reactions (Streilein et al., 1980b; Maguire, 1980), and their lymphocytes are able to induce graft versus host reactions in vivo (Streilein and Billingham, 1970a,b) and participate in mixed lymphocyte reactions in vitro (Duncan and Streilein, 1977). All of these forms of reactivity have been shown to be mediated by T lymphocytes in other species and provide prima facie evidence for the existence of similar cells in the hamster. In mice, the Thy-1 cell surface marker has come to be equated with T lymphocytes. A Thy-1 homolog also exists in hamsters and was originally though to identify T cells in the species (Blasecki and Houston, 1978). However, Witte and Streilin (1983b) have clearly shown that the great majority of lymphocytes in lymph nodes, spleen, and thymus of hamsters express the Thy-1 molecule. Since a significant proportion of these cells also express surface immunoglobulin, it is clear that the Thy-1 molecule must also be present on B cells and therefore cannot be used to distinguish T cells from other lymphoreticular cells in this species.

The most definitive analysis of T lymphocytes, their subpopulations, and distinctive surface molecules has recently been accomplished by Dr. P. Witte (1984, Witte et al., 1985; Witte and Streilein, 1986). This investigator has raised monoclonal antibodies in mice against hamster lymphoid cells and has described two relevant sets of cell surface molecules on putative T cells. The first set, identified by monoclonal antibodies (MAb) nos. 20 and 110, appears to be present on all T lymphocytes, both in the thymus and in the periphery. A second set of surface molecules, identified by MAb no. 38, recognizes a subset of T cells both within the thymus and in peripheral tissues. Peripheral T cells that lack the MAb 38 molecule on their surface resemble, in functional terms, the helper/inducer set of T cells found in other species. These cells are able to proliferate in response to alloantigenic determinants, they produce interleukin 2 (IL-2) in response to antigenic challenge, they provide helper function to B cells, and they mediate both delayed-type and contact hypersensitivity in adoptive transfer assays. By contrast, T cells bearing the MAb 38-positive molecule fail to perform any of the functional properties just described, but they are capable of responding to exogenous IL-2 with a proliferative response.

Attempts to identify functional cytotoxic T lymphocytes in the Syrian hamster have been unrewarding until very recently. Studies on the nature of the MHC in this species (see above) suggested that hamsters either did not express class I MHC molecules, or that there was little or no polymorphism. In an effort to gain functional evidence of polymorphic class I molecules, Nelles and Streilein (1980b) attempted to demonstrate genetic restriction of cytotoxic activity to vacciniavirus in hamsters. While hamsters were able to produce virus-specific cytotoxic activity, it was absolutely not genetically restricted; in fact, no participation of T cells was observed in any cytotoxic reactions (Nelles and Streilein, 1980a). However, other investigators claimed that T cells were able to effect cytotoxicity in hamsters (Henderson, 1979; Kimmel et al., 1982; Cremer et al., 1982; O'Hara and Duncan, 1984). The availability of the monoclonal antibodies produced by Witte made it possible to test the hypothesis directly. Surprisingly, MAb 38-positive T cells harvested from the lungs of influenza A-infected hamsters appear to be cytotoxic T cells in that they are able to kill influenza-infected target cells with antigen specificity (Stein-Streilein et al., 1985). However, the vast majority of cytotoxic cells that hamsters possess endogenously or produce in response to immunogenic challenge, are MAb 38-negative and are undoubtedly not of the cytotoxic T-cell lineage. Thus, it would appear that at least two populations of T lymphocytes exist in the Syrian hamster. One is the hamster homolog of the helper/inducer/DTH subset, and is class II MHC oriented. The other population bears some resemblance to cytotoxic T cells, although more work needs to be done in order to confirm this finding.

To date no satisfactory assay exists for the detection of suppressor T lymphocytes in hamsters (Lause and Streilein, 1977, 1978; Streilein et al., 1981). Consequently, it is not known whether T-suppressor cells are included in either or both of the subpopulations (nos. 20, 110, 38) just described. The monoclonal antibodies have been used in an attempt to document the pathway of ontogenetic maturation of hamster T cells through the thymus (Witte, 1984; Witte and Streilein, 1986). While much remains to be accomplished in this regard, it does appear that a peripheral population of mature T cells may exist without ever having gone through an intrathymic stage. This is an important area of investigation, not only in hamsters, but in other species as well.

B. B Lymphocytes

Much less attention has been devoted to hamster B cells. As in other species, hamster B lymphocytes can be identified by virtue of the presence of surface immunoglobulin. Witte and Streilein (1983a) have shown that surface expression of class II MHC molecules also distinguishes B cells from T lymphocytes. At the present, there are no unique surface molecules that have been identified on hamster B cells, which would allow for a description of stages in the sequence of maturation. To the present, a hamster equivalent of IgD has not been unequivocally demonstrated.

Hamster B lymphocytes appear to undergo the same program of maturation as that described in the B cells of other species. B Cells that respond to nominal antigen only in the presence of T cells exist, as do those B cells which can respond to antigen in the absence of T-cell influence (Ahmed et al., 1978). There is, however, some question about the capacity of these latter cells. Zwilling et al. (1982) have reported that there may be a defect in T-independent B lymphocytes that is revealed on stimulation with bacterial lipopolysaccharide (LPS). The intriguing suggestion has been made that some hamster B lymphocytes may fail to pass beyond a certain stage in the maturation process, and that a significant portion of B cells in adult hamsters may be functionally immature (Ahmed et al., 1978).

The relative proportions of T and B lymphocytes in various hematopoietic and lymphoid compartments of the hamster is fairly typical of that found in other rodents (Streilein, 1979). However, the lack of reagents able to identify either distinct subpopulations of B cells or more than one subpopulation of T cells precludes any statements about the relative distributions and contributions of lymphocyte subsets in these various lymphoid compartments.

C. Natural Killer Cells

Although the existence of cytotoxic T lymphocytes in the hamster is open to considerable question, there seems to be little question that this species is well endowed with cells of the natural killer (NK) variety (Datta et al., 1979; Rees et al., 1980). This subject has been amply reviewed by Trentin and Datta (1980). NK-Cell activity in the hamster has been obtained from spleen cells (Haddada and de Vaux Saint Cyr, 1980; Haddada et al., 1983), from peritoneal exudate cells (Yang et al., 1982, 1983), from bronchoalveolar lavage (Stein-Streilein and Guffee, 1985), and from peripheral lymph nodes (Gee et al., 1979; Nelles et al., 1981). NK-Cell activity in this species is able to kill tumor targets and virus-infected targets. Cell surface determinants specific for NK cells have yet to be identified in hamsters. It has been well demonstrated that asialo-G_{M1} is a cell surface molecule on hamster NK cells (Stein-Streilein and Guffee, 1985). Stein-Streilein is currently evaluating five newly prepared monoclonal antibodies that appear to be specific for NK cells and not only lyse these cells in the presence of complement, but can block their effector function, suggesting that they bind to the surface molecule used for attachment/recognition (J. Stein-Streilein, personal communication). Not surprisingly, it has been demonstrated that some peritoneally derived NK cells, especially those obtained following vaccinia infection, can acquire the capacity to express the Thy-1 molecule (Yang et al., 1983). As in other species, the contribution NK cells make to immunity and protection against viruses and neoplasms remains to be worked out. However, evidence does exist in hamsters that NK-cell activity can participate in protection against susceptibility to virus infections. Specifically, a role for NK cells has been documented in infection with Pichinde virus (Gee et al., 1979, 1981), vaccinia (Nelles et al., 1981), and influenza A (Stein-Streilein and Guffee, 1985). In addition, NK cells have been implicated in host defenses against adeno 12-transformed hamster cell lines (Cook and Lewis, 1983) and other types of neoplasms. One gets the sense that in this species an inordinately large proportion of the cytotoxic cell activity is contributed by NK cells, rather than cytotoxic T cells.

D. Macrophages and Dendritic Cells

Macrophages also participate in cytotoxic activity in the Syrian hamster. Virus-induced cytotoxic activity within the peritoneal cavity has been shown to be at least partly the responsibility of activated macrophages (Chapes and Tompkins, 1979, 1981), and it has been demonstrated that the cytotoxic cell responsible for protection against SV40-transformed cells in the hamster is the macrophage (Cook and Lewis, 1983). Finally, the pulmonary macrophages of hamsters possess cytolytic capability, and this is readily inducible by LPS (Panke et al., 1982).

Dendritic cells are a relatively newly described member of the lymphoreticular cell group (Steinman and Nussenzweig, 1980). These nonphagocytic cells of hematopoietic origin appear to function primarily as antigen-presenting cells (Austyn et al., 1983). While little has been examined about dendritic cells in the hamster, some work has been done describing the epidermal Langerhans cell in this species (Bergstresser et al., 1980). It is thought that the epidermal Langerhans cell is the cutaneous representation of the dendritic cell line (Streilein and Bergstresser, 1981). It has been shown that Langerhans cells, when harvested from hamster epidermis (see above), function both in vivo and in vitro as effective antigen-processing and -presenting cells (Sullivan et al., 1985).

VI. RECENT ADVANCES IN HAMSTER IMMUNOBIOLOGY

A. Contact Hypersensitivity

The capacity to manifest and study contact hypersensitivity in mice has made it possible for significant inroads to be made in our understanding of the cellular basis of immune reactivities. Since 1980 two laboratories (Streilein, 1980; Streilein et al., 1980b; Maguire, 1980) have adapted the study of contact hypersensitivity to the Syrian hamster. These studies have made it possible to demonstrate that hamsters are capable of mounting typical contact hypersensitivity reactions, as measured by the ear swelling response. Immunization procedures with hapten which lead to contact hypersensitivity evoke hapten-specific antibody formation (Streilein, 1980). By using adoptive transfer assays of lymphoid cells, it has been possible to show that the contact-hypersensitive response in this species is mediated by cells of the T-lymphocyte lineage. Hamster epidermis contains Langerhans cells (Bergstresser et al., 1980; Streilein and Bergstresser, 1981), which resemble those found in mice and humans. In an effort to demonstrate that these cells play a critical antigen-presenting role in the induction of contact hypersensitivity, Sullivan et al. (1985) have isolated relatively purified Langerhan cells from hamster epidermis. When these cells are derivatized with hapten (trinitrophenol, TNP) and injected by either the intravenous or subcutaneous routes, they induce contact hypersensitivity. Thus, the first formal proof that Langerhan cells function as antigen-presenting cells in vivo came from studies in the Syrian hamsters.

As mentioned previously, questions have been raised concerning the quality and quantity of T-lymphocyte function in Syrian hamsters. The capacity to study contact hypersensitivity has made it possible to examine down-regulation or suppression of this response: the study of down-regulation of contact hypersensitivity in mice has been crucial to our understanding of the cascade of interacting suppressor T cells that regulate the response. Inoculation of hapten-derivatized lymphoid cells (or free hapten) intravenously into Syrian hamsters induces an unresponsive state, which is revealed by the relative failure of treated hamsters to mount hypersensitivity reactions subsequently when immunized in the conventional manner (Streilein et al., 1980b). Adoptive transfer of down-regulation with lymphoid cells from unresponsive hamsters has been accomplished. However, the precise phenotype of the cells in the adoptive transfer inoculum that are responsible for down-regulation remains obscure. Preliminary evidence indicates that it is not a typical T cell, although critical studies with the monoclonal antibodies recently produced by P. Witte have not been conducted.

B. Pulmonary Immunity

Histological analysis of the hamster lung reveals a relatively clean tissue, similar to that of human. Endogenous pulmonary infection with microorganisms is not common in this species, in contrast to other rodents used for experimental analysis (Toolan, 1977). Thus, Stein-Streilein and colleagues have exploited immune responses initiated in hamster lungs as a means of studying pulmonary immunology without the complicating features of endogenous infections. By inoculating immunocompetent lymphocytes intratracheally into semiallogeneic recipients, it was found that local graft versus host reactions developed not only within alveolar spaces, but in the draining hilar lymph nodes (Stein-Streilein et al., 1981b). Since no comparable reactions were elicited by intravenous inoculation of similar lymphoid cells, it was concluded that pathways exist for the traffic of lymphocytes across the alveolar walls into the interstitium and its lymphatic drainage. Intratracheal inoculations of SRBC produced evidence of both local and systemic immune responses (Stein-Streilein et al., 1981a). It is significant that, depending on the dose of inoculated antigen, a completely local (intrapulmonary) response could be achieved without evidence of systemic participation. More recently, TNP in the form of trinitrobenzenesulfonate (TNBS) has been inoculated into the hamster pulmonary tree. Hamsters treated in this way developed typical systemic contact hypersensitivity, as revealed when their ears are challenged with a direct application of dilute hapten (Stein-Streilein, 1983). Moreover, and perhaps important with regard to the pathogenesis of certain human inflammatory and fibrotic diseases of the lung, hamsters that are first rendered contact hypersensitive to hapten by skin painting and are then challenged intrapulmonarily with hapten, develop an acute inflammatory response which bears some resemblance to the fibrotic reaction found in idiopathic fibrosis in humans. Thus, the hamster may prove to be an important model for the study of inflammatory diseases of the lung with a potential immunological pathogenesis.

Stein-Streilein and collaborators have used intrapulmonary inoculation of antigen in hamster to study the immunobiology of various lymphoid cells. For example, following intratracheal installation of TNBS, the bronchoalveolar lavage fluid contains populations of cells with the capacity to transfer adoptively contact hypersensitivity to the hapten (Stein-Streilein, 1983). In related experiments, influenza A virus inoculated intratracheally into the hamster lung produces an inflammatory lesion that gradually resolves (Stein-Streilein et al., 1985). Once again, bronchoalveolar lavage fluid contains cells which have the capacity to mediate DTH reactions and cells which are cytotoxic for virus-infected target cells. Importantly, while some of the cells that are cytotoxic belong to the NK-cell lineage (Stein-Streilein and Guffee, 1985), Stein-Streilein has clearly shown that the virus-specific cytotoxic cells in the lavage fluid are of the T-cell lineage. This represents the first unambiguous description of cytotoxic T cells in this species. It is of particular interest that while cytotoxic T cells can be harvested from bronchoalveolar spaces of influenza-infected hamsters, no comparable cells can be identified from hilar lymph nodes nor from the spleen in similarly infected hamsters, indicating that, in hamsters, special local conditions are required in order for cytotoxic T cells to be activated.

C. Tumor Immunity

Hamsters stand alone among rodent species in an important way! Only in this species can cells that have been transformed *in vitro* by SV40 infection be shown to grow and form tumors following subcutaneous injection *in vivo*. This unique property has been exploited by numerous laboratories in an effort to understand the nature of the hamster immune response to these neoplastic cells.

De Vaux Saint Cyr and her colleagues have studied for many years the responses of hamsters to an SV40-induced fibrosarcoma (de Vaux Saint Cyr et al., 1977; Loisillier et al., 1977; Zuinghedau et al., 1979; Heimer and de Vaux Saint Cyr, 1979; de Vaux Saint Cyr and Loisillier, 1978; Haddada and de Vaux Saint Cyr, 1980; Haddada et al., 1983). When this tumor is inoculated subcutaneously into adult hamsters, it forms progressively growing local tumors. During the early growth of the tumor, antibodies reactive with tumor-specific antigens begin to appear and to accumulate in the peripheral blood. Histopathological analysis reveals that there is a peritumoral accumulation of plasma cells. Circulating immune complexes form and progressively increase in the blood; they deposit in the kidneys and contribute to renal failure. The spleens of these animals become enlarged, in part due to metastases from the subcutaneous tumor, but also because of the accumulation within the spleen of immunoreactive cells. Two types of these cells have been studied most recently: The first appears to be a suppressor cell which is revealed when cocultured with normal cells being stimulated with the T-cell mitogen, concanavalin A (Con A). The second is a non-T, non-B lymphocyte which has the capacity to inhibit tumor cell growth *in vitro*. These findings suggest that hamsters are capable of making lymphoid cells that inhibit tumor growth, but in these experiments, the cells are not of the immune, cytotoxic T-cell variety. In fact, specific immunizations of hamsters by repeated inoculations of tumor cells did lead to effector cells which could destroy the tumor. Phenotyping of these cells placed them in the NK-cell, macrophage- and/or antibody-dependent, cell-mediated cytotoxicity category. No cytotoxic T cells were in evidence.

Lewis and Cook have also collaborated on a series of interesting studies examining the response of hamsters to *in vitro*-transformed SV40 cells, as compared to cells transformed with human adenoviruses (Cook and Lewis, 1979, 1983, 1984; Cook et al., 1980; Lewis and Cook, 1982). The SV40-transformed cells grow equally well in syngeneic and allogeneic hamster hosts before the age of 3 weeks. Subsequently, the tumor cells are rejected but only by older allogeneic hosts. These workers have demonstrated that the cell responsible for rejecting the tumors is an activated macrophage and that macrophage activation is an ontogenetically dependent event which gradually takes place during the first 3 weeks of extrauterine life. By contrast, adeno 2-transformed hamster cells fail to grow in allogeneic hosts, irrespective of age. But they do grow in syngeneic hosts. Alternatively, adeno 12-infected cells fail to grow even in syngeneic hamsters. The work of these investigators offers hope that a link can be made between resistance to tumors, expression of histocompatibility markers, and ontogenetic acquisition of adaptive and nonadaptive host defense mechanisms.

REFERENCES

Adner, M. M., Sherman, J. D., and Dameshek, W. (1965). The normal development of the lymphoid mass in the golden hamster and its relationship to the effects of thymectomy. *Blood* **25**, 511–521.

Ahmed, A., Hare, J. A., and Sell, K. W. (1978). Morphological and functional characteristics of lymphoid cells from LHC hamsters. *Fed. Proc., Fed. Am. Soc. Exp. Biol.* **37**, 2045–2046.

Atherton, S. S., Streilein, R. D., and Streilein, J. W. (1984). Lack of polymorphism for C-type retrovirus sequences in the Syrian hamster. *ICSU Short Rep.* **1**, 128–129.

Austyn, J. M., Steinman, R. M., Weinstein, D. E., Granelli-Piperno, A., and Palladino, M. A. (1983). Dendritic cells initiate a two-stage mechanism for T lymphocyte proliferation. *J. Exp. Med.* **157**, 1101–1115.

Barta, O., Oyekan, P. P., Malone, J. B., and Klei, T. R. (1980). Hemolytic complement and its components in Syrian hamsters: A study of five strains uninfected and infected with *Brugia pahangi*. *Adv. Exp. Med. Biol.* **134**, 103–110.

Bergstresser, P. R., Fletcher, C. R., and Streilein, J. W. (1980). Surface densities of Langerhans cells in relation to rodent epidermal sites with special immunologic properties. *J. Invest. Dermatol.* **74**, 77–80.

Billingham, R. E. (1978). Concerning the laboratory career of *Mesocricetus auratus* with special reference to transplantation. *Fed. Proc., Fed. Am. Soc. Exp. Biol.* **37**, 2024–2027.

Billingham, R. E., and Hildemann, W. H. (1958). Studies on the immunological responses of hamsters to skin homografts. *Proc. R. Soc. London, Ser. B.* **148**, 216–232.

Billingham, R. E., and Silvers, W. K. (1964). Syrian hamsters and transplantation immunity. *Plast. Reconstr. Surg.* **34**, 329–340.

Billingham, R. E., Sawchuck, G. H., and Silvers, W. K. (1960). Studies on the histocompatibility genes of the Syrian hamster. *Proc. Natl. Acad. Sci. U.S.A.* **46**, 1079–1090.

Blasecki, J. W., and Houston, K. J. (1978). Identification, functional characterization and partial purification of thymus-derived lymphocytes in inbred hamsters. *Immunology* **35**, 1–11.

Bollin, E., Jr., and Sulkowski, E. (1981). Partial purification and characterization of Syrian hamster interferon. *J. Gen. Virol.* **52**, 227–233.

Chapes, S. K., and Tompkins, W. A. F. (1979). Cytotoxic macrophages induced in hamsters by vaccinia virus. *J. Immunol.* **123**, 303–308.

Chapes, S. K., and Tompkins, W. A. (1981). Distribution of macrophage cytotoxic and macrophage helper functions on BSA discontinuous gradients. *J. Reticuloendothel. Soc.* **30**, 517–530.

Coe, J. E. (1978). Humoral immunity and serum proteins in the Syrian hamsters. *Fed. Proc., Fed. Am. Soc. Exp. Biol.* **37**, 2030–2031.

Coe, J. E. (1980). Comparative immunology of old world hamsters—Cricetinae. *Adv. Exp. Med. Biol.* **134**, 95–101.

Coe, J. E., and Ross, M. J. (1985). Hamster female protein, a sex-limited pentraxin, is a constituent of Syrian hamster amyloid. *J. Clin. Invest.* **76**, 66–74.

Coggin, J. H., Jr., Thomas, K. V., and Huebner, R. (1978). Horizontally transmitted lymphomas of Syrian hamsters. *Fed. Proc., Fed. Am. Soc. Exp. Biol.* **37**, 2086–2088.

Coggin, J. H., Jr., Bellomy, B. B., Thomas, K. V., and Pollock, W. J. (1983). B-Cell and T-cell lymphomas and other associated diseases induced by an infectious DNA viroid-like agent in hamsters. *Am. J. Pathol.* **110**, 254–266.

Cook, J. L., and Lewis, A. M., Jr. (1979). Host response to adenovirus 2-transformed hamster embryo cells. *Cancer Res.* **39**, 1455–1461.

Cook, J. L., and Lewis, A. M., Jr., (1983). Differential NK cell and macrophage killing of hamster cells infected with nononcogenic or oncogenic adenovirus. *Science* **224**, 612–615.

Cook, J. L., Lewis, A. M., Jr., and Kirkpatrick, C. H. (1979). Age-related and thymus-dependent rejection of adenovirus 2-transformed cell tumors in the Syrian hamster. *Cancer Res.* **39**, 3335–3340.

Cook, J. L., Hibbs, J. B., Jr., and Lewis, A. M., Jr. (1980). Resistance of simian virus 40-transformed hamster cells to the cytolytic effect of activated macrophages: A possible factor in species-specific viral oncogenicity. *Immunology* **77**, 6773–6777.

Cremer, N. E., O'Keefe, B., Hagens, S. J., and Diggs, J. (1982). Cell-mediated cytotoxicity toward measles virus-infected target cells in randomly bred Syrian hamsters. *Infect. Immun.* **38**, 580–587.

Darden, A. G., and Streilein, J. W. (1985). Syrian hamsters express two monomorphic class I major histocompatibility complex molecules. *Immunogenetics* **17**, 980–981.

Datta, S. K., Gallagher, M. T., and Trentin, J. J. (1979). Natural cell-mediated cytotoxicity in hamsters. *Int. J. Cancer* **23**, 728–734.

de Vaux Saint Cyr, C., and Louisillier, F. (1978). Renal deposition of soluble immune complexes in hamsters bearing an SV40-induced tumour. *Ann. Immunol. (Paris)* **129**, 429–438.

de Vaux Saint Cyr, C., Louisillier, F., and Zuinghedau, J. (1977). Humoral and cellular immune response during the growth of an SV40-induced tumour in hamsters. *Ann. Microbiol. (Paris)* **128**, 385–398.

DiPaolo, J. A., Evans, C. H., DeMarinis, A. J., and Doniger, J. (1984). Inhibition of radiation-initiated and prompted transformation of Syrian hamster embryo cells by lymphotoxin. *Cancer Res.* **44**, 1465–1471.

Duncan, W. R., and Streilein, J. W. (1977). Analysis of the major histocompatibility complex in Syrian hamster. III. Cellular and humoral immunity to alloantigens. *J. Immunol.* **118**, 832–839.

Duncan, W. R., and Streilein, J. W. (1978a). Analysis of the major histocompatibility complex in Syrian hamsters. I. Skin graft rejection, GvH reactions, mixed lymphocyte reactions, and immune response gene in inbred strains. *Transplantation* **25**, 12–16.

Duncan, W. R., and Streilein, J. W. (1978b). Analysis of the major histocompatibility complex in Syrian hamsters. II. Linkage studies. *Transplantation* **25**, 17–22.

Duncan, W. R., and Streilein, J. W. (1981). Genetic analyses of alloreactions between recently wild and classical inbred strains of Syrian hamsters: Evidence in favor of a major histocompatibility complex. *Immunogenetics* **13**, 393–403.

Evans, C. H., and DiPaolo, J. A. (1981). Lymphotoxin: An anticarcinogenic lymphokine as measured by inhibition of chemical carcinogen or ultraviolet-irradiation-induced transformation of Syrian hamster cells. *Int. J. Cancer* **27**, 45–49.

Gee, S. R., Clark, D. A., and Rawls, W. E. (1979). Differences between Syrian hamster strains in natural killer cell activity induced by infection with Pichinde virus. *J. Immunol.* **123**, 2618–2626.

Gee, S. R., Chan, M. A., Clark, D. A., and Rawls, W. E. (1981). Role of natural killer cells in Pichinde virus infection of Syrian hamsters. *Infect. Immun.* **31**, 919–928.

Gorer, P. A. (1937). The genetic and antigenic basis of tumour transplantation. *J. Pathol. Bacterol.* **44**, 691–697.

Gotze, E. S. (1977). "The Major Histocompatibility System in Man and Animals." Springer-Verlag, Berlin and New York.

Haddada, H., and de Vaux Saint Cyr, C. (1980). Suppressive and cytostatic activities in the spleen of tumor-bearing hamsters. *Eur. J. Cancer* **16**, 841–848.

Haddada, H., de Vaux Saint Cyr, C., and Duthu, A. (1983). In vivo studies of spleen lymphoid cells implicated in antitumor immunity in hamsters. *Cancer Immunol. Immunother.* **15**, 96–100.

Heimer, R., and de Vaux Saint Cyr, C. (1979). Circulating immune complexes in sera of hamsters bearing simian virus 40-induced tumors. *Cancer Res.* **39**, 2919–2922.

Henderson, F. W. (1979). Pulmonary cell-mediated cytotoxicity in hamsters with parainfluenza virus type 3 pneumonia. *Am. Rev. Respir. Dis.* **120**, 41–47.

Jaenisch, R., and Berns, A. (1977). Tumor virus expression during mammalian embryogenesis. *In* "Concepts in Mammalian Embryogenesis" (M. Sherman, ed.), pp. 267–314. MIT Press, Cambridge, Massachusetts.

Johnson, K. P., Feldman, E. G., and Byington, D. P. (1975). Effect of neonatal thymectomy on experimental subacute sclerosing panencephalitis in adult hamsters. *Infect. Immun.* **12**, 1464–1469.

Jordan, R. K. (1974). Ultrastructure studies on cells containing secretory granules in the early embryonic thymus. *In* "Biological Activity of Thymic Hormones" (D. W. van Bekkum, ed.), p. 69. Kooyker Sci. Publ., New York.

Kimmel, K. A., Wyde, P. R., and Glezen, W. P. (1982). Evidence of a T-cell-mediated cytotoxic response to parainfluenza virus type 3 pneumonia in hamsters. *RES: J. Reticuloendothel. Soc.* **31**, 71–83.

Klein, J., Juretic, A., Baxevanis, C. N., and Nagy, Z. A. (1981). The traditional and a new version of the mouse H-2 complex. *Nature (London)* **291**, 455–460.

Lause, D., and Streilein, J. W. (1977). Hamsters, unlike rats, fail to regulate the response of immunocompetent cells to alloantigens. *Transplant. Proc.* **10**, 819–822.

Lause, D., and Streilein, J. W. (1978). Lymphoid cell interactions in allograft immunity: The failure of hamsters to generate suppressor cells upon specific immunization. *Transplantation* **26**, 80–83.

Lewis, A. M., Jr., and Cook, J. L. (1982). Spectrum of tumorigenic phenotypes among adenovirus 2-, adenovirus 12-, and simian virus 40-transformed Syrian hamster cells defined by host cellular immune–tumor cell interactions. *Cancer Res.* **42**, 939–944.

Linna, T. J. (1968). Cell migration from the thymus to other lymphoid organs in hamsters of different ages. *Blood* **31**, 727–746.

Loisillier, F., Zuinghedau, J., and de Vaux Saint Cyr, C. (1977). An anatomy-pathological study of the lymphoid system in hamsters during the growth of an SV40-induced tumour. *Br. J. Exp. Pathol.* **58**, 533–540.

Maguire, H. C., Jr. (1980). Allergic contact dermatitis in the hamster. *J. Invest. Dermatol.* **75**, 166–169.

McGuire, K. (1985). Molecular characterization of immunologically related loci of the Syrian hamster. Ph.D. Dissertation, University of Texas Health Science Center at Dallas.

McGuire, K. L., Duncan, W. R., and Tucker, P. W. (1986). Structure of a class I gene from Syrian hamster. *J. Immunol.* **137**, 366–372.

Murphy, M. R. (1971). Natural history of the Syrian golden hamster—a reconnaissance expedition. *Am. Zool.* **11**, 632.

Nelles, M. J., and Streilein, J. W. (1980a). Hamster T cells participate in MHC alloimmune reactions but do not effect virus-induced cytotoxic activity. *Immunogenetics* **11**, 75–86.

Nelles, M. J., and Streilein, J. W. (1980b). Immune response to acute virus infection in the Syrian hamster. I. Studies on genetic restriction of cell-mediated cytotoxicity. *Immunogenetics* **10**, 185–199.

Nelles, M. J., Duncan, W. R., and Streilein, J. W. (1981). Immune response to acute virus infection in the Syrian hamster. II. Studies on the identity of virus-induced cytotoxic effector cells. *J. Immunol.* **126**, 214–218.

O'Hara, R. M., Jr., and Duncan, W. R. (1984). Antigen-specific xenogeneic CTL activity in the Syrian hamster. *J. Immunol.* **133**, 1163–1167.

Ohno, S., and Wallace, W. B. (1983). Polymorphism and monomorphism in class-I MHC antigens. *Immunol. Today* **4**, 320–322.

Palm, J., Silvers, W. K., and Billingham, R. E. (1967). The problem of histocompatibility in wild hamsters. *J. Hered.* **58**, 41–43.

Panke, E. S., Zwilling, B. S., Somers, S. D., Campolito, L. B., and Packer, B. J. (1982). The effects of lipopolysaccharide, BCG-immune T lymphocytes, and lymphokines on generation of tumoricidal pulmonary macrophages in Syrian hamsters. *Exp. Lung Res.* **3**, 81–90.

Phillips, J. T., Streilein, J. W., and Duncan, W. R. (1979). The biochemical characterization of Syrian hamster cell surface alloantigens. I. Analysis of allogeneic differences between recently wild and highly inbred hamsters. *Immunogenetics* **7**, 445–455.

Phillips, J. T., Duncan, W. R., and Streilein, J. W. (1981). The biochemical characterization of Syrian hamster cell-surface alloantigens. II. Immunochemical relationships between cell-surface alloantigens and class II MHC homologues. *Immunogenetics* **12**, 485–496.

Rees, R. C., Hassan, Z. M., and Potter, C. W. (1980). Detection of natural cytotoxicity in Syrian hamsters. *Br. J. Cancer* **41**, 485–488.

Reimann, J., and Miller, R. G. (1983). Polymorphism and MHC gene function. *Dev. Comp. Immunol.* **7**, 403–412.

Riesenfeld, I., and Alm, G. V. (1977). Spontaneous and induced appearance of murine leukemia virus antigen containing cells in organ cultures of embryonic mouse thymus. *Int. J. Cancer* **20**, 309–317.

Robles, C. P., Proia, R. L., Hart, D. A., and Eidels, L. (1980). Immunoglobulins on the surfaces of hamster lymphocytes. *Adv. Exp. Med. Biol.* **134**, 87–94.

Silvers, W. K., Gasser, D. L., and Murphy, M. R. (1975). Number of histocompatibility loci in Syrian hamsters. *J. Immunol.* **115**, 1309–1311.

Solomon, J. B. (1978). Immunological milestones in the ontogeny of hamster, guinea pig, sheep, and man. *Fed. Proc., Fed. Am. Soc. Exp. Biol.* **37**, 2028–2030.

Solomon, J. B., Leiper, J., and Reid, T. M. S. (1972). Ontogeny of haemolytic plaque-forming cells in the hamster: The response to sheep and mouse erythrocytes. *Immunology* **22**, 63–67.

Steinman, R. M., and Nussenzweig, M. C. (1980). Dendritic cells: Features and functions. *Immunol. Rev.* **53**, 127–165.

Stein-Streilein, J. (1983). Allergic contact dermatitis induced by intratracheal administration of hapten. *J. Immunol.* **131**, 1748–1753.

Stein-Streilein, J., and Guffee, J. (1986). In-vivo treatment of mice or hamsters with antibodies to asialo-G_{M1}, increasing morbidity and mortality to pulmonary influenza infection. *J. Immunol.* **136**, 1435–1441.

Stein-Streilein, J., Gross, G. N., and Hart, D. A. (1981a). Comparison of intratracheal and intravenous inoculation of sheep erythrocytes in the induction of local and systemic immune responses. *Transplantation* **24**, 145–150.

Stein-Streilein, J., Lipscomb, M. F., Hart, D. A., and Darden, A. (1981b). Graft-versus-host reaction in the lung. *Transplantation* **32**, 38–44.

Stein-Streilein, J., Witte, P. L., Streilein, J. W., and Guffee, J. (1985). Local cellular defenses in influenza-infected lungs. *Cell. Immunol.* **95**, 234–246.

Streilein, J. W. (1978). Symposium: Hamster immune responses: Experimental models linking immunogenetics, oncogenesis and viral immunity. *Fed. Proc., Fed. Am. Soc. Exp. Biol.* **37**, 2023–2109.

Streilein, J. W. (1979). III. Hamster. *In* "Inbred and Genetically Defined Strains of Laboratory Animals" (P. L. Altman and D. D. Katz, eds.), Part 2, pp. 425–504. Fed. Am. Soc. Exp. Biol., Bethesda, Maryland.

Streilein, J. W. (1980). Contact hypersensitivity, humoral immunity and specific unresponsiveness can be induced in Syrian hamsters with simple haptens. *J. Immunol.* **124**, 577–585.

Streilein, J. W. (1981). Symposium report. Hamster immune responsiveness and experimental models of infectious and oncologic diseases. *Fed. Proc., Fed. Am. Soc. Exp. Biol.* **40**, 2343–2352.

Streilein, J. W., and Bergstresser, P. R. (1981). Langerhans cell function dictates induction of contact hypersensitivity or unresponsiveness to DNFB in Syrian hamsters. *J. Invest. Dermatol.* **77,** 272–277.

Streilein, J. W., and Billingham, R. E. (1970a). An analysis of graft-versus-host disease in Syrian hamsters. I. The epidermolytic syndrome: Description and studies on its procurement. *J. Exp. Med.* **132,** 163–180.

Streilein, J. W., and Billingham, R. E. (1970b). An analysis of graft-versus-host disease in Syrian hamsters. II. The epidermolytic syndrome: Studies on its pathogenesis. *J. Exp. Med.* **132,** 181–197.

Streilein, J. W., and Duncan, W. R. (1979). Alloimmune reactions among recently wild Syrian hamsters and classical inbred strains include alloantibody production. *Immunogenetics* **9,** 563–574.

Streilein, J. W., and Duncan, W. R. (1983). On the anomalous nature of the major histocompatibility complex in Syrian hamsters, Hm-1. *Transplant. Proc.* **15,** 1540–1545.

Streilein, J. W., Duncan, W. R., and Homburger, F. (1980a). Immunogenetic relationships among genetically defined, inbred domestic Syrian hamster strains. *Transplantation* **30,** 358–361.

Streilein, J. W., Witte, P., Burnham, K., and Bergstresser, P. R. (1980b). Induction and regulation of contact hypersensitivity in Syrian hamsters. *Adv. Exp. Med. Biol.* **134,** 43–57.

Streilein, J. W., Phillips, J. T., Stein-Streilein, J., Proia, D. A., and Duncan, W. R. (1981). Biochemical characterization of Syrian hamster cell surface alloantigens. III. Hamster alloantisera immunoprecipitate class II-like, but not class I-like molecules. *Transplantation* **32,** 106–110.

Streilein, J. W., Gerboth-Darden, A., and Phillips, J. T. (1984). Primordial MHC function may be best served by monomorphism. *Immunol. Today* **5,** 87–88.

Sullivan, S., Bergstresser, P. R., and Streilein, J. W. (1985). Intravenously injected, TNP-derivatized, Langerhans cell-enriched epidermal cells induce contact hypersensitivity in Syrian hamsters. *J. Invest. Dermatol.* **84,** 249–252.

Sullivan, T. J., Hart, D. A., and Streilein, J. W. (1980). IgE-Dependent release of inflammatory mediators from hamster mast cells *in vitro*. *Adv. Exp. Med. Biol.* **134,** 33–41.

Sung, E., Duncan, W. R., Streilein, J. W., and Jones, P. P. (1982). Detection of two distinct class II alpha:beta:Ii complexes in the Syrian hamster. *Immunogenetics* **16,** 425–433.

Toolan, H. W. (1977). Susceptibility of the Syrian hamster to virus infection. *Fed. Proc., Fed. Am. Soc. Exp. Biol.* **37,** 2065–2068.

Trentin, J. J., and Datta, S. K. (1980). Natural killer (NK) cells in hamster and their modulation in tumorigenesis. *Adv. Exp. Med. Biol.* **134,** 153–163.

Witte, P. L. (1984). Description of T lymphocyte heterogeneity in the Syrian hamster. Ph.D. Dissertation, University of Texas Health Science Center at Dallas.

Witte, P. L., and Streilein, J. W. (1983a). Monoclonal antibodies to hamster Class II MHC molecules distinguish T and B cells. *J. Immunol.* **130,** 2282–2286.

Witte, P. L., and Streilein, J. W. (1983b). Thy-1 antigen is present on B and T lymphocytes of the Syrian hamster. *J. Immunol.* **131,** 2903–2907.

Witte, P. L., and Streilein, J. W. (1986). Development and ontogeny of hamster T cell subpopulations. *J. Immunol.* **137,** 45–54.

Witte, P. L., Streilein, J. W., and Shannon, W. A., Jr. (1980). C-Type particles in thymic development: A correlation with thymus function. *Adv. Exp. Med. Biol.* **134,** 455–462.

Witte, P. L., Stein-Streilein, J., and Streilein, J. W. (1985). Description of phenotypically distinct T-lymphocyte subsets which mediate helper/DTH and cytotoxic functions in the Syrian hamster. *J. Immunol.* **134,** 2908–2915.

Yang, H., Cain, C. A., Woan, C., and Tompkins, W. A. (1982). Evaluation of hamster natural cytotoxic cells and vaccinia-induced cytotoxic cells for Thy-1.2 homologue by using a mouse monoclonal anti-Thy-1.2 antibody. *J. Immunol.* **129,** 2239–2243.

Yang, H., Cain, C. A., and Tompkins, W. A. F. (1983). Induction of Thy-1.2 homologue-positive and Thy-1.2 homologue-negative nonspecific cytotoxic lymphocytes in the hamster. *J. Immunol.* **131,** 622–627.

Young, R. L., Ward, H., Hartshorn, D., and Block, M. (1963). The relationship between antibody formation and the appearance of plasma cells in newborn hamsters. *J. Infect. Dis.* **112,** 67–76.

Zuinghedau, J., Duthu, A., and de Vaux St Cyr, C. (1979). Presence and significance of tumour cells in the spleen of tumour-bearing hamsters. *Br. J. Cancer* **39,** 594–597.

Zwilling, B. S., Koegwel, M., and Campolito, L. B. (1982). Lack of lipopolysaccharide response of inbred strains of Syrian hamsters. *Dev. Comp. Immunol.* **6,** 349–358.

Chapter 14

Experimental Biology: Use in Infectious Disease Research

J. K. Frenkel

I.	Hamsters as Unique Infection Models	228
II.	Infections with Disseminated Lesions	229
	A. Syphilis	229
	B. Toxoplasmosis	229
	C. Other Protozoan Infections	229
	D. Mycobacterial Infection	229
	E. Leprosy	231
	F. Melioidosis	231
	G. Fungal Infections	231
	H. Viral Infections	231
	I. Helminths	233
III.	Infections with Cerebral and Ocular Lesions	233
	A. Measles and Subacute Sclerosing Panencephalitis (SSPE)	233
	B. Scrapie and Creutzfeldt–Jakob Disease (CJD)	234
	C. Other Viruses	234
	D. Toxoplasmosis and Besnoitiosis	235
	E. *Encephalitozoon cuniculi*	235
	F. Malaria	235
	G. Bacteria	235
IV.	Infections with Muscle Lesions	235
V.	Infections with Respiratory Tract Lesions	235
	A. *Mycoplasma* Infection	236
	B. Legionellosis	236
	C. Histoplasmosis	236
	D. Virus Infection	236
VI.	Infections with Kidney Lesions	236
	A. Leptospirosis	236
	B. Intravascular Coagulation with Glomerular Thrombosis	237
VII.	Infections with Liver Lesions	237
	A. Equine Rhinopneumonitis (Herpesvirus)	237
	B. Leptospirosis	237
	C. Amebiasis	237
	D. Trematode Infections	237
	E. Nematodes	238
	F. Bacterial Infection	238

        VIII. Infections with Gastrointestinal Lesions 238
              A. Antibiotic-Associated Colitis due to *Clostridium difficile* 238
              B. Other Bacterial Enteritides 238
              C. Helminthic Infection 239
         IX. Infections with Reticuloendothelial and Blood Vascular Lesions 239
              A. Leishmaniasis 239
              B. Histoplasmosis 239
              C. Babesiosis 240
          X. Infections with Lesions in Endocrine Organs 240
              A. Adrenal Cortex 240
              B. Pineal Gland 241
              C. Thyroid Gland 241
              D. Placenta 241
         XI. Infections with Skin and Other Lesions 241
              A. Staphylococcal Infection 241
              B. Impetigo Model 241
              C. Gonorrhea 242
              D. Demodectic Mange 242
              E. Mite Infection 242
        XII. Hamsters as Sentinels of Infection 242
              A. Sentinels in Nature 242
              B. Indicators of Arthropod Infection 242
              C. Sentinels of Contamination 242
              D. Hamsters as Filters of Infection 242
       XIII. Hamsters as Models of Cellular Immunity against Infection 242
        XIV. Conclusions 243
              References 243

I. HAMSTERS AS UNIQUE INFECTION MODELS

Soon after its introduction to the laboratory, *Mesocricetus auratus* was found to be useful for the study of leishmaniasis (Adler and Theodor, 1931), yellow fever (Findley, 1934), and leprosy (Adler, 1937). The utility of hamsters in infectious disease research can be considered a major early impetus for their use in experimental biology.

Hamsters, which are more difficult to handle and more costly than mice, have generally been used for their special species attributes. Their cheek pouch, apparently devoid of lymphoid drainage (Barker and Billingham, 1971) (see also Chapter 2), permitted experiments to study localized infection. Their capacity to hibernate (see Chapter 16) make hamsters useful for studies of infection during hypothermia (Bessemans *et al.*, 1956; Chute, 1961). Their tendency toward cannibalism was used to compare oral and parenteral transmission of scrapie, perhaps applied as a model for unconventional virus infection of humans (Prusiner *et al.*, 1985). Hamsters are the only known animals in which progressive adrenal lesions develop during certain chronic infections. As in humans, the principal adrenocortical secretory product was found to be cortisol. This has helped in analyzing the pathogenesis of infectious adrenal necrosis leading to Addison's disease in humans (Frenkel, 1956, 1961a). Their relatively greater corticoid sensitivity, compared to rats, mice, and guinea pigs, makes hamsters unique for the study of corticoid effects on infection (Frenkel and Havenhill, 1963). Their relative genetic uniformity (see Chapter 13) has facilitated skin transplantation among some groups of hamsters; however, for the transplantation of lymphoid cells for the study of infection immunity (Frenkel and Wilson, 1972), isologous strains, tolerant or irradiated recipients are necessary. The availability of inbred strains of hamsters (see Chapter 15), suggests that some strains might be useful as unique models of infectious disease. Only a few such models have been developed so far (Schell *et al.*, 1979, 1980b). More appropriate models are needed for many infections. Such models can be found by serendipity or by trial. However, our understanding of pathogenesis is generally too limited to predict what strains of hamsters might furnish a new or needed model of infection. Little can be gained by agreeing to continue using models that represent pathogenesis in only a limited manner.

While there are numerous papers dealing with infection in hamsters, this chapter concentrates on those studies where hamsters were shown to be of special model value, citing representative studies.

II. INFECTIONS WITH DISSEMINATED LESIONS

A. Syphilis

Hamsters develop skin lesions after intradermal infection with *Treponema pallidum bosnia A,* the agent of endemic syphilis, and with *Treponema pertenue,* the cause of yaws or frambesia. Unlike rabbits, historically the more popular experimental host, hamsters have no spontaneous treponemal infection that interferes with experimental studies (Schell and Musher, 1983). Skin ulcers develop during the fourth week at the intracutaneous infection sites, and an erythematous scaling lesion spreads peripherally, persisting for 16 weeks, when healing begins. The regional lymph nodes enlarge during the first 12 weeks of infection with *T. pallidum* and contain large numbers of treponemes which can be seen by dark-field microscopy or after staining with Dieterle or Warthin–Starry silver stains. Homologous and cross-immunity is demonstrable by inhibition of skin lesions and a reduced number of spirochetes in lymph nodes, starting about 10 weeks after primary infection. Cross-immunity between *T. pallidum* and *T. pertenue* was also demonstrated (Schell and Musher, 1983).

Only the primary and secondary syphilitic infections are modeled in hamsters (and rabbits), and although inapparent chronic infection persists for months or years, no tertiary lesions have been described. Neither was congenital syphilis observed in offspring of such hamsters. The hamster model was used to determine the importance of cellular immune mechanisms by using adoptive transfer of lymphoid cells, making use of an inbred line (Schell *et al.,* 1980a,b, 1982).

B. Toxoplasmosis

Hamsters develop generalized toxoplasmosis after oral or parenteral infection, permitting study of acute and chronic infections. Asymptomatic to fatal infections, during the acute, subacute, or chronic stage have been observed with different strains of *Toxoplasma*. Death, after 7–10 days is usually due to pneumonia, in the third week or later, due to encephalitis. Retinochoroiditis, followed by trophic cataracts, develop in some hamsters. Most of the retinal lesions follow cyst rupture; because the liberated bradyzoites are destroyed, lesions are interpreted on a delayed-hypersensitivity basis (type IV). A few instances of retinitis with proliferating tachyzoites have also been seen (Frenkel, 1958) (Figs. 1,2). When chronically infected hamsters are treated with pharmacological doses of corticosteroids (Frenkel, 1960b), toxoplasmosis reactivates with pneumonia and encephalitis, leading to death. Both lesions start focally, presumably from ruptured cysts, and expand with innumerable tachyzoites destroying the cells

Fig. 1. Focal retinochoroiditis in hamster 10 months after subcutaneous infection with *Toxoplasma*. Proliferating tachyzoites (arrow) are associated with this lesion. Magnification: ×133 PASH (periodic acid, Schiff, hematoxylin). [Reprinted with permission from Frenkel (1955); © by Ophthalmic Publ. Co.]

(Frenkel *et al.,* 1975). Adrenal lesions with proliferating tachyzoites were found during otherwise inactive chronic infection, and have been linked to the immunosuppressive effect of endogenous cortisol (Frenkel, 1956) (Figs. 3–5). Although immunity to *Toxoplasma* has often been attributed to chronic persisting infection, also called premunition or concomitant immunity, immunity in the absence of chronic infection has been studied in hamsters employing strain *ts-4,* a special nonpersisting mutant (Elwell and Frenkel, 1984a,b).

C. Other Protozoan Infections

Hamsters were used by Rubio (1959) to study natural and acquired immunity to Chagas' disease, an important disease of humans and dogs. Use of hamster in malaria is described in Section III,F.

D. Mycobacterial Infection

Hamsters are uniquely susceptible to infection with several mycobacteria including bacillus Calmette-Guérin (BCG) vaccine. The production of lesions in hamsters with a clinical isolate led Buhler and Pollack (1953) to attribute human pathogenicity to a photochromogenic "yellow bacillus" which was later named *Mycobacterium kansasii*. Because prior isolates were found nonpathogenic to guinea pigs, they had been regarded as saprophytes without primary pathogenic roles in humans. Investigation of the pathogenic potential of atypical acid-fast bacilli, none of which are pathogenic to guinea pigs, was appreciably furthered by the availability of hamsters as

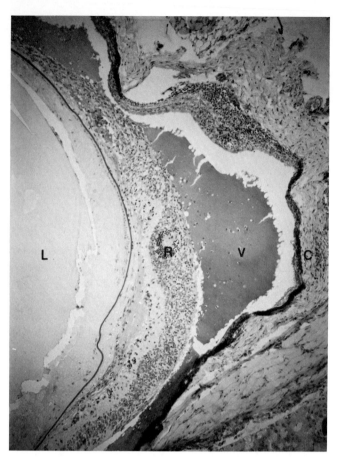
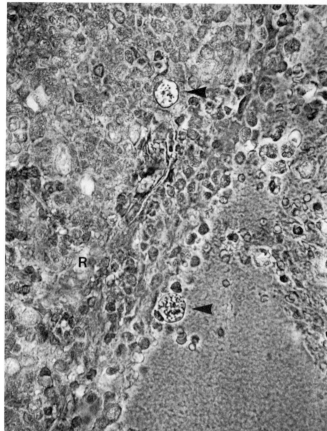
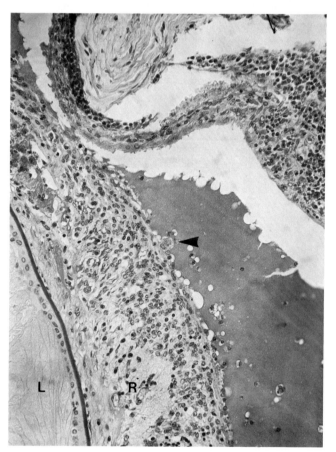

Fig. 2. Chronic *Toxoplasma* retinochoroiditis in hamster 4 months after subcutaneous infection. The loss of retinal rod and cone layer is accompanied by gliosis, inflammatory reaction, and separation of the retina (R) from choroid (C), forming an inflammatory vesicle (V). The retina has become adherent to the lens (L). *Toxoplasma* cysts (arrows) are present in the retina with cyst wall shown by silver impregnation. Cyst rupture is the likely pathogenic mechanism because no tachyzoites are present. Magnification: (A) 80×; (B) 200× PASH; (C) 500×. Wilder's reticulum impregnation. [Reprinted with permission from Frenkel (1961b); © by Williams & Wilkins.]

sensitive experimental hosts. Disseminated infections, often accompanied by a sarcoidlike granulomatous reaction, are easily produced after intraperitoneal or even subcutaneous inoculation (Frenkel, 1958, 1961a).

The sensitivity of hamsters to human BCG vaccine (*Mycobacterium bovis* BCG) was discovered by Hauduroy and Rosset (1951). Numerous bacilli may be present which give rise to granulomatous lesions in many organs, particularly the liver, lymph nodes, and spleen, and the infection may even be fatal. Because even large doses of killed BCG or tuberculin alone can be fatal to hamsters, a million organisms intraperitoneally were toxic, whereas smaller numbers showed pathogenic effects from progressive infection. Hamsters have been used to compare the pathogenicity of various strains of the BCG vaccine (Jespersen and Bentzon, 1964), using 10^6–10^8 bacteria. Many hamsters die from such infections in less than a year, but in some the infections persisted for over 1 year.

Tubercle bacilli of human pathogenicity are, of course, also pathogenic to hamsters, including strains that had become resistant to isoniazid and which have lost their pathogenicity for guinea pigs (Saenz, 1958a,b). A laboratory epidemic of tuberculosis has been described in hamsters (Chute *et al.*, 1954).

Schaumann bodies, such as are found in cases of sarcoidosis, are readily formed in the epithelioid and giant cells of hamsters infected with various mycobacteriae (Saenz, 1956). Ultrastructurally they are composed of concentric protein layers, containing an apatite, similar to bone, with traces of iron, which forms around the mycobacteria (Rasmussen and Caulfield, 1960; Dumont and Sheldon, 1965). The hormonal effects on the sarcoidlike granulomatous reaction and Schaumann bodies were studied (Frenkel, 1958), after finding that granulomas in the hamster adrenal were dissimilar and underwent necrosis (Frenkel, 1961a). It was found that exogenous and adrenal cortisol increased the number of mycobacteria and decreased the number of sarcoidlike granulomas and Schaumann bodies and favored the development of necrosis (Frenkel, 1972).

E. Leprosy

Leprosy bacilli were first transferred to hamsters in 1937 by Adler (1937, 1948). Because *Mycobacterium leprae* could not be grown on synthetic media and there were no other known susceptible animals, his success was an important impetus toward the use of hamsters as an experimental animal. The proliferation of leprosy bacilli in hamsters with a production of histiocytic granulomas has been studied extensively, especially by Binford (1959). The lesions resemble the lepromatous form of leprosy. More recently, leprosy bacilli have been grown in cell cultures, in the tail and footpads of mice and the viscera of the armadillo, and the role of hamsters as an experimental animal has declined.

F. Melioidosis

This rare infection, caused by *Malleomyces pseudomallei*, was studied in hamsters with emphasis on pathogenic and low-pathogenic strains and vaccination (Dannenberg and Scott, 1960).

G. Fungal Infections

Hamsters are susceptible to most disseminating deep fungal infections. Histoplasmosis has been studied extensively, with hamsters more sensitive to smaller inocula than mice (Drouhet and Schwarz, 1956; Schwarz *et al.*, 1957). After intraperitoneal inoculation, most of the yeast forms grow in phagocytic cells of the spleen, lymph nodes, and liver. A model for the study of disseminated histoplasmosis has also been developed after intrapulmonary inoculation (Bauman and Chick, 1969). Conchoid Schaumann bodies often develop as in mycobacterial and leishmanial infection (Okudaira *et al.*, 1961). The phagocytosis of *Histoplasma* by macrophages was studied ultrastructurally by Dumont and Robert (1970). Hamsters develop immunity after infection with smaller inocula than mice, and hamsters were useful to study strains not infecting mice and experimental chemotherapy. Immunity can be transferred adoptively with lymphoid cells of immune donors, but not with antiserum (Frenkel, 1972).

Cryptococcosis of the central nervous system (CNS), extending from posterior pharyngeal lesions, was observed after intranasal inoculation (Herrold, 1965). After intraperitoneal inoculation, a large amount of gelatinous capsular material may accumulate in the peritoneum.

Blastomycosis (Korij *et al.*, 1984), paracoccidioidomycosis (Peracoli *et al.*, 1982), coccidioidomycosis (Drouhet, 1961), *Nocardia asteroides* and *Nocardia brasiliensis* (Macotela-Ruiz and Mariat, 1963), and sporotrichosis (Mariat and Drouhet, 1954) have all been studied in hamsters. Hypercorticism interferes with the development of immunity, not only in hamsters with histoplasmosis (Frenkel, 1962), but also with cryptococcosis, blastomycosis, sporotrichosis, and coccidioidomycosis (Frenkel, 1972).

H. Viral Infections

Many viruses will infect the hamster or its cells in culture, but most of these viruses can be studied equally well in mice. The infectivity and oncogenicity of viruses in hamsters has been reviewed by Eddy (1972), Toolan (1978), Marsh and Hanson (1978), Johnson (1978), and in Chapter 12. This section concentrates on special uses of hamsters in virological research.

Generalized infections are produced by certain arboviruses. Several strains of Venezuelan equine encephalitis (VEE) with differing pathogenicity for humans were compared in hamsters for the rate of clearance from blood, the virus titers reached in several organs, and the ability to kill hamsters (Jahrling and Scherer, 1973a). Also, 100 clones of virus were compared for hamster pathogenicity (Jahrling and Scherer, 1973b,c).

The histopathology and virus growth curves of avirulent and virulent VEE strains were compared in adult hamsters, showing effects mainly on hematopoietic cells (Austin and Scherer, 1971). The possibility that encephalitogenic viruses may also give rise to extraneural lesions in humans was successfully studied in baby hamsters with St. Louis encephalitis virus injected subcutaneously. The animals generally became moribund within 2 days. Extraneural lesions were not revealed by light microscopy, but immunofluorescence and electron microscopy showed widespread involvement, including the exocrine and endocrine pancreas, myocardium, smooth muscle, vascular endothelium, endocardium, and lamina propria of the intestine (Harrison et al., 1980, 1982).

Susceptibility to Pichinde virus infections (Arenavirus) is genetically determined in hamsters. Intraperitoneal inoculation was fatal to adult MHA hamsters, tut not to adult LSH hamsters, or outbred LVG/LAK hamsters (Buchmeier and Rawls, 1977). Both inbred strains survived footpad inoculations, after which only LSH hamsters developed an inflammatory response (Gee et al., 1981). Immunosuppression by cyclophosphamide rendered adult LVG animals susceptible to lethal infection, and virus grew to high titers (Buchmeier and Rawls, 1977).

A putative interferon effect was studied in hamsters infected with encephalomyocarditis (EMC) virus. Such infection ordinarily is fatal, but protection was induced by poly(I:C) (Round and Stebbings, 1981).

EMC viruses give rise to generalized infection in hamsters with variable myocardial and neural lesions (Jungeblut, 1958); and possibly the β cells of the pancreas may have also been affected. Coxsackie viruses may also be pathogenic, mainly in baby hamsters with widespread lesions (Lu et al., 1961).

Colorado tick fever virus, a member of the orbiviruses, was isolated and studied experimentally in hamsters and separated from dengue virus infection. Hamsters are the only animal model available (Florio et al., 1946).

Cellular immunity and nonspecific resistance were compared in hamsters (Frenkel and Caldwell, 1975). These studies used Oriboca and Ossa viruses (bunyavirus group C) and equine herpesvirus, which gave rise to fatal infections, and yellow fever (17D vaccine strain), vesicular stomatitis, and Newcastle disease virus, which produce asymptomatic infections.

The protective role of defective interfering virus particles and of synthetic polyribonucleotide interferon inducers, were studied in a fatal generalized vesicular somatitis infection of LSH and MHA hamsters (Fultz et al., 1982). Protection was shown by serotype-specific and nonspecific interfering particles, and by the interferon inducer, but this did not always correlate with serum levels of putative hamster interferon. Lymphoreticular cells appeared principally affected in hamsters dying after 3 days (Fultz et al., 1981; see reference 10 in Fultz et al., 1982).

Studies on the genetic restriction of cell-mediated cytotoxicity were conducted with nonfatal vaccinia and lymphocytic choriomeningitis virus infection (Nelles and Streilein, 1980). A chronic virus infection model is available with vesicular stomatitis virus (Fulz et al., 1982). These models are useful for studies of viral persistence, pathogenesis, and immunity. Chronic, asymptomatic lymphocytic choriomeningitis infection has been observed also spontaneously, giving rise to confounding data (Baum et al., 1966; Thacker et al., 1982).

Infant hamsters appear to be the most sensitive laboratory animal for the isolation of Machupo and Junin arenaviruses, agents of Bolivian and Argentine hemorrhagic fever. Hamsters develop antibody, but shed virus in the throat and urine for several months (Justines and Johnson, 1968). Congenitally infected Calomys, the natural host, often do not develop antibody, while shedding virus (Justines and Johnson, 1969).

Placental involvement was studied in hamsters with herpesvirus infection (probably type I). This agent generally does not produce disease after peripheral injection, but in pregnant hamsters adrenocortical and placental infection with necrosis was observed, although little virus was transferred to the offspring (Ferm and Low, 1965). Newborns suffered fatal infection when injected directly, whereas adults generally developed immunity, except when hypercorticoid.

The related equine rhinopneumonitis (herpes)virus is highly pathogenic for adult hamsters which die between 1 and 7 days after infection. A single infectious dose was generally fatal, and formalinized vaccine was protective (Frenkel and Caldwell, 1975). Immunity could be adoptively transferred with lymphoid cells from immune donors and with serum (Frenkel, 1967). The pathology of the infection in hamsters was studied by McCollum et al. (1962).

Acyclovir given intramuscularly prevented fatal CNS infection of hamsters with a strain of herpes, which after intracutaneous inoculation spread to the CNS. Acyclovir was effective even when given 48 hr after infection at a time when hamsters showed early paresis (Van Ekdom and Versteeg, 1982). Virus and drug titers were measured.

Measles virus gives rise to generalized infection, which, however, is essentially asymptomatic even in newborn hamsters (McKendall et al., 1981).

Parvovirus infection (H-1) of fetal and newborn hamsters was discovered and reviewed by Toolan (1972) and is discussed in Chapter 6. Hamsters that had been injected at birth developed a head deformity with flattened face, protruding tongue, abnormal teeth, bone fragility, remained dwarfed, and sometimes died. Newborn rats and mice were refractory. The virus was contained in several human tumors; other similar viruses were isolated from rats. The virus infects many tissues, multiplies in the placenta, and may give rise to fetal death or deformity in hamsters. The virus persists in a latent form and may need a helper virus to proliferate. However, in rapidly growing tissue, whether fetal, neoplastic, inflammatory, or reparative (callus), the virus can proliferate without helper. In the fetal hamster, infection leads to cerebellar hypoplasia with ataxia, and necrosis of odontoblasts and osteoblasts, which produces malformation of teeth and bones. Intranuclear inclusions are found in the germinal layers of the cerebellum before it becomes hypoplastic. The H-1 virus may inhibit its helper virus, and the number of tumors produced by some helper viruses, such as adenovirus type 12.

Virus infections that are oncogenic in hamsters are described by Trentin (Chapter 12).

A viroidlike agent gave rise to malignant lymphomas when young hamsters were exposed, by injection or even contact. The agent could be reisolated from the lymphomas. For further detail see Chapter 9. Hamsters are a convenient small animal in which the natural spread and pathogenesis of a communicable neoplasm can be studied (Coggin *et al.*, 1981, 1983).

I. Helminths

Hamsters have been useful to investigate many helminthic infections, but it is not always clear why hamsters are better models than other animal hosts. Variability of susceptibility to *Schistosoma mansoni* was tentatively interpreted as a polygene-controlled phenomenon (Yong *et al.*, 1983). A degree of cross-immunity between *S. mansoni* and *Schistosoma hematobium* was uncovered when studying double infections (Mansour *et al.*, 1984). The clinical observation that schistosome-infected patients may develop a particularly intractable *Salmonella paratyphi* infection was reproduced in hamsters (Mikhail *et al.*, 1981, 1982). Chronic *Salmonella* infections were produced in schistosome-infected hamsters, whereas in normal hamsters intracardially administered *Salmonella* was eliminated in 4 days. The *Salmonella* were intimately associated with the adult worms. *In vitro* tests have indicated that such association appears to favor the growth of *Salmonella*.

Immunosuppression, as reflected by the number of microfilaria in the blood of hamsters, was studied with *Dipetalonema vitae* (Neilson, 1978). In normal hamsters, the circulation of microfilaria is transient, lasting only about 4 months. However, after becoming free of microfilaria, they reappeared after 4–7 days in hamsters that were injected with methylprednisolone or cyclophosphamide. Inbred hamsters were employed as hosts for *D. vitae* by Neilson (1978) and for *Brugia malayi* by Crandall *et al.* (1982). *Brugia pahangi* infection produces lymphangitis with granulomas (Malone *et al.*, 1976).

III. INFECTIONS WITH CEREBRAL AND OCULAR LESIONS

Intracerebral inoculation with herpes, measles, rabies, and various arboviruses gives rise to encephalitis which is similar to that produced by these same agents in mice. Of special interest are models of CNS involvement that follows extraneural inoculations of hamsters, and generalized infection as is seen in animal and human diseases. Often, newborn animals must be used to yield such a semblance, the model value of which is being investigated. This subject has been reviewed by Johnson (1978) and Toolan (1978), and in the case of "slow virus" infections by Marsh and Hanson (1978).

A. Measles and Subacute Sclerosing Panencephalitis (SSPE)

Efforts to produce generalized asymptomatic measles infection followed by encephalitis were successful in only 6 of 26 newborn hamsters that were inoculated intraperitoneally with a SSPE strain of measles virus adapted to the hamster brain (Carrigan *et al.*, 1978). Neither viremia nor viral spread to parenchymal organs were sufficient to establish CNS disease, and cyclophosphamide did not increase the rate of CNS invasion. Perivascular inflammation, rare giant cells in the hippocampus, rare cytoplasmic inclusions, and cell necrosis were seen histologically. Virus was recovered from brain, lungs, spleen, and kidney of certain animals.

Defective measles virus was isolated in hamsters from patients with SSPE by Johnson and Norrby (1974). Previously, a chronic progressive infection had been produced in weanling and adult hamsters after intracerebral inoculation of a hamster-adapted virus (Byington and Johnson, 1972). Measles antigen could be found by immunofluorescence in the brain of weanling hamsters for 55 days, and defective, cell-associated virus for 81 days, using cocultivation with monkey kidney cells. Intranuclear inclusions, inflammatory cells, and astrocytosis were found histologically. In adult hamsters, the CNS involvement ended by 21 days; however, after treatment with antilymphocyte serum, more than 50% died showing measles antigen, intranuclear inclusions, and encephalitis (Byington and Johnson, 1975). This suggested that the abortive infection of adult hamsters was due to an effective immune response which could be inhibited. The immune response of these hamsters was studied (Tobler *et al.*, 1982) using a rosetting response of lymphocytes with cells persistently infected with measles virus and by hemagglutination inhibition titers.

Chronic relapsing myelitis was observed in hamsters following experimental measles infection (Carrigan and Johnson,

1980), suggesting a similarity to multiple sclerosis, in which a measles etiology had been suspected (Norrby, 1978).

Retinopathy was found in hamsters infected intracerebrally with measles virus by Parhad et al. (1980). Retinal lesions consisted of inclusion-bearing cells, necrosis, and gliosis. Retinal virus titers were similar to those in the brain. The lesions simulated those of human SSPE and measles retinopathy.

B. Scrapie and Creutzfeldt–Jakob Disease (CJD)

Scrapie is an infectious disease of sheep and goats with an incubation period of over a year. Mice were the first laboratory animals to which it was transmitted. Later, scrapie was reproduced in hamsters, with an incubation period of as short as 2 months (Kimberlin and Walker, 1977). CJD, a progressive dementia of humans with cerebellar impairment, extrapyramidal signs, myoclonus, and blindness, was first transmitted to chimpanzees, squirrel monkeys, and later to smaller animals, including hamsters (Manuelidis et al., 1978; Brown 1984; Moreau-Dubois et al., 1982). However, infections in hamsters appeared less predictable than those in mice and guinea pigs. Both scrapie and CJD are histologically characterized by intracellular vacuoles ("spongiform encephalopathy") and gliosis. The distribution of these lesions was studied by conventional histology and monoclonal antibodies for fibrous astrocytes, comparing intracerebrally inoculated hamsters, mice, and squirrel monkeys (Masters et al., 1984). It was found that the scrapie agent reproduced exponentially in hamsters, reaching 10^{11} infectious units per gram of wet brain after about 80 days. Gliosis also increased progressively during infection and appeared useful as an indicator of the presence of scrapie infection, when other causes of diffuse encephalopathy could be excluded. However, the vacuolar change did not progress after 2 months, which is the time when clinical symptoms appeared. Hamsters generally died after 3–4 months. Thus, hamsters appear to be the most useful small animal for the study of scrapie infection because of the high titers and the sensitive histological indicators of progression. In guinea pigs, the virus titer in the end stage of disease was only 10^5.

The cannibalistic tendencies of hamsters were used to study the transmissibility of scrapie by ingestion (Prusiner et al., 1985). Hamsters developed ataxia and tremors and died after 100–160 days with similar lesions after inoculation. This hamster model shed light on the presumed cannibalistic transmission of kuru in humans. Hamsters are also useful to compare susceptibilities to enteral and parenteral transmission of scrapie and the related CJD. Electroencepalographic changes were also found (Gourmelon et al., 1983).

Retinal degeneration begins prior to the onset of CNS signs in scrapie-infected hamsters, with retinal viral titers paralleling those present in the brain (Hogan et al., 1981; Buyukmihci et al., 1982a,b). These retinal changes, which were limited to the photoreceptors, began in the inner segment and progressed to the outer segment of the rod and cone layer. Optic nerve degeneration began later in the disease (Buyukmihci et al., 1982b). Although not stated, these lesions may be visible with the ophthalmoscope because of a slow pupillary reflex after dark adaptation (Frenkel, 1955).

Mink encephalopathy agent was compared with scrapie in hamsters by Marsh and Kimberlin (1975), who found similar lesions.

The decontamination of tissues and materials contaminated with scrapie and other slow viruses was studied in hamsters using pooled brain suspensions with a titer of over 10^{10} (Brown et al., 1982). The greatest reduction was achieved by autoclaving (7.5 logs) followed by 1:10 dilution of sodium hypochlorite (3–4 logs); it was suggested that a combination of exposure to chemicals and autoclaving may be necessary to sterilize high-titer scrapie virus-infected tissues. It was determined that formalin fixation, followed by routine histological processing left about 6.8 logs of the original infection of 9.6 logs LD_{50} per gram. Masters and Gajdusek (1982) recommended adding 10% phenol after fixation in 10% formalin.

C. Other Viruses

Hydrocephalus in newborn hamsters was produced by mumps, reovirus 1, and polyomavirus (Johnson and Johnson, 1968; Margolis and Kilham, 1969; Nielsen and Baringer, 1972). The viruses multiplied in the ependymal cells, and obstruction of the aqueduct of Sylvius followed. However, with human respiratory syncytial virus infection in hamsters, hydrocephalus was not accompanied by aqueductal stenosis (Lagace-Simard et al., 1982).

Rabies gives rise to encephalitis after intramuscular injection in hamsters and to infection of other tissues including the brown fat, skeletal muscle, kidneys, and salivary glands; hamsters were significantly more susceptible to experimental infection with rabies virus than two species of bats (Sulkin et al., 1959). Treatment with corticosteroids interferes with the development of immunity (Enright et al., 1970). The antibody response was studied by Coe and Bell (1977), and immunization with modified live virus by Koprowski (1952a).

Poliovirus has been studied in hamsters treated with corticosteroids; only certain strains were pathogenic (Aronson and Shwartzman, 1956).

Persistent bornavirus infection in adult hamsters was studied by Anzil et al. (1973).

The vervet monkey (Marburg) agent, a large RNA virus, gives rise to neuronal necrosis, splenic and liver lesions in newborn hamsters (Zlotnik and Simpson, 1969; Siegert, 1971).

D. Toxoplasmosis and Besnoitiosis

Toxoplasmosis causes development of fatal encephalitis in hypercorticoid hamsters, with renewed proliferation of tachyzoites, as was described in Section II,B. Encephalomyelitis develops also in eucorticoid hamsters during chronic infection with several strains. The hamsters become hyperactive, circle, and often hold their head to one side (Frenkel, 1953, 1956). During chronic infection, Toxoplasma cysts are present in the brain and their disintegration gives rise to inflammatory lesions about 200 μm in diameter, which are followed by gliosis. Most of the liberated intracystic bradyzoites are killed by immune-mediated action, and the lesions are those of type IV hypersensitivity. In the presence of many cysts, clinically significant lesions develop over a period of months in the late stage of the disease, at which time posterior paralysis develops, due to the superimposed spinal cord lesions. This is slowly ascending until the hamster dies with a secondary infection and loss of bladder and bowel control. These lesions in eucorticoid hamsters are distinct from lesions in hypercorticoid hamsters because of the absence of proliferating tachyzoites.

Chronic retinochoroiditis often develops after several months of chronic toxoplasmosis in hamsters. Most of the lesions result from cyst rupture, and a few are from proliferating tachyzoites (Figs. 1,2) (Frenkel, 1955, 1961b). The pupillary reflex of hamsters is rather slow, and when approached in the dark, one can examine the fundus for a minute or so before the pupil constricts. Although Toxoplasma cysts are too small to be seen with the ophthalmoscope, the larger Besnoitia jellisoni cyst is visible, especially when it projects into the vitreous. Infection of the retina occurs in close to half of the hamsters even after subcutaneous infection with Besnoitia jellisoni, followed by prophylaxis with sulfadiazine for a week until generalized immunity is developed. The diagnosis is made by periodic fundus examination. After the retinochoroiditis has progressed, trophic cataracts often develop (Frenkel, 1958a; 1961b).

E. Encephalitozoon cuniculi

This small, low-pathogenic microsporidian organism infects mice and other laboratory animals, including hamsters (Frenkel, 1971; Meiser et al., 1972). While its main importance lies in being an unrecognized and unsuspected confounding influence of other experiments, CNS infection with formation of pseudocysts containing spores may follow generalized infections. Goodpasture's carbol fuchsin or Gram's stain demonstrate the 2×3 μm organisms, whereas they are easily overlooked with hematoxylin and eosin or Giesma stains. The organism can be definitively identified from its ultrastructure detail (Pakes et al., 1975), especially the presence of a polar filament and a single nucleus. The unfamiliarity of many experimenters with this organism has led to erroneous claims of virus isolation (Frenkel, 1971).

F. Malaria

A model for cerebral falciparum malaria was developed in hamsters and mice (Rest, 1983). Plasmodium berghei gave rise to fulminant infection in the majority of hamsters, producing more gametocytes after mosquito infection, than in mice and rats (Yoeli and Most, 1964). Pulmonary and hepatic macrophage dynamics in relation to phagocytosis of parasitized red cells were analyzed by MacCallum (1969a,b).

G. Bacteria

Spiroplasma was pathogenic to newborn hamsters after intracerebral injection (Kirchhoff et al., 1981). After a prolonged illness characterized by spasms, muscular tremors, disturbances in motor control, and inability to feed, death often ensued. The bacteria could be isolated from brain, liver, and spleen, and ocular involvement was common.

IV. INFECTIONS WITH MUSCLE LESIONS

Trichinella spiralis infects hamsters that were fed infected meat; the LD_{50} is about 850 larvae (Sadun and Norman, 1956). Death occurred during the period of muscle invasion at about 30 days. When hamsters are infected with doses over 4000 larvae, they die during the second week during the intestinal phase. Electrocardiographic changes were studied by Bernard and Sudak (1960) in hamsters infected with 225 larvae. Atrioventricular block was observed in two hamsters after 2 and 4 weeks of infection; intraventricular block was found in 7 of 62 hamsters with older infections. Metabolic changes, especially of uric acid and creatine excretion, were studied by Bernard (1961).

V. INFECTIONS WITH RESPIRATORY TRACT LESIONS

Except for infections with Sendai virus, hamsters appear to be relatively free from spontaneous pulmonary virus infections. For experimental inoculation of the respiratory tract, two techniques have been used for accurate delivery of infectious doses. Lightly anesthetized hamsters are placed on a board, their upper and lower incisors immobilized, the tongue extended, and the inocula delivered under the glottis into the trachea as described by Barile et al. (1981). Another technique consists of inserting a needle into the trachea, transcutaneously, as described by Bauman and Chick (1969). Either technique is suitable for quantitative work with infectious agents, whereas intranasal inoculation is not.

A. *Mycoplasma* Infection

Hamsters were instrumental in isolating *Mycoplasma pneumoniae*, the causative agent of primary atypical pneumonia (Eaton et al., 1944); serological and immunological responses to these agents were also studied by Dajani et al. (1965). Hamster lungs support the growth of *M. pneumoniae* and have been used to isolate strains from patients with primary atypical pneumonia (Collier and Clyde, 1974). Hamster trachea in organ culture (Collier and Baseman, 1973) and ciliary respiratory epithelium grown in monolayer cultures (Gabridge et al., 1978) have also been used. Attachment of tritiated *Mycoplasma* to such tracheal cells was compared with attachment to other cells (Chandler et al., 1982). Killed *Mycoplasma* vaccines were tested in hamsters which were challenged intratracheally with virulent *M. pneumoniae* and scored histologically and by culture (Barile et al., 1981). Two or more doses of inactivated vaccine were required to protect hamsters. The nature and duration of immunity after vaccination was studied by Hayatsu et al. (1981). Protection was measured by clearance of the challenge inocula after 2 weeks from trachea and lung. Protection persisted for 6 months, and a single booster extended it to 10 months. Serum antibody titers reached a peak 2 weeks after the last vaccination and appeared to protect after passive transfer. Metabolism-inhibiting antibodies were detected in the bronchial washings of animals showing resistance. No cell-mediated immune responses (migration inhibition test, footpad reaction) were observed. Double infections of *M. pneumoniae* and pneumococci and the effect of vaccination were studied by Liu et al. (1970).

B. Legionellosis

A hamster model was devised by means of intratracheal inoculation with *Legionella pneumophila*, serogroup 1, grown on charcoal yeast extract agar for 3 days (Richards et al., 1984). The LD_{50} was consistently $10^{8.4} \pm 0.1$ organisms. Histologically, a fibrinopurulent pneumonia developed in these hamsters which was said to be morphologically similar to that seen in human Legionnaires' disease. A mixed neutrophil and alveolar macrophage response was maximal after 4 days. IgG and IgM antibody responses reached a maximum after 2 weeks, and diminished to approximately preinoculation levels by 6 weeks. Inoculation of preopsonized or of unopsonized bacteria yielded the same LD_{50}. By comparing the infection in nude mice and their hirsute littermates, the importance of cellular immunity was shown by Drutz et al. (1984). This aspect has not yet been studied in hamsters.

C. Histoplasmosis

Hamsters were employed to study extrapulmonary dissemination of *Histoplasma capsulatum* to the spleen following intratracheal infection (Bauman and Chick, 1969). Four weeks after infection, the spleens were homogenized and cultured quantitatively. Between 10^4 and 10^5 years organisms had to be injected intratracheally in order to infect the spleens of half the hamsters. This simulates the pathogenesis of human histoplasmosis, where pulmonary infection by itself is usually self-limited, and where illness is generally produced by dissemination. This model was employed to assess the effectiveness of vaccines and should be useful for strain comparisons, chemotherapy, and other experimental approaches requiring reproducible lesions.

D. Virus Infection

Influenza A viruses were grown in the nasal turbinates and the lungs of hamsters (Ali et al., 1982). Strains of influenza A reported as virulent for humans tended to grow to higher titers in nose and lung, were more infectious to hamsters by the intranasal route, and showed a high incidence of spread to cage contacts. This model may be useful in testing of vaccines and potential chemotherapeutic agents.

VI. INFECTIONS WITH KIDNEY LESIONS

A. Leptospirosis

Hamsters are highly susceptible to many *Leptospira* species, and more so than young guinea pigs traditionally employed. Inadvertent infections with *Leptospira ballum* have been observed in hamsters injected with peritoneal washings, or kidney or brain suspensions from inapparently infected mice (Frenkel, 1972). The infection is asymptomatic in mice, but gave rise to fatal illness in hamsters after 4–6 days. Intense jaundice was present from hemolysis and hepatitis. The urine was dark brownish red. The kidneys were swollen and dark green, with the tubules containing hemoglobin-stained protein casts. In hamsters that were on sulfadiazine prophylaxis (60–120 mg/dl in drinking water), death from leptospirosis was delayed for from 2 to 3 weeks, and was due to uremia with prominent signs of encephalopathy. Small, pale kidneys with regenerating tubules, mostly without lumens were found in these animals. The renal lesions produced by *L. interrogans* were studied ultrastructurally (Badiola *et al.*, 1983). The illness in hamsters could be prevented and *L. ballum* eliminated from the chronically infected mice by use of chlortetracycline (50 mg/100 gm food or water) (Stoenner *et al.*, 1958); penicillin was ineffective.

Hamsters did not develop chronic infection with *L. ballum*; however, they became carriers of *L. canicola* after penicillin treatment. The carrier state could be terminated with streptomycin (Brunner and Meyer, 1950), the tetracyclines, and erythromycin (Weber *et al.*, 1956).

Vaccination with lyophilized cultures or with an avirulent high-passage culture protected hamsters against fatal *L. canicola, L. icterohemorrhagiae,* (Brunner and Meyer, 1950), or *L. pomona* infection (Stahlheim, 1968). The duration of immunity conferred by commercial killed vaccines, used for dogs in Belgium, was tested in hamsters (Desmecht, 1982). Ten and 18 months after two subcutaneous doses of vaccine, most hamsters resisted challenge, which was uniformly fatal to the controls.

B. Intravascular Coagulation with Glomerular Thrombosis

Cortical thrombi of both kidneys were observed in pregnant hamsters after superior mesenteric artery occlusion and the release of bacterial products, or after the injection of colchicine. Thus the hamster has been suggested as a model for the study of the generalized Schwartzman reaction (Galton, 1964, 1965, 1966; Galton and Slater, 1965).

VII. INFECTIONS WITH LIVER LESIONS

A. Equine Rhinopneumonitis (Herpesvirus)

This virus is highly pathogenic to adult hamsters, one infectious dose being equivalent to one fatal dose. Duration of illness is dose related. Death occurs 1–5 days after intraperitoneal inoculation. Many liver cells contain intranuclear inclusions. Formalinized live virus was shown to be an effective vaccine (Frenkel and Caldwell, 1975; see also Section II,H).

B. Leptospirosis

The hepatitis accompanying *Leptospira* infection in hamsters may be severe and contributes to the fatal outcome (see Section VI,A). The livers of pregnant hamsters and fetuses infected with *L. canicola* were studied by light and electron microscopy, and showed severe changes (Sapp *et al.*, 1980).

C. Amebiasis

Liver abscesses have been produced in hamsters with several strains of *Entamoeba histolytica*. Even axenic trophozoites when injected intrahepatically or intraperitoneally gave rise to liver abscesses, and sometimes metastatic abscesses are produced (Ghadirian and Meerovitch, 1978). However, after intracecal inoculation, only three of six hamsters developed local abscesses. Some of the liver abscesses were absorbed in the third week of infection. Complement depletion of the hamsters with cobra venom factor increases the frequency and severity of lesions (Capin *et al.*, 1980a; Ghadirian and Meerovitch, 1982). Immunosuppression by neonatal thymectomy and anti-T-cell sera increases the size of primary and metastatic abscesses (Ghadirian and Meerovitch, 1981). The capacity of hamsters to eliminate *Candida albicans* was reduced, as amebic infection progressed (Capin *et al.*, 1980b). Vaccines against amebiasis have been tested in hamsters (Gold *et al.*, 1982).

D. Trematode Infections

Extramedullary hematopoiesis, especially of eosinophils, accompanied infection with *Schistosoma mansoni, Schistosoma mekongi,* and *Schistosoma japonicum* in hamsters and mice (Byram *et al.*, 1978). The liver is the principal focus of dense oviposition, accompanied by developing eosinophils. The humoral immune response to *S. mansoni* was studied in hamsters by Young *et al.* (1983).

Opisthorchis viverrini, a common liver fluke of humans in Thailand, can also infect hamsters; metacercarias found in cyprinid fish are the communicable stage. The flukes live in the bile ducts of humans, chronic infections are common and can lead to cholangitis and cholangiocarcinoma in humans. The effects of high- and low-protein diets on infected hamsters was studied by Flavell *et al.* (1980), who observed marked bile duct proliferation with low-protein diets; however, no tumors were produced.

Thamavit *et al.* (1978) had shown earlier that dimethylnitrosamine added to an infection with 100 metacercariae produced cholangiofibrosis with cholangiocarcinoma in 100% of hamsters, whereas groups receiving either the infection or the carcinogen alone did not show any carcinoma. It was postulated that cholangiocarcinoma in these animals arose because of the carcinogen's greater effect on the proliferating bile duct epithelium stimulated by the parasite, whereas resting epithelium was not susceptible. Carcinogenic agents, particularly nitrosamines, are contained in favored food dishes in Thailand (Juttijudata *et al.*, 1982).

Collagen synthesis and degradation in hamsters infected with *Opisthorchis* was greater in early than late infection (Hutadilok and Ruenwongsa, 1983). Attempts to induce protective immunity in hamsters against reinfection with *Opisthorchis* were unsuccessful; however, prior infection resulted generally in a reduced egg output by the worms (Sirisinha *et al.*, 1983).

Dicrocoelium dendriticum (a sheep parasite) was transferred to hamsters and used for studies of the antibody response (Bode and Geyer, 1981).

Immunity to *Fasciola hepatica* cercarial infection was studied by Hillyer *et al.* (1977).

E. Nematodes

Cirrhosis developed in hamsters infected with the filaria *Dipetalonema vitae* (Neilson, 1978; Van Marck *et al.*, 1983).

F. Bacterial Infection

Tyzzer's disease due to *Bacillus piliformis,* usually found in mice, has also been studied in hamsters (Zook *et al.*, 1977).

VIII. INFECTIONS WITH GASTROINTESTINAL LESIONS

A. Antibiotic-Associated Colitis due to *Clostridium difficile*

Hamsters were essential in unraveling what appears now to be the principal cause of antibiotic-associated colitis (AAC) in humans. Small (1968) described the fatal enterocolitis in hamsters given lincomycin. The illness began with anorexia, ruffled fur, diarrhea, and led to death with a fluid-distended ileum, cecum, and colon. The penicillin and lincomycin hamster models were studied in the early and mid-1970s by several investigators (Frenkel, 1972; Humphrey *et al.,* 1974; Green, 1974). But not until 1977 was it shown that fecal extracts of hamsters with clindamycin-induced colitis were toxic, and that they could be neutralized with antitoxin from *Clostridium sordellii* (Rifkin *et al.,* 1977), that vancomycin protected hamsters against clindamycin-induced toxicity (Browne *et al.,* 1977; Bartlett *et al.,* 1977a); and that a toxigenic *Clostridium* species, resistant to clindamycin, could be cultured from hamster gut content (Bartlett *et al.,* 1977b). Bacteria-free filtrates from feces of clindamycin-treated hamsters were fatal in as little as 1 day after intraperitoneal injection and polyvalent *Clostridium* antitoxin neutralized its toxicity. Lusk *et al.* (1978) illustrated the lesions by light and scanning electron microscopy. Further microscopic and ultrastructural studies by Humphrey *et al.* (1979) showed pseudomembranous, ulcerative, and hemorrhagic lesions consistent with toxic effect. There are only minor differences between the lesions produced in humans and hamsters. The hamster lesions occur mainly in the cecum, whereas in the human the distal colon is involved.

By 1979, *Clostridium difficile,* a newly described human pathogen, was recognized by Fekety *et al.* (1979) and Bartlett (1979) as a low-level constituent of most hamster and some human intestinal flora which is inhibited by elements of the normal flora. Clindamycin and some other antibiotics, by reducing the prevalent anerobes such as bacteroides and coliforms, permits *C. difficile* to grow to larger numbers and to secrete an appreciable amount of toxin which has general and local effects on the gut epithelium. A large number of antimicrobial agents were tested in hamsters, and the colitis (typhlitis) produced by a combination of agents was characterized. These studies suggested that suppression of normal flora, overgrowth, suppression, and recolonization with clostridia, played a role in different experiments. Although vancomycin inhibited *C. difficile,* sometimes hamsters died after the end of a prolonged period of administering vancomycin (Fekety *et al.,* 1979). Even non-antibiotic-treated hamsters may develop fatal colitis after contact infections (Rehg and Lu, 1982). By ion-exchange chromatography, two separate toxins were identified: toxin A, which is more lethal to mice and gives rise to water loss in the cecum of hamsters, but is less cytotoxic than toxin B, which gives rise to hemorrhage and inflammation in the cecum (Libby *et al.,* 1982). Hamsters needed to be immune to both toxins in order to survive clindamycin administration. The suppression of *C. difficile* by normal hamster cecal flora and the prevention of AAC was shown by Wilson *et al.* (1981). Thus by investigating the paradoxical antibiotic toxicity in hamsters, the cause, mechanism, and treatment of pseudomembranous colitis associated with antibiotic use in humans was clarified. These studies are further detailed in Chapter 11.

B. Other Bacterial Enteritides

"Wet-tail," proliferative ileitis, atypical ileal hyperplasia, and enzootic intestinal adenocarcinoma, refers to a form of spontaneous ileocecitis of hamsters discussed in detail in Chapter 7. With its cause uncertain, it cannot yet be used as an experimental model. However, it may confound other experimental results.

A transmissible, non-Hodgkin's lymphoma, apparently caused by a viroid, has already been referred to in Section II,F, and is further discussed in Chapter 9. It is associated with a multifocal, necrotizing, ulcerative bowel disease involving the entire small intestine and occurring 2–3 weeks after exposure to the causative agent. Because it gives rise to diarrhea and ulceration, and to intestinal perforation and sepsis, these bowel lesions are often fatal. They can serve as an early clinical marker for exposure of hamsters to the viroid lymphomagenic agent (Manci *et al.,* 1984).

The intestinal microflora of hamsters has been studied

(Hagen et al., 1965). The anaerobic flora was examined with more recent techniques during the investigation of antibiotic-associated cecitis (Lusk et al., 1978). Attempted artificial infection of LSH and outbred hamsters with *Clostridium perfringens* failed (Goldman et al., 1972).

Campylobacter fetus was isolated from normal hamsters and those with proliferative ileitis (Fox et al., 1981). However, culture-inoculated hamsters did not develop ideal hyperplasia, although they had become colonized (Lentsch et al., 1982). From 40 to 80% of pet hamsters were found to be carriers of *Campylobacter jejuni* (Fox and Hering, 1983). This infection is discussed in Chapter 7.

Enteropathogenic *Escherichia coli* gave rise to acute enteritis in weanling hamsters, invaded intestinal epithelial cells and mesenteric lymph nodes, and gave rise to some fatalities (Frisk et al., 1981).

Elimination from the hamster gut of selected gram-negative bacteria which had been previously introduced was successful using a 4-week treatment with dihydrostreptomycin (Angulo et al., 1978).

C. Helminthic Infection

Certain worms of mice, such as *Hymenolepsis nana* and *Syphacia obvelata*, are transmissible to hamsters (Wantland, 1968); they are discussed in Chapter 8.

The nematode *Angiostrongylus cantonensis* infected hamsters which could serve as definitive host if infected with a small number of larvae (Ishii et al., 1980). Large numbers of larvae were fatal. The cattle lungworm *Dictiocaulus viviparus* was studied in hamsters and other laboratory rodents (Wade et al., 1960).

Hamsters can serve as definitive hosts for *Taenia solium* and *Taenia saginata* of humans (Verster, 1974).

Cestode larvae from the stomach wall of Alaskan haddock developed to adulthood in hamsters (Hasegawa et al., 1980).

The acanthocephalid *Moniliformis dubius* was studied in concurrent infection with *Hymenolepis diminuta* (Holmes, 1962).

The trematode of cats, *Opisthorchis felineus*, gave rise to intestinal infection in hamsters. These adults from 10 and 25 infecting cercariae yielded more eggs in hamsters than in other rodents, and even cats (Lämmler et al., 1968).

Schistosomiasis has been studied extensively in hamsters using all three or four species found in humans (Mostofi, 1967). These infections have been used for chemotherapeutic and immunological studies. Whereas the WO outbred strain developed immunity, the inbred LGN strain was less able to do so (Smith and Clegg, 1976).

Hamsters were found to be better laboratory hosts than mice for *Plagiorchis proximus*, a trematode; from 40 to 60 adults were accommodated in the small intestine, and egg production was good (Rachford, 1970).

IX. INFECTIONS WITH RETICULOENDOTHELIAL AND BLOOD VASCULAR LESIONS

A. Leishmaniasis

Visceral leishmaniasis from *Leishmania infantum* was the first experimental infection that was studied in hamsters (Adler and Theodor, 1931). Sometimes large inocula are used, yielding rapidly progressive infections fatal in 2 weeks, such as for chemotherapeutic assays (Kinnamon et al., 1980; Germuth et al., 1950; Stauber, 1966). The histiocytes of the spleen and liver become heavily parasitized; if the infection persists for over a month, amyloidosis tends to develop and the hamsters eventually die with renal failure (Gellhorn et al., 1946). A proliferative glomerulonephritis was also described by Agu et al. (1981).

For biological studies smaller inocula, or a range of inocula, are useful. From 1 to 1000 infectious doses give rise to a granulomatous response with epithelioid and giant cells, with amastigotes persisting for a number of months to be ultimately enclosed in conchoid bodies (Frenkel, 1972). Polyclonal B-cell activation was compared for visceral and cutaneous leishmaniasis (Campos-Neto and Bunn-Moreno, 1982). Visceral involvement with *L. donovani* led to antibodies and hypergammaglobulinemia due to polyclonal activation of B cells. However, after *L. braziliensis* and *L. mexicana amazonensis* infections limited to the skin, antibodies were stimulated without either an increase of Ig concentration or of numbers of antibody-forming cells per spleen. The role of antigenic load and of species differences must be considered. Because of the availability of many isologous strains of mice, most immunity studies have recently been performed in mice. *Leishmania mexicanum*, *L. braziliensis*, and *L. tropica* grow best in the cooler parts of the body, giving rise to subcutaneous lesions in the ear, tail, and feet. Amyloidosis also commonly develops.

Subcutaneous nodules that formed after infections with *L. mexicana*, *L. braziliensis*, and *L. garnhami* showed similar histological responses (Bretana et al., 1983). The macrophagic granuloma was accompanied by neutrophils, some eosinophils, and plasma cells, and by increasing numbers of fibroblast in regressing lesions. Amastigotes occurred within the phagolysosomes of macrophages, within lacunar cells, and interstitially.

B. Histoplasmosis

The predominantly reticuloendothelial location of *Histoplasma capsulatum* should be mentioned here, although the generalized infection was discussed in Sections II,G, and V,C. The intracellular parasitism in macrophages was analyzed ultrastructurally by Dumont (1972). Some fungi were destroyed in Schaumann bodies and others in phagolysosomes. Sensitized macrophages developed complex nuclear bodies.

C. Babesiosis

This protozoan parasite of red cells has been studied extensively in hamsters, which are more susceptible than many other animals. *Babesia microti*, occurring in the northeastern United States, was adapted to hamsters by Western *et al.* (1970), giving rise to anemia and reticulocytosis (Roth *et al.*, 1981). *Babesia rodhaini* produces fatal infection in hamsters (Han *et al.*, 1982). Hamsters were treated successfully with clindamycin–vancomycin alone or together with oral quinine (Rowin *et al.*, 1982), establishing the first effective and safe treatment of babesiosis.

X. INFECTIONS WITH LESIONS IN ENDOCRINE ORGANS

A. Adrenal Cortex

Hamsters are unique among small laboratory animals in secreting principally cortisol rather than corticosterone, and in their immunity being highly corticosteroid sensitive (Frenkel and Havenhill, 1963; Frenkel *et al.*, 1965). In the course of studies of chronic infection with two related protozoans *Toxoplasma gondii* and *Besnoitia jellisoni*, adrenal necrosis was observed (Frenkel, 1956). Either organism actively proliferated in the adrenal, destroying cells, whereas other tissues were immune. The adrenal immune defect resulted from local hypercorticism. Adrenal immunity could be maintained or restored when corticosteroid levels were lowered in the adrenals, either by replacement doses of corticoids causing feedback inhibition or by hypophysectomy. However, tachyzoites of *Besnoitia* often proliferated around the subcutaneous deposits of cortisol in these immune hamsters. Adrenal infection (Figs. 3–5) involves both cortex and medulla, and the cortisol-rich blood percolates through the medulla. Adrenal vein blood of hamsters contains 5- to 10-fold higher corticoid levels than the peripheral venous blood (Frenkel *et al.*, 1965). Cortisol did not affect *Besnoitia* proliferation in cell culture or in nonimmune hamsters. The adrenal lesions can be studied during two periods of time (Fig. 3). During the third week of infection, when infection regresses in most organs due to the development of immunity, this lags behind in the adrenals where focal necrotic lesions develop, which are easily reproducible and which lend themselves to experimental manipulation, especially with *Besnoitia* infection (Frenkel, 1956) (Fig. 5). Again, after several months of chronic infection with either *Toxoplasma* or *Besnoitia*, one sporadically sees hamsters lose weight and become weak, developing the equivalent of Addison's disease

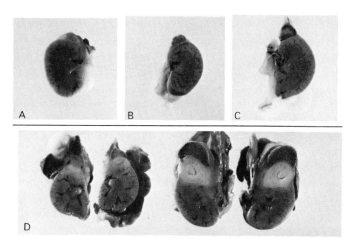

Fig. 3. Kidneys with attached adrenals from hamsters with *Besnoitia* or *Toxoplasma* infection. (A) normal; (B) *Besnoitia* infection 11 days with small focal necrosis (light) and hemorrhage (dark); (C) *Toxoplasma* infection of 1 year's duration with partial necrosis and hyperplasia; (D) *Besnoitia* infection 175 days with marked enlargement due to necrosis and compensatory hyperplasia. The liver and spleen are adherent above. [Reprinted from Frenkel (1956) by copyright permission of The Rockefeller University Press.]

from bilateral adrenocortical infection and necrosis, which exceeded the regenerative capacity of the cortex. This late lesion is not easily produced or manipulated (Figs. 3C, D, and 4).

Disseminated *Mycobacterium kansasii* infection in hamsters, which elicits a pure granulomatous reaction in most tissues, but not in the adrenals, provides an indication of the anti-inflammatory activity of cortisol in addition to its immunosuppression (Frenkel, 1972). The latter is suggested by the greater number of mycobacteria in the adrenal than in the liver.

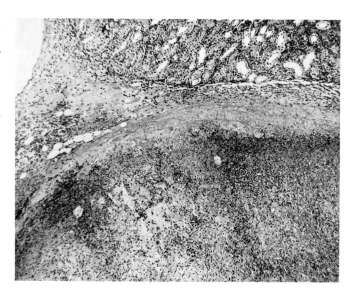

Fig. 4. Chronic *Besnoitia* infection of hamster. The infection is actively progressing in the adrenals, whereas the adjacent kidney is unaffected. Hematoxylin–eosin stain. Magnification: 60×.

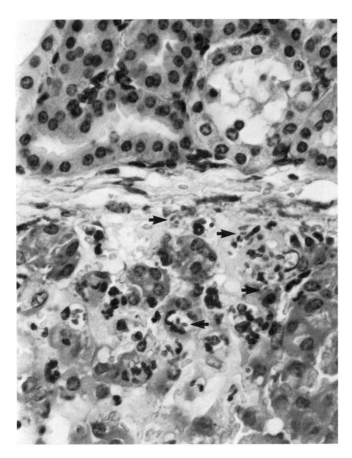

Fig. 5. Acute *Besnoitia* infection with adrenal involvement but sparing of the adjacent kidney. *Besnoitia* tachyzoites (arrows) are seen within and between cortical cells. Hematoxylin–eosin stain. Magnification: 532×. [Reprinted from Frenkel (1956) by copyright permission of The Rockefeller University Press.]

Recognition of local hypercorticism as the basis of infectious adrenal necrosis by *Besnoitia, Toxoplasma,* and *Mycobacterium kansasii,* suggested application of this hypothesis to humans (Frenkel, 1960a, 1961a). Infectious adrenal necrosis in humans results from tuberculosis, histoplasmosis and other fungal infections, syphilis, cytomegalovirus, and possibly other viral infections. In the adrenals, organisms are more numerous, the granulomatous reaction is depressed, and there is more necrosis than in other involved organs of the same patients. Clarification of the pathogenesis of adrenal necrosis leading to Addison's disease in humans depended on the hamster–*Besnoitia* model and an understanding of antiinflammatory and immunity-depressing corticosteroid effects.

B. Pineal Gland

During corticoid-produced relapse of chronic toxoplasmic infection in hamsters, pineal necrosis due to proliferating organisms is common, as is encephalitis and pneumonia (Frenkel *et al.,* 1975). The predisposing causes of pineal involvement are unknown and have not been studied systematically.

C. Thyroid Gland

During a study of disseminated mycobacterial infection (*M. kansasii*) in hamsters, it was noted that there were fewer bacteria in the thyroid, that Schaumann's bodies were fewer, but that epithelioid and giant-cell reactions were larger and that paraamyloid deposits were plentiful. This was duplicated in other organs by feeding 1% powdered thyroid in the diet (Frenkel, 1958, 1972).

D. Placenta

Placental infections may be asymptomatic and without lesions as with mumps and rat virus (Ferm and Kilham, 1963a,b); whereas mumps does not pass into the fetus, the RV strain of parvovirus does so frequently.

Herpes infection of pregnant hamsters gives rise to severe adrenal and placental necrosis and to inflammation and inclusion bodies; although traces of virus passed the placenta, no lesions were found in near-term fetuses except when they were injected directly (Ferm and Low, 1965). *Escherichia coli* endotoxin injected intravenously gave rise to placental necrosis within 24–48 hr (Lanning and Hilbelink, 1983).

XI. INFECTIONS WITH SKIN AND OTHER LESIONS

A. Staphylococcal Infection

Injections of *Staphylococcus aureus* into the cheek pouch led to abscess formation at the injection site and some dissemination. Lesions subsided after 3 days (Young, 1954).

B. Impetigo Model

Hamsters were explored as a model for human impetigo, usually caused by group A β-hemolytic streptococci and *Staphylococcus aureus* (Cushing and Mortimer, 1970; Dajani and Wannamaker, 1970). Bacteriocins from *Staphylococcus epidermidis* or *S. aureus* observed *in vitro* were also effective *in vivo*. Probably in part because of this, the number of streptococci found associated with staphylococci in patients with impetigo varies (Dajani and Wannamaker, 1971). Several chemotherapeutic regimens were assessed in hamsters (Dajani and Wannamaker, 1971).

C. Gonorrhea

Gonorrheal infection and immunity was studied in hamsters by Arko (1974).

D. Demodectic Mange

Hamsters are infected with two species of hair follicle mites. Biotin deficiency decreases the population of *Demodex criceti* and *D. aurati* (Nutting and Rauch, 1963). Distribution and biology of *D. aurati* was described by Nutting and Rauch (1963), and demodectic mange by Estes *et al.* (1971).

E. Mite Infection

The tropical rat mite *Liponyssus bacoti* feeds well on hamsters, which can be used as experimental hosts involving this arthropod (Scott, 1958).

XII. HAMSTERS AS SENTINELS OF INFECTION

A. Sentinels in Nature

Because of their great susceptibility to certain arboviruses, their attraction to the insect vectors, and their self-sufficiency, hamsters have been exposed in mesh cages in areas where certain infections were suspected. *Leishmania* has been isolated from such hamsters exposed above ground level in Peru, Russia, and Panama (Herrer *et al.*, 1973, 1980; Herrer, 1982). Viral isolation or antibody response was used to indicate the presence of arbovirus (togavirus)-infected mosquitos at treetop or other levels (Dickerman *et al.*, 1980; Srihongse *et al.*, 1967; Sanmartin *et al.*, 1971).

B. Indicators of Arthropod Infection

Mosquitos, ticks, and other arthropods collected in nature can be triturated and injected into hamsters to isolate any of the organisms hamsters are highly susceptible to. For example, *Babesia microti* was isolated from ticks (Walter, 1981).

C. Sentinels of Contamination

Leptospirosis, as discussed in Sections III, VI, and VII, can be a latent infection in mice and rats. Inocula passaged in these hosts may get contaminated and then give rise to spurious results. Their susceptibility to leptospirosis makes hamsters a good model to detect such contamination.

D. Hamsters as Filters of Infection

Haemobartonella, *Eperythrozoon*, and similar erythrocytic parasites of rats and mice can contaminate other blood parasites, such as *Plasmodium* and *Babesia*, passaged in mice or rats. Because hamsters are resistant to *Haemobartonella* but susceptible to rodent *Plasmodium* and *Babesia*, repeated passage of the latter may free it from its contaminants (Hsu and Geiman, 1952).

XIII. HAMSTERS AS MODELS OF CELLULAR IMMUNITY AGAINST INFECTION

The chance observations of adrenal necrosis in a hamster chronically infected with *Toxoplasma* (Frenkel, 1956) and of relapsing toxoplasmosis in a patient treated with corticosteroids suggested the investigation of corticoid sensitivity of acquired immunity in hamsters, the nature of cell-mediated immunity, and the pathogenesis of infectious adrenal necrosis. Because of their greater sensitivity to corticosteroids than mice and rats (Frenkel and Havenhill, 1963), hamsters were useful to compare the immunity-depressing potencies of cortisol and potentiated analogs in producing relapse of chronic *Besnoitia* infections (Frenkel, 1960a). However, corticoids did not enhance acute infections in hamsters or in cell cultures, indicating that corticosteroids were not infection enhancing, but immunodepressing. As established antibody levels were not decreased by corticosteroids and passive transfer of antibody was not protective, it was suspected that corticoids affected cellular immunity. The role of lymphoid cells in protective cell-mediated immunity was shown by the adoptive transfer of spleen and lymph node cells from hamsters immune to *Toxoplasma* and *Besnoitia* (Frenkel, 1967). This adoptive transplantation of lymphoid cells was contingent on isogenicity of donor and recipient hamsters. Allogeneic hamsters had to be irradiated (500–600 rem) or made tolerant to the donor cells. The LD_{50} for such hamsters was 900–925 rem. The characteristics of cellular immunity in hamsters were delineated by means of total-body irradiation; transplanted cellular immunity was inactivated by 50–100 rem to the donor, and by 800 rem to the recipient, while about 400–500 rem inhibited active immunization, and 1500–1800 rem in divided doses were necessary to abolish established immunity (Frenkel and Wilson, 1972). The effects of several cytostatics, antimetabolites, and alkylating agents were compared with antilymphocyte serum and 600 rem during the period of immunogenesis (Wilson and Frenkel, 1971).

The specificity of cellular immunity was shown in intact hamsters by titrated challenge (Frenkel and Caldwell, 1975).

The roles of lymphocytes and other inflammatory cells were analyzed after specific and nonspecific intraperitoneal challenge of hamsters, and the role of macrophages *in vitro*, indicating that specificity resided in lymphocytes of hamsters, whereas macrophages were effector cells (Hoff and Frenkel, 1974). The corticoid sensitivity of these lymphocytes and of macrophages before and after specific sensitization was characterized and compared to the level of cortisol in the circulation of hamsters and in the adrenal vein blood (Lindberg and Frenkel, 1977). Because *Toxoplasma* infects mainly fibroblasts, heptocytes, neurons, and myocardial cells of hamsters, the expression of acquired immunity on such "somatic cells," as opposed to macrophages, was investigated. It was found that infected fibroblasts and kidney cells of hamsters could be protected with specifically sensitized lymphocytes (Chinchilla and Frenkel, 1978). The first specific mediators of protective cellular immunity were described from hamsters as a 4000- to 5000-Da polypeptide (Chinchilla and Frenkel, 1978). While these mediators were specific for *Toxoplasma* and *Besnoitia* as used for immunization, a nonspecific lymphokine measuring greater than 43,000 Da resembling γ-interferon was also described. The dynamics of *Toxoplasma* infection in naive and immunized hamsters was correlated with the appearance of mediators, antibody, and clinical immunity (Reyes and Frenkel, 1987). Whereas the latter two appeared between 7 and 16 days, delayed-type hypersensitivity measured by skin test was present 4 days after infection (Elwell and Frenkel, 1984a,b). Thus a wealth of information about protective cellular immunity was developed in hamsters, including data on corticosteroid and radiation sensitivity of the mediator lymphocyte, the characteristics of lymphokines secreted, and the role of corticosteroids in the pathogenesis of adrenal infection. Similar models, of importance in medicine, are not available in rats, mice, or guinea pigs.

XIV. CONCLUSIONS

Hamsters have been useful to study numerous infectious diseases. They have been *uniquely useful or essential* in clarifying the pathogenesis of ocular, chronic progressive toxoplasmosis, and relapsing toxoplasmosis in the hypercorticoid host, in the study of *Mycobacterium kansasii* and other atypical mycobacterial infections, in the clarification of the pathogenesis of adrenal necrosis leading to Addison's disease in humans, in demonstrating protective cellular immunity to intracellular infection, in a study of hormonal effects on sarcoidlike granulomas and Schaumann bodies, and in gaining an understanding of the pathogenesis of parvovirus infection, primary atypical pneumonia, leptospirosis, antibiotic-associated colitis, and babesiosis.

Hamsters have been *valuable* in the study of the pathogenesis of syphilis, leprosy, visceral fungal infections, scrapie, legionellosis, histoplasmosis, influenza, and several arbovirus infections, particularly as sentinels.

Hamsters present *problems* because lymphocytic choriomeningitis virus and *Encephalitozoon* infections may be present subclinically, which may confound other experimental endeavors.

REFERENCES

Adler, S. (1937). Inoculation of human leprosy into Syrian hamster. *Lancet* **233**, 714–715.

Adler, S. (1948). Origin of the golden hamster, *Cricetus auratus*, as a laboratory animal. *Nature (London)* **162**, 256–257.

Adler, S., and Theodor, O. (1931). Investigations on Mediterranean Kala-azar. II. *Leishmania infantum*. *Proc. R. Soc. London, Ser. B* **108**, 453–463.

Agu, W. E., Farrell, J. P., and Soulsby, E. J. L. (1981). Proliferative glomerulonephritis in experimental *Leishmania donovani* infection of the golden hamster. *Comp. Immunol. Microbiol. Infect. Dis.* **4**, 353–368.

Ali, M. J., Teh, C. Z., Jennings, R., and Potter, C. W. (1982). Transmissibility of influenza viruses in hamsters. *Arch. Virol.* **72**, 187–197.

Angulo, A. F., Spaans, J., Zemmouchi, L., and Van Der Waaij, D. (1978). Selective decontamination of the digestive tract of Syrian hamsters. *Lab. Anim.* **12**, 157–158.

Anzil, A. P., Blinzinger, K., and Mayr, A. (1973). Persistent *Borna* virus infection in adult hamsters. *Arch. Gesamte Virusforsch.* **40**, 52–57.

Arko, R. J. (1974). An immunologic model in laboratory animals for the study of *Neisseria gonorrhoeae*. *J. Infect. Dis.* **129**, 451–455.

Aronson, S. M., and Shwartzman, G. (1956). The histopathology of brown fat in experimental poliomyelitis. *Am. J. Pathol.* **32**, 315.

Austin, F. J., and Scherer, W. F. (1971). Studies of viral virulence. I. Growth and histopathology of virulent and attenuated strains of Venezuelan encephalitis virus in hamsters. *Am. J. Pathol.* **62**, 195–210.

Badiola, J., Thierman, A. B., and Cheville, N. F. (1983). Pathologic features of leptospirosis in hamsters caused by *Leptospira interrogans* serovars *hardjo* and *szwajizak*. *Am. J. Vet. Res.* **44**, 91–99.

Barile, M. F., Chandler, D. K. F., Yoshida, H., Grabowski, M. W., Harasawa, R., and Ahmed, O. A. (1981). Hamster challenge potency assay for evaluation of *Mycoplasma pneumoniae* vaccines. *Isr. J. Med. Sci.* **17**, 682–686.

Barker, C. F., and Billingham, R. E. (1971). The lymphatic status of hamster cheek pouch tissue in relation to its properties as a graft and as a graft site. *J. Exp. Med.* **133**, 620–639.

Bartlett, J. G. (1979). Antibiotic-associated pseudomembranous colitis. *Rev. Infect. Dis.* **1**, 530–539.

Bartlett, J. G., Onderdonk, A. B., and Cisneros, R. L. (1977a). Clindamycin-associated colitis in hamsters: Protection with vancomycin. *Gastroenterology* **73**, 772–776.

Bartlett, J. G., Onderdonk, A. B., Cisneros, R. L., and Kasper, D. L. (1977b). Clindamycin-associated colitis due to a toxin-producing species of *Clostridium* in hamsters. *J. Infect. Dis.* **136**, 701–705.

Baum, S. G., Lewis, A. M., Rowe, W. P., and Huebner, R. J. (1966). Epidemic nonmeningitic lymphocytic-choriomeningitis-virus infection. An outbreak in a population of laboratory personnel. *N. Engl. J. Med.* **274**, 934–936.

Bauman, D. S., and Chick, E. W. (1969). An experimental model for studying extrapulmonary dissemination of *Histoplasma capsulatum* in hamsters. *Am. Rev. Respir. Dis.* **100**, 79–81.

Bernard, G. R. (1961). Experimental trichinosis of the golden hamster. III. Urinary changes with observations upon the urinary uric acid/creatinine ratio. *J. Infect. Dis.* **108**, 1–11.

Bernard, G. R., and Sudak, F. N. (1960). Experimental trichinosis in the golden hamster. II. Electrocardiographic changes. *Am. Heart J.* **60**, 89–93.

Bessemans, A., Vanderputte, J., and Baert, H. (1956). Evolution, pendant l'hibernation, de la syphilis, la trypanosomiase et la tuberculose chez le hamster doré de Syrie. *Rev. Belge Pathol. Med. Exp.* **25**, 491–497.

Binford, C. H. (1959). Histocytic granulomatous mycobacterial lesions produced in the golden hamster (*Cricetus auratus*) inoculated with human leprosy. Negative results in experiments using other animals. *Lab. Invest.* **8**, 901–923.

Bode, L., and Geyer, E. (1981). Experimental dicrocoeliasis. The humoral immune response of golden hamsters and rabbits to primary infection with *Dicrocoelium dendriticum*. *Z. Parasitenkd.* **66**, 167–178.

Bretana, A., Avila, J. L., Lizardo, G., Convit, J., and Rondon, A. J. (1983). *Leishmania* species comparative ultrastructure of experimental nodules and diffuse human cutaneous lesions in American leishmaniasis. *Exp. Parasitol.* **55**, 377–385.

Brown, P. (1984). Biological and chemotherapeutic forays into the field of unconventional viruses. *In* "Targets for the Design of Antiviral Agents" (E. De Clercq and R. T. Walker, eds.), pp. 131–157. Plenum, New York.

Brown, P., Rohwer, R. G., Moreau-Dubois, M. C., Green, E. M., and Gajdusek, D. C. (1981). Use of the golden Syrian hamster in the study of scrapie virus. *In* "Hamster Immune Responses in Infectious and Oncologic Diseases" (J. W. Streilein, D. A. Hart, J. Stein-Streilein, W. R. Duncan, and R. E. Billingham, eds.), pp. 365–373. Plenum, New York.

Brown, P., Rohwer, R. G., Green, E. M., and Gajdusek, D. C. (1982). Effect of chemicals, heat, and histopathologic processing on high-infectivity hamster-adapted scrapie virus. *J. Infect. Dis.* **145**, 683–687.

Browne, R. A., Fekety, R., Silva, J. Jr., Boyd, D. I., Work, C. O., and Abrams, G. D. (1977). The protective effect of vancomycin on clindamycin-induced colitis in hamsters. *Johns Hopkins Med. J.* **141**, 183–192.

Brunner, K. T., and Meyer, K. F. (1950). Immunization of hamsters and dogs against experimental leptospirosis. *J. Immunol.* **64**, 365–372.

Buchmeier, M. J., and Rawls, W. E. (1977). Variation between strains of hamsters in the lethality of Pichinde virus infections. *Infect. Immun.* **16**, 413–421.

Buhler, V. B., and Pollack, A. (1953). Human infection with atypical acid-fast organisms. Report of 2 cases with pathologic findings. *Am. J. Clin. Pathol.* **23**, 363.

Buyukmihci, N., Goehring-Harmon, F., and Marsh, R. S. (1982a). Retinal degeneration during clinical scrapie encephalopathy in hamsters. *J. Comp. Neurol.* **205**, 153–160.

Buyukmihci, N., Goehring-Harmon, F., and Marsh, R. S. (1982b). Optic nerve degeneration in hamsters experimentally infected with scrapie. *Exp. Neurol.* **78**, 780–785.

Byington, D. P., and Johnson, K. P. (1972). Experimental subacute sclerosing panencephalitis in the hamster: Correlation of age with chronic inclusion-cell encephalitis. *J. Infect. Dis.* **126**, 18–26.

Byington, D. P., and Johnson, K. P. (1975). Subacute sclerosing panencephalitis virus in immunosuppressed adult hamsters. *Lab. Invest.* **32**, 91–97.

Byram, J. E., Imohiosen, E. A. E., and von Lichtenberg, F. (1978). Tissue eosinophil proliferation and maturation in schistosome-infected mice and hamsters. *Am. J. Trop. Med. Hyg.* **27**, 267–270.

Campos-Neto, A., and Bunn-Moreno, M. M. (1982). Polyclonal B cell activation in hamsters *Mesocricetus auratus* infected with parasites of the genus *Leishmania*. *Infect. Immun.* **38**, 871–876.

Capin, R., Capin, N. R., Carmona, M., and Ortiz-Ortiz, L. (1980a). Effect of complement depletion on the induction of amoebic liver abscess in the hamster *Mesocricetus auratus*. *Arch. Invest. Med.* **11** (Sp. 1), 173–180.

Capin, R., Gonzalez-Mendoza, A., and Ortiz-Ortiz, L. (1980b). Diminution of the activity of the mononuclear phagocytic system in hamsters *Mesocricetus auratus* infected with *Entamoeba histolytica*. *Arch. Invest. Med.* **11** (Sp. 1), 235–240.

Carrigan, D. R., and Johnson, K. P. (1980). Chronic relapsing myelitis in hamsters associated with experimental measles virus infection. *Proc. Nat. Acad. Sci. U.S.A.* **77**, 4297–4300.

Carrigan, D. R., McKendall, R. R., and Johnson, K. P. (1978). CNS disease following dissemination of SSPE measles virus from intraperitoneal inoculation of suckling hamsters. *J. Med. Virol.* **2**, 347–357.

Chandler, D. K. F., Collier, A. M., and Barile, M. F. (1982). Attachment of *Mycoplasma pneumoniae* to hamster tracheal organ cultures, tracheal outgrowth monolayers, human erythrocytes, and WiDr human tissue culture cells. *Infect. Immun.* **35**, 937–942.

Chinchilla, M., and Frenkel, J. K. (1978). Mediation of immunity to intracellular infection (*Toxoplasma* and *Besnoitia*) within somatic cells. *Infect. Immun.* **19**, 999–1012.

Chute, R. M. (1961). Infections of *Trichinella spiralis* in hibernating hamsters. *J. Parasitol.* **47**, 25–29.

Chute, R. N., Kenton, H. B., and Sommers, S. C. (1954). A laboratory epidemic of human-type tuberculosis in hamsters. *Am. J. Clin. Pathol.* **24**, 223–226.

Coe, J. E., and Bell, J. F. (1977). Antibody response to rabies virus in Syrian hamsters. *Infect. Immun.* **16**, 915–919.

Coggin, J. H., Oakes, J. E., Huebner, R. J., and Gilden, R. (1981). Unusual filterable oncogenic agent isolated from horizontally transmitted Syrian hamster lymphomas. *Nature (London)* **290**, 336–338.

Coggin, J. H., Bellomy, B. B., Thomas, K. V., and Pollock, W. J. (1983). B-Cell and T-cell lymphomas and other associated diseases induced by an infectious DNA viroid-like agent in hamsters (*Mesocricetus auratus*). *Am. J. Pathol.* **110**, 254–266.

Collier, A. M., and Baseman, J. B. (1973). Organ culture techniques with mycoplasmas. *Ann. N.Y. Acad. Sci.* **225**, 277–289.

Collier, A. M., and Clyde, W. A., Jr. (1974). Appearance of *Mycoplasma pneumoniae* in lungs of experimentally infected hamsters and sputum from patients with natural disease. *Am. Rev. Respir. Dis.* **110**, 765–773.

Crandall, C. A., Neilson, J. T., and Crandall, R. B. (1982). Evaluation of inbred strains of hamsters as hosts for *Brugia malayi*. *Trans. R. Soc. Trop. Med. Hyg.* **76**, 277.

Cushing, A. H., and Mortimer, E. A. (1970). A hamster model for streptococcal impetigo. *J. Infect. Dis.* **122**(3), 224–226.

Dajani, A. S., and Wannamaker, L. W. (1970). Experimental infection of the skin in the hamster simulating human impetigo. I. Natural history of the infection. *J. Infect. Dis.* **122**, 196.

Dajani, A. S., and Wannamaker, L. W. (1971). Experimental infection of the skin in the hamster simulating human impetigo. III. Interaction between staphylococci and Group A staphylococci. *J. Exp. Med.* **134**, 588–599.

Dajani, A. S., Clyde, A., and Denny, F. W. (1965). Experimental infection with *Mycoplasma pneumoniae* (Eaton's agent). *J. Exp. Med.* **121**, 1071–1086.

Dannenberg, A. M., Jr., and Scott, E. M. (1960). Melioidosis: Pathogenesis and immunity in mice and hamsters. III. Effect of vaccination with avirulent strains of *Pseudomonas pseudomallei* on the resistance to the establishment and the resistance to progress of respiratory melioidosis caused by virulent strains; all-or-none aspects of this disease. *J. Immunol.* **84**, 233–246.

Desmecht, M. (1982). Durée de l'immunité conférée par le vaccin de la leptospirose canine chez le hamster. *Ann. Med. Vet.* **126**, 55–58.

Dickerman, R. W., Pinheiro, F. P., Oliva, O. F., Travassos, da Rosa, J. F., and Calisher, C. H. (1980). Eastern encephalitis virus from virgin forests of Northern Brazil. *Bull.—Pan Am. Health Organ.* **14,** 15–21.

Drouhet, E. (1961). Coccidiomycose d'importation observée en France. *Bull. Soc. Pathol. Exot. Ses Fil.* **54,** 1002–1007.

Drouhet, E., and Schwarz, J. (1956). Comparative studies with 18 strains of *Histoplasma*. *J. Lab. Clin. Med.* **47,** 128–139.

Drutz, D. J., DeMarsh, P., Edelstein, J., Owens, R. W., and Finegold, S. (1984). *Legionella pneumophila* pneumonia in athymic nude mice. In "*Legionella*" (C. Thornberry, A. Balows, J. C. Feeley, and W. Jakubowski, eds.), pp. 134–135. Am. Soc. Microbiol., Washington, D.C.

Dumont, A. (1972). Ultrastructural aspects of phagocytosis of facultative intracellular parasites by hamster peritoneal macrophages. *RES, J. Reticuloendothel. Soc.* **11,** 469–491.

Dumont, A., and Robert, A. (1970). Electron microscopic study of phagocytosis of *Histoplasma capsulatum* by hamster peritoneal macrophages. *Lab. Invest.* **23,** 278–286.

Dumont, A., and Sheldon, H. (1965). Changes in the fine structure of macrophages in experimentally produced tuberculous granulomas in hamsters. *Lab. Invest.* **14,** 2034–2055.

Eaton, M. D., Meiklejohn, G., and van Herick, W. (1944). Studies on the etiology of primary atypical pneumonia. A filterable agent transmissible to cotton rats, hamsters and check embryos. *J. Exp. Med.* **79,** 649–668.

Eddy, B. E. (1972). Viral infectivity of the Syrian hamster. *Prog. Exp. Tumor Res.* **16,** 454–496.

Elwell, M. R., and Frenkel, J. K. (1984a). Immunity to toxoplasmosis in hamsters. *Am. J. Vet. Res.* **43,** 2668–2674.

Elwell, M. R., and Frenkel, J. K. (1984b). Acute toxoplasmosis in hamsters and mice: Measurement of pathogenicity by fever and weight loss. *Am. J. Vet. Res.* **45,** 2663–2667.

Enright, J. B., Franti, C. E., Frye, F. L., and Behymer, D. E. (1970). Effects of corticosteroids on rabies in mice. *Can. J. Microbiol.* **16,** 667–675.

Estes, P. C., Richter, C. B., and Franklin, J. A. (1971). Demodectic mange in the golden hamster. *Lab. Anim. Sci.* **21,** 825–828.

Fekety, R., Silva, J., Toshniwal, R., Allo, M., Armstrong, J., Browne, R., Ebright, J., and Rifkin, G. (1979). Antibiotic-associated colitis: Effects of antibiotics on *Clostridium difficile* and the disease in hamsters. *Rev. Infect. Dis.* **1,** 386–397.

Ferm, V. H., and Kilham, L. (1963a). Mumps virus infection of the pregnant hamster. *J. Embryol. Exp. Morphol.* **11** (Part 4), 659–665.

Ferm, V. H., and Kilham, L. (1963b). Rat virus (RV) infection in fetal and pregnant hamsters. *Proc. Soc. Exp. Biol. Med.* **112,** 623–626.

Ferm, V. H., and Low, R. J. (1965). Herpes simplex virus infection in the pregnant hamster. *J. Pathol. Bacteriol.* **89,** 295–300.

Findley, G. M. (1934). The infectivity of neurotropic yellow fever virus for animals. *J. Pathol. Bacteriol.* **38,** 1–6.

Flavell, D. J., Pattanapanyasat, K., Lucas, S. B., and Vongsangnak, V. (1980). *Opisthorchis-viverrini* liver changes in golden hamsters maintained on high and low protein diets. *Acta Trop.* **37,** 337–350.

Florio, L., Hammon, W. M., Laurent, A., and Stewart, M. O. (1946). Colorado tick fever and dengue. An experimental immunological and clinical comparison. *J. Exp. Med.* **83,** 295–301.

Fox, J. G., and Hering, A. M. (1983). The pet hamster as a potential reservoir of human campylobacteriosis. *J. Infect. Dis.* **147,** 784.

Fox, J. G., Zanotti, S., and Jordan, H. V. (1981). The hamster as a reservoir of *Campylobacter fetus* subspecies *jejuni*. *J. Infect. Dis.* **143,** 856.

Frenkel, J. K. (1953). Host, strain and treatment variation as factors in the pathogenesis of toxoplasmosis. *Am. J. Trop. Med. Hyg.* **2,** 390–415.

Frenkel, J. K. (1955). Ocular lesions in hamsters with chronic *Toxoplasma* and *Besnoitia* infection. *Am. J. Ophthalmol.* **39,** 202–225.

Frenkel, J. K. (1956). Effects of hormones on the adrenal necrosis produced by *Besnoitia jellisoni* in golden hamsters. *J. Exp. Med.* **103,** 375–398.

Frenkel, J. K. (1958). Hormonal effects on a sarcoid-like response with Schaumann bodies and amyloid in golden hamsters infected with photochromogenic mycobacteria. *Am. J. Pathol.* **34,** 586–587.

Frenkel, J. K. (1960a). Pathogenesis of infections of the adrenal gland, leading to Addison's disease in man. Role of corticoids in adrenal and generalized infection. *Ann. N.Y. Acad. Sci.* **94,** 391–440.

Frenkel, J. K. (1960b). Evaluation of infection-enhancing activity of effects of anti-modified corticoids. *Proc. Soc. Exp. Biol. Med.* **103,** 552–555.

Frenkel, J. K. (1961a). Infections involving the adrenal cortex. In "The Adrenal Cortex" (H. D. Moon, ed.), pp. 201–219. Harper (Hoeber), New York.

Frenkel, J. K. (1961b). Pathogenesis of toxoplasmosis with a consideration of cyst rupture in *Besnoitia* infection. *Surv. Ophthalmol.* **6,** 799–825.

Frenkel, J. K. (1962). Role of corticosteroids as predisposing factors in fungal diseases. *Lab. Invest.* **11,** 1192–1208.

Frenkel, J. K. (1967). Adoptive immunity to intracellular infection. *J. Immunol.* **98,** 1309–1319.

Frenkel, J. K. (1971). Protozoal diseases of laboratory animals. In "Pathology of Protozoal and Helminthic Diseases with Clinical Correlations" (R. A. Marcial-Rojas, ed.), pp. 318–369. Williams & Wilkins, Baltimore, Maryland.

Frenkel, J. K. (1972). Infection and immunity in hamsters. *Prog. Exp. Tumor Res.* **16,** 326–367.

Frenkel, J. K., and Caldwell, S. A. (1975). Specific immunity and nonspecific resistance to infection: Listeria, protozoa, and viruses in mice and hamsters. *J. Infect. Dis.* **131,** 201–209.

Frenkel, J. K., and Havenhill, M. A. (1963). The corticoid sensitivity of golden hamsters, rats, and mice. Effects of dose, time and route of administration. *Lab. Invest.* **12,** 1204–1220.

Frenkel, J. K., and Jacobs, L. (1958). Ocular toxoplasmosis: Pathogenesis, diagnosis and treatment. *AMA Arch. Ophthalmol.* **59,** 260–279.

Frenkel, J. K., and Wilson, H. R. (1972). Effects of radiation on specific cellular immunities: Besnoitiosis and a herpesvirus infection of hamsters. *J. Infect. Dis.* **125,** 216–230.

Frenkel, J. K., Cook, K., Grady, H. J., and Pendleton, S. K. (1965). Effects of hormones on adrenocortical secretion of golden hamsters. *Lab. Invest.* **14,** 142–156.

Frenkel, J. K., Nelson, B. M., and Arias-Stella, J. (1975). Immunosuppression and toxoplasmic encephalitis: Clinical and experimental aspects. *Hum. Pathol.* **6,** 97–111.

Frisk, C. S., Wagner, J. E., and Owens, D. R. (1981). Hamster (*Mesocricetus auratus*) enteritis caused by epithelial cell-invasive *Escherichia coli*. *Infect. Immun.* **31,** 1232–1238.

Fultz, P. N., Shadduck, J. A., Kang, C. Y., and Streilein, J. W. (1981). Involvement of cells' hematopoietic origin in genetically determined resistance of Syrian hamsters to vesicular stomatitis virus. *Infect. Immun.* **34,** 540–549.

Fultz, P. N., Shadduck, J. A., Yang, C. Y., and Streilein, J. W. (1982). Mediators of protection against lethal systemic vesicular stomatitis virus infection in hamsters: Defective interfering particles, polyinosinate-polycytidylate, and interferon. *Infect. Immun.* **37,** 679–686.

Gabridge, M. G. H., Gunderson, H., Schaeffer, S. L., and Dee Barden-Stahl, Y. (1978). Ciliated respiratory epithelial monolayers: New model for *Mycoplasma pneumoniae* infection. *Infect. Immun.* **21,** 333–336.

Galton, M. (1964). Studies of the generalized Schwartzman reaction in the pregnant golden hamster. *Am. J. Pathol.* **44,** 613–627.

Galton, M. (1965). Particle retention by the renal glomerular capillaries in the generalized Schwartzman reaction. *Proc. Soc. Exp. Biol. Med.* **119,** 1139–1141.

Galton, M. (1966). Thrombosis in the pregnant Syrian hamster. *Trans. N. Y. Acad. Sci.* [2] **28,** 423–438.

Galton, M., and Slater, S. M. (1965). Naturally occurring fatal disease of the pregnant golden hamster. *Proc. Soc. Exp. Biol. Med.* **120,** 873–876.

Gee, S. R., Chan, M. A., Clark, D. A., and Rawls, W. E. (1981). Role of natural killer cells in Pichinde virus infection of Syrian hamsters. *Infect. Immun.* **31,** 919–928.

Gellhorn, A., Van Dyke, H. B., Pyles, W. J., and Tupikova, N. A. (1946). Amyloidosis in hamsters with leishmaniasis. *Proc. Soc. Exp. Biol. Med.* **61,** 25–30.

Germuth, F. G., Eagle, H., and Oyama, V. (1950). An evaluation of the criteria of cure in experimental leishmaniasis of the golden hamster. *Am. J. Trop. Med.* **30,** 377–385.

Ghadirian, E., and Meerovitch, E. (1978). Behavior of axenic IP-106 strain of *Entamoeba histolytica* in the golden hamster. *Am. J. Trop. Med.* **27,** 241–247.

Ghadirian, E., and Meerovitch, E. (1981). Effect of immunosuppression on the size and metastasis of amoebic liver abscesses in hamsters. *Parasite Immunol.* **3,** 329–338.

Ghadirian, E., and Meerovitch, E. (1982). Effect of complement depletion of hepatic amebiasis in hamsters. *Clin. Immunol.* **24,** 315–319.

Ginovker, N. G., and Konovalova, L. A. (1981). Circadian activity of the golden hamster liver in the acute stage of opisthorchiasis. *Med. Parazitol. Parazit. Bolezni* **50,** 35–38.

Gold, D., Diamantstein, T., and Hahn, H. (1982). Attempts to immunize golden hamsters against *Entamoeba histolytica*. *Ann. Trop. Med. Parasitol.* **76,** 367–370.

Goldman, P. M., Andrews, E. J., and Lang, C. M. (1972). A preliminary evaluation of *Clostridium* sp. in the etiology of hamster enteritis. *Lab. Anim. Sci.* **22,** 721–724.

Gourmelon, P., Court, L., and Gibbs, C. J. (1983). La tremblante expérimentale du hamster: Analyse de l'activité électrique cérébrale. *Trav. Sci.* **4,** 195–197.

Green, R. H. (1974). The association of viral activation with penicillin toxicity in guinea pigs and hamsters. *Yale J. Biol. Med.* **47,** 629–642.

Hagen, C. A., Shefner, A. M., and Ehrlich, R. (1965). Intestinal microflora of normal hamsters. *Lab. Anim. Care* **15,** 185–193.

Han, S. S., Saeki, H., Hiyama, M., and Ishii, T. (1982). Susceptibility of small laboratory animals to *Babesia rodhaini* infections. *Jpn. J. Vet. Sci.* **44,** 497–502.

Harrison, A. K., Murphy, F. A., Gardner, J. J., and Bauer, S. P. (1980). Myocardial and pancreatic necrosis induced by Rocio virus, a new flavivirus. *Exp. Mol. Pathol.* **32,** 102–113.

Harrison, A. K., Murphy, F. A., and Gardner, J. J. (1982). Visceral target organs in systemic St. Louis encephalitis virus infection of hamsters. *Exp. Mol. Pathol.* **37,** 292–304.

Hasegawa, H., Hotta, T., and Otsura, M. (1980). Diphyllobothriid cestodes in Northern Japan. 4. Plerocercoids recovered from the Alaskan haddock *Theragra chalcogramma* and adults developed in the golden hamster. *Jpn. J. Parasitol.* **29,** 473–482.

Hauduroy, P., and Rosset, W. (1951). BCG et hamster. *Rev. Tuberc. (Paris)* **15,** 1071.

Hayatsu, E., Kawakubo, Y., Yayoshi, M., Araake, M., Wakai, M., Yoshida, A., Yoshioka, M., and Nishiyama, Y. (1981). Immunologic responses of hamsters in the acquired immune state of *Mycoplasma pneumoniae* infection. *Microbiol. Immunol.* **25,** 1255–1263.

Herrer, A. (1982). Use of the golden hamster as sentinel animal in areas where uta (cutaneous leishmaniasis) is endemic. *Rev. Inst. Med. Trop. Sao Paulo* **24,** 162–167.

Herrer, A., Christensen, H. A., and Beumer, R. J. (1973). Detection of leishmanial activity in nature by means of sentinel animals. *Trans. R. Soc. Trop. Med. Hyg.* **67,** 870–879.

Herrer, A., Hidalgo, V., and Meneses, O. (1980). Cutaneous leishmaniasis uta in Peru. Use of contact insecticides and reactivation of the disease during the last years. *Rev. Inst. Med. Trop. Sao Paulo* **22,** 203–206.

Herrold, K. M. (1965). *Cryptococcus neoformans*: Pathogenesis of the disease in Syrian hamsters. Pt. 1. *Fed. Proc., Fed. Am. Soc. Exp. Biol.* **24,** 492.

Hillyer, G. V., Del Llano De Diaz, A., and Reyes, C. N. (1977). *Schistosoma mansoni:* Acquired immunity in mice and hamsters using antigens of *Fasciola hepatica*. *Exp. Parasitol.* **42,** 348–355.

Hoff, R. L., and Frenkel, J. K. (1974). Cell-mediated immunity against *Besnoitia* and *Toxoplasma* in specifically and cross-immunized hamsters and in cultures. *J. Exp. Med.* **139,** 560–580.

Hogan, R. N., Baringer, J. R., and Prusiner, S. B. (1981). Progressive retinal degeneration in scrapie-infected hamsters. *Lab. Invest.* **44,** 34–42.

Holmes, J. C. (1962). Effects of concurrent infections on *Hymenolepis diminuta* (Cestoda) and *Moniliformis dubius* (Acanthocephala). III. Effects in hamsters. *J. Parasitol.* **48,** 97–100.

Hsu, D. Y. M., and Geiman, W. M. (1952). Synergistic effect of *Haemobartonella muris* on *Plasmodium berghei* in white rats. *Am. J. Trop. Med. Hyg.* **1,** 747–749.

Humphrey, C. D., Pittman, J. C., and Pittman, F. E. (1974). A hamster model for lincomycin colitis. *Gastroenterology* **66,** 847.

Humphrey, C. D., Lushbaugh, W. B., Condon, C. W., Pittman, J. C., and Pittman, F. E. (1979). Light and electron microscopic studies of antibiotic-associated colitis in the hamster. **20,** 6–15.

Hutadilok, N., and Ruenwongsa, P. (1983). Liver collagen turnover in hamsters during infection by the human liver fluke, *Opisthorchis viverrini*. *Mol. Biochem. Parasitol.* **8,** 71–77.

Ishii, A. I., Kind, H., Hayashi, M., Fujio, Y., and Sand, M. (1980). Experimental light infection of *Angiostrongylus cantonensis* in hamsters *Mesocricetus auratus*. *Int. J. Zoonoses* **7,** 120–124.

Jahrling, P. B., and Scherer, W. F. (1973a). Homogeneity of Venezuelan encephalitis virion populations of hamster-virulent and benign strains, including the attenuated TC83 vaccine. *Infect. Immun.* **7,** 905–910.

Jahrling, P. B., and Scherer, W. F. (1973b). Growth curves and clearance rates of virulent and benign Venezuelan encephalitis viruses in hamsters. *Infect. Immun.* **8,** 456–462.

Jahrling, P. B., and Scherer, W. F. (1973c). Histopathology and distribution of viral antigens in hamsters infected with virulent and benign Venezuelan encephalitis viruses. *Am. J. Pathol.* **72,** 25–38.

Jespersen, A., and Bentzon, M. W. (1964). The virulence of various strains of BCG determined on the golden hamster. *Acta Tuberc. Scand.* **44,** 222–249.

Johnson, K. P. (1978). Viral infections of the hamster central nervous system. *Fed. Proc., Fed. Am. Soc. Exp. Biol.* **37,** 2074–2075.

Johnson, K. P., and Norrby, E. (1974). Subacute sclerosing panencephalitis (SSPE) agent in hamsters. III. Induction of defective measles infection in hamster brain. *Exp. Mol. Pathol.* **21,** 166–178.

Johnson, R. T., and Johnson, K. P. (1968). Hydrocephalus following viral infection: The pathology of aqueductal stenosis developing after experimental mumps virus infection. *J. Neuropathol. Exp. Neurol.* **27,** 591–606.

Jungeblut, C. W. (1958). Columbia SK group of viruses. In "Handbuch der Virusforschung" (R. Doerr, ed.), pp. 459–580. Springer-Verlag, Berlin and New York.

Justines, G., and Johnson, K. M. (1968). Use of oral swabs for detection of Machupo virus infection in rodents. *Am. J. Trop. Med. Hyg.* **17,** 788–790.

Justines, G., and Johnson, K. M. (1979). Immune tolerance in *Calomys callosus* infected with Machupo virus. *Nature (London)* **222,** 1090–1091.

Juttijudata, P., Chiemchaisri, C., Palavatana, C., and Churnratanakul, S. (1982). A clinical study of cholangiocarcinoma caused cholestasis in Thailand. *Surg., Gynecol. Obstet.* **155,** 373–376.

Kimberlin, R. H., and Walker, C. A. (1977). Characteristics of a short incubation model of scrapie in the golden hamster. *J. Gen. Virol.* **34,** 295.

Kinnamon, K. E., Steck, E. A., Loizeaux, P. S., Chapman, W. L., Jr., Waits, V. B., and Hanson, W. L. (1980). Leishmaniasis: In search of a new chemotherapeutic agent. *Am. J. Vet. Res.* **41,** 405–407.

Kirchhoff, H., Kuwabara, T., and Barile, M. F. (1981). Pathogenicity of *Spiroplasma* sp. strain SMCA in Syrian hamsters: Clinical, microbiological, and histological aspects. *Infect. Immun.* **31,** 445–452.

Koprowski, H. (1952a). Immunization with modified living virus with particular reference to rabies and hog cholera. *Vet. Med. (Kansas City, Mo.)* **47**, 144–150.

Koprowski, H. (1952b). Latent or dormant viral infections. *Ann. N. Y. Acad. Sci.* **54**, 963–976.

Korij, W., Soltani, K., Chyu, J. Y. H., and Rippon, J. W. (1984). Immunohistopathological studies of blastomycosis. The use of labeled specific antigens in a hamster model of disease. *Mycopathology* **85**, 17–20.

Lagace-Simard, J., Descoteaux, J. P., and Lussier, G. (1982). Experimental pneumovirus infections. 2. Hydrocephalus of hamsters and mice due to infection with human respiratory syncytial virus (RS). *Am. J. Pathol.* **107**, 36–40.

Lämmler, G., Zahner, H., and Texdorf, I. (1968). Infektionsversuche mit Darmnematoden, Cestoden und Trematoden bei *Mastomys natalensis* (Smith, 1834). *Z. Parasitenkol.* **31**, 166–202.

Lanning, J. C., and Hilbelink, D. R. (1983). Effects of endotoxin on placental labyrinth development in the hamster. *Teratology* **27**, A60.

Lentsch, R. H., McLaughlin, R. M., Wagner, J. E., and Day, T. J. (1982). *Campylobacter fetus* subspecies *jejuni* isolated from Syrian hamsters with proliferative ileitis. *Lab. Anim. Sci.* **32**, 511–514.

Libby, J. M., Jortner, B. S., and Wilkins, T. D. (1982). Effects of the two toxins of *Clostridium difficile* in antibiotic-associated cecitis in hamsters. *Infect. Immun.* **36**, 822–829.

Lindberg, R. E., and Frenkel, J. K. (1977). Cellular immunity to *Toxoplasma* and *Besnoitia* in hamsters: Specificity and the effects of cortisol. *Infect. Immun.* **15**, 855–862.

Liu, C., Jayanetra, P., and Voth, D. W. (1970). Effect of combined *Mycoplasma pneumoniae* and pneumococcal infections in hamsters. *Ann. N. Y. Acad. Sci.* **174**, 828–834.

Lu, T. Y., Wenner, H. A., and Kamitsuka, P. S. (1961). Experimental infections with Coxsackie viruses. II. Myocarditis in cynomolgus monkeys infected with B4 virus. *Arch. Gesamte Virusforsch.* **10**, 451–464.

Lusk, R. H., Fekety, R., Silva, J., Browne, R. A., Ringler, D. H., and Abrams, G. D. (1978). Clindamycin-induced enterocolitis in hamsters. *J. Infect. Dis.* **137**, 464–475.

MacCallum, D. K. (1969a). Time sequence study on the hepatic system of macrophages in malaria-infected hamsters. *RES, J. Reticuloendothel. Soc.* **6**, 232–252.

MacCallum, D. K. (1969b). A study of macrophage–pulmonary vascular bed interactions in malaria-infected hamsters. *RES, J. Reticuloendothel. Soc.* **6**, 253–270.

McCollum, W. H., Doll, E. R., Wilson, J. C., and Johnson, C. B. (1962). Isolation and propagation of equine rhinopneumonitis virus in primary monolayer kidney cell cultures of domestic animals. *Cornell Vet.* **52**, 164–173.

McKendall, R. R., Carrigan, D. R., and Johnson, K. P. (1981). Lymphoid cell infection by measles virus in newborn hamsters. *J. Neuroimmunol.* **1**, 261–274.

Macotela-Ruiz, E., and Mariat, F. (1963). Sur le production de mycétomes expérimentaux par *N. asteroides* et *N. brasiliensis*. *Bull. Soc. Pathol. Exot. Ses Fil.* **56**, 46–54.

Malone, J. B., Leininger, J. R., and Chapman, W. L. (1976). *Brugia pahangi* histopathological study of golden hamsters. *Exp. Parasitol.* **40**, 62–73.

Manci, E. A., Heath, L. S., Leinbach, S. S., and Coggin, J. H. (1984). Lymphoma associated ulcerative bowel disease in the hamsters (*Mesocricetus auratus*) induced by an unusual agent. *Am. J. Pathol.* **116**, 1–8.

Mansour, N. S., Soliman, G. N., and Elassal, F. M. (1984). Studies on experimental mixed infections of *Schistosoma mansoni* and *Schistosoma haematobium* in hamsters. *Z. Parasitenkd.* **70**, 345–357.

Manuelidis, E. E., Gorgacz, E. J., and Manuelidis, L. (1978). Viremia in experimental Creutzfeldt–Jakob disease. *Science* **200**, 1069.

Margolis, G., and Kilham, L. (1969). Hydrocephalus in hamsters, ferrets, rats, and mice following inoculations with Reovirus type I. II. Pathologic studies. *Lab. Invest.* **21**, 189–198.

Mariat, F., and Drouhet, E. (1954). Sporotrichose expérimentale due hamster. Observation de formes astéroïdes de *Sporotrichum*. *Ann. Inst. Pasteur, Paris* **86**, 485–492.

Marsh, R. F., and Hanson, R. P. (1978). The Syrian hamster as a model for the study of slow virus diseases caused by unconventional agents. *Fed. Proc., Fed. Am. Soc. Exp. Biol.* **37**, 2076–2078.

Marsh, R. F., and Kimberlin, R. H. (1975). Comparison of scrapie and transmissible mink encephalopathy in hamsters. I. Clinical signs, pathology and pathogenesis. *J. Infect. Dis.* **131**, 104–110.

Masters, C. L., and Gajdusek, D. C. (1982). The spectrum of Creutzfeldt–Jakob disease and the virus-induced subacute spongiform encephalopathies. *In* "Recent Advances in Neuropathology" (W. T. Smith and J. B. Cavanagh, eds.), pp. 139–163. Churchill-Livingstone, Edinburgh and New York.

Masters, C. L., Rohwer, R. G., Franko, M. C., Brown, P., and Gajdusek, D. C. (1984). The sequential development of spongiform change and gliosis of scrapie in the golden Syrian hamsters. *J. Neuropathol. Exp. Neurol.* **43**, 242–252.

Meiser, J., Kinzel, V., and Jirovec, O. (1972). Nosematosis as an accompanying infection of plasmacytoma ascites in Syrian golden hamsters. *Pathol. Microbiol.* **37**, 249–260.

Mikhail, I. A., Higashi, G. I., Mansour, N. S., Edman, D. C., and Elwan, S. H. (1981). *Salmonella paratyphi-A* in hamsters *Mesocricetus auratus* concurrently infected with *Schistosoma mansoni*. *Am. J. Trop. Med. Hyg.* **30**, 385–393.

Mikhail, I. A., Higashi, G. I., Edman, D. C., and Elwan, S. H. (1982). Interaction of *Salmonella paratyphi* A and *Schistosoma mansoni* in hamsters *Mesocricetus auratus*. *Am. J. Trop. Med. Hyg.* **31**, 328–334.

Moreau-Dubois, M. C., Brown, P., Rohwer, R. G., Masters, C. L., Franko, M., and Gajdusek, D. C. (1982). Experimental scrapie in golden Syrian hamsters: Temporal comparison of *in vitro* cell-fusing activity with brain infectivity an histopathological changes. *Infect. Immun.* **37**, 195.

Mostofi, F. K. (1967). "Bilharziasis." Springer, New York.

Neilsen, S. L., and Baringer, J. R. (1972). Reovirus-induced aqueductal stenosis in hamsters. *Lab. Invest.* **27**, 531–537.

Neilson, J. T. M. (1978). Alteration of amicrofilaremia in *Dipetalonema viteae*-infected hamsters with immunosuppressive drugs. *Acta Trop.* **35**, 57–61.

Nelles, M. J., and Streilein, W. J. (1980). Immune response to acute virus infection in the Syrian hamster. I. Studies on genetic restriction of cell-mediated cytotoxicity. *Immunogenetics* **10**, 185–199.

Norrby, E. (1978). Viral antibodies in multiple sclerosis. *Prog. Med. Virol.* **24**, 1–39.

Nutting, W. B., and Rauch, H. (1963). Distribution of *Demodex aurati* in the host (*Mesocricetus auratus*) skin complex. *J. Parasitol.* **49**, 323–329.

Okudaira, M., Schwarz, J., and Adriano, S. M. (1961). Experimental production of Schaumann bodies by heterogeneous microbial agents in the golden hamster. *Lab. Invest.* **10**, 968–982.

Pakes, S. P., Shadduck, J. A., and Cali, A. (1975). Fine structure of *Encephalitozoon cuniculi* from rabbits, mice and hamsters. *J. Protozool.* **22**, 481–488.

Parhad, I. M., Johnson, K. P., and Wolinsky, J. S. (1980). Measles retinopathy. A hamster model of acute and chronic lesions. *Lab. Invest.* **43**, 52–60.

Peracoli, M. T. S., Mota, N. G. S., and Montenegro, M. R. (1982). Experimental paracoccidioidomycosis in the Syrian hamster. Morphology and correlation of lesions with humoral and cell-mediated immunity. *Mycopathologia* **79**, 7–17.

Prusiner, S. B., Cochran, S. P., and Alpers, M. P. (1985). Transmission of scrapie in hamsters. *J. Infect. Dis.* **152**, 971–978.

Rachford, F. W. (1970). The hamster as a laboratory host for *Plagiorchis proximus* (Barker, 1915). *J. Parasitol.* **56**, 1137.

Rasmussen, P., and Caulfield, J. B. (1960). The ultrastructure of Schaumann bodies in the golden hamster. *Lab. Invest.* **9**, 330–338.

Rehg, J. E., and Lu, Y. S. (1982). *Clostridium difficile* typhlitis in hamsters not associated with antibiotic therapy. *J. Am. Vet. Med. Assoc.* **181**, 1422–1423.

Rest, J. R. (1983). Pathogenesis of cerebral malaria in golden hamsters *Mesocricetus auratus* and inbred mice. *Contrib. Microbiol. Immunol.* **7**, 139–146.

Reyes, L., and Frenkel, J. K. (1987). Specific and nonspecific mediation of protective immunity to *Toxoplasma gondii*. *Infect. Immun.* **55**(4), 856–863.

Richards, S. W., Peterson, P. K., Niewohner, D. E., and Hoidal, J. N. (1984). Legionnaires disease: A hamster model. *In* "*Legionella*" (C. Thornberry, A. Balows, J. C. Feeley, and W. Jakubowski, eds.), pp. 133–134. Am. Soc. Microbiol., Washington, D.C.

Rifkin, G. D., Fekety, F. R., Silva, J., Jr., and Sack, R. B. (1977). Antibiotic-induced colitis. Implication of a toxin neutralised by *Clostridium sordellii* antitoxin. *Lancet* **2**, 1103–1106.

Rifkin, G. D., Silva, J., Jr., and Fekety, R. (1978). Gastrointestinal and systemic toxicity of fecal extracts from hamsters with clindamycin-induced colitis. *Gastroenterology* **74**, 52–57.

Roth, E. F., Tanowitz, H., Wittner, M., Ueda, Y., Hsieh, H. S., Neumann, G., and Nagel, R. L. (1981). *Babesia microti* biochemistry and function of hamster erythrocytes infected from a human source. *Exp. Parasitol.* **51**, 116–123.

Round, E. M., and Stebbings, N. (1981). A putative interferon induced in hamsters by poly(C), poly(C). *J. Interferon Res.* **1**, 451–456.

Rowin, K. S., Tanowitz, H. B., and Wittner, M. (1982). Therapy of experimental babesiosis. *Ann. Intern. Med.* **97**, 556–558.

Rubio, M. (1959). Natural and acquired immunity against *Trypanosoma cruzi* in the hamster (*Cricetus auratus*). *Biol. Trab. Inst. Biol. "Juan Noe" Fac. Med. Univ. Chile* **27/28**, 95–116.

Sadun, E. H., and Norman, L. (1956). Effect of single inocula, of varied size, on the resistance of hamsters to *Trichinella spiralis*. *J. Parasitol.* **42**, 608–612.

Saenz, A. (1956). Corps de Metchnikoff-Schaumann chez le hamster doré inoculé avec du BCG. *Ann. Inst. Pasteur, Paris* **90**, 751–762.

Saenz, A. (1958a). La tuberculose expérimentale du hamster doré. *Ann. Inst. Pasteur, Paris* **95**, 534–556.

Saenz, A. (1958b). La tuberculose expérimentale du hamster doré. *Ann. Inst. Pasteur, Paris* **95**, 694–720.

Sanmartin, C., Trapido, H., Barreto, P., and Lesmes, C. I. (1971). Isolations of Venezuelan and eastern equine encephalomyelitis viruses from sentinel hamsters exposed in the Pacific lowlands of Colombia. *Am. J. Trop. Med. Hyg.* **20**, 469–473.

Sapp, W. J., Siddique, I. H., Williams, C. S., and Graham, T. (1980). Histopathologic evaluation of livers of pregnant hamsters infected with *Leptospira canicola*. *Am. J. Vet. Res.* **41**, 1288–1292.

Schell, R. F., and Musher, D. N. (1983). Pathogenesis and immunology of treponemal infection. *Immunol. Ser.* **20**, 121–135.

Schell, R. F., LeFrock, J. L., Babu, J. P., and Chan, J. K. (1979). Use of CB hamster in the study of *Treponema pertenue*. *Br. J. Vener. Dis.* **55**, 316–319.

Schell, R. F., Chan, J. K., LeFrock, J. L., and Bagasra, O. (1980a). Endemic syphilis: Transfer of resistance to *Treponema pallidum* strain Bosnia A in hamsters with a cell suspension enriched in thymus-derived cells. *J. Infect. Dis.* **141**, 752–758.

Schell, R. F., LeFrock, J. L., Chan, J. K., and Bagasra, O. (1980b). LSH hamster model of syphilitic infection. *Infect. Immun.* **28**, 909–913.

Schell, R. F., Azadegan, A. A., Nitskansky, S. G., and LeFrock, J. L. (1982). Acquired resistance of hamsters to challenge with homologous and heterologous virulent treponemes. *Infect. Immun.* **37**, 617–621.

Schwarz, J., Baum, G. L., Wang, J. C., and Rubel, H. (1957). Comparative susceptibility to histoplasmosis of hamsters and mice and evaluation of their value for laboratory diagnosis. *Lab. Invest.* **6**, 547–550.

Scott, H. G. (1958). Control of mites on hamsters. *J. Econ. Entomol.* **51**, 412.

Siegert, R. (1971). "Marburg-Virus-Krankheit." Springer-Verlag, Berlin and New York.

Sirisinha, S., Tuti, S., Tawatsin, A., Vichasri, S., Upatham, E. S., and Bunnag, D. (1983). Attempts to induce protective immunity in hamsters against infection by a liver fluke of man (*Opisthorchis viverrini*). *Parasitology* **86**, 127–136.

Small, J. D. (1968). Fatal enterocolitis in hamsters given lincomycin hydrochloride. *Lab. Anim. Care* **18**, 411–420.

Smith, M. A., and Clegg, J. A. (1976). Different levels of acquired immunity to *Schistosoma mansoni* in 2 strains of hamsters. *Parasitology* **73**, 47–52.

Srihongse, S., and Johnson, K. M. (1969). Hemagglutinin production and infectivity patterns in adult hamsters inoculated with group C and other new world arboviruses. *Am. J. Trop. Med. Hyg.* **18**, 273–279.

Srihongse, S., Scherer, W. F., and Galindo, P. (1967). Detection of arboviruses by sentinel hamsters during the low period of transmission. *Am. J. Trop. Med. Hyg.* **16**, 519–524.

Stahlheim, O. H. (1968). Vaccination of hamsters, swine and cattle with viable avirulent *Leptospira pomona*. *Am. J. Vet. Res.* **29**, 1463–1468.

Stauber, L. A. (1966). Characterization of strains of *Leishmania donovani*. *Exp. Parasitol.* **18**, 1–11.

Stoenner, H., Grimes, E., Thrailkill, F., and Davis, E. (1958). Elimination of *Leptospira ballum* from a colony of Swiss albino mice by use of chlortetracycline hydrochloride. *Am. J. Trop. Med. Hyg.* **7**, 423–426.

Sulkin, S. E., Krutzsch, P. H., Allen, R., and Wallis, C. (1959). Studies of the pathogenesis of rabies in insectivorous bats [and hamsters]. *J. Exp. Med.* **110**, 369–388.

Thacker, W. L., Lewis, V. J., Shadduck, J. H., and Winkler, W. G. (1982). Infection of Syrian hamsters with lymphocytic choriomeningitis virus: Comparison of detection methods. *Am. J. Vet. Res.* **43**, 1500–1502.

Thamavit, W., Bhamarapravati, N., Sahaphong, S., Vajrasthira, S., and Angsubhakorn, S. (1978). Effects of dimethylnitrosamine on induction of cholangiocarcinoma in *Opisthorchis viverrini*-infected Syrian golden hamsters. *Cancer Res.* **38**, 4634–4639.

Tobler, L. H., Johnson, K. P., and Buehring, G. C. (1982). Immune response of hamsters to experimental central nervous infection with measles virus. *J. Neuroimmunol.* **2**, 307–320.

Toolan, H. W. (1972). Parvovirus-induced infection of Syrian hamsters. *Prog. Exp. Tumor Res.* **16**, 411–426.

Toolan, H. W. (1978). Susceptibility of the Syrian hamster to virus infection. *Fed. Proc., Fed. Am. Soc. Exp. Biol.* **37**, 2065–2068.

Van Ekdom, L. T. S., and Versteeg, J. (1982). Prevention and curative effects of acyclovir on central nervous system infections in hamsters inoculated with herpes simplex virus. *Am. J. Med.* **73**(1A), 161–164.

Van Marck, E. A. E., Haque, A., and Gigase, P. L. J. (1983). Liver cirrhosis in hamsters *Mesocricetus auratus* infected with *Dipetalonema viteae*. *Contrib. Microbiol. Immunol.* **7**, 245–250.

Verster, A. (1974). The golden hamster as a definitive host of *Taenia solium* and *Taenia saginata*. *Onderstepoort J. Vet. Res.* **41**, 23–28.

Wade, A. E., Fox, L. E., and Swanson, L. E. (1960). Studies on infection and immunity with the cattle lungworm, *Dictyocaulus viviparus* (Bloch). I. Infection in laboratory animals. *J. Vet. Res.* **21**, 753–575.

Walter, G. (1981). Isolation of *Babesia microti* from free-living nymphs of *Ixodes ricinus*. *Acta Trop.* **38**, 187–188.

Wantland, W. W. (1968). Parasitology. *In* "The Golden Hamster—Its Biology and Use in Medical Research" (R. A. Hoffman, P. F. Robinson, and H. Magalhaes, eds.), pp. 171–184. Iowa State Univ. Press, Ames.

Weber, W. J., Creamer, H. R., and Bohl, E. H. (1956). Chemotherapy in hamsters chronically infected with *Leptospira canicola*. *J. Am. Vet. Med. Assoc.* **129,** 271–273.

Western, K. A., Benson, G. D., Gleason, N. N., Healy, G. R., and Schultz, M. G. (1970). Babesiosis in a Massachusetts resident. *N. Engl. J. Med.* **283,** 854–856.

Wilson, H. R., and Frenkel, J. K. (1971). Immunosuppressive agents in intracellular infection: Besnoitiosis in hamsters. *Infect. Immun.* **3,** 756–761.

Wilson, H. R., and Lollini, L. O. (1980). *Leishmania braziliensis braziliensis*: Metastatic infection in a golden hamster. *Trans. R. Soc. Trop. Med. Hyg.* **74,** 833.

Wilson, K. R., Silva, J., Jr., and Fekety, F. R. (1981). Suppression of *Clostridium difficile* by normal hamster cecal flora and prevention of antibiotic-associated cecitis. *Infect. Immun.* **34,** 626–268.

Yoeli, M., and Most, H. (1964). A study of *Plasmodium berghei* in *Thamnomys surdaster* and in other experimental hosts. *Am. J. Trop. Med. Hyg.* **13,** 659–663.

Yong, W. K., Das, P. K., and Dachlan, Y. P. (1983). *Schistosoma mansoni* infection of Syrian golden hamsters. The host humoral immune response in relation to the adult worm burdens after primary infection. *Z. Parasitenkd.* **69,** 41–51.

Young, G. (1954). Experimental staphylococcus infection in the hamster cheek pouch: The process of localization. *J. Exp. Med.* **99,** 299–306.

Zlotnik, I., and Simpson, D. I. H. (1969). The pathology of experimental vervet monkey disease in hamsters. *Br. J. Exp. Pathol.* **50,** 393–399.

Zook, B. C., Huang, K., and Rhorer, R. G. (1977). Tyzzer's disease in Syrian hamsters. *J. Am. Vet. Med. Assoc.* **171,** 833–836.

Chapter 15

Experimental Biology: Genetic Models in Biomedical Research

F. Homburger and Jeanne Peterson

I. Introduction 251
    A. Hamster Genetics prior to 1960 251
    B. Disease Models prior to 1960 252
    C. Concept of Disease Models 252
II. Specific Disease Models 253
    A. Coagulopathies: Thrombosis 253
    B. Amyloidosis 254
    C. "Quaking" or Parkinsonism 255
    D. Spontaneous Seizures: Epilepsy 255
    E. Hydrocephalus 256
    F. Hindleg Paralysis: Model for Demyelinizing Neuropathies 256
    G. Obesity 257
    H. Prostatic Hyperplasia 257
    I. Muscular Dystrophy 258
    J. Carcinogen Susceptibility 259
    K. Slow Acetylation 260
    References 261

I. INTRODUCTION

Certain diseases of Syrian hamsters occur spontaneously. They appear to be inherited or have been proven to have a genetic basis, and have pathological manifestations resembling those of humans (Homburger and Bajusz, 1970).

Such naturally occurring anomalies may be used as models of human disease. These models are described in this chapter, along with certain inherited traits which mimic human susceptibility or resistance toward particular noxious environmental factors such as carcinogens, or toward drugs and other chemicals, such as inherited susceptibility to carcinogens or variable rate of acetylation (Homburger et al., 1972).

A. Hamster Genetics prior to 1960

During the first three decades following the introduction of hamsters into the laboratory in 1931 (Yerganian, 1972), only 11 clearly defined genetic traits were described, and all of these affected mainly the coat color or dermal pigmentation (see Table I). Two of these mutations displayed additional anatomical abnormalities, as explained below.

B. Disease Models prior to 1960

Animals homozygous for the piebald gene, first reported in 1949, have coats with irregular white patches. In addition, 20–30% of these animals show urogenital abnormalities, as discussed by Robinson (1968).

Animals homozygous for the anophthalmia gene, first described in 1958, have white fur and are born without eyes.

In addition, it has been shown that certain strains of hamsters were susceptible to dental caries induction, while other strains were resistant. However, according to Keyes (Robinson, 1968), in these studies it was not possible to demonstrate that these differences were due to genetic differences alone.

As is customary with any new species of animal entering the laboratory, efforts were made to give it the rank of an "inbred laboratory specimen" (Yerganian, 1972). The study of disease models in Syrian hamsters intensified when inbred hamsters became available in substantial numbers and underwent examination for phenotypic deviations from the norm. Since 1960, 23 new genetic mutations have been described in hamsters, and 5 of these show severe abnormalities which are well suited for study as models of human disease. These mutations are shown in Table I.

C. Concept of Disease Models

Although the concept "spontaneous animal model of human disease" is today generally accepted and understood, the terms "animal model" or "animal model of human disease" are not

Table I

Hereditary Traits in Syrian Hamsters

Mutation	Gene symbol	Disease	Hair coat	Reference
Piebald	s	x	x	Foote (1949)
Mottled white	Mo		x	Magalhaes (1954a)
Tawny	T		x	Magalhaes (1954b)
Cream	c		x	Robinson (1955)
Ruby eye	ru		x	Robinson (1955)
Acromelanic white	c^d		x	Robinson (1957)
Anophthalmic white	Wh	x	x	Knapp and Polivanov (1958)
Dermal pigmentation	—		x	Robinson (1959)
White band	Ba		x	Robinson (1960c)
Brown	b*		x	Robinson (1960b)
Light undercolor	—		x	Robinson (1960a)
Cardiomyopathy	cm	x		Homburger et al. (1962)
Frost	f		x	Whitney (1963)
Rust	r†		x	Whitney et al. (1964)
Tortoise-shell	To		x	Robinson (1966)
Dark gray	dg		x	Nixon and Connelly (1967b)
Lethal gray	Lg		x	Nixon and Connelly (1967b)
Hindleg paralysis	pa	x		Nixon and Connelly (1967a)
Dominant spot	Ds		x	Nixon et al. (1969)
Hydrocephalus	hy	x		Yoon and Slaney (1972)
Hairless	hr		x	Nixon (1972)
Naked	N		x	Festing and Wright (1972)
Satin	Sa		x	Robinson (1972)
Longhair	l		x	Schimke et al. (1973)
Rex	rx		x	Whitney and Nixon (1973)
Jute	J		x	Peterson and Yoon (1975)
Seizure	sz	x		Yoon et al. (1976)
Umbrous	U		x	Robinson (1977)
Pinto	pi		x	Nixon and Connelly (1977)
Fur loss	fs		x	Peterson et al. (1981)
Fur deficiency	fd		x	Peterson et al. (1981)
Juvenile gray	jg		x	Peterson et al. (1981)
Ashen	A		x	Peterson et al. (1981)
Quaking	q	x		Peterson et al. (1981)

*Now pink-eyed dilation, p.
†Now brown, b.

found in medical dictionaries (Dorland, 1965; Stedman, 1957; Veillon-Nobel, 1969), nor are they discussed in the 1961 edition of Webster's dictionary in its lengthy definition of "model" as "a person or thing that exactly resembles another," nor does it appear in Webster's 1979 edition. The preface to "Spontaneous Animal Models of Human Disease" (Andrews et al., 1979) carefully avoids defining the concept.

In the broadest sense, disease models could include manifestations of pathology in lower organisms, as well as in mammals. For example, Mendel's studies on sweet peas are pertinent to the understanding of human heritable disease. More strictly speaking, an animal model is a complex of pathological manifestations in animals that resemble similar disease manifestations known to occur in humans. This broad definition includes the pathological manifestations of infections, as well as the pathological sequelae of exposure to oncogenic, toxic, and nutritional agents which can induce the same disease manifestations in animals as are provoked in humans. A narrower definition of spontaneous disease model, applied in this chapter, excludes such disorders induced by extraneous agents, except for susceptibility or resistance to carcinogens and certain chemicals.

In this chapter, the only hamster diseases to be discussed are those which may serve as models of human disease, are observed to be without apparent extraneous cause, and are either demonstrably inherited or shown by exclusion of other causes to be probably due to genetic factors.

II. SPECIFIC DISEASE MODELS

A. Coagulopathies: Thrombosis

1. Human Disease

Thrombosis plays a role in a variety of human vascular diseases, including diseases of peripheral arteries (iliac and femoral) and cerebral and coronary arteries, superficial veins and deep veins (superficial thrombophlebitis and deep venous thrombosis, the latter difficult to diagnose and often ending in pulmonary embolism). Mural, intracardiac thrombi occur in the elderly and in patients of any age with dilated cardiomyopathy, or as a consequence of valvular stenosis. Intraluminal thrombi also occur in the aorta.

Thrombosis represents "hemostasis in the wrong place" and its mechanism is poorly understood, even though it has been studied for 100 years. It remains questionable whether thrombosis is a consequence of intrinsic abnormalities of the hemostatic mechanism or occurs as a normal response to abnormal stimuli (Weatherall et al., 1983; Robbins et al., 1984).

2. Animal Models

Except for the hamster model of atrial thrombosis, there seems to be no spontaneous animal model of thrombosis. A variety of stimuli have been used to cause platelet microthrombi. These include electric current, burning by lasers, or direct application of adenosine diphosphate (ADP). Insertion of cannulas into large arteries has also been used. Inhalation of cigarette smoke, or injection of bacterial endotoxin or thrombin, are other means to induce thrombosis. The infusion of ADP into the coronaries of pigs has been used to cause coronary thrombosis. Vascular (venous) stasis, massive trauma, and other means have been employed to induce experimental thrombosis (Friedberg, 1966; Jordan et al., 1951).

3. Model of Thrombosis in Hamsters

Auricular thrombosis in the Syrian hamster has been described by McMartin and Dodds (1982). Its frequency increases with age, and at the time 75% of the animals have died spontaneously (117 weeks for males and 85 weeks for females), 73% of the surviving animals showed atrial thrombosis. Thrombi were predominantly in the right atrium. The left auricle and the ventricles were also involved in 13% of the animals. Noncardiac thrombi were rare. Thrombosis was often accompanied by bilateral ventricular hypertrophy. Usually there was myocardial degeneration, especially in the oldest hamsters. Bernfeld et al. (1974), using BIO* 15.16 male hamsters allowed to live from 35 to 80 weeks, found atrial thrombosis in 24% of their animals.

A study on two birth cohorts of approximately 100 male and 100 female hamsters each of the BIO F_1D Alexander hybrids (male BIO 15.16 × female BIO 87.20), allowed to live out their natural life span (defined as that time at which there remain 25% survivors) of 85 weeks for females and 117 weeks for males, showed auricular thrombosis in 26.4% and 57.8% of the males and in 32.3% and 41.1% of the females (Homburger et al., 1983).

Schmidt et al. (1983) studied the pathology of 750 male hamsters used as controls in carcinogenesis studies and allowed to live up to 120 weeks. The conditions of husbandry were similar to those described in the other studies cited above. The first incidences of atrial thrombosis were seen in hamsters dying at 52 weeks, and the incidence increased with age. There were 14.6% of the animals with atrial thrombosis and 6.7% with cardiomyopathy, showing that thrombosis does occur without cardiac degeneration. The frequency of this association does not permit conclusions as to its etiological significance. There are considerable variations in the incidence of atrial thrombosis, even though the food, bedding, and type of

---

*Bio-Research Consultants, Inc., Cambridge, Massachusetts (now Bio-Breeders, Inc., Fitchburg, Massachusetts 01420).

cage used were comparable in most of these studies. Just how important may be the role of heredity in this variability is not clear. Age does play a definite, though poorly understood role, and the frequency of this disorder clearly increases with advancing age.

McMartin and Dodds (1982) appear to be the only investigators who have studied coagulation and fibrinolysis in this syndrome. They found changes consistent with concomitant consumption coagulopathy, namely, reduced factor II, IV, VIII, and X activity, reduced plasminogen, elevated fibrinogen–fibrin split products, and thrombocytopenia. However, similar coagulopathies were also noted in dying hamsters without atrial thrombosis.

4. Discussion of the Hamster Model

McMartin and Dodds (1982) point out the similarities that exist between some forms of human thrombosis and the hamster model of atrial thrombosis. In humans, atrial thrombosis is a frequent complication of right heart failure and thrombi occur in the right atrium. Increasing age is clearly a factor in deep vein thrombosis of humans, as it is in the atrial thrombosis of the hamster. Strains of hamsters with a high frequency of spontaneous age-related atrial thrombosis clearly are a good model for studies of the mechanism(s) of an often fatal human disorder.

B. Amyloidosis

1. Human Disease

In humans, amyloidosis occurs in various forms which are classified according to type: amyloid of immunoglobulin origin (AIO), or amyloid B, and amyloid of unknown origin (AUO), or amyloid A. Amyloidosis is also classified with respect to distribution of amyloid within the body, the immunochemical nature of the amyloid, and the presence or absence of underlying disease (Robbins and Cotran, 1979; Weatherall et al., 1983).

According to these criteria, the following classification might be offered:
a. *Primary amyloidosis*
   Type of amyloid: AIO (resulting from plasma cell dyscrasia)
   Amyloid in serum, Bence–Jones protein in urine
   No other disease present
b. *Secondary amyloidosis*
   Type of amyloid: AUO
   Amyloid deposited in the tissues (liver, kidney, spleen, adrenals)
   Accompanies any one of many underlying diseases usually characterized by protracted cellular disintegration (as in tuberculosis, leprosy, osteomyelitis, bronchiectasis, rheumatoid arthritis, Hodgkin's disease, cancer, familial Mediterranean fever—an autosomal recessive)
c. *Hereditofamilial amyloidosis*
   Amyloid mostly AUO
   Associated disorders with organ-specific autosomal-dominant inheritance (e.g., neuropathies, nephropathies, cardiopathies)
d. *Localized amyloidosis*
   Amyloid of both types, deposited in lungs, larynx, skin, urinary bladder, tongue, eye, or microscopic amyloid deposits associated with tumors of the pancreas, thyroid gland, pheochromocytomas, gastric cancer; also in the islands of Langerhans in diabetics
e. *Amyloid of aging*
   In the eighth and ninth decades in humans, amyloid found in the heart, brain, pancreas, and spleen

Little is known about factors modifying human amyloidosis. Colchicine has been observed to prevent the deposition of amyloid in Mediterranean fever (Cohen et al., 1981). The literature contains no information on whether the incidence of amyloidosis in humans differs in males and females. Cohen (1984), through personal communication, stated that in a series of 300 patients with amyloidosis (1984), he found no significant difference of incidence between men and women.

2. Amyloidosis in Hamsters

Amyloidosis, which occurs in many aging hamsters, has been described in Chapter 10. Several of its features make hamster amyloidosis an attractive system by which to study mechanisms of the human disease.

Amyloidosis is subject to genetic regulation, as great variation in amyloid deposition is observed among aging animals of different inbred strains.

For example, Burek et al. (1984), in a 2-year inhalation study of Ela:Eng(Syr) hamsters (Engle Laboratory Animals, Inc., Farmersburg, Indiana) kept in wire-bottomed cages and fed Purina Laboratory Chow, observed "amyloid in some organ(s) in excess of 50% of the control males and 90% of the control females." Bernfeld et al. (1974), on the other hand, found far less frequent amyloidosis, although there was some renal amyloidosis in 86% of BIO 15.16 inbred hamsters and in 69% of BIO 87.20 inbred hamsters. In a later study (Bernfeld et al., 1979), 102 BIO 15.16 control males exhibited only sporadic infiltration of the liver with amyloid and adrenal amyloidosis in 5–10% of the animals. In 4% of the kidneys, some scattered amyloid deposits were observed. In Bernfeld's studies, the hamsters were kept in shoebox-type cages and were fed Purina Laboratory Chow.

Homburger et al. (1983) studied the histopathology of two birth cohorts of about 100 male and 100 female BIO F_1D Alexander hybrid hamsters (a cross between BIO 15.16 sires and BIO 87.20 dams), also kept in solid-bottomed cages and fed Purina Laboratory Chow. In the first cohort, 5% of the males and 51.7% of the females had amyloidosis. In the second cohort, the respective figures were 3.3% and 15.6%. In later studies involving over 200 animals of the BIO F_1D Alexander hybrid type, the incidence of amyloidosis has remained low (4.2% in females and 0.5% in males). The only clear difference in the conditions of these later experiments was slightly larger cages, which resulted in animals of slightly lower body weight. It appears that there is not only genetically determined variability of the frequency and intensity of amyloidosis, but that sex is an important modifier and that other, perhaps environmental, factors may modulate the end result.

Coe (1977, 1983; Coe and Ross, 1983, 1984; Coe et al., 1981) has recently studied the relationships that exist between amyloid and the female protein, a sex-limited protein of Syrian hamsters with serum levels in females 100 times greater than those of normal males. He observed that female protein of hamsters is structurally homologous with human C-reactive protein and, especially, with amyloid P component.

3. Characterization of the Hamster Model

The little which is known about hamster amyloid structure indicates resemblance to the human substance. It has recently been shown that the amyloid found in aging hamsters is associated with both proteins AA and AP, characteristic of secondary amyloidosis (Brandwein et al., 1984). The organ distribution of amyloid in aging hamsters is much as that seen in humans. In both species, there are clear indications that hereditary factors play significant roles in bringing about deposition of amyloid in certain individuals more than in others. Clarification of the mechanisms of amyloid synthesis, transport, and deposition in one species undoubtedly would contribute to the understanding of the disease process in the other.

C. "Quaking" or Parkinsonism

1. Human Disease

The average age of onset of Parkinson's disease is about 55 years. Onset under 40 years of age is rare. The disease is slightly more common in men than in women. The incidence rises exponentially with age, reaching 1 in 200 in the elderly. The onset is insidious, and in 70% of the patients the first symptom is tremor. The other cardinal signs are rigidity, akinesia (scarceness or absence of motion), and postural changes. There often is observed the classical, expressionless, rigid facies of Parkinsonism, sometimes with dribbling of saliva. The disorder is usually progressive but can be controlled to some extent with modern (levodopa) treatment.

The disease mechanism and the pathology are well studied and understood. There is dopamine deficiency in the striate body consequent to death of the substantia nigra. Loss of dopamine and cell death in the substantia nigra are closely correlated with the severity of akinesia. Loss of about 80–85% of nigral neurons and depletion of about a similar percentage of the dopamine content are necessary for symptoms of parkinsonism to appear.

The therapeutic agent levopeda is converted into dopamine in the brain by the enzyme dopa decarboxylase, thereby restoring striatal dopamine action (Weatherall et al., 1983).

2. Animal Models

Tremor and dyskinasia have been described in certain inbred strains of mice referred to as "tremblers," a dominant mutation resulting in the triad of signs, spastic paralysis, action tremor, and frequent convulsions, especially in young mice (Green, 1966). Evidence appears to be lacking that this complex of signs is generated by mechanisms resembling the pathogenesis of Parkinson's disease.

3. Hamster Model of Parkinsonism

Hamsters lacking motor coordination were found in 1978 among the offspring of one mating of the BIO 53.58 inbred line* (Peterson et al., 1981). Affected animals show severe coarse tremor, especially when they attempt to move. When placed on their backs they are capable of righting themselves, but with some difficulty. They rarely reproduce, although a very small number of successful matings has been observed.

From matings between normal animals with quaking siblings, it could be deduced that the condition is a simple recessive trait (Peterson et al., 1981).

No neurohistological studies or biochemical investigations are available on this model, which in its manifestations closely resembles human parkinsonism.

D. Spontaneous Seizures: Epilepsy

1. Human Disease

Epilepsy is defined as a continuing tendency to epileptic seizures. The basic event common to all seizures is a paroxysmal discharge of cerebral neurons. Not all such discharges result in overt events, and between seizures, electroencephalograms of epileptics may show spikes which represent paroxysmal discharges of neurons (Weatherall et al., 1983).

One theory holds that there exists a genetically determined seizure proneness and intrinsic trigger. Audiogenic seizures are inducible in certain mice by noise, and seizures can be

---

*Then maintained at TELACO, at that time a division of Bio-Research Consultants.

induced in genetically epilepsy-prone rats by various stimuli (Green, 1966).

2. Hamster Model of Epilepsy

The hamster model of epilepsy as observed in the inbred BIO 86.93 strain possesses both genetically determined seizure proneness and an intrinsic trigger, which gives these animals a continuing tendency to epileptic seizures, at least during part of their lives (Yoon *et al.*, 1976). The condition is transmitted as a simple recessive.

Between the ages of 30 and 60 days, seizures lasting from 2 to 5 hr occur spontaneously and frequently. They can also be induced by any mild stress, but not by noise, buzzer, or painful stimuli.

The description of a typical seizure, taken from the original paper (Yoon *et al.*, 1976), follows:

> When placed in a new cage, it is typical of hamsters to move rapidly around the cage, exploring, digging, and trying to escape. This activity continues for some time in normal animals. However, in seizure hamsters, a reduction in normal activity, as well as obvious clumsiness, can be noted after the first few minutes. The ears are held back tightly against the head and they walk with their heads and bodies held much lower to the cage floor than do normal animals. At this time, close observation reveals subtle and yet definite facial contortions: the lower lip quivers somewhat and is pulled downward, exposing the roots of the lower incisors in a very unnatural manner. The activities comprise stage 1 and last approximately 10 to 20 minutes.
> Then, the animals' clumsiness gives way to serious locomotor difficulty. They frequently fall and completely roll over, especially when trying to stand up on their hind legs. After this, the hind legs curl up and remain so for some time. Then the hind legs become extended outward in a tonic spasm (stage 2). At this time, however, the front legs are still functional and the animals often crawl around the cage dragging the hind legs. This stage may last 1 to 2 hours or even longer.
> At the third stage, both front and hind legs are in a rigid, quivering state, both front and hind legs are extended forward, with the body twisted, the tail curled upward, the ears held back, and the eyes rolled back and glassy. This third stage lasts for 20 to 40 minutes. Then the animals recover and usually revert back to stage 2, where they may remain for 1–2 hours more, finally falling asleep.
> It should be noted here that, at weaning, seizure hamsters usually are smaller and thinner than their normal littermates. However, they outgrow this condition by about day 60, when seizure susceptibility disappears.
> In some rare instances, animals have been observed to have seizures after day 60, or even at day 90. This usually occurs in females when delivering a litter.

3. Discussion

In the inbred hamster lines, spontaneous seizures are predictably frequent in all animals between the ages of 30 and 60 days, and in addition, easily induced by mild stress, such as the placing of animals in a new cage.

The hamster epilepsy model would appear to mimic a number of features of human epilepsy, but for unknown reasons it has not found wide use in pharmacological or physiopathological studies.

E. Hydrocephalus

1. Human Disease

Hydrocephalus designates an increase in the volume of cerebrospinal fluid in the head. Compensatory hydrocephalus occurs in any degenerative disease which results in shrinkage of the brain tissue. Clinically significant hydrocephalus can result from increased secretion of cerebrospinal fluid or from interference with its flow and absorption. In obstructive hydrocephalus, the fluid is prevented from leaving the ventricles. In communicating hydrocephalus, the obstruction lies between the foramina of the fourth ventricle and the arachnoid villi on the convexity of the brain. The incidence of hydrocephalus is about 1 in 1000 live births, with a predilection of 3:2 of males over females (Weatherall *et al.*, 1983). There is at least one form of obstructive hydrocephalus known in humans (Brickers and Adams, 1949).

2. Animal Models of Hydrocephalus

Hereditary hydrocephalus has been reported in mice, rats, rabbits, dogs, swine, and cattle (Shannon and Handler, 1968).

3. Hamster Model of Hydrocephalus

Yoon and Slaney (1972) described a hydrocephalus in mutant hamsters which they showed to be due to a simple recessive gene. Since the cerebral aqueduct was normal during early stages of hydrocephalic development, it appears that the narrowing of the aqueduct observed in later stages of ventricular distention is a secondary phenomenon and that this form of hydrocephalus is due to excessive secretion of cerebrospinal fluid.

F. Hindleg Paralysis: Model for Demyelinizing Neuropathies

1. Human Disease

Peripheral neuropathies in humans take many forms, depending on which nerves are involved. Conditions that lead to death of the neuron as a whole are referred to as neuronopathies; conditions that have selective effects on the axons are known as axonopathies. Axonal interruption leads to Wallerian degeneration below the site of injury. Other neuronopathies affect the myelin either directly or by interference with Schwann cell function. As a consequence, there occurs selective demyelination with relative preservation of axonal integrity (Weatherall et al., 1983).

2. Animal Models of Neuropathies

Experimental models of neuronal and axonal degeneration involve the surgical or traumatic interruption of axons or destruction of neurons by invasive means. We are not aware of spontaneous animal models of peripheral neuropathy in animals other than hamsters.

3. Hamster Model of Peripheral Neuropathy

Nixon and Connelly (1967a) described a progressive hindleg paralysis in Syrian hamsters due to a sex-linked recessive gene. This mode of inheritance entails the predictable involvement of all males in an inbred line (BIO 12.14). Early neuropathological studies by Nauta and by Manuelidis (personal communications) have been summarized by Homburger and Bajusz (1970). These observations suggested a diminution of the neuronal population dorsal to the anterior horn motor cells or fiber degeneration bilaterally in the lateral column of the spinal cord, and evidence of axon terminal degeneration in the spinal dorsal horn and the nuclei gracilis and cunnatus.

In detailed studies of fine structure, Hirano (1977, 1978a,b; Hirano and Dembitzer, 1976) defined the morphological changes found in the nervous system of hamsters with hindleg paralysis. He described eosinophilic rodlike structures in myelinated fibers of spinal roots of normal as well as paralyzed hamsters, although these structures were more frequent in the paralyzed hamsters (Hirano and Dembitzer, 1976). Hirano also found alterations in the myelin sheets, but failed to find changes compatible with amyotropic lateral sclerosis (1977), and noted lamellar alterations of the endoplasmic reticulum (1978b). The significance of these observations remains obscure. The syndrome of inherited hindleg paralysis appears to be a spontaneous peripheral neuropathy.

G. Obesity

1. Human Disease

Obesity is a generally recognized risk factor for cardiovascular disease and other metabolic disorders (diabetes, gout, etc.), accompanied by a host of potentially life-threatening complications and by itself an undesirable condition.

2. Animal Models of Obesity

A number of animal models of obesity exist, such as the obese rat and the obese mouse.

3. Hamster Model of Obesity

The hamster model of obesity is of historical interest only, since the inbred line of obese hamsters described by Homburger et al. (1964) is extinct.

Characteristics of the inbred BIO 4.24 line were described by Homburger and Bajusz (1970), and the enzymatic mechanisms of this type of obesity were studied by Bernfeld (1979). In this inbred line of hamsters, all females were markedly obese after the age of 9 months. The food consumption of the obese animals was the same as that of normal controls. The fat content of the carcasses of the obese animals was significantly higher than in nonobese controls, and there were marked increases in the levels of glycerol kinase in the fats of certain body sites of the obese animals.

Unusually frequent endocrine anomalies were observed in the females of the BIO 4.24 line, but their relationship to the obesity was not clear. There was a high incidence (average 23%) of benign adenomas of the adrenal cortex in the animals of this line with the frequency twice as great in females as in males. Nearly one-fifth of the females had pituitary microadenomas and/or intermediary lobe hyperplasia, which were never noted in nonobese lines. The presence of such lesions was, however, not correlated with the severity of the obesity. These endocrine anomalies, combined with the obesity of the females, led to the extinction of this line before the interrelationship of endocrine lesions and obesity could be clarified.

H. Prostatic Hyperplasia

1. Human Disease

In men over the age of 70, prostatic hyperplasia is so common that it may be regarded as a normal aging process. It does occur, however, in 50–60% of men between the ages of 40 and 59 and may require surgical intervention for relief of urinary obstruction in 5–10% of those men whose prostate is enlarged.

Histologically, many patterns of prostatic nodular hyperplasia may be discerned. There may be mainly glandular proliferation and dilatation, or fibrous and muscular proliferation of the stroma. All of these elements are usually involved in proportions that vary from case to case. Usually, the epithelial element predominates in the form of aggregations of small to large, to cystically dilated glands, lined by a single layer of tall columnar or flattened cuboidal epithelium, based upon an intact basement membrane (Robbins et al., 1984).

The etiology of prostatic hyperplasia is still unclear. Reasonable speculations assume that an imbalance between male and female sex hormones must be involved.

2. Animal Models of Prostatic Hyperplasia

Spontaneous models of benign human prostatic hypertrophy have been described in dogs and mastomys. In other species, including mice and rats and guinea pigs, hormonal manipulations are required to induce prostatic hypertrophy (Andrews *et al.*, 1979).

3. Hamster Model of Prostatic Hyperplasia

Prostatic hyperplasia was observed in two inbred lines of Syrian hamsters (Homburger and Nixon, 1970), where it occurred in all males over the age of 150 days. It was not observed in any other inbred lines (except in small numbers of very old animals) and has not been described in an aging study on approximately 750 control hamsters allowed to live out their life span (Schmidt *et al.*, 1983). While its genetic mechanism is unknown, it may be assumed that the prostatic hypertrophy in the BIO 87.20 and BIO 2.4 inbred lines is somehow determined genetically.

Anatomically, the hyperplasia is of a glandular epithelial, cystic type, and thus is representative of only one morphological component seen in human prostatic hyperplasia. This makes it a useful model for the study of mechanisms that control the glandular aspects of prostatic hyperplasia. Studies by Schaffner (1981; Wang and Schaffner, 1976) have implicated cholesterol metabolism as a possible etiological factor in this disorder. This may be of special interest, since it was believed at one time that diabetics had unusually high incidences of prostatic hyperplasia and it is still thought that the disorder is more prevalent in hypertensives than in normotensive men.

I. Muscular Dystrophy

1. Human Disease

The human muscular dystrophies comprise a very important group of disease entities affecting approximately 250,000 people in the United States alone. Progressive muscular dystrophy is characterized by progressive atrophy of skeletal muscles without demonstrable lesions of the spinal cord. Weakness and visible muscular wasting make up its clinical manifestations. There are several recognized forms of the disease, which are either categorized by eponyms, such as Thomsen's disease or Duchenne's disease, or by more generic terminology, such as progressive muscular dystrophy of childhood, autosomal limb–girdle muscular dystrophy, ocular dystrophy, pseudohypertrophic muscular dystrophy, myotonic dystrophy, severe generalized familial muscle dystrophy, and mild restricted muscular dystrophy. In general, the early signs of the disease are muscular weakness, especially during physical effort, and significantly elevated levels of serum phosphocreatine kinase. The progressive course of the disease varies, and as the names above imply, can be slow and mild or rapid and fatal.

A variety of causes has been suggested. There is considerable evidence that the muscular dystrophies are truly myogenic in origin; however, neurogenic causation is supported by some investigators. A membrane defect, hormonal basis, persistent immaturity, immunological defect, contractile protein defect, and oxygen deficiency or incapability of utilizing oxygen have all been proposed as causes or essential characteristics of the pathogenesis of muscular dystrophy of humans.

Many affected individuals have a history of familial involvement, and a sex-linked recessive gene has been incriminated in some clinical types (e.g., Duchenne). Recently, cases have been described where an autosomal-recessive gene is responsible (Weatherall *et al.*, 1983). The classification of human dystrophies is illustrated in Scheme 1.

1. X-Linked muscular dystrophy
   Severe (Duchenne type)
   Benign (Becker type)
2. Autosomal-recessive muscular dystrophy
   Limb–girdle type
   Childhood dystrophy (except Duchenne type)
   Congenital dystrophies
3. Autosomal-dominant muscular dystrophy
   Fascioscapulohumeral dystrophy
4. Distal muscular dystrophy
5. Ocular muscular dystrophy
6. Ocular pharyngeal muscular dystrophy

SCHEME 1. The classification of human dystrophies.

The pathology of the skeletal muscle shows moderate to severe hypertrophy of muscle fibers, with great variability of fiber diameter, occasional prominent nuclei, sarcoplasmic basophilia, myolysis, and regeneration.

Muscular dystrophy in humans is a genetically determined primary degenerative myopathy. Its pathogenesis is as yet unknown, no deficiency of any single enzyme nor any deficiency of a single muscle protein having been demonstrated. The possibility of a primary defect of the plasma membrane of muscle cells exists, but has not been proven. No specific marker for the dystrophy gene has been identified in other cells, so that the antenatal diagnosis of dystrophy is not possible.

The pathological process is basically the same for all types of dystrophy, although the distribution and rate of involvement of different muscle groups vary from type to type, as does the rate at which enzymes such as creatine kinase leak into the serum. Marked variation in fiber size and central nuclei of muscle fibers are the outstanding histopathological features (Weatherall *et al.*, 1983).

In human dystrophy, involvement of the heart is rare. The histopathology of the cardiomyopathies is different from that of muscular dystrophy.

2. Hamster Model of Muscular Dystrophy

A description and historical review of inherited muscular dystrophy in Syrian hamsters was published in 1979 (Homburger, 1979) and included illustrations of key features of the histopathological findings.

Inbred strains carrying the autosomal-recessive gene at that time numbered four, two of which have since been lost. At present, the surviving homozygotes are the BIO 14.6 inbred line and the BIO 53.58 inbred line perpetuated at the University of Toronto by Dr. Michael Sole.

The earliest manifestations of dystrophy which permit its diagnosis by means other than muscle biopsy are tongue changes (Handler et al., 1975) evident by inspection, early hindleg fatigue revealed by forced swimming (Bajusz, 1966), and elevated serum phosphocreatine kinase (Eppenberger et al., 1964).

In all homozygous animals (100% of the inbred line), there is gradual progression of the muscular disease leading to death in 6–12 months, depending on the strain studied. The cause of death is usually heart failure with generalized edema and ascites caused by cardiac myopathy, which always accompanies the skeletal muscular dystrophy.

Histologically, there is disappearance of muscle fibers through atrophy and fragmentation, with the remaining fibers varying greatly in size. The sarcolemmal nuclei are increased in number, enlarged, and of greatly variable shape. Hyalin and granular degeneration and myoblasts appear in affected muscle. Bands of young connective tissue and fat cells may separate the atrophic muscle fibers. There may be some connective tissue reaction around foci of necrosis.

The earliest change noted by light microscopy consists of perinuclear halos which gradually become larger and fuse, forming pale eosinophilic sleeves around rows of nuclei.

In the heart, focal myolysis in the absence of any significant cellular infiltration is the prominent feature. While this process is focal, it also is progressive and becomes locally diffuse. Eventually, focal fibrosis and calcification develop. In some strains, there was marked compensatory hypertrophy, dilatation, and heart failure. In other strains, dilatation and hypertrophy did not occur (Mohr and Lossnitzer, 1974).

3. Discussion

The myopathic, cardiomyopathic Syrian hamster closely imitates certain forms of human muscular dystrophy, both by its autosomal-recessive mode of genetic transmission and its morphological manifestation of muscular dystrophy.

Since all animals of an inbred strain are predictably affected, this model is ideal for studies of mechanisms at all levels, from molecular biology studied in subcellular structures to physiopathological studies in the isolated muscle or the intact animal.

The model is also well suited to studies of cardiomyopathy and of heart failure due to cardiac muscle impairment.

The hamster model is being used intensively for these purposes, as attested by a continuously expanding literature.

J. Carcinogen Susceptibility

1. Human Disease

Cancer is a complex of diseases with variable manifestations, but having in common the characteristic of uncontrolled tumor growth, which results in the interference with vital functions and invasion of various organs by tumor growth. Since the cause of most cancer is unknown, many assumptions as to its nature have been made, among them that carcinogens in the environment, both chemical and physical (such as radiation), can lead to cancer. Various chemicals have been indicted, based on their association with cancer as demonstrated by epidemiological studies, and animal tests have been developed with the aim of demonstrating, if possible, which of the innumerable chemicals surrounding us might be potential carcinogens.

2. Animal Models for Carcinogenesis

Customarily, mice and rats are used in animal bioassays for carcinogenic potency of chemicals. These rodents have the advantage of being small, easily housed and fed, relatively short-lived, and susceptible to known carcinogens.

Their principal drawback is a high incidence of "spontaneous" cancers that do occur in untreated control animals and resemble those known to be induced by chemical carcinogens. In addition, there is a tendency toward the development of liver cancer in mice given chemicals known not to be carcinogenic in other species.

3. Hamster Model for Carcinogenesis

The adequacy of Syrian hamsters for long-term animal bioassay has been reviewed by Homburger and Adams (1987). They concluded that hamsters are eminently suitable for long-term carcinogenesis bioassay and offer opportunities to improve the standard long-term carcinogenesis bioassay, especially as regards compounds which are demonstrably metabolized by hamsters as they are by humans. The hamster could well replace the mouse as a second species and eliminate some of the confusion which results from high spontaneous tumor incidences in mice, and from a tendency of their liver cells to be altered by various mechanisms, perhaps unrelated to carcinogenesis (Homburger et al., 1983).

Experience with studies of chemical carcinogenesis in hamsters (Homburger, 1968, 1969, 1972) has suggested that many apparent contradictions resulted from the genetically controlled variability of the hamster's susceptibility to carcinogens. Consequently, studies were initiated to test the susceptibility of inbred hamsters to various carcinogens.

Eleven inbred lines of hamsters and one random-bred group were tested for their response to the subcutaneous administration of 500 µg of benzo[a]pyrene, methylcholanthrene, or benzanthracene in 0.1 ml of tricaprylin. Other groups of animals received a total dose of 250 mg of methylcholanthrene in corn oil by stomach tube.

The subcutaneous injections of benzo[a]pyrene and methylcholanthrene caused local sarcomas in all studied strains, with varying tumor incidences and times of latency. The range of tumor incidence was from 56 to 100% for methylcholanthrene and from 12 to 64% for benzo[a]pyrene.

The gastric gavage with methylcholanthrene resulted in gastrointestinal tumors with varying incidences. The BIO 87.20 strain, which was most susceptible to subcutaneous injection of methylcholanthrene (100% tumors, average latency of 2 weeks), was also the only tested strain susceptible to the induction of gastric cancer. It also showed a high incidence of intestinal cancer, as did another inbred line, BIO 15.16. Other tested lines were resistant to this type of carcinogenesis.

The gastric administration of methylcholanthrene was later repeated in 10 additional inbred lines (Adams et al., 1979), with special attention paid to the development of mammary cancer. Again, tumor incidence varied between strains from 40 to 95%. The incidence of metastases in various strains varied from 18 to 80%.

In studies on the carcinogenicity of cigarette smoke, in which the two strains most susceptible to the induction of subcutaneous cancer by methylcholanthrene were used (BIO 15.16 and BIO 87.20), there were significant differences in susceptibility to cigarette smoke-induced larynx cancer. Invasive carcinoma of the larynx was nearly five times more frequent in BIO 15.16 hamsters than in BIO 87.20 animals.

4. Inherited Difference in Carcinogenesis— Susceptibility in Animals Other than Hamsters

Greenstein (1954) tabulated the variability of carcinogen susceptibility in various inbred strains of mice, and in the case of methylcholanthrene injected subcutaneously, this ranged from low (20–40% of the animals with induced tumors) to very high (80–100% of the animals with induced tumors).

Characteristics of inbred strains of mice, rats, guinea pigs, rabbits, chickens, amphibia, and fish have been reviewed by Festing (1979).

5. Discussion

The dependence of cancer incidence on heredity in humans and animals is generally recognized, although the specific role of genetics in the initiation, promotion, or suppression of cancer is not fully understood. In rodents, susceptibility to chemical carcinogens is demonstrably linked to genetic background. The hamster is a case in point and has the additional property, not common to rats and mice, that in spite of high susceptibility to the induction of cancer by chemical carcinogens, the incidence of spontaneous tumors may be extremely low. In some cases, the types of tumors induced in such hamsters by a chemical (such as methylcholanthrene) are completely absent in untreated control animals. This unique characteristic of hamsters could be used to improve methods for in vivo carcinogenesis bioassay (Homburger et al., 1983).

K. Slow Acetylation

1. Human Disorder

Slow and fast acetylators represent a form of genetic polymorphism, that is, "a type of variation in which individuals with clearly distinct qualities exist together in a freely interbreeding single population" (Weatherall et al., 1983). A polymorphism of drug metabolism is demonstrable in human populations by study of the acetylation by N-acetyltransferase of commonly used compounds. Such studies permit the identification of slow or fast acetylators, as determined by simple autosomal Mendelian genetics. As a result of differences in acetylation rates, the responses of fast versus slow acetylators to administered compounds will differ.

In slow acetylators, for example, the adverse effects of salicylazosulfapyridine are increased and adverse hematological reactions to dapsone are also enhanced. They are more prone to develop peripheral neuropathy when treated with isoniazid for tuberculosis.

Rapid acetylators respond less rapidly to isoniazid, because the drug is rapidly excreted and they are more prone to develop isoniazid hepatitis; they require higher doses of hydralazine for control of hypertension.

2. Animal Models of Slow/Fast Acetylation

Hein et al. (1982) reviewed the literature and noted that the acetylation polymorphism had been described in inbred mice and rabbits and among noninbred hamsters.

3. N-Acetylation Polymorphism in Hamsters

Hein et al. (1982) described N-acetylation polymorphism using in vitro liver N-acetyltransferase preparations from 26 strains of inbred hamsters. Six arylamine substrates were used: isoniazid, p-aminobenzoic acid, p-aminosalicylic acid, sulfamethazine, procainamide, and 2-aminofluorene. There was monomorphic expression, with only slight variability for the N-acetylation of isoniazid, sulfamethazine, procainamide, and 2-aminofluorene. N-Acetylation of p-aminobenzoic acid and

p-aminosalicylic acid showed polymorphism. The inbred lines BIO 1.5 and BIO 82.73 had over 400-fold lower p-aminobenzoic acid N-acetylation and over 20-fold lower p-aminosalicylic acid N-acetylation than did the other inbred strains. This represents a unique pattern of N-acetylation polymorphism.

4. Discussion

The pharmacogenetic expression of the N-acetylation polymorphism in the hamster is qualitatively opposite of that in humans and rabbits. Each species has particular attributes as an animal model to aid in the understanding of N-acetylation polymorphism in humans. Studies in mice may aid especially in mapping gene locus for N-acetylation in the mouse genome. Studies in hamsters may be helpful in furthering the understanding of the biochemical nature of the process.

ACKNOWLEDGMENT

Grateful acknowledgment is made for the continuous efforts of Dr. C. G. Van Dongen, vice president for production, of Bio-Research Consultants, Inc., without whose skills and care many of the disease models discussed herein could not have been perpetuated and made available to the scientific community. Dr. Van Dongen is now president of Bio-Breeders, Inc., Fitchburg, Massachusetts 01420.

REFERENCES

Adams, R. A., Homburger, F., Russfield, A. B., and Soto, E. (1979). Methylcholanthrene-induced metastatic mammary carcinoma in several inbred hamster strains. *Prog. Exp. Tumor Res.* **24**, 408–413.

Andrews, E. J., Ward, B. C., and Altman, N. H., eds. (1979). "Spontaneous Animal Models of Human Disease." Academic Press, New York.

Bajusz, E., ed. (1966). "Experimental Primary Myopathies and Their Relationship to Human Muscle Disease," Proc. Conf., 1965, *Ann. N.Y. Acad. Sci.*, Vol. 138, pp. 1–366. N.Y. Acad. Sci., New York.

Bernfeld, P. (1979). Glycerokinase levels in adipose tissue of obese hamsters. *Prog. Exp. Tumor Res.* **24**, 139–144.

Bernfeld, P., Homburger, F., and Russfield, A. B. (1974). Strain differences in the response of inbred hamsters to cigarette smoke inhalation. *J. Natl. Cancer Inst. (U.S.)* **53**, 1141–1157.

Bernfeld, P., Homburger, F., Soto, E., and Pai, K. J. (1979). Cigarette smoke inhalation studies in inbred Syrian golden hamsters. *JNCI, J. Natl. Cancer Inst.* **63**, 675–689.

Brandwein, S. R., Skinner, M., and Cohen, A. S. (1984). Isolation and characterization of spontaneously occurring amyloid fibrils in aged Syrian golden hamsters. *Clin. Res.*, Abstr. No. 3182.

Brickers, D. S., and Adams, R. D. (1949). Hereditary stenosis of the aqueduct of Silvius as a cause of congenital hydrocephalus. *Brain* **72**, 246–262.

Burek, J. D., Nitschke, K. D., Bell, T. J., Wackerle, D. L., Childs, R. C., Beyer, J. E., Dittenber, D. A., Rampy, L. W., and McKenna, M. D. (1984). Methylene chloride: A two-year inhalation toxicity and oncogenicity study in rats and hamsters. *Fundam. Appl. Toxicol.* **4**, 30–47.

Coe, J. E. (1977). A sex-limited serum protein of Syrian hamsters: Definition of female protein and regulation by testosterone. *Proc. Natl. Acad. Sci. U.S.A.* **74**, 730–733.

Coe, J. E. (1983). Homologs of CRP: A diverse family of proteins with similar structure. *Contemp. Top. Mol. Immunol.* **9**, 211–238.

Coe, J. E., and Ross, M. J. (1983). Hamster female protein: A divergent acute phase protein in male and female Syrian hamsters. *J. Exp. Med.* **157**, 1421–1433.

Coe, J. E., and Ross, M. J. (1984). Metabolism of hamster female protein is influenced by gender and by presence of amyloid. *Fed. Proc., Fed. Am. Soc. Exp. Biol.* **43**, 1859, Abstr. No. 2585.

Coe, J. E., Margossian, S. S., Slayter, H. S., and Sogn, J. A. (1981). Hamster female protein: A new pentraxin structurally and functionally similar to C-reactive protein and amyloid P component. *J. Exp. Med.* **153**, 977–991.

Cohen, A. S. (1967). Amyloidosis. *N. Engl. J. Med.* **277**, 522–628.

Cohen, A. S. (1984). Lack of sex difference in incidence of amyloidosis. Personal communication.

Cohen, A. S., Rubenow, A., Kayne, H., and Libbey, C. A. (1981). Colchicine therapy in primary amyloidosis: A preliminary report. *Rev. Rhem., Spec. No., Int. Congr. Rheumatol., 15th 1981*, Abstr. No. 1171.

Dorland (1965). "Illustrated Medical Dictionary," 24th ed. Saunders, Philadelphia, Pennsylvania.

Eppenberger, M., Nixon, C. W., Baker, J. R., *et al.* (1964). Serum phosphocreatine kinase in hereditary muscular dystrophy and cardiac necrosis of Syrian golden hamsters. *Proc. Soc. Exp. Biol. Med.* **117**, 465–468.

Festing, M. F. W. (1979). "Inbred Strains in Biomedical Research." Oxford Univ. Press, London and New York.

Festing, M. F. W., and Wright, M. K. (1972). New semi-dominant mutation in the Syrian hamster. *Nature (London)* **236**, 81–82.

Foote, C. L. (1949). A mutation in the golden hamster. *J. Hered.* **40**, 100–101.

Friedberg, C. K. (1966). "Diseases of the Heart," 3rd ed. Saunders, Philadelphia, Pennsylvania.

Glenner, G. G. (1980). Amyloid deposits and amyloidosis. *N. Engl. J. Med.* **302**, 1333–1343.

Green, E. L., ed. (1966). "The Biology of the Laboratory Mouse." McGraw-Hill, New York.

Greenstein, J. P. (1954). "Biochemistry of Cancer," 2nd ed., p. 148. Academic Press, New York.

Handler, A. H., Russfield, A. B., and Homburger, F. (1975). Tongue lesions specific for diagnosis of myopathy in inbred Syrian hamsters. *Proc. Soc. Exp. Biol. Med.* **148**, 573–577.

Hein, D. W., Omichinski, J. G., Brewer, J. A., and Weber, W. W. (1982). A unique pharmacogenetic expression of the N-acetylation polymorphism in the inbred hamster. *J. Pharmacol. Exp. Ther.* **220**, 8–15.

Hirano, A. (1977). Fine structural changes in the mutant hamster with hindleg paralysis. *Acta Neuropathol.* **39**, 225–230.

Hirano, A. (1978a). A possible mechanism of demyelination in the Syrian hamster with hindleg paralysis. *Lab. Invest.* **38**, 115–121.

Hirano, A. (1978b). Changes of the neuronic endoplasmic reticulum in the peripheral nervous system in mutant hamsters with hindleg paralysis and normal controls. *J. Neuropathol. Exp. Neurol.* **37**, 75–84.

Hirano, A., and Dembitzer, H. M. (1976). Eosinophilic rod-like structures in myelinated fibers of hamster spinal roots. *Neuropathol. Appl. Neurobiol.* **2**, 225–232.

Homburger, F. (1968). Chemical carcinogenesis in Syrian hamsters. *Prog. Exp. Tumor Res.* **10**, 163–237.

Homburger, F. (1969). Chemical carcinogenesis in the Syrian hamster: A review. *Cancer (Philadelphia)* **23**, 313–338.

Homburger, F. (1972). Chemical carcinogenesis in the Syrian hamster. *Prog. Exp. Tumor Res.* **10**, 152–175.

Homburger, F. (1979). Myopathy of hamster dystrophy: History and morphologic aspects. *Ann. N.Y. Acad. Sci.* **317**, 2–17.

Homburger, F., and Adams, R. A. (1987). Adequacy of Syrian hamsters for long-term animal bioassays. In "Handbook of Carcinogen Testing" (H. A. Milman and E. K. Weisburger, eds.). Noyes Data Corp., Park Ridge, New Jersey.

Homburger, F., and Bajusz, E. (1970). New models of human disease in Syrian hamsters. *JAMA, J. Am. Med. Assoc.* **212**, 604–610.

Homburger, F., and Nixon, C. W. (1970). Cystic prostatic hypertrophy in two inbred lines of Syrian hamsters. *Proc. Soc. Exp. Biol. Med.* **134**, 284–286.

Homburger, F., Baker, J. R., Nixon, C. W., and Whitney, R. (1962). Primary generalized polymyopathy and cardiac necrosis in an inbred line of Syrian hamsters. *Med. Exp.* **6**, 339–345.

Homburger, F., Nixon, C. W., and Baker, J. (1964). Some anatomical characteristics of eight inbred strains of *Mesocricetus auratus auratus*. *J. Genet.* **59**, 1–6.

Homburger, F., Hsueh, S. S., Kerr, C. S., and Russfield, A. B. (1972). Inherited susceptibility of inbred strains of Syrian hamsters to induction of subcutaneous sarcomas and mammary and gastrointestinal carcinomas by subcutaneous and gastric administration of polynuclear hydrocarbons. *Cancer Res.* **32**, 360–366.

Homburger, F., Adams, R. A., Bernfeld, P., Van Dongen, C. G., and Soto, E. (1983). A new first-generation hybrid Syrian hamster BIO F1D Alxander for *in vivo* carcinogenesis bioassay, as a third species or to replace the mouse. *Surv. Synth. Pathol. Res.* **1**, 125–133.

Jordan, R. A., Scheifley, C. H., and Edwards, J. E. (1951). Mitral thrombosis and atrial embolism in mitral stenosis. *Circulation* **3**, 363–367.

Knapp, B. H., and Polivanov, S. (1958). Anophthalmic albino: A new mutation in the Syrian hamster. *Cricetus (Mesocricetus) auratus. Am. Nat.* **92**, 317–318.

Laird, H. E., II, Dailey, J. W., and Jobe, P. C. (1984). Neurotransmitter abnormalities in rodents. *Fed. Proc., Fed. Am. Soc. Exp. Biol.* **43**, 2505–2509.

McMartin, D. N., and Dodds, W. J. (1982). Atrial thrombosis in aging Syrian hamsters: An animal model of human disease. *Am. J. Pathol.* **107**, 277–279.

Magalhaes, H. (1954a). Mottled white, a sex-linked lethal mutation in the golden hamster. *Anat. Rec.* **120**, Abstr. No. 112, 752.

Magalhaes, H. (1954b). Cream and tawny, coat color mutations in the golden hamster. *Anat. Rec.* **120**, Abstr. No. 113, 752.

Mohr, U., and Lossnitzer, K. (1974). Morphological investigation in hamsters of strain BIO 82.62 with hereditary myopathy and cardiomyopathy. *Bietr. Pathol. Anat. Allg. Pathol.* **153**, 178–193.

Nixon, C. W. (1972). Hereditary hairlessness in the Syrian golden hamster. *J. Hered.* **63**, 215–217.

Nixon, C. W., and Connelly, M. E. (1967a). Hind-leg paralysis: A new sex-linked mutation in the Syrian hamster. *J. Hered.* **59**, 276–278.

Nixon, C. W., and Connelly, M. E. (1967b). Dark gray and lethal gray: Two new coat mutations in Syrian hamsters. *J. Hered.* **58**, 295–296.

Nixon, C. W., and Connelly, M. E. (1977). Pinto—a new coat patterning factor in Syrian hamsters. *J. Hered.* **68**, 399–402.

Nixon, C. W., Whitney, R., Beaumont, J. H., and Connelly, M. E. (1969). Dominant spotting: A new mutation in the Syrian hamster. *J. Hered.* **60**, 299–300.

Peterson, J. S., and Yoon, C. H. (1975). Jute: A new coat color mutation in the Syrian hamster. *J. Hered.* **66**, 246–247.

Peterson, J. S., de Groot, C. T., and Yoon, C. H. (1981). Five new mutations in the Syrian hamster: Fur loss, fur deficiency, juvenile gray, ashen, and quaking. *J. Hered.* **72**, 445–446.

Robbins, S. L., Cotran, R. S., and Kumar, V. (1984). "Pathologic Basis of Disease." Saunders, Philadelphia, Pennsylvania.

Robinson, R. (1955). Two new mutations in the Syrian hamster. *Nature (London)* **176**, 353–354.

Robinson, R. (1957). Partial albinism in the Syrian hamster. *Nature (London)* **180**, 443–444.

Robinson, R. (1959). Genetic studies of the Syrian hamster. III. Variation of dermal pigmentation. *Genetica (The Hague)* **30**, 393–411.

Robinson, R. (1960a). Light undercolor in the Syrian hamster. *J. Hered.* **51**, 111–115.

Robinson, R. (1960b). Occurrence of a brown mutation in the Syrian hamster. *Nature (London)* **187**, 170–171.

Robinson, R. (1960c). White band, a new spotting mutation in the Syrian hamster. *Nature (London)* **188**, 764–765.

Robinson, R. (1966). Sex-linked yellow in the Syrian hamster. *Nature (London)* **212**, 824–825.

Robinson, R., (1968). In "The Golden Hamster—Its Biology and Use in Medical Research" (R. A. Hoffman, P. F. Robinson, and H. Magalhaes, eds.), pp. 41–72. Iowa State Univ. Press, Ames.

Robinson, R. (1972). Satin—a new coat mutation in the Syrian hamster. *J. Hered.* **63**, 52.

Robinson, R. (1977). Umbrous: A dominant darkening gene in the Syrian hamster. *J. Hered.* **68**, 328.

Schaffner, C. (1981). Prostatic cholesterol metabolism: Regulation and alteration. In "The Prostatic Cell: Structure and Function," pp. 279–324. Alan R. Liss, Inc., New York.

Schimke, D. J., Nixon, C. W., and Connelly, M. E. (1973). Longhair: A new mutation in the Syrian hamster. *J. Hered.* **64**, 236–237.

Schmidt, R. E., Eason, R. L., Hubbard, G. B., Young, J. T., and Eisenbrandt, D. L. (1983). "Pathology of Aging Syrian Hamsters." CRC Press, Boca Raton, Florida.

Shannon, M. W., and Handler, H. L. (1968). X-Linked hydrocephalus. *J. Med. Genet.* **5**, 326–328.

Stedman (1957). "Medical Dictionary," 19th ed. Williams & Wilkins, Baltimore, Maryland.

Veillon-Nobel (1969). "Medical Dictionary." Springer-Verlag, Berlin and New York.

Wang, G. M., and Schaffner, C. P. (1976). Effect of candicin and colestipol on the testes and prostate glands of BIO 87.20 hamsters. *Invest. Urol.* **14**, 66–71.

Weatherall, D. J., Ledingham, J. G. G., and Warrell, D. A., eds. (1983). "Oxford Textbook of Medicine." Oxford Univ. Press, London and New York.

Webster, N., ed. (1961). "Third New International Dictionary of the English Language." C. C. Merriam Co., Springfield, Massachusetts.

Webster's New Twentieth Century Dictionary, 2nd ed. (1979). William Collins Publishers, Inc., New York.

Whitney, R. (1963). Principles of breeding and management. In "Animals for Research" (W. Lane-Petter, ed.), pp. 265–392. Academic Press, London.

Whitney, R., and Nixon, C. W. (1973). Rex coat: A new mutation in the Syrian hamster. *J. Hered.* **64**, 239.

Whitney, R., Burns, G., and Nixon, C. W. (1964). Rust, a new mutation in Syrian hamsters. *Am. Nat.* **98**, 121–122.

World Health Organization (WHO) (1973). "Pharmacogenetics, Report of a WHO Scientific Group," W.H.O. Tech. Rep. Ser., No. 524. W.H.O., Geneva.

Yerganian, G. (1972). History and cytogenetics of hamsters. *Prog. Exp. Tumor Res.* **16**, 2–34.

Yoon, C. H., and Slaney, J. (1972). Hydrocephalus: A new mutation in the Syrian hamster. *J. Hered.* **63**, 344–346.

Yoon, C. H., Peterson, J. S., and Corrow, D. (1976). Spontaneous seizures: A new mutation in Syrian golden hamsters. *J. Hered.* **67**, 115–16.

Chapter 16

Experimental Biology: Other Research Uses of Syrian Hamsters

Christian E. Newcomer, Douglas A. Fitts, Bruce D. Goldman, Michael R. Murphy, Ghanta N. Rao, Gerald Shklar, and Joel L. Schwartz

I.	Introduction	264
II.	Physiological Regulation of Ingestive Behavior	264
	A. Fluid Intake and Salt Preference	264
	B. Ethanol Intake	266
	C. Food Intake	266
	D. Conclusion	267
III.	Behavior and Neuroscience Research	267
	A. Sexual Behavior	267
	B. Activity Rhythms	268
	C. Other Behaviors	268
	D. Neuroscience	268
	E. Special Support Requirements	269
IV.	Reproductive Physiology	269
	A. Estrous Cycle	269
	B. Circadian Timing of Hormone Secretion and Ovulation	269
	C. Gestation and Early Development	270
	D. Photoperiodism and Pineal Physiology	270
	E. Measurement of Pituitary Hormones	272
	F. Other Reproductive Endocrine Influences	272
V.	Hibernation and Cold Adaptation Research	273
	A. Criteria for Hibernation	273
	B. Environmental Factors in Hibernation	273
	C. Neuroendocrinological Aspects of Hibernation	274
	D. Cellular Alterations in Hibernation	275
VI.	Teratology	275
	A. Physical Factors	276
	B. Pharmacological Compounds	276
	C. Heavy-Metal Compounds	276
	D. Other Teratogenic Compounds	276
VII.	Toxicology	277
	A. General Considerations	277
	B. General Toxicology Studies	277
	C. Toxicology of Pharmacological Compounds	277
	D. Inhalation Studies	278

VIII. Radiobiological Research 278
  A. Radiation Syndromes 278
  B. Other Radiation Studies 281
IX. Oral Pathology: Chemical Carcinogenesis in the Hamster
    Cheek Pouch 281
  A. Induction and Development of Cheek Pouch Tumors 281
  B. Factors Contributing to Carcinogenesis 282
  C. Immunological and Pharmacological Modulation of
     Cheek Pouch Neoplasia 284
  D. Conclusion 285
X. Chemical Carcinogenesis at Sites Other than Cheek Pouch .. 285
  A. Carcinogen Susceptibility 285
  B. Mean Life Span 286
  C. Incidence of Spontaneous Tumors 286
  D. Incidence of Nonneoplastic Lesions 286
  E. Conclusions 287
XI. Dental Caries and Periodontal Disease 288
  A. Dental Caries 288
  B. Periodontal Disease 288
    References 289

I. INTRODUCTION

Although the Syrian hamster has not been used as extensively as rats and mice in many areas of biomedical research, use of the hamster has been instrumental to the progress of research in numerous disciplines. The purpose of this chapter is to discuss the major areas of investigation not covered in other chapters of this volume which have benefited from the use of Syrian hamsters. Occasional reference to other hamster species is included for comparative purposes. Within the topics chosen for presentation, the authors have emphasized studies of particular importance to the expansion and enrichment of a field of study. This approach has provided the reader with an opportunity to review quickly the dominant themes of several key areas of research at the expense of a thorough explication of other areas. However, extensive bibliographical documentation is provided for the reader interested in additional information.

II. PHYSIOLOGICAL REGULATION OF INGESTIVE BEHAVIOR

Water and energy are efficiently regulated in the body through systems that detect errors as they occur, and adaptive responses bring the regulated substance back to its optimal value. Physiological components of this homeostatic mechanism alter the normal functioning of the body during emergencies to protect against an exacerbation of the existing problem. Behavioral mechanisms, such as eating and drinking, then replace the needed substance, and in many cases these behaviors have evolved with great specificity. The forms these compensatory behaviors take during emergency situations depend heavily on the choices available in the habitat. Preferences for certain foods and salts evolve according to a commodity's relative abundance or scarcity, and a study of species differences both in the preference for an item and in the behavioral response to emergency situations can yield important information about the evolution of regulatory systems in those species Accordingly, the following section describes research on the control of ingestion in golden Syrian hamsters, including their intake of water, salt, ethanol, and food.

A. Fluid Intake and Salt Preference

Golden hamsters have both renal and respiratory mechanisms for water conservation which are typical of desert-adapted rodents, but, unlike some desert rodents, they cannot survive on metabolic water alone (Schmidt-Nielsen and Schmidt-Nielsen, 1950, 1951). Even when eating a high-carbohydrate diet at 90% relative humidity, hamsters need to ingest preformed water to prevent weight loss and debilitation (Schmidt-Nielsen and Schmidt-Nielsen, 1951). The few reports on water and salt balance in the hamster have tended to be at odds with the literature on hydromineral regulations in rats, and have highlighted the diverse strategies by which rodents from differing habitats maintain an optimal volume and distribution of water in the body (Kutscher, 1968, 1969; Carpenter, 1956).

1. Water Intake

During water deprivation, hamsters maintained on dry rodent chow lose carcass and plasma water in direct proportion to the loss of weight (Fitts and Wright, 1980; Kutscher, 1968); food intake also declines (Fitts and Wright, 1979). When water is returned, dehydrated hamsters drink immediately, and eat a meal within the first hour of rehydration (Fitts and Wright, 1979, 1980).

Experimental dehydration of the intracellular fluid space by a subcutaneous injection of 1 ml of 10% NaCl produces robust water drinking by otherwise normally hydrated hamsters (Lowy and Yim, 1982; Ritter and Bach, 1978). Water intake by hamsters dilute this 1.67 M NaCl load to a calculated 0.26 M after 2 hr (Ritter and Bach, 1978) and 0.19 M after 4 hr (Lowy and Yim, 1982), nearly restoring this highly hypertonic fluid load to isotonicity (0.15 M). Other rodents also exhibit the ability to respond to osmotic stimuli with drinking. This drinking response to either water deprivation or relative dehydration with 10% NaCl is progressively reduced in hamsters pretreated with 1 or 10 mg/kg of naltrexone, an opiate antagonist (Lowy and Yim, 1982), suggesting an opiate-sensitive link in the hamster's behavioral response to osmotic dehydration analogous to that implicated in other species.

In many species, drinking results from a reduction of plasma volume, even without a change in osmotic pressure or intracellular fluid volume. This is accomplished experimentally either by injection of a diuretic or by dialysis with a hyperoncotic colloid, such as polyethylene glycol (PEG). Treatment with PEG failed to elicit water consumption by hamsters, however, despite a measured 30% decline in plasma and blood volumes (Fitts *et al.*, 1982). By comparison, water deprivation for 72 hr produced only 15% depletion of plasma volume in hamsters (Kutscher, 1968; Fitts and Wright, 1980), so "extracellular thirst" probably does not contribute to drinking following water deprivation in this species.

One suggested mechanism that would account for the drinking seen in rats after depletion of extracellular volume is the formation of angiotensin II (A-II) in the blood. Many phylogenetically diverse species respond to either intravenous or cerebroventricular A-II injections by drinking. Hamsters do increase water intake following large intracerebroventricular (50–200 ng) and intraperitoneal doses of A-II (Kobayashi *et al.*, 1979; Miceli and Malsbury, 1983). Receptors for A-II have been identified in virtually every area of the hamster brain (Harding *et al.*, 1981), but specific functions for these receptors remain to be determined. Why hamsters drink in response to A-II injections, but not to PEG, a physiological stimulus for A-II formation, is unknown. Careful investigations using intravenous administration of physiological doses of A-II appear warranted to evaluate the possible role of A-II in osmotic thirst in hamsters.

Starvation, as well as water deprivation, can influence water intake, although the direction of influence depends on the species. Much of the daily water consumption by laboratory rodents occurs within a few minutes before or after consumption of a meal of dry chow. This food-related drinking is absent when food is withheld, and the daily intake of many species declines accordingly. However, many desert-adapted rodent species, including hamsters, drink more water during extended starvation (Kutscher, 1968, 1969). Whereas kangaroo rats and gerbils progressively increase water intake up to 500%, beginning on the first day of food deprivation (Wright, 1976), hamsters at first decrease intake for about 2 days before developing this polydipsic pattern (Kutscher, 1969).

Food deprivation polydipsia is not presently explainable. It is not secondary to any change in the renal response to vasopressin, because exogenous vasopressin elicits excellent water retention in all desert species tested during starvation (Kutscher, 1968). The polydipsia is of questionable survival value, because gerbils given a ration of water equal to their normal baseline show extended survival compared to animals allowed to develop the polydipsia (Kutscher, 1973). In fact, it is difficult to conceive of a situation in the desert habitat when water would be plentiful while food was scarce. Consequently, the phenomenon observed in the laboratory may never occur in the wild. Nevertheless, its presence in so many desert rodents with a wide geographic distribution suggests that some adaptation to a desert habitat consistently interacts with the imposed conditions to generate this behavior.

2. Sodium Intake

Carpenter (1956) first observed that hamsters lacked a preference for NaCl solutions in two-bottle tests, even in dilute concentrations that rats find highly palatable; subsequent reports have confirmed that hamsters avoid sodium salt solutions at all tested concentrations (Fitts *et al.*, 1982, 1983; Nickerson and Molteni, 1971; Salber and Zucker, 1974; Zucker *et al.*, 1972). This finding was helpful in the interpretation of studies showing extremely poor survival by adrenalectomized (ADX) hamsters if they were given 1% NaCl to drink without hormonal replacement (Snyder and Wyman, 1951). Rats treated in this way compensate for natriuresis by drinking copious quantities of salt solution, but ADX hamsters drink little saline solution, lose weight rapidly, and die (Snyder and Wyman, 1951).

Addition of either sucrose or saccharin to a saline solution greatly increases the acceptability of salt to ADX hamsters, promoting greater survival (Nickerson and Molteni, 1971; Salber and Zucker, 1974). These studies suggest that sodium is indeed the critical factor in mortality of ADX hamsters, although confusion existed for years following reports that neither the adrenal glands nor mineralocorticoid hormones appeared to affect sodium status of hamsters (Gaunt *et al.*, 1971; Snyder and Syman, 1951). Recent studies have verified that adrenalectomy and mineralocorticoid (DOCA) have profound effects on the sodium status of hamsters (Fitts *et al.*, 1983). The time delay for detection of the DOCA-induced sodium retention is longer than the usual 4–5 hr routinely observed in rats, and the reason for this delay is unknown. A further, complicating problem in ADX hamsters is a progressive anorexia, especially in male hamsters, which undoubtedly hastens the demise of untreated animals after surgery (Fitts *et al.*, 1983).

The poor survival of hamsters following adrenalectomy led to speculation that they lacked a mechanism for sodium appetite. Rats respond to sodium deficits by voraciously drinking sodium salt solutions which were previously considered unpalatable. Early tests for salt appetite in hamsters using adrenalectomy, diuretics, and DOCA injections failed to demonstrate such an appetite (Salber and Zucker, 1974; Wong and Jones, 1978; Zucker *et al.*, 1972). Later studies succeeded in eliciting salt appetite by offering much lower concentrations (e.g., 0.15 M) of saline after injections of furosemide, PEG, or DOCA (Fitts *et al.*, 1982, 1983; Wong, 1981). Results in ADX hamsters were less convincing (Fitts *et al.*, 1983; Zucker *et al.*, 1972), possibly because of the rapidly developing debilitation and anorexia produced by this procedure.

B. Ethanol Intake

Another striking difference between hamsters and other commonly used laboratory animals is their preference for ethanol. When hamsters are offered varying concentrations of ethanol or water to drink, they show an impressive intake of the alcohol (Fitts and St. Dennis, 1981; Kulkosky, 1978; Kulkosky and Cornell, 1979; Slighter, 1970). Hamsters drink significant quantities of ethanol in concentrations up to 70% (w/v) under free-choice conditions, and drink virtually all of their daily water ration as ethanol at concentrations below 15% (Kulkosky and Cornell, 1979). By contrast, rats drink ethanol only in very dilute concentrations and reject it altogether at concentrations exceeding 12–20% when water is also available. Furthermore, rats avoid even the more dilute solutions of ethanol when other solutions containing calories, such as carbohydrates or fats, are also offered. Hamsters continue consuming ethanol to some degree when an isocaloric solution of dextrose and water is available as an alternative, and may even prefer ethanol to the carbohydrate at some concentrations (30% v/v) (Fitts and St. Dennis, 1981).

Daily dosages of ethanol during free-choice consumption by hamsters range from 10 to 16 gm/kg/day (Kulkosky, 1978; Kulkosky and Cornell, 1979). During "forced choice," when only a 20 or 40% (v/v) ethanol solution is available, their intake can exceed 20–25 gm/kg/day (St. Dennis, 1981). These dosages would be sufficient to induce organ damage in rats, but as yet no signs of ethanol withdrawal or organ disorder have been demonstrated following prolonged consumption by hamsters (Harris *et al.*, 1979; Kulkosky, 1978; McMillan *et al.*, 1977; St. Dennis, 1981). This failure has frustrated attempts to use hamsters as a model of alcohol-induced diseases, and appears to result from an extremely rapid and efficient clearance of ethanol from the blood (St. Dennis, 1981; Thurman *et al.*, 1978). Alcohol dehydrogenase, the principle enzyme responsible for metabolizing ethanol to acetaldehyde, exhibits an activity in hamster liver homogenates of about two to three times the activity of similar preparations from rat livers (Kulkosky and Cornell, 1979; St. Dennis, 1981). Thus, hamsters metabolize ethanol at a rate that precludes a significant and prolonged elevation of blood ethanol, even considering their high intakes. However, the avid preference of hamsters for ethanol requires more explanation than an indication of their high metabolic capacity. Perhaps the preference for and remarkable metabolism of ethanol both evolved from a necessity for eating fermented grains hoarded in the hamsters' damp burrows. Unfortunately, no naturalistic information of sufficient quality exists to allow an informed judgment on the evolution of ethanol metabolism in hamsters.

C. Food Intake

While examining morphological and enzymatic changes of the digestive system during fasting, Simek (1967) first noticed that hamsters adapted very poorly to a regimen of alternating days of feeding and fasting. Subsequent research in many laboratories has confirmed that hamsters fail to increase food consumption after fasts varying from a few hours to 4 days (Borer *et al.*, 1979; DiBattista, 1983; Fitts and Wright, 1980; Kutscher, 1969; Rowland, 1982; Silverman and Zucker, 1976). Under normal feeding conditions, hamsters eat a meal about every 2 hr and show no circadian variations in food intake (Borer *et al.*, 1979). Following a period of fasting, the probability of a hamster eating a meal on presentation of food is very high, but the size of the meal is normal, and the daily intake thereafter is no greater than the usual daily food consumption (Borer *et al.*, 1979; Silverman and Zucker, 1976). If food is available for only a certain time period each day, or only on alternate days, hamsters will eat exactly as much as they would ordinarily eat during that time, and will consequently lose weight (Silverman and Zucker, 1976). Indefinite limitation of the feeding schedule in this fashion leads to debilitation and death (Silverman and Zucker, 1976). This behavior contrasts sharply with that of the rat, which oridinarily compensates for a period of fasting by increasing food consumption immediately, both in the first meal and during the first day.

This finding brings into question the ability of hamsters to regulate food intake. Nonfasted hamsters apparently regulate intake quite well over the long term, as can readily be seen by the constancy of daily food intake and body weight change under steady-state conditions (Borer and Kooi, 1975; Borer *et al.*, 1979; Silverman and Zucker, 1976). Their failure to increase consumption following fasting suggests that hamsters either fail to detect the short-term deficit in consumption, or are anatomically or metabolically limited in their rate of food consumption. At present, attempts to identify such a limiting factor have failed. For example, diluting the diet with bulk or water produces precise adjustments by hamsters, so the absolute quantity of material ingested is not a limitation (Silverman and Zucker, 1976). Furthermore, certain conditions, such as lactation and cold stress, do lead to increases in food consumption by hamsters, so there is not an inherent inability to eat more food (Fleming and Miceli, 1983; Lyman, 1954). These considerations have strengthened the notion that hamsters simply fail to respond to some types of energy deficits, and that this failure must somehow be reflected in the metabolism of the animal.

Borer *et al.* (1979) examined serum concentrations of glucose, insulin, free fatty acids, and ketones in hamsters following different degrees of weight loss induced by restricted feeding, and found changes which were comparable to those seen in rats. That is, a shift from the utilization of glucose to free fatty acids was marked by a decline in plasma glucose and insulin and an increase in free fatty acids and ketones. Despite the fact that hamsters in this study failed to increase consumption of chow after fasts of varying duration, hamsters did regain weight at rates which were proportional to the duration of fasting, suggesting that compensation was metabolic rather than behavioral.

Several studies have found that hamsters fail to respond to the glucose analogs, 2-deoxy-D-glucose and 5-thioglucose, with an increase in feeding, although both acute and chronic insulin treatments do elevate food intake and weight gain (DiBattista, 1982, 1983; Ritter and Bach, 1978; Rowland, 1978, 1983; Sclafani and Eisenstadt, 1980; Silverman, 1978). The glucose analogs and insulin are all effective stimuli to feeding in rats, so this species difference in response to the glucose analogs may provide a clue to the failure of hamsters to compensate for a period of fasting.

Investigations into the suppressive effects of several treatments on appetite in hamsters have also shown mixed results. Feeding is effectively suppressed by osmotic stimulation, such as water deprivation or hypertonic saline injection (Fitts and Wright, 1979), in both hamsters and rats. The gut peptide cholecystokinin suppresses food intake in hamsters, whether it is administered peripherally or directly into the cerebral ventricles (Miceli and Malsbury, 1983); this effect is elicited only following peripheral administration in rats. Finally, the long-lasting opiate antagonist, naltrexone, failed to suppress feeding in hamsters, although it potently reduced feeding, drinking, and body weight gain in rats (Lowy and Yim, 1982).

It is apparent that the regulatory systems controlling feeding are quite different in hamsters and rats. It should be pointed out, however, that because of the hamster's vigorous and diligent hoarding behavior, there may have been little selective pressure for a mechanism supporting extraordinary food intake following fasting. Hoarding behavior does increase following fasting, even if feeding does not (Anderson and Shettleworth, 1977), so if food is at all available, it will soon be contained in the hamster's hoard.

D. Conclusion

The rather unusual responses of hamsters with respect to eating after fasting, drinking following extracellular fluid depletion, and the respective preference–aversion functions for salt and ethanol, all seem consistent with the hamsters' evolution as a burrowing, hoarding, hibernating desert rodent. Many authors have invoked these adaptations as possible origins for the behavioral "anomalies" they observe. Yet, there is a paucity of confirming field studies with hamsters, which might determine critical factors such as the average salinity of available drinking water, dietary preferences, seasonal variations in food availability, and the degree of fermentation that takes place in the hoard. Such studies would greatly improve the case for speculations about these behaviors in terms of the natural history and evolution of hamsters.

III. BEHAVIOR AND NEUROSCIENCE RESEARCH

By 1978, over 200 journal articles had been written on hamster behavior and at least 400 have been published since. The purpose of this section is to highlight certain dominant themes in the use of hamsters for research in the fields of behavior and neuroscience and provide references for further study (in particular, see Murphy, 1979; Siegel, 1984).

The popularity of hamsters for behavioral and brain research derives primarily from three factors: (1) The hamster readily displays its natural behavior patterns in the presence of human observers, under the controlled and artificial conditions of the laboratory, thereby conveniently yielding great quantities of reliable and relevant data; (2) the hamster possesses many behavioral characteristics that make it well suited as an example or model of particular natural phenomena; (3) the suitability of hamsters for research in both neuroscience and behavior has provided a positive-feedback situation in which work in each field stimulates work in the other; in addition, the extensive endocrinological data base on hamsters provides support for studies on hormonal influences on behavior.

A. Sexual Behavior

The sexual behavior of hamsters is easily elicited and measured in the laboratory. Early research concentrated on describing the vigorous copulatory pattern of the male (Bunnell *et al.,* 1976), contrasting it with the rigid lordotic, receptive posture of the female (Carter, 1973; Noble, 1979). More recent work has also attended to courtship and proceptive behavior (Lisk *et al.,* 1983; Murphy, 1980; Steel, 1981). The reproductive roles of ultrasonic auditory communication (Floody, 1979; Miceli and Malsbury, 1982) and olfactory signaling via scent marking (Johnston, 1979) also have received considerable attention. The hamster has been used to study drug effects on sexual behavior. For example, Murphy (1981) found a dose-dependent reduction of sexual performance and motivation in methadone-treated male hamsters, while Turley and Floody (1981) reported the stimulation of receptive and proceptive sexual behaviors in tetrahydrocannabinol-treated females.

Hamster sexual behavior has provided an excellent model for studies on the hormonal control of the development, and adult expression, of a genetically constituted behavior pattern. The sexual behavior of both sexes is virtually eliminated by gonadectomy but can be restored by hormone replacement therapy (Lisk and Heimann, 1980; Lisk and Reuter, 1980; Stetson and Tate-Ostroff, 1981). The differentiation of male versus female sexual patterns depends heavily on the levels of reproductive steroid hormones during development (Etgen, 1981; Vomachka et al., 1981) and has provided a model for research in neurotoxicology (Gray, 1982). The influence of endrogens and estrogens on specific central neural sites has been correlated with various sexual behaviors including receptivity, lordosis, and copulation (DeBold et al., 1982; Lisk and Greenwald, 1983; Vomachka et al., 1982).

The discovery that the removal of the main and accessory olfactory bulbs completely eliminates sexual behavior in the male hamster (Murphy and Schneider, 1970) spearheaded an area of research using hamsters that is still among the most active. Later work confirmed that sexual arousal in the male hamster is totally dependent on chemosensory stimulation from the olfactory and vomeronasal systems (Winans and Powers, 1977). The neuropathways mediating the olfactory control of sexual behavior have been traced from the olfactory bulb into the hypothalamus (Lehman et al., 1983). The anatomy and function of the vomeronasal organ have been analyzed (Meredith, 1983). Further basic limbic neuroanatomy elucidating the efferent projections of the olfactory amygdala has been conducted in support of behavioral work (Kevetter and Winans, 1981). In a parallel area of inquiry, the search for the pheromones involved in stimulating male hamster sexual arousal has led to an analysis of the function of hamster vaginal secretions (Johnston, 1981; Macrides et al., 1974; Murphy, 1973) and the identification of one of its attractant components, dimethyl disulfide (Singer et al., 1976).

B. Activity Rhythms

Because of the hamster's regular circadian activity patterns in a running wheel, it has become a popular model in circadian rhythm research. An early study by Elliott et al. (1972) determined that the position of a light interval relative to the circadian locomotor activity influenced the photoperiodic testicular response of the hamster. Recent research has been greatly influenced by the discovery that hamster rhythmic activity patterns are eliminated by lesions of the suprachiasmatic nuclei (SCN) (Stetson and Watson-Whitmyre, 1976); currently, the hamster is the most widely used animal in studies on the SCN (Davis et al., 1983; Davis and Gorski, 1984; Rusak and Groos, 1982; Stephan et al., 1982). Other recent uses of hamsters in circadian activity research using hamsters include determining if there are one or multiple circadian oscillators (Earnest and Turek, 1982, 1983), and relating circadian activity to other cyclic phenomena, such as the 4-day estrous cycle (Carmichael et al., 1981).

C. Other Behaviors

In addition to sexual behavior, other social behaviors of the hamster that have been studied in detail have been aggression (Murphy, 1976; Potegal et al., 1981) and maternal behavior (Siegel and Rosenblatt, 1980). The development of olfactory preferences (Leonard, 1982a) and behavioral thermoregulation (Leonard, 1982b) also have received considerable attention.

Unique characteristics of the hamster's eating and drinking behavior have resulted in its use in studies on the behavioral regulation of body fluids described earlier (Section II) (Fitts et al., 1983), food intake (Rowland, 1982), alcohol consumption (McCoy et al., 1981), and somatic growth (Borer et al., 1983). The hamster is also used in learning research (Shettleworth and Juergensen, 1980).

D. Neuroscience

The extensive neuroanatomical, neurochemical, neurophysiological, and neurobehavioral research using hamsters stems in part from the male hamster's critical dependence on the olfactory and vomeronasal sense for so much of its natural behavior, an attribute which was discussed earlier. An even more significant influence on the hamster's current popularity in neuroscience is its use in some early fundamental work on the functional organization of the visual system and the development of the visual system following neonatal lesions (Schneider, 1973). Recent work on vision using hamsters has focused on the neuroanatomy (Crain and Hall, 1981; Frost, 1981) and electrophysiology (McHaffie and Stein, 1982; Rhoades, 1981) of several components of the visual system including the superior colliculus, lateral posterior nucleus, and visual cortex. Investigations combining approaches such as anatomy and behavior or neurophysiology have yielded information on the function of aberrant retinotectal projections (Finlay and Cairns, 1981) and of the role of the lateral posterior nucleus in the hamster visual system (Mooney et al., 1984). Three recent papers provide an overview of much of the research on the hamster visual system (Chalupa, 1981; Finlay and Sengelaub, 1981; Stein, 1981).

In addition to the work in vision, other sensory systems of the hamster that have been studied include the olfactory, which may be suitable for studies of neuronal plasticity (Davis and

Macrides, 1983; Kream *et al.*, 1984; Small and Leonard, 1983), gustatory (Smith *et al.*, 1983; Whitehead and Frank, 1983), somatosensory (Jacquin *et al.*, 1984), and auditory (Schweitzer and Cant, 1984) systems. Particular attention has also been directed toward the hippocampus (Chiaia *et al.*, 1983), the cerebellum (LaVelle and LaVelle, 1983), the pyramidal tract (Reh and Kalil, 1982), and the neocortex (Reep and Winans, 1982).

Hamsters are used as neurological models in the study of the aging process (Buschmann and LaVelle, 1983) and diseases such as scrapie (Buyukmihci *et al.*, 1983), muscular dystrophy (Boegman and Wood, 1981), and measles (Tobler *et al.*, 1983).

E. Special Support Requirements

Three aspects of research with hamsters in the fields of behavior and neuroscience demand special consideration from support personnel:

1. Since hamsters are nocturnal animals, their most natural and vigorous behavioral responses are shown during the dark part of their 24-hr cycle. Therefore, it may be necessary to house hamsters on a reversed light–dark cycle in order to accommodate the schedules of research personnel using hamsters in behavioral studies.
2. Since social interactions extraneous to the research being conducted may affect the behaviors being studied, hamsters used for behavioral research should be housed one to a cage when possible. In addition, hamsters are believed to be solitary in the wild.
3. The time investment and other considerations that go into animal training, surgery, and long-term maintenance can result in individual hamsters being worth as much as a thousand times the cost of the original animal. Therefore, careful attention to treating individual cases of injury or disease as well as preventive medicine on a population basis is often justified.

IV. REPRODUCTIVE PHYSIOLOGY

The Syrian hamster has been widely used for research in various aspects of reproductive physiology. This species offers some distinct features which are advantageous in reproductive studies.

A. Estrous Cycle

Perhaps the most useful feature for the reproductive physiologist is the remarkable regularity of the hamster estrous cycle (Orsini, 1961). Unlike the laboratory rat, which shows both 4-day and 5-day cycles, the female hamster ovulates every 4 days. (See Chapter 4 for further details.)

The days of the hamster estrous cycle are generally called metestrus (or diestrus 1), diestrus (or diestrus 2), proestrus, and estrus. Ovulation occurs in the very early morning (i.e., 1–2 AM) of estrus. The important events of the estrous cycle have been investigated by examining changes in the titers of circulating reproductive hormones and associated ovarian changes on each day of the cycle (Bex and Goldman, 1975; Bast and Greenwald, 1974; Baranczuk and Greenwald, 1973; Leavitt and Blaha, 1970).

In general, the estrous cycle patterns of secretion of the various reproductive hormones in Syrian hamsters are very similar to those reported for the laboratory rat. Ovarian follicular development appears to be initiated by a rise in serum follicle-stimulating hormone (FSH) concentration which occurs early on the day of estrus (Bex and Goldman, 1975; Bast and Greenwald, 1974). After several hours the serum FSH concentration decreases 2- to 4-fold, and follicular maturation continues in the presence of this lower FSH level. The low level of luteinizing hormone (LH) present during this time is probably also necessary for follicular maturation (Bex and Goldman, 1975).

As several of the follicles approach maturity, they begin to secrete estradiol in increased amounts. If one ovary is removed at any time before the evening of diestrus, the remaining ovary compensates by producing about twice its usual number of mature follicles (Greenwald, 1962). In the hemiovariectomized hamster, the recruitment of the extra follicles in the remaining ovary appears to be stimulated by an increase in serum FSH which occurs shortly following the removal of one ovary and which lasts for several hours (Bex and Goldman, 1975; Bast and Greenwald, 1977). Thus, it seems that the ovaries—most likely the maturing follicles—secrete hormones which are involved in the regulation of FSH secretion. Estradiol and inhibin, a hormone originating from the gonads, have been suggested as probable candidates for such feedback inhibitors of pituitary FSH secretion (Chappel, 1979; Grady *et al.*, 1982). It appears that this feedback system between the ovaries and the FSH-secretory apparatus provides for the regulation of the number of maturing follicles. In the Syrian hamster, this ultimately results in the ovulation of 7–12 ova at the end of each cycle (Goldman and Porter, 1970; Bast and Greenwald, 1977).

B. Circadian Timing of Hormone Secretion and Ovulation

The circadian timing of preovulatory LH release and subsequent ovulation is very precise in the Syrian hamster. As in the rat, it appears that LH release in the hamster is timed by a neural signal originating from the suprachiasmatic nuclei (Legan and Karsch, 1975; Stetson and Watson-Whitmyre,

1976). These nuclei, located just above the optic chiasm, are believed to act as a circadian pacemaker, or clock, and their phase of oscillation is entrained to the light–dark cycle (Rusak and Zucker, 1978). Thus on natural day length, LH release in the female hamster occurs during the afternoon within an approximately 2-hr "window" of time. Ovulation occurs about 10–12 hr after the LH surge, resulting in the release of the ova during the night (Goldman and Porter, 1970). This is probably an adaptive characteristic, since the hamster is a nocturnal animal and would be most likely to encounter a potential mate at night. To ensure further that mating will occur at about the time of ovulation, the female hamster becomes sexually receptive in response to the sequential increases in circulating levels of estradiol and progesterone which occur prior to ovulation (Goldman and Sheridan, 1974). Estradiol secretion increases markedly a full day prior to preovulatory LH release, but progesterone secretion begins only in response to the LH surge (Leavitt and Blaha, 1970; Bosley and Leavitt, 1972; Baranczuk and Greenwald, 1973; Lukaszewska and Greenwald, 1970; Norman et al., 1973). Therefore, the female hamster becomes sexually active shortly following the release of LH and subsequent progesterone secretion. Sexual receptivity continues for only a few hours and overlaps the time of ovulation (Goldman and Sheridan, 1974).

C. Gestation and Early Development

The hamster is easy to breed and raise in the laboratory, and the gestation period is only 16 days—the shortest known for any eutherian mammal. The short gestation results in the birth of pups at a relatively early developmental stage, and this may facilitate certain types of studies of early development. For example, hamsters have been fairly widely used in studies of the influence of hormones on sexual differentiation during perinatal life. As with the rat, perinatal exposure to androgen or estrogen leads to a permanent failure of ovulatory cycles, presumably due to masculinization of brain centers which are involved in regulating the cyclic release of pituitary gonadotropic hormones in normal females (Swanson, 1966; McEwen et al., 1977; Goldman and Brown, 1979). Unlike the rat, however, if female hamsters or neonatally castrated males are treated with androgen at birth, they remain capable of displaying feminine sexual behavior in adulthood (Tiefer and Johnson, 1975; Whitsett and Vandenberg, 1975; Coniglio et al., 1973). The failure of this feminine sexual response to be strongly inhibited by early exposure to androgen is correlated with the observation that adult male hamsters frequently exhibit female sexual behavior following castration and treatment with female sex hormones (Tiefer and Johnson, 1971). This type of behavioral response is rarely observed in male rats (Barraclough and Gorski, 1962).

Hamsters are also notably different from laboratory rats and mice with respect to sexual differences in aggressive behavior. In the Syrian hamster, females tend to be somewhat more aggressive than males, and perinatal exposure to androgen does not result in greatly enhanced aggressiveness in adulthood (Payne and Swanson, 1970, 1971). In rats and mice, males are more aggressive than females, largely as a result of the action of androgen on the brain during perinatal development (Bronson and Desjardins, 1970; Barr et al., 1976).

In laboratory rats and mice, it has been suggested that α-fetoprotein may be involved in early sexual differentiation (Plapinger and McEwen, 1978). The α-fetoproteins of these species have been shown to bind estrogen (Aussel et al., 1973; Raynaud, 1973). Relatively large amounts of estrogen are present in the blood of female rat pups during neonatal life (Weisz and Gunsalus, 1973), and estrogen is capable of "masculinizing" the developing brain. Indeed, there is evidence that the normal action of testosterone to masculinize the brain in the developing male rat is dependent on conversion of the androgen to estrogen within the brain cells. It has been proposed that α-fetoprotein may "protect" neonatal female rats from their own circulating estrogens by binding the steroid and preventing its uptake by brain cells (Plapinger and McEwen, 1978). Estrogen-binding activity was not detected in the serum of newborn Syrian, Djungarian, or Turkish hamsters (J. Terkel and B. Goldman, unpublished data); therefore, the α-fetoproteins of these species, like those of humans and rabbits, probably do not bind estrogen.

D. Photoperiodism and Pineal Physiology

The Syrian hamster is a highly photoperiodic species and has been perhaps the most widely used mammalian model for studying the mechanisms by which changes in day length regulate seasonal reproductive changes (Elliott and Goldman, 1981). When hamsters are exposed to day lengths with 12.5 hr of illumination or less each day, the animals cease reproductive activity after a conditioning exposure to short days for 5–10 weeks. On longer day lengths, hamsters remain reproductively active throughout the year. In male hamsters, exposure to short days leads to decreased plasma concentrations of LH, FSH, and prolactin, and a subsequent involution of the testes (Goldman et al., 1982). In females, short-day exposure leads to decreased baseline LH and FSH levels and the occurrence of daily afternoon surges in these hormones (Seegal and Goldman, 1975; Bridges and Goldman, 1975). Follicular maturation and ovulation do not occur during this period (Reiter, 1968). At least two apparently separate mechanisms are responsible for the changes in pituitary gonadotropin secretion during exposure to short days. First, an increased sensitivity to the inhibitory feedback effects of sex steroid hormones leads to the suppression of gonadotropin secretion by relatively small amounts of gonadal steroids, so that circulating gonadotropin levels remain low even after gonadal regression occurs (Turek, 1977). Further, it has been shown that exposure to short days results in an inhibition of pituitary gonadotropin secretion even in gonadectomized hamsters; thus, a pathway for regulating gonadotropin secretion which is independent of steroid feedback must also be affected by day length (Turek and Ellis, 1981).

All the effects of photoperiod which are described above appear to be mediated by the pineal gland (Fig. 1). Pinealectomy renders the animal unresponsive to changes in the photoperiod. Pinealectomized Syrian hamsters remain reproductively active regardless of changes in day length, although other species show somewhat different responses to removal of the pineal (Goldman *et al.*, 1982). For example, the Turkish hamster (*Mesocricetus brandti*), which is closely related to the Syrian hamster, undergoes testicular involution after pinealectomy (Carter *et al.*, 1982). Sheep and ferrets continue to show alternating patterns of reproductive activity and anestrus after removal of the pineal (Herbert, 1972; Karsch *et al.*, 1984), but the pinealectomized animals are unable to respond appropriately to changes in the photoperiod (Goldman, 1983).

The pineal hormone, melatonin, has been shown to be involved in the mediation of photoperiodic effects in several species, most notably in Syrian and Djungarian hamsters, white-footed mice, ferrets, and sheep (Goldman, 1983). Pineal melatonin secretion is largely regulated by neural circadian oscillators, the suprachiasmatic nuclei, which receive input from the retina and which are responsive to light (Moore and

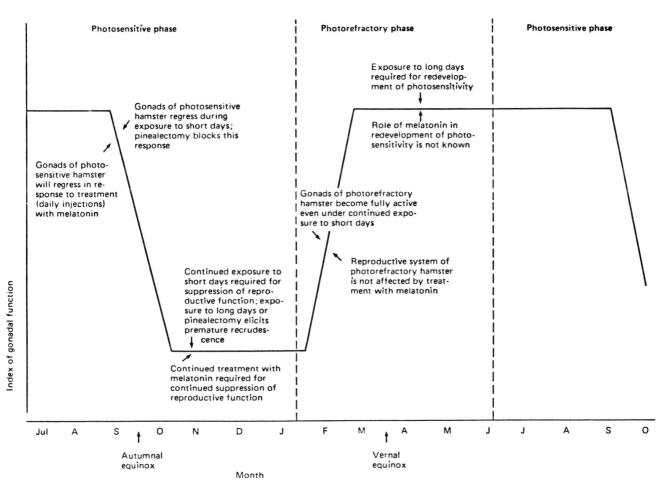

Fig. 1. Schematic representation of components of the annual reproductive cycle in the Syrian hamster. Gonadal activity decreases in the autumn in response to a decrease in the day length. In this species, the critical day length is approximately 12.5 hr. Several months after the photoperiodically induced regression of the gonads, a "spontaneous" recrudescence of the reproductive system occurs. This event does not require a change in day length; rather, it occurs prior to the advent of "long days." The timing of spontaneous recrudescence appears to be established by an endogenous timing mechanism. The photosensitive and photorefractory phases of the annual cycle are indicated in this scheme. In the photorefractory hamster, exposure to several weeks of long (summer) days is required to reestablish the photosensitive state. The effects of day length, melatonin, and pinealectomy on gonadal function vary depending on the phase of the annual cycle, as indicated in the diagram. From Goldman (1983).

Klein, 1974). The pathway from the suprachiasmatic nuclei to the pineal is multisynaptic (Klein et al., 1983), and lesions of either the suprachiasmatic nuclei, paraventricular nuclei, or superior cervical ganglia lead to a loss of responsiveness to photoperiod changes, similar to the results of pinealectomy (Reiter, 1980; Pickard and Turek, 1983).

The pattern of melatonin secretion changes with alterations in the photoperiod. Plasma melatonin concentrations are elevated during a portion of the night, and in Djungarian hamsters and sheep the duration of this nocturnal elevation in melatonin increases as day length decreases. Thus, long-duration melatonin peaks "signify" short days and elicit the appropriate reproductive responses in these species (Darrow and Goldman, 1983; Goldman, 1983; Karsch et al., 1984). A similar mechanism may be operative in Syrian hamsters, but the data for this species are not yet conclusive (Goldman et al., 1982; Stetson and Watson-Whitmyre, 1984). Nevertheless, numerous studies have revealed that the effects of exposure to a short-day photoperiod can be mimicked by administering daily injections of melatonin to Syrian hamsters maintained in a long photoperiod. In these experiments, the time of day at which melatonin is injected is often very important, particularly when the test animal has an intact pineal (Goldman et al., 1982). Following pinealectomy, treatment with daily injections of melatonin can still induce gonadal atrophy, but in the pinealectomized animals multiple daily injections were required to obtain large effects and the time of day of treatment appeared to be less important than with the pineal intact animals (Stetson and Watson-Whitmyre, 1984).

The inhibitory effects of short days on hamster reproductive activity are not permanent. After 5–6 months of short-day exposure the pituitary–gonadal axis becomes fully functional again (Fig. 1). The physiological mechanism which triggers the spontaneous recrudescence of the gonads at this time is not known, but it probably involves some type of endogenous timing system. In any event, the return to reproductive function after long-term exposure to short days is believed to provide a means by which hamsters can begin to breed early in the spring, following the short-photoperiod environment of fall and winter (Reiter, 1973). Photorefractory hamsters (i.e., animals which have recovered gonadal function following long-term exposure to short days) do not undergo gonadal regression in response to treatment with exogenous melatonin (Fig. 1). Thus the state of photorefractoriness in this species includes not only a failure of reproductive system to be inhibited by continued exposure to short photoperiod, but also a refractoriness to melatonin (Goldman et al., 1982).

E. Measurement of Pituitary Hormones

Studies of hamster reproductive endocrinology have been greatly facilitated by the development and availability of radioimmunoassays for the three pituitary hormones most intimately involved with reproduction: LH, FSH, and prolactin. Two heterologous assays have been employed to measure serum and pituitary LH concentrations in Syrian hamsters. One of these employs an antiserum raised against ovine LH (Niswender et al., 1968; Goldman et al., 1971; Blake et al., 1973), while the other uses antisera formed against rat LH. The latter antisera are available from the National Hormone and Pituitary Program (NHPP).* A heterologous assay for hamster FSH has been developed using antisera to rat FSH which are available from the NHPP (Bex and Goldman, 1975). Several suitable rat FSH antisera have been available; however, it is important to check these carefully, since the first antiserum to rat FSH which was made available by the NHPP (i.e., anti-rat FSH-1) showed little or no cross-reaction with hamster FSH (B. D. Goldman, unpublished data).

Two homologous hamster prolactin radioimmunoassays have been developed which are highly sensitive (Borer et al., 1982; Soares et al., 1983). Hamster prolactin has also been measured by using an antibody formed against rat prolactin (Reiter et al., 1975; Goldman et al., 1981), but this method is inferior to the homologous assays previously mentioned. While the LH, FSH, and prolactin assays described here have been used mainly in the Syrian hamster, they generally have also been found suitable for use in Djungarian and Turkish hamsters (Goldman, 1980; Carter and Goldman, 1983b; Yellon and Goldman, 1984). However, the rat LH antiserum is not suitable for use in the Djungarian hamster (B. D. Goldman, unpublished data).

F. Other Reproductive Endocrine Influences

As an adaptation to the harsh winter environment in their native northern USSR habitat, Djungarian hamsters are capable of daily torpor, a thermoregulatory phenomenon in which the body temperature is decreased to approximately 19°C for a few hours during the day. This response, along with the decreased body mass displayed by these hamsters during the winter months, allows the animal to conserve energy which would otherwise be required to maintain a higher body temperature. Also, the Djungarian hamster molts from its gray summer pelage to develop a white winter coat, which provides for camouflage against a snow cover. Both the development of the winter pelage and the exhibition of daily torpor are elicited by exposure to short photoperiod, and these responses to short days are prevented by pinealectomy (Vitale et al., 1985).

The endocrine regulation of fur pigmentation has been further investigated by both *in vivo* and *in vitro* approaches. It has been reported that serum prolactin concentrations become markedly decreased in short days (Duncan and Goldman,

---

*Suite 501, 210 Fayette Avenue, Baltimore, Maryland 21201.

1985). This inhibition of prolactin secretion appears to be caused by the increased duration of nocturnal pineal melatonin secretion characteristic of short-day exposed animals (Carter and Goldman, 1983a; Goldman et al., 1984). When new fur grows under conditions of low circulating prolactin levels, the hairs are unpigmented. Thus, the decrease in prolactin in short days is involved in the production of the white winter pelage (Duncan and Goldman, 1984, 1985). Additional studies have revealed that melatonin may exert a direct effect on the hair follicles, inhibiting the biosynthesis of melanin (Logan and Weatherhead, 1980). This would suggest a dual role of pineal melatonin in the production of winter pelage; both a direct effect on hair pigmentation and an indirect effect resulting from the ability of melatonin to inhibit pituitary prolactin secretion. It should be noted that circulating prolactin levels are also decreased in short days in Syrian and Turkish hamsters (Carter et al., 1982: Goldman et al., 1982), yet these species do not undergo a change in pelage color. It has been suggested that changes in prolactin levels may be involved in the regulation of changes in pelage only in species which normally undergo changes in the coat at specific times (Duncan, 1984).

V. HIBERNATION AND COLD ADAPTATION RESEARCH

Several hamster species have been useful in studies of major thermoregulatory adaptations including hibernation. Syrian, Turkish, and European hamsters are all capable of hibernation, while the Djungarian hamster exhibits daily torpor. Although hibernation is not invariably produced in the Syrian hamster maintained under presumably favorable conditions (Lyman, 1948; Hall and Goldman, 1980), the Syrian hamster has become established as an important model in this area of research and is used more frequently than other hamster species. The reader should refer to an excellent review of hibernation and the effects of low temperature in the hamster which thoroughly details the early descriptive studies in this species (Hoffman, 1968); many of these studies have not been included in this section. Recent research in hibernation and related areas has consisted of finely controlled experimental studies which have begun to differentiate the complex effects of hibernation and hibernatory preparation from the physiological and biochemical responses concomitant with cold adaptation in the hamster.

A. Criteria for Hibernation

One of the problems complicating the quantitative investigation of hamster hibernating capability is the lack of standardized criteria of hibernation in this species. However, criteria have been outlined recently for hibernation in the Turkish hamster (*Mesocricetus brandti*) (Lyman and O'Brien, 1977), and similar criteria for the Syrian hamster have also been formulated (Jansky et al., 1980). Briefly, the period of time between cold exposure and hibernation (i.e., prehibernation period) and the proportion of time spent in hibernation relative to the period of cold exposure were considered informative criteria in the studies of both species; according to Jansky et al. (1980), the latter proved to be the superior indicator under diverse experimental conditions. Several additional criteria were applied to *M. brandti,* including the period of continuous hibernation, the number of animals in the group which hibernate during cold exposure, and the length of the hibernating season. Additional efforts to develop uniform criteria of hibernation in the hamster appear to be warranted to facilitate interstudy comparisons of experimental data.

B. Environmental Factors in Hibernation

The exact role of various environmental influences, including seasonality, photoperiod, ambient temperature, availability of food, and isolation as factors predisposing to hibernation in the Syrian hamster has been reviewed previously (Hoffman, 1968). Most investigators agree that the availability of food stores (Lyman, 1954) and isolation of animals (Hoffman, 1968; Jansky et al., 1980) should be routinely incorporated into experimental designs to facilitate successful hibernation. However, neither of these conditions can be considered absolute requirements of hibernation. For example, hamsters which had aroused from hibernation were observed to reenter hibernation in the absence of a food hoard (Lyman and Leduc, 1958), and hamsters fasted for 22 hr selected an 8°C box rather than a 24°C box in a T-maze apparatus and exhibited quickened hibernation when introduced into a 5°C cold room (Gumma et al., 1967).

Although hibernation has been observed in hamsters kept at room temperature (Hoffman, 1968; Jansky et al., 1980), cold exposure is regarded as an important factor capable of inducing hibernation. A hypothermic hamster model has been developed by the use of extreme cold exposure (Musacchia, 1972, 1976) and has proven to be useful in the investigation of carbohydrate metabolism (Musacchia and Deavers, 1980) and renal physiology in hypothermia (Tempel et al., 1977). Hypothermia is induced in hamsters taken from room temperature (21°–24°C) and placed in a $He-O_2$ (80:20) atmosphere at −8°C. In approximately 4 hr, when the hamster becomes torpid and its rectal temperature falls to 7°–8°C, the hamster is placed in a room at 6°–7°C, where it can survive for slightly more than 24 hr. The use of this model may afford the investigator some opportunity to isolate the effects of hypothermia from those of hibernation in which complex neuroendocrinological, neurophysiological, and behavioral factors are confounding variables.

A seasonal influence in the ability of the hamster to hibernate evidenced by a variation in the prehibernatory phase has been proposed (Smit-Vis and Smit, 1963). The requirement for cold exposure during the prehibernatory phase was noted to decrease progressively from November to February, whereas cold exposure for at least 3 months was necessary at other times of the year. This observation has recently been challenged by Jansky et al. (1980), who concluded that there were no prerequisite qualitative changes for the occurrence of hibernation at different seasons of the year. Although some seasonal differences in the frequency of hibernation and the length of the prehibernatory phase were noted by Jansky et al. (1980), they were attributed to the effect of environmental temperature or the physiological status of the animals during the experiment. A lack of a seasonal tendency to hibernate has also been documented in the Turkish hamster (*M. brandti*) (Lyman and O'Brien, 1977).

C. Neuroendocrinological Aspects of Hibernation

The role of the pineal gland in the photoperiodic reproductive changes of the hamster has been discussed previously (Section IV,D). Gonadal atrophy has been recognized as essential for successful prolonged hibernation in the hamster (Smit-Vis, 1972: Smit-Vis and Smit, 1970). Gonadectomy is facilitatory to hibernation, whereas testosterone supplementation is inhibitory to hibernation or torpor in the four hamster species previously mentioned (Fig. 2) (Smit-Vis and Smit, 1970; Jansky et al., 1980; Hall and Goldman, 1980; Vitale et al., 1985). The administration of exogenous melatonin to hamsters kept in continuous light induces gonadal suppression and increases the frequency of hibernation (Jansky et al., 1980), confirming the role of the pineal gland in this mechanism. The influence of exogenous melatonin on other endocrine systems in the hamster has also been documented (Vriend and Reiter, 1977; Vaughan et al., 1982).

Serotonin, or 5-hydroxytryptamine (5-HT), has been hypothesized to be important in the control of hibernation of the hamster through the modulation of the pineal gland, hypothalamus, or other central sites (Jansky, 1978; Jansky et al., 1980). Brain stem turnover of 5-HT increases significantly at the onset of hibernation and again during hibernation (Novotna et al., 1975). The occurrence of hibernation was also increased by feeding a diet high in tryptophan (Jansky, 1978), the precursor of 5-HT production, and by the administration of a drug combination that simulates 5-HT production and inhibits nor-

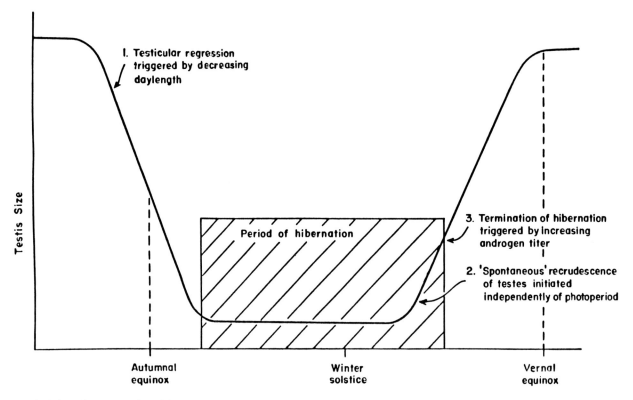

Fig. 2. Schematic representation of the hypothetical role of the testes in regulation of the annual cycle of hibernation. In European and Turkish hamsters testicular regression occurs in late summer–autumn under the influence of decreased day length. Hibernation begins after testicular regression has taken place. During the later stages of hibernation, testicular recrudescence begins, accompanied by increasing blood titers of testosterone. Increased circulating testosterone is associated with the termination of hibernation (Hall and Goldman, 1980); conversely, castration prolongs the hibernation "season" (Hall et al., 1982). This provides an endocrine mechanism for coordination of the termination of hibernation with the onset of the breeding season.

epinephrine metabolism (Jansky *et al.*, 1980). Plasma tryptophan levels are depressed during this process (Duncan and Trickleband, 1978). Increased availability of tryptophan for brain 5-HT metabolism may be related to declining liver tryptophan pyrrolase (tryptophan 2,3-dioxygenase) activity resulting from diminished plasma corticosteroid levels. However, this mechanism occurs in temperature stress, including hyperthermia, as well as in hibernation (Novotna and Civin, 1979; Novotna, 1980). Pharmacological inhibition of the synthesis of norepinephrine has also been reported to prevent the arousal of hibernating hamsters (Hudson and Wang, 1979). The effector pathways by which the neurohumors 5-HT and norepinephrine have an effect on hibernation have not yet been elucidated.

Several other endocrinological alterations of hibernation have been reported, but in most instances insufficient evidence has accumulated to allow the differentiation of changes which occur prior to and promote hibernation versus those which occur secondary to the hibernation process (Hudson and Wang, 1979). The endocrinological interactions which regulate thermogenesis are of central importance in successful hibernation of the hamster and have been extensively investigated. Norepinephrine, in conjunction with glucocorticoids and thyroid hormones, is the primary regulator of brown adipose tissue (Landsberg and Young, 1983; Himms-Hagen, 1983). The hamster has a relatively high capacity for nonshivering thermogenesis (NST) induced by norepinephrine (Vybiral and Jansky, 1972; Jansky, 1973), which is dependent on a specific brown fat mitochondrial protein, "thermogenin" (Cannon *et al.*, 1980). Brown fat triglyceride utilization accounts for all the energy necessary for arousal from hibernation in the hamster, but the relative contributions of norepinephrine-induced NST and of the release of free fatty acids from brown adipose tissue has not yet been determined (Cannon *et al.*, 1980). Hamsters do not develop large fat reserves prior to hibernation, and generally lose weight due to the depletion of glycogen stores during the prehibernatory phase (Lyman, 1948). This suggests that gluconeogenesis and glycogenolysis are also essential for thermoregulation during hibernation. Hamsters remain normoglycemic during hibernation and will not enter hibernation if adrenalectomized (Musacchia and Deavers, 1980; Jansky *et al.*, 1980). When hypothermia was induced rapidly by exposure to extreme cold, hamsters utilized intravenously infused glucose to sustain their survival, and glucocorticoid supplements enabled these animals to arouse spontaneously from hypothermia (Musacchia and Deavers, 1980). Potential roles for glucocorticoids in hamster hibernation other than gluconeogenesis have also been discussed (Musacchia and Deavers, 1980).

The hamster maintains normal thyroid function prior to and during hibernation, in contrast to some other hibernating species (Tashima, 1965; Bauman *et al.*, 1969; Hudson, 1980). This adaptation may reflect the requirement for an active thyroid for the continuous metabolism of food eaten during the animal's periodic arousal (Hudson and Wang, 1979). Thyroidectomy in the hamster reduced the tendency to hibernation, whereas reverse triiodothyronine was demonstrated to increase the number of hibernating animals but not affect the prehibernatory phase (Jansky *et al.*, 1980). However, a role has not been ascribed to thyroid hormones in the regulation of hibernation.

D. Cellular Alterations in Hibernation

In other hibernating species, decreased thyroid activity during hibernation was associated with increased cellular membrane fluidity, a compensatory change thought to confer a cellular resistance to cold (Willis, 1978; Augee *et al.*, 1979; Hudson, 1980). While changes in thyroid function cannot be correlated with specific changes in membrane fluidity in the hamster during hibernation, several plasma membrane alterations compatible with increased fluidity have been reported (Goldman, 1975; Blaker and Moscatelli, 1978). Position-specific changes in the fatty acid chain composition of ethanolamine glycerophospholipids in synaptosomal, brain mitochondrial, and myelin membrane lipids have been characterized in hibernating hamsters (Blaker and Moscatelli, 1979; Demediuk and Moscatelli, 1983). These changes were not detected in cold-acclimated hamsters, indicating that these animals and hibernating hamsters represent chemically distinct populations (Demediuk and Moscatelli, 1983). Increased polarity of brain gangliosides has also been described in hibernating hamsters and may allow enhanced calcium binding at reduced body temperature (Hilbig and Rahmann, 1979; Rahmann and Hilbig, 1980).

Shifts of the breakpoints in Arrhenius plots of select liver plasma membrane enzymes have been documented in the hibernating hamster (Houslay and Palmer, 1978) and may be considered indirect evidence of increased membrane fluidity at reduced temperatures. However, in a comprehensive discussion of Arrhenius plot data of numerous enzymes derived from ectotherm, endotherm, and hibernating species, including the hamster, it was concluded that differences in the Arrhenius plots of control and hibernating animals were not demonstrable (Charnock, 1978). In addition, differences in the bulk membrane fluidity between active and hibernating hamsters were not observed with the sensitive steady-state fluorescence polarization technique (Cossins and Wilkinson, 1982). These findings tend to indicate that homoioviscous adaptation may not be important in hibernation in this species.

VI. TERATOLOGY

Although the use of the hamster undoubtedly will never surpass that of the rat and the mouse in experimental teratology, it has secured a substantial role in this discipline. Several features of the hamster conducive to its use in experimental teratology are the ease of obtaining timed matings, the rapid embryonic development, and the low spontaneous malformation rate (Wilson et al., 1978). The hamster has been utilized in the investigation of a wide variety of teratogens. The following review emphasizes the compounds that have been the most actively investigated in the hamster.

A. Physical Factors

Two physical factors that have been shown to be teratogenic in hamsters are environmental heat stress and extreme whole-body hypothermia. Smith (1958) showed that reducing the pregnant dam's core body temperature to 0°C for up to 45 min on one occasion was teratogenic or lethal for hamster pups between gestation days 1.5 and 8.5, or days 9.5–11.5, respectively, but was harmless to the pups after day 12.5 of gestation. Heat exposure for two 20-min periods at 40°C was also teratogenic between days 7 and 13 of gestation, and the day of exposure influenced the teratological effects produced (Umpierre and Dukelow, 1977).

B. Pharmacological Compounds

The influence of the day of gestation that a treatment is administered on the teratological manifestations produced is evident for many types of compounds in the hamster and other animal species. In conjunction with a teratological study of retinoic acid (vitamin A_1), Shenefelt (1972) reviewed in detail the normal embryonic development of the hamster and correlated the critical period in gestation for various malformations with these embryonic events. Retinoic acid, a rapidly excreted derivative of vitamin A, is a potent teratogen which alters fetal morphology within 6 hr of administration and produces many different anomalies (Shenefelt, 1972). Vertebral defects in hamster pups exposed to retinoic acid on days 7, 8, and 9 of pregnancy resulted from a disruption of somatogenesis induced by vascular damage (Wiley, 1983). Ocular and orbital developmental abnormalities due to retinoic acid have also been reported in hamsters (Cauwenbergs et al., 1978). Vitamin A is a less potent teratogen than retinoic acid but also produces alterations in the cranial somites, resulting in cranioschisis (Marin-Padilla and Ferm, 1965).

The teratogenic potential of many other pharmacological compounds has been evaluated in the hamster. Several inbred strains of hamsters are sensitive to the teratogenic effects of thalidomide (Homburger et al., 1965). The corticosteroid compounds, triamcinolone acetonide and hydrocortisone, have been demonstrated to be teratogens resulting in meningeal defects and partial cleft palates (Kelly, 1980; Chaudhry and Shah, 1973). Hydrocortisone is most effective when administered on day 11 of gestation, and lower doses produced a higher incidence of partial cleft palates than did higher doses of this compound. Some analgesic compounds or their metabolic products have been shown to be teratogens in the hamster (LaPointe and Harvey, 1964; Rutkowski and Ferm, 1982). The metabolic product of acetaminophen, p-aminophenol, produced a dose-dependent effect in the pups of dams treated on day 8 of gestation, resulting in many types of abnormalities without compromising maternal health. The o-aminophenol derivative was also teratogenic but m-aminophenol was not. Vincristine, vinblastine, colchicine, actinomycin D, 5-bromodeoxyuridine, and 6-aminonicotinamide are teratogens in the hamster through various nucleotoxic mechansims (Ferm, 1963; Elis and DiPaolo, 1970; Ruffalo and Ferm, 1965; Turbow et al. 1971). Colchicine, vincristine, and vinblastine arrest cellular mitotic activity and produce various skeletal, neural, and ophthalmic abnormalities in the pups of dams treated at day 8 of gestation (Ferm, 1963). Actinomycin D is both feticidal and teratogenic but through different mechanisms. Embryo lethality is an indirect effect through the hormonal alteration of the dam, whereas the teratogenic alterations result from a direct effect on the embryos and can be ameliorated by the concomitant administration of nucleic acids to the dam (Elis and DiPaolo, 1970). Tryptophan may also be teratogenic in the hamster. Tryptophan appears to be embryotoxic in the hamster by elevating serotonin levels which result in placental vasculature damage (Meier and Wilson, 1983).

C. Heavy-Metal Compounds

Many heavy-metal compounds are teratogens in the hamster, including indium nitrate, germanium tetroxide, gallium sulfate, nickel carbonyl, cadmium compounds, and organic and inorganic mercury compounds (Ferm and Carpenter, 1970; Sunderman et al., 1981; Tassinari and Long, 1982; Reuhl et al., 1981a). The effects of cadmium compounds were determined to be highly specific for the anterior neural segment when given on day 8 of gestation (Tassinari and Long, 1982; Ferm, 1971). The hamster placenta is permeable to organic and inorganic mercury compounds (Gale and Hanlon, 1976). Both inorganic and organic mercury compounds are teratogenic and embryotoxic and result in many different developmental abnormalities (Harris et al., 1972; Gale and Germ, 1971). Hamster pups that were exposed in utero to methyl mercury developed cerebellar lesions: neural lesions were most pronounced during the first 15 postpartum days and persisted into adulthood (Reuhl et al., 1981a,b).

D. Other Teratogenic Compounds

Other substances which have been evaluated as teratogens in the hamster include plant-borne compounds, as well as herbicides and pesticides. The pesticides and herbicides reported to be teratogenic include phthalimide derivatives, thiuram derivatives, carbaryl, diazinon, 2,4,5-trichlorophenoxyacetic acid (2,4,5-T), 2,4-dichlorophenoxyacetic acid (2,4-D), and dioxin (Robens, 1969, 1970; Collins and Williams, 1971). Dioxin contamination of 2,4,5-T significantly increased the incidence and severity of the abnormalities noted (Collins and Williams, 1971); however, 2,4,5-T and 2,4-D alone were also teratogenic in hamsters. Acetonitrile, a plant-associated compound, presumably yields cyanide during metabolism, thereby producing its teratogenic effects (Willhite, 1983). Aflatoxin B_1 and β-aminopropionitrile, a lathyrogen, are other plant-associated compounds that are known to be teratogenic for hamsters (Wiley and Joneja, 1978; DiPaolo et al., 1969).

VII. TOXICOLOGY

The hamster has been utilized in the toxicological evaluation of many compounds, including drugs, hormones, food additives and contaminants, cigarette smoke and its components, industrial chemicals, heavy metals, and other environmental contaminants. Many of these compounds are carcinogenic in the hamster; for a discussion of this aspect of toxicology refer to Section X of this chapter. The hamster is suited to a variety of toxicology bioassays both *in vivo* and *in vitro* but generally is used less frequently than the rat and mouse for which larger biological data bases exist (Gak et al., 1976; Arnold and Grice, 1979; Shubik, 1972). However, in certain bioassays (e.g., inhalational studies) and in the investigation of particular compounds, hamsters are used extensively and may have become the preferred species through the development of inbred strains (see Chapter 15) (Homburger et al., 1979; Shubik, 1972).

A. General Considerations

Both the source and level of dietary protein have been demonstrated to have a marked effect on hepatic microsomal drug metabolism in the hamster, and should be considered in toxicological studies using this species (Birt et al., 1982; Birt and Schuldt, 1982). Generally, microsomal protein, microsomal P-450 content, and the activities of the detoxifying enzymes, aniline hydroxylase and aryl hydrocarbon hydroxylase, increased in hamsters as dietary protein increased up to 20 gm/100 gm diet. Some differences in these parameters were noted between the sexes in the response to various levels of dietary protein. Consideration should also be given to the effect of gut bacteria on drug metabolism in the planning of a long-term bioassay in the hamster (Arnold and Grice, 1979).

B. General Toxicology Studies

The toxicology of several dietary components or contaminants have been evaluated in the hamster including selenium, monosodium glutamate, and mycotoxins. Renal lesions consisting of tubular epithelial necrosis and tubular calcification were attributed to the toxic effects of the mycotoxin, citrinin (Jordan et al., 1978). The mycotoxin patulin produced gastrointestinal tract lesions that were attributed to antimicrobial activity, rather than a direct toxic effect, of this compound (McKinley and Carlton, 1980). Aflatoxin B_1 produced hepatic necrosis, but the solvent used as a carrier, N,N-dimethylformamide, also induced hepatic lesions (Ungar et al., 1976). Hamsters appeared more susceptible to the solvent poisoning than did rats, rabbits, and guinea pigs which had been administered this solvent in aflatoxin studies. Monosodium glutamate administered parenterally to hamsters during the neonatal period had a selective neurotoxic effect and resulted in alterations of the reproductive neuroendocrine axis (Lamperti and Blaha, 1980; Pickard et al., 1982). Dietary selenium, when fed as sodium selenite above 10 ppm, becomes toxic for hamsters possibly by producing hepatic atrophy (Julius et al., 1983).

DDT and numerous other organochlorine compounds, dioxin, paraquat, and maleic hydrazide are among the insecticides and herbicides which have been examined in the hamster (Gak et al., 1976; Cabral and Shubik, 1977; Carbal et al., 1977; Hinkle, 1973; Gray et al., 1979; Olson et al., 1980a,b; Kim et al., 1980, 1981; Iwasaki et al., 1981). The hamster is uniquely tolerant of the acute toxic effects of DDT and its two intermediate metabolites, DDD and DDE, compared with the rat and the mouse (Gak et al., 1976), but for other organochlorine compounds, the toxicity in the rat, mouse, and hamster is comparable. The hamster is also much less sensitive to the lethal effects of 2,3,7,8-tetrachorodibenzo-p-dioxin (TCDD) than other animal species in which this has been examined (Olson et al., 1980a), possibly due to its enhanced metabolism and excretion in this species (Olson et al. 1980b).

C. Toxicology of Pharmacological Compounds

The toxicology of several experimental and clinically useful pharmacological compounds has been studied in hamsters. Gak et al. (1976) noted that the hamster has a peculiar resistance to the action of several pharmacological compounds, including barbiturates, morphine, and colchicine. Morphine does not induce a narcotic reaction in the hamster even when the LD_{50} (1250 mg/kg) is administered (Houchin, 1943). The hamster is also extremely resistant to the toxic effects of colchicine: the LD_{50} (600 mg/kg) is approximately 600 times the dose known to be lethal to a healthy man (Orsini and Pansky, 1952). The comparative metabolism and cardiotoxicity of the anthracycline antibiotics adriamycin and daunorubicin has been reported

(Bachur et al., 1973). The antineoplastic compound, daunorubicin, produces clinical signs and lesions in the hamster which can be ameliorated by the administration of 1,2-bis-3,5-dioxopiperazinyl-1-ylpropane (1CRF-187), a compound which alters free-radical toxicity (Herman et al., 1983). Administration of 1CRF-187 also reduces acetaminophen hepatotoxicity in the hamster (El-Hage et al., 1983). The antineoplastic compound, 5-fluorouracil produces dermal and peripheral blood toxicity in humans; these features of 5-fluorouracil toxicity have also been examined in hamsters (Cherrick, and Weissman, 1974; Cherrick and McKelvy, 1975). Several retinoids, a class of antineoplastic and dermatological drugs, have been shown to produce testicular lesions and sperm abnormalities in hamsters (Stinson et al., 1980). Reserpine toxicity in the hamster produces ultrastructural hepatocellular changes comparable in many respects to those produced by starvation (Winborn and Seelig, 1970).

Bleomycin produces pulmonary interstitial fibrosis in the hamster analogous to that observed in a small percentage of patients treated parenterally with this antineoplastic compound. The hamster model is produced by the direct intratracheal instillation of bleomycin and has proved useful in the evaluation of oxygen therapy and toxicity, and for the correlation of respiratory function tests with disease and biochemical composition of the lungs (Goldstein et al., 1979; Rinaldo et al., 1982; Tryka et al., 1982). A method for the production of pulmonary emphysema in the hamster has also been described which entailed the instillation of porcine pancreatic elastase into the lung (Lucey et al., 1980). These models permit the influence of preexisting pulmonary disease on the toxicity of a compound to be examined in inhalation and intratracheal instillation studies.

D. Inhalation Studies

Inhalation studies, and to a lesser extent, intratracheal instillation studies, constitute the predominant use of the hamster in toxicological research. A summary of some of the key research with an emphasis on recent investigations in inhalational toxicology is presented in Table I. However, the enormous amount of literature which has accumulated in inhalational toxicology precludes a complete description of this area. Hamsters are regarded as particularly useful in these studies because they are susceptible to respiratory carcinogens but do not have a high incidence of spontaneous pulmonary neoplasia, and because they are resistant to chronic pulmonary infections, specifically, murine mycoplasmosis (Van Hoosier et al., 1971; Werner et al., 1979). In addition, the histopathological characteristics of tracheal tumors in the hamster induced by carcinogens are similar to bronchogenic carcinoma, the most common primary lung tumor of humans (Saffiotti et al., 1968; Grubbs et al. 1979). Characterization of the anatomical, histological, morphometric, and ultrastructural features of the hamster respiratory tract as it relates to this model of human respiratory carcinogenesis has been reported (Kennedy and Little, 1979).

The method of preparation of in vitro organ cultures from hamster tracheal explants also has been described, and these organ culture systems have proved to be useful in both toxicological and infectious disease investigations (Mossman and Craighead, 1979; Gabridge, 1979). Other in vitro systems have been developed using hamster tissues for the evaluation of airborne pollutants but have not received widespread acceptance in this area of research (Samuelsen et al., 1978).

VIII. RADIOBIOLOGICAL RESEARCH

The hamster has been utilized in a wide variety of radiobiological investigations and, along with the Chinese hamster (Cricetus griseus), is among the most radioresistant animals ever studied.

A. Radiation Syndromes

According to Eddy and Casarett (1972), the plot of survival time versus radiation exposure level for the Syrian hamster has a conformation comparable to that of other species, but the dose levels required to produce the three recognized radiation syndromes (i.e., hematopoietic, intestinal, and central nervous system) are increased significantly. Radioresistance may be related to the pattern of recovery and proliferative capacities of specific organ systems in the hamster (Eddy and Casarett, 1972). The median lethal dose (50% survival) 30 days postirradiation ($LD_{50/30}$) required to produce injury to the hematopoietic system in the hamster was approximately 900–1000 roentgen(R), which was several hundred roentgen higher than that necessary in some other species. Resistance of the hematopoietic system of the hamster was attributed to the early regenerative activity and hyperplasia evident in the bone marrow following irradiation. Hamsters irradiated with a higher dose developed necrosis of the intestinal epithelium, but the intestinal epithelium was maintained for a longer period and the regenerative response of the epithelium was more vigorous than that observed in other species. Consequently, an LD_{50} estimated at an 8-day survival interval was considered a more appropriate parameter for the hamster than the LD_{50} at a 5-day survival normally applied to other species. Hamsters also survive longer than most other species when irradiated with 8000–10,000 R sufficient to produce central nervous system disturbances and death.

In addition to the classical radiation syndromes, a newly recognized radiation syndrome has been described in the ham-

Table I
Inhalation and Intratracheal Instillation Studies in the Syrian Hamster

Method of administration	Material administered	Biological parameters examined	Comments	Reference
Instillation	9,10-Dimethyl-1,2-benzanthracene	Tumor induction	Early model of respiratory neoplasia	Della Porta et al. (1958)
Instillation	Benzo[a]pyrene and ferric oxide	Pulmonary histopathology	Respiratory neoplasia induced	Saffiotti et al. (1968)
Instillation	Benzo[a]pyrene and ferric oxide	Epithelial ultrastructure	Histogenesis of squamous-cell carcinoma characterized	Harris et al. (1971)
Instillation	Benzo[a]pyrene	Tumor	Synergism with furfural studied	Feron (1972)
Instillation	1-Methyl-1-nitrosourea	Incidence and histopathology	Model to evaluate antineoplastic compounds	Grubbs et al. (1979)
Instillation	N-Nitroso-N-methylurea	Histopathology	Dose dependency and topography of neoplasm	Nettesheim and Tsutomu (1979)
Instillation	Polyethylene glycol and p-isooctylphenyl ether (Triton X-100)	Thymidine uptake and autoradiography	Tissue repair studied	Hackett et al. (1980)
Inhalation/ instillation	Triton X-100	Histopathology	Effect of route of exposure on histopathology studied	Damon et al. (1982)
Instillation	$CdCl_2$	Morphology	Histopathological response mediated by lathrogen treatment	Niewoehner and Hoidal (1982)
Instillation	As_2O_3	Histopathology	Carcinogenicity of compound determined	Ishinishi et al. (1983)
Instillation	^{210}Po	Autoradiography	Distribution in lung following dosing	Kennedy et al. (1977)
Instillation	^{210}Po	Histopathology	Progression of histopathological changes studied	Kennedy et al. (1978)
Instillation	^{210}Po	Histopathology	Instillation contrasted with other routes of lung α-irradiation exposure	Anderson et al. (1979)
Instillation	^{210}Po, Benzo[a]pyrene	Histopathology, autoradiography	Cellular proliferation potentiates carcinogenesis	Shami et al. (1982a)
Inhalation	^{90}Y on fused clay particles	Biochemistry, respiratory physiology, histopathology	Pulmonary fibrosis	Pickrell et al. (1976)
Inhalation	$^{144}CeO_2$	Histopathology	Influence of hamster age and frequency of dosing on tumor development examined	Lundgren et al. (1982)
Inhalation	None	Morphometric analysis	Mathematical modeling of particle inhalation, translocation, and clearance in tissue	Desrosiers et al. (1978)
Inhalation	Cigarette smoke (various types)	Histopathology and numerous other parameters	Synergistic effect of 7,12-dimethylbenz[a]anthracene diethylnitrosamine evaluated and asbestos	Dontenwill et al. (1973)
Inhalation	Cigarette smoke	Biochemical and hematological data	—	Dontenwill et al. (1974)
Inhalation	Cigarette smoke	Histopathology, mortality, organ and body weights, serum constituents	Model system different from that of Dontewill et al. (1973)	Bernfeld et al. (1974)
Inhalation	Cigarette smoke	Histopathology	Lung parenchymal response to smoke	Ketkar et al. (1977)
Inhalation	Cigarette smoke	Histopathology	Effect of smoke on nasal mucosa evaluated	Basrur and Harada (1979)
Inhalation	Cigarette smoke and vitamin A	Histopathology	Effect of vitamin A depletion and supplementation on carcinogenesis evaluated	Meade et al. (1979)

(continued)

Table I (*Continued*)

Method of administration	Material administered	Biological parameters examined	Comments	Reference
Inhalation/ instillation	Acetaldehyde	Mortality data and histopathology	Acetaldehyde evaluated as cofactor in carcinogenesis	Feron (1979)
Inhalation	Cigarette smoke	Histopathology	Various N-nitrosamines of tobacco smoke examined as carcinogens	Hoffman et al (1979) Hecht et al. (1983)
Inhalation	Cigarette smoke, chrysotile asbestos, NiO, CoO, talc, baby powder, radon daughters, uranium ore dust, diesel engine exhaust, asbestos, cement, fly ash, plutonium dioxides	Mortality data, histopathology	Numerous compound synergisms studied	Werner et al. (1979)
Inhalation	Various consumer products	Mortality data	Comparative value of lavage in acute toxicity testing discussed	Damon et al. (1979)
Inhalation	Cigarette smoke	Histopathology and ultrastructure	Relation of nicotine content to lesions	Reznik-Schuller (1980)
Inhalation	Cigarette smoke	Scanning electron microscopy	Conjunctival epithelial response	Basrur and Basu (1980)
Inhalation	Commercial hair spray	Histopathology	—	Gupta and Drew (1976)
Inhalation	N_2O	Biochemical	Effect on elastin and collagen content of lung	Kleinerman (1979)
Inhalation	$CdCl_2$ and $CrCl_3$	Biochemical and cytological	Validation of techniques to evaluate lung injury	Henderson et al. (1979)
Inhalation	N_2O	Biochemical and cytological	Correlation of biochemical and cytological indicators of tissue damage in lung	Denicola et al. (1981)
Inhalation	Ozone	Morphological and cytokinetic analysis	Regenerative capacity of respiratory epithelium	Shami et al. (1982b)
Inhalation	Silica and fly ash	Immunological indices	Differential effect of silica and fly ash on alveolar macrophage function	Burns and Zarkower (1982)
Inhalation	$^{238}PuO_2$ and $^{239}PuO_2$	Histopathology, mortality data, radioactive retention	—	Saunders (1977)
Inhalation	$^{238}PuO_2/ZrO_2$ microspheres	Cell growth, chromosome analysis, and autoradiography	Chromosomal aberrations induced	Stroud (1977)
Inhalation	^{137}Cs-Labeled aluminosilicate particles	Radioactive retention	Regional differences in lung examined	Thomas and Raabe (1978)
Inhalation	$^{238}PuO_2$	Autoradiography	Calculation of local dosimetry of α-irradiation	Diel (1978)
Inhalation	$^{238}PuO_2/ZrO_2$ microspheres	Histopathology	—	Thomas and Smith (1979)
Inhalation	^{137}Cs-Labeled aluminosilicate particles	Retention of radioactivity	Effect of induce emphysema on particle clearance	Hahn and Hobbs (1979)
Inhalation	^{90}Y on fused clay particles	Biochemistry, respiratory physiology, histopathology	Pulmonary fibrosis	Pickrell et al. (1976)
Inhalation	$^{238}PuO_2$	Radioactive retention	Effect of time and dose on regional lung clearance of radioactivity	Diel et al. (1981)
Inhalation	$^{239}PuO_2$	Histopathology	Tumor incidence	Thomas et al. (1981)
Inhalation	$^{239}PuO_2$	Histopathology	Radiation dose effect examined	Lundgren et al. (1983)
Inhalation	$^{238}PuO_2$	Pulmonary physiology, histopathology	Correlation of pathophysiology and histopathology findings	Pickrell et al. (1983)
Inhalation	^{222}Rn daughters	Mortality data, histopathology	Hamster not deemed appropriate animal model for study of uranium ore	Cross et al. (1981)
Inhalation	$^{144}CeO_2$	Histopathology	Influence of hamster age and frequency of dosing on tumor development examined	Lundgren et al. (1982)
Inhalation	None	Morphometric analysis	Mathematical modeling of particle inhalation, translocation, and clearance in tissue	Desrosiers et al. (1978)

ster. The syndrome is correlated with a new plateau in the dose survival time response occurring 40–57 hr after doses of 30,000–60,000 R (Tsubouchi and Matsuzawa, 1981). In this dose range, hamsters developed necrosis of the cells in the islets of Langerhans, resulting in hyperglycemia, hyperkalemia, ketonemia, and acidosis which progresses until death (Tsubouchi et al., 1981). Insulin treatment reduced the hyperglycemia and extended the survival period to 5 days, which is comparable to survival of hamsters with the gastrointestinal syndrome.

Eddy and Casarett (1972) have demonstrated that the hamster exhibits a biphasic pattern in the rate of recovery from radiation injury using the split-dose lethality technique. This response in the hamster is unlike that of other small mammals but has been observed in some larger species including dogs and nonhuman primates (Holloway et al., 1968). Split-dose irradiation also has been used to characterize radiation-induced anorexia in the hamster (Kindt et al., 1980). Immediately following total or partial irradiation, hamsters developed anorexia related to decreased gastric emptying, perhaps mediated through radiation-sensitive target sites in the head or abdomen. The hamsters developed a more severe anorexia 6–10 days later, presumably associated with the ensuing hematopoietic syndrome; this period of anorexia is prolonged and more pronounced in nonsurvivors.

B. Other Radiation Studies

One of the main uses of hamsters in radiobiological research has been the induction of neoplasia either by whole-body irradiation or by intrapulmonary exposure to radioactive compounds. A summary of inhalation and intratracheal instillation studies using radioactive compounds is included in Table I, and the reader should refer to Chapter 12 for information on radiation-induced neoplasia. Other aspects of radiobiological research in the hamster that relate to neoplasia also have been investigated. Low-dose localized irradiation of the head and neck region of hamsters with 20 R at weekly intervals for 15–17 weeks has been shown to increase the incidence of tumorogenesis induced chemically by 7,12-dimethylbenz[a]anthracene in a variety of oral tissue (Lurie and Rippey, 1978; Lurie and Cutler, 1979). Repeated low-dose irradiation of hamster tumor-bearing cheek pouch also had increased blood volume and vascular perfusion (Lurie and Rippey, 1978). This response of tumor vasculature to irradiation was evaluated as a therapeutic modality in the hamster and found not to afford any tumor control alone. However, radiation and hyperthermia treatment in combination presumably alter the internal environment of the tumor to produce a greater degree of tumor control than can be ascribed to the summation of direct cell killing by these therapeutic modalities (Eddy and Chmielewski, 1982).

Hamsters also have been utilized in the investigation of antedotes for radioactive contamination including ^{252}Cf, ^{241}Am, ^{238}Pu, and ^{239}Pu compounds. The biodistribution and metabolism of ^{252}Cf, ^{241}Am, ^{238}Pu, and ^{239}Pu has been reviewed and related to the antedotal efficacy of the chelating agent, diethylenetriaminepentaacetate (DTPA) (Seidel, 1977, 1978; Stradling et al., 1981; Winter and Seidel, 1981, 1982), and of the radioprotectant, iron sorbital citrate. According to Seidel (1978), DTPA removed about 50% of the ^{252}CF activity from the skeleton of the hamster, but its effectiveness in removing ^{252}CF from the liver was considerably lower. Stradling et al. (1981) studied the efficacy of DTPA and its lipophilic derivative Puchel* on the removal of ^{239}PuO$_2$ from lung and extrapulmonary sites of the hamster and concluded that bronchopulmonary lavage removed all the lung deposits. Extrapulmonary ^{239}Pu deposits became stable 2 days after exposure and were much more difficult to remove, requiring high doses of supplemental intravenous Puchel. Stradling et al. (1981) considered the efficacy of Puchel superior to that of DTPA in removing ^{239}Pu deposits and developed recommendations for the treatment of humans based on the hamster model.

IX. ORAL PATHOLOGY: CHEMICAL CARCINOGENESIS IN THE HAMSTER CHEEK POUCH

The primary aims of any experimental model of oral cancer are to gain an understanding of the mechanisms of neoplasia and gather data that will help the clinician to provide more effective treatment of patients. The parameters of the model must duplicate as closely as possible the human disease. The hamster oral mucous membrane tumor model for experimental epidermoid carcinoma is a widely used and extensively studied model. Originally developed by Salley (1954), it was later studied in great detail by Morris (1961) and others (Shklar, 1965, 1968). Among its many advantages, this model simulates human oral carcinoma in many aspects. As in other chemical carcinogenesis systems, there is a relatively long latent period; the tumors are slowly growing and can be carefully followed, counted, and measured.

A. Induction and Development of Cheek Pouch Tumors

Hamster cheek pouch tumors, as in human oral malignancy, are epidermoid (squamous-cell) carcinomas, rather than sarcomas of underlying connective tissue. The hosts are outbred, and passenger viruses are found very rarely. Furthermore, the hamster does not develop spontaneous tumors of the buccal

---

*Puchel, National Radiological Protection Board, Harwell, England.

pouch to complicate the interpretation of the numbers of tumors induced by the chemical carcinogen. The chemically induced epidermoid carcinomas of hamster buccal pouch are preceded by a precancerous state similar to human oral leukoplakia, so that not only the tumors can be studied, but also the initial transitional lesions leading to frank carcinomas. Since a clear sequence of tumor development exists in this excellent model system, it becomes possible to study a variety of local and systemic influences that may be capable of augmenting, inhibiting, or, in other ways, modifying the sequence of neoplastic transformation and development.

Epidermoid carcinoma is produced in the pouch through the topical application by brush of a 0.5% solution of 7,12-dimethylbenz[a]anthracene (9,10-dimethyl-1,2-benzathracene) (DMBA) in mineral oil, applied three times weekly, using a number 3 or 4 sable brush.

The sequence of tumor development as viewed grossly and histologically is as follows: hyperkeratosis and chronic inflammation occurring at 4–6 weeks, hyperkeratosis and dysplasia occurring at 6–8 weeks, carcinoma in situ at 8–10 weeks, papillary carcinoma and frank invasion of the dermis at 10–12 weeks, and extensive tumor with invasion and surface necrosis at 12–14 weeks (Figs. 3–7). Santis et al. (1964) showed that epidermoid carcinoma of the hamster pouch is preceded by a hyperkeratotic and dysplastic lesion comparable to human leukoplakia of the dysplastic or precancerous variety.

DMBA carcinogenic activity has been shown to occur through the aryl hydrocarbon monooxygenase system (Slaga, et al., 1979a). Therefore, it is probable that the production of diol epoxides derived from this system stimulates transformation of epithelial cells (Slaga et al., 1979b).

The buccal pouch mucosa is an excellent tissue for the study of potential carcinogenicity of various substances. Potential

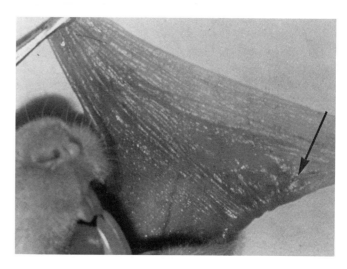

Fig. 4. Left buccal pouch of hamster after 6 weeks of DMBA application. Numerous white leukoplakic patches are evident (arrow).

oral carcinogens can be applied to the pouch for periods up to 6–12 months. In addition, the histological picture will not be complicated by appendages such as sebaceous glands or hair follicles in this model. The histological picture will demonstrate whether a certain agent has produced dysplasia or early neoplasia.

B. Factors Contributing to Carcinogenesis

Carcinogenesis of the oral mucosa of the hamster requires that the carcinogen be in contact with the epithelial cell surface. When dimethyl sulfoxide (DMSO) was applied together with DMBA, epithelial carcinogenesis was found to be retarded and sarcomas were produced by the DMBA being car-

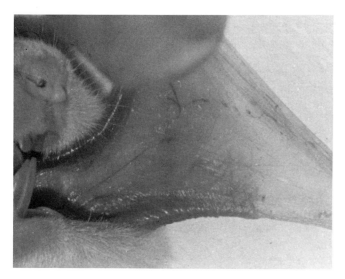

Fig. 3. Normal left buccal pouch of Syrian hamster.

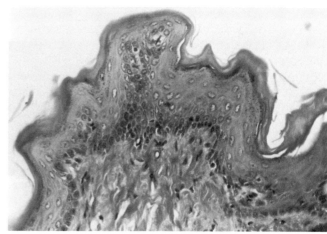

Fig. 5. Leukoplakia of hamster buccal pouch, showing microscopic features of hyperkeratosis and mild dysplasia. Magnification: ×82.5.

Fig. 6. Extensive tumor development after 16 weeks of DMBA application.

ried into the underlying connective tissue (Shklar *et al.*, 1969). Following radiation damage, the buccal pouch was also found to be less susceptible to hydrocarbon-stimulated malignancy than normal epithelium (Shklar *et al.*, 1970).

Solt and Shklar (1982) have shown that DMBA application to the hamster buccal pouch will produce γ-glutamyltransferase (GGT)-positive foci. GGT is an enzyme bound to the cell membrane, and is associated with the production of glutathione. Glutathione is a principal scavenger of free radicals that are involved in the autoperoxidation of the cell. Specifically, it was demonstrated that following three applications of DMBA (0.5%), isolated clones of GGT-positive cells were observed histologically, and thought to be initiated clones that could be promoted to malignant cells.

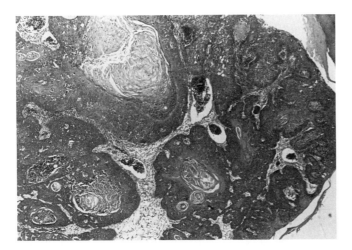

Fig. 7. Epidermoid carcinoma of hamster buccal pouch showing extensive areas of keratin formation. Magnification: ×79.5.

The kinetics of pouch carcinogenesis has been studied by Reiskin and Berry (1968), who found an increase in cell turnover following carcinogen application. Chronic irritation, produced either by a chemical agent such as croton oil (Silberman and Shklar, 1963) or by mechanical damage (Renstrup *et al.*, 1962), enhances buccal pouch carcinogenesis, presumably by acting in a cocarcinogenic manner. Buccal pouch irritation induced by topically applied retinoids has also been shown to enhance buccal pouch carcinogenesis (Levij and Polliack, 1969; Levij *et al.*, 1970; Gilmore and Giunta, 1981).

Safour *et al.* (1984) have shown that surgical manipulation of a pouch epidermoid carcinoma led to metastasis of malignant cells to cervical lymph nodes. These findings were in contrast to an earlier study by Shklar (1968), who found that manipulation and incision of the developing carcinoma did not result in deeper invasion of these lesions or metastatic spread.

An association between alcohol intake and oral cancer has been well established (Rothman and Keller, 1972). In hamsters given alcohol, DMBA carcinogenesis of the buccal pouch was significantly augmented and may be related to the hepatic damage induced by alcohol (Friedman and Shklar, 1978). Protzel *et al.* (1964) showed that in mice, liver damage by alcohol or carbon tetrachloride enhanced tumor formation by carcinogenic agents.

Smoking is a major cause of oral cancer in humans (Rothman and Keller, 1972). Products associated with cigarette smoke—for example, condensates of tobacco smoke—have been used to induce buccal pouch tumors (Kendrick, 1964; Moore and Miller, 1958). Tobacco in the form of snuff does not appear to produce preneoplastic or neoplastic change (Homburger, 1971). Treatment with snuff does impair the ability of peritoneal-derived macrophages to lyse cells of an orally derived squamous-cell carcinoma of the hamster cheek pouch (Antoniades *et al.*, 1984a). The carcinogen benzo[*a*]pyrene (3,4-benzpyrene), which is found in tobacco smoke condensate, has not been shown to induce buccal pouch tumors when used in a manner similar to DMBA or methylcholanthrene.

The concept of two phases in carcinogenesis, as originally suggested by Berenblum (1941), has remained untested until recently in the hamster cheek pouch model. Odukoya and Shklar (1982), have reported on two-phase carcinogenesis in the hamster cheek pouch. DMBA was used both as the initiator and promoter. Tissues painted with a 0.1% solution of DMBA in mineral oil for 10 weeks did not develop tumors if left untreated for a subsequent 6 weeks, but if they were then painted for an additional 4 weeks with an 0.5% solution of DMBA there was a rapid development of epidermoid carcinoma. The second course of DMBA painting purportedly acted as a promoter, since 4 weeks of DMBA application does not result in rapid tumor development without prior initiation. In a second experiment, Odukoya and Shklar (1984) treated hamsters for 10 weeks, three times a week, with a 0.1% DMBA solution; then following a period of 6 weeks of no treatment, treated for 6 weeks with 40% benzoyl peroxide. Carcinomas subsequently developed only in the experimental group, demonstrating that a noncarcinogenic agent, benzoyl peroxide, can promote tumor formation in the cheek pouch.

C. Immunological and Pharmacological Modulation of Cheek Pouch Neoplasia

Several aspects of the immune system response in relation to chemical carcinogenesis of the hamster pouch have been investigated. The buccal pouch of the hamster has been described as an immunologically privileged site (Billingham and Silvers, 1964). Evidence from studies of immunosuppressive drugs such as cortisone (Shklar, 1966, 1967) and methotrexate (Shklar et al., 1966) disputes this concept, however, since these agents enhanced carcinogenesis. Furthermore, specific antilymphocyte serum was found to enhance the rapidity of tumor development by depressing the cell-mediated immune response (Giunta and Shklar, 1971; Woods, 1969). Cortisone accelerated tumor development, whereas antilymphocyte serum and antimetabolites such as methotrexate increased the anaplastic and invasive characteristics of epidermoid carcinomas. Antimetabolite drugs, when used after tumor development, however, caused the regression of tumors by destroying the epithelial cells within the tumor. One such drug, azathioprine, was shown to be particularly effective against tumors of the cheek pouch (Sheehan et al., 1971).

Immunoenhancing agents such as bacillus Calmette-Guérin (BCG) (Giunta et al., 1974) and levamisole (Eisenberg and Shklar, 1977; Shklar et al., 1979) have been found to inhibit the development of buccal pouch tumors and may prove to be useful in the management of human oral cancer.

Sonis and Shklar (1981) have shown that there was an increase in cell-mediated immune activity associated with 13-cis-retinoic acid treatment. Guinta et al. (1975) have shown that the cheek pouch lacks lymph vessels and that macrophages mediate the drainage of particulate matter (e.g., India ink particles or fluoresceinated latex beads) from the pouch to the submandibular lymph nodes. Antoniades et al. (1984b) have shown that there was a significant inhibition of cytolysis of epidermoid carcinoma target cells by macrophages derived from hamsters with cheek pouch tumors. Langerhans cells associated with antigen processing (Silberberg-Sinakin et al., 1976) and delayed-hypersensitivity reactions (Stingl et al., 1978) have been demonstrated to be present in the cheek pouch (Fig. 8). These cells, stained for ATPase activity, are markedly reduced in number following repeated treatment of the pouch with 0.5% DMBA for 10 weeks (Schwartz et al., 1981); however, a single, double, or triple application of DMBA 0.5% resulted in an increase of ATPase-positive Langerhans cells in the pouch (Hassan et al., 1984).

A number of chemoprophylactic agents that are either non-

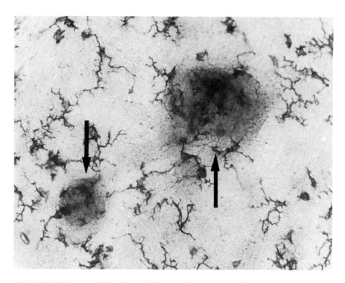

Fig. 8. ATPase-positive Langerhans cells (arrows) of hamster buccal pouch epithelium as seen in a whole-mount preparation.

toxic or minimally toxic have been studied experimentally in the hamster in an attempt to identify compounds efficacious in the prevention or inhibition of oral cancer. Retinoids, such as 13-cis-retinoic acid (Shklar et al., 1980) and retinyl acetate (Burge-Bottenbley and Shklar, 1983), given systematically resulted in a decrease in the number and size of tumors of the cheek pouch. α-Tocopherol (vitamin E) has also been shown to prevent tumor formation in the pouch, used either systemically or topically (Shklar, 1982; Weerapradist and Shklar, 1982; Odukoya et al., 1984). Vitamin E, at a concentration equivalent to 200 IU, had an in vitro proliferative effect on cell cultures derived from hamster cheek pouch squamous carcinoma. However, if two or three times this dose was used, an inhibition occurred 5 and 10 days following initial plating of the cells.

Other substances such as chlorpromazine (Polliack and Levij, 1972), ibuprofen (Cornwall et al., 1983), and indomethacin and aspirin (Perkins and Shklar, 1982), have produced an inhibition of oral carcinogenesis of the hamster cheek pouch (Fig. 9). The inhibition by these agents may be through an alteration in vascular responses to a growing cheek pouch tumor. Lurie et al. (1983) has recently demonstrated that following the application of 0.5% DMBA for 11 weeks, changes in vascular volume, dilation of capillaries, and capillary proliferation occurred in association with dysplastic and neoplastic epithelial changes.

The hamster buccal pouch carcinoma has been transplanted directly to the peritoneal cavity of inbred and neonatal hamsters (Merk et al., 1979). The transplanted tumors become more anaplastic and grow more rapidly than primary tumors. These transplanted tumors can then be reimplanted to the cheek pouch and become a model for a more anaplastic epidermoid carcinoma (Meng et al., 1982). In addition, the hamster

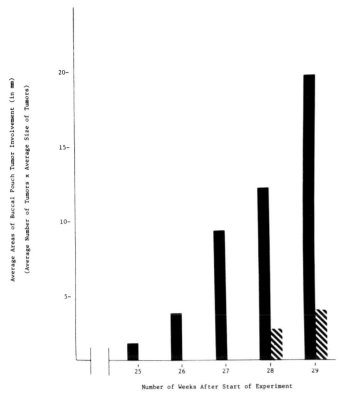

Fig. 9. Graph illustrating the prevention and inhibition of buccal pouch carcinogenesis by ibuprofen. Solid bars, DMBA only; hatched bars, DMBA plus ibuprofen.

cheek pouch carcinoma can, as discussed above, be grown in culture (Fig. 10), and recently, a cell line derived from a primary hamster epidermoid carcinoma has been developed (Odukoya *et al.*, 1983). The culture of normal epithelial cells derived from the cheek pouch can also be performed (Chieli *et al.*, 1979).

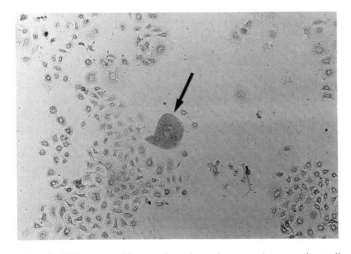

Fig. 10. Cell culture of hamster buccal pouch tumor. A tumor giant cell stains positively for GGT (arrow).

D. Conclusion

Although the hamster buccal pouch carcinoma model has proved extremely useful in a variety of studies, it has certain deficiencies when related to human oral cancer. The carcinomas tend to be well differentiated and papillary, whereas human oral cancer tends to be anaplastic and more aggressive clinically. Human oral cancer often metastasizes to regional nodes and occasionally to major organs. Metastasis in the hamster model is dependent on tumor size and mechanical manipulation. The pouch is a unique anatomical structure not found in the human.

Regardless of the model's shortcomings, it gives investigators the opportunity to complement *in vivo* studies with information obtained *in vitro* from tissue cultures derived from normal or neoplastic cheek pouch, thereby strengthening its role in oral carcinogenesis investigations.

X. CHEMICAL CARCINOGENESIS AT SITES OTHER THAN CHEEK POUCH

Rats, mice, and Syrian hamsters are the species recommended for chemical carcinogenicity studies (International Agency for Research on Cancer, 1980; Grice, 1984). The major reason for selection of these rodents is that they can be conveniently housed in large numbers in a controlled laboratory environment, thus providing a sufficient number of animals for statistical evaluation of the results. In selecting a rodent species for carcinogenicity studies, the susceptibility of various organs to carcinogenic activity of chemicals, a mean life span of at least 18 months (preferably 24 months), a low incidence of spontaneous tumors, and a low incidence of nonneoplastic lesions are important. When the metabolism of the chemical to be tested for carcinogenicity is known for humans, it is preferable to select the rodent species that metabolizes the chemical quantitatively more similar to the human.

A. Carcinogen Susceptibility

The hamster is susceptible to carcinogenic activity of various classes of chemicals. Arnold and Grice (1979) reviewed 15 volumes of the International Agency for Research on Cancer (IARC) monographs on the evaluation of carcinogenic activity of 304 chemicals. Of these compounds, 130 chemicals were classified as carcinogens. However, only 35 of these carcinogenic chemicals were tested in both the hamster and the mouse, and only 38 were studied in both the hamster and the rat. The results were similar in the hamster and the rat for 84% of the chemicals tested, and 86% of the chemicals tested caused similar neoplastic changes in the hamster and the mouse. They concluded that where comparison of species was possible, the hamster was generally neither more nor less sensitive to carcinogenic affects than the rat or the mouse. The target organs for carcinogenicity of a given chemical in the hamster may be different from that of the rat or the mouse. Homburger and Adams (1985) reviewed carcinogenicity studies in Syrian hamsters by various routes of administration and concluded that the Syrian hamster is not merely adequate but eminently suitable for chemical carcinogenicity studies.

In a review by Shubik (1972), the Syrian hamster was the recommended species for testing of aromatic amines that may cause tumors in the urinary bladder. Aromatic amines also caused intestinal cancer in hamsters with atypical proliferative enteritis (Williams et al., 1981). Susceptibility of the Syrian hamster for the carcinogenic activity of chlorinated hydrocarbons was considered equivocal (Shubik, 1972). Li et al. (1983) produced renal carcinomas in the castrated male hamster with synthetic and natural estrogens, showing that the hamster is susceptible to renal carcinogenesis by chemicals. This tumor model may provide unique opportunity to evaluate hormonal factors involved in renal carcinogenesis and for investigation of therapeutics.

Lijinsky (1984) reviewed the carcinogenic activity of nitrosamines in hamsters and rats. Nitrosamines caused tumors in the forestomach, liver, pancreas, nasal cavity, lung, trachea, and occasionally in the esophagus of the hamster. The esophagus is the most common site for tumor induction by nitrosamines in the rat but not in the hamster. However, there is no indication that the hamster is generally more or less susceptible than the rat for tumor development by nitrosamines. Some nitrosamines are more potent carcinogens in the hamster than in the rat, and vice versa. Ten or more nitrosamines induced pancreatic tumors in hamsters, but none of the 350 nitrosamines reviewed induced this type of tumor in the rat. Most of the nitrosamines causing pancreatic tumors in the hamster also induced cholangiocarcinomas, hepatocellular tumors, or hemangiosarcomas of the liver. The hamster is the recommended model for induction of ductal tumors of the pancreas by nitrosamines. Pancreatic tumors of the hamster are morphologically more similar to the human tumors than the pancreatic tumors of the rat and the mouse (Pour et al., 1976). Since pancreatic tumors are common human tumors, the Syrian hamster is the animal model of choice for induction of pancreatic tumors (Mohr, 1979).

Aflatoxin B_1 induced cholangiocarcinomas and hepatocellular tumors in the hamster as well as in the rat (Moore et al., 1982). Mori et al. (1984) reported that some azo dyes caused neoplastic lesions in the liver of the rat but not in the hamster. The antithyroid drug aminotriazole (amitrole) when fed in the diet caused thyroid and pituitary tumors in the rat but not in the hamster or the mouse (Steinhoff et al., 1983). The antioxidant butylated hydroxyanisole (BHA), a common food additive, caused papillomas of the forestomach in the hamster and the rat but not in the mouse (Ito et al., 1983). The forestomach of the hamster is considered to be more susceptible than the rat to the carcinogenic activity of BHA. Dichloromethane is not carcinogenic in the hamster but caused benign mammary tumors in the female rat (Burek et al., 1984).

Inhalation and intratracheal instillation studies with hamsters are reviewed and discussed in Section VII and Table I.

B. Mean Life Span

The mean life span of outbred, inbred, and hybrid strains of hamsters is markedly less than the rat and the mouse stocks and strains routinely used for chemical carcinogenicity studies (Grice, 1984). Various studies [Redman et al., 1979; Sher, 1982; National Toxicology Program (NTP), 1983a,b] reported a mean life span of 10–23 months for the outbred Syrian hamster. Bernfeld (1979) discussed several inbred strains with a mean life span of 14–25 months. Homburger et al. (1983) reported that a hybrid Syrian hamster strain designated as BIO F_1D Alexander with a mean life span of 20–24 months may be suitable for long-term chemical carcinogenicity studies.

C. Incidence of Spontaneous Tumors

The incidence of major tumors in viral antibody-free outbred hamsters (Charles River Laboratories) reviewed in six long-term carcinogenicity studies done for the United States National Institute of Occupational Safety and Health (NIOSH). The purpose of the review was to aid in selection of hamster strains for the National Toxicology Program (NTP, 1985) chemical carcinogenicity studies. These studies used 3040 male and 1740 female hamsters and were continued until the animals were 20–25 months of age. The approximate mean life span of male and female hamsters in these studies was 20 months. The most common tumor reported was adrenal adenoma, with an incidence of 22% in the males and 10% in the females. Tumors with a 1–5% incidence included lymphoma/leukemia of the spleen and lymph nodes and tumors of the thyroid and parathyroid. Tumors with a 0.1–1% incidence included blood vessel and cholangial tumors of the liver, stomach tumors, adenomas of the pituitary, granulosa cell tumors of the ovary, and tumors of the uterus. For additional information on neoplastic diseases of hamsters, refer to Chapter 9.

D. Incidence of Nonneoplastic Lesions

Spontaneous nonneoplastic lesions are frequently seen in Syrian hamsters as early as 5 months of age. Pour et al. (1976) described various lesions with sites and incidences. A very common lesion is amyloidosis with an incidence as high as 86%. Homburger et al. (1983) reported a low incidence of this lesion in the BIO F₁D Alexander hybrid. Common lesions observed in the six NIOSH studies reviewed by the NTP (1985) are compared with the lesions of the BIO F₁D Alexander hybrid in Table II. For additional information on amyloidosis and other nonneoplastic lesions, refer to Chapter 10.

E. Conclusions

The Syrian hamster is an appropriately sized rodent to be housed in large numbers in a controlled environment and to provide adequate samples of blood and tissues for various clinical and histopathological evaluations. These characteristics make it desirable for toxicological and carcinogenicity testing of chemicals. The hamster is susceptible to tumor development in the liver, kidney, pancreas, stomach, nasal cavity, tracheobronchial tree, lungs, gallbladder, urinary bladder, and small and large intestines. The incidence of spontaneous tumors is low except in the adrenal and the lymphoid tissue. Because of these advantages, the Syrian hamster is considered a good animal model for chemical carcinogenicity studies and often considered for use in place of the mouse or the rat.

A characteristic that detracts from its desirability for use in long-term carcinogenicity studies is that the mean life span of the Syrian hamster is much shorter than that of the rat or the mouse. Furthermore, the high incidence of spontaneous lesions such as cysts in the liver, thrombi in the heart, amyloidosis of major organs such as the liver, kidney, spleen, adrenal, and thyroid, and nephrosis are major disadvantages for use of the hamster in long-term studies. In addition, the nutrient requirements of adult hamsters are not adequately determined (Committee on Animal Nutrition, 1978), and the optimum husbandry procedures for long-term holding of this rodent are not well established. Improper nutrient concentrations of the diet may be contributing to the nephrosis and amyloidosis, resulting in shorter life span of the hamster (Feldman *et al.*, 1982). If the breeding practices in outbred hamster colonies are not well controlled, there may be genetic variability between production colonies, and within a production colony over time. Genetic characteristics of the parental inbred strains of the BIO F₁D Alexander hybrid are not well established, and it is not known whether these parental strains are closely related or are diverse enough to provide broad genetic makeup of the hamster species to this hybrid.

The Syrian hamster will be useful for carcinogenicity studies with special classes of chemicals where the metabolism and target organ for tumor induction are similar to that of the human. This rodent will also be useful in elucidating mechanisms of carcinogenesis at special sites such as the pancreas, lung, and skin. However, until outbred stocks and hybrid stains with low incidence of nonneoplastic lesions and long lifespan similar to that of the rat and the mouse are developed, and the nutrient requirements and husbandry procedures are well established, the hamster may not be the rodent of choice for long-term (≥24 months) carcinogenicity studies with many chemicals.

Table II

Incidence of Major Spontaneous Nonneoplastic Lesions[a]

	Outbred						Hybrid			
	Males			Females			Males[b]		Females	
Lesions	Mean (%)	Range[c] (%)	(N)	Mean (%)	Range (%)	(N)	Mean (%)	Range[d] (%)	Mean (%)	Range (%)
Amyloidosis							4	3–5	34	16–52
Liver	17	6–33	(236)	72	49–87	(188)				
Kidney	26	17–42	(236)	81	77–86	(188)				
Spleen	15	6–24	(228)	70	60–80	(187)				
Adrenal	20	9–33	(230)	67	57–87	(185)				
Thyroid	18	3–38	(217)	57	46–64	(183)				
Liver cysts	1	0–4	(236)	19	0–31	(188)	41	33–50	36	23–49
Nephrosis	46	37–60	(236)	16	3–37	(188)	65	59–72	50	44–56
Heart thrombi	10	6–17	(236)	37	29–51	(186)	42	26–58	32	28–37
Ileal hyperplasia	13	8–23	(232)	5	0–8	(188)	<1		<1	

[a]For details on the sources of the information see Section X, C and D.
[b]N = 177.
[c]Range of six studies with 30–48 animals per study.
[d]Range of two studies with 87 and 90 animals.

XI. DENTAL CARIES AND PERIODONTAL DISEASE

The Syrian hamster has emerged as a convenient and relevant animal model in dental caries and periodontal disease research following its recognition as a caries-susceptible animal in the early 1940s (Arnold, 1942). Dental caries has been depicted as a multifactorial disease resulting from the interaction of three primary contributing factors—host, diet, and oral microflora—each of which encompasses a number of subfactors (Keyes, 1968). Through the use of the hamster, many basic research and applied investigations of dental caries have come to fruition and have been thoroughly reviewed elsewhere (Keyes, 1968; Jordan and Van Houte, 1972; Gustafson et al., 1968). The various methodologies employed in the production, quantitation, and pathology of dental caries and periodontal disease have also been detailed and evaluated in other works (Keyes, 1968; Navia, 1977). The salient features of the hamster model and it recent uses will be considered here.

A. Dental Caries

The infectious and transmissible nature of dental caries and periodontal disease has been ascertained in the hamster (Keyes, 1968; Jordan and Van Houte, 1972). Keyes (1960) speculated that dental caries was infectious and transmissible in rodents based on differences in the patterns of disease among different colonies. Subsequently, *Streptococcus mutans* was identified as the etiological agent of coronal caries in the hamster (Fitzgerald and Keyes, 1960) several years prior to the renewed interest in the role of this organism in human dental caries.

The availability of hamsters which do not harbor endogenous *S. mutans* in the oral cavity has allowed the investigation of oral colonization and caries production in hamsters by human oral streptococcal isolates (Jordan and Van Houte, 1972; Zinner et al., 1965). Hamsters pretreated with antibiotics to suppress their endogenous bacterial populations and subsequently inoculated with an antibiotic-resistant *S. mutans* strain, have also been used in this type of investigation (Keyes, 1968). Following inoculation, examination of the hamster's dentition should be conducted periodically to obtain a comprehensive picture of the disease as it proceeds through the establishment of the organism, plaque development, enamel decalcification, and loss of tooth structure. Examination has routinely included sampling of the bacterial plaque and photography to record the disease at various stages (Jordan and Van Houte, 1972).

Many human strains of *S. mutans* have been determined to be cariogenic in the hamster and produce severe coronal lesions comparable to those produced by *S. mutans* endogenous to the hamster (Zinner et al., 1965; Krasse, 1965; Jordan and Van Houte, 1972). Human *S. salivarius* also becomes implanted and produces caries in hamsters, but its cariogenic activity is milder than that of *S. mutans* (Jordan and Van Houte, 1972). *Streptococcus sanguis*, the predominant streptococci of human dental plaque, does not naturally occur in the oral cavity of hamsters but can be established there if the sucrose-responsive plaque microflora of the hamster is suppressed with an antibiotic (Krasse and Carlsson, 1970; Jordan and Van Houte, 1972).

Diet #2000,* a caries-conducive diet, has been formulated to sustain the health and growth of the hamster as well as provide the substrates to promote cariogenic activity (Keyes and Jordan, 1964). Sucrose, the main component of this diet, appears to be of critical importance of the establishment of *S. mutans* in the hamster and human (Krasse, 1965; Krasse et al., 1967). Sucrose is necessary for the elaboration of particular extracellular polymers by oral bacteria, and these polymers play a role in the maintenance and metabolism of dental plaque (Fitzgerald and Jordan, 1968). Recently, it has been shown that the pathogenesis of dental caries in the hamster is augmented by *S. mutans* strains which elaborate water-insoluble 1,3-α-glycans in addition to water-soluble 1,6-α-dextrans (Simonson et al., 1983). Daily administration of 1,3-α-D-glucanase in the drinking water significantly reduced the mean carious area scores and total caries score in hamsters maintained on diet #2000, demonstrating the utility of the hamster in the examination of new modalities of dental caries prevention.

The hamster model has contributed in many other areas of dental therapy. Fitted vinyl mouth applicators have been developed to allow the application of dental medicaments directly to the dentition of the hamster (Horii and Keyes, 1964). The beneficial effects of fluoride treatment in hamsters were identified using this approach and resulted in the initiation of clinical tests in humans using a similar technique (Keyes et al., 1966; Englander and Keyes, 1966). The suppression or exacerbation of dental caries by various antibiotics also has been evaluated in the *S. mutans*-infected hamster model (Keyes, 1968); however, the poor tolerance of the hamster for some antibiotics imposes limitations in this area of research (Chapter 11).

The hamster caries model has also been used to examine the protective effects of a dental pit and tissue sealant. Unlike humans, however, the retention of the sealant on the teeth of hamsters afforded no protective effect against caries, suggesting that hamsters may not be an appropriate model for the evaluation of dental sealants (Kandelman et al., 1979).

B. Periodontal Disease

The study of periodontal disease in the hamster has evolved in parallel with the investigation of dental caries in this species (Jordan and Van Houte, 1972). Several aspects of periodontal disease in the hamster have favored the utilization of this animal model: the influence of dietary factors, the infectious etiol-

---

*Diet #901283, ICN Nutritional Biochemicals, Cleveland, Ohio 44128.

ogy, and the ease of reproducibility in this relatively inexpensive host system. The early studies of periodontal disease in the hamster have been reviewed previously (Keyes, 1968).

Periodontal disease in the hamster is promoted by high-carbohydrate rations, and the physical consistency of these rations may be an important etiological factor (Keyes, 1968). Hamsters fed diet #2000 or a cereal-base, starch-laden diet will develop periodontal disease; however, if a commercially available laboratory rodent diet is fed in its pelleted form, the gross plaque deposits do not form and subgingival plaque deposits diminish (Keyes, 1968). Periodontal infections do not appear to result from a unique interaction of a specific dietary component with the etiological agent; rather, a variety of carbohydrates will support the development of periodontal disease (Keyes, 1968).

Periodontal disease in the hamster has been associated with a single etiological agent, *Actinomyces viscosus*, that has been extensively studied (Jordan and Keyes, 1964; Howell *et al.*, 1965; Howell and Jordan, 1963). Studies indicate that *A. viscosus* of the hamster represents a distinctive serotype not found among the variety of catalase-positive and -negative *Actinomyces* and related diphtheroidal types prevalent in the natural flora of the human (Slack and Gerenczer, 1970; Jordan and Van Houte, 1972). Other *Actinomyces* species also accompany *A. viscosus* in the hamster flora, but their ability to induce periodontal disease in hamsters is unknown (Jordan and Van Houte, 1972).

Several pathological features of periodontal disease in the hamster are at variance with the pathology of the disease in humans and have contributed to the decline in use of this animal model (Keyes, 1968; Miller and Ripley, 1975). In the pathogenesis of peridontal disease, hamsters develop changes in the level and shape of the gingival tissues and do not develop the epithelial ulceration and vascular changes evident in humans prior to the accumulation of inflammatory exudates. In contrast, in the hamster inflammatory cells migrate through intact gingival sulcal epithelium forming a supraepithelial chronic exudate as a barrier against irritants in the plaque (Keyes, 1968; Miler and Ripley, 1975). According to Miller and Ripley (1975), other aspects limiting the applicability of this model in the study of the human disease included the lack of bone involvement, the type of bone and marrow relationship, the structure of the temporomandibular joint and the resulting pattern and forces of mastication, and the localization of maximal disease at interproximal M_1 and M_2 rather than at mesial M_1.

REFERENCES

Anderson, E. C., Holland, L. M., Prine, J. R., and Smith, D. M. (1979). Tumorigenic hazard of particulate alpha activity in Syrian hamster lungs. *Radiat. Res.* **78**, 82–97.

Anderson, M. C., and Shettleworth, S. T. (1977). Behavioral adaptation to a fixed-interval and fixed-time food delivery in golden hamsters. *J. Exp. Anal. Behav.* **27**, 33–49.

Antoniades, D., Niukian, K., Schwartz, J. L., and Shklar, G. (1984a). Effects of smokeless tobacco on the immune system. *J. Oral Med.* **39**, 136–141.

Antoniades, D., Schwartz, J. L., and Shklar, G. (1984b). The effect of chemically induced oral carcinogenesis on peritoneal macrophages. *J. Clin. Lab. Immunol.* **14**, 19–22.

Arnold, D. L., and Grice, H. C. (1979). The use of the Syrian hamster in toxicology assays with emphasis on carcinogenesis bioassay. *Prog. Exp. Tumor Res.* **24**, 222–234.

Arnold, F. A., Jr. (1942). The production of carious lesions in the molar teeth of hamsters (*C. auratus*). *Public Health Rep.* **57**, 1599–1604.

Augee, M. L., Raison, J. K., and Hulbert, A. J. (1979). Seasonal changes in membrane lipid transitions and thyroid function in the hedgehog. *Am. J. Physiol* **236**, 589–593.

Aussel, C., Uriel, J., and Mercier-Bodard, C. (1973). Rat alpha-fetoprotein: Isolation, characterization and estrogen-binding properties. *Biochimie* **55**, 1431–1437.

Bachur, N. R., Egorin, M. J., and Hildebrand, R. C. (1973). Daunorubicin and adriamycin metabolism in the golden Syrian hamster. *Biochem. Med.* **8**, 352–361.

Baranczuk, R., and Greenwald, G. S. (1973). Peripheral levels of estrogen in the cyclic hamster. *Endocrinology (Baltimore)* **92**, 805–812.

Barr, G. A., Gibbons, J. L., and Moyer, K. E. (1976). Male–female differences and the influence of neonatal and adult testosterone on intraspecies aggression in rats. *J. Comp. Physiol. Psychol.* **90**, 1169–1183.

Barraclough, C. A., and Gorski, R. A. (1962). Studies of mating behavior in the androgen-sterilized female rat in relation to the hypothalamic regulation of sexual behavior. *J. Endocrinol.* **25**, 175–182.

Basrur, P. K., and Basu, P. K. (1980). Effect of cigarette smoke on the surface structure of the conjunctival epithelium. *Can. J. Ophthalmol.* **15**, 20–23.

Basrur, P. K., and Harada, T. (1979). Alterations in the nasal mucosa of Syrian golden hamsters exposed to cigarette smoke. *Prog. Exp. Tumor Res.* **24**, 283–301.

Bast, J. D., and Greenwald, G. S. (1974). Serum profiles of follicle stimulating hormone, luteinizing hormone, and prolactin during the estrous cycle of the hamster. *Endocrinology (Baltimore)* **94**, 1295–1299.

Bast, J. D., and Greenwald, G. S. (1977). Acute and chronic elevations in serum levels FSH after unilateral ovariectomy in the cyclic hamster. *Endocrinology (Baltimore)* **100**, 955–966.

Bauman, T. R., Anderson, R. R., and Turner, C. W. (1969). Thyroid hormone secretion rates and food consumption of the hamster (*Mesocricetus auratus*) at 25.5° and 4.5°. *Gen. Comp. Endocrinol.* **10**, 92–98.

Berenblum, I. (1941). The mechanism of carcinogenesis: A study of the significance of cocarcinogenic action and related phenomena. *Cancer Res.* **1**, 807–814.

Bernfeld, P. (1979). Longevity of the Syrian hamster. *Prog. Exp. Tumor Res.* **24**, 118–126.

Bernfeld, P., Homburger, F., and Russfield, A. B. (1974). Strain differences in the response of inbred Syrian hamsters to cigarette smoke inhalation. *J. Natl. Cancer Inst. (U.S.)* **53**, 1141–1157.

Bex, F. J., and Goldman, B. D. (1975). Serum gonadotropins and follicular development in the Syrian hamster. *Endocrinology (Baltimore)* **96**, 923–933.

Billingham, R. L., and Silvers, W. K. (1964). Syrian hamster and transplantation immunity. *Plast. Reconstr. Surg.* **34**, 329–353.

Birt, D. F., and Schuldt, G. H. (1982). Dietary amino-acids and hepatic microsomal drug metabolism in Syrian hamsters. *Drug Nutr. Interact.* **1**, 177–188.

Birt, D. F., Hruza, D. S., and Baker, P. V. (1982). Effects of dietary protein level on hepatic microsomal mixed function oxidase systems during aging in 2 generations of Syrian hamsters. *Toxicol. Appl. Pharmacol.* **68**, 77–86.

Blake, C. A., Norman, R. L., and Sawyer, C. H. (1973). Validation of an ovine:ovine LH radioimmunoassay for use in the hamster. *Biol. Reprod.* **8**, 299–304.

Blaker, W. D., and Moscatelli, E. A. (1978). The effect of hibernation on the lipids of brain myelin and microsomes in Syrian hamsters. *J. Neurochem.* **31**, 1513–1518.

Blaker, W. D., and Moscatelli, E. A. (1979). The effect of hibernation on the positional distribution of ethanolamine glycerophospholipid fatty-acids in hamster (*Mesocricetus auratus*) brain membranes. *Lipids* **14**, 1027–1031.

Boegman, R. J., and Wood, P. L. (1981). Axonal transport in dystropic hamsters. *Can. J. Physiol. Pharmacol.* **59**, 202–204.

Borer, K. T., and Kooi, A. A. (1975). Regulatory defense of the exercise-induced weight elevation in hamsters. *Behav. Biol.* **13**, 301–310.

Borer, K. T., Rowland, N., Mirow, A., Borer, R. C., Jr., and Kelch, R. P. (1979). Physiological and behavioral responses to starvation in the golden hamster. *Am. J. Physiol.* **236**, E105–E112.

Borer, K. T., Kelch, R. P., and Corley, K. (1982). Hamster prolactin: Physiological changes in blood and pituitary concentrations as measured by a homologous radioimmunoassay. *Neuroendocrinology* **35**, 13–21.

Borer, K. T., Potter, C. D., and Fileccia, N. (1983). Basis for the hypoactivity that accompanies rapid weight gain in hamsters. *Physiol. Behav.* **30**, 389–397.

Bosley, C. G., and Leavitt, W. W. (1972). Dependence of preovulatory progesterone on critical period in the cyclic hamster. *Am. J. Physiol.* **222**, 129–133.

Bridges, R. S., and Goldman, B. D. (1975). Diurnal rhythms in gonadotropins and progesterone in lactating and photoperiod induced acyclic hamsters. *Biol. Reprod.* **13**, 617–622.

Bronson, F. H., and Desjardins, C. (1970). Neonatal androgen administration and adult aggressiveness in female mice. *Gen. Comp. Endocrinol.* **15**, 320–325.

Bunnell, B. N., Boland, B. D., and Dewbury, D. A. (1976). Copulatory behavior of golden hamsters (*Mesocricetus auratus*). *Behaviour* **61**, 180–206.

Burek, J. D., Nitschke, K. D., Bell, T. J., Wacherle, D. L., Childs, R. C., Beyer, J. E., Dittenber, D. A., Rampy, L. W., and McKenna, M. J. (1984). Methylene chloride: A two-year inhalation toxicity and oncogenicity study in rats and hamsters. *Fundam. Appl. Toxicol.* **4**, 30–47.

Burge-Bottenbley, A., and Shklar, G. (1983). Retardation of experimental oral cancer development by retinyl acetate. *Nutr. Cancer* **5**, 121–129.

Burns, C. A., and Zarkower, A. (1982). The effects of silica and fly ash dust inhalation on alveolar macrophage effector cell function. *RES: J. Reticuloendothel. Soc.* **32**, 449–460.

Buschmann, M. B. T., and LaVelle, A. (1983). Morphometry of nuclei, nuclear envelopes and nucleoli in aging hamster cerebrum. *Neurobiol. Aging* **4**, 197–202.

Buyukmihci, N., Goehring-Harmon, F., and Marsh, R. F. (1983). Neural pathogenesis of experimental scrapie after intraocular inoculation of hamsters. *Exp. Neurol.* **81**, 396–406.

Cabral, J. R. P., and Shubik, P. (1977). Lack of carcinogenicity of DDT in hamsters. *Fed. Proc., Fed. Am. Soc. Exp. Biol.* **36**, 1079.

Cabral, J. R. P., Shubik, P., Mollner, T., and Raitano, F. (1977). Carcinogenic activity of hexachlorobenzene in hamsters. *Nature (London)* **269**, 510–511.

Cannon, B., Nedergaard, J., and Sundin, U. (1980). Thermogenesis, brown fat and thermogenin. *In* "Survival in the Cold: Hibernation and Other Adaptations" (X. J. Musacchia and L. Jansky, eds.), pp. 99–119. Elsevier/North-Holland, New York.

Carmichael, M. S., Nelson, R. J., and Zucker, I. (1981). Hamster activity and estrous cycles: Control by a single versus multiple circadian oscillators. *Proc. Natl. Acad. Sci. U.S.A.* **78**, 7830–7834.

Carpenter, J. A. (1956). Species differences in taste preferences. *J. Comp. Physiol. Psychol.* **49**, 139–143.

Carter, C. S. (1973). Stimuli contributing to the decrement in sexual receptivity of female golden hamsters (*Mesocricetus auratus*). *Anim. Behav.* **21**, 827–834.

Carter, D. S., and Goldman, B. D. (1983a). Antigonadal effects of timed melatonin infusion in pinealectomized male Djungarian hamsters (*Phodopus sungorus sungorus*): Duration is the critical parameter. *Endocrinology (Baltimore)* **113**, 1261–1267.

Carter, D. S., and Goldman, B. D. (1983b). Progonadal role of the pineal in the Djungarian hamster (*Phodopus sungorus sungorus*): Mediation by melatonin. *Endocrinology (Baltimore)* **113**, 1268–1273.

Carter, D. S., Hall, V. D., Tamarkin, L., and Goldman, B. D. (1982). Pineal is required for testicular maintenance in the Turkish hamster (*Mesocricetus brandti*). *Endocrinology (Baltimore)* **111**, 863–871.

Cauwenbergs, P., Wiley, M. J., and Taylor, I. M. (1978). Retinoic acid-induced developmental abnormalities of the orbit and eye. *Can. Fed. Biol. Soc. Proc.* **21**, 157 (abstr.).

Chalupa, L. M. (1981). Some observations on the functional organization of the golden hamster's visual systems. *Behav. Brain Res.* **3**, 189–200.

Chappel, S. C. (1979). Cyclic fluctuations in ovarian FSH-inhibiting material in golden hamsters. *Biol. Reprod.* **21**, 447–453.

Charnock, J. S. (1978). Membrane lipid phase transitions: A possible biological response to hibernation? *In* "Strategies in Cold: Natural Torpidity and Thermogenesis" (L. C. H. Wang and J. W. Hudson, eds.), pp. 417–452. Academic Press, New York.

Chaudhry, A. P., and Shah, R. M. (1973). Estimation of hydrocortisone dose and optimal gestation period for cleft palate induction in golden hamsters. *Teratology* **8**, 139–142.

Cherrick, H. M., and McKelvy, B. D. (1975). Peripheral blood toxicity of topical 5-fluorouracil on the Syrian hamster cheek pouch. *J. Dent. Res.* **54**, 522–526.

Cherrick, H. M., and Weissman, D. (1974). Effects of topically applied 5-fluorouracil in the Syrian hamster. *J. Invest. Dermatol.* **63**, 284–286.

Chiaia, N., Foy, M., and Teyler, T. J. (1983). The hamster hippocampal slice. II. Neuroendocrine modulation. *Behav. Neurosci.* **97**, 839–843.

Chieli, E., Malvaldi, G., and Tongiani, R. (1979). A quantitative cytochemical study of isolated epithelial cells of the hamster cheek pouch. *Acta Histochem.* **6**, 104–129.

Collins, T. F. Y., and Williams, C. H. (1971). Teratogenic studies with 2,4,5-T and 2,4-D in the hamster. *Bull. Environ. Contam. Toxicol.* **6**, 559–567.

Committee on Animal Nutrition (1978). Nutrient requirements of the hamster. *In* "Nutrient Requirements of Laboratory Animals," 3rd rev. ed., No. 10, pp. 71–79. Natl. Acad. Sci., Washington, D.C.

Coniglio, L. P., Paup, D. C., and Clemens, L. G. (1973). Hormonal factors controlling the development of sexual behavior in the male golden hamster. *Physiol. Behav.* **10**, 1087–1094.

Cornwall, H., Odukoya, O., and Shklar, G. (1983). Oral mucosal tumor inhibition by ibuprofen. *J. Oral Maxillofac. Surg.* **41**, 795–800.

Cossins, A. R., and Wilkinson, H. L. (1982). The role of homeoviscous adaptation in mammalian hibernation. *J. Therm. Biol.* **7**, 107–110.

Crain, B. J., and Hall, W. C. (1981). The normal organization of the lateral posterior nucleus in the golden hamster and its reorganization after neonatal superior colliculus lesions. *Behav. Brain Res.* **3**, 223–238.

Cross, F. T., Palmer, R. F., Busch, R. H., Stuart, B. O., and Filipy, R. E. (1981). Development of lesions in Syrian golden hamsters following exposure to radon daughters and uranium ore dust. *Health Phys.* **41**, 135–154.

Damon, E. G., Phipps, D. R., Henderson, T. R., and Jones, R. K. (1979). Bronchopulmonary lavage in the assessment of relative acute pulmonary toxicity of pressurized consumer products. *Toxicol. Appl. Pharmacol.* **49**, 497–504.

Damon, E. G., Halliwell, W. H., Henderson, T. R., Mokler, B. V., and Jones, R. K. (1982). Acute toxicity of polyethylene glycol-P-isooctyphenyl ether in Syrian hamsters exposed by inhalation or bronchopulmonary lavage. *Toxicol. Appl. Pharmacol.* **63,** 53–61.

Darrow, J. M., and Goldman, B. D. (1983). The pineal gland and mammalian photoperiodism. *Neuroendocrinology* **37,** 386–396.

Davis, B. J., and Macrides, F. (1983). Tyrosine hydroxylase immunoreactive neurons and fibers in the olfactory system of the hamster. *J. Comp. Neurol.* **214,** 427–440.

Davis, F. C., and Gorski, R. A. (1984). Unilateral lesions of the hamster suprachiasmatic nuclei: Evidence for redundant control of circadian rhythms. *J. Comp. Physiol. A* **154,** 221–232.

Davis, F. C., Darrow, J. M., and Menaker, M. (1983). Sex differences in the circadian control of hamster wheel-running activity. *Am. J. Physiol.* **244,** R93–R105.

DeBold, J. F., Malsbury, C. W., Harris, V. S., and Malenka, R. (1982). Sexual receptivity: Brain sites of estrogen action in female hamsters. *Physiol. Behav.* **29,** 589–593.

Della Porta, G., Kolb, L., and Shubik, P. (1958). Induction of tracheal carcinomas in the Syrian golden hamster. *Cancer Res.* **18,** 592–597.

Demediuk, P., and Moscatelli, E. A. (1983). Synaptosomal and brain mitochondrial lipids in hibernating and cold acclimated golden hamsters (*Mesocricetus auratus*). *J. Neurochem.* **40,** 1100–1105.

Denicola, D. B., Rebar, A. H., and Henderson, R. F. (1981). Early damage indicators in the lung. Biochemical and cytological response to nitrogen dioxide inhalation. *Toxicol. Appl. Pharmacol.* **60,** 301–312.

Desrosiers, A., Kennedy, A., and Little, J. B. (1978). Radon-222 daughter dosimetry in the Syrian golden hamster lung. *Health Phys.* **35,** 607–627.

DiBattista, D. (1982). Effects of 5-Thioglucose on feeding and glycemia in the hamster. *Physiol. Behav.* **29,** 803–806.

DiBattista, D. (1983). Food deprivation and insulin-induced feeding in the hamster. *Physiol. Behav.* **30,** 683–687.

Diel, J. H. (1978). Local dose to lung tissue from inhaled plutonium-238 dioxide particles. *Radiat. Res.* **75,** 348–372.

Diel, J. H., Mewhinney, J. A., and Snipes, M. B. (1981). Distribution of inhaled plutonium-238 dioxide particles in Syrian hamster lungs. *Radiat. Res.* **88,** 299–312.

DiPaolo, J. A., Elis, J., and Erwin, H. (1969). Teratogenic response by hamsters, rats and mice to aflatoxin B_1. *Nature (London)* **215,** 638–639.

Dontenwill, W., Chevalier, H. T., Harke, H. R., Lafrenz, U., Rechzeh, G., and Schneider, B. (1973). Investigations of the effect of chronic cigarette smoke inhalation in Syrian golden hamsters. *J. Natl. Cancer Inst. (U.S.)* **51,** 1781–1832.

Dontenwill, W., Chevalier, H. J., Harke, H. P., Lafrenz, U., Reckzeh, G., and Leuschner, F. (1974). Biochemical and hematological investigations in the Syrian golden hamster after cigarette smoke inhalation. *Lab. Anim.* **8,** 217–235.

Duncan, M. J. (1984). The annual pelage color cycle in the Djungarian hamster (*Phodopus sungorus*): Regulation by circulating prolactin levels. Ph.D. Thesis, Worcester Polytechnic Institute, Worcester, Massachusetts.

Duncan, M. J., and Goldman, B. D. (1984). Hormonal regulation of the annual pelage color cycle in the Djungarian hamster (*Phodopus sungorus*). II. Role of prolactin. *J. Exp. Zool.* **230,** 97–103.

Duncan, M. J., and Goldman, B. D. (1985). Physiological doses of prolactin stimulate the pelage pigmentation in Djungarian hamster (*Phodopus sungorus*). *Am. J. Physiol.* **248,** R664–R667.

Duncan, R. J. S., and Trickleband, M. D. (1978). On the stimulation of the rate of hydroxylation of tryptophan in the brain of hamsters during hibernation. *J. Neurochem.* **31,** 553–556.

Earnest, D. J., and Turek, F. W. (1982). Splitting of the circadian rhythm of activity in hamsters: Effects of exposure to constant darkness and subsequent re-exposure to constant light. *J. Comp. Physiol.* **145,** 405–411.

Earnest, D. J., and Turek, F. W. (1983). Role for acetylcholine in mediating effects of light on reproduction. *Science* **219,** 77–79.

Eddy, H. A., and Casarett, G. W. (1972). Pathology of radiation syndromes in the hamster. *Prog. Exp. Tumor Res.* **16,** 98–119.

Eddy, H. A., and Chmielewski, G. (1982). Effect of hyperthermia radiation and adriamycin combinations on tumor vascular function. *Int. J. Radiat. Oncol., Biol. Phys.* **8,** 1167–1176.

Eisenberg, E., and Shklar, G. (1977). Levamisole and hamster pouch carcinogenesis. *Oral Surg., Oral Med. Oral Pathol.* **43,** 562–574.

El-Hage, A. N., Herman, E. H., and Ferrans, V. J. (1983). Examination of the protective effect of 1CRF-187, 1,2-bis(3,5-dioxopiperazin-1-yl)-propane and dimethylsulfoxide against acetaminophen-induced hepatotoxicity in Syrian golden hamsters. *Toxicology,* **16,** 295–304.

Elis, J., and DiPaolo, J. A. (1970). The alteration of actinomycin D teratogenicity by hormones and nucleic acids. *Teratology* **3,** 33–38.

Elliott, J. A., and Goldman, B. D. (1981). Seasonal reproduction: Photoperiodism and biological clocks. *In* "Neuroendocrinology of Reproduction" (N. T. Adler, ed.), pp. 377–423. Plenum, New York.

Elliott, J. A., Stetson, M. H., and Menaker, M. (1972). Regulation of testis function in golden hamsters: A circadian clock measures photoperiodic time. *Science* **178,** 771–773.

Englander, H. R., and Keyes, P. H. (1966). The prevention of dental caries in the Syrian hamster following repeated topical application of sodium fluoride gels. *J. Am. Dent. Assoc.* **73,** 1342.

Etgen, A. M. (1981). Differential effects of two antagonists on the development of masculine and feminine sexual behavior in hamsters. *Horm. Behav.* **15,** 299–311.

Feldman, D. B., McConnell, E. E., and Knapka, J. J. (1982). Growth, kidney disease, and longevity of Syrian hamsters (*Mesocricetus auratus*) fed varying levels of protein. *Lab. Anim. Sci.* **32,** 613–618.

Ferm, V. (1963). Congenital malformations in hamster embryos after treatment with vinblastine and vincristine. *Science* **141,** 426.

Ferm, V. (1971). Developmental malformations induced by cadmium. *Biol. Neonate* **19,** 101–107.

Ferm, V., and Carpenter, S. (1970). Teratogenic and embryopathic effects of indium, gallium and germanium. *Toxicol. Appl. Pharmacol.* **16,** 166–170.

Feron, V. J. (1972). Respiratory tract tumors in hamsters after intratracheal instillations of benzo[a]pyrene alone and with furfural. *Cancer Res.* **32,** 28–36.

Feron, V. J. (1979). Effects of exposure to acetaldehyde in Syrian hamsters simultaneously treated with benzo[a]pyrene or diethylnitrosamine. *Prog. Exp. Tumor Res.* **24,** 162–176.

Finlay, B. L., and Cairns, S. J. (1981). Relationship of aberrant retinotectal projections to visual orienting after neonatal tectal damage in hamster. *Exp. Neurol.* **72,** 308–317.

Finlay, B. L., and Sengelaub, D. R. (1981). Toward a neuroethology of mammalian vision: Ecology and anatomy of rodent visuomotor behavior. *Behav. Brain Res.* **3,** 133–149.

Fitts, D. A., and St. Dennis, C. (1981). Ethanol and dextrose preferences in hamsters. *J. Stud. Alcohol* **42,** 901–907.

Fitts, D. A., and Wright, J. W. (1979). Control of feeding during saline consumption and fluid deprivation in hamsters. *Physiol. Behav.* **22,** 963–969.

Fitts, D. A., and Wright, J. W. (1980). Control of body fluids and saline intake in hamsters. *Behav. Neural Biol.* **30,** 68–79.

Fitts, D. A., Corp, E. S., and Simpson, J. B. (1982). Salt appetite and intravascular volume depletion following colloid dialysis in hamsters. *Behav. Neural Biol.* **32,** 75–88.

Fitts, D. A., Yang, O. O., Corp, E. S., and Simpson, J. B. (1983). Sodium retention and salt appetite following deoxycorticosterone in hamsters. *Am. J. Physiol.* **244,** R78–R83.

Fitzgerald, J. R., and Jordan, H. V. (1968). Polysaccharide-producing bacteria and caries. *In* "The Art and Science of Dental Caries Research" (R. S. Harris, ed.), pp. 79–86. Academic Press, New York.

Fitzgerald, J. R., and Keyes, P. H. (1960). Demonstration of the etiologic role of streptococci in experimental caries in the hamster. *J. Am. Dent. Assoc.* **61**, 9–19.

Fleming, A. S., and Miceli, M. (1983). Effects of diet on feeding and body weight regulation during pregnancy and lactation in the golden hamster (*Mesocricetus auratus*). *Behav. Neurosci.* **97**, 246–254.

Floody, O. R. (1979). Behavioral and physiological analyses of ultrasound production by female hamsters (*Mesocricetus auratus*). *Am. Zool.* **19**, 443–455.

Friedman, A., and Shklar, G. (1978). Alcohol and hamster buccal pouch carcinogenesis. *Oral Surg., Oral Med. Oral Pathol.* **46**, 774–810.

Frost, D. O. (1981). Orderly anomalous retinal projections to the medial geniculate, ventrobasal and lateral posterior nucleus of the hamster. *J. Comp. Neurol.* **203**, 227–256.

Gabridge, M. G. (1979). Hamster tracheal organ cultures as models for infection and toxicology studies. *Prog. Exp. Tumor Res.* **24**, 85–95.

Gak, J. C., Graillot, C., and Truhaut, R. (1976). Use of the golden hamster in toxicology. *Lab. Anim. Sci.* **26**, 274–280.

Gale, T. F., and Ferm, V. (1971). Embryopathic effects of mercuric salts. *Life Sci.* **10**, 1341–1347.

Gale, T. F., and Hanlon, D. P. (1976). The permeability of the Syrian hamster placenta to mercury. *Environ. Res.* **12**, 26–31.

Gaunt, R., Gisoldi, E., and Smith, N. (1971). Refractoriness to renal effects of aldosterone in the golden hamster. *Endocrinology (Baltimore)* **89**, 63–69.

Gilmore, W., and Giunta, J. L. (1981). The effect of 13-*cis*-retinoic acid on hamster buccal pouch carcinogenesis. *Oral Surg., Oral Med. Oral Pathol.* **51**, 256–265.

Giunta, J., and Shklar, G. (1971). The effect of antilymphocyte serum on experimental hamster buccal pouch carcinogenesis. *Oral Surg., Oral Med. Oral Pathol.* **31**, 344–355.

Giunta, J., Reif, A. E., and Shklar, G. (1974). Bacillus Calmette-Guérin and antilymphocyte serum in carcinogenesis. *Arch. Pathol.* **98**, 237–240.

Giunta, J. L., Schwartz, J. L., and Antoniadis, D. V. (1985). Studies on the vascular drainage system of the hamster buccal pouch. *J. Oral Pathol.* **14**, 263–267.

Goldman, B. D. (1980). Seasonal cycles in testis function in two hamster species: Relation to photoperiod and hibernation. *In* "Testicular Development, Structure, and Function" (A. Steinburger and E. Steinburger, eds.), pp. 401–409. Raven Press, New York.

Goldman, B. D. (1983). The physiology of melatonin in mammals. *In* "Pineal Research Reviews" (R. J. Reiter, ed.), pp. 145–182. Alan R. Liss, Inc., New York.

Goldman, B. D., and Brown, S. (1979). Sex differences in serum LH and FSH patterns in hamsters exposed to short photoperiod. *J. Steroid Biochem.* **11**, 531–535.

Goldman, B. D., and Porter, J. C. (1970). Serum LH levels in intact and castrated golden hamsters. *Endocrinology (Baltimore)* **87**, 676–679.

Goldman, B. D., and Sheridan, P. J. (1974). The ovulatory surges of gonadotropin and sexual receptivity in the female golden hamster. *Physiol. Behav.* **12**, 991–995.

Goldman, B. D., Mahesh, V. B., and Porter, J. C. (1971). The role of the ovary in control of cyclic LH release in the hamster, *Mesocricetus auratus*. *Biol. Reprod.* **4**, 57–65.

Goldman, B. D., Matt, K. S., Roychoudhury, P., and Stetson, M. H. (1981). Prolactin release in golden hamsters: Photoperiod and gonadal influences. *Biol. Reprod.* **24**, 287–292.

Goldman, B. D., Carter, D. S., Hall, V. D., Roychoudhury, P., and Yellon, S. M. (1982). Physiology of pineal melatonin in three hamster species. *In* "Melatonin Rhythm Generating System: Developmental Aspects" (D. C. Klein, ed.), Steamboat Springs Symp., pp. 210–231. Karger, Basel.

Goldman, B. D., Darrow, J. M., and Yogev, L. (1984). Effects of timed melatonin infusions on reproductive development in the Djungarian hamster (*Phodopus sungorus*). *Endocrinology (Baltimore)* **114**, 2074–2083.

Goldman, S. S. (1975). Cold resistance of the brain during hibernation. III. Evidence of a lipid adaptation. *Am. J. Physiol.* **228**, 834–838.

Goldstein, R. H., Lucey, E. C., Franzblau, C., and Snider, G. L. (1979). Failure of mechanical properties to parallel changes in lung connective tissue composition in bleomycin induced pulmonary fibrosis in hamsters (*Mesocricetus auratus*). *Am. Rev. Respir. Dis.* **120**, 67–74.

Grady, R. R., Charlesworth, M. C., and Schwartz, N. B. (1982). Characterization of the FSH suppressing activity in follicular fluid. *Recent Prog. Horm. Res.* **38**, 409.

Gray, L. E., Jr. (1982). Neonatal chlordecone exposure alters behavioral sex differentiation in female hamsters. *Neurotoxicology* **3**, 67–80.

Gray, L. E., Jr., Kavlock, R., Chernoff, N., Lawton, D., and Gray, J. (1979). The effects of endrin administration during gestation of the behavior of the golden hamster (*Mesocrietus auratus*). *Toxicol. Appl. Pharmacol.* **48**, A200 (abstr.).

Greenwald, G. S. (1962). Temporal relationship between unilateral ovariectomy and the ovulatory response of the remaining ovary. *Endocrinology (Baltimore)* **71**, 664–668.

Grice, H. C. (1984). "Current Issues in Toxicology." Springer-Verlag, Berlin and New York.

Grubbs, C. J., Moon, R. C., Norikane, K., Thompson, H. J., and Becci, P. J. (1979). 1-Methyl-1-nitrosourea induction of cancer in a localized area of the Syrian golden hamster trachea. *Prog. Exp. Tumor Res.* **24**, 348–355.

Gumma, M. R., South, F. E., and Allen, J. N. (1967). Temperature preference in golden hamsters. *Anim. Behav.* **15**, 534–537.

Gupta, B. N., and Drew, R. T. (1976). The effect of aerosol hair spray inhalation in the hamster. *Am. Ind. Hyg. Assoc. J.* **37**, 357–360.

Gustafson, G., Steiling, E., and Brunius, E. (1968). The use of animals in dental research. *In* "Methods of Animal Experimentation" (W. I. Gay, ed.), Vol. 3, pp. 263–321. Academic Press, New York.

Hackett, N. A., Henderson, R. F., and Rebar, A. H. (1980). Kinetics of lung cell labeling after lung lavage assay. *Toxicol. Appl. Pharmacol.* **52**, 169–176.

Hahn, F. F., and Hobbs, C. H. (1979). The effect of enzyme-induced pulmonary emphysema in Syrian hamsters on the deposition and long term retention of inhaled particles. *Arch. Environ. Health* **34**, 203–211.

Hall, V. D., and Goldman, B. D. (1980). Effects of gonadal steroid hormones on hibernation in the Turkish hamster (*Mesocricetus brandti*). *J. Comp. Physiol.* **135**, 107–114.

Hall, V. D., Bartke, A., and Goldman, B. D. (1982). Role of the testis in regulating the duration of hibernation in the Turkish hamster, *Mesocricetus brandti*. *Biol. Reprod.* **27**, 802–810.

Harding, J. W., Stone, L. P., and Wright, J. W. (1981). The distribution of angiotensin II binding sites in rodent brain. *Brain Res.* **205**, 265–274.

Harris, C. C., Sporn, M. B., Kaufman, D. O., Smith, J. M., Baker, M. S., and Saffiotti, U. (1971). Acute ultrastructural effects of benzo[*a*]pyrene and ferric oxide on the hamster tracheobronchial epithelium. *Cancer Res.* **31**, 1977–1989.

Harris, R. A., Krause, W., Goh, E., and Case, J. (1979). Behavioral and biochemical effects of chronic consumption of ethanol by hamsters. *Pharmacol., Biochem. Behav.* **10**, 343–347.

Harris, S. B., Wilson, J. G., and Printz, R. R. (1972). Embryotoxicity of methyl mercuric chloride in golden hamsters. *Teratology* **6**, 139–142.

Hassan, M. M. A., Schwartz, J. L., and Shklar, G. (1984). Acute effect of DMBA application on Langerhans cells of the hamster buccal pouch mucosa. *Oral Surg., Oral Med. Oral Pathol.* **58**, 191–198.

Hecht, S. S., Adams, J. D., Numoto, S., and Hoffman, D. (1983). Induction of respiratory tract tumors in Syrian golden hamsters by a single dose of 4-methylnitrosoamino-1-3-pyridyl-1 butanone and the effect of smoke inhalation. *Carcinogenesis (London)* **4**, 1287–1290.

Henderson, R. F., Rebar, A. H., Pickrell, J. A., and Newton, G. J. (1979). Early damage indicators in the lung. 3. Biochemical and cytological response of the lung to inhaled metal salts. *Toxicol. Appl. Pharmacol.* **50,** 123–136.

Herbert, J. (1972). Initial observations on pinealectomized ferrets kept for long periods in either daylight or artificial illumination. *J. Endocrinol.* **55,** 591–597.

Herman, E. H., El-Hage, A. N., Ferrans, V. J., and Witiak, D. T. (1983). Reduction by 1CRF-187 racemic (1,2-bis(3,5-dioxopiperazan-1-yl-propane) of acute daunorubicin toxicity in Syrian golden hamsters. *Res. Commun. Chem. Pathol. Pharmacol.* **40,** 217–232.

Hilbig, R., and Rahmann, H. (1979). Changes in brain ganglioside composition of normothermic and hibernating golden hamsters (*Mesocricetus auratus*). *Comp. Biochem. Physiol.* **64,** 527–532.

Himms-Hagen, J. (1983). Thyroid hormones and thermogenesis. *In* "Mammalian Thermogenesis" (L. Girardier and M. J. Stock, eds.), pp. 141–177. Chapman & Hall, New York.

Hinkle, D. K. (1973). Fetotoxic effects on pentachlorophenol in the golden Syrian hamster. *Toxicol. Appl. Pharmacol.* **25,** 455 (abstr.).

Hoffman, D., Rivenson, A., Hecht, S. S., Hilfrich, J., Kobayashi, N., and Wynder, E. L. (1979). Model studies in tobacco carcinogenesis with the Syrian golden hamster. *Prog. Exp. Tumor Res.* **24,** 370–390.

Hoffman, R. A. (1968). Hibernation and the effects of low temperature. *In* "The Golden Hamster—Its Biology and Use in Biomedical Research" (R. A. Hoffman, P. F. Robinson, and H. Mugalhaes, eds.), pp. 25–40. Iowa State Univ. Press, Ames.

Holloway, R. J., Leong, G. F., Ainsworth, E. J., Albright, M. L., and Baum, S. J. (1968). Recovery from radiation injury in the hamster as evaluated by the split-dose technique. *Radiat. Res.* **33,** 37–49.

Homburger, F. (1971). Response of the oral mucosa, cheek pouch and facial skin of Syrian hamsters to chronic irritation, polycyclic hydrocarbons and chewing tobacco. *Toxicol. Appl. Pharmacol.* **19,** 41.

Homburger, F., and Adams, R. A. (1985). Adequacy of the Syrian hamster for long-term animal bioassays. *In* "Handbook of Carcinogen Testing" (H. A. Melman and E. K. Weisburger, eds.), pp. 326–344. Noyes Data Corp. Park Ridge, New Jersey.

Homburger, F., Chaube, S., Eppenberger, M., Bogdonoff, P. D., and Nixon, C. W. (1965). Susceptibility of certain inbred strains of hamsters to teratogenic effects of thalidomide. *Toxicol. Appl. Pharmacol.* **7,** 686–693.

Homburger, F., Adams, R. A., Soto, E., and Van Dongen, C. G. (1979). Susceptibility and resistance to chemical carcinogens by inbred Syrian hamsters. *Prog. Exp. Tumor Res.* **24,** 215–221.

Homburger, F., Adams, R. A., Bernfeld, P., Van Dongen, C. G., and Soto, E. (1983). A new first generation hybrid Syrian hamster, BIO F1D Alexander for *in vivo* carcinogenesis bioassay, as a third species or to replace the mouse. *Surv. Synth. Pathol. Res.* **1,** 125–133.

Horii, A. A., and Keyes, P. H. (1964). A vinyl applicator for assessing drugs in the treatment of caries and periodontal disease in the hamster. *J. Dent. Res.* **43,** 152.

Houchin, O. B. (1943). Toxic levels of morphine for the hamster. *Proc. Soc. Exp. Biol. Med.* **54,** 339–340.

Houslay, M. D., and Palmer, R. W. (1978). Changes in the forms of Arrhenius plots of the activity of glucogen-stimulated adenylate cyclase and other hamster liver plasma membrane enzymes occurring on hibernation. *Biochem. J.* **174,** 909–920.

Howell, A., Jr., and Jordan, H. V. (1963). A filamentous microorganism isolated from periodontal plaque in hamsters. II. Physiological and biochemical characteristics. *Sabouradia* **3,** 93–105.

Howell, A., Jr., Jordan, H. V., Georg, L., and Pine, L. (1965). *Odontomyces viscosus*, gen. nov., spec. nov., a filamentous microorganism isolated from periodontal plaque in hamsters. *Sabouradia* **4,** 65–68.

Hudson, J. W. (1980). The role of the endocrine glands with spatial reference to the thyroid glands. *In* "Survival in the Cold: Hibernation and Other Adaptations" (X. J. Musacchia and L. Jansky, eds.), pp. 33–54. Elsevier/North-Holland, New York.

Hudson, J. W., and Wang, L. C. H. (1979). Hibernation: Endocrinologic aspects. *Annu. Rev. Physiol.* **41,** 287–303.

International Agency for Research on Cancer (IARC) (1980). "Long-term and Short-Term Screening Assays for Carcinogens: A Critical Appraisal," IARC Monogr. Suppl. 2, p. 30. IARC, Lyon.

Ishinishi, N., Yamamoto, A., Hisanaga, A., and Imamasu, J. (1983). Tumorigenicity of arsenic trioxide to the lung in the Syrian golden hamster by intermittent instillation. *Cancer Lett.* **21,** 141–148.

Ito, N., Fukushima, S., Imaida, K., Sakata, T., and Masui, T. (1983). Induction of papillomas in the forestomach of hamsters by butylated hydroxyanisole. *Gann* **74,** 459–461.

Iwasaki, M., Miyaoka, T., Tsuda, S., Shirasu, Y., and Harada, T. (1981). Effects of maleic hydrazide on cigarette smoke inhalation toxicity in Syrian golden hamsters. *J. Pestic. Sci.* **6,** 17–24.

Jacquin, M. F., Mooney, R. D., and Rhoades, R. W. (1984). Altered somatosensory receptive fields in hamster colliculus after infraorbital nerve section and xylocaine injection. *J. Physiol. (London)* **348,** 471–492.

Jansky, L. (1973). Nonshivering thermogenesis and its thermoregulatory importance. *Biol. Rev. Cambridge Philos. Soc.* **48,** 85–132.

Jansky, L. (1978). Time sequence of physiological changes during hibernation. The significance of the serotonergic pathways. *In* "Strategies in Cold: Natural Torpidity and Thermogenesis" (L. C. H. Wang and J. W. Hudson, eds.), pp. 299–325. Academic Press, New York.

Jansky, L., Kahlerova, Z., Nedoma, J., and Andrews, J. F. (1980). Humoral control of hibernation in golden hamsters. *In* "Survival in the Cold: Hibernation and Other Adaptations" (X. J. Musacchia and L. Jansky, eds.), pp. 13–31. Elsevier/North-Holland, New York.

Johnston, R. E. (1979). Olfactory preferences, scent marking and "proceptivity" in female hamsters. *Horm. Behav.* **13,** 21–39.

Johnston, R. E. (1981). Attraction to odors in hamsters: An evaluation of methods. *J. Comp. Physiol.* **95,** 951–960.

Jordan, H. V., and Keyes, P. H. (1964). Aerobic, gram-positive, filamentous bacteria as etiologic agents of experimental periodontal disease in hamsters. *Arch. Oral Biol.* **9,** 401–414.

Jordan, H. V., and Van Houte, J. (1972). The hamster as a model for odontopathic infections. *Prog. Exp. Tumor Res.* **16,** 539–556.

Jordan, W. H., Carlton, W. W., and Sansing, G. A. (1978). Citrinin mycotoxicosis in the Syrian hamster. *Food Cosmet. Toxicol.* **16,** 355–364.

Julius, A. D., Davies, M. J., and Birt, D. F. (1983). Toxic effects of dietary selenium in the Syrian hamster. *Ann. Nutr. Metab.* **27,** 296–305.

Kandelman, D., Lepage, Y., and Bélanger, Y. (1979). Evaluation of a bisgma pit and fissure sealant experimentation on golden Syrian hamsters. *J. Dent. Res.* **58,** Spec. Issue A, 297 (abstr.).

Karsch, F. J., Bittman, E. L., Foster, D. L., Goodman, R. L., Legan, S. J., and Robinson, J. E. (1984). Neuroendocrine basis of seasonal reproduction. *Recent Prog. Horm. Res.* **40,** 185–232.

Kelly, A. (1980). Embryonic meningeal development in the golden hamster (*Mesocricetus auratus*) and the effect of triamcinolone acetonide on this development. *Anat. Rec.* **196,** 94–95 (abstr.).

Kendrick, F. J. (1964). Some effects of chemical carcinogen and of cigarette smoke condensate upon hamster cheek pouch mucosa. *Health Sci.* **24,** 3698–3716.

Kennedy, A. R., and Little, J. B. (1979). Respiratory system differences relevant to lung carcinogenesis between Syrian hamsters and other species. *Prog. Exp. Tumor Res.* **24,** 302–314.

Kennedy, A. R., Worcester, J., and Little, J. B. (1977). Deposition and localization of polonium-210 intratracheally instilled in the hamster lung as determined by autoradiography of freeze dried sections. *Radiat. Res.* **69,** 553–572.

Kennedy, A. R., McGandy, R. B., and Little, J. B. (1978). Serial sacrifice study of pathogenesis of polonium-210 induced lung tumors in Syrian golden hamsters. *Cancer Res.* **38,** 1127–1135.

Ketkar, M. B., Reznik, G., and Mohr, U. (1977). Pathological alterations in

Syrian golden hamster lungs after passive exposure to cigarette smoke. *Toxicology* **7**, 265–274.

Kevetter, G. A., and Winans, S. S. (1981). Connections of the corticomedial amygdala in the golden hamster. II. Efferents of the "olfactory amygdala." *J. Comp. Neurol.* **197**, 99–111.

Keyes, P. H. (1960). The infectious and transmissible nature of experimental dental caries. *Arch. Oral Biol.* **1**, 304–320.

Keyes, P. H. (1968). Odontopathic infections. In "The Golden Hamster—Its Biology and Use in Biomedical Research" (R. A. Hoffman, P. F. Robinson, and H. Mugalhaes, eds.), pp. 253–284. Iowa State Univ. Press, Ames.

Keyes, P. H., and Jordan, H. V. (1964). Peridontal lesions in the Syrian hamster. III. Findings related to an infectious and transmissible component. *Arch. Oral Biol.* **9**, 377–400.

Keyes, P. H., Rowberry, S. A., Englander, H. R., and Fitzgerald, R. J. (1966). Bio-assays of medicaments for the control of dentobacterial plaque, dental caries, and periodontal lesions in Syrian hamsters. *J. Oral Ther. Pharmacol.* **3**, 157.

Kim, S. J., Roberts, J. F., and Koo, J. O. (1980). Effect of paraquat on erythrocytic 2,3-diphosphoglycerate in hamsters (*Mesocricetus auratus*). *Pharmacologist* **22**, 172 (abstr.).

Kim, S. J., Roberts, J. F., and Koo, J. O. (1981). Effects of paraquat on angiotensin-converting enzyme activity in hamsters (*Mesocricetus auratus*) lungs. *Am. Rev. Respir. Dis.* **123**, 141 (abstr.).

Kindt, A., Sattler, E. L., and Schraub, A. (1980). Radiation-induced anorexia in Syrian hamsters. *Radiat. Environ. Biophys.* **18**, 149–155.

Klein, D. C., Smoot, R., Weller, J. L., Higa, S., Markey, S. P., Creed, G. J., and Jacobowitz, D. M. (1983). Lesions of the paraventricular nucleus area of the hypothalamus disrupt the suprachiasmatic–spinal cord circuit in the melatonin rhythm-generating system. *Brain Res. Bull.* **10**, 647–652.

Kleinerman, J. (1979). Effects of nitrogen dioxide on elastin and collagen contents of lung. *Arch. Environ. Health* **34**, 228–232.

Kobayashi, H., Uemura, H., Wada, M., and Takei, Y. (1979). Ecological adaptation of angiotensin-induced thirst mechanism in tetrapods. *Gen. Comp. Endocrinol.* **38**, 93–104.

Krasse, B. (1965). The effect of diet on the implantation of caries-inducing streptococci in hamsters. *Arch. Oral Biol.* **10**, 215–221.

Krasse, B., and Carlsson, J. (1970). Various types of streptococci and experimental caries in hamsters. *Arch. Oral Biol.* **15**, 25–32.

Krasse, B., Edwardsson, S., Swensson, I., and Trell, L. (1967). Implantation of caries-inducing streptococci in the human oral cavity. *Arch. Oral Biol.* **12**, 231–236.

Kream, R. M., Davis, B. J., Kawano, T., Margolis, F. L., and Macrides, F. (1984). Substance P and catecholaminergic expression in neurons of the hamster main olfactory bulb. *J. Comp. Neurol.* **222**, 140–154.

Kulkosky, P. J. (1978). Free-selection ethanol intake of the golden hamster (*Mesocricetus auratus*). *Physiol Psychol.* **64**, 505–509.

Kulkosky, P. J., and Cornell, N. W. (1979). Free-choice ethanol intake and ethanol metabolism in the hamster and rat. *Pharmacol., Biochem. Behav.* **11**, 439–444.

Kutscher, C. L. (1968). Plasma volume change during water deprivation in gerbils, hamsters, guinea pigs and rats. *Comp. Biochem. Physiol.* **25**, 929–936.

Kutscher, C. L. (1969). Species differences in the interaction of feeding and drinking. *Ann. N.Y. Acad. Sci.* **157**, 539–552.

Kutscher, C. L. (1973). Interaction of food and water deprivation on drinking: Effect of body water losses and characteristics of solution offered. *Physiol. Behav.* **9**, 753–758.

Lamperti, A., and Blaha, G. (1980). Further observations on the effects of neonatally administered monosodium glutamate on the reproductive axis of hamsters (*Mesocricetus auratus*). *Biol. Reprod.* **22**, 687–694.

Landsberg, L., and Young, J. B. (1983). Automatic regulation of thermogenesis. In "Mammalian Thermogenesis" (L. Girardier and M. J. Stock, eds.), pp. 99–140. Chapman & Hall, New York.

LaPointe, R., and Harvey, G. (1964). Salicylamide-induced anomalies in hamster embryos. *J. Exp. Zool.* **156**, 197–199.

LaVelle, F. W., and LaVelle, A. (1983). A difference in the nuclei of Purkinje cells of the nodular lobe of the hamster cerebellum. *Brain Res.* **265**, 119–124.

Leavitt, W. W., and Blaha, G. G. (1970). Circulating progesterone levels in the golden hamster during the estrous cycle, pregnancy and lactation. *Biol. Reprod.* **3**, 353–361.

Legan, S. J., and Karsch, F. J. (1975). A daily signal for the LH surge in the rat. *Endocrinology (Baltimore)* **95**, 57–62.

Lehman, M. N., Powers, J. B., and Winans, S. S. (1983). Stria terminalis lesions alter the temporal pattern of copulatory behavior in the male golden hamster. *Behav. Brain Res.* **8**, 109–128.

Leonard, C. M. (1982a). Some speculations concerning neurological mechanisms for early olfactory recognition. In "Development of Perception" (R. N. Aslin, J. R. Alberts, and M. R. Peterson, eds.), Vol. 1, pp. 383–405. Academic Press, New York.

Leonard, C. M. (1982b). Shifting strategies for behavioral thermoregulation in developing golden hamsters. *J. Comp. Physiol. Psychol.* **96**, 234–243.

Levij, I. S., and Polliack, A. (1969). Lymphoma-like hamster cheek pouch with topical vitamin A palmitate. *Pathol. Microbiol.* **34**, 288–394.

Levij, I. S., Rwomushana, J. W., and Polliack A. (1970). Effect of topical cyclophosphamide, methotrexate, and vinblastine on 9,10-dimethyl-1,2-benzanthracene (DMBA) carcinogenesis in the hamster cheek pouch. *Eur. J. Cancer* **6**, 187–193.

Li, J. J., Li, S. A., Klicka, J. K., Parsons, J. A., and Lam, L. K. T. (1983). Relative carcinogenic activity of various synthetic and natural estrogens in the Syrian hamster kidney. *Cancer Res.* **43**, 5200–5204.

Lijinsky, W. (1984). Species differences in nitrosamine carcinogenesis. *Cancer Res. Clin. Oncol.* **108**, 46–55.

Lisk, R. D., and Greenwald, D. P. (1983). Central plus peripheral stimulation by androgen is necessary for complete restoration of copulatory behavior in the male hamster. *Neuroendocrinology* **36**, 211–217.

Lisk, R. D., and Heimann, J. (1980). The effects of sexual experience and frequency of testing on retention of copulatory behavior following castration in the hamster. *Behav. Neural Biol.* **28**, 156–171.

Lisk, R. D., and Reuter, L. A. (1980). Relative contributions of estradiol and progesterone to the maintenance of sexual receptivity in mated female hamsters. *J. Endocrinol.* **87**, 175–183.

Lisk, R. D., Ciaccio, L. A., and Catanzaro, C. (1983). Mating behaviour of the golden hamster under seminatural conditions. *Anim. Behav.* **31**, 659–666.

Logan, A., and Weatherhead, B. (1980). Post-tyrosine inhibition of melanogenesis by melatonin in hair follicles in vitro. *J. Invest. Dermatol.* **74**, 47–50.

Lowy, M. T., and Yim, G. K. W. (1982). Drinking, but not feeding, is opiate-sensitive in hamsters. *Life Sci.* **30**, 1639–1644.

Lucey, E. C., O'Brien, J. J., Jr., Pereira, W., Jr., and Snider, G. L. (1980). Arterial blood gas values in emphysematous hamsters (*Mesocricetus auratus*). *Am. Rev. Respir. Dis.* **121**, 83–90.

Lukaszewska, J. H., and Greenwald, G. S. (1970). Progesterone levels in the cyclic and pregnant hamster. *Endocrinology (Baltimore)* **86**, 1–9.

Lundgren, D. L., Hahn, F. F., and McClellan, R. O. (1982). Effects of single and repeated inhalation exposure of Syrian hamsters to aerosols of cerium-144 dioxide. *Radiat. Res.* **90**, 374–394.

Lundgren, D. L., Hahn, F. F., Rebar, A. H., and McClellan, R. O. (1983). Effects of the single or repeated inhalation exposure of Syrian hamsters to aerosols of plutonium-239-dioxide. *Int. J. Radiat. Biol. Relat. Stud. Phys., Chem. Med.* **43**, 1–18.

Lurie, A. G., and Cutler, L. S. (1979). Effects of low-level x-radiation on

7,12-dimethylbenz[a]anthracene-induced lingual tumors in Syrian golden hamsters. *JNCI, J. Natl. Cancer Inst.* **63,** 147–152.

Lurie, A. G., and Rippey, R. M. (1978). Low level-x-radiation-induced vascular alterations in normal and 7,12-dimethylbenz[a]anthracene-treated tumor-bearing cheek pouch epithelium of Syrian hamsters. *Radiat. Res.* **76,** 127–138.

Lurie, A. G., Tatemattsu, M., Nakatszk, T., Rippey, R. M., and Nobuyuki, I. (1983). Anatomical and functional vascular changes in hamster cheek pouch during carcinogenesis induced by 7,12-dimethylbenz-[a]antharacene. *Cancer Res.* **43,** 5486–5494.

Lyman, C. P. (1948). The oxygen consumption and temperature regulation of hibernating hamsters. *J. Exp. Zool.* **109,** 55–78.

Lyman, C. P. (1954). Activity, food consumption and hoarding in hibernators. *J. Mammal.* **35,** 545–552.

Lyman, C. P., and Leduc, E. H. (1958). Changes in blood sugar and tissue glycogen during arousal from hibernation. *J. Cell. Comp. Physiol.* **41,** 471–488.

Lyman, C. P., and O'Brien, R.C. (1977). A laboratory study of the Turkish hamster (*Mesocricetus brandti*). *Breviora Mus. Comp. Zool.* **442,** 1–27.

McCoy, G. D., Haisley, A. D., Powchik, P., and Tambone, P. C. (1981). Ethanol consumption by Syrian golden hamsters. *J. Stud. Alcohol* **42,** 508–513.

McEwen, B. S., Lieberburg, I., Chaptal, C., and Krey, L. C. (1977). Aromatization: Important for sexual differentiation of the neonatal rat brain. *Horm. Behav.* **9,** 249–255.

McHaffie, J. G., and Stein, B. E. (1982). Eye movements evoked by electrical stimulation in the superior colliculus of rats and hamsters. *Brain Res.* **247,** 243–253.

McKinley, E. R., and Carlton, W. W. (1980). Patulin mycotoxicosis in the Syrian hamster. *Food Cosmet. Toxicol.* **18,** 173–180.

McMillan, D. E., Ellis, F. W., Frye, G. D., and Pick, J. R. (1977). Failure of signs of physical dependence to develop in hamsters after prolonged consumption of large doses of ethanol. *Pharmacol., Biochem. Behav.* **7,** 55–57.

Macrides, F., Bartke, A., Fernandez, F., and D'Angelo, W. (1974). Effects of exposure to vaginal odor and receptive females on plasma testosterone in the male hamster. *Neuroendocrinology* **15,** 355–364.

Marin-Padilla, M., and Ferm, V. (1965). Somite necrosis and developmental malformations induced by vitamin A in the golden hamster. *J. Embryol. Exp. Morphol.* **13,** 1–8.

Meade, P. D., Tamashiro, S., Harada, T., and Basrur, P. K. (1979). Influence of vitamin A on the laryngeal response of hamsters exposed to cigarette smoke. *Prog. Exp. Tumor Res.* **24,** 320–329.

Meier, A. H., and Wilson, A. M. (1983). Tryptophan feeding adversely influences pregnancy. *Life Sci.* **32,** 1193–1196.

Meng, C. L., Shklar, G., and Albright, J. (1982). A transplantable anaplasticoral cancer model. *Oral Surg., Oral Med. Oral Pathol.* **53,** 179–187.

Meredith, M. (1983). Sensory physiology of pheromone communication. In "Pheromones and Reproduction in Mammals" (J. G. Vandenbergh, ed.), pp. 199–252. Academic Press, New York.

Merk, L., Shklar, G., and Albright, J. (1979). Transplantation of hamster buccal pouch carcinoma to neonatal hamsters. *Oral Surg., Oral Med. Oral Pathol.* **47,** 533–541.

Ficeli, M. O., and Malsbury, C. W. (1982). Sagittal knife cuts in the near and far lateral preoptic area-hypothalamus reduce ultrasonic vocalizations in female hamsters. *Physiol. Behav.* **29,** 953–956.

Miceli, M. O., and Malsbury, C. W. (1983). Feeding and drinking responses in the golden hamster following treatment with cholecystokinin and angiotensin II. *Peptides (Fayetteville, N.Y.)* **4,** 103–106.

Miller, W. A., and Ripley, J. F. (1975). Early periodontal disease in the Syrian hamster. *J. Periodontol.* **46,** 368–374.

Mohr, U. (1979). The Syrian golden hamster as a model in cancer research. *Prog. Exp. Tumor Res.* **24,** 245–252.

Mooney, R. D., Fish, S. E., and Rhoades, R. W. (1984). Anatomical and functional organization of pathway from superior colliculus to lateral posterior nucleus in hamster. *J. Neurophysiol.* **51,** 407–431.

Moore, C., and Miller, A. J. (1958). Effect of cigarette smoke tar on hamster cheek pouch. *Arch. Surg. (Chicago)* **76,** 786–794.

Moore, M. R., Pitot, H. C., Miller, E., and Miller, J. A. (1982). Cholangiocellular carcinomas induced in Syrian golden hamsters administered aflatoxin B_1 in large doses. *JNCI, J. Natl. Cancer Inst.* **68,** 271–277.

Moore, R. Y., and Klein, D. C. (1974). Visual pathways and the central neural control of the circadian rhythm in pineal serotonin N-acetyltransferase activity. *Brain Res.* **71,** 17–33.

Mori, Y., Niwa, T., Toyoshi, K., Nagai, H., Koda, A., Kawada, K., Ojima, A., and Konishi, Y. (1984). Carcinogenic activities of hydroxymethyl derivatives of 4-dimethylamino)azobenzene in the liver of rats, mice and hamsters. *Exp. Pathol.* **26,** 15–19.

Morris, A. L. (1961). Factors influencing experimental carcinogenesis in the hamster cheek pouch. *J. Dent. Res.* **40,** 3–15.

Mossman, B. T., and Craighead, J. E. (1979). Use of hamster tracheal organ cultures for assessing the cocarcinogen effects of inorganic particulars on respiratory epithelium. *Prog. Exp. Tumor Res.* **24,** 37–47.

Murphy, M. R. (1973). Effects of female hamster vaginal discharge on the behavior of male hamsters. *Behav. Biol.* **9,** 367–375.

Murphy, M. R. (1976). Olfactory stimulation and olfactory bulb removal: Effects on territorial aggression in male Syrian golden hamsters. *Brain Res.* **113,** 95–110.

Murphy, M. R. (1979). Behavior patterns: Hamster. Part I. Development and control. In "Inbred and Genetically Defined Strains of Laboratory Animals" (P. L. Altman and D. D. Katz, eds.), Part 2, pp. 455–465. Fed. Am. Soc. Exp. Biol., Bethesda, Maryland.

Murphy, M. R. (1980). Sexual preferences of male hamsters: Importance of preweaning and adult experience, vaginal secretion, and olfactory or vomeronasal sensation. *Behav. Neural Biol.* **30,** 323–340.

Murphy, M. R. (1981). Methadone reduces sexual performance and sexual motivation in the male Syrian golden hamster. *Pharmacol., Biochem. Behav.* **14,** 561–567.

Murphy, M. R., and Schneider, G. E. (1970). Olfactory bulb removal eliminates mating behavior in the male golden hamster. *Science* **167,** 302–303.

Musacchia, X. J. (1972). Heat and cold acclimation in helium-cold hypothermia in the hamster. *Am. J. Physiol.* **222,** 495–498.

Musacchia, X. J. (1976). Helium-cold hypothermia, an approach to depressed metabolism and thermoregulation. In "Regulation of Depressed Metabolism and Thermogenesis" (L. Jansky and X.J. Musacchia, eds.), pp. 137–157. Thomas, Springfield, Illinois.

Musacchia, X. J., and Deavers, D. R. (1980). The regulation of carbohydrate metabolism in hibernators. In "Survival in the Cold: Hibernation and Other Adaptations" (X. J. Musacchia and L. Jansky, eds.), pp. 55–75. Elsevier/North-Holland, New York.

National Toxicology Program (NTP) (1983a). "Lifetime Carcinogenesis Studies of Chrysotile Asbestos in Syrian Golden Hamsters (Feed Studies)," NTP Tech. Rep. Ser. No. 246. NTP, Washington, D.C.

National Toxicology Program (NTP) (1983b). "Lifetime Carcinogenesis Studies of Amosite Asbestos in Syrian Golden Hamsters (Feed Studies)," NTP Tech. Rep. Ser. No. 249. NTP, Washington, D.C.

National Toxicology Program (NTP) (1985). A summary of data compiled for a meeting to discuss the "Selection of Hamster Strain for Chemical Toxicity and Carcinogenicity Studies" (unpublished data).

Navia, J. M. (1977). "Animal Models in Dental Research." Univ. of Alabama Press, University.

Nettesheim, P., and Tsutomu, Y. (1979). Studies with a new experimental model in respiratory carcinogenesis. *Prog. Exp. Tumor Res.* **24,** 330–344.

Nickerson, P. A., and Molteni, A. (1971). Survival of bilaterally adrenalectomized hamsters drinking saline sucrose. *J. Appl. Physiol.* **31,** 675–678.

Niewoehner, D. E., and Hoidal, J. R. (1982). Lung fibrosis and emphysema: Divergent responses to a common injury? *Science* **217,** 359–360.

Niswender, G. D., Midgley, A. R., Monroe, S. E., and Reichert, L. E. (1968). Radioimmunoassay for rat luteinizing hormone with anti-ovine LH serum and ovine LH-^{131}I. *Proc. Soc. Exp. Biol. Med.* **128,** 807–811.

Noble, R. G. (1979). The sexual responses of the female hamster: A descriptive analysis. *Physiol. Behav.* **23,** 1001–1005.

Norman, R. C., Blake, C. A., and Sawyer, C. H. (1973). Estrogen-dependent twenty-four-hour periodicity in pituitary LH release in the female hamster. *Endocrinology (Baltimore)* **93,** 965–970.

Novotna, R. (1980). The metabolism of 5-hydroxytryptamine in the hypothalamus and the liver tryptophane pyrrolase activity of a hibernator, *Mesocricetus auratus,* in stress. *Physiol. Bohemoslov.* **29,** 243–252.

Novotna, R., and Civin, J. (1979). The relationship between tryptophan pyrrolase activity and 5-hydroxytryptamine metabolism in the brain and induction of hibernation. *Physiol. Bohemoslov.* **28,** 339–346.

Novotna, R., Jansky, L., and Drahota, Z. (1975). Effect of hibernation on serotonin metabolism in the brainstem of the golden hamster (*Mesocricetus auratus*). *Gen. Pharmacol.* **6,** 23–26.

Odukoya, O., and Shklar, G. (1982). Two phase carcinogenesis in hamster buccal pouch. *Oral Surg., Oral Med. Oral Pathol.* **54,** 547–552.

Odukoya, O., and Shklar, G. (1984). Initiation and promotion in experimental oral carcinogenesis. *Oral Surg., Oral Med. Oral Pathol.* **58,** 315–320.

Odukoya, O., Schwartz, J. L., Weichselbaum, R., and Shklar, G. (1983). An epidermoid carcinoma cell line derived from hamster DMBA-induced buccal pouch tumors. *JNCI, J. Natl. Cancer Inst.* **71,** 1253–1264.

Odukoya, O., Hawash, F., and Shklar, G. (1984). Retardation of experimental oral cancer by topical vitamin E. *Nutr. Cancer* **6,** 98–104.

Olson, J. R., Holscher, M. A., and Neal R. A. (1980a). Toxicity of 2,3,7,9-tetrachlorodibenzo-*p*-dioxin in the golden Syrian hamster. *Toxicol. Appl. Pharmacol.* **55,** 67–78.

Olson, J. R., Gasiewicz, T. A., and Neal, R. A. (1980b). Tissue distribution excretion and metabolism of 2,3,7,8-tetrachlorodibenzo-*p*-dioxin in the golden Syrian hamster. *Toxicol. Appl. Pharmacol.* **56,** 78–85.

Orsini, M. W. (1961). The external phenomena characterizing the stages of the estrous cycle, pregnancy, pseudopregnancy, lactation and the anestrous hamster, *Mesocricetus auratus Waterhouse. Proc. Anim. Care Panel* **61,** 193–206.

Orsini, N. W., and Pansky, B. (1952). The natural resistance of the hamster to colchicine. *Science* **115,** 88–89.

Payne, A. P., and Swanson, H. H. (1970). Agonistic behaviour between pairs of hamsters of the same and opposite sex in a neutral observation area. *Behaviour* **36,** 259–269.

Payne, A. P., and Swanson, H. (1971). Hormonal control of aggressive dominance in the female hamster. *Physiol. Behav.* **6,** 355–357.

Perkins, T. M., and Shklar, G. (1982). Delay in hamster buccal pouch carcinogenesis by aspirin and indomethacin. *Oral Surg., Oral Med. Oral Pathol.* **53,** 170–178.

Pickard, G. E., and Turek, F. W. (1983). The hypothalamic paraventricular nucleus mediates photoperiodic control of reproduction but not the effects of light on the circadian rhythm of activity. *Neurosci. Lett.* **43,** 67–72.

Pickard, G. E., Turek, F. W., Lamperti, A. A., and Silverman, A. J. (1982). The effect of neonatally administered monosodium glutamate on the development of retinofugal projections and the entrainment of the circadian locomotor activity. *Behav. Neural Biol.* **34,** 433–444.

Pickrell, J. A., Harris, D. V., Benjamin, S. A., Cuddihy, R. G., Pfleger, R. C., and Mauderly, J. L. (1976). Pulmonary collagen metabolism after lung injury from inhaled yttrium-90 in fused clay particles. *Exp. Mol. Pathol.* **25,** 70–81.

Pickrell, J. A., Diel, J. H., Slauson, D. O., Halliwell, W. H., and Mauderly, J. L. (1983). Radiation-induced pulmonary fibrosis resolves spontaneously if dense scars are not formed. *Exp. Mol. Pathol.* **38,** 22–32.

Plapinger, L., and McEwen, B. (1978). Gonadal steroid–brain interactions in sexual differentiation. *In* "Biological Determinants of Sexual Behavior" (J. B. Hutchison, ed.), pp. 153–218. Wiley, New York.

Polliack, A., and Levij, I. S. (1972). Antineoplastic effect of chlorpromazine in chemical carcinogenesis in the hamster cheek pouch. *Cancer Res.* **32,** 1912–1917.

Potegal, M., Blau, A., and Glusman, M. (1981). Effects of anteroventral septal lesions on intraspecific aggression in male hamsters. *Physiol. Behav.* **26,** 407–412.

Pour, P., Moch, N., Greiser, E., Mohr, U., Althoff, J., and Cardesa, A. (1976). Spontaneous tumors and common diseases in two colonies of Syrian hamsters. I. Incidence and sites. *J. Natl. Cancer Inst. (U.S.)* **56,** 931–967.

Protzel, M., Giardiana, A. C., and Albano, E. H. (1964). The effect of liver imbalance in the development of oral tumor in mice following the application of benzpyrene or tobacco tar. *Oral Surg., Oral Med. Oral Pathol.* **18,** 622–635.

Rahmann, H., and Hilbig, R. (1980). Involvement of neuronal gangliosides and thermal adaptation. *In* "Survival in the Cold: Hibernation and Other Adaptations" (X. J. Musacchia and L. Jansky, eds.), pp. 177–189. Elsevier/North-Holland, New York.

Raynaud, J. P. (1973). Influence of rat estradiol-binding plasma protein (EBP) on uterotrophic activity. *Steroids,* pp. 249–258.

Redman, H. C., Hobbs, C. H., and Rebar, A. H. (1979). Survival distribution of Syrian hamsters. *Prog. Exp. Tumor Res.* **24,** 108–117.

Reep, R. L., and Winans, S. A. (1982). Efferent connections of dorsal and ventral angular insular cortex in the hamster, *Mesocricetus auratus. Neuroscience* **7,** 2609–2635.

Reh, T., and Kalil, K. (1982). Development of the pyramidal tract in the hamster. II. An electron microscopic study. *J. Comp. Neurol.* **205,** 77–88.

Reiskin, A. P., and Berry, R. J. (1968). Cell proliferation and carcinogenesis in the hamster cheek pouch. *Cancer Res.* **28,** 898–905.

Reiter, R. J. (1968). Changes in the reproductive organs of cold-exposed and light-deprived female hamsters (*Mesocricetus auratus*). *J. Reprod. Fertil.* **16,** 217–222.

Reiter, R. J. (1973). Pineal control of a seasonal reproductive rhythm in male golden hamsters exposed to natural daylight and temperature. *Endocrinology (Baltimore)* **92,** 423–430.

Reiter, R. J. (1980). The pineal gland and its hormones in the control of reproduction in mammals. *Endocr. Rev.* **1,** 109–131.

Reiter, R. J., Vaughan, M. K., Rudeen, P. K., Vaughan, G. M., and Waring, P. J. (1975). Melatonin–pineal relationships in female golden hamster. *Proc. Soc. Exp. Biol. Med.* **149,** 290–293.

Renstrup, G., Smulow, J., and Glickman, I. (1962). Effect of chronic mechanical irritation on chemically induced carcinogenesis in the hamster cheek pouch. *Am. Dent.* **64,** 770.

Reuhl, K. R., Chang, L. W., and Townsend, J. W. (1981a). Pathological effects of in utero methyl mercury exposure on the cerebellum of the golden Syrian hamster (*Mesocricetus auratus*). 1. Early effects on the neonatal cerebellar cortex. *Environ. Res.* **26,** 281–306.

Reuhl, K. R., Chang, L. W., and Townsend, J. W. (1981b). Pathological effects of in utero methyl mercury exposure on the cerebellum of the golden Syrian hamster (*Mesocricetus auratus*). 2. Residual effects on the adult cerebellum. *Environ. Res.* **26,** 307–327.

Reznik-Schuller, H. M. (1980). Acute effects of cigarette smoke inhalation on the Syrian hamster lungs. *J. Environ. Pathol. Toxicol.* **4,** 285–292.

Rhoades, R. W. (1981). Expansion of the ipsilateral visual corticotectal projection in hamsters subjected to partial lesions of the visual cortex during infancy: Electrophysiological experiments. *J. Comp. Neurol.* **197,** 447–450.

Rinaldo, J., Goldstein, R. H., and Snider, G. L. (1982). Modification of oxygen toxicity after lung injury by bleomycin in hamsters (*Mesocricetus auratus*). *Am. Rev. Respir. Dis.* **126,** 1030–1033.

Ritter, R. C., and Bach, O. K. (1978). Feeding in response to insulin but not 2-deoxy-D-glucose in the hamster. *Am. J. Physiol.* **234,** E20–E24.

Robens, J. F. (1969). Teratogenic studies of carbaryl, diazinon, norea, disulfram, and thiram in small laboratory animals. *Toxicol. Appl. Pharmacol.* **15,** 152–163.

Robens, J. F. (1970). Teratogenic activity of several phthalimide derivatives in the golden hamster. *Toxicol. Appl. Pharmacol.* **16,** 24–34.

Rothman, K., and Keller, A. (1972). The effect of joint exposure to alcohol and tobacco on risk of cancer of the mouth and pharynx. *J. Chronic Dis.* **25,** 711–719.

Rowe, N. H., and Gorlin, R. J. (1959). The effect of vitamin A deficiency upon experimental carcinogenesis. *J. Dent. Res.* **38,** 72–83.

Rowland, N. (1978). Effects of insulin and 2-deoxy-D-glucose on feeding in hamsters and gerbils. *Physiol. Behav.* **21,** 291–294.

Rowland, N. (1982). Failure of deprived hamsters to increase their food intake: Some behavioral and physiological determinants. *J. Comp. Physiol. Psychol.* **96,** 591–603.

Rowland, N. (1983). Physiological and behavioral responses to glucoprivation in the golden hamster. *Physiol. Behav.* **30,** 743–747.

Ruffalo, P., and Ferm, V. (1965). The teratogenicity of 5-bromodeoxyuridine in the pregnant Syrian hamster. *Life Sci.* **4,** 633–637.

Rusak, B., and Groos, G. (1982). Suprachiasmatic stimulation phase shifts rodent circadian rhythms. *Science* **215,** 1407–1409.

Rusak, B., and Zucker, I. (1978). Neural regulation of circadian rhythms. *Physiol. Rev.* **59,** 449–526.

Rutkowski, J. V., and Ferm, V. H. (1982). Comparison of the teratogenic effects of the isometric forms of aminophenol in the Syrian golden hamster. *Toxicol. Appl. Pharmacol.* **63,** 264–269.

Saffiotti, U., Cefis, F., and Kolb, L. H. (1968). A method for experimental induction of bronchogenic carcinoma. *Cancer Res.* **20,** 857–864.

Safour, I. M., Wood, N. K., Tsiklakis, K., Doemung, D. B., and Joseph, G. (1984). Incisional biopsy and seeding in hamster cheek pouch carcinoma. *J. Dent. Res.* **63,** 1116–1120.

St. Dennis, C. D. (1981). The male golden Syrian hamster (*Mesocricetus auratus*); a unique laboratory model for studies that involve the semivoluntary consumption of ethanol. Ph. D. Dissertation, Washington State University, Pullman.

Salber, P., and Zucker, I. (1974). Absence of salt appetite in adrenalectomized and DOCA-treated hamsters. *Behav. Biol.* **10,** 295–311.

Salley, J. J. (1954). Experimental carcinogenesis in the cheek pouch of the Syrian hamster. *J. Dent. Res.* **33,** 253–262.

Samuelsen, G. S., Rasmussen, R. E., Nair, B. K., and Crocker, T. T. (1978). Novel culture and exposure system for measurement of effects of airborne pollutants on mammalian cells. *Environ. Sci. Technol.* **12,** 426–429.

Santis, H., Shklar, G., and Chauncey, H. H. (1964). Histochemistry of experimentally induced leukoplakia and carcinoma of the hamster buccal pouch. *Oral Surg., Oral Med. Oral Pathol.* **17,** 84–95.

Saunders, C. L. (1977). Inhalation toxicology of plutonium-238 dioxide and plutonium-239 dioxide in Syrian golden hamsters. *Radiat. Res.* **70,** 334–344.

Schmidt-Nielsen, B., and Schmidt-Nielsen, K. (1950). Pulmonary water loss in desert rodents. *Am. J. Physiol.* **162,** 31–36.

Schmidt-Nielsen, B., and Schmidt-Nielsen, K. (1951). A complete account of the water metabolism in kangaroo rats and an experimental verification. *J. Cell. Comp. Physiol.* **38,** 165–181.

Schneider, G. E. (1973). Early lesions of the superior colliculus: Factors affecting the formation of abnormal retinal projections. *Brain Behav. Evol.* **8,** 73–109.

Schwartz, J., Solt, D. B., Pappo, J., and Weischselbaum, R. (1981). Distribution of Langerhans cells in normal and carcinogen-treated mucosa of buccal pouches of hamsters. *J. Dermatol. Surg. Oncol.* **7,** 1005–1010.

Schweitzer, L., and Cant, N. B. (1984). Development of the cochlear innervation of the dorsal cochlear nucleus of the hamster. *J. Comp. Neurol.* **225,** 228–243.

Sclafani, A., and Eisenstadt, D. (1980). 2-deoxy-D-glucose fails to induce feeding in hamsters fed a preferred diet. *Physiol. Behav.* **24,** 641–643.

Seegal, R. F., and Goldman, B. D. (1975). Effects of photoperiod on cyclicity and serum gonadotropins in the Syrian hamster. *Biol. Reprod.* **12,** 223–231.

Seidel, A. (1977). The deposition of americanium-241 and californium-252 in the skeleton of Chinese hamster, Syrian hamster and the rat. *Health Phys.* **33,** 83–86.

Seidel, A. (1978). Excorporation efficacy with calcium diethylenetriamine pentaacetic acid of americanium-241 and californium-252 in the skeleton, liver and kidney of the rat and Syrian hamster and Chinese hamster; Lack of correlation with biological half-times. *Radiat. Res.* **76,** 60–69.

Shami, S. G., Thibodeau, L. A., Kennedy, A. E., and Little, J. B. (1982a). Proliferation and morphological changes in the pulmonary epithelium of the Syrian golden hamster during carcinogenesis initiated by polonium-210 alpha radiation. *Cancer Res.* **42,** 1405–1411.

Shami, S. G., Thibodeau, L. A., Kennedy, A. R., and Little, J. B. (1982b). Recovery from ozone induced injury in the lungs of the Syrian golden hamster. *Exp. Mol. Pathol.* **36,** 57–71.

Sheehan, R., Shklar, G., and Tennenbaum, R. (1971). Azathioprine effect on experimental buccal pouch tumors. *Arch. Pathol.* **21,** 264–270.

Shenefelt, R. (1972). Morphogenesis of malformations in hamsters caused by retinoic acid. *Teratology* **5,** 103–118.

Sher, S. P. (1982). Tumors in control hamsters, rats, and mice: Literature tabulation. *CRC Crit. Rev. Toxicol.* **10,** 49–79.

Shettleworth, S. J., and Juergensen, M. R. (1980). Reinforcement and the organization of behavior in golden hamsters: Brain stimulation reinforcement for seven action patterns. *J. Exp. Psychol. Anim. Behav. Processes* **6,** 352–375.

Shklar, G. (1965). Metabolic characteristics of experimental hamster pouch carcinomas. *Oral Surg., Oral Med. Oral Pathol.* **20,** 336–339.

Shklar, G. (1966). Cortisone and hamster buccal pouch carcinogenesis. *Cancer Res.* **26,** 2461–2463.

Shklar, G. (1967). The effect of cortisone on the induction and development of hamster buccal pouch carcinomas. *Oral Surg., Oral Med. Oral Pathol.* **23,** 241–249.

Shklar, G. (1968). The effect of manipulation and incision on experimental carcinoma of hamster buccal pouch. *Cancer Res.* **28,** 2180–2182.

Shklar, G. (1982). Oral mucosal carcinogenesis in hamsters inhibition by vitamin E. *JNCI, J. Natl. Cancer Inst.* **68,** 791–797.

Shklar, G., Cataldo, E., and Fitzgerald, A. L. (1966). The effect of methotrexate on chemical carcinogenesis of hamster cheek pouch. *Cancer Res.* **26,** 2218–2224.

Shklar, G., Turbiner, S., and Siegel, W. (1969). Chemical carcinogenesis of hamster mucosa. Reactions to dimethylsulfoxide. *Arch. Pathol.* **87,** 637–642.

Shklar, G., Meyer, I., and Stevens, W. (1970). A variance in hamster pouch carcinogenesis in tissues irradiated with orthovoltage and cobalt 60. *Oral Surg., Oral Med. Oral Pathol.* **30,** 431–438.

Shklar, G., Eisenberg, E., and Flynn, E. (1979). Immunoenhancing agents and experimental leukoplakia and carcinoma of the hamster buccal pouch. *Prog. Exp. Tumor Res.* **24,** 269–282.

Shklar, G., Schwartz, J. L., Grau, D., Trickler, D. B., and Wallace, K. D. (1980). Inhibition of hamster pouch carcinogenesis by 13-*cis*-retinoic acid *Oral Surg., Oral Med. Oral Pathol.* **50,** 45–53.

Shubik, P. (1972). The use of the Syrian golden hamster in chronic toxicity testing. *Prog. Exp. Tumor Res.* **16,** 176–184.

Siegel, H. I. (1984). "The Hamster: Reproduction and Behavior." Plenum, New York.

Siegel, H. I., and Rosenblatt, J. S. (1980). Hormonal and behavioral aspects of maternal care in the hamster: A review. *Neurosci. Biobehav. Rev.* **4**, 17–26.

Silberberg-Sinakin, I., Thorbecke, G. J., and Baer, R. L. (1976). Antigen-bearing Langerhans cells in the skin, dermal lymphatics, and in lymph nodes. *Cell Immunol.* **25**, 137–151.

Silberman, S., and Shklar, G. (1963). The effect of a carcinogen (DMBA) applied to the hamster's buccal pouch in combination with croton oil. *Oral Surg., Oral Med. Oral Pathol.* **16**, 1344–1360.

Silverman, H. J. (1978). Failure of 2-deoxy-D-glucose to increase feeding in the golden hamster. *Physiol. Behav.* **21**, 859–864.

Silverman, H. J., and Zucker, I. (1976). Absence of post-fast food compensation in the golden hamster (*Mesocricetus auratus*). *Physiol. Behav.* **17**, 271–285.

Simek, V. (1967). Influence of intermittent fasting on morphological changes of the digestive system and on the activity of some enzymes in the golden hamster (*Mesocricetus auratus*). *Acta Soc. Zool. Bohemoslov.* **32**, 89–95.

Simonson, L. G., Lamberts, B. L. Reiher, D. A., and Walter, R. G. (1983). Prevention of dental caries in hamsters (*Mesocricetus auratus*) by an endo-1,3-alpha-d-glucanase. *J. Dent. Res.* **62**, 395–397.

Singer, A. G., Agosta, W. C., O'Connell, R. J., Pfaffmann, C., Bowen, D. V., and Field, F. H. (1976). Dimethyl disulfide: An attractant pheromone in hamster vaginal secretion. *Science* **191**, 948–950.

Slack, J. M., and Gerenczer, M. A. (1970). Two new serological groups of Actinomyces. *J. Bacteriol.* **103**, 266–267.

Slaga, T. J., Huberman, E., and Digiovanni, J. (1979a). The importance of the "Bay Region" diol-epoxide in 7,12-dimethylbenz[*a*]anthracene skin tumor initiation and mutagenesis. *Cancer Lett.* **6**, 213–220.

Slaga, T. J., Gleason, and Digiovanni, J. (1979b). Potent tumor initiating activity of the 3,4-dihydrodiol of 7,12-dimethylbenz[*a*]anthracene in mouse skin. *Cancer Res.* **39**, 1934–1936.

Slighter, R. G., Jr. (1970). Alcohol selection and position selection in hamsters caged singly and in groups. *Q. J. Stud. Alcohol* **31**, 20–27.

Small, R. K., and Leonard, C. M. (1983). Rapid fiber reorganization after early olfactory tract section and bulbectomy in the hamster. *J. Comp. Neurol.* **214**, 353–369.

Smith, A. N. (1958). The effects of foetal development of freezing pregnant hamsters (*Mesocricetus auratus*). *J. Embryol. Exp. Morphol.* **5**, 311–323.

Smith, D. V., Van Buskirk, R. L., Travers, J. B., and Bieber, S. L. (1983). Gustatory neuron types in hamster brain stem. *J. Neurophysiol.* **50**, 522–540.

Smit-Vis, J. H. (1972). The effect of pinealectomy and of testosterone administration on the occurrence of hibernation in adult male golden hamsters. *Acta Morphol. Neurol. Scand.* **10**, 269–281.

Smit-Vis, J. H., and Smit, G. J. (1963). Occurrence of hibernation in the golden hamster, *Mesocricetus auratus* Waterhouse. *Experimentia* **19**, 363–364.

Smit-Vis, J. H., and Smit, G. J. (1970). Hibernation and testes activity in golden hamster. *Neth. J. Zool.* **20**, 502–506.

Snyder, J. G., and Wyman, L. C. (1951). Sodium and potassium of blood and urine in adrenalectomized golden hamsters. *Am. J. Physiol.* **167**, 328–332.

Soares, M. J., Colosi, P., and Talamantes, F. (1983). Development of a homologous radioimmunoassay for secreted hamster prolactin. *Proc. Soc. Exp. Biol. Med.* **172**, 379–381.

Solt, D. B., and Shklar, G. (1982). Rapid induction of (gamma)-glutamyl transpeptidase-rich intraepithelial clones in 7,12-dimethylbenz[*a*]anthracene treated hamster buccal pouch. *Cancer Res.* **42**, 285–295.

Sonis, S., and Shklar, G. (1981). Preliminary immunologic studies on retinoid inhibition of experimental carcinogenesis. *J. Oral Med.* **36**, 117–119.

Steel, E. (1981). Control of proceptive and receptive behavior by ovarian hormones in the Syrian hamster (*Mesocricetus auratus*). *Horm. Behav.* **15**, 141–156.

Stein, B. E. (1981). Organization of the rodent superior colliculus: Some comparisons with other mammals. *Behav. Brain Res.* **3**, 175–188.

Steinhoff, D., Weber, H., Mohr, V., and Boehme, K. (1983). Evaluation of Amitrole (aminotriazole) for potential carcinogenicity in orally dosed rats, mice and golden hamsters. *Toxicol. Appl. Pharmacol.* **69**, 161–169.

Stephan, F. K., Donaldson, J. A., and Gellert, J. (1982). Retinohypothalamic tract symmetry and phase shifts of circadian rhythms in rats and hamsters. *Physiol. Behav.* **29**, 1153–1159.

Stetson, M. H., and Tate-Ostroff, B. (1981). Hormonal regulation of the annual reproductive cycle of golden hamsters. *Gen. Comp. Endocrinol.* **45**, 329–344.

Stetson, M. H., and Watson-Whitmyre, M. (1976). Nucleus suprachiamaticus: The biological clock in the hamster? *Science* **191**, 197–199.

Stetson, M. H., and Watson-Whitmyre, M. (1984). The physiology of the pineal and its hormone melatonin in annual reproduction in rodents. In "The Pineal Gland" (R. J. Reiter, ed.), pp. 109–153. Raven Press, New York.

Stingl, G., Katz, S. I., Shevach, E. M., Rosenthal, A. S., and Green, I. (1978). Analogous functions of macrophages and Langerhans cells in the initiation of the immune response. *J. Invest. Dermatol.* **71**, 59–64.

Stinson, S. F., Reznik-Schueller, H., Reznik, G., and Donahoe, R. (1980). Atrophy induced in the tubules of the testes of Syrian hamsters by 2 retinoids. *Toxicology* **17**, 343–354.

Stradling, G. N., Stather, J. W., Ham, S. E., and Summer, S. A. (1981). The use of Puchel® and diethylenetriamine pentaacetic-acid for removing plutonium-238 dioxide from the lungs of Syrian hamsters. *Health Physics* **41**, 387–391.

Stroud, A. N. (1977). Chromosome aberrations induced in Syrian hamster lung cells by inhaled plutonium-238-oxide zirconium oxide particles. *Radiat. Res.* **69**, 583–590.

Sunderman, F. W., Shen, S. K., Reid, M. C., and Allpass, P. R. (1981). Teratogenicity and embryotoxicity of nickel carbonyl in Syrian hamsters. *Ann. Clin. Lab. Sci.* **11**, 84 (abstr.).

Swanson, H. H. (1966). Modification of the reproductive tracts of hamsters of both sexes by neonatal administration of androgen or oestrogen. *J. Endocrinol.* **36**, 327–328.

Tashima, L. S. (1965). The effects of cold exposure and hibernation on the thyroidal activity of *Mesocricetus auratus*. *Gen Comp. Endocrinol.* **5**, 267–277.

Tassinari, M. S., and Long, S. Y. (1982). Normal and abnormal midfacial development in the cadmium-treated hamster. *Teratology* **25**, 101–114.

Tempel, G. E., Musacchia, X. J., and Jones, S. B. (1977). Mechanisms responsible for decreased glomerular filtration in hibernation and hypothermia. *J. Appl. Physiol.* **42**, 420–425.

Thomas, R. G., and Smith, D. M. (1979). Lung tumors from plutonium dioxide zirconium dioxide aerosol particles in Syrian hamster. *Int. J. Cancer* **27**, 594–599.

Thurman, R. G., Bleyman, M. A., and McMillan, D. E. (1978). Rapid blood ethanol elimination and withdrawal resistance in the Syrian golden hamster. *Pharmacologist* **20**, 160.

Thomas, R. G., Drake, G. A., London, J. E., Anderson, E. C., Prine, J. R., and Smith, D. M. (1981). Pulmonary tumors in Syrian hamsters following inhalation of plutonium-239 labeled plutonium oxide. *Int. J. Radiat. Biol. Relat. Stud. Phys., Chem. Med.* **40**, 605–612.

Thomas, R. L., and Raabe, O. G. (1978). Regional deposition of inhaled cesium-137 labeled monodisperse and polydisperse aluminosilicate aerosols in Syrian hamsters. *Am. Ind. Hyg. Assoc. J.* **39**, 1009–1018.

Tiefer, L., and Johnson, W. (1971). Female sexual behavior in male golden hamsters. *J. Endocrinol.* **51**, 615–620.

Tiefer, L., and Johnson, W. (1975). Neonatal androstenedione and adult sexual behavior in the golden hamster. *J. Comp. Physiol. Psychol.* **88,** 239–247.

Tobler, L. H., Johnson, K. P., and Buehring, G. C. (1983). Immune response of hamsters to experimental central nervous system infection with measles virus. *J. Neuroimmunol.* **2,** 307–320.

Tryka, A. F., Godleski, J. J., Skornik, W. A., and Brain, J. D. (1982). Bleomycin and 70 percent oxygen exposure: Early and late effects. *Am. Rev. Respir. Dis.* **125,** 92 (abstr.).

Tsubouchi, S., and Matsuzawa, T. (1981). A new plateau in the dose survival time response of the golden hamster (*Mesocricetus auratus*) from whole-body irradiation. *Int. J. Radiat. Biol. Relat. Stud. Phys., Chem. Med.* **40,** 87–94.

Tsubouchi, S., Suzuki, H., Ariyoshi, H., and Matsuzawa, T. (1981). Radiation-induced acute necrosis of the pancreatic islet and the diabetic syndrome in the golden hamster (*Mesocricetus auratus*). *Int. J. Radiat. Biol. Relat. Stud. Phys., Chem. Med.* **40,** 95–106.

Turbow, M., Clark, W., and DiPaolo, J. (1971). Embryonic abnormalities in hamsters following intrauterine injection of 6-aminonicotinamide. *Teratology* **4,** 427–432.

Turek, F. W. (1977). The interaction of photoperiod and testosterone in regulating serum gonadotropin levels in castrated male hamsters. *Endocrinology (Baltimore)* **101,** 1210–1215.

Turek, F. W., and Ellis, G. B. (1981). Steroid-dependent and steroid-independent aspects of the photoperiodic control of seasonal reproductive cycles in male hamsters. *In* "Biological Clocks in Seasonal Reproductive Cycles" (B. K. Follett and D. E. Follett, eds.), pp. 251–260. Wright, Bristol, England.

Turley, W. A., and Floody, O. R. (1981). Delta-9-tetrahydrocannabinol stimulates receptive and proceptive sexual behaviors in female hamsters. *Pharmacol., Biochem. Behav.* **14,** 745–747.

Umpierre, C. C., and Dukelow, W. R. (1977). Environmental heat stress effects in the hamster. *Teratology* **16,** 155–158.

Ungar, H., Sullman, S. F., and Zuckerman, A. J. (1976). Acute and protracted changes in the liver of Syrian hamsters induced by a single dose of aflatoxin B_1: Observations on pathologic effects of the solvent dimethylformamide. *Br. J. Exp. Pathol.* **57,** 157–164.

Van Hoosier, G. L., Jr., Spjut, H. J., and Trentin, J. J. (1971). Spontaneous tumors of the Syrian hamster: Observations in a closed breeding colony and a review of the literature. *In* "Defining the Laboratory Animal," pp. 450–473. Natl. Acad. Sci., Washington, D.C.

Vaughan, M. K., Richardson, B. A., Craft, C. M., Rowanda, M. C., and Reiter, R. J. (1982). Interaction of aging, photoperiod and melatonin on plasma thyroid hormones and cholesterol levels in female Syrian hamsters (*Mesocricetus auratus*). *Gerontology* **28,** 345–353.

Vitale, P. M., Darrow, J. M., Duncan, M. J., Shustak, C. A., and Goldman, B. D. (1985). Effects of photoperiod, pinealectomy, and castration on body weight and daily torpor in Djungarian hamsters. (*Phodopus sungorus*). *J. Endocrinol.* **106,** 367–375.

Vomachka, A. J., Ruppert, P. H., Clemens, L. G., and Greenwald, G. S. (1981). Adult sexual behavior deficits and altered hormone levels in male hamsters given steroids during development. *Physiol. Behav.* **26,** 461–466.

Vomachka, A. J., Richards, N. R., II, and Lisk, R. D. (1982). Effects of septal lesions on lordosis in female hamsters. *Physiol. Behav.* **29,** 1131–1135.

Vriend, J., and Reiter, R. J. (1977). Free thyroxin index in normal, melatonin-treated and blind hamsters. *Horm. Metab. Res.* **9,** 231–234.

Vybiral, S., and Jansky, L. (1972). Thermoregulatory significance of catecholamine thermogenesis in golden hamsters. *Physiol. Bohemoslov.* **21,** 121–122.

Weerapradist, W., and Shklar, G. (1982). Vitamin E inhibition of hamster buccal pouch carcinogenesis. *Oral Surg., Oral Med. Oral Pathol.* **54,** 304–312.

Weisz, J., and Gunsalus, P. (1973). Estrogen levels in immature female rats: True or Spurious—ovarian or adrenal? *Endocrinology (Baltimore)* **93,** 1057–1065.

Werner, A. P., Stuart, B. O., and Sanders, C. L. (1979). Inhalation studies with Syrian golden hamsters. *Prog. Exp. Tumor Res.* **24,** 177–198.

Whitehead, M. C., and Frank, M. E. (1983). Anatomy of the gustatory system in the hamster: Central projections of the chorda tympani and the lingual nerve. *J. Comp. Neurol.* **220,** 378–395.

Whitsett, J. M., and Vandenberg, J. G. (1975). Influence of testosterone propionate administered neonatally on puberty and bisexual behavior in female hamsters. *J. Comp. Physiol. Psychol.* **88,** 248–255.

Wiley, M. J. (1983). The pathogenesis of retinoic acid-induced vertebral abnormalities in golden Syrian hamster fetuses. *Teratology* **28,** 341–354.

Wiley, M. J., and Joneja, M.G. (1978). The morphogenesis of beta-aminopropionitrile-induced rib malformation in fetal golden Syrian hamsters. *Teratology* **18,** 173–186.

Willhite, C. C. (1983). Developmental toxicology of acetonitrite in the Syrian golden hamster. *Teratology* **27,** 313–326.

Williams, G. M., Chandrasekawan, V., Katayam, S., and Weisburger, J. H. (1981). Carcinogenicity of 3-methyl-2-naphthylamine and 3,2'-demethyl-4-amino-biphenyl to the bladder and gastrointestinal tract-of-the Syrian golden hamster with atypical proliferative enteritis. *JNCI, J. Natl. Cancer Inst.* **67,** 481–486.

Willis, J. S. (1978). Cold tolerance of mammalian cells: Prevalence and properties. *In* "Strategies in Cold: Natural Torpidity and Thermogenesis" (L. C. H. Wang and J. W. Hudson, eds.), pp. 377–415. Academic Press, New York.

Wilson, J. G., Kalter, H., Palmer, A. K., Hoar, R. M., Shenefelt, R. E., Beck, F., Earl, F. L., Nelson, N. S., Berman, E., Stara, J. F., Selby, L. A., and Scott, W. J., Jr. (1978). Developmental abnormalities. *In* "Pathology of Laboratory Animals" (C. K. Benirschke, F. M. Garner, and T. C. Jones, eds.), Vol. 2, pp. 1818–1945. Springer-Verlag, Berlin and New York.

Winans, S. S., and Powers, J. B. (1977). Olfactory and vomeronasal deafferentation of male hamsters: Histological and behavioral analyses. *Brain Res.* **126,** 325–344.

Winborn, W. B., and Seelig, L. L., Jr. (1970). Cytologic effects of reserpine on hepatocyte: An ultrastructural study of drug toxicity. *Lab. Invest.* **23,** 216–229.

Winter, R., and Seidel, A. (1981). The influence of an iron sorbitol citrate complex on the deposition of monomeric plutonium-239 in 4 rodent species. *Health Phys.* **40,** 100–104.

Winter, R., and Seidel, A. (1982). Comparison of the subcellular distribution of monomeric plutonium-239 and iron-59 in the liver of the rat, mouse, Syrian hamster and Chinese hamster. *Radiat. Res.* **89,** 113–123.

Wong, R. (1981). Maintenance diet and the effects of furosemide on hamsters. *Am. J. Physiol.* **94,** 339–354.

Wong, R., and Jones, W. (1978). Saline intake in hamsters. *Behav. Biol.* **24,** 474–480.

Woods, D. A. (1969). Influence of antilymphocyte serum on DMBA induction of oral carcinoma. *Nature (London)* **224,** 276–277.

Wright, J. W. (1976). Effect of hunger on the drinking behaviour of rodents adapted for mesic and xeric environments. *Anim. Behav.* **24,** 300–304.

Yellon, S. M., and Goldman, B. D. (1984). Photoperiod control of reproductive development in the male Djungarian hamster (*Phodopus sungorus*). *Endocrinology (Baltimore)* **114,** 664–670.

Zinner, D. D., Aran, A. P., Jablon, J. J., Brust, B., and Saslaw, M. S. (1965). Experimental caries induced in animals by streptococci of human origin. *Proc. Soc. Exp. Biol. Med.* **118,** 776–770.

Zucker, I., Wade, G. N., and Ziegler, R. (1972). Sexual and hormonal influences on eating, taste preferences, and body weight of hamsters. *Physiol. Behav.* **8,** 101–111.

Appendix

Selected Normative Data for the Syrian Golden Hamster

Charles W. McPherson

Parameter	Value	Reference[a]
Adult weight		4
Male	85–130 gm	4
Female	95–150 gm	4
Body surface area	260 cm²:125 gm	4
Life span	18–24 months	4
Diploid chromosome number	44	4
Birth weight	2.4–3.0 gm	2
Eyes open	15 days	2
Weaning age	21 days (30–40 gm)	2
Puberty		
Male	42 days	2
Female	42 days	2
Reproductive life		
Male	10 months	2
Female	10 months	2
Gestation	15–16 days	4
Litter size	4–16 pups	2
Body temperature	37°–38°C	4
Respiratory rate	76 breaths/min	5
Tidal volume	0.66 ml	5
Minute volume	50 ml	5
Heart rate	250–500 beats/min	4
Blood pressure (mean)	97.8 mm Hg	3
Systolic	109.5 mm Hg	3
Stroke volume	0.047 ml/beat	1
Blood volume	7.2 ± 1.5 ml/100 gm	8
Blood		
pH	7.48	8
P_aO_2	71.8 ± 4.9 mm Hg	8
P_vO_2	30.4 ± 3.3 mm Hg	9
P_aCO_2	41.1 ± 2.4 mm Hg	8
P_vCO_2	55.5 ± 4.1 mm Hg	9
Red blood cells		
Males	7.5 ± 1.4 $10^6/\mu l$	6
Females	7.0 ± 1.5 $10^6/\mu l$	6
Packed cell volume		
Males	52.0 ± 2.3 ml%	6
Females	49.0 ± 4.9 ml%	6
Hemoglobin		
Males	16.8 ± 1.2 gm/dl	6
Females	16.0 ± 1.5 gm/dl	6
Platelets		
Males	410 ± 75 $10^3/\mu l$	6
Females	360 ± 121 $10^3/\mu l$	6
Leukocytes (total)		
Males	7.6 ± 1.3 $10^3/\mu l$	6
Females	8.6 ± 1.5 $10^3/\mu l$	6
Neutrophils		
Males	21.1 ± 2.5%	6
Females	29.0 ± 3.1%	6
Eosinophils		
Males	0.9 ± 0.32%	6
Females	0.7 ± 0.24%	6
Basophils		
Males	1.0 ± 2.0%	6
Females	0.5 ± 0.7%	6
Lymphocytes		
Males	73.5 ± 9.4%	6
Females	67.9 ± 8.5%	6
Monocytes		
Males	2.5 ± 0.8%	6
Females	2.4 ± 1.0%	6
Food consumption	10–12 gm/100 gm body wt/day	4

(continued)

Parameter	Value	Reference[a]
Water consumption	8–10 ml/100 gm body wt/day	4
Nutrient requirements	Amount in diet	7
Protein	15%	
Fat	5%	
Digestible energy	4.2 kcal/gm	
Amino acids		7
Arginine	0.75%	
Histidine	0.40%	
Isoleucine	1.39%	
Lysine	1.20%	
Methionine	0.32%	
Phenylalanine	0.83%	
Threonine	0.70%	
Tryptophan	0.34%	
Tyrosine	0.57%	
Valine	0.91%	
Minerals		7
Calcium	0.59%	
Magnesium	0.06%	
Phosphorus	0.30%	
Potassium	0.61%	
Sodium	0.15%	
Cobalt	1.1 mg/kg	
Copper	1.6 mg/kg	
Fluoride	0.024 mg/kg	
Iodine	1.6 mg/kg	
Iron	140.0 mg/kg	
Manganese	3.65 mg/kg	
Selenium	0.1 mg/kg	
Zinc	9.2 mg/kg	
Vitamins		7
A (retinyl palmitate)	2.0 mg/kg	
D	2484 IU/kg	
E	3.0 mg/kg	
K_1	4.0 mg/kg	
Biotin	0.6 mg/kg	
Choline	2000 mg/kg	
Folic acid	2.0 mg/kg	
Inositol	100.0 mg/kg	
Niacin	90.0 mg/kg	
Pantothenate (Ca)	40.0 mg/kg	
Riboflavin	15.0 mg/kg	
Thiamin	20.0 mg/kg	
B_6	6.0 mg/kg	
B_{12}	10.0 mg/kg	

[a]REFERENCES: (1) Abelmann W., Jeffrey, F., and Wagner, F. (1972). Circulatory dynamics in heart failure of Syrian hamsters. *Prog. Exp. Tumor Res.* **16,** 261–273. (2) Balk, M. W., and Slater, G. (1987). Care and management of Syrian hamsters. In "Laboratory Hamsters" (G. L. Van Hoosier and C. W. McPherson, eds.), Chapter 4. Academic Press, Orlando, Florida. (3) Callahan, A. B., Degelman, J., and Lutz, B. R. (1959). Systolic blood pressure in the hamster as determined by a strain-gauge transducer. *J. Appl. Physiol.* **14,** 1051–1052. (4) Harkness, J. E., and Wagner, J. E. (1983). Biology and husbandry—The hamster. In "The Biology and Medicine of Rodents and Rabbits," 2nd ed. pp. 24–31. Lea & Febiger, Philadelphia, Pennsylvania. (5) Mauderly, J. L., Tesarek, J. E., and Sifford, L. J. (1979). Respiratory measurements of unsedated small laboratory mammals using non-rebreathing valves. *Lab. Anim. Sci.* **29,** 323–329. (6) Mitruka, B. M., and Rawnsley, H. M. (1977). "Clinical Biochemical and Hematological Reference Values in Normal Experimental Animals." Masson, New York. (7) National Research Council (Committee on Animal Nutrition) (1978). Nutrient requirements of the hamster. In "Nutrient Requirements of Laboratory Animals," pp. 70–79. N.R.C., Natl. Acad. Sci., Washington, D. C. (8) O'Brien, J. J. Jr., Lucey, E. C., and Snider, G. L. (1979). Arterial blood gases in normal hamsters at rest and during exercise. *J. Appl. Physiol.* **56,** 806–810. (9) Volkert, W. A., and Musacchia, X. J. (1970). Blood gases in hamsters during hypothermia by exposure to $He-O_2$ mixture and cold. *Am. J. Physiol.* **219,** 919–922.

Part II

The Striped or Chinese Hamster

Chapter 17

Biology and Care

Albert Chang, Arthur Diani, and Mark Connell

I. History	305
II. Description and Behavior	306
III. Biology	306
A. Cytogenetics and Fetal Development	306
B. Metabolism	307
C. Enzyme Studies	308
D. Hormones	308
E. Morphology of Organs and Tissues	309
F. Control Effects of Cyclic Nucleotides on Cell Proliferation	313
IV. Care and Husbandry	313
A. Housing	313
B. Environmental Requirements	314
C. Breeding	314
D. Feeding	315
References	316

I. HISTORY

The laboratory hamster from China is traditionally classified as *Cricetulus griseus* (Honacki *et al.*, 1982) but some authorities (Nowak and Paradiso, 1983) do not recognize this species and include these animals in *Cricetulus barabensis*. The Chinese hamster, also known as the striped-back hamster or gray hamster, originated from Beijing (Peking), China, where it was captured from the street and sold as a pet. The first documented report of Chinese hamsters as laboratory animals appeared in 1919 when Hsieh used them to identify the pneumococcal types in an attempt to treat his patients at the Peking Union Medical College (Hsieh, 1919). Later, because of its availability, it gained acceptance as the most widely used host to *Leishmania donovani*, the causative agent for kala-azar (black fever), in studies on the cyclic transmission of the parasite (Yerganian, 1972). In the mid-1920s, the source of captured Chinese hamsters for use in the laboratory was extended to other parts of China such as Shantung Province. Although breeding stocks were shipped from China to other parts of the world such as India, Near East, and England, attempts to establish colonies for laboratory use had invariably failed (Smith, 1957). The pugnacious habits of the animals and the irregularities of the estrous cycle of captured females made them extremely difficult to breed. In China, breeding Chinese hamsters for laboratory use appeared to be successful, however (C. Y. Chang and Wu, 1938).

Outside China, the first successful domestication of the Chinese hamster was reported by Schwentker (1957) and Yerganian (1958) at Harvard Medical School, Boston, Massachusetts. The original breeding stock of 10 male and 10 female captured animals were shipped from China on the eve of the founding of the People's Republic of China in December 1948. A vivid account of the tumultuous event was described in a letter to Yerganian from Watson of Rockefeller Institute (Yerganian, 1972). Although the colony at Harvard Medical School has since been extinct, the discovery of hereditary diabetes in the colony (Meier and Yerganian, 1959) led to the establishment of two diabetic Chinese hamster colonies in North America in the 1960s, one at The Upjohn Company, Kalamazoo, Michigan (Dulin and Gerritsen, 1967), and another at the C. H. Best Institute, Toronto, Ontario (Sirek and Sirek, 1967). In the following two decades, a large body of literature on the characterization of the diabetic syndrome in the Upjohn colony of Chinese hamsters was accumulated, and it is discussed in Chapter 19 of this text. This chapter concentrates on the biology and care of the Chinese hamster. Currently, a number of Chinese hamster colonies exist throughout the world in countries such as the United States, Japan, Scotland, Switzerland, Germany, and Sweden.

II. DESCRIPTION AND BEHAVIOR

The Chinese hamsters are small rodents with a grayish black coat of very soft fur. The adult males weigh 30–35 gm and the females are about 10% smaller than the males. Their daily food intake is about 7–10 gm/100 gm body weight per day and water intake about 12 ml/100 gm body weight per day, consumed mainly between 8 PM and 8 AM (Thompson, 1971). However, food intake varies with caloric density.

Chinese hamsters, in particular the females, are belligerent in nature and fight with each other at first encounter. It is essential that they are housed in individual cages. Fights erupt most frequently in encounters between females and least often between males. In fact, the presence of females usually suppresses the fighting level of males, whereas the presence of males appears to have no effect on the aggressive nature of the females. After they stop fighting, which usually lasts about 20 min, the animals separate and never engage in contact-promoting behaviors such as nasonasal and nasoanal investigation or mutual grooming. The winner of these fighting encounters is usually resident of the nest site (Skirrow and Rysan, 1976).

The juvenile female Chinese hamsters show little preference between an empty nest site and a site occupied by an adult male, but they avoid nest sites of adult females. The juvenile males prefer an empty nest site to a site of an adult male, and they also avoid nest sites of adult females. The pregnant young females become aggressive toward any adult male littermates and force them out of the shared nest site before they give birth to their litter. During their subsequent pregnancies, the adult females always change their nest sites between litters and they do not tolerate older pups or strange young animals near their new nest sites (Dasser, 1981).

The marking pattern most frequently used by the Chinese hamsters is scratching the flank gland vigorously with a hindfoot, followed immediately by a perineal drag which consists of depressing the anogenital region on the substrate. Such movements spread the scent-bearing secretion of the flank gland onto the substrate. Other sexual behaviors are characterized by brief lordosis, rapid intermount pursuit, brief postejaculatory lock, and one to three thrusts per mount (Daly, 1976). In female–female encounters, marking frequency appears to correlate with behavior dominance, whereas in mixed-sex encounters the males mark more frequently, although the females are invariably dominant in fights and chases. Interestingly, marking which is the only presumably threatening behavior in male–male and female–female encounters succeeds rather than precedes the initial fighting. The lack of warning behaviors and the presence of a high level of food-caching activities indicate that, in its natural habitat, the Chinese hamster must live a solitary and highly dispersed existence with relatively infrequent encounters with its conspecifics (Skirrow and Rysan, 1976).

A study on the behavioral ontogeny of captive Chinese hamsters (Stolba and Kummer, 1972) showed that behavioral correlates of suckling are fully developed at birth. As the hamster matures, ineffective limb movements develop into components of walking and self-grooming. The emergence of mature behaviors follows this sequence with some temporal overlaps: grooming, sniffing, feeding, digging, escape, aggression, and hoarding (Stolba *et al.*, 1978).

III. BIOLOGY

A. Cytogenetics and Fetal Development

The Chinese hamster is very often the species of choice to study chromosome abnormalities because it has a small number of chromosomes ($2n = 22$), which can be divided into four morphologically distinguishable subgroups. A study on the effect of aging showed an increase in the number of terminal chiasmata and univalents and a decrease in chiasma frequency in the first meiotic chromosomes of the oocytes obtained from aged female Chinese hamsters. However, no correlation between the univalents and first meiotic nondisjunction was found (Sugawara and Mikamo, 1983). The Chinese hamster was also used to study the gametic and zygotic selection of genome imbalance by direct analyses of autosomal reciprocal translocation in preimplantation embryos. The data suggested that the cleavage of embryos with chromosome abnormalities was retarded as compared with karyotypically normal zygotes by day 4 of gestation (Sonta et al., 1984). A survey of 226 preimplantation embryos of the Chinese hamster showed only 0.9% genome mutations, consisting of one autosomal trisomy and monosomy each (Binkert and Schmid, 1977). This low spontaneous chromosome aberration rate makes the Chinese hamster a valuable animal for mutagen testing. A search for aneuploidy and polyploidy in the germline cells of the males showed 15 aneuploid out of 392 second metaphases and 14 tetraploid, 4 hexaploid, and 2 octaploid out of 225 spermatogonial metaphases (Hulten et al., 1970). In the Chinese hamsters, giant oocytes and zygotes, twice the normal size, also occur with marked frequency, making up 0.4–0.5% of oviductal eggs (Funaki and Mikamo, 1980; Funaki, 1981). These giant eggs undergo normal meiosis and fertilization and develop normally as digynic triploids at least up to the second cleavage metaphase. Their mitotic activity is normal or even higher than their diploid siblings.

The midpoint temperature (T_m) of thermal denaturation of reassociated DNA from Chinese hamster cells was determined to be 72.5°C (Rake, 1974). According to Kit et al. (1959), the Chinese hamster has a genome size of 2.7×10^{-12} gm/haploid cell, corresponding to 2.48×10^9 nucleotide pairs. The intranuclear distribution of various types of DNA in Chinese hamster cells appears to differ from those of mouse, guinea pig, horse, chicken, and Japanese quail. A study determining the buoyant density, satellite composition, and renaturation kinetics of the DNA of whole nuclei, euchromatin, heterochromatin, and nucleoli showed the presence of a tiny satellite in nucleolar DNA of the Chinese hamster (Comings and Mattoccia, 1972). The genome of Chinese hamster also lacks a class of repetitive sequences with a repetition frequency of about 1000 or more which are commonly found in the DNA of higher organisms. Further, 80% of the single-copy DNA (~60% of the total genome) in Chinese hamster is interspersed with repetitive DNA sequences. The fact that its genome contains relatively few moderately repetitive sequences (9% of the genome) renders this observation particularly striking (Leipoldt et al., 1982). A survey of 26 genetic loci by means of electrophoretic variability showed a high degree of heterozygosity, suggesting that the Chinese hamster species is highly polymorphic (Csaikl, 1984a,b).

A detailed study on the characterization of the gametes and fertilization in vivo of the Chinese hamsters has been reported (Yanagimachi et al., 1983). Fertilization began at 2–3 hr after ovulation and was completed within the next 4–5 hr. At 20- to 26-hr intervals after ovulation, the first three cleavages occurred. Four days postovulation, eggs reached the blastocyst stage and entered the uterus. Implantation occurred between days 5 and 6 (Pickworth et al., 1968). The embryonic and fetal development of the Chinese hamsters has been fully characterized and subdivided into a number of developmental stages.

Its prenatal development mimics closely but lies 2 days behind that of the mouse (ten-Donkelaar et al., 1979a).

B. Metabolism

1. Lipid Metabolism

The Chinese hamster shows plasma cholesterol levels in a range of 130–140 mg/dl with half-lives of 2.1 and 12.0 days for pool A and B, respectively. Cholesterol turnover, pool size, tissue exchange, and tissue content in the Chinese hamster have also been documented in the same report (Chobanian et al., 1974b; Feingold et al., 1984). The plasma levels of prostaglandin $F_{2\alpha}$ in the female Chinese hamster show no age-related changes, in contrast to that in other rodent species such as mice and rats (Parkening et al., 1982). The Chinese hamster shows inordinately high levels of free fatty acids in the serum, ranging from 2.05 to 5.11 mEq/liter which is about five to nine times greater than the levels in rat, dog, and human. At least two factors contribute to this phenomenon. First, both the epididymal and perirenal adipose tissues of Chinese hamster show elevated lipolytic activity as evidenced by their high free fatty acid-releasing activity in vitro. Second, there is a relatively large amount of highly active epididymal fat tissue per unit body weight in the Chinese hamster as compared with other animal species such as the rat. The Chinese hamster adipose tissues show a response to glucose and insulin similar to rat tissues in the suppression of free fatty acid release (Campbell and Green, 1964).

2. Carbohydrate Metabolism

Aortic glucose metabolism in the Chinese hamster has been documented (Chobanian et al., 1974a). When radiolabeled glucose was incubated with aorta in vitro, a large portion of the radioactivity was found in aortic lipids, predominantly in the phospholipid and triglyceride fractions. The incorporation was greatly enhanced by the presence of insulin. In contrast to most mammals in which the principal sugar of semen is usually free fructose, the accessory organs of the Chinese hamster are devoid of fructose and sorbitol but contain some glucose which serves as the major nutrient for seminal metabolism. In addition, an appreciable fraction of a sialic acid-like substance is present in the Chinese hamster vesicular secretion (Fouquet, 1971).

The glucose tolerance curve of the Chinese hamster in response to an intraperitoneal injection of 2 gm/kg glucose is comparable to those obtained from other mammals including humans (Boquist, 1967c). Administration of glucagon, epinephrine, hydrocortisone, or cobalt chloride elicited transient hyperglycemia. The Chinese hamster responds to bovine insulin with high sensitivity, but porcine growth hormone shows no effect on its blood glucose level.

3. Water Metabolism

When the Chinese hamster is kept in an environment with severely limited drinking water supply, its rectal temperature decreases. Its mean maximum urine concentration also increases to a value as high as that of desert rodents, and its maximum urine osmotic concentration shows a direct relationship to the relative renal medullary thickness. When it is allowed water *ad libitum*, the Chinese hamster consumes 90% of the specific water intake on the basis of its body mass (Trojan, 1979).

4. Metal Ion Metabolism

The Chinese hamster has been used to study the metabolic behavior of hafnium, a tetravalent metallic element discovered in 1923, in mammals (Taylor *et al.*, 1983). The organ distribution of hafnium was found to be highest in the skeleton and lowest in the liver at 7 days postinjection and it was bound mainly to transferrin, the iron-transport protein. Thionein, another metalloprotein, has also been investigated and found to be present in high concentration in the livers of newborn Chinese hamsters (Bakka and Webb, 1981). It is essentially a Zn metalloprotein, whereas in other species such as the Syrian hamster it contains more Cu than Zn. Zn, which plays a pivotal role in the storage and secretion of insulin, was found to be present in high content in the Chinese hamster islet β cells (Havu *et al.*, 1977). In animals fed a Zn-deficient diet, decreased granulation of the β cells was seen but the α_1, α_2, and agranular cells showed no morphological deviations. These animals also appeared to be glucose intolerant, but their plasma glucose and insulin levels remained normal (Boquist and Lernmark, 1969).

5. Metabolism in the Pancreatic Islets

Since the detection of spontaneous diabetes in the Chinese hamster, a great deal of research effort has been spent on islet metabolism, and the subject is covered in Chapter 19 of this text. In addition to a large body of data accumulated on cAMP (Rabinovitch *et al.*, 1976) and calcium metabolism (Siegel *et al.*, 1979), the level of monoamine oxidase activity was also measured in the islets as well as in the acinar pancreas and the liver (Feldman *et al.*, 1980). The monoamine oxidase activity may play a role in the regulation of serotonin, which is also present in appreciable levels in the Chinese hamster islets.

C. Enzyme Studies

Isolation, purification, and partial characterization of the following enzymes have been carried out in the Chinese hamster liver: galactokinase (Talbot and Thirion, 1982), alcohol dehydrogenase (Talbot *et al.*, 1981), glutamine synthetase (glutamate–ammonia ligase) (Tiemeier and Milman, 1972), and purine-nucleoside phosphorylase (Milman *et al.*, 1976). An assay and purification procedure of Chinese hamster brain hypoxanthine phosphoribosyltransferase is also available (Olsen and Milman, 1978). Electrophoretic characteristics of the isozymes of lactate dehydrogenase (Chang *et al.*, 1977), alcohol dehydrogenase (Talbot *et al.*, 1981), galactokinase (Nichols *et al.*, 1974; Sun *et al.*, 1975), guanylate kinase (Jamil and Fisher, 1977), pyrimidine-5'-nucleotidase (Swallow *et al.*, 1983), and amylase (Dawson and Huang, 1981) have been described for those isolated from various Chinese hamster tissues. The female Chinese hamsters showed greater renal α- and β-galactosidase activity than the male animals (Chang, 1978a,b). The adrenomedullary dopamine β-hydroxylase (β-monooxygenase) activity in the male Chinese hamster was found to be twice as high in the 30-month as in the 4-month-old animals, contributing to an age-dependent increase in circulating catecholamine levels (Banerji *et al.*, 1984). Chinese hamster tissue levels of the following enzymes have also been well documented: renal acid glycohydrolases (Chang and Greenberg, 1978), renal glucosyltransferase (Chang *et al.*, 1980), plasma, hepatic and renal α-mannosidase (Chang and Perry, 1980), and hepatic and renal gluconeogenic, glycolytic, and pentose phosphate shunt enzymes (Chang and Schneider, 1970a,b).

D. Hormones

The discovery of hereditary diabetes led to extensive studies on the pancreatic islet hormones in the Chinese hamster. The subject is discussed in Chapter 19 of this text. In addition to insulin and glucagon, somatostatin has also been studied in the pancreas, stomach, and hypothalamus, and found to be present in appreciable quantity (Petersson *et al.*, 1977). Plasma and pituitary concentrations of luteinizing hormone (LH), follicle-stimulating hormone (FSH), and prolactin have been measured in female Chinese hamsters during various stages of estrous cycle and age periods (Parkening *et al.*, 1980, 1981). The pattern appears to be similar to that found in the Syrian hamster but different from those observed in the rat and the mouse. However, LH secretion in the male Chinese hamster is unaffected by aging, in contrast to the Syrian hamster and the rat (Parkening *et al.*, 1983). In the Chinese hamster, plasma progesterone concentration fluctuates in a biphasic pattern during the 4-day estrous cycle, with peak values at diestrus and proestrus. Its level increases 3-fold from day 5 to day 8 of the 20-day gestation period, remains elevated until parturition and then falls precipitously during lactation to pregestation level (Sato *et al.*, 1984). Similar to other seasonally breeding rodents, the male Chinese hamster shows a transient but time-dependent decrease in testicular and seminal vesicle weights when transferred from a long (light/dark = 14:10) to a short (light/dark = 5:19) photoperiod (Bartka and Parkening, 1981). The male Chinese hamster also shows a black pigmented spot on its abdomen which appears to be an accessory sex characteristic under endocrine control. The pattern does not occur in males castrated prior to puberty, but it develops when 2-month-old female Chinese hamsters are treated with testosterone (Belcic, 1971).

In a white mutant (*Ws*) with a dark dorsal stripe, the male is infertile but the female reproduces normally. The testicular weight of the white male is only about a quarter of that of the wild-type male, and no spermatogenic cells are found in the seminiferous tubules of the sexually mature mutant male. The infertility of the male is caused by a bilateral hypoplasia of the testes. The inheritance is dominant and pleiotropic to the *Ws* gene (Geyer *et al.*, 1975).

E. Morphology of Organs and Tissues

The purpose of this section is to review the available literature concerning the normal topographic and microscopic anatomy of organs and tissues of the Chinese hamster. A partial description can also be found in Chapter 19 where a morphological comparison between diabetic and nondiabetic animals is made.

1. Topographic Antaomy

A thorough atlas which displays elegant illustrations of the head of the Chinese hamster and its associated tissues and organs has been published (Horber *et al.*, 1974). The cranial bones, dentition, salivary glands, blood vessels, nerves, lymphatic vessels, orbital contents, and major anatomical areas of the brain and its associated vasculature, were identified on detailed sketches. The orbital bleeding technique via puncture of the retrobulbar vein by a microcapillary tube and the parapharyngeal method of hypophysectomy were also discussed. The topographic anatomy of the neck and thoracic cavity has been reported (Geyer, 1973). The glands (including thyroid and parathyroid), musculature, and blood vessels of the neck, epiglottis, trachea, lobes of the lung and the heart, and associated great vessels were delineated on figures and compared with those in the golden hamster. Bronchograms have provided an excellent representation of the tracheal bifurcation as well as the divisions (bronchi, lobar bronchi, and segmental bronchi) of the bronchial tree (Eckel *et al.*, 1974). The entire contents of the peritoneal cavity of the Chinese hamster have been described through a series of exquisite photographs and drawings (Geyer *et al.*, 1972). The lobes of the liver, gallbladder, pancreas, and components of the gastrointestinal tube were intricately defined. The liver biopsy technique and partial removal of the pancreas were also elucidated. The gross anatomy of the urogenital system has been outlined on a series of line drawings which illustrate the kidneys, ureters, urinary bladder, urethra, and organs of the male and female reproductive tract (Geyer, 1972). A surgical procedure for removal of the testes and adrenal gland was also explained. Since the Chinese hamster has been used extensively for the evaluation of toxicity associated with experimental drugs (see Chapter 19), a comprehensive list of the weight and dimensions of the carcass and most organs of this animal has been published (Reznik *et al.*, 1973). This publication serves as a definitive standard against which the body and organ weights of drug-treated Chinese hamsters may be compared.

2. Microscopic Anatomy

a. Pancreatic Islets. Due to the fact that the diabetic Chinese hamster has been widely studied as a model of type I (insulin-deficient) diabetes mellitus (see Chapter 19), numerous investigations have been conducted to characterize the normal morphology of the pancreatic islets of the nondiabetic animal. Under light microscopy, aldehyde–fuchsin stained islets of adult Chinese hamsters consisted of a core of heavily granulated β cells with a peripheral mantle of α cells (Like *et al.*, 1974a). Autoradiography revealed minimal [^{3}H]thymidine labeling of β cells, suggestive of the fact that β-cell replication is uncommon in the islets of adult Chinese hamsters (Like *et al.*, 1974a). The pancreatic islets have also been evaluated by qualitative electron microscopy (Boquist 1967a,b; Luse *et al.*, 1967; Like *et al.*, 1974b). Endocrine cells in conjunction with fenestrated capillaries and small bundles of collagen comprised a typical islet. Numerous desmosomes and nexuses connected the endocrine cells to the one another. Peripheral, unmyelinated nerves and cholinergic and adrenergic terminals juxtaposed to the endocrine cells were also observable. The β cells displayed numerous round storage granules enclosed by loose-fitting membranes. The granules were divided into two main populations based on size and electron density. The more lucent granules were greater in size and filled the majority of the space enclosed by the limiting membrane in contrast to a population of diminutive, densely stained granules. Bar-shaped insulin granules were atypical. Although most of the organelle population was inconspicuous, numerous lysosomal structures were evident in the β cells. The peripheral α cells possessed larger and more heavily stained secretory granules with tight limiting membranes. The rough endoplasmic reticulum, Golgi apparatus, irregular-shaped and electron-dense lysosomes, and bundles of perinuclear microfilaments were also characteristic features of α cells. The peripheral δ cells displayed circular, oval and teardrop-shaped granules which were smaller in size and less dense in comparison with the α-cell granules.

A recent histochemical study demonstrated the presence of heavy acetylcholinesterase and adrenergic activity within the islets of the Chinese hamster (Diani *et al.*, 1983). Unfortunately, the specific endocrine cell populations associated with these activities were not established.

Morphometric analysis of pancreatic islets has been performed to determine islet volume and β-cell mass of 22-month-old Chinese hamsters (Carpenter et al., 1967). The average islet volume was shown to occupy slightly less than 2% of the total pancreatic volume, and approximately 74% of the total islet volume was composed of β cells. Thus, the β-cell mass represented slightly more than 1.2% of the total pancreatic volume. In a subsequent study, histometric evaluation of pancreatic islets was done on 15-day, 3- to 6-month, and 8- to 19-month-old Chinese hamsters (Carpenter et al., 1970). It was found that islet volume accounted for about 0.7% of the total pancreatic tissue of the combined age groups. Furthermore, the islets of all three age groups were composed of an average of 75% β cells, 18% α cells, and 7% blood vessels. It was also observed that neither islet to pancreas volume ratio nor the islet components changed appreciably over the 22-month period. This suggests that islet morphology remains relatively constant over the first 22 months of life in the Chinese hamster.

b. Retina. The layers and cell types of the normal Chinese hamster retina have been identified to serve as baseline data for comparison with that of the diabetic animal (Soret et al., 1974). In general, the layers and cell types of the retina of the normal animal are compatible with those of the human, with one prominent exception. The receptor cell population in the Chinese hamster is composed solely of rods which are situated in the external aspect of the retina.

The retinal vasculature of the Chinese hamster has been examined by fundus photography, fluorescein angiography, and trypsin-digested flat mounts (Federman and Gerritsen, 1970). The large retinal vessels, which vary from 8 to 12, emanated from the optic disk in either a straight or tortuous manner. The optic disk was visualized as a raised structure with pigmentation and a conical configuration. The main retinal vessels traversed the nerve fiber and ganglion cell layers, where they ramified into smaller vessels. Extensive capillary beds were observed in the inner plexiform, inner nuclear, and outer plexiform layers. Although the capillaries of the posterior pole achieved the highest density, structural differences were not apparent among capillaries of the various areas of the retina. Unlike the human retina, a macula densa was not detectable in the Chinese hamster. An average ratio of 2:1 was morphometrically determined to exist between the retinal capillary endothelial cells and intramural pericytes. This ratio is higher compared with the 1:1 ratio in the normal human retina.

Qualitative electron microscopy has been utilized to describe the structural components of the retinal capillaries (Ghosh et al., 1970). In general, two to three endothelial cells, connected by junctional complexes, enveloped the capillary lumen. The capillary was surrounded in its entirety by a thick trilayered basement membrane. The central and thickest portion of the basement membrane was sandwiched between internal and external laminae. In association with the endothelial cells, one or more intramural pericytes, which projected away from the capillary lumen, also surrounded the capillary and were engulfed by the basement membrane. Between the endothelial cell and pericyte, the basement membrane was reduced to a very thin lamina which became discontinuous and allowed the two cell types to contact each other.

c. Kidney. The microscopic anatomy of the kidney of the Chinese hamster has received considerable attention, with much of the effort focused on the glomerulus (Lawe, 1962; Shirai et al., 1967; Soret et al., 1974). Under light microscopy, the structure of the glomerulus, glomerular capillary loops, and afferent and efferent arterioles was virtually identical to that of other rodents. Advance in age seemed to produce modest thickening and some mild degenerative changes of the mesangium and limited dilation of the capillary loops. Ultrastructural analysis of the glomerulus revealed a typical filtration membrane composed of porous endothelial processes, a uniformly thin basement membrane, and podocyte pedicles. Aging effects such as moderate thickening of the basement membrane and mesangium, and occasional vacuolization, fusion of pedicles, and adhesion of podocytes were noted. Systematic morphometric evaluation confirmed that thickening of the glomerular basement membrane was positively correlated with the aging process (Diani et al., 1986). At 11–15 months of age, the thickness averaged 99 nm, which expanded to a mean of 119 nm in 19- to 23-month-old animals. In a recent study, angiotensin II-like activity was localized by immunocytochemistry in the juxtaglomerular apparatus and also in the smooth muscle cells of the arcuate and interlobular arteries and afferent arterioles of the kidney (Taugner et al., 1983). This finding implies that angiotensin II may have a unique but undefined intracellular function in the smooth musculature of resistance vessels.

d. Brain. Although the adult structure of the Chinese hamster brain has been studied in a fragmentary manner, the morphogenesis of certain regions of the brain has been thoroughly described. A series of detailed reports with elaborate line drawings derived from photomicrographs of conventionally stained and autoradiographic ([^{3}H]thymidine) tissue sections have characterized the embryonic development of the diencephalon (Keyser, 1972), basal forebrain (ten-Donkelaar and Dederen, 1979; Lammers et al., 1980), and amygdaloid nuclear complex (ten-Donkelaar et al., 1979b). The adult morphology of the cerebral cortex of the Chinese hamster was provided by ultrastructural analysis of the vasculature, neurophil, and neuronal processes (Luse et al., 1970). In general, the glial tissue as well as axons and dendrites were devoid of necrotic changes, and the vascular basement membranes were not thickened.

A very limited description of the microscopic structure of the hypothalamus has been put forth (Bestetti and Rossi, 1982). The location and morphology of the cell bodies and processes of the α- and β-tanycytes as well as the composition of the arcuate nucleus and median eminence were reported.

Due to its rather complex anatomy, the pineal body of the Chinese hamster has been the focal point of intensive research. The pineal organ is composed of a sac and superficial and deep components that are connected by a stalk (Gregorek et al., 1977). Pinealocytes are most concentrated in the deep portion but also extend into the stalk and superficial region. The pineal sac overhangs the stalk and deep pineal and resembles the morphology of the choroid plexus (Gregorek et al., 1977). Morphometry has shown a striking correlation between pinealocyte structure and diurnal rhythm (Matsushima and Morisawa, 1981). In the initial phase of the dark period, the pinealocytes were reduced in size and the number of synaptic ribbons was elevated. Since reserpine and propranolol prevented the diminution of cell size but did not interfere with the increase in number of synaptic ribbons, it is likely that these two phenomena are controlled by dissimilar mechanisms.

The pinealocytes and glial cells of the superficial pineal have been carefully scrutinized under the electron microscope (Matsushima and Morisawa, 1982). Unique features of the pinealocytes were mitochondria of various shapes and sizes with bizarre cristae, stacks of flattened cisternae juxtaposed to smooth and rough endoplasmic reticulum, and 100-nm and 300-nm granulated vesicles. The glial cells were typified by dense concentrations of cytoplasmic fat droplets and sporadic pigmentation. The synaptic ribbons of pinealocytes in the superficial pineal were grouped into three categories (types 1, 2, and 3) on the basis of their configuration (Matsushima et al., 1983b). The type 1 and 3 synaptic ribbons also displayed variation in shape during a 24-hr light–dark cycle (Matsushima et al., 1983a).

e. Pituitary Gland. The adenohypophysis of the Chinese hamster has been examined under the electron microscope, and distinct morphological criteria for the differentiation of seven cell types were generated. Mammotropes (common in females, infrequent in males) were elongated cells with 600-nm pleomorphic granules. The most numerous cell in the distal lobe of the pituitary was the somatotrope, which displayed storage and synthesis stages. The granules, which became less numerous in the synthesis phase, were intensely stained and averaged 300 nm in size. The number of somatotropes was calculated to be approximately 17 cells per 10,000 μm^2 of adenohypophysis (Deslex and Rossi, 1976). Corticotropes, confined primarily to the pars intermedia, were irregular in shape with long processes and electronlucent 160-nm granules. Two types of gonadotropes were distinguishable. The first type was a rounded cell with 190- to 320-nm granules of variable electron density, whereas the second type was round or polygonal with 120- to 220-nm granules. Thyrotropes were uncommon and characterized as polygonal cells with extended cellular processes and 120- to 160-nm granules. The remaining cell type was the chromophobe cell, which was devoid of granules but possessed copious amounts of smooth endoplasmic reticulum (Deslex et al., 1976).

f. Peripheral and Autonomic Nerves. Teased-fiber wholemount preparations and electron microscopy were employed to investigate the distal peripheral tibial nerve of the Chinese hamster (Schlaepfer et al., 1974). A linear proportional correlation was observed between myelinated fiber diameter and internode length. Myelin internodes displayed uniform length and thickness along the fibers. Under electron microscopy, the myelinated and unmyelinated fibers and the endoneural blood vessels resembled those of most rodent species. The distal motor and sensory nerve fibers and the hindlimb of the Chinese hamster were subjected to quantitative microscopic assessment (Kennedy et al., 1982). The diameter of the sensory and motor myelinated fibers ranged from approximately 1 to 7 μm with a mean of 3 μm. Myelin thickness comprised about 19% of the total fiber diameter. The number of myelinated fibers in the sensory nerves (average, 233) was considerably greater than those of the motor nerves (average, 12). Mixed, motor, and sensory conduction velocities were also tabulated for the hindlimb nerves (Kennedy et al., 1982).

Morphometry was performed on the infradiaphragmatic portion of the sympathetic trunk (Diani et al., 1981a) and the ventral abdominal division of the vagus nerve (Diani et al., 1984). A bimodal distribution of myelinated fibers with population peaks at 3–4 μm and 9–10 μm and a single population peak of 1.2–1.4 μm for unmyelinated fibers were obtained for the sympathetic trunk. The majority of myelinated fibers spanned a range of 4–6 μm, whereas the highest percentage of unmyelinated fibers fell in the 0.8- to 1.0-μm range for the vagus nerve. Numerical and volume densities were also calculated for the myelinated and unmyelinated nerve fibers of these autonomic nerves.

g. Aorta. Length and diameter measurements along with qualitative ultrastructural analysis of the aorta have been garnered from newborn through adult Chinese hamsters (Soret et al., 1976a). The length of the aorta varied from 18 mm in young pups to 50 mm in adults. Although the diameter of the aorta increased with the aging process, it decreased from the level of the heart to the abdomen within a given age group. The tunica intima was composed of a thin layer of endothelial cells situated upon an internal elastic membrane and subendothelial pockets of reticular and collagen fibrils. Alternating lamellae of smooth muscle cells and elastic membranes constituted the tunica media. Bridges between the intima and media were observed to traverse the fenestrations of the internal elastic membrane (Soret et al., 1976b). Four types of bridges were defined on the basis of their constituents: (1) smooth muscle cells, (2) endothelial cells, (3) smooth muscle and endothelial cells, and (4) collagen fibers. Scattered fibroblasts and vast arrays of collagen fibers were the primary components of the tunica adventitia.

h. Reproductive Tract. Much of the literature concerning spermatogenesis and oogenesis in the Chinese hamster has been reviewed in Chapter 19, Section VI. It should be pointed out that a very comprehensive review of the process of spermatogenesis has been generated by light-microscopic autoradiography (Oud and de Rooij, 1977; Lok *et al.*, 1982, 1983, 1984; Lok and de Rooij, 1983a,b). Various nonspecific and nucleoside phosphatase enzymes have been localized in specific subcellular structures during spermatogenesis (J. P. Chang *et al.*, 1974). One notable feature was the identification of glucose-6-phosphatase, which appears to be unique to the spermatid of the Chinese hamster in comparison with humans (Yokoyama and Chang, 1977).

The motility and structure of some of the components of mature spermatozoa have been investigated in the Chinese hamster. The sperm tend to beat in an asymmetric manner which induces a circular swimming motion (Phillips, 1972a). Forward movement in a straight line was provided by rotation around a central axis. Silver staining has been very beneficial in the gross differentiation of various substructures of spermatozoa such as the acrosome, subacrosomal region, perforatorium, postacrosomal sheath, neck, midpiece, annulus, principal piece, and endpiece (Elder and Hsu, 1981). Critically point-dried replicas of the head of Chinese hamster sperm demonstrated that a unique array of tubules and vesicles are distributed on the apical aspect of the acrosome (Phillips, 1975). These tubules and vesicles displayed an adhesive quality which may play a critical role in the adherence of sperm to the ovum during fertilization (Phillips, 1972b). The acrosomal zonule was detected as a ringlike structure juxtaposed between the inner acrosomal membrane and the nuclear membrane during all stages of development through spermiation (Maxwell, 1982). Its function has remained obscure. The midpiece of Chinese hamster spermatozoa is composed of several spiral strands of mitochondria (Phillips, 1977). The strands are oriented in a staggered fashion but form a very precise pattern. The nuclear ring was defined as a segmented troughlike structure which surrounds the nucleus of the spermatid (Rattner and Olson, 1973). Although its purpose is unclear, it may regulate the microtubules of the manchette.

A few scattered reports have been published on other cell types in the male reproductive tract. Conventional ultrastructure techniques and ultracytochemistry were employed to differentiate the ciliated and nonciliated cells in the efferent ductules (Yokoyama and Chang, 1971). On the basis of specific enzyme activities, the ciliated cells were functionally responsible for sperm transport, whereas the nonciliated cells controlled fluid composition of the ductules. Two nonpathological paracrystalline inclusions have been described in the Leydig cells of sexually mature Chinese hamsters (Payer and Parkening, 1983a,b). The type A inclusion was localized in the nucleoplasm and cytoplasm and was characterized by serrated microfilaments. The type B inclusion was restricted to the cytoplasm and exhibited straight microfilaments.

Glycogen content of oocytes from 20-day-old female Chinese hamsters has been examined (Takeuchi and Sonta, 1983). Glycogen was absent from oocytes which were less than 20 μm in diameter. Numerous cytoplasmic aggregates of two to four β-glycogen particles were observed in oocytes which achieved a diameter of greater than 40 μm.

i. Oral Cavity. The parotid and mandibular salivary glands of the Chinese hamster have been described under light and electron microscopy. The parotid gland was characterized by the presence of light, dark, and specific light acinar cells, as well as intercalated and striated ducts (Suzuki *et al.*, 1981). Nerve endings were observed near acinar cells, but myoepithelial cells were completely absent from the parotid gland. The acinar cells of the mandibular gland were predominantly light cells which displayed sexual dimorphism (Suzuki *et al.*, 1982). As in the parotid gland, intercalated and striated ducts were common but the latter displayed secretory and striated areas.

Detailed description of the postnatal development of the teeth and the wearing process of the molars has been published (Habermehl, 1971). Newborn animals were found to be equipped with incisors which were hypothesized to assist in the nursing process. A reasonable estimate of the age of Chinese hamsters was also derived from the extent of wear on the molars. With freeze–fracture replicas, the enamel of incisors showed an unusual configuration (Warshawsky and Nanci, 1982). The enamel crystallites were slender, flat, tortuous ribbons with variable length and similar widths.

j. Gastrointestinal Tube. Detailed accounts of the microscopic structure of the gastrointestinal tube of the Chinese hamster have been published during the last few years. At the light-microscopic level, the morphology of the mucosa, submucosa, muscularis, and serosa of the small intestine was elucidated (Diani *et al.*, 1976). Quantification of the number and size of villi per cross section of gut was also conducted. At the ultrastructural level, morphometric evaluation of true mucosal capillaries indicated that the basement membrane thickness averaged 274 Å and was positively correlated with age (Diani *et al.*, 1981b). Barium radiology was utilized to measure the rate of total gastric emptying (200 min), initial barium stool formation (100 min), and initial fecal passage (210 min) (Diani *et al.*, 1979). The unmyelinated axons of Auerbach's plexus in the small intestine maintained a uniform diameter of 0.6 μm and displayed a few scattered glycogen particles and neurofilaments (Diani *et al.*, 1979). Histochemistry has documented the presence of pseudoisocyanine-positive (presumably insulin-containing) (von-Dorsche et al., 1977) and neurotensin-immunoreactive (Helmstaedter et al., 1977) cells in the intestine of the Chinese hamster.

k. Musculoskeletal System. The morphology of neuromuscular terminals and the muscle spindle have undergone extensive analysis by electron microscopy. Specializations of the neuromuscular junction of the Chinese hamster included thin, ramified nerve endings which terminated on an oval region of the skeletal muscle fiber surface and synaptic depressions that possessed an irregular contour and random placement of junctional folds (Desaki and Uehara, 1981). The intrafusal fibers of the neuromuscular spindle of the Chinese hamster displayed some unique characteristics such as markedly thick filaments (Desaki, 1978), underdeveloped or absent infoldings of the sarcolemma, and a pleated basal lamina (Desaki and Uehara, 1983).

The hip joint and intervertebral discs are the only areas of the skeletal system that have been examined. In general, the morphology of the hip joint was remarkably similar to that of other age-matched rodents, particularly mice (Silberberg et al., 1976). Aging alterations, representation of spondylosis, and disc prolapse were observed in geriatric animals (Silberberg and Gerritsen, 1976).

3. Hemogram

The hemogram of the Chinese hamster is shown in Table I.

F. Control Effects of Cyclic Nucleotides on Cell Proliferation

The Chinese hamster has been used as an animal model to study the role of cyclic nucleotides in the control of cellular processes (Kovar and Fremuth, 1976). Specifically, the proliferation activity of bone marrow cells was measured *in vivo*. cAMP was found to inhibit cell proliferation at doses of 3–30 mg/kg, and the predicted stimulation effect of cGMP was not detected. In fact, cGMP inhibited the entrance of bone marrow cells into mitosis *in vivo* at doses of 0.1–100 mg/kg. In the same study, cAMP also decreased the radiosensitivity of the Chinese hamster when given intraperitoneally 10 min prior to radiation exposure, whereas cGMP reversed the protective effect of cAMP. cGMP also induced radiosensitivity of the animal. There appeared to be a correlation between radiosensitivity and bone marrow cell proliferation activity (Kovar and Fremuth, 1976).

IV. CARE AND HUSBANDRY

The information contained in this section was derived from a literature review and practical experience associated with the maintenance of the Upjohn colony of Chinese hamsters for more than 20 years. The colony consists of inbred lines which are either essentially 100% spontaneously glycosuric or 100% aglycosuric animals. They have been inbred for over 30 generations.

A. Housing

From weaning (i.e., 21 days postpartum) to 3 months of age, the animals are housed in plastic boxes, 17 × 7 × 5 in. in size, each of which has a stainless-steel top with a drop-in feeder and a water bottle. Small holes are drilled in each corner of the box to allow moisture to escape in the event of a "leaky" bottle or from excessive dampness caused by polyuria from the diabetic animals, and the bedding used in the boxes is finely chopped wood chips. The hamsters are housed with a maximum of four per cage, and only animals of the same sex are housed within an individual box. This is critical to the operation because of the record keeping with accompanies the inbred breeding program. At 3 months of age, the animals are either put into breeding or designated for research. Each of these dispositions has slightly different housing requirements.

Males used in breeding are housed in stainless-steel suspended cages with a stainless-steel mouse-sized mesh covering the bottom and front of the cage. These cages are 10 × 10 × 5 inches in size, which allows the male to have a rather large area for his domain without the threat of other males nearby. They are thus able to concentrate on reproductive activity without having to worry about territorial rights. Females used in breeding are also housed individually but are kept in plastic boxes identical to the ones used for the weaned hamsters discussed earlier. It is extremely important that these boxes have several small holes drilled into each corner to allow excess moisture to escape. If the bedding becomes too wet and the pups are allowed to remain damp for any length of time, neonatal mortality will be excessive.

Table I

Hemogram of the Chinese Hamster[a]

Parameter	Mean ± SD
Erythrocytes ($10^6/\mu$l)	7.1 ± 0.01
Packed cell volume (%)	42.1 ± 5.6
Hemoglobin (gm/dl)	12.4
Total leukocytes ($10^3/\mu$l)	5.5
Neutrophils (%)	19.3 ± 2.2
Monocytes (%)	76.1 ± 7.8
Eosinophils (%)	1.7 ± 0.7
Basophils (%)	0.1 ± 0.04
Sedimentation rate (mm/hr)	3.5 ± 1.7

[a]From Moore (1966).

Animals that are designated for research at 3 months of age are housed in the stainless-steel suspended cages identical to those used for the male breeders. These cages, however, are equipped with a solid stainless-steel divider which splits each cage into two equally sized compartments. The dividers must fit the floor, back, and front of the cage snugly in order to assure that the animals cannot pass through or bite each other. If two hamsters happen to occupy the same compartment, fighting will erupt and inevitably lead to the death of at least one hamster.

B. Environmental Requirements

The most crucial environmental requirement for the success of a Chinese hamster colony is the control of light cycles and noise level, especially in the breeding area. Noise is the most severe environmental deterrent to reproduction for the Chinese hamsters. If noise is not kept to a minimal level, the female breeders will become anestrous. Routine animal care activities should therefore be carried out either prior to the dark cycle or after females are removed from the males cage.

Lights in the breeding room are on a reverse day–night cycle because males form copulation plugs 1–2 hr after initiation of the dark cycle. With the lights off from 6 AM to 6 PM, females are then placed with males at a very opportune time for mating to occur. The breeding room needs to be relatively light-tight, so that illumination from outside the breeding area does not enter the breeding room during the dark cycle. The preferred temperature range for the Chinese hamster is 72°–74°F, with a relative humidity of approximately 50%.

C. Breeding

1. Methods

The early attempts to breed Chinese hamsters invariably failed due to the aggressiveness of the animals, especially the females. If an adult female and male were left together, the male would be killed. Therefore, females with nonagressive behavior (Calland, 1984) or sexually aggressive males (Moore, 1965) were selected as breeders. The modified test-mating and monogamous mating have been used to breed the Chinese hamster (Camden *et al.*, 1968, 1969). The first system involves placing a female with a male for 7 days or longer, but separating them before a litter is born. The second system involves leaving a male and female together during their entire reproductive life. Both systems produced acceptable numbers of litters. In monogamous matings, mate killing was highest in nonsibling mating and decreased with each successive generation of sibling mating. Also, a decrease in the average litter size weaned was observed and attributed to losses of entire litters prior to weaning (Camden *et al.*, 1969; Porter and Lacey, 1969; Festing, 1970). Different approaches to propagate the Chinese hamsters, namely, hand-mating and artificial insemination, have been used with moderate success (Avery, 1968). A practical and more economical method to produce Chinese hamsters on a commercial scale was also reported (Cisar *et al.*, 1972). In this scheme, three males were placed in a breeding cage with three to five female littermates at weaning, and the authors reported a 3.5% loss of males per month.

The contents of vaginal smears of 4-day cyclic females were examined every 3 hr for 5 days. A light–dark cycle of 14 and 10 hr was used, with the lights turned on at 6 AM. The proestrous phase started at about time 0 for day 1; the day of the proestrous phase was designated as day 1 of the estrous cycle. It was concluded that copulation occurs between 6 PM and midnight of day 1 (Kita *et al.*, 1979a). The vaginal mucous discharge and cell components were examined twice a day throughout the gestation period in the females. Plug in vagina and sperms in the smear were clearly observed on day 0 of pregnancy. Aqueous mucinlike vaginal discharge was found in 82.6% on day 14. The hemorrhagic mucous discharge lasting for 2 or 4 days was followed by the sticky mucin in most cases. This was found to be a reliable means to diagnose pregnancy, which can be detected as early as day 10 after copulation in the Chinese hamster (Kita *et al.*, 1979b). After the third litter, the female breeders show a progressive decline in litter size. The mean number of litters produced is 4.9 ± 0.5 and the mean age of female breeders at the time of delivering the last litter is 16.4 ± 0.5 months (Parkening, 1982).

2. Breeding Management in the Upjohn Colony

The animals are placed into breeding at 3 months of age. Females are bred through 1 year of age, while males continue to breed until 15 months of age. The following criteria are used to decide whether an animal will be placed into the breeding program.

a. 90-Day Nonfasting Blood Glucose (NFBG). Each hamster has a blood glucose measurement taken at 90 days of age. The preferred blood glucose range for diabetic animals is 250–350 mg/dl. Values above this range tend to make an animal totally infertile or infertile at a younger age. Another problem encountered when using breeding pairs that have high blood glucose levels is that their offspring generally are prediabetic for a very short time after weaning. This is a definite factor for consideration for those investigators that are interested in doing research with a prediabetic animal. Conversely, hamsters with NFBG values below 250 mg/dl generally do not produce pups that will maintain a steady diabetic state with glycosuria and a very high blood glucose level.

b. Number of Siblings in the Litter. The number of siblings in a litter is important to the management of a breeding colony because large litters are a very desirable trait that needs to be selected for whenever possible. This is a continual area of concern, because selection of possible breeders from a litter of three is much more restrictive than from a litter of five or six pups.

The optimum breeding ratio is two or three females for every male. To keep males sexually active, it is necessary to place a female with them two to three times per week. For this reason a number of females are kept on hand for random breeding. These females are bred to any male, and their offspring are primarily targeted for research rather than breeding.

Juvenile hamsters mate faster initially if they are placed with an experienced male. To accomplish this within the breeding management scheme young females are inbred back to their fathers when possible for their first litter. While the females are being bred to their sires, it gives the young males an opportunity to breed with older and experienced females used for random breeding.

c. Familial History. Familial history is used extensively in the breeding program to ensure the inbred nature of the sublines. The history is also used in selection of breeding hamsters by means of viewing parental records of glycosuria. If a dam or sire of one litter was aglycosuric or only mildly glycosuric, the pups from that litter would not be selected. Pups from a litter where the dam and sire were strongly glycosuric would be preferentially selected. Familial history is also useful in selecting breeders whose families have had a history of large litters.

d. Current Subline Population. The number of females and age spread are extremely important for the continuance of a genetic subline. If there are fewer than 15 females in a given subline, it is extremely difficult to be able to produce adequate hamsters for proper selection, and the subline can run into extinction in a very brief period.

Likewise, if the hamsters that are in a breeding program are too old or too young, serious reproductive difficulties may emerge. Therefore, it is critical to maintain a continuous flow of hamsters into and out of the breeding colony, so that the subline reflects a range of ages from 3 to 12 months.

e. Overall Appearance. Prior to selecting a hamster for the breeding program, one must pay attention to such things as the general health, size and weight of an animal, and whether it has missing appendages that might hamper its reproductive performance.

Chinese hamsters in the Upjohn colony are "hand-bred," that is, females are checked daily for signs of estrus at 7 AM (i.e., 1 hr after the start of the dark period). When a female is determined to be in proestrus or estrus by visual inspection (i.e., pink, swollen vulva), she is placed in the appropriate male's cage. They are left together until 1 PM, and the female's vagina is checked for evidence of a copulation plug. If a plug is present, it is termed a "probable" mating. If there are copulation plugs present on the paper beneath a male's cage, but none found in the vagina, it is termed a "possible" mating. All females are returned to their respective cages following examination. Females in which no evidence of mating occurred are put back into their cages and checked daily for estrus until a mating takes place.

Pregnancy determination can be made at 15 days after mating. If a female is pregnant, a small spot of blood will appear on the vaginal orifice. Pups are born 21 days after a successful mating. Probability of pregnancy from a female with a "probable" mating is 60–70% and with a "possible" mating, 30–40%. Pups are weaned at 21 days of age, and the dams are checked for estrus the day after weaning.

Litter sizes in the Upjohn colony average four pups per litter, and a female generally has two or three litters while in the breeding colony. It has been found that one way of greatly increasing the number of litters that a female can produce is by fostering her newborn pups to another dam. For fostering, a specific line that has a history of producing abundant, large litters is used. Pups are fostered at 3–4 days of age. Beyond that time there is a greatly increased risk of rejection by the foster mother. Dams whose litters have been fostered show estrus within days after removal of their pups and can be placed into breeding as soon as that occurs. Fostering of litters has saved several sublines from extinction and is a very useful and productive technique.

D. Feeding

The Chinese hamster is a nocturnally feeding rodent, and it displays circadian variation in feeding (Billington *et al.*, 1984). However, unlike other rodents, opiate antagonists such as naloxone and butorphanol have no effect on Chinese hamster feeding. The following peptides, which are known to have satiety effects in rats, show no effect in the Chinese hamster: cholecystokinin, bombesin, somatostatin, and pancreatic polypeptide. Thus, the opiateric and peptidergic influences on feeding are very different between the Chinese hamster and other rodents. The Chinese hamster shows an impressive preference for 10% alcohol to water (Goas *et al.*, 1979).

In the Upjohn colony hamsters are fed Purina Mouse Breeder Chow 5015 *ad libitum,* and feeders are kept low enough so that fresh food can be added weekly. This prevents the food from becoming stale while sitting in the open feeders on the cages. The Purina 5015 diet is able to meet the requirements for most of the hamsters, the exception being lactating females. Newly lactating females have two problems. The first is that when a female has just given birth to a litter of pups she is most likely to cannibalize them. To diminish the amount of cannibalization, the dam's diet is supplemented with wheat germ and a fresh slice of apple daily. These dietary supplements seem to satisfy most dams, making them much less likely to destroy their pups. If a particular female or a particular subline of hamsters is more likely to cannibalize their young pups, then the dam's diet is enriched with wheat germ a week prior to the date she is expected to deliver.

The second problem area is lactation, primarily with the diabetic animals. Chinese hamsters do not seem to produce an excessive amount of milk for their litters. The dietary supplements described above also enable the dam to receive a concentrated supply of calories to keep her in lactation and to enable her to produce more milk without wasting away. Pups are also able to feed on the supplements as soon as they can move about the cage. This enables them to get at an additional food source if the dam is having problems producing an adequate milk supply. The apple slice also serves as an important source of water for pups that are too small to reach the water spigots.

Water delivery to Chinese hamsters can be either through the use of individual water bottles or an automatic watering system. The automatic watering system has been found to be extremely beneficial in reducing the animal care workforce. The only problem encountered with the automatic watering system was the loss of 10% of the older animals when they were switched from individual water bottles. These animals refused to look at the back of their cages for water instead of the front in spite of enticements placed on the new water spigots to lure them to the water. Automatic watering systems are available for racks as well as individual boxes.

ACKNOWLEDGMENTS

The authors wish to thank Dr. George Gerritsen of the Diabetes and Gastrointestinal Diseases Research Unit of The Upjohn Company for assistance with the literature search and advice concerning the content of this chapter. Gratitude is extended to Linda Rogers of the Diabetes and Gastrointestinal Diseases Research Unit of The Upjohn Company for highly professional and expedient secretarial contributions. The authors are also indebted to Shelia Graber, Susan Fierke, and the staff of the Upjohn Company Technical Library for collection and reproduction of literature utilized in this chapter.

REFERENCES

Avery, T. L. (1968). Observations on the propagation of Chinese hamsters. *Lab. Anim. Care* **18**, 151–159.

Bakka, A., and Webb, M. (1981). Metabolism of zinc and copper in the neonate: Changes in the concentrations and contents of thionein-bound Zn and Cu with age in the livers of the newborn of various mammalian species. *Biochem. Pharmacol.* **30**, 721–725.

Banerji, T. K., Parkening, T. A., and Collins, T. J. (1984). Adrenomedullary catecholaminergic activity increases with age in male laboratory rodents. *J. Gerontol.* **39**, 264–268.

Bartka, A., and Parkening, T. A. (1981). Effects of short photoperiod on pituitary and testicular function in the Chinese hamster, *Cricetulus griseus*. *Biol. Reprod.* **25**, 958–962.

Belcic, I. (1971). Ein hormonal gesteuerter Geschlechtsdimorphismus in Fellmaster des chinesischen Hamsters. *Z. Versuchstierkd.* **13**, 193–196.

Bestetti, G., and Rossi, G. L. (1982). Hypothalamic changes in diabetic Chinese hamsters: A semiquantitative, light and electron microscopic study. *Lab. Invest.* **47**, 516–522.

Billington, C. J., Morley, J. E., Levine, A. S., and Gerritsen, G. C. (1984). Feeding systems in Chinese hamsters. *Am. J. Physiol.* **247**, R405–R411.

Binkert, F., and Schmid, W. (1977). Preimplantation embryos of Chinese hamster. I. Incidence of karyotype anomalies in 226 control embryos. *Mutat. Res.* **46**, 63–76.

Boquist, L. (1967a). Morphology of the pancreatic islets of the non-diabetic adult Chinese hamster, *Cricetulus griseus:* Light microscopical findings. *Acta Soc. Med. Ups.* **72**, 331–344.

Boquist, L. (1967b). Morphology of the pancreatic islets of the non-diabetic adult Chinese hamster, *Cricetulus griseus:* Ultrastructural findings. *Acta Soc. Med. Ups.* **72**, 345–357.

Boquist, L. (1967c). Some aspects on the blood glucose regulation and the glutathione content of the nondiabetic adult Chinese hamster, *Cricetulus griseus*. *Acta Soc. Med. Ups.* **72**, 358–375.

Boquist, L., and Lernmark, A. (1969). Effects on the endocrine pancreas in Chinese hamsters fed zinc-deficient diets. *Acta Pathol. Microbiol. Scand.* **76**, 215–228.

Calland, C. J. (1984). Establishment of a Chinese hamster breeding colony. *Lab. Anim. Sci.* **34**, 519.

Camden, R. W., Poole, C. M., and Flynn, R. J. (1968). Mating systems for Chinese hamster production. *U.S. A.E.C. Argonne Natl. Lab.*, December, pp. 190–191.

Camden, R. W., Poole, C. M., and Flynn, R. J. (1969). Monogamous mating of Chinese hamster. *U.S. A.E.C. Argonne Natl. Lab.*, December, pp. 62–63.

Campbell, J., and Green, G. R. (1964). Free fatty acid metabolism in Chinese hamsters. *Can. J. Physiol. Pharmacol.* **44**, 47–57.

Carpenter, A. M., Gerritsen, G. C., Dulin, W. E., and Lazarow, A. (1967). Islet and beta cell volumes in diabetic Chinese hamsters and their nondiabetic siblings. *Diabetologia* **3**, 92–96.

Carpenter, A. M., Gerritsen, G. C., Dulin, W. E., and Lazarow, A. (1970). Islet and beta cell volumes in offspring of severely diabetic (ketotic) Chinese hamsters. *Diabetologia* **6**, 168–176.

Chang, A. Y. (1978a). Difference in renal α-galactosidase levels in male and female Chinese hamsters (*Cricetulus griseus*). *Comp. Biochem. Physiol. B* **61B**, 133–137.

Chang, A. Y. (1978b). Sexually differentiated activity and isozymes of renal β-galactosidase in the Chinese hamster. *Int. J. Biochem.* **9**, 567–572.

Chang, A. Y., and Greenberg, H. S. (1978). Acid glycohydrolase in Chinese hamster with spontaneous diabetes. III. Line-dependent variance. *Biochim. Biophys. Acta* **525**, 134–141.

Chang, A. Y., and Perry, C. S. (1980). Acid glycohydrolase in Chinese hamster (*Cricetulus griseus*) with spontaneous diabetes. VI. Diabetes-dependent differences in α-D-mannosidase activities in plasma, liver and kidney. *Comp. Biochem. Physiol. B* **65B**, 489–495.

Chang, A. Y., and Schneider, D. I. (1970a). Metabolic abnormalities in the pancreatic islets and livers of the diabetic Chinese hamster. *Diabetologia* **6**, 180–185.

Chang, A. Y., and Schneider, D. I. (1970b). Rate of gluconeogenesis and levels of gluconeogenic enzymes in liver and kidney of diabetic and normal Chinese hamsters. *Biochim. Biophys. Acta* **222**, 587–592.

Chang, A. Y., Noble, R. E., and Greenberg, H. S. (1977). Variance in LDH isozyme patterns in a Chinese hamster (*Cricetulus griseus*) colony. *Comp. Biochem. Physiol. B* **58B**, 119–123.

Chang, A. Y., Noble, R. E., Perry, C. S., and Greenberg, H. S. (1980). Renal glucosyltransferase activity in highly inbred spontaneously diabetic Chinese hamsters. *Diabetologia* **19**, 40–44.

Chang, C. Y., and Wu, H. (1938). Growth and reproduction of laboratory bred hamsters, *Cricetulus griseus*. *Chin. J. Physiol.* **13**, 109–118.

Chang, J. P., Yokoyama, M., Brinkley, B. R., and Mayahara, H. (1974). Electron microscopic cytochemical study of phosphatases during spermiogenesis in Chinese hamster. *Biol. Reprod.* **11**, 601–610.

Chobanian, A. V., Gerritsen, G. C., Brecher, P. I., and McCombs, L. (1974a). Aortic glucose metabolism in the diabetic Chinese hamster. *Diabetologia* **10**, 589–593.

Chobanian, A. V., Gerritsen, G. C., Brecher, P. I., and Kessler, M. (1974b). Cholesterol metabolism in the diabetic Chinese hamster. *Diabetologia* **10**, 595–600.

Cisar, C. F., Gumperz, E. P., Nicholson, F. S., and Moore, W., Jr. (1972). A practical method for production breeding of Chinese hamsters (*Cricetulus griseus*). *Lab. Anim. Sci.* **22**, 725–727.

Comings, D. E., and Mattoccia, E. (1972). DNA of mammalian and avian heterochromatics. *Exp. Cell Res.* **71**, 113–131.

Csaikl, F. (1984a). Electrophoretic comparison of Syrian and Chinese hamster species. *Hereditary* **52**, 141–144.

Csaikl, F. (1984b). Fragment comparison of hamster mitochondrial DNA: General conclusions about the evolution of mitochondrial DNA. *Comp. Biochem. Physiol. B* **78B**, 325–329.

Daly, M. (1976). Behavioral development in three hamster species. *Dev. Psychobiol.* **9**, 315–323.

Dasser, V. (1981). Dispersal tendencies and social behavior of young Chinese hamsters, *Cricetulus griseus*. *Behaviour* **78**, 1–20.

Dawson, W. D., and Huang, L. L. (1981). Comparative genetics of hamster amylases. *Biochem. Genet.* **19**, 623–633.

Desaki, J. (1978). Unusually thick filaments in intrafusal muscle fibers of the Chinese hamster. *J. Electron Microsc.* **27**, 313–315.

Desaki, J., and Uehara, Y. (1981). The overall morphology of neuromuscular junctions as revealed by scanning electron microscopy. *J. Neurocytol.* **10**, 101–110.

Desaki, J., and Uehara, Y. (1983). A fine structural study of the termination of intrafusal muscle fibres in the Chinese hamster. *Cell Tissue Res.* **234**, 723–733.

Deslex, P., and Rossi, G. L. (1976). Quantitative evaluation of somatotrophic cells (SC) in the adenohypophysis of normal and diabetic male Chinese hamsters. *Diabetologia* **12**, 489–493.

Deslex, P., Rossi, G. L., and Probst, D. (1976). Ultrastructural study of the adenohypophysis of the Chinese hamster. *Acta Anat.* **96**, 35–54.

Diani, A. R., Gerritsen, G. C., Stromsta, S., and Murray, P. (1976). A study of the morphological changes in the small intestine of the spontaneously diabetic Chinese hamster. *Diabetologia* **19**, 101–109.

Diani, A. R., Grogan, D. M., Yates, M. E., Risinger, D. L. and Gerritsen, G. C. (1979). Radiologic abnormalities and autonomic neuropathology in the digestive tract of the ketonuric diabetic Chinese hamster. *Diabetologia* **17**, 33–40.

Diani, A. R., Davis, D. E., Fix, J. D., Swartzman, J., and Gerritsen, G. C. (1981a). Morphometric analysis of autonomic neuropathology in the abdominal sympathetic trunk of the ketonuric diabetic Chinese hamster. *Acta Neuropathol.* **53**, 293–298.

Diani, A. R., Weaver, E. A., and Gerritsen, G. C. (1981b). Capillary basement membrane thickening associated with the small intestine of the ketonuric diabetic Chinese hamster. *Lab. Invest.* **44**, 388–391.

Diani, A. R., Peterson, T., and Gilchrist, B. J. (1983). Islet innervation of nondiabetic and diabetic Chinese hamsters. I. Acetylcholinesterase histochemistry and norepinephrine fluorescence. *J. Neural Transm.* **56**, 223–238.

Diani, A. R., West, C., Vidmar, T. J., Peterson, T., and Gerritsen, G. C. (1984). Morphometric analysis of the vagus nerve in nondiabetic and ketonuric diabetic Chinese hamsters. *J. Comp. Pathol.* **94**, 495–504.

Diani, A. R., Sawada, G. A., Peterson, T., Wyse, B. M., Blanks, M. C., Vidmar, T. J., and Gerritsen, G. C. (1986). Systematic evaluation of microangiopathy in diabetic Chinese hamsters. I. Morphometric analysis of minimal glomerular basement membrane thickness in 11–15 and 19–23 month old Chinese hamsters. *Microvasc. Res.* **31**, 306–316.

Dulin, W. E., and Gerritsen, G. C. (1967). Summary of biochemical, physiological and morphological changes associated with diabetes in Chinese hamster. *Int. Congr. Ser.—Excerpta Med.* **172**, 806–812.

Eckel, H., Reznik, G., Reznik-Schuller, H., and Mohr, U. (1974). Bronchographic studies of the European hamster (*Cricetus cricetus* L), the Syrian golden hamster (*Mesocricetus auratus* W) and the Chinese hamster (Cricetulus griseus M). *Z. Versuchstierkd.* **16**, 322–328.

Elder, F. F. B., and Hsu, T. C. (1981). Silver staining patterns of mammalian epididymal spermatozoa. *Cytogenet. Cell Genet.* **30**, 157–167.

Federman, J. L., and Gerritsen, G. C. (1970). The retinal vasculature of the Chinese hamster: A preliminary study. *Diabetologia* **6**, 186–191.

Feingold, K. R., Lear, S. R., and Moser, A. H. (1984). De novo cholesterol synthesis in three different animal models of diabetes. *Diabetologia* **26**, 234–29.

Feldman, J. M., White-Owen, C., and Klatt, C. (1980). Golden hamster pancreatic islets: A tissue rich in monoamine oxidase. *Endocrinology (Baltimore)* **107**, 1504–1511.

Festing, M. (1970). Breeding Chinese hamsters (*Cricetulus griseus*) in monogamous pairs. *Z. Versuchstierkd.* **12**, 89–90.

Fouquet, J. P. (1971). Secretion of free glucose and related carbohydrates in the male accessory organs of rodents. *Comp. Biochem. Physiol. A* **40A**, 305–317.

Funaki, K. (1981). Active development in preimplantation stages of giant digynic triploids in the Chinese hamster. *Proc. Jpn. Acad., Ser. B* **57**, 18–22.

Funaki, K., and Mikamo, K. (1980). Cytogenetic evidence of the giant diploid oocytes capable of developing into digynic triploids in the Chinese hamster. *Proc. Jpn. Acad., Ser. B* **56**, 134–140.

Geyer, H. (1972). Anatomische Untersuchungen am Harn und Geschlechtsapparat des chinesischen Zwerghamsters. *Z. Versuchstierkd.* **14**, 107–123.

Geyer, H. (1973). Zur topographischen Anatomie der Brusthohle und des Halses beim chinesischen Zwerghamster. *Z. Versuchstierkd.* **15**, 34–49.

Geyer, H., Habermehl, K. H., Wissdorf, H., and Belcic, I. (1972). Die Topographie der Bauchorgane des chinesischen Zwerghamsters. *Z. Versuchstierkd.* **14**, 50–64.

Geyer, H., Bertschinger, H., Strittmatter, J., and Morel, J. (1975). Erbliche Unfruchtbarkeit bei männlichen weissen Chinesenhamstern (*Cricetulus griseus* Milne Edwards 1867). *Z. Versuchstierkd.* **17**, 78–90.

Ghosh, M., Hausler, H. R., Basu, P. K., and Stachowska, B. (1970). An electron microscopic study of retinal capillaries of normal and spontaneously diabetic Chinese hamsters. *Can. J. Ophthalmol.* **5**, 187–193.

Goas, J. A., Pelham, R. W. and Lippa, A. S. (1979). Endocrine factors contributing to the ethanol preferences of rodents. *Pharmacol., Biochem. Behav.* **10**, 557–560.

Gregorek, J. C., Seibel, H. R., and Reiter, R. J. (1977). The pineal complex and its relationship to other epithalamic structures. *Acta Anat.* **99**, 425–434.

Habermehl, K. H. (1971). Gebißentwicklung, Backenzahnabnutzung und Zahnalterschätzung beim chinesischen Zwerghamster. *Schweiz. Arch. Tierheilkd.* **113**, 278–286.

Havu, N., Lundgren, G., and Falkmer, S. (1977). Zinc and manganese contents of microdissected pancreatic islets of some rodents. *Acta Endocrinol. (Copenhagen)* **86**, 570–577.

Helmstaedter, V., Taugner, C., Feurle, G. E., and Forssmann, W. G. (1977). Localization of neurotensin-immunoreactive cells in the small intestine of man and various mammals. *Histochemistry* **53**, 35–41.

Honacki, J. H., Kinman, K. E., and Koeppl, J. W. (1982). "Mammal Species of the World: A Taxonomic and Geographic Reference." Allen Press and Association of Systematics Collections, Lawrence, Kansas.

Horber, P. J., Geyer, H., and Habermehl, K. H. (1974). Topographisch-anatomische und histologische Untersuchungen am Kopf des chinesischen Zwerghamsters mit besonderer Berücksichtigung der Blutentnahme und der Hypophysektomie. *Z. Versuchstierkd.* **16**, 214–238.

Hsieh, E. T. (1919). A new laboratory animal (*Cricetulus griseus*). *Natl. Med. J. China* **5**, 20–24.

Hulten, M., Karlman, A., Lindstea, J., and Tiepolo, L. (1970). Aneuploidy and polyploidy in germ-line cells of the male Chinese hamster (*Cricetulus griseus*). *Hereditas* **65**, 197–202.

Jamil, T., and Fisher, R. A. (1977). An investigation of the homology of guanylate kinase isozymes in mammals and further evidence for multiple GUK gene loci. *Biochem. Genet.* **15**, 847–858.

Kennedy, W. R., Quick, D. C., Miyoshi, T., and Gerritsen, G. C. (1982). Peripheral neurology of the diabetic Chinese hamster. *Diabetologia* **23**, 445–451.

Keyser, A. (1972). The development of the diencephalon of the Chinese hamster. An investigation of the validity of the criteria of subdivision of the brain. *Acta Anat.* **59**, 1–178.

Kit, S., Fiscus, J., Ragland, R. S., Graham, O. L., and Gross, A. L. (1959). Biochemical studies on the Chinese hamster (22 chromosomes) and the Syrian hamster (44 chromosomes). *Exp. Cell Res.* **16**, 411.

Kita, M., Kobayashi, H., Ino, T., and Nakata, K. (1979a). Vaginal smear cycle in 4-day cyclic Chinese hamsters, *Cricetulus griseus*. *Exp. Anim.* **28**, 11–19.

Kita, M., Nishikawa, T., Ino, T., and Yamashita, K. (1979b). Changes in the mucous discharge and the vaginal smear during pregnancy in the Chinese hamster (*Cricetulus griseus*). *Exp. Anim.* **28**, 365–372.

Kovar, J., and Fremuth, F. (1976). Control effects of cyclic nucleotides on the proliferation activity of the bone-marrow cells of the Chinese hamster *in vivo* and their relation to the radiosensitivity. *Acta Univ. Carol., Med.* **22**, 329–366.

Lammers, G. J., Gribnau, A. A., and ten-Donkelaar, H. J. (1980). Neurogenesis in the basal forebrain in the Chinese hamster (*Cricetulus griseus*). II. Site of neuron origin: Morphogenesis of the ventricular ridges. *Anat. Embryol.* **158**, 193–211.

Lawe, J. E. (1962). Renal changes in hamster with hereditary diabetes mellitus. *Arch. Pathol.* **73**, 88–96.

Leipoldt, M., Eckhardt, R., and Schmid, M. (1982). Comparative DNA/DNA reassociation kinetics in three hamster species. *Comp. Biochem. Physiol. B* **72B**, 385–391.

Like, A. A., Gerritsen, G. C., Dulin, W. E., and Gaudreau, P. (1974a). Studies in the diabetic Chinese hamster: Light microscopy and autoradiography of pancreatic islets. *Diabetologia* **10**, 501–508.

Like, A. A., Gerritsen, G. C., Dulin, W. E., and Gaudreau, P. (1974b). Studies in the diabetic Chinese hamster: Electron microscopy of pancreatic islets. *Diabetologia* **10**, 509–520.

Lok, D., and de Rooji, D. G. (1983a). Spermatogonial multiplication in the Chinese hamster. I. Cell cycle properties and synchronization of differentiating spermatogonia. *Cell Tissue Kinet.* **16**, 7–18.

Lok, D., and de Rooji, D. G. (1983b). Spermatogonial multiplication in the Chinese hamster. III. Labelling indices of undifferentiated spermatogonia throughout the cycle of the seminiferous epithelium. *Cell Tissue Kinet.* **16**, 31–40.

Lok, D., Weenk, D., and de Rooij, D. G. (1982). Morphology, proliferation, and differentiation of undifferentiated spermatogonia in the Chinese hamster and the ram. *Ant. Rec.* **203**, 83–99.

Lok, D., Jansen, M. T., and de Rooij, D. G. (1983). Spermatogonial multiplication in the Chinese hamster. II. Cell cycle properties of undifferentiated spermatogonia. *Cell Tissue Kinet.* **16**, 19–29.

Lok, D., Jansen, M. T., and de Rooij, D. G. (1984). Spermatogonial multiplication in the Chinese hamster. IV. Search for long cycling stem cells. *Cell Tissue Kinet.* **17**, 135–143.

Luse, S. A., Caramia, F., Gerritsen, G. C., and Dulin, W. E. (1967). Spontaneous diabetes mellitus in the Chinese hamster: An electron microscopic study of the islets of Langerhans. *Diabetologia* **3**, 97–108.

Luse, S. A., Gerritsen, G. C., and Dulin, W. E. (1970). Cerebral abnormalities in diabetes mellitus: An ultrastructural study of the brain in early onset diabetes mellitus in the Chinese hamster. *Diabetologia* **6**, 192–198.

Matsushima, S. A., and Morisawa, Y. (1981). Quantitative morphological studies on pinealocytes of the Chinese hamster, *Cricetulus griseus*. *Jikeikai Med. J.* **28**, 29–34.

Matsushima, S. A., and Morisawa, Y. (1982). Ultrastructural observations on the pineal gland of the Chinese hamster, *Cricetulus griseus*. 1. The superficial pineal. *Cell Tissue Res.* **222**, 531–546.

Matsushima, S. A., Morisawa, Y., Aida, I., and Abe, K. (1983a). Circadian variations in pinealocytes of the Chinese hamster, *Cricetulus griseus*. A quantitative electron microscopic study. *Cell Tissue Res.* **228**, 231–244.

Matsushima, S. A., Saki, Y., and Aida, I. (1983b). Effects of melatonin on synaptic ribbons in pinealocytes of the Chinese hamster, *Cricetulus griseus*. A quantitative electron-microscopic study. *Cell Tissue Res.* **233**, 59–67.

Maxwell, W. L. (1982). The acrosomal zonule. *Tissue Cell* **14**, 283–288.

Meier, H., and Yerganian, G. (1959). Spontaneous diabetes mellitus in the Chinese hamster (*Cricetulus griseus*). I. Pathological findings. *Proc. Soc. Exp. Biol. Med.* **100**, 810–815.

Milman, G., Anton, D. L., and Weber, J. L. (1976). Chinese hamster purine-nucleotide phosphorylase: Purification, structural, and catalytic properties. *Biochemistry* **15**, 4967–4973.

Moore, W., Jr. (1965). Observations on the breeding and care of the Chinese hamster, *Cricetulus griseus*. *Lab. Anim. Care* **15**, 94–101.

Moore, W., Jr. (1966). Hemogram of the Chinese hamster. *Am. J. Vet. Res.* **27**, 608–610.

Nichols, E. A., Elsevier, S. M., and Ruddle, F. H. (1974). A new electrophoretic technique for mouse, human, and Chinese hamster galactokinase. *Cytogenet. Cell Genet.* **13**, 275–278.

Nowak, R. M., and Paradiso, J. L. (1983). "Walker's Mammals of The World." Johns Hopkins Univ. Press, Baltimore, Maryland.

Olsen, A. S., and Milman, G. (1978). Hypoxanthine phosphoribosyltransferase from Chinese hamster brain and human erythrocytes. *In* "Methods in Enzymology" (P. A. Hoffee and M. E. Jones, eds.), Vol. 51, pp. 543–549. Academic Press, New York.

Oud, J. L., and de Rooij, D. G. (1977). Spermatogenesis in the Chinese hamster. *Anat. Rec.* **187**, 113–123.

Parkening, T. A. (1982). Reproductive senescence in the Chinese hamster (*Cricetulus griseus*). *J. Gerontol.* **37**, 283–287.

Parkening, T. A., Collins, T. J., and Smith, E. R. (1980). A comparative study of prolactin levels in five species of aged female laboratory rodents. *Biol. Reprod.* **22**, 513–518.

Parkening, T. A., Calcote, R. D., and Collins, T. J. (1981). Plasma and pituitary concentrations of LH, FSH, and prolactin in reproductively senescent Chinese hamsters during various stages of the estrous cycle. *Biol. Reprod.* **25**, 825–831.

Parkening, T. A., Brouhard, B. H., and LaGrone, L. F. (1982). Plasma concentrations of $PGF_{2\alpha}$ in aging mice and Chinese hamsters. *IRCS Med. Sci.: Libr. Compend.* **10**, 151–152.

Parkening, T. A., Collins, T. J., and Smith, E. R. (1983). Measurement of plasma LH concentrations in aged male rodents by a radioimmunoassay and a radioreceptor assay. *J. Reprod. Fertil.* **69**, 717–722.

Payer, A. F., and Parkening, T. A. (1983a). Two types of paracrystalline inclusions in the Leydig cell of the Chinese hamster (*Cricetulus griseus*). *J. Ultrastruct. Res.* **83**, 161–167.

Payer, A. F., and Parkening. T. A. (1983b). Membrane-bound intranuclear inclusions in the Leydig cell of the Chinese hamster (*Cricetulus griseus*). *J. Ultrastruct. Res.* **84**, 317–325.

Petersson, B., Elde, R., Efendic, S., Hokfelt, T., Johansson, O., Luft, R., Cerasi, E., and Hellerström, C. (1977). Somatostatin in the pancreas, stomach and hypothalamus of the diabetic Chinese hamster. *Diabetologia* **13**, 463–466.

Phillips, D. M. (1972a). Comparative analysis of mammalian sperm motility. *J. Cell Biol.* **53**, 561–573.

Phillips, D. M. (1972b). Substructure of the mammalian acrosome. *J. Ultrastruct. Res.* **38**, 591–604.

Phillips, D. M. (1975). Cell surface structure of rodent sperm heads. *J. Exp. Zool.* **191**, 1–8.

Phillips, D. M. (1977). Mitochondrial disposition in mammalian spermatozoa. *J. Ultrastruct. Res.* **58**, 144–154.

Pickworth, S., Yerganian, G., and Chang, M. C. (1968). Fertilization and early development in the Chinese hamster, *Cricetulus griseus*. *Anta. Rec.* **162**, 197–208.

Porter, G., and Lacey, A. (1969). Breeding the Chinese hamster (*Cricetulus griseus*) in monogamous pairs. *Lab. Anim.* **3**, 65–68.

Rabinovitch, A., Renold, A. E., and Cerasi, E. (1976). Decreased cyclic AMP and insulin responses to glucose in pancreatic islets of diabetic Chinese hamsters. *Diabetologia* **12**, 581–587.

Rake, A. V. (1974). DNA reassociation kinetics of closely related species. *Biochem. Genet.* **11**, 261–277.

Rattner, J. B., and Olson, G. (1973). Observations on the fine structure of the nuclear ring of the mammalian spermatid. *J. Ultrastruct. Res.* **43**, 438–444.

Reznik, G., Reznik-Schueller, H., and Mohr, U. (1973). Comparative studies of organs in the European hamster (*Cricetus cricetus* L.) the Syrian golden hamster (*Mesocricetus auratus* W.) and the Chinese hamster (Cricetulus griseus M). *Z. Versuchstierkd.* **15**, 272–282.

Sato, T., Komeda, K., and Shirama, K. (1984). Plasma progesterone concentrations during the estrous cycle, pregnancy, and lactation in the Chinese hamster, *Circetulus griseus*. *Exp. Anim.* **33**, 501–508.

Schlaepfer, W. W., Gerritsen, G. C., and Dulin, W. E. (1974). Segmental demyelination in the distal peripheral nerves of chronically diabetic Chinese hamsters. *Diabetologia* **10**, 541–548.

Schwentker, V. (1957). The Chinese hamster. *In* "The U.F.A.W. Handbook on the Care and Management of Laboratory Animals" (A. N. Worden, ed.), 2nd ed. Williams & Wilkins, Baltimore, Maryland.

Shirai, T., Welsh, G. W., and Sims, E. A. H. (1967). Diabetes mellitus in the Chinese hamster. II. The evolution of renal glomerulopathy. *Diabetologia* **3**, 266–286.

Siegel, E. G., Wollheim, C. B., Sharp, G. W. G., Herberg, L., and Renold, A. E. (1979). Defective calcium handling and insulin release in islets from diabetic Chinese hamsters. *Biochem. J.* **180**, 233–236.

Silberberg, R., and Gerritsen, G. (1976). Aging changes in intervertebral discs and spondylosis in Chinese hamsters. *Diabetes* **25**, 477–483.

Silberberg, R., Gerritsen, G., and Hasler, M. (1976). Articular cartilage of diabetic Chinese hamsters. *Arch. Pathol. Lab. Med.* **100**, 50–54.

Sirek, O. V., and Sirek, A. (1967). The colony of Chinese hamsters of the C. H. Best Institute. A review of experimental work. *Diabetologia* **3**, 65–73.

Skirrow, M. H., and Rysan, M. (1976). Observations on the social behavior of the Chinese hamster, *Cricetulus griseus*. *Can. J. Zool.* **54**, 361–368.

Smith, C. (1957). The introduction and breeding of the Chinese striped hamster (*Cricetulus griseus*) in Great Britain. *J. Anim. Tech. Assoc.* **7**, 59–60.

Sonta, S., Fukui, K., and Yamamura, H. (1984). Selective elimination of chromosomally unbalanced zygotes at the two-cell stage in the Chinese hamster. *Cytogenet. Cell Genet.* **38**, 5–13.

Soret, M. G., Dulin, W. E., Mathews, J., and Gerritsen, G. C. (1974). Morphologic abnormalities observed in retina, pancreas and kidney of diabetic Chinese hamsters. *Diabetologia* **10**, 567–579.

Soret, M. G., Peterson, T., and Block, E. M. (1976a). Electron microscopy of the aorta in young and adult Chinese hamsters. *Artery (Fulton, Mich.)* **2**, 109–128.

Soret, M. G., Peterson, T., and Block, E. M. (1976b). Ultrastructural study of cytoplasmic bridges through the membrana elastica interna in the aorta of the Chinese hamster, *Cricetulus griseus*. *Artery (Fulton, Mich.)* **2**, 451–466.

Stolba, A. (1978). Sozialsystem adulter chinesischer Zwerghamster (*Cricetulus griseus*). *Z. Tierpsychol.* **45**, 389–413.

Stolba, A., and Kummer, H. (1972). Zur Verhaltensontogenese des chinesischen Zwerghamsters. *Cricetulus griseus* (Milne-Edwards, 1867). *Rev. Suisse Zool.* **79**, 89–101.

Sugawara, S., and Mikamo, K. (1983). Absence of correlation between univalent formation and meiotic nondisjunction in aged female Chinese hamsters. *Cytogenet. Cell Genet.* **35**, 34–40.

Sun, N. C., Chang, C. C., and Chu, E. H. Y. (1975). A rapid electrophoretic technique for human and Chinese hamster galactokinase. *Humangenetik* **29**, 351–353.

Suzuki, S., Nishinakagawa, H., and Otsuka, J. (1981). Fine structure of the parotid gland of Chinese hamster, *Cricetulus griseus*. *Exp. Anim.* **30**, 241–250.

Suzuki, S., Nishinakagawa, H., and Otsuka, J. (1982). Fine structure of the mandibular gland of Chinese hamster, *Cricetulus griseus*. *Exp. Anim.* **31**, 97–106.

Swallow, D. M., Turner, V. S., and Hopkinson, D. A. (1983). Isozymes of rodent 5'-nucleotidase: Evidence for two independent structural loci *Umph-1* and *Umph-2*. *Ann. Hum. Genet.* **47**, 9–17.

Takeuchi, I. K., and Sonta, S. I. (1983). Electron microscopic study on glycogen particles in ovarian oocytes of the Chinese hamster. *Annot. Zool. Jpn.* **56**, 167–173.

Talbot, B. G., and Thirion, J. P. (1982). Isolation, purification and partial characterization of galactokinase from Chinese hamster liver. *Int. Biochem.* **14**, 719–726.

Talbot, B. G., Qureshi, A. A., Cohen, R., and Thirion, J. P. (1981). Purification and properties of two distinct groups of ADH isozymes from Chinese hamster liver. *Biochem. Genet.* **19**, 813–829.

Taugner, R., Buhrle, C. P., Ganten, D., Hackenthal, E., Hardegg, C., Hardegg, G., Nobiling, R., and Unger, T. (1983). Angiotensin-like activity in resistance vessels. Immunocytochemical study in Chinese hamsters. *Histochemistry* **78**, 61–70.

Taylor, D. M., Lehmann, M., Planas-Bohne, F., and Seidel, A. (1983). The metabolism of radiohafnium in rats and hamsters: A possible analog of plutonium for metabolic studies. *Radiat. Res.* **95**, 339–358.

ten-Donkelaar, H. J., and Dederen, P. J. (1979). Neurogenesis in the basal forebrain of the Chinese hamster (*Cricetulus griseus*). I. Time of neuron origin. *Anat. Embryol.* **156**, 331–348.

ten-Donkelaar, H. J., Geysbarts, L. G. M., and Dederen, P. J. W. (1979a). Stages in the prenatal development of the Chinese hamster (*Cricetulus griseus*). *Anat. Embryol.* **156**, 1–28.

ten-Donkelaar, H. J., Lammers, G. J., and Gribnau, A. A. (1979b). Neurogenesis in the amygaloid nuclear complex in a rodent (the chinese hamster). *Brain Res.* **165**, 348–353.

Thompson, R. (1971). The water consumption and drinking habits of a few species and strains of laboratory animals. *J. Inst. Anim. Technicians* **22**, 29–36.

Tiemeier, D. C., and Milman, G. (1972). Chinese hamster liver glutamine synthetase. *J. Biol. Chem.* **247**, 2272–2277.

Trojan, M. (1979). Vergleichende Untersuchungen über den Wasserhaushalt und die Nierenfunktion der palaarktischen hamster *Cricetus cricetus* (Leske, 1779), *Mesocricetus auratus* (Waterhouse, 1839), *Cricetulus griseus* (Milne-Edwards, 1867) and *Phodopus sungorus* (Pallas, 1770). *Zool. J. Physiol.* **83**, 192–223.

von-Dorsche, H. H., Hartlelt, E., Fehrmann, P., and Krause, R. (1977). Occurrence of pseudoisocyanine-positive endocrine cells in the midgut of the hamster (*Cricetulus griseus*). *Acta Histochem.* **59**, 168–171.

Warshawsky, H., and Nanci, A. (1982). Stereo electron microscopy of enamel crystallites. *J. Dent. Res.* **61**, 1504–1514.

Yanagimachi, R., Kamiguchi, Y., Sugawara, S., and Mikamo, K. (1983). Gametes and fertilization in the Chinese hamster. *Gamete Res.* **8**, 97–117.

Yerganian, G. (1958). The striped-back or Chinese hamster (*Cricetulus griseus*). *J. Natl. Cancer Inst. (U.S.)* **20**, 705–727.

Yerganian, G. (1972). History and cytogenetics of hamsters. *Prog. Exp. Tumor Res.* **16**, 2–41.

Yokoyama, M., and Chang, J. P. (1971). An ultracytochemical and ultrastructural study of epithelial cells in ductuli efferentes of Chinese hamster. *J. Histochem. Cytochem.* **19**, 766–774.

Yokoyama, M., and Chang, J. P. (1977). Cytochemical study of glucose-6-phosphatase in Chinese hamster testis. *Biol. Reprod.* **17**, 265–268.

Chapter 18

Diseases

Warren C. Ladiges

    I. Infectious Diseases 321
        A. Introduction 321
        B. Viral Diseases 322
        C. Bacterial Diseases 322
        D. Parasitic Diseases 323
   II. Neoplastic Diseases 324
        A. Hepatocellular Hyperplasia and Adenoma 324
        B. Uterine Tumors 324
        C. Myeloproliferative Disease and Leukemia 325
        D. Other Neoplasms 325
  III. Metabolic and Genetic Conditions 325
        A. Spontaneous Diabetes Mellitus 325
        B. Primary Chromosomal Anomalies 325
  IV. Traumatic Disorders 326
   V. Reproductive Disorders 326
        A. Infertility 326
        B. Dystocia 326
        C. Uterine Fistula 326
        D. Testicular Degeneration and Atrophy 327
  VI. Miscellaneous Diseases 327
        A. Cerebral Hemorrhage 327
        B. Degenerative Renal Disease 327
        C. Pulmonary Granulomas 327
        D. Periodontitis 328
        E. Spondylosis 328
        References 328

I. INFECTIOUS DISEASES

A. Introduction

Very little has been reported concerning spontaneously occurring infectious diseases in Chinese hamsters, and they apparently have relatively few spontaneous diseases as compared to other laboratory rodents. They are, however, susceptible to a large number of experimental infections and have been used extensively in viral and parasitological disease investigations. Table I is a representative list of infectious disease agents which have been experimentally studied in the Chinese hamster.

Table I

Susceptibility of the Chinese Hamster to Representative Experimental Infectious Diseases

Disease agent	Comments	Reference
Scrapie virus	Short incubation period	Chandler and Turfrey (1972)
Rabies virus		Yen (1936)
Influenza virus		Yen (1940)
Viral encephalitis virus		Chang et al. (1951)
Rous sarcoma virus	Retrovirus transformation of CHI fibroblasts	Hillova et al. (1985)
Pneumococcal bacteria	Pneumonia	Hsieh (1919)
Corynebacterium (C. diphtheriae)	Diphtheria	Lu and Zia (1935)
Mycobacterium (M. tuberculosis)	Tuberculosis	Wang (1951)
Leptospirosae	Seven of nine strains lethal	Plesko (1977)
Toxoplasm (T. gondii)	Highly sensitive	Fujii et al. (1983)
Leishmania	Leishmaniasis	Hindle et al. (1926)
Babesia (B. microti)		Krampitz and Baumler (1978)
Monila	Monoliasis	Kurotchkin and Lin (1930)
Schistosoma (S. mansoni)	Cheek pouch	Crabtree and Wilson (1984)
Spirometra (S. erinacei)	Growth-promoting effect	Hirai et al. (1983)
Hymenolepis (H. microstoma)	Cortisone dependent	Ritterson (1971)
Trichinella (T. spiralis)	Methotrexate increases susceptibility	Ritterson (1968)

B. Viral Diseases

1. Antibody Prevalence Studies

Schiff *et al.* (1973) evaluated 35 Chinese hamsters, 2, 3, and 4 months of age, from a colony of 200, and found evidence of hemagglutination inhibition (HI) titers to PVM, Kilham's rat virus, GDVII, Toolan's H-1, and reovirus type 3. None of the 35 animals examined in the three age groups had evidence of HI titers to Sendai virus or polyomavirus. Using the complement fixation (CF) test, no animals were found to have detectable antibody levels to mouse adenovirus, mouse hepatitis virus, or lymphocytic choriomeningitis virus (LCM).

Bowen *et al.* (1975) found a positive CF titer to LCM in 1 of 25 Chinese hamsters exposed to infected golden Syrian hamsters. However, no virus isolations were made from autopsy material (pooled liver, spleen, kidney, lung, brain) or serum blood clots in 15 of the animals.

2. Endogenous Chinese Hamster Virus

Tihon and Hellman (1976) observed type C viral particles in electron micrographs in the CHO-K1 line of Chinese hamster ovary cells. They found that the production of the virus was greatly enhanced by treating the cells with dibutyryl-cAMP plus testosterone propionate, as well as with bromodeoxyuridine. The extracellular viral particles, which sedimented at a density of approximately 1.16 gm/ml, contained 70S RNA and RNA-dependent DNA polymerase. The enzyme was found to differ antigenically from that described for Raucher leukemia virus (RLV). Using immunodiffusion analysis and a radioimmunoassay, they found some cross-reactivity between Chinese hamster viral proteins and RLV, but not with type C primate or feline viruses.

C. Bacterial Diseases

1. Intestinal Microflora

Schiff *et al.* (1973) evaluated 30 Chinese hamsters aged 2, 3, and 4 months from a group of 200 for intestinal bacteria. Quantitative recovery of major groups is shown (Table II). There were high numbers of lactobacilli with $1.3-1.4 \times 10^9$ viable bacteria per gram of cecal material. The low obligate anaerobe count at 2 and 4 months of age may have been due to overgrowth by other organisms. Bacterial genera isolated from the ceca of the same animals are shown in Table III.

2. *Mycoplasma*

a. History. Hill (1983) described a new *Mycoplasma* species which he isolated from the conjunctiva and nasopharynx of Chinese hamsters from one colony.

Table II

Quantitative Recovery of Major Groups of Bacteria from the Intestinal Tract of Chinese Hamsters[a]

Bacterial groups isolated	Age (months)		
	2	3	4
Total aerobes	5×10^{6b}	1.3×10^5	3.5×10^5
Obligate anaerobes	$<10^6$	1.7×10^9	$<10^6$
Enterics	9.5×10^4	7.6×10^2	2.5×10^5
Micrococcaeae	1.8×10^6	1.3×10^4	9.7×10^4
Lactobacillus	1.3×10^9	1.3×10^9	1.4×10^9
Streptococceae	2×10^6	4.2×10^4	1.4×10^4

[a] From Schiff *et al.* (1973); reprinted with permission *from Laboratory Animal Science.*
[b] Mean recovery expressed as log numbers of viable bacteria per gram of cecal material.

Table III

Frequency of Bacterial Genera Isolation from Cecal Contents of Chinese Hamsters[a]

Organisms isolated	Age (months)		
	2	3	4
Bacillus	10/10[b]	9/10	10/10
Bacterioides	10/10	10/10	10/10
Gaffkya	7/10	8/10	9/10
Escherichia	10/10	10/10	10/10
Lactobacillus	10/10	10/10	10/10
Micrococcus	10/10	10/10	10/10
Proteus	4/10	2/10	2/10
Pseudomonas	7/10	0/10	0/10
Staphylococcus	1/10	0/10	8/10
Streptococcus	9/10	10/10	10/10

[a]From Schiff *et al.* (1973); reprinted with permission from *Laboratory Animal Science.*

[b]Number of animals with positive culture/number of animals examined.

b. Etiology. The new *Mycoplasma* strain was named *M. cricetula.* It was serologically distinct from 67 recognized *Mycoplasma* species and designated as strain CH (WCTC10190). Growth characteristics were similar to other species. Sheep erythrocytes were lysed but not absorbed, and electrophoretic protein patterns were distinct from other rodent mycoplasms. Little is yet known about the presence of the organism in other animals. Hill (1974) was unable to infect gerbils or lemmings experimentally by inoculation into the nasopharynx or conjunctiva.

c. Epizootiology. The organism was isolated from the conjunctiva of all 55 animals examined from one colony. Mycoplasma was found in the nasopharynx of 47 of the animals, and 3 had positive lung cultures. No isolates were recovered from eight vaginal swabs. Four animals from an unrelated colony were found to have serologically identical *Mycoplasma* species from the conjunctiva. Prevalence and mode of transmission are not known and will require further investigation.

d. Clinical Signs and Pathology. The organisms appear to have a preference for infecting the conjunctiva. There is no evidence that it is associated with clinical disease, but only a small number of animals have been evaluated thus far.

e. Significance. Further studies are needed to determine the pathogenic significance of this organism for the Chinese hamster.

3. Tyzzer's Disease

a. Case History. Zook *et al.* (1977) diagnosed Tyzzer's disease in 10 of 30 female Chinese hamsters 30–60 days of age. They were purchased from a vendor and placed in a quarantine room. One of 18 animals which was removed on the fourth day and experimentally killed 5 days later showed pathological lesions. Four hamsters were found dead in the quarantine room on the seventh day and two more died on day 9, with an additional sick animal being killed. Two more animals died within the next several days.

b. Clinical Signs and Pathology. Affected animals were lethargic, had rough hair coats, evidence of diarrhea, and were in a humpback posture. Gross necropsy lesions consisted of multiple disseminated ill-defined white foci in the liver and distention of the large intestine with semiliquid yellow feces. Microscopically, there was focal coagulation necrosis of hepatic cells with no significant cellular inflammatory response. Giemsa, toluidine blue, and methenamine silver stains revealed long pleomorphic organisms grouped in interlacing bundles, starburst clusters, and parallel formations within the cytoplasm of hepatic cells adjacent to necrotic foci. No lesions or bacilli were observed in other representative tissues examined.

c. Diagnosis. Histopathological lesions were consistent with a diagnosis of Tyzzer's disease caused by the organism *Bacillus piliformis.* It was suspected that stress may have been involved in the disease outbreak in these recently shipped animals, especially since some of the animals may have had difficulty in reaching the water bottle sipper tubes. The disease needs to be differentiated from septicemic or other systemic disease conditions. Diagnosis is easily made by observing typical organisms in liver sections stained by the Giemsa, methenamine–silver, or toluidine blue techniques.

d. Control and Prevention. Efforts should be made to decrease stressful conditions whenever possible. It is important that all animals have a regular source of clean water and adequate food supply. Affected animals should be eliminated.

e. Significance. Affected animals could invalidate research by producing erroneous data or because of termination of an experiment.

D. Parasitic Diseases

Very little has been reported concerning spontaneous parasitic diseases in the Chinese hamster. Schiff *et al.* (1973) consistently found *Trichomonas* sp. in 2-, 3-, and 4-month-old animals from a group of 200. Benjamin and Brooks (1977) found histopathological evidence of demodectic mange in 2 of 157 animals. Krampitz and Baumler (1978) were able to infect 5 of 5 Chinese hamsters with the blood parasite *Babesia microti.* They used intraperitoneal injections of parasitized blood obtained from the common field vole, which is one of several

Bavarian rodent natural hosts for the organism. These findings have significance because of potential cross-contamination between infected voles and Chinese hamsters in the same animal facility.

II. NEOPLASTIC DISEASES

A. Hepatocellular Hyperplasia and Adenoma

1. Epizootiology

Kohn and Guttman (1964) found hepatomas in 5 of 112 nonirradiated control animals, but did not describe any nonneoplastic lesions of the liver. Ward and Moore (1969) reported hepatomas in 18 of 50 nonirradiated control animals and also found a similar incidence in animals receiving ^{131}I. They also described hepatocellular hyperplasia in some of the livers and suggested they might be preneoplastic lesions. Benjamin and Brooks (1977) described hepatic nodular hyperplasia in 111 of 157 aging animals but found evidence of adenoma in only 1 of 157. Brooks et al. (1983), in a later ^{239}Pu irradiation study, described hepatocellular malignancy in 2 of 93 nonirradiated control animals, while more than 90% had some degree of hyperplastic change. Hepatocellular hyperplasia and adenoma are discussed together because of their common origin, although they appear to be separate, albeit possibly related, entities.

2. Pathology

Grossly the two conditions were very similar, often appearing as multiple nodules from less than 1 mm to more than 1 cm in diameter. They were usually lighter in color than the surrounding tissue. Histologically, Benjamin and Brooks (1977) described the nodular hyperplastic lesions as varying from small nodules with normal structural patterns of sinusoids and central vein to coalescing masses with disruption of architectural formations. Parenchymal cells were usually enlarged due to an increase of eosinophilic and sometimes vacuolated cytoplasm. Nuclei were enlarged but retained smooth, benign appearances. There was compression of surrounding parenchyma by all nodules.

Benjamin and Brooks (1977) found evidence of hepatocellular adenoma in only 1 of 157 animals examined. They distinguished the condition from nodular hyperplasia by the histological appearance of enlarged hepatic cells with abundant hyperchromatic nuclei containing multiple nucleoli. Ward and Moore (1969) also distinguished between nodular hyperplasia and hepatomas but found a much higher incidence of hepatomas. They described the hepatoma histology as varying from small nodules of well-differentiated hepatic cells with a loss of normal lobular orientation and compressing surrounding normal parenchyma to large growth of anaplastic cells resembling liver cells with large round or oval nuclei, prominent nucleoli, and abundant eosinophilic cytoplasm. The tumor cells often grew in sheets, with degenerative changes often seen centrally. Mitotic figures were rarely seen and no evidence of metastasis was observed. They also saw hepatocellular hyperplasia in some of the livers and suggested they might be preneoplastic lesions.

3. Significance

Since the Chinese hamster is used in toxicological research, these spontaneous lesions must be distinguished from experimentally induced lesions.

B. Uterine Tumors

1. Epizootiology

Ward and Moore (1969) found adenocarcinomas in 30 of 120 females examined, while Benjamin and Brooks (1977) described malignant uterine neoplasms in 13 of 77 female hamsters. Brownstein and Brooks (1980) then characterized the incidence and classification of endomyometrial neoplasms in 21 of 93 multiparous noninbred animals 3 to 4 years of age.

2. Clinical Signs and Pathology

Animals with malignant uterine neoplasms were frequently observed with vaginal hemorrhage. Gross necropsy examination often showed hemorrhage into the peritoneal cavity, with the tumor spreading throughout. The tumors directly involved other organs by local invasion and often formed nodular peritoneal implants with involvement of local abdominal lymph nodes. The tumor tissue was usually white to yellow with a hard nodular surface.

A histological classification of endomyometrial neoplasms (Table IV) was formulated by Brownstein and Brooks (1980). The most commonly occurring type was adenocarcinoma of the endometrium, ranging from simple acinar to papillary growths. They were usually well differentiated, but the more anaplastic types consisted of polymorphic cytoplasm and nuclei, frequent mitotic figures, and prominent fibrous stromal proliferation. Local vascular invasion was usually observed. Extraperitoneal metastasis to the lungs occurred in only one animal with a mixed endometrial carcinosarcoma.

3. Significance

Spontaneous endomyometrial cancers of Chinese hamsters are useful models for studying endomyometrial neoplasms of

18. DISEASES

Table IV

Classification of Endomyometrial Neoplasms in 21 Chinese Hamsters[a]

Site	Type	Occurrence[b]	Metastasis
Endometrium	Adenocarcinoma	13	Abdominal spread (10)
	Mixed adeno-squamous carcinoma	3	Abdominal spread (3)
	Mixed müllerian carcinosarcoma	2	Abdominal spread (1), lung metastasis (1)
	Mixed mesodermal	1	Abdominal spread
Myometrium	Leiomyoma	1	
	Leiomyosarcoma	1	Abdominal spread

[a] From Brownstein and Brooks (1980); reprinted with permission from the *Journal of the National Cancer Institute*.
[b] From a group of 93 nulliparous noninbred females with a median survival time of 1040 days.

women, since the classification and relative frequency are similar.

C. Myeloproliferative Disease and Leukemia

1. Epizootiology

Benjamin and Brooks (1977) found evidence of hematopoietic hyperplasia and leukemia in bone marrow and spleen in 49 of 157 Chinese hamsters examined. Bone marrow hyperplasia, usually of the granulocytic type, occurred in 36 animals and in about half was associated with degenerative or inflammatory changes in some other organ system, and especially with other neoplastic conditions. In the other half no association could be made with any obvious underlying lesions in other tissues. Of the 49, 2 had lesions consistent with myelofibrosis, while an additional 3 were diagnosed as having myelogenous leukemia.

2. Pathology

Splenomegaly and pale bone marrow were usually seen grossly in all of the conditions. Microscopically, the marrow was hypercellular with replacement of erythroid cells by granulocytic cells. A diagnosis of leukemia was made when cells consisted mainly of myelocytes and myeloblasts, with leukemic infiltrates sometimes found in the liver, kidney, brain, and spinal cord, and routine involvement of the spleen.

3. Significance

The leukemic condition in Chinese hamsters could potentially be useful for studying myelogenous leukemia, especially if further studies of other colonies confirm the occurrence of the disease. Adequate care must be taken to distinguish leukemoid reactions from malignancy.

D. Other Neoplasms

Table V lists other reports of malignancies in the Chinese hamster.

III. METABOLIC AND GENETIC CONDITIONS

A. Spontaneous Diabetes Mellitus

Spontaneous diabetes mellitus in Chinese hamsters will be discussed in detail in Chapter 19, since it has been found to be an adequate model for studying several of the aspects of the human disease. It should be emphasized that since it is a spontaneous disease, it could occur in animals being used in research protocols unrelated to diabetes. It would be important to define this condition especially in cytogenetic studies, where the abnormal cellular metabolism could produce adverse, unreliable data.

B. Primary Chromosomal Anomalies

Mikamo and Kamiguchi (1983) studied the primary incidence of spontaneous chromosomal aberrations in oocytes and

Table V

Infrequently Occurring Spontaneous Malignancies Reported in the Chinese Hamster

Tumor type	Occurrence	Reference
Adrenal cortical carcinoma	1/157	Benjamin and Brooks (1977)
	4/112	Kohn and Guttman (1964)
Pancreatic adenocarcinoma	2/3	Poel and Yerganian (1961)
Sweat gland adenocarcinoma	1/157	Benjamin and Brooks (1977)
Squamous-cell carcinoma	1/157	Benjamin and Brooks (1977)
Subcutaneous fibrosarcoma	1/157	Benjamin and Brooks (1977)
Undifferentiated sarcoma	1/157	Benjamin and Brooks (1977)
Gastric adenocarcinoma	1/112	Kohn and Guttman (1964)
Bile duct carcinoma	1/112	Kohn and Guttman (1964)
Ovarian adenoma[a]	1/77	Benjamin and Brooks (1977)

[a] Of interest is the report by Kohn and Guttman (1964) of the occurrence of ovarian tumors in 18 of 74 irradiated females, but none in 69 control animals.

Table VI

Primary Incidences of Spontaneous Chromosomal Anomalies, Classified According to their Origins and Causal Mechanisms in the Chinese Hamster[a]

Chromosomal anomaly	Origin and causal mechanism	Incidence (%)
Aneuploidy	Nondisjunction and anaphase lagging	
	During oogenesis	2.1
	During spermatogenesis	0.7
Mosaic	Nondisjunction and anaphase lagging	
	During first somatic division	0.5
Triploidy	Diandry	0.9
	Digyny	0.3
	Uncertain	0.3
Haploidy	Parthenogenesis	0.5
Structural anomaly	Breakage during oogenesis	1.3
	Breakage during spermatogenesis	1.4
Total		8

[a]From Mikamo and Kamiguchi (1983); reprinted with permission from Elsevier Biomedical Press.

early zygotes of Chinese hamsters and classified the lesions according to their causal mechanisms (Table VI). The overall incidence of chromosomal anomalies was 8%. Abnormal chromosomal segregation constituted 3.3% of the conditions, while ploidy anomalies accounted for 2.0%. Structural defects occurring during oogenesis or spermatogenesis made up the remaining 2.7% chromosomal anomalies. It was concluded that the 8% incidence was much lower than that estimated for the human population, possibly due to the lack of influence of mutagens in a controlled environment. It is important to appreciate the spontaneous chromosomal defects in the Chinese hamster, since this species is used extensively as an animal model for cytogenetic studies.

IV. TRAUMATIC DISORDERS

Female littermates often become quite aggressive as they reach maturity. Severe bite wounds can be inflicted, especially in the neck and tail areas, and death is a frequent consequence. In addition, females can become quite aggressive following mating, inflicting bite wounds which often result in death of the male. The obvious prevention is to separate litters before fighting occurs, and to provide some means of removing the male following mating. Some groups have selectively bred for nonaggressive females in an attempt to reduce this problem (Yerganian, 1958).

V. REPRODUCTIVE DISORDERS

A. Infertility

1. Etiology

Infertility has been reported in young females in open-bottom cages (Avery, 1968). The cause was thought to be an excess growth of hair around the vulva, preventing penile penetration during copulation attempts. Infertility has been associated with diabetic females, especially inbred animals. Aging has also been studied as a cause of poor reproductive performance (Parkening, 1982). Aged female Chinese hamsters, 17–20 months old, have decreased ovulation rates compared to 5- to 6-month-old animals.

2. Clinical Signs

Animals fail to conceive, or show decreased conception rates. Older animals will show relatively normal estrous cycles but fail to ovulate either spontaneously or after mating. They may conceive and deliver live litters but with significantly decreased litter size.

3. Treatment

A change from an open-bottom to a solid-bottom caging system is indicated if young disease-free females are not conceiving. Progesterone has been used to increase conception, both in inbred and/or diabetic breeders and aged animals, immediately following mating, with some success (Avery, 1968; Parkening, 1982).

4. Significance

The Chinese hamster female may be a good rodent model to study the effects of aging on the reproductive system, since, unlike the aging rat or mouse, it tends to maintain a relatively normal estrous cycle until late in life.

B. Dystocia

Dystocia may occur as a result of fetal wedging in the proximal portion of the vagina during parturition attempts (Yerganian, 1958). Surgical removal of the fetuses can be performed in order to save valuable animals.

C. Uterine Fistula

Two females from a breeding colony were diagnosed as having a uterine fistula (van der Gulden, 1981). They were initially observed with an enlarged abdomen and purulent drainage from the flank area. Necropsy showed a purulent exudate and a retained fetus in the uterine horn.

D. Testicular Degeneration and Atrophy

Benjamin and Brooks (1977) reported 14 of 80 males with testicular degeneration and atrophy in an aging colony of outbred hamsters.

VI. MISCELLANEOUS DISEASES

A. Cerebral Hemorrhage

1. Epizootiology

Cerebral hemorrhage was found in 20% of both control and experimental animals in a ^{131}I chronic toxicity study (Ward and Moore, 1969). Deaths usually occurred sporadically when animals were 1–2 years old. The cause could not be specifically ascertained but was considered to be a degenerative or inflammatory process of the anterior cerebral artery.

2. Clinical Signs and Pathology

Deaths were sporadic over a span of 1–2 years. Grossly, the hemorrhage was most evident between the cerebral hemispheres, with blood often in the lateral ventricles. In some aged hamsters, a small brownish depression was observed on the medial aspect of the cerebral hemispheres just caudal to the superior cerebral vein. When cut transversely, this lesion was a well-circumscribed 1- to 2-cm mass.

Histologically, the lesions were found to be restricted to the anterior cerebral artery, which in the Chinese hamster branches from the internal carotid artery and unites with the corresponding artery from the other side. The lesions occurred where the combined trunk passes caudad and upward between the cerebral hemispheres. The vascular changes in acute cases included an increase in diameter of the lumen and thickening of the arterial walls with occasional inflammatory cells. An amorphous eosinophilic material, which was periodic acid–Schiff (PAS) positive but fibrin negative, was usually observed within the media. In chronic cases the arterial wall was greatly thickened, and numerous macrophages were observed surrounding the lesions. No vascular changes as described were found in other sites of the brain or within other body organs.

3. Significance

The incidence of cerebral hemorrhage in Chinese hamsters is not known, since it has not been reported in other colonies.

Studies which include careful evaluation of the Chinese hamster brain may help to determine the animal model potential.

B. Degenerative Renal Disease

1. Intercapillary Glomerulosclerosis

Intercapillary glomerulosclerosis has been described in Chinese hamsters in association with diabetes mellitus (Meier and Yerganian, 1961). It has also been reported as a progressive condition associated with aging (Guttman and Kohn, 1960; Kohn and Guttman, 1964). Nephropathology associated with diabetes mellitus is discussed further in Chapter 19.

2. Nephrosclerosis

a. Epizootiology. Benjamin and Brooks (1977) described nephrosclerosis in 46 of 157 animals which were older than 600 days.

b. Pathology. Severely affected kidneys were small in size, pale tan in color, and with a pitted texture. Microscopically, early changes consisted of degeneration of tubules, mild interstitial fibrosis, and triangular atrophic foci in the cortex. Progressive lesions consisted of hyaline sclerosis of glomeruli, more severe interstitial fibrosis, and severe tubular degeneration and atrophy.

c. Significance. Degenerative renal disease in Chinese hamsters may be useful as model systems in the study of aging or diabetes mellitus.

C. Pulmonary Granulomas

1. Epizootiology

Pulmonary granulomas were observed in 54 of 157 aging Chinese hamsters (Benjamin and Brooks, 1977). The cause could not be determined, but an infectious agent was suggested, since there was no difference in incidence of lesions observed in animals on contact bedding or in suspended wire cages.

2. Pathology

Grossly, the lesions appeared as subpleural 1- to 3-mm yellowish gray foci with variable involvement of the lungs from one to two lesions in some animals up to entire lobes in other animals. Microscopically, mild lesions consisted of alveolar collections of lipid-filled macrophages, while more severe lesions showed well-developed granulomas with lipid-filled macrophages, mixed inflammatory cells, focal septal fibrosis, and occasionally, cholesterol clefts. Severe alveolar and bronchiolar epithelial hyperplasia were seen in several animals.

3. Significance

Even though lesions are usually mild, they should be distinguished from experimentally induced conditions.

D. Periodontitis

Periodontitis was found in a strain of Chinese hamster with hereditary diabetes mellitus (Cohen *et al.*, 1961). It corresponds closely to periodontitis in humans with diabetes mellitus.

E. Spondylosis

The incidence of spondylosis was increased in hamsters with spontaneous diabetes mellitus compared to nondiabetic control animals (Silberberg and Gerritsen, 1976).

REFERENCES

Avery, T. I. (1968). Observations on the propagation of Chinese hamsters. *Lab. Anim. Care* **18**(2), 151–159.

Benjamin, S. A., and Brooks, A. L. (1977). Spontaneous lesions in Chinese hamsters. *Vet. Pathol.* **14**, 449–462.

Bowen, G. S., Calisher, C. H., Winkler, W. G., Kraus, A. L., Fowler, E. H., Garman, R. H., Fraser, D. W., and Hinman, A. L. (1975). Laboratory studies of a lymphocytic choriomeningitis virus outbreak in man and laboratory animals. *Am. J. Epidemiol.* **102**(3), 233–240.

Brooks, A. L., Benjamin, S. A., Hahn, F. F., Brownstein, D. G., Griffith, W. C., and McClellan, R. D. (1983). The induction of liver tumors by ^{239}Pu citrate or ^{239}PuO$_2$ particles in the Chinese hamster. *Radiat. Res.* **96**, 135–151.

Brownstein, D. G., and Brooks, A. L. (1980). Spontaneous endomyometrial neoplasms in aging Chinese hamsters. *JNCI, J. Natl. Cancer Inst.* **64**(5), 1209–1214.

Chandler, R. L., and Turfrey, B. A. (1972). Inoculation of voles, Chinese hamsters, gerbils and guinea pigs with scrapie brain material. *Res. Vet. Sci.* **13**(3), 219–224.

Chang, N., Liu, Y., and Lin, F. (1951). Experimental encephalitis virus infection in Chinese hamsters. *Chin. Med. J.* **69**, 420–426.

Cohen, M. M., Shklar, G., and Yerganian, G. (1961). Periodontal pathology in a strain of Chinese hamster, *Cricetulus griseus*, with hereditary diabetes mellitus. *Am. J. Med.* **31**(6), 864–867.

Crabtree, J. E., and Wilson, R. A. (1984). *Schistosoma mansoni*, cellular reactions to challenge infections in the cheek pouch skin of chronically infected Chinese hamsters. *Parasitology* **84**(1), 59–69.

Fujii, H., Kamiyama, T., and Hagiwara, T. (1983). Species and strain differences in sensitivity to *Toxoplasma* infection among laboratory rodents. *Jpn. J. Med. Sci. Biol.* **36**(6), 343–346.

Guttman, P. H., and Kohn, H. I. (1960). Progressive intercapillary glomerulosclerosis in the mouse, rat, and Chinese hamster, associated with aging and x-ray exposure. *Am. J. Pathol.* **37**, 293–308.

Hill, A. (1974). Mycoplasmas of small animal hosts. *Colloq.—Inst. Natl. Sante Rech. Med.* **33**, 311–316.

Hill, A. (1983). *Mycoplasma cricetuli*, a new species from the conjunctivas of Chinese hamsters. *Int. J. Syst. Bacteriol.* **33**(1), 113–117.

Hillova, J., Hiu, M., Mariages, R., and Belehrad, J. (1985). RSV provirus with some flanking sequences is formed on different size classes of Chinese hamster chromosomes. *Intervirology* **23**(1), 29–43.

Hindle, E., Hon, P. C., and Patton, W. S. (1926). Serological studies on Chinese Kala-azar. *Proc. R. Soc. London, Ser. B* **100**, 368–373.

Hirai, K., Shimaku, K., Tsuboi, T., Torii, M., Nishida, H., and Yamane, Y. (1983). Biological effects of *Spirometra erinacei* plerocercoids in several species of rodents. *Z. Parasitenkd.* **69**, 489–499.

Hsieh, E. T. (1919). A new laboratory animal, *Cricetulus griseus*. *Natl. Med. J. China* **5**, 20–24.

Kohn, H. I., and Guttman, P. H. (1964). Life span, tumor incidence and intercapillary glomerulosclerosis in the Chinese hamster (*Cricetulus griseus*) after whole-body and partial body exposure to x-rays. *Radiat. Res.* **21**, 622–443.

Krampitz, H. E., and Baumler, W. (1978). Occurrence, host range and seasonal prevalence of *Babesia microti* (Fran Ca, 1912) in rodents of southern Germany. *Z. Parasitenkd.* **58**, 15–33.

Kurotchkin, T. J., and Lin, L. E. (1930). The effect of animal passage upon pathogenic monilia. *Natl. Med. J. China* **16**, 337.

Lu, K. J., and Zia, S. H. (1935). Use of the Chinese hamster for testing the virulence of *C. diptheriae*. *Proc. Soc. Exp. Biol. Med.* **33**, 334–337.

Meier, H., and Yerganian, G. (1961). Spontaneous diabetes mellitus in the Chinese hamster (*Cricetulus griseus*). II. Findings in the offspring of diabetic parents. *J. Am. Diabetes Assoc.* **10**(1), 12–18.

Mikamo, K., and Kamiguchi, Y. (1983). Primary incidences of spontaneous chromosomal anomalies and their origins and causal mechanisms in the Chinese hamster. *Mutat. Res.* **108**, 265–278.

Parkening, T. A. (1982). Reproductive senescence in the Chinese hamster (*Cricetulus griseus*). *J. Gerontol.* **37**(3), 283–287.

Plesko, I. (1977). Experimental leptospirosis in Chinese hamsters. *Biologia (Bratislava)* **32**(3), 164–172.

Poel, W. E., and Yerganian, G. (1961). Adenocarcinoma of the pancreas in diabetes-prone Chinese hamsters. *Am. J. Med.* **31**(6), 861–863.

Ritterson, A. L. (1968). Effect of immunosuppressive drugs (6-mercaptopurine and methotrexate) on resistance of Chinese hamsters to the tissue phase of *Trichinella spiralis*. *J. Infect. Dis.* **118**(4), 365–369.

Ritterson, A. L. (1971). Resistance of Chinese hamsters to *Hymenolepis microstoma* and its reversal by immunosuppression. *J. Parasitol.* **57**(6), 1247–1250.

Schiff, L. F., Shefren, A. M., Barbera, P. W., and Poiley, S. M. (1973). Microbial flora and viral contact status of Chinese hamsters (*Cricetus griseus*). *Lab. Anim. Sci.* **23**(6), 899–902.

Silberberg, R., and Gerritsen, G. (1976). Aging changes in intervertebral discs and spondylosis in Chinese hamsters. *Diabetes* **25**(6), 477–483.

Tihon, C., and Hellman, A. (1976). Characterization of an endogenous Chinese hamster virus. *Fed. Proc., Fed. Am. Soc. Exp. Biol.* **35**(7), 1612.

van der Gulden, W. J. I. (1981). Uterus fistula in Chinese hamsters (*Cricetulus griseus*). *Z. Versuchstierkd.* **23**, 346–347.

Wang, U. F. L. (1951). The use of Chinese hamsters (*Cricetulus griseus*) in a study of *Mycobacterium tuberculosis*. II. For differentiation of human and bovine types. *Chin. Med. J.* **69**, 155–159.

Ward, B. C., and Moore, W. (1969). Spontaneous lesions in a colony of Chinese hamsters. *Lab. Anim. Care* **19**(4), 516–521.

Yen, A. C. H. (1936). Experimental virus infection in the Chinese hamster. II. Susceptibility to street rabies. *Proc. Soc. Exp. Biol. Med.* **34**, 648–651.

Yen, A. C. H. (1940). Susceptibility of Chinese hamsters to the mouse passage virus of human influenza. *Chin. Med. J., Suppl.* **3**, 342–348.

Yerganian, G. (1958). The striped-back or Chinese hamster, *Cricetulus griseus*. *J. Natl. Cancer Inst. (U.S.)* **20**(4), 705–727.

Zook, B. C., Albert, E. W., and Rhorer, R. G. (1977). Tyzzer's disease in the Chinese hamster, *Cricetulus griseus*. *Lab. Anim. Sci.* **27**(6), 1033–1035.

Chapter 19

Use in Research

Arthur Diani and George Gerritsen

I. Introduction	330
II. Spontaneous Diabetes Mellitus	330
A. Characterization	330
B. Therapy	334
III. Chemically Induced Diabetes Mellitus	335
A. Streptozotocin and *N*-Nitrosomethylurea	335
B. Alloxan	336
C. Monosodium Glutamate	336
D. Cortisone	336
E. Zinc Deficiency	336
IV. General Toxicology	336
A. Radiation	337
B. Environmental Pollutants	337
C. Therapeutic Drugs	337
D. Pesticides	338
E. Nitroso Compounds	338
F. Other Agents	338
V. Genetic Toxicology	338
A. Radiation	338
B. Environmental Pollutants	339
C. Therapeutic Drugs	339
D. Pesticides	339
E. Artificial Sweeteners	339
F. Mycotoxins	340
G. Vitamins	340
H. Inhalation Anesthetics	340
I. Hallucinogenic Drugs	340
J. Nitroso Compounds	340
K. Other Agents	340
VI. Cytogenetics of Germ and Somatic Cells	341
VII. Cell Lines Derived from Chinese Hamster Tissues and Organs	341
References	341

Copyright © 1987 by Academic Press, Inc.
All rights of reproduction in any form reserved.

I. INTRODUCTION

The purpose of this chapter is to review the principal types of biomedical research for which the Chinese hamster (*Cricetulus griseus*) (also known as *Cricetulus barabensis*) has been utilized. These categories of investigation encompass spontaneous and chemically induced diabetes mellitus, general and genetic toxicology, cytogenetics, and culture of cell lines derived from organs and tissues.

II. SPONTANEOUS DIABETES MELLITUS

Numerous detailed reviews on the spontaneously diabetic Chinese hamster have previously been published (Sirek and Sirek, 1967, 1971; Wappler and Fiedler, 1972; Hunt et al., 1976; Chang, 1978; Grodsky and Frankel, 1981; Gerritsen, 1982; Bell and Hye, 1983). Due to the thorough coverage of the earlier literature on this diabetic animal model, the purpose of this section is to emphasize the more recent publications and provide a brief survey of the historical data.

A. Characterization

1. Physiology

Most of the earlier metabolic studies have shown that the genetically diabetic Chinese hamster is a lean animal which displays glucose intolerance, mild to severe hyperglycemia, polyuria, glycosuria, hypoinsulinemia, occasional ketonuria, and high free fatty acid levels (Dulin and Gerritsen, 1967; Gerritsen and Dulin, 1967; Dulin et al., 1970a,b; Gerritsen and Blanks, 1974). Furthermore, these metabolic disturbances appeared to be influenced by an interaction of genotype with environmental factors such as amount of food consumed, dietary composition, and uterine environment of the dam (Dulin and Gerritsen, 1967, 1972; Gerritsen and Dulin, 1967, 1972; Gerritsen et al., 1970, 1974a, 1976, 1981; Dulin et al., 1970b; Gerritsen and Blanks, 1974, 1979; Gerritsen, 1975; Schmidt et al., 1976).

More recently, expansion of the Chinese hamster colony has resulted in greater breeding capability and opportunity for selection along with more predictable onset and characterization of diabetes. For example, at 90 days of age, all of the animals were found to be glycosuric and a majority displayed nonfasting blood glucose values above 300 mg/dl (Gerritsen et al., 1984).

The plasma insulin and glucagon levels are markedly deranged in chronically diabetic Chinese hamsters. Although there has been considerable phenotypic heterogeneity due to genetic and environmental influences (Dulin and Gerritsen, 1972), most established diabetic hamsters, in general, have low nonfasting plasma insulin levels (Gerritsen and Dulin, 1967; Sims and Landau, 1967; Gerritsen and Blanks, 1974; Gerritsen, 1982). In contrast, the plasma insulin of newly diagnosed diabetic Chinese hamsters tends to be within normal limits or slightly elevated (Gerritsen and Blanks, 1974; Gerritsen, 1982). Plasma glucagon levels are usually elevated in the long-term diabetic state (Chang et al., 1977a). Although the diabetic Chinese hamster occasionally displays transient insulin resistance in the early stages of the disease, the chronic absolute deficiency of plasma insulin and high plasma glucagon suggest that this animal model is most representative of the type I human diabetic.

Concentrations of pancreatic hormones are drastically altered in the chronically diabetic state. A decrease in pancreatic insulin has been a very consistent observation in the islets of Langerhans of diabetic Chinese hamsters (Gerritsen and Dulin, 1967; Sims and Landau, 1967; Chang et al., 1977a; Gerritsen, 1982). In contrast, pancreatic glucagon levels are generally elevated, particularly with long-term disease (Grodsky et al., 1974; Frankel et al., 1975; Chang et al., 1977a). Pancreatic somatostatin concentration is markedly depressed in the diabetic pancreas (Petersson et al., 1977).

Several *in vitro* studies have documented that islet hormone secretions of diabetic Chinese hamsters are abnormal. Insulin secretion was reduced during perfusion of the whole pancreas with glucose, arginine, or potassium ions (Frankel et al., 1974, 1975, 1982; Grodsky et al., 1974), but was normal in response to theophylline (Frankel et al., 1975). Isolated islets or fragments of pancreas also displayed a subnormal capacity to release insulin (Malaisse et al., 1967; Chang and Schneider, 1970a; Chang, 1970; Rabinovitch et al., 1976). This finding may be related to defective calcium ion handling by the diabetic β cells (Siegel et al., 1979, 1981). In general, glucose suppression of glucagon secretion from the perfused islets of diabetics was deranged (Frankel et al., 1974, 1982; Grodsky et al., 1974).

The physiological characteristics of the prediabetic Chinese hamster have been adequately examined. The prediabetic hamster is defined as an animal which carries the diabetic genotype but does not express the typical metabolic lesions of diabetes until later in life. Many of the characteristics of the prediabetic state have been summarized recently (Dulin et al., 1984). Blood glucose levels of prediabetics usually fall within normal limits, and glycosuria is not detectable (Gerritsen, 1982). Plasma insulin was reported to be normal in prediabetics (Gerritsen and Dulin, 1972; Gerritsen et al., 1974b; Gerritsen, 1975), except for a modest elevation just prior to development of hyperglycemia (Gerritsen and Blanks, 1979; Gerritsen et al., 1981). Pancreatic insulin content was normal, whereas pancreatic glucagon was slightly elevated during prediabetes (Gerritsen and Dulin, 1967; Gerritsen *et al.,* 1974b; Grodsky *et al.,* 1974; Gerritsen, 1982; Dulin *et al.,* 1984). Plasma glucagon was unremarkable when compared with age-matched nondiabetics (Dulin *et al.,* 1984). *In vitro* perfusion of the prediabetic pancreas demonstrated that release of insulin, glucagon, and somatostatin in response to a glucose load plus arginine stimulus was normal (Frankel *et al.,* 1984).

2. Morphology

Due to the primary involvement of the pancreas in the diabetic syndrome of the Chinese hamster, numerous investigations have focused on the structural deformities in the islets of Langerhans. The major morphological derangement in the diabetic islet has been observed in the β cells, which degranulate with duration of the disease (Malaisse *et al.,* 1967; Boquist, 1969; Like *et al.,* 1974a). The defective β cell in the diabetic Chinese hamster is characterized by proliferative rough endoplasmic reticulum, expansion of the Golgi apparatus, and glycogen deposition (Luse *et al.,* 1967; Boquist, 1969; Soret *et al.,* 1974; Like *et al.,* 1974b). Morphometric analysis has revealed that the islets of diabetic hamsters are diminutive as a result of β-cell death (Carpenter *et al.,* 1967, 1970). Variable alterations have been observed in α cells, but the most consistent encompass proliferation of lysosomes and dilatation of the rough endoplasmic reticulum (Orci *et al.,* 1970; Soret *et al.,* 1974; Like *et al.,* 1974b). The nuclear membranes of both α and β cells were reported to contain an elevated quantity of pores (Orci *et al.,* 1974). Although the δ cells (somatostatin-producing) were devoid of structural derangement (Boquist, 1969; Like *et al.,* 1974b), the number decreased in concert with depressed islet size (Petersson *et al.,* 1977). At the light-microscopic level, the cholinergic and adrenergic innervation of the diabetic islet was significantly diminished under conditions of elevated plasma glucose and ketone levels (Diani *et al.,* 1983). Unfortunately, the specific cell types which were deficiently innervated were not elucidated in this study.

The development of lesions in the islets of prediabetic Chinese hamsters has been evaluated. In the earliest stages of prediabetes, the islets were hypertrophic, increased in number, and composed primarily of hypergranulated β cells (Meier and Yerganian, 1961). These changes were subsequently followed by degranulation and sporadic glycogen deposition within β cells of diabetic hamsters (Meier and Yerganian, 1961; Carpenter *et al.,* 1967). It should be pointed out that despite the occasional presence of lymphocytes within the islets of prediabetic Chinese hamsters, the sporadic nature of these observations suggests that autoimmunity probably is not the cause of diabetes in this animal model.

Some of the peripheral tissues and organs of the diabetic Chinese hamster have been studied by light and/or electron microscopy. Unfortunately, the lack of systematic investigations has generated only fragmentary data, which has led to an incomplete understanding of some of the complications in this animal model. Due to limitations imposed by small numbers of animals and incomplete characterization of the diabetic state, the kidney has been thoroughly examined only recently. Glycogen deposition in the renal tubules seemed to be the most consistent derangement in the kidney (Soret *et al.,* 1973, 1974). Mesangial expansion, glomerular basement membrane thickening, and other minor structural deformities have been reported on a more sporadic basis (Shirai *et al.,* 1967; Conforti, 1972; Soret *et al.,* 1974). Recently, systematic morphometric analysis confirmed that glomerular basement membrane thickness is significantly increased in 11- to 15-month-old and 19- to 23-month-old diabetic animals and positively correlated with age and severity of hyperglycemia (Diani *et al.,* 1985a). Overall, the renal lesions in the Chinese hamster develop slowly and may correspond to the very early and mild stages of human diabetic nephropathy.

Examination of the eye of the diabetic Chinese hamsters has documented the presence of lesions in the retina. Microaneurysms similar to those in the diabetic human retina have not been observed in the hamster. However, thrombuslike abnormalities were associated with the retinal capillary beds (Soret *et al.,* 1973, 1974). The most consistent abnormality involved massive glycogen aggregates, which probably were deposited within the Muller cells of the outer nuclear layer (Soret *et al.,* 1973, 1974). Quantitative evaluation of trypsin-digested flat mounts of the retinal vasculature revealed that the endothelial cell/pericyte ratio was inflated in the capillaries of diabetics (Federman and Gerritsen, 1970). Thickening, lamination, vacuolization of the retinal capillary basement membrane were confirmed by qualitative electron microscopy (Ghosh *et al.,* 1970). A recent investigation showed that the level of the neuropeptide vasoactive intestinal peptide (VIP) was significantly elevated in the choroidal vasculature of prediabetic and diabetic Chinese hamsters (Diani *et al.,* 1985b). The disruption of normal VIP levels in the prediabetic state suggests that impairment of this neuropeptide may be one of the causative factors underlying ocular complications in this animal model.

Morphological aberrations have been identified in the central, peripheral, and autonomic nervous systems of diabetic Chinese hamsters. In the cerebral cortex, the basement membranes of capillaries were markedly thickened (Luse *et al.,* 1970). Neuronal pathology in the brain was characterized by degenerate processes, abnormal myelin and synapses, enlarged mitochondria in dendrites, and accumulation of neurofibrils within the axons (Luse *et al.,* 1970). The hypothalamus of diabetic hamsters displayed aggregates of lipid droplets in the β-tanycyte processes of the median eminence and impaired growth of cell bodies in the arcuate nucleus (Bestetti and Rossi, 1982). Furthermore, the number of somatotrophs in the adenohypophysis was increased (Deslex and Rossi, 1976). These findings imply that the aberrant hypothalamic–hypophyseal axis may be an important etiological agent underlying the development of the diabetic syndrome in this animal model.

The peripheral nervous system of the Chinese hamster has been examined from morphological and functional viewpoints. In the chronically diabetic animal with ketonuria, segmental demyelination and Wallerian degeneration were recognized in the distal aspect of the tibial nerve (Schlaepfer et al., 1974). In a subsequent study with mildly diabetic Chinese hamsters, sensory and motor conduction velocities were decreased in the tibial nerve without evidence of anatomical pathology (Kennedy et al., 1982). The presence of functional abnormalities in peripheral nerves without morphological derangement is not compatible with results obtained in diabetic humans. Therefore, the Chinese hamster may be useful as a model for only certain aspects of diabetic neuropathy.

Extensive qualitative and morphometric analyses of the autonomic nervous system have been conducted in the diabetic Chinese hamster. Decreased acetylcholinesterase staining, aberrant myelination, and axonal degeneration of the pelvic visceral nerves were linked to urinary bladder distention (Dail et al., 1977), which resembled the neurogenic bladder syndrome of diabetic humans. More recently, it has been demonstrated that high levels of the neuropeptide VIP in the bladder of prediabetics and diabetics may also be implicated in the cystopathy of this animal model (Diani et al., 1985b). The abdominal sympathetic trunk has been evaluated by detailed morphometry which showed that both myelinated and unmyelinated fibers were reduced in size (Diani et al., 1981a). Furthermore, numerical density was impaired in both types of fibers. Similar morphometric alterations were also identified in the ventral abdominal vagus nerve (Diani et al., 1984). These structural alterations in the autonomic nervous system are thought to be critical factors underlying gastrointestinal malfunction in the diabetic Chinese hamster.

The gastrointestinal tube and its innervation have been systematically investigated by morphological methods. Using light microscopy, damaged villi, blood vascular lesions, lymphocyte aggregates, and reduced quantity of Auerbach's plexuses were noted (Diani et al., 1976). At the electron-microscopic level, the axons of Auerbach's plexuses were enlarged with heavy glycogen deposition, aggregates of neurofilaments, and accumulation of lamellar bodies (Diani et al., 1979). These damaged nerves were believed to be the cause of the atony, distention, delayed motility (Diani et al., 1979), and overpopulation with aerobic bacteria (Diani et al., 1981b) in the gut of the diabetic Chinese hamster. Capillary basement membrane thickening was also evident in the mucosal capillaries, particularly in hamsters with severe ketonuria (Diani et al., 1981c).

The aorta of the diabetic and hypercholesterolemic Chinese hamster was reported to display some very mild atherosclerotic alterations at the ultrastructural level. Endothelial cells with aggregates of lipid droplets, smooth muscle cell invasion of the intima, fragmentation of the internal elastic membrane, and calcium deposits of the media were observed (McCombs et al., 1974). However, well-developed atherosclerotic lesions were not observed in a subsequent study of diet-induced hypercholesterolemia in the mildly diabetic Chinese hamster (Soret et al., 1976). Therefore, on the basis of very limited observations, it seems that the Chinese hamster may not be a good model for the advanced stages of diabetic atherosclerosis.

The testicular tissue of the male diabetic Chinese hamster was characterized primarily by a diminutive germinal epithelium and a reduced quantity of Leydig cells (Schoffling et al., 1967; Sirek and Sirek, 1971). Though somewhat variable, impaired spermatogenesis was also evident (Schoffling et al., 1967). These observations, concomitant with pelvic neuropathy (Dail et al., 1977), may offer a partial explanation for the poor reproductive capacity of this animal model. Meaningful studies of the reproductive tract of the female have not been done.

The skeletal system of diabetic Chinese hamsters has received limited attention. In the hip joint, the chondrocytes in the head of the femur possessed atrophic rough endoplasmic reticulum, aberrant mitochondria, and large lipid inclusions (Silberberg et al., 1976). The occurrence and onset of spondylosis of the vertebrae were accelerated in diabetic animals (Silberberg and Gerritsen, 1976).

The ability of the diabetic Chinese hamster to respond to dermal wounds of the ear has been explored. A reduction of fibroblast and polymorphonuclear infiltration into the wound along with considerable edema was observed (Weringer and Arquilla, 1981).

3. Biochemistry

The biochemical changes in the spontaneously diabetic Chinese hamster have been previously reviewed (Chang, 1981; Gerritsen, 1982). Much of the research has focused on the pancreatic islets, liver, kidney, adipose tissue, diaphragm, and plasma, with scattered reports on other tissues.

The pancreatic islet content and secretion of insulin, glucagon, and somatostatin in spontaneously diabetic hamsters were previously discussed in Section II,A,1 of this chapter. Insulin synthesis was shown to be deficient in isolated islets of diabetic Chinese hamsters. This conclusion was derived from the fact that radiolabeled leucine was incorporated into insulin at a reduced rate in diabetics (Chang, 1970). This impairment of insulin synthesis is believed to be a major factor underlying diminished insulin content and secretion in the islets of diabetic Chinese hamsters.

The malfunction of the hepatic tissue was deemed to be a serious impediment to metabolic activity in the diabetic Chinese hamster. After incubation of the liver, *in vitro*, with radiolabeled molecules, the incorporation of glucose and pyruvate into CO_2, glycogen, and fatty acid was depressed in diabetic hamsters (Sims and Landau, 1967). In general, the rate of hepatic gluconeogenesis was moderately enhanced (Chang and Schneider, 1970b). This condition was exacerbated in ketonuric diabetic animals where the levels of all four key gluconeogenic enzymes (i.e., phosphoenolpyruvate carboxykinase, glucose-6-phosphatase, fructose-1,6-bisphosphatase, and pyruvate carboxylase) were elevated (Chang and Schneider, 1970a). Although unchanged in nonketonuric diabetics, the activity of hepatic glycolytic and pentose phosphate shunt enzymes were reduced in ketonuric diabetics (Chang and Schneider, 1970a; Chang et al., 1977b). Lactate dehydrogenase activity and isozyme patterns in the liver were variable but seemed to be more closely correlated with inheritance factors other than diabetes (Chang et al., 1977c). Hepatic protein synthesis was unremarkable in both diabetic and ketonuric diabetic animals (Chang, 1974). In an apparent effort to compensate for severe insulinopenia, the plasma membranes of hepatocytes displayed an increased insulin-binding capacity in diabetics (Hepp et al., 1975). The levels of two hepatic acid glycohydrolases, N-acetyl-β-D-hexosaminidase and α-D-mannosidase, were increased in diabetic hamsters (Chang, 1981).

Evidence of functional impairment also was documented in the kidney of the diabetic Chinese hamster. Incorporation of the C-2 of pyruvate into carbon dioxide was depressed in kidneys of severely diabetic animals (Sims and Landau, 1967). Renal gluconeogenesis and phosphoenolpyruvate carboxykinase activity were elevated in diabetic hamsters (Chang and Schneider, 1970b). Protein synthesis, however, was reported to be within normal limits in renal tissue (Chang, 1974). The acid glycohydrolases have been intensively evaluated in the kidney because of their regulation of glycoprotein metabolism and possible mediation of capillary basement membrane thickening. Despite some variability, N-acetyl-β-D-hexosaminidase, α-D-mannosidase, α-D-galactosidase, β-D-galactosidase, and β-D-glucuronidase activities were below normal in kidneys from diabetics (Chang, 1981). A significant increase in hyaluronic acid and heparan sulfate content and decreased quantities of chondroitin sulfate have been detected in renal cell lines derived from diabetic Chinese hamsters (Ginsberg et al., 1981).

There was considerable variation, probably due to inheritance factors, in the activity of acid glycohydrolases in the plasma of diabetic Chinese hamsters (Chang, 1981). The most consistent results indicated that activities of N-acetyl-β-D-hexosaminidase and β-D-galactosidase were elevated in diabetics (Chang, 1981). The levels of plasma glycoproteins of diabetic Chinese hamsters were within normal limits with the exception of elevated hexosamines (Copeland et al., 1982). The electrophoretic profile of plasma glycoproteins indicated that α_2-macroglobulin was displaced on the gel in diabetic compared with nondiabetic animals (Copeland and Ginsberg, 1982).

Although the diabetic Chinese hamster is a nonobese animal model, the adipose tissue has received some attention. In general, most studies have documented that the adipose tissue of the diabetic hamster is unremarkable with respect to enzyme activity (Chang et al., 1977a,b), glucose oxidation to carbon dioxide or conversion of glucose into fatty acids (Gerritsen and Dulin, 1967; Sims and Landau, 1967), and insulin sensitivity (Dulin and Gerritsen, 1967).

Skeletal muscle of the diabetic Chinese hamster was almost totally devoid of biochemical abnormalities. With the exception of lowered glucose oxidation (Sims and Landau, 1967), the muscle of the diaphragm displayed normal insulin sensitivity and incorporation of amino acids into protein (Sims and Landau, 1967; Dulin and Gerritsen, 1967).

Various other organs of the diabetic Chinese hamster have been evaluated by biochemical techniques on the nonsystematic basis. Absorption of free fatty acids in the gut was greater in diabetics, probably as a result of intestinal hypertrophy (Parkinson, 1973). Synthesis of cholesterol was significantly increased in the small and large intestine of diabetics whether expressed per whole organ or per gram of weight (Feingold et al., 1984). In diabetic hypercholesterolemic hamsters, cholesterol content was elevated in the liver, aorta, and adipose tissue. Furthermore, the turnover rate of plasma cholesterol and the size of the rapidly exchanging pool were increased (Chobanian et al., 1974b). Sections of aorta from diabetic hamsters have been incubated in vitro to measure arterial glucose metabolism. Overall, glucose utilization and conversion of glucose to lactic acid, carbon dioxide, lipid, glycogen, and protein were depressed, particularly when ketonuria was present also (Chobanian et al., 1974a). High doses of insulin, in vivo, reversed these aortic perturbations. Accelerated polyol pathway activity leading to excessive levels of fructose and sorbitol was demonstrated in the sciatic nerve and lens of diabetic Chinese hamsters (Holcomb et al., 1974). Unfortunately, the potential relationship between the increased activity of the polyol pathway and diabetic complications has not been explored.

4. Genetics

The inheritance factors which mediate diabetes mellitus in the Chinese hamster have been investigated for some time; however, limitations imposed by inadequate numbers of animals have made it difficult to examine the genetic composition of diabetic Chinese hamsters in detail. Recent expansion of the Chinese hamster colony has increased breeding and selection to the point where the epidemiology of blood glucose levels and onset and incidence of glycosuria can be effectively studied. Furthermore, the increased reproductive capacity has provided for enhancement of experimental crossing and genotypic analysis.

The original hypothesis for inheritance of diabetes in the

Chinese hamster was proposed by Yerganian. The genetic data, based on limited numbers of animals, inferred that the diabetic syndrome in this animal model was controlled by a single recessive gene and certain modifier genes (Meier and Yerganian, 1959, 1961; Yerganian, 1964). The single-gene concept was modified later by Butler, who examined epidemiological data from other colonies. On the basis of glycosuria data and examination of the phenotypes of offspring from various matings, it was concluded that inheritance of diabetes in the Chinese hamster was extremely complex. Therefore, Butler proposed the polygenic hypothesis, which states that there are a minimum of four recessive diabetogenic genes in the Chinese hamster (Butler, 1967; Butler and Gerritsen, 1970). Any two of these genes in the homozygous state can generate a glycosuric hamster. Further studies showed that inbreeding for homozygosity produced an earlier onset and increased incidence of glycosuria (Schmidt et al., 1970; Gerritsen et al., 1974a) but did not provide conclusive support for the polygenic hypothesis. It was not until recently that sufficient numbers of animals from pure inbred genetic lines became available to allow accurate analysis of genotype in the diabetic Chinese hamster. Extensive genetic experiments including backcrosses have revealed that certain sublines of Chinese hamsters possess a single recessive gene for diabetes, whereas others possess two recessive genes (Gerritsen et al., 1984). Moreover, the onset of glycosuria in the diabetics is regulated by a multiple-gene system. The inbred nondiabetic Chinese hamster appears to be totally devoid of any recessive genes for diabetes.

Due to the inbreeding of diabetic Chinese hamsters for over 40 generations and the advanced knowledge of inheritance factors, this animal model was appropriate for the study of genetic markers of diabetes. Sodium dodecyl sulfate (SDS) two-dimensional gel electrophoresis revealed that a specific protein is recognizable in the liver tissue of inbred prediabetic and diabetic but not nondiabetic hamsters (Dulin et al., 1984). Although this important preliminary observation needs further study, it infers that gene products may serve as potentially useful markers for the early identification of diabetic individuals.

5. Immunology

The immune system of the diabetic Chinese hamster has received minimal attention until recently. After primary and secondary challenge with sheep erythrocytes, total antibody production and switchover to IgG were considerably less than that of age-matched nondiabetics (Fletcher-McGruder and Gerritsen, 1984). Hyperglycemia per se was not the principal defect underlying immune deficiency in the spontaneously diabetic animal, since streptozotocin-diabetic Chinese hamsters displayed a normal immune response. Despite the occasional presence of lymphocytes in the islets of prediabetic hamsters, several lines of evidence suggest that the diabetic syndrome in the Chinese hamster is not an autoimmune disorder. Suppression of the immune system of prediabetic hamsters with cyclophosphamide did not prevent or delay the onset of hyperglycemia or glucosuria (Gishizky et al., 1984). In addition, the administration of antilymphocyte serum to or neonatal thymectomy of prediabetics did not affect the subsequent development of diabetes (Fletcher-McGruder and Gerritsen, 1985). Finally, passive transfer of lymphocytes from diabetic Chinese hamsters did not elicit any manifestations of diabetes in nude mice (Fletcher-McGruder and Gerritsen, 1985).

6. Teratogenesis

The effect of maternal spontaneous diabetes on the prenatal and postnatal development of the offspring has been sparsely evaluated in genetically diabetic Chinese hamsters. In an early investigation, perinatal mortality of litters from diabetic dams was not found to be increased. The only significant alteration was an increase in the weight of pups at birth and during the initial 5 weeks of life (Heisig and Schall, 1971). Teratogenic effects associated with diabetes were recognized in the Chinese hamster colony maintained at the Asahikawa Medical College (Japan). Although no abnormalities were observed during preimplantation, a higher incidence of embryonic mortality, gross anomalies, and retardation of fetal growth were found during the postimplantation period in diabetic offspring (Funaki and Mikamo, 1983).

7. Appetite Control Systems

Although the diabetic Chinese hamster has been reported to display hyperphagia, few detailed studies have been conducted to examine the appetite control systems in this animal model. Recently it has been shown that diabetic and nondiabetic Chinese hamsters exhibit a similar natural circadian rhythm for feeding within a 24-hr period despite excessive consumption of food by the diabetics (van Sickle et al., 1981). It was also documented that various opiates and putative satiety peptides have minimal effect on modulation of food intake in diabetic Chinese hamsters (Billington et al., 1984). Therefore, the factors underlying hyperphagia in this genetically diabetic animal model remain to be elucidated.

B. Therapy

A variety of therapeutic regimens have been utilized on prediabetic and diabetic Chinese hamsters either to prevent or to ameliorate the metabolic disturbances of diabetes. Dietary modification has been shown to have a pronounced effect on diabetes in this animal model. Since the prediabetic hamster often exhibits hyperphagia, food consumption was limited to that of a nondiabetic hamster, which resulted in delayed onset of glycosuria and prevention of urinary ketone spillage (Gerritsen and Dulin, 1972; Gerritsen et al., 1974b; Gerritsen, 1975). In another study, dietary restriction also retarded the development of glycosuria and hyperglycemia and eliminated hyperinsulinemia in nonhyperphagic prediabetic hamsters (Gerritsen et al., 1981). Alteration of dietary composition also appeared to play a major role in the prevention of severe metabolic complications of diabetes. A reduction of fat and replacement of animal fat with vegetable fat in the diet either ameliorated or prevented ketonuria (Grodsky et al., 1974; Gerritsen et al., 1981). Simple fasting or decreased caloric intake have elicited a modest depression of hyperglycemia in animals with established diabetes (Gerritsen and Dulin, 1967; Dulin and Gerritsen, 1972).

Since the Chinese hamster is an insulin-deficient model of diabetes mellitus, treatment with exogenous insulin has had beneficial effects on metabolic aberrations. A substantial reduction in glycosuria was observed in severely diabetic animals after injection with NPH insulin (Gerritsen and Dulin, 1966). Decreased glycosuria was noted also in mildly diabetic hamsters following administration of the sulfonylurea tolbutamide (Gerritsen and Dulin, 1966). Continuous insulin infusion with a minipump for 7 or 14 days generated a significant improvement in blood glucose levels and endogenous insulin release from the perfused pancreas of diabetic animals (Frankel et al., 1979; Bringer et al., 1981). Treatment of prediabetic Chinese hamsters with continuous low-dose insulin via a minipump for 4 weeks increased plasma insulin levels but did not retard the onset or severity of glycosuria or hyperglycemia (Frankel and Grodsky, 1979). Thus, insulin treatment of prediabetic animals does not seem to prevent the development of diabetes and the associated metabolic alterations.

A few experimental drugs have been tested in the diabetic Chinese hamster in an attempt to alleviate the metabolic disturbances. Chronic dosing with an inhibitor of gluconeogenesis reduced blood glucose levels and hepatic glycogen but also produced an acceleration in mortality (Schillinger and Loge, 1976). Centpiperalone, a recognized hypoglycemic agent, was not efficacious in diabetic Chinese hamsters (Chatterjee et al., 1980). Ciglitazone, a novel hypoglycemic agent in obese, insulin-resistant animal models, only lowered plasma lipids in the insulinopenic diabetic hamsters (Chang et al., 1983).

III. CHEMICALLY INDUCED DIABETES MELLITUS

Numerous chemical agents have been utilized to induce all or some of the metabolic and morphological complications of diabetes mellitus on a permanent or transient basis in Chinese hamsters which do not appear to carry genes for diabetes. Some of the more pertinent abnormalities exhibited by the hamster are glycosuria, impaired glucose tolerance, hyperglycemia, abnormal plasma insulin, and β-cell abnormality. These alterations are similar to those manifested by chemically induced diabetic mice and rats. The substances which are known to promote these disturbances in this animal include streptozotocin, N-nitrosomethylurea, alloxan, monosodium glutamate, cortisone, zinc deficiency, and a variety of other chemicals. The purpose of this section will be to review the metabolic and morphological alterations associated with administration of the various diabetogenic agents in the Chinese hamster.

A. Streptozotocin and N-Nitrosomethylurea

Streptozotocin, an antibiotic derived from *Streptomyces achromogenes*, has been the principal diabetogenic agent in the Chinese hamster due to its cytotoxic effect on the islet β cell. A single intravenous injection of 75 mg/kg of streptozotocin into fasted animals caused insulinopenia, hyperglycemia, increased serum free fatty acids, and elevated hepatic glucose-6-phosphatase concomitant with permanent degranulation and degeneration of the islet β cells (Losert et al., 1971; Richter et al., 1971). In contrast, a 25-mg dose of the drug injected once a month for 4 consecutive months, together with daily NPH insulin therapy, produced a transient glycosuria of 1–2 months in adult hamsters (Berman et al., 1973).

Detailed morphological and metabolic studies have been performed to document the acute and chronic effects of a single 175- or 200-mg/kg dose of streptozotocin in adult animals. Within 24 hr, a triphasic blood glucose response, hyperglycemia, glycosuria, and severe degeneration of the β cells, characterized by nuclear pyknosis, cytoplasmic vacuolization, and necrosis, were present (Wilander and Boquist, 1972). Dilatation of the endoplasmic reticulum was demonstrable in the α cells (Dulin and Soret, 1977). At 24 hr islets were deranged and almost indistinguishable, along with subnormal amounts of pancreatic insulin and glucagon (Dulin and Soret, 1977). Several weeks after streptozotocin treatment, evidence of limited regeneration of islet tissue such as growth of tubular structures and buds was found. Massive islets with hypertrophied cells, hydropic degeneration, and necrosis were also observed (Wilander, 1974). At 3 months after streptozotocin administration, enlarged islets and regeneration were not apparent and only diminutive islets with degranulated β cells and a few scattered α cells were detected (Wilander, 1974). Furthermore, hyperglycemia and glycosuria were persistent and led to a high mortality rate (Wilander, 1975). Morphometric analysis confirmed that after 3 months of streptozotocin diabetes, islet volume and diameter were markedly decreased in Chinese hamsters (Wilander, 1975). In another study enzyme activities as well as organ weights were found to be abnormal in the liver, kidney, and epididymal fatpads (Chang *et al.*, 1977d). Glycohydrolase activities also have been reported to be impaired in the plasma, kidney, and liver of streptozotocin-diabetic hamsters (Chang *et al.*, 1979).

Prevention or control of streptozotocin diabetes in the Chinese hamster has been attempted. The level of nicotinamide within β cells is depleted by streptozotocin, and complete protection from the diabetogenic effects of streptozotocin can be achieved by pretreatment of Chinese hamsters with 500 mg/kg of nicotinamide (Wilander, 1975). Normoglycemia has been maintained in streptozotocin-diabetic Chinese hamsters by means of isograft or xenograft implantation of cultured pancreatic islet cells into the peritoneal cavity (Archer *et al.*, 1980).

The streptozotocin molecule is composed of a *N*-nitrosomethylurea residue which is conjugated to the second carbon of 2-deoxy-D-glucose. Several studies have indicated that the *N*-nitrosomethylurea moiety by itself is diabetogenic in Chinese hamsters. A single dose of 50 mg/kg *N*-nitrosomethylurea promoted an initial triphasic blood glucose curve and moderate levels of hyperglycemia. Nuclear pyknosis of islet cells was observed within 3 hr. At 24 hr, degeneration and disintegration of most islet cells had occurred (Wilander and Gunnarson, 1975; Wilander and Tjalve, 1975a). Radiolabeled *N*-nitrosomethylurea, at a dose of 50 mg/kg, has been shown to accumulate selectively in the islets of Chinese hamster within 1 hr of administration. This specific uptake by the islet endocrine tissue appears to explain the marked diabetogenicity of this compound (Wilander and Tjalve, 1975b).

B. Alloxan

The pyrimidine alloxan is another compound which has been extensively used to induce diabetes through its specific cytotoxic effect on the β cells. Intramuscular or intraperitoneal injection of a single 300-mg/kg dose of alloxan into Chinese hamsters induced a triphasic blood glucose curve and a temporary, modest degree of hyperglycemia (Boquist, 1967). Alloxan also was reported to induce β-cell necrosis, which was followed by widespread regeneration of the β cells and normalization of blood glucose within a few weeks after induction of diabetes. Thus, alloxan was considered to be a transitory diabetogenic agent in the Chinese hamster (Boquist, 1968a,b; Boquist and Falkmer, 1970).

C. Monosodium Glutamate

A single subcutaneous injection of monosodium glutamate at a dose of 4 mg/gm into newborn Chinese hamsters provoked remarkable morphological changes in the hypothalamus. Within 6 hr after injection, severe necrosis, cell death, and loss of nerve cell bodies were evident in the arcuate and ventromedial hypothalamic nuclei (Komeda *et al.*, 1980). The appearance of early lesions and cell death in the arcuate and ventromedial nuclei imply that the hypothalamus may play a central role in the development of monosodium glutamate-induced diabetes (Komeda *et al.*, 1980). In a subsequent experiment, subcutaneous injections of monosodium glutamate at a dose of 4 mg/gm were given to Chinese hamsters on each of the first 3 days after birth (Komeda *et al.*, 1980). Glycosuria occurred at 4 weeks of age, along with fasting hyperglycemia and abnormal glucose tolerance. Although the animals did not develop obesity, food and water consumption and urine volume were substantially increased. The quantity of pancreatic islets was decreased, and the β cells were degranulated with heavy glycogen deposition. Mild alleviation of hyperglycemia was obtained in monosodium glutamate-induced diabetic Chinese hamsters when a synthetic trypsin inhibitor was incorporated into the diet (Sakai *et al.*, 1981).

D. Cortisone

Subcutaneous injections of cortisone for each of 14 consecutive days at 1, 3, or 12 mg/day produced a significant elevation of serum insulin, glucose, and glycosuria in a small percentage of Chinese hamsters. If a combination of cortisone and growth hormone were administered, the levels of hyperinsulinemia and hyperglycemia were reduced (Campbell *et al.*, 1966). Daily intramuscular injection of hydrocortisone for 8 days at a dose of 50 mg/kg elicited moderate hyperglycemia which normalized within 4 days of cessation of treatment (Boquist, 1967).

E. Zinc Deficiency

Consumption of a zinc-deficient diet by Chinese hamsters for up to 4 months led to impaired glucose tolerance without glycosuria, hyperglycemia, or hyperinsulinemia. Moderate degranulation of the β cells also was detectable (Boquist and Lernmark, 1969).

IV. GENERAL TOXICOLOGY

The nondiabetic Chinese hamster has received considerable attention as a rodent model for the evaluation of potential toxicity of a wide range of agents. It is generally agreed that the Chinese hamster's usefulness in toxicology studies emanates from its small size (30–35 gm), relatively long life span of 3–4 years, and ease of handling, care, and breeding. An abundance of literature has accumulated concerning the general toxic effects of agents on various organs of the Chinese hamster. The purpose of this section will be to review the toxicity of such diverse agents as radiation, environmental pollutants, therapeutic drugs, pesticides, nitrosocompounds, and various other agents.

A. Radiation

The predominant forms of radiation which have been tested for toxicity in the Chinese hamster include radioisotopes, X irradiation, and microwaves. Intraperitoneal injection of ^{144}Ce into Chinese hamsters resulted in dissemination, primarily to liver and bone. This radioisotope shortened the life span of the animals and induced liver tumors such as hemangiosarcomas, fibrosarcomas, and other neoplasms (Benjamin et al., 1976). It also was shown that the frequency of ^{144}Ce-induced liver tumors could be increased by stimulation of hepatocyte mitosis through partial hepatectomy (Brooks et al., 1982). Plutonium-239 had comparable effects to ^{144}Ce, with additional complications in bone, such as osteosarcomas (Benjamin et al., 1976). Intravenous injection of [^{239}Pu]citrate or ^{239}PuO$_2$ particles demonstrated that the former was more evenly distributed in the liver tissue and yielded a higher frequency of tumors (Brooks et al., 1983). It also has been clearly documented that ^{239}Pu becomes selectively bound to the lysosomes of hepatocytes (Sütterlin et al., 1984). This observation provided a reasonable explanation for the prolonged retention of ^{239}Pu and high incidence of tumors in the liver of Chinese hamsters. Intraperitoneal injection of ^{90}Sr shortened the life expectancy of Chinese hamsters but generated minimal tumor development in bone (Benjamin et al., 1976). ^{241}Am was injected intraperitoneally into Chinese hamsters and collected primarily in the liver, skeleton, and kidney. Life expectancy was reduced by this radioisotope also (McKay et al., 1969).

Whole-body X irradiation was determined to be extremely lethal to Chinese hamsters. The rate of mortality showed a positive correlation with age up to about 100 days of age and then leveled off (Ward et al., 1972). Other parameters, such as the rate of mitosis of white blood cells in the marrow and the peripheral hematocrit, were deleteriously affected by X irradiation also. However, the results fluctuated according to the 24-hr circadian rhythm of the animals (Lappenbusch, 1972). After exposure of neonatal Chinese hamsters to X irradiation for up to 14 days after birth, it was observed that massive oocyte death occurred in the ovary on day 4 (pachytene) and days 12–14 (dictyate) (Tateno and Mikamo, 1984).

Microwave irradiation induced focal lesions in the small vessels and neuropil of the brains of Chinese hamsters. These localized disturbances provoked a short-term and reversible, leaky blood–brain barrier, as evidenced by increased amounts of the tracer horseradish peroxidase in endothelial vesicles. Accumulation of horseradish peroxidase in the endothelial-cell cytoplasm, accumulation of platelets, and perivascular edema were also transient findings (Albert, 1979; Albert and Kerns, 1981). X Irradiation followed by microwave exposure had a partial protective effect on Chinese hamsters. The X-ray LD$_{50}$ was increased, and the peripheral blood count was less severely depleted (Lappenbusch et al., 1973).

B. Environmental Pollutants

A wide range of environmental contaminants, particularly those associated with the manufacture of various polymers, have been tested for toxic effects in Chinese hamsters. Rapid turnover, representative of cell death and renewal of pneumocytes, alveolar macrophages, and epithelial cells of the smaller bronchi and alveoli was observed after a 24-hr exposure to nitrogen dioxide (Hackett, 1979). Styrene (vinylbenzene), a monomer used in the synthesis of resins and rubbers, was markedly toxic in Chinese hamsters. This pollutant depleted glutathione in the liver and kidney and impaired enzyme activity in the liver of adult animals (Heinonen and Vainio, 1980). Accelerated embryonic death also was noted in pregnant Chinese hamsters after inhalation of styrene but malformation of the fetuses was nonexistent (Kankaanpaa et al., 1980). Inhalation of vinylidene chloride, used in manufacture of copolymers, also had minor toxic effects on enzyme activities in the kidney and liver (Oesch et al., 1983). Intratracheal instillation of benzo[a]pyrene, a potent aromatic hydrocarbon, has been reported to accentuate aryl hydrocarbon hydroxylase activity in the lungs, kidney, and liver of Chinese hamsters. This finding is considered to be critical because this microsomal enzyme may transform benzo[a]pyrene to a carcinogenic compound (Mitchell, 1980). Six months of exposure to diesel exhaust evoked severe changes in pulmonary capability of Chinese hamsters. Lung weight was increased whereas vital capacity and carbon monoxide transfer were depressed, probably due to emphysema (Vinegar et al., 1981).

C. Therapeutic Drugs

There have been limited studies concerning the potential toxicity of therapeutic compounds such as the antitumor agent cyclophosphamide and the synthetic estrogen diethylstilbestrol, for therapy of prostatic cancer. Acute oral administration of cyclophosphamide caused listlessness, decreased muscle tone, hyperreflexia, impaired gait, irregular pulmonary function, roughening of the hair, and advanced mortality in Chinese hamsters. Furthermore, the variation in LD$_{50}$ of cyclophosphamide was related to the time of the year in which it was administered (Pericin, 1981). The toxic effects of cyclophosphamide on Chinese hamster cell cultures were found to be relieved by additional therapy with mesna (Millar et al., 1983).

Reduction of cyclophosphamide toxicity was also observed in intestinal stem cells after treatment with lucanthone (Milligan et al., 1982). Diethylstilbestrol generated a dose-related jaundice with histological lesions in the liver. Although the mortality rates of males and females were equal, hyperbilirubinemia was more severe in the females (Coe et al., 1983).

D. Pesticides

The phenoxyherbicides, 2,4-dichlorophenoxyacetic acid and 4-chloro-2-methylphenoxyacetic acid, have been given by gavage to Chinese hamsters. Both compounds stimulated an increase in size and number of peroxisomes, usually predictive of neoplasia in hepatic cells. Furthermore, clofibrate, a phenoxy acid derivative, displayed an even greater capacity to induce proliferation of peroxisomes than either of the herbicides (Vainio et al., 1982).

E. Nitroso Compounds

The nitroso compounds are well recognized to be potent tumorigenic agents, particularly in the liver and to a lesser degree, in the kidney and gastrointestinal tube. The carcinogenicity of several of these nitroso compounds has been assessed in the Chinese hamster. Dibutylnitrosamine, chronically administered to Chinese hamsters, generated the formation of tumors in the urinary bladder, forestomach, and oral and nasal cavities, along with an increased mortality rate (Althoff et al., 1971). Long-term treatment with diethylnitrosamine was responsible for induction of cancerous tumors in the esophagus, stomach, and liver of Chinese hamsters (Baker et al., 1974). Shortened life span and various types of tumors have also been observed in Chinese hamsters treated with dimethylnitrosamine (Reznik et al., 1976a), N-nitrosomethylurea (Reznik et al., 1976b), and N-nitrosomorpholine and N-nitrosopiperidine (Reznik et al., 1976c).

F. Other Agents

A few miscellaneous agents have been verified to induce toxicity in various organs of Chinese hamsters. Caffeine, given chronically to potential sires, produce a shift in the sex ratio of their offspring to females (Weathersbee et al., 1975). A single intraperitoneal injection of mineral or pristane oil into Chinese hamsters generated an inflammatory process in the pancreas characterized by lymphoplasmacytosis and hyperplasia or metaplasia of the acinar, ductal, and endothelial cells. These structural aberrations were followed by granuloma formation which involved most of the exocrine pancreas (Yerganian et al., 1979).

V. GENETIC TOXICOLOGY

The Chinese hamster has served as an ideal model for genetic toxicology studies, primarily because of its large chromosomes and low chromosome number ($2N = 22$). A voluminous quantity of literature exists concerning the effects of a variety of insults on chromosome composition in certain cells of the Chinese hamster. These insults include radiation, environmental pollutants, therapeutic drugs, pesticides, artificial sweeteners, mycotoxins, vitamins, inhalation anesthetics, hallucinogenic drugs, nitroso compounds, ethanol, and other miscellaneous agents.

A. Radiation

X irradiation has been found to exert a profound deleterious effect on germ-cell chromosomes in the Chinese hamster. A high frequency of chromosome aberrations has been observed in oocytes at various stages during meiosis (Mikamo et al., 1981; Koishi, 1983). Likewise, chromatid breaks and reciprocal translocations were common manifestations of irradiated spermatogonia (Brewen and Preston, 1973).

With respect to somatic cell types, the incidence of chromosome damage to thyroid cells was closely correlated with the dose of X irradiation (Moore and Colvin, 1968a). Furthermore, it has been shown that X rays in combination with other agents exacerbate chromosome abnormalities in blood cells. Exposure to X rays and ozone or isoniazid greatly increased the number of chromosome breaks and modified the division of bone marrow cells (Miltenburger and Korte, 1976). Limited protection against X-irradiation-induced chromosome damage and alteration of mitotic rates in bone marrow cells was obtained by treatment with S-2-aminoethylisothiouronium bromide hydrobromide (Barta and Fremuth, 1976) and cyclic AMP (Fremuth and Kovar, 1978).

A wide variety of radioisotopes have induced chromosomal aberrations in Chinese hamsters. Cobalt-60 produced chromatid breaks, premature chromosome condensation, or abnormal mitoses in testes, bone marrow (Brooks and Lengemann, 1967), and liver (Brooks et al., 1971). Plutonium-239 elicited genetic damage to liver cells (Brooks et al., 1976a) and lymphocytes (Brooks et al., 1976b). Chromatid deletions and exchanges, isochromatid deletions and chromosome exchanges, and other abnormalities were common occurrences in the bone marrow of Chinese hamsters after exposure to ^{90}Sr and ^{90}Sr-^{90}Y (Brooks and McClellan, 1968, 1969). Thyroid cells displayed chromosome lesions after injury with ^{131}I (Moore and Colvin, 1968b). Chromosome aberrations in hepatic cells were detectable after injection with ^{144}Ce (Brooks et al., 1972a), ^{252}Cf (Brooks et al., 1972b), or [^{241}Am]citrate (McKay et al., 1972).

Genetic disruption due to microwave radiation also has been

explored in Chinese hamsters. "Chromosome stickiness" without chromatid aberrations was noted in bone marrow cells (Janes et al., 1969). Chromosome breaks in the corneal epithelial cells and lens opacity were found in Chinese hamsters after brief exposure to microwave radiation (Yao, 1978).

B. Environmental Pollutants

The number of environmental mutagens that have undergone genetic analysis in the Chinese hamster is enormous. Since detailed description of the genetic toxicology associated with each of these environmental contaminants goes far beyond the scope of this chapter, a list of these agents and the appropriate literature citations are shown in Table I.

C. Therapeutic Drugs

A vast array of therapeutic drugs has been carefully analyzed for genetic toxicity in the Chinese hamster. A list of these compounds and the most pertinent literature sources are provided in Table II.

D. Pesticides

A limited number of pesticides have been administered to Chinese hamsters to determine their ability to produce genetic anomalies. The herbicide 2,4,5-trichlorophenoxyacetic acid (2,4,5-T) was not genotoxic even at the highest dose (Herbold et al., 1982). Ethoxyquin, utilized in the control of scald on fruit trees, induced a depressed rate of micronucleus formation in polychromatic erythrocytes and severe chromosome injury in marrow cells (Renner, 1984). Dichlorvos, an organophosphorus insecticide, displayed no genetic side effects in either bone marrow cells or spermatogonia of Chinese hamsters (Dean and Thorpe, 1972).

E. Artificial Sweeteners

Since artificial sweeteners have been linked to disorders such as bladder cancer, both saccharin and cyclamate have been intensively studied for their ability to cause genetic damage. In general, saccharin administration in the Chinese hamster did not generate chromosome damage to bone marrow cells (van Went-de Vries and Kragten, 1975) or spermatogonia (Machemer and Lorke, 1975). However, an extremely high dose, 7.5 gm/kg, of this sweetener proved to be weakly mutagenic by causing sister chromatid exchange (Renner, 1979). Sodium cyclamate failed to induce chromosome aberrations in spermatogonia of Chinese hamsters (Machemer and Lorke, 1975). Although cyclamate per se did not cause genetic complications in the Chinese hamster, one of its metabolic by-products,

Table I

Environmental Pollutants Which Have Been Tested for Their Ability to Produce Genetic Aberrations in Chinese Hamsters

Pollutant	Genetic aberration(s)	Reference
Benzene	None	Siou et al. (1981)
3,4-Benzopyrene	Sister chromatid exchange in bone marrow cells	Bayer et al. (1981)
n-Butyl acrylate	None	Engelhardt and Klimisch (1983)
4-Chloromethyl-biphenyl	Sister chromatid exchange and dicentric chromosomes in bone marrow cells	Albanese (1982)
Chrysene	None	Basler et al. (1977)
Cigarette smoke	None	Madle et al. (1981)
Diesel exhaust	Increase of micronuclei in polychromatic erythrocytes and decrease in mitotic index of bone marrow cells	Pereira et al. (1981)
7,12-Dimethylbenzanthracene	Chromosome breakage in bone marrow cells	Kato et al. (1969)
Ethanol and cigarette smoke	Increase in mitotic index of bone marrow cells	Korte et al. (1981)
Formaldehyde	Chromatid gaps, breaks and exchanges of ovary cells	Natarajan et al. (1983)
Fluorescent whitening agents	None	Muller et al. (1975)
o-Phenylenediamine	Increase of micronuclei in polychromatic erythrocytes	Wild et al. (1980)
Phenanthrene	Sister chromatid exchange in bone marrow cells	Bayer (1978)
Smoke flavors	Sister chromatid exchange in bone marrow cells	Pool et al. (1983)
Sodium selenite	Chromatid breaks and sister chromatid exchange in bone marrow cells	Norppa et al. (1980a)
Styrene	None	Norppa et al. (1980b)
Vinyl chloride	Chromatid breaks, fragments, gaps, deletions and sister chromatid exchange in bone marrow cells	Basler and Rohrborn (1980)

Table II

Therapeutic Drugs Which Have Been Evaluated for Their Ability to Cause Genetic Damage in Chinese Hamsters

Drug	Genetic damage	Reference
Acyclovir	Chromosome breakage in bone marrow cells	Clive et al. (1983)
Benzodiazepine tranquilizers	None	Schmid and Staiger (1969)
Busulfan	Chromosome gaps in intestinal and bone marrow cells	Miltenburger et al. (1980)
3-Chinuclidylbenzilate and 3-chinuclidinol	Chromosome gaps and breaks in bone marrow cells	Sram (1975)
Cyclophosphamide	Chromosome gaps and exchanges in bone marrow cells and spermatogonia	Miltenburger et al. (1981)
Cytembena	Chromosome breaks and increase of micronuclei in polychromatic erythrocytes	Goetz et al. (1976)
Isoniazid	None	Adler et al. (1978)
6-Mercaptopurine	Chromosome breaks and gaps in bone marrow cells	Frohberg and Schencking (1975)
Phenylbutazone	None	Muller and Strasser (1971)
Pirenzepine	None	Baumeister (1980)
Povidone-iodine	None	Merkle and Zeller (1979)
Praziquantel	None	Machemer and Lorke (1978)
Propafenone hydrochloride	None	Rohrborn et al. (1980)
Quinine	None	Munzner and Renner (1983)
Sulfonylureas	Sister chromatid exchange in bone marrow cells	Renner and Munzner (1980)
Trenimon	Chromosome deletions, dicentrics, translocations, inversions and complex rearrangements in embryos	Binkert and Schmid (1977)
Triaziquone	Chromatid breaks, deletions and exchanges in oocytes	Hansmann et al. (1974)

cyclohexamine, was responsible for chromatid exchange and various types of chromatid breaks in lymphocytes (van Went-de Vries et al., 1975). In contrast, the bone marrow cells and spermatogonia were determined to be free of cyclohexamine-induced chromosome aberrations (Brewen et al., 1971; Machemer and Lorke, 1976).

F. Mycotoxins

Mycotoxins, the metabolites of fungi, have been identified as major food contaminants which can cause liver cancer and other serious health problems in animals and humans. Several of the mycotoxins such as aflatoxin B_1, aflatoxin G_1, and platulin were shown to produce severe chromosome and chromatid breaks in marrow cells of Chinese hamsters (Korte, 1980; Barta et al., 1984). Mycotoxin T-2 not only increased chromosome injury but also reduced the rate of division of marrow cells (Norppa et al., 1980c).

G. Vitamins

The only vitamin which has been tested for mutagenic properties in the Chinese hamster is ascorbic acid. Although sister chromatid exchanges were observed in fibroblasts in vitro, vitamin C had no remarkable genetic effects on bone marrow cells of Chinese hamsters (Speit et al., 1980).

H. Inhalation Anesthetics

Due to the apparent increased risk of carcinogenesis and other serious side effects with exposure to inhalation anesthetics such as halothane, these compounds have been evaluated in rodent models. Halothane-related chromosome damage was completely absent in Chinese hamster bone marrow after prolonged exposure (Basler and Rohrborn, 1981).

I. Hallucinogenic Drugs

Potential chromosome damage from the abuse of hallucinogenic drugs has been a serious concern for the human population. In order to evaluate the genotoxicity of these drugs, LSD was chronically administered to Chinese hamsters. Numerous chromosome breaks were found, but upon withdrawal of the drug a reduced frequency of abnormalities was observed (Moorthy and Mitra, 1978).

J. Nitroso Compounds

Although nitroso compounds appear to display antitumor activity, they also possess genotoxic capabilities. Intraperitoneal injection of N-methyl-N-nitrosourea to Chinese hamsters produced sister chromatid exchange in bone marrow cells. However, this genetic damage was prevented by pretreatment with n-alkanols (Stahl and Bayer, 1983).

K. Other Agents

The genotoxicity of a few other agents has been assessed in the Chinese hamster. Caffeine (Basler et al., 1979) and cocoa

(Renner and Munzner, 1982) elevated the rate of sister chromatid exchange in bone marrow cells. The flame retardants, tris(2,3-dibromopropyl) phosphate and tris(2-chlorethyl) orthophosphate displayed a dose-dependent increase of micronuclei in polychromatic erythrocytes of the bone marrow (Sala *et al.*, 1982). Single cell protein, a food supplement, was completely devoid of mutagenicity as determined by a series of genetic toxicology tests (Renner and Munzner, 1978).

VI. CYTOGENETICS OF GERM AND SOMATIC CELLS

A massive volume of literature concerning cytogenetic studies in the Chinese hamster has accrued primarily because of the small number of chromosomes ($2N = 22$) in this rodent species. This small number of chromosomes compared with humans and other vertebrates has allowed for in-depth microscopic analysis of banding patterns and systematic evaluation of mitosis, meiosis, spermatogenesis, and oogenesis. Since the literature concerning this field of investigation in the Chinese hamster is diffuse, Tables III and IV have been utilized to summarize the pertinent cytogenetic studies on the germ and somatic cells of this animal.

VII. CELL LINES DERIVED FROM CHINESE HAMSTER TISSUES AND ORGANS

Although the primary purpose of this chapter is to review the *in vivo* research involving the Chinese hamster, it is important to point out that a copious number of publications on cell lines derived from Chinese hamster tissues and organs has accumulated. These cell lines have been utilized for *in vitro* experimentation involving such diverse areas as cytogenetics, cytotoxicity, cell physiology, genotoxicity, and immunology. According to the literature, the two most prominent cell lines obtained from the Chinese hamster are the ovary (CHO) fibroblasts and V-79 fibroblasts (Bradley *et al.*, 1981). Other cell lines which have received less attention include lymphocytes, embryo fibroblasts, lung fibroblasts, spleen cells, peritoneal cells, and epithelial cells from kidney and seminiferous tubules. Numerous hybrid cell lines involving crosses between Chinese hamster and human or mouse cells along with mutants, cybrids, and spheroids have also been established. The sources and pertinent characteristics of some of the major Chinese hamster cell lines can be procured from the most recent catalog of the American Type Culture Collection of cell lines, viruses, and antisera.

Table III
Cytogenetic Studies of Male and Female Germ Cells in Chinese Hamsters

Study	Reference(s)
Chromosome activities during meiosis of male germ cells	Jhanwar *et al.* (1981)
Chromosome activities during meiosis of female germ cells	Mikamo and Sugawara (1980)
Chromosome anomalies of oocytes	Mikamo and Kamiguchi (1983)
Chromosome activities during spermatogenesis	Murkherjee and Ghosal (1969); Barcellona and Brinkley (1973)
Chromosome banding patterns of sperm	Utakoji (1966); Pathak *et al.* (1976); Vistorin *et al.* (1977)
Cytochemistry and behavior of dense bodies in sperm	Takanari *et al.* (1982)
Karotyping of synaptonemal complex in sperm	Dresser and Moses, 1980

Table IV
Cytogenetic Studies of Somatic Cells in The Chinese Hamster

Study	Reference(s)
Banding patterns of chromosomes	Raicu *et al.* (1970); Kakati and Sinha (1972); Ray and Mohandas (1976); Gamperl *et al.* (1978)
Chromosome activities during mitosis	Bregman (1972); Vig and Miltenburger (1976); Jhanwar and Chaganti (1981)
Chromosome analysis of Rous sarcoma viruses	Kato (1968)
Cytotaxonomy	Yerganian (1972)
DNA replication	Utakoji and Hsu (1965)
Effects of bromodeoxyuridine on chromosomes	Lin *et al.* (1984)

ACKNOWLEDGMENTS

The authors wish to thank Sheila Graber, Susan Fierke, and the staff of the Upjohn Company Technical Library for their excellent and expedient assistance with the literature search. The authors also are indebted to Linda Rogers of the Diabetes and Gastrointestinal Diseases Research Unit of the Upjohn Company for her outstanding and highly professional secretarial assistance.

REFERENCES

Adler, I. D., Schmaltz, A., Rathenberg, R., Muller, D., Strasser, F. F., and Perret, R. (1978). Cytogenetic study in spermatocytes of mice and Chinese hamsters after treatment with isoniazid (INH). *Hum. Genet.* **42**, 50–54.

Albanese, R. (1982). 4-Chloromethylbiphenyl (4CMB)—an *in vivo* cytogenetic study in the Chinese hamster (*Cricetulus griseus*). *Mutat. Res.* **100**, 309–12.

Albert, E. N. (1979). Current status of microwave effects on the blood brain barrier. *J. Microwave Power* **14**, 281–286.

Albert, E. N., and Kerns, J. M. (1981). Reversible microwave effects on the blood brain barrier. *Brain Res.* **230**, 153–164.

Althoff, J., Kruger, F. W., Mohr, U., and Schmahl, D. (1971). Dibutylnitrosamine carcinogenesis in Syrian, Golden and Chinese hamsters. *Proc. Soc. Exp. Biol. Med.* **136,** 168–173.

Archer, J., Kaye, R., and Mutter, G. (1980). Control of streptozotocin diabetes in Chinese hamsters by cultured mouse islet cells without immunosuppression: A preliminary report. *J. Surg. Res.* **28,** 77 85.

Baker, J. R., Mason, M. M., Yerganian, G., Weisburger, E. K., and Weisburger, J. H. (1974). Induction of tumors of the stomach and esophagus in inbred Chinese hamsters by oral diethylnitrosamine. *Proc. Soc. Exp. Biol. Med.* **146,** 291–293.

Barcellona, W. J., and Brinkley, B. R. (1973). Effects of actinomycin D on spermatogenesis in the Chinese hamster. *Biol. Reprod.* **8,** 335–349.

Barta, I., and Fremuth, F. (1976). Effect of chemical radioprotection on radiation injury of chromosomes. *Acta Univ. Carol., Med.* **22,** 263–288.

Barta, I., Adamkova, M., Markarjan, D., Adzigitov, F., and Prokes, K. (1984). The mutagenic activity of aflatoxin B_1 in the *Cricetulus griseus* hamster and *Macaca mulatta* monkey. *J. Hyg., Epidemiol., Microbiol., Immunol.* **28,** 149–59.

Basler, A., and Rohrborn, G. (1980). Vinyl chloride: An example for evaluating mutagenic effects in mammals *in vivo* after exposure to inhalation. *Arch. Toxicol.* **45,** 1–7.

Basler, A., and Rohrborn, G. (1981). Lack of mutagenic effects of halothane in mammals *in vivo*. *Anesthesiology* **55,** 143–147.

Basler, A., Herbold, B., Peter, S., and Rohrborn, G. (1977). Mutagenicity of polycyclic hydrocarbons. II. Monitoring genetical hazards of chrysene *in vitro* and *vivo*. *Mutat. Res.* **48,** 249–254.

Basler, A., Bachmann, U., Roszinsky-Koecher, G., and Rohrborn, G. (1979). Effects of caffeine on sister chromatid exchanges *in vivo*. *Mutat. Res.* **59,** 209–214.

Baumeister, M. (1980). Comparative studies on the mutagenicity of pirenzepine with submammalian bacterial and mammalian systems. *Toxicol. Appl. Pharmacol.* **54,** 384–391.

Bayer, U. (1978). *In vivo* induction of sister chromatid exchanges by 3 polyaromatic hydrocarbons. *Carcinog.—Compr. Surv.* **3,** 423–428.

Bayer, U., Siegers, C. P., and Younes, M. (1981). Enhancement of benzo[a]pyrene-induced sister chromatid exchanges as a consequence of glutathione depletion *in vivo*. *Toxicol. Lett.* **9,** 339–344.

Bell, R., and Hye, R. J. (1983). Animal models of diabetes mellitus: Physiology and pathology. *J. Surg. Res.* **35,** 433–460.

Benjamin, S. A., Brooks, A. L., and McClellan, R. O. (1976). Biological effectiveness of ^{239}Pu, ^{144}Ce and ^{90}Sr citrate in producing chromosome damage, bone-related tumours, liver tumours and life shortening in the Chinese hamster. *In* "Biological and Environmental Effects of Low-Level Radiation" (M. Lewis, ed.), Vol. 1, pp. 143–152. IAEA, Vienna.

Berman, L. D., Hayes, J. A., and Sibay, T. M. (1973). Effect of streptozotocin in the Chinese hamster (*Cricetulus griseus*). *J. Natl. Cancer Inst. (U.S.)* **51,** 1287–1294.

Bestetti, G., and Rossi, G. L. (1982). Hypothalamic changes in diabetic Chinese hamsters. A semiquantitative, light and electron microscopic study. *Lab. Invest.* **47,** 516–522.

Billington, C. J., Morley, J. E., Levine, A. S., and Gerritsen, G. C. (1984). Feeding systems in Chinese hamsters. *Am. J. Physiol.* **247,** 405–411.

Binkert, F., and Schmid, W. (1977). Pre-implantation embryos of Chinese hamster. II. Incidence and type of karyotype anomalies after treatment of the paternal post-meiotic germ cells with an alkylating mutagen. *Mutat. Res.* **46,** 77–86.

Boquist, L. (1967). Some aspects on the blood glucose regulation and the glutathione content of the non-diabetic adult Chinese hamster, *Cricetulus griseus*. *Acta Soc. Med. Ups.* **72,** 358–375.

Boquist, L. (1968a). Alloxan administration in the Chinese hamster. *Virchows Arch. B* **1,** 157–168.

Boquist, L. (1968b). Alloxan administration in the Chinese hamster. II. Ultrastructural study of degeneration and subsequent regeneration of the pancreatic islet tissue. *Virchows Arch. B* **1,** 169–181.

Boquist, L. (1969). Pancreatic islet morphology in diabetic Chinese hamsters. A light microscopic and electron microscopic study. *Acta Pathol. Microbiol. Scand.* **75,** 399–414.

Boquist, L., and Falkmer, S. (1970). Morphologic changes in the pancreatic islets of the Chinese hamster in spontaneous diabetes and some experimental conditions. *Z. Versuchstierkd.* **12,** 96–99.

Boquist, L., and Lernmark, A. (1969). Effects on the endocrine pancreas in Chinese hamsters fed zinc-deficient diets. *Acta Pathol. Microbiol. Scand.* **76,** 215–228.

Bradley, M. A., Bhuyan, B., Francis, M. C., Langenbach, R., Peterson, A., and Huberman, E. (1981). Mutagenesis by chemical agents in V79 Chinese hamster cells: A review and analysis of the literature. *Mutat. Res.* **87,** 81–142.

Bregman, A. A. (1972). Asynchronous centromere division in the Chinese hamster. *Cytogenetics* **11,** 102–112.

Brewen, J. G., and Preston, R. J. (1973). Chromosomal interchanges induced by radiation in spermatogonial cells and leukocytes of mouse and Chinese hamster. *Nature (London), New Biol.* **244,** 111–113.

Brewen, J. G., Pearson, F. G., Jones, K. P., and Luippold, H. E. (1971). Cytogenetic effects of cyclohexylamine and N-OH-cyclohexylamine on human leucocytes and Chinese hamster bone marrow. *Nature (London), New Biol.* **230,** 15–16.

Bringer, J., Heldt, A., and Grodsky, G. M. (1981). Prevention of insulin aggregation by dicarboxylic amino acids during prolonged infusion. *Diabetes* **30,** 83–85.

Brooks, A. L., and Lengemann, F. W. (1967). Comparison of radiation-induced chromatid aberrations in the testes and bone marrow of the Chinese hamster. *Radiat. Res.* **32,** 587–595.

Brooks, A. L., and McClellan, R. O. (1968). Cytogenetic effects of strontium-90 on the bone marrow of the Chinese hamster. *Nature (London)* **219,** 761–763.

Brooks, A. L., and McClellan, R. O. (1969). Chromosome aberrations and other effects produced by ^{90}Sr-^{90}Y in Chinese hamsters. *Wenner-Gren Cent. Int. J. Radiat. Biol.* **16,** 545–561.

Brooks, A. L., Mead, D. K., and Peters, R. F. (1971). Effect of chronic exposure to ^{60}Co on the frequency of metaphase chromosome aberrations in the liver cells of the Chinese hamster *in vivo*. *Int. J. Radiat. Biol.* **20,** 599–604.

Brooks, A. L., McClellan, R. O., and Benjamin, S. A. (1972a). The effects of ^{144}Ce-^{144}Pr on the metaphase chromosomes of the Chinese hamster liver cells *in vivo*. *Radiat. Res.* **52,** 481–498.

Brooks, A. L., Mewhinney, J. A., and McClellan, R. O. (1972b). The *in vivo* cytogenetic effects of ^{252}Cf on liver and bone marrow of the Chinese hamster. *Health Phys.* **22,** 701–706.

Brooks, A. L., McClellan, R. O., Peters, R. F., and Mead, D. K. (1976a). Effect of size and alpha flux of plutonium-239 oxide particles on production of chromosome aberrations in liver of Chinese hamster. *In* "Biological and Environmental Effects of Low-Level Radiation" (M. Lewis, ed.), Vol. 1, pp. 131–142. IAEA, Vienna.

Brooks, A. L., LaBauve, R. J., McClellan, R. O., and Jensen, D. A. (1976b). Chromosome aberration frequency in blood lymphocytes of animals with plutonium-239 lung burdens. *ERDA Symp. Ser.* **37,** 106–111.

Brooks, A. L., Benjamin, S. A., Jones, R. K., and McClellan, R. O. (1982). Interaction of ^{144}Ce and partial hepatectomy in the production of liver neoplasms in the Chinese hamster. *Radiat. Res.* **91,** 573–588.

Brooks, A. L., Benjamin, S. A., Hahn, F. F., Brownstein, D. G., Griffith, W. C., and McClellan, R. O. (1983). The induction of liver tumors by plutonium-239 citrate or plutonium-239 dioxide particles in the Chinese hamster. *Radiat. Res.* **96,** 135–150.

Butler, L. (1967). The inheritance of diabetes in the Chinese hamster. *Diabetologia* **3,** 124–129.

Butler, L., and Gerritsen, G. C. (1970). A comparison of the modes of inheritance of diabetes in the Chinese hamster and the KK mouse. *Diabetologia* **6,** 163–167.

Campbell, J., Rastogi, K. S., and Hausler, H. R. (1966). Hyperinsulinemia with diabetes induced by cortisone, and the influence of growth hormone in the Chinese hamster. *Endocrinology (Baltimore)* **79**, 749–756.

Carpenter, A. M., Gerritsen, G. C., Dulin, W. E., and Lazarow, A. (1967). Islet and beta cell volumes in offspring of diabetic Chinese hamsters and their nondiabetic siblings. *Diabetologia* **3**, 92–96.

Carpenter, A. M., Gerritsen, G. C., Dulin, W. E., and Lazarow, A. (1970). Islet and beta cell volumes in offspring of severely diabetic ketotic Chinese hamsters. *Diabetologia* **6**, 168–176.

Chang, A. Y. (1970). Insulin synthesis and secretion by isolated islets of spontaneously diabetic Chinese hamsters. *Int. Symp. Ser.* **16**, 515–552.

Chang, A. Y. (1974). Hepatic and renal protein synthesis in normal diabetic and ketotic Chinese hamsters. *Diabetologia* **10**, 555–558.

Chang, A. Y. (1978). Spontaneous diabetes in animals. *Gen. Pharmacol.* **9**, 447–450.

Chang, A. Y. (1981). Biochemical abnormalities in the Chinese hamster, *Cricetulus griseus* with spontaneous diabetes. *Int. J. Biochem.* **13**, 41–44.

Chang, A. Y., and Schneider, D. I. (1970a). Metabolic abnormalities in the pancreatic islets and livers of the diabetic Chinese hamster. *Diabetologia* **6**, 180–185.

Chang, A. Y., and Schneider, D. I. (1970b). Rate of gluconeogenesis and levels of gluconeogenic enzymes in liver and kidney of diabetic and normal Chinese hamsters. *Biochim. Biophys. Acta* **222**, 587–592.

Chang, A. Y., Noble, R. E., and Wyse, B. M. (1977a). Spontaneous diabetes in the Chinese hamster: Comparative studies in M and L lines. *Int. Congr. Ser.—Excepta Med.* **413**, 691–702.

Chang, A. Y., Noble, R. E., and Wyse, B. M. (1977b). Comparison of highly inbred diabetic and nondiabetic lines in the Upjohn colony of Chinese hamsters. *Diabetes* **26**, 1063–1071.

Chang. A. Y., Noble, R. E., and Greenberg, H. S. (1977c). Variance in LDH isoenzyme patterns in a Chinese hamster (*Cricetulus griseus*) colony (EN). *Comp. Biochem. Physiol. B* **58B**, 119–123.

Chang, A. Y., Noble, R. E., and Wyse, B. M. (1977d). Streptozotocin induced diabetes in the Chinese hamster: Biochemical and endocrine disorders. *Diabetologia* **13**, 595–602.

Chang, A. Y., Perry, C. S., and Wyse, B. M. (1979). Alterations in glycohydrolase activities in streptozotocin diabetic Chinese hamsters (*Cricetulus griseus*). *Comp. Biochem. Physiol. B* **63B**, 341–344.

Chang, A. Y., Wyse, B. M., Gilchrist, B. J., Peterson, T., and Diani, A. R. (1983). Ciglitazone, a new hypoglycemic agent. 1. Studies in ob/ob and db/db mice, diabetic Chinese hamsters, and normal and streptozotocin-diabetic rats. *Diabetes* **32**, 830–838.

Chatterjee, A. K., Murthi, P. S. R., and Murkherjee, S. K. (1980). Effect of centpiperalone in insulin-deficient diabetes. *Indian J. Exp. Biol.* **18**, 1005–1008.

Chobanian, A. V., Gerritsen, G. C., Brecher, P. I., and McCombs, L. (1974a). Aortic glucose metabolism in the diabetic Chinese hamster. *Diabetologia* **10**, 589–593.

Chobanian, A. V., Gerritsen, G. C., Brecher, P. I., and Kessler, M. (1974b). Cholesterol metabolism in the diabetic Chinese hamster. *Diabetologia* **10**, 595–600.

Clive, D., Turner, N. T., Hozier, J., Batson, A. G., and Tucker, W. E. (1983). Preclinical toxicology studies with acyclovir: Genetic toxicity tests. *Fundam. Appl. Toxicol.* **3**, 587–602.

Coe, J. E., Ishak, K. G., and Ross, M. J. (1983). Diethylstilbestrol-induced jaundice in the Chinese and Armenian hamster. *Hepatology* **3**, 489–496.

Conforti, A. (1972). Ultrastructural changes in the kidney in spontaneous diabetes of the Chinese hamster. *Cricetulus griseus. Acta Diabetol. Lat.* **9**, 655–668.

Copeland, E. J., and Ginsberg, L. C. (1982). Major plasma glycoproteins of diabetic and nondiabetic Chinese hamsters. *J. Hered.* **73**, 311–313.

Copeland, J., Blashfield, K., Bauer, B., Gerritsen, G. C., and Ginsberg, L. C. (1982). Plasma glycoproteins of diabetic and normal Chinese hamsters. *Experientia* **38**, 301–302.

Dail, W. G., Evan, A. P., Gerritsen, G. C., and Dulin, W. E. (1977).Abnormalities in pelvic visceral nerves: A basis for neurogenic bladder in the diabetic Chinese hamster. *Invest. Urol.* **15**, 161–166.

Dean, B. J., and Thorpe, E. (1972). Cytogenetic studies with dichlorvos in mice and Chinese hamsters. *Arch. Toxikol.* **30**, 39–49.

Deslex, P., and Rossi, G. L. (1976). Quantitative evaluation of somatotropic cells in the adenohypophysis of normal and diabetic male Chinese hamsters. *Diabetologia* **12**, 489–493.

Diani, A. R., Gerritsen, G. C., Stromsta, S., and Murray, P. (1976). A study of the morphological changes in the small intestine of the spontaneously diabetic Chinese hamster. *Diabetologia* **12**, 101–109.

Diani, A. R., Grogan, D. M., Yates, M. E., Risinger, D. L., and Gerritsen, G. C. (1979). Radiologic abnormalities and autonomic neuropathology in the digestive tract of the ketonuric diabetic Chinese hamster. *Diabetologia* **17**, 33–40.

Diani, A. R., Davis, D. E., Fix, J. D., Swartzman, J., and Gerritsen, G. C. (1981a). Morphometric analysis of autonomic neuropathology in the abdominal sympathetic trunk of the ketonuric diabetic Chinese hamster. *Acta Neuropathol.* **53**, 293–298.

Diani, A. R., Stutsman, S. G., Eldridge, D. W., and Gerritsen, G. C. (1981b). Alteration of aerobic microflora populations in the intestine of the ketonuric diabetic Chinese hamster. *Microbios Lett.* **15**, 135–140.

Diani, A. R., Weaver, E. A., and Gerritsen, G. C. (1981c). Capillary basement membrane thickening associated with the small intestine of the ketonuric diabetic Chinese hamster. *Lab. Invest.* **44**, 388–391.

Diani, A. R., Peterson, T., and Gilchrist, B. J. (1983). Islet innervation of nondiabetic and diabetic Chinese hamsters. 1. Acetylcholinesterase histochemistry and norepinephrine fluorescence. *J. Neural Transm.* **56**, 223–238.

Diani, A. R., West, C., Vidmar, T. J., Peterson, T., and Gerritsen, G. C. (1984). Morphometric analysis of the vagus nerve in nondiabetic and ketonuric diabetic Chinese hamsters. *J. Comp. Pathol.* **94**, 495–504.

Diani, A. R., Sawada, G. A., Peterson, T., Wyse, B. M., Blanks, M. C., Vidmar, T. J., and Gerritsen, G. C. (1986a). Systematic evaluation of microangiopathy in diabetic Chinese hamsters. I. Morphometric analysis of minimal glomerular basement membrane thickness in 11–15 and 19–23 month old Chinese hamsters. *Microvasc. Res.* **31**, 306–316.

Diani, A. R., Peterson, T., Sawada, G. A., Wyse, B. M., Blanks, M. C., Gerritsen, G. C., Terenghi, G., Varndell, I., Polak, J. M., Blank, M. A., and Bloom, S. R. (1985b). Elevated levels of vasoactive intestinal peptide in the eye and urinary bladder of diabetic and prediabetic Chinese hamsters. *Diabetologia* **28**, 302–307.

Dresser, M. E., and Moses, M. J. (1980). Synaptonemal complex karyotyping in spermatocytes of the Chinese hamster (*Cricetulus griseus*). IV. Light and electron microscopy of synapsis and nucleolar development by silver staining. *Chromosoma* **76**, 1–22.

Dulin, W. E., and Gerritsen, G. C. (1967). Summary of biochemical, physiological and morphological changes associated with diabetes in the Chinese hamster. *Int. Congr. Ser.—Excerpta Med.* **172**, 806–812.

Dulin, W. E., and Gerritsen, G. C. (1972). Interaction of genetics and environment on diabetes in the Chinese hamster as compared with human and other diabetic animal species. *Acta Diabetol. Lat.* **9**, 48–84.

Dulin, W. E., and Soret, M. G. (1977). Chemically and hormonally induced diabetes. *In* "The Diabetic Pancreas" (B. W. Volk and K. F. Wellman, eds.), pp. 425–465. Plenum, New York.

Dulin, W. E., Chang, A. Y., and Gerritsen, G. C. (1970a). Comparison of diabetes in the Chinese hamster, KK mouse and db mouse. *Int. Congr. Ser.—Excerpta Med.* **231**, 868–800.

Dulin, W. E., Chang, A. Y., and Gerritsen, G. C. (1970b). Spontaneous diabetes in the Chinese hamster. *Colloq. Int. C.N.R.S.* **924**, 133–153.

Dulin, W. E., Gerritsen, G. C., Chang, A. Y., Anderson, N. L., Tollaksen, S., and Sammons, D. W. (1984). Characteristics of the prediabetic state in the Chinese hamster and evaluation of environmental factors that affect expression of the diabetic genes. *In* "Lessons from Animal Diabetes" (E.

Shafrir and A. E. Renold, eds.), pp. 199–209. John Libbey, London and Paris.

Engelhardt, G., and Klimisch, H. J. (1983). n-Butyl acrylate: Cytogenetic investigations in the bone marrow of Chinese hamsters and rats after 4-day inhalation. *Fundam. Appl. Toxicol.* **3,** 640–641.

Federman, J. L., and Gerritsen, G. C. (1970). The retinal vasculature of the Chinese hamster: A preliminary study. *Diabetologia* **6,** 186–191.

Feingold, K. R., Lear, S. R., and Moser, A. H. (1984). De novo cholesterol synthesis in three different animal models of diabetes. *Diabetologia* **26,** 234–239.

Fletcher-McGruder, B. L., and Gerritsen, G. C. (1984). Deficient humoral antibody response of the spontaneously diabetic Chinese hamster. *Proc. Soc. Exp. Biol. Med.* **175,** 74–78.

Fletcher-McGruder, B. L., and Gerritsen, G. C. (1985). Effect of anti-lymphocyte serum and neonatal thymectomy upon onset of diabetes in the Chinese hamster. *Proc. Soc. Exp. Biol. Med.* **180,** 92–97.

Frankel, B. J., and Grodsky, G. M. (1979). Effect of continuous low-dose insulin treatment on subsequent incidence of diabetes in genetically prediabetic Chinese hamsters. *Diabetes* **28,** 544–547.

Frankel, B. J., Gerich, J. E., Hagura, R., Fanska, R. E., Gerritsen, G. C., and Grodsky, G. M. (1974). Abnormal secretion of insulin and glucagon by the *in vitro* perfused pancreas of the genetically diabetic Chinese hamster. *J. Clin. Invest.* **53,** 1637–1646.

Frankel, B. J., Gerich, J. E., Fanska, R. E., Gerritsen, G. C., and Grodsky, G. M. (1975). Responses to arginine of the perfused pancreas of the genetically diabetic Chinese hamster. *Diabetes* **24,** 273–279.

Frankel, B. J., Schmid, F. G., and Grodsky, G. M. (1979). Effect of continuous insulin infusion with an implantable seven-day minipump in the diabetic Chinese hamster. *Endocrinology (Baltimore)* **104,** 1532–1539.

Frankel, B. J., Heldt, A. M., and Grodsky, G. M. (1982). Insulin and glucagon release in the diabetic Chinese hamster: Differences among inbred sublines. *Diabetologia* **22,** 292–295.

Frankel, B. J., Heldt, A. M., Gerritsen, G. C., and Grodsky, G. M. (1984). Insulin, glucagon and somatostatin release from the prediabetic Chinese hamster. *Diabetologia* **27,** 387–391.

Fremuth, F., and Kovar, J. (1978). Effects of cyclic nucleotides on radiosensitivity and proliferation *in vivo*. *Stud. Biophys.* **68,** 47–60.

Frohberg, H., and Schencking, M. S. (1975). *In vivo* cytogenetic investigations in bone marrow cells of rats, Chinese hamsters and mice treated with 6-mercaptopurine. *Arch. Toxicol.* **33,** 209–224.

Funaki, K., and Mikamo, K. (1983). Developmental stage-dependent teratogenic effects of maternal spontaneous diabetes in the Chinese hamster. *Diabetes* **32,** 637–643.

Gamperl, R., Vistorin, G., and Rosenkranz, W. (1978). Comparison of chromosome banding patterns in 5 members of Cricetinae with comments on possible relationships. *Caryologia* **31,** 343–354.

Gerritsen, G. C. (1975). Experimental prevention of diabetes in prediabetics: Studies of Chinese hamsters. *Compr. Ther.* **1,** 25–29.

Gerritsen, G. C. (1982). The Chinese hamster as a model for the study of diabetes mellitus. *Diabetes* **31,** 14–23.

Gerritsen, G. C., and Blanks, M. C. (1974). Characterization of Chinese hamsters by metabolic balance, glucose tolerance and insulin secretion. *Diabetologia* **10,** 493–500.

Gerritsen, G. C., and Blanks, M. C. (1979). Effect of diet limitation on development of diabetes in nonhyperphagic prediabetic Chinese hamsters. *Adv. Exp. Med. Biol.* **119,** 237–242.

Gerritsen, G. C., and Dulin, W. E. (1966). Serum proteins of Chinese hamsters and response of diabetics to tolbutamide and insulin. *Diabetes* **15,** 331–335.

Gerritsen, G. C., and Dulin, W. E. (1967). Characterization of diabetes in the Chinese hamster. *Diabetologia* **3,** 74–84.

Gerritsen, G. C., and Dulin, W. E. (1972). Effect of diet restriction on onset of development of diabetes in prediabetic Chinese hamsters. *Acta Diabetol. Lat.* **9,** 597–613.

Gerritsen, G. C., Needham, L. B., Schmidt, F. L., and Dulin, W. E. (1970). Studies on the prediction and development of diabetes in offspring of diabetic Chinese hamsters. *Diabetologia* **6,** 158–162.

Gerritsen, G. C., Johnson, M. A., Soret, M. G., and Schultz, J. R. (1974a). Epidemiology of Chinese hamsters and preliminary evidence for genetic heterogeneity of diabetes. *Diabetologia* **10,** 581–588.

Gerritsen, G. C., Blanks, M. C., Miller, R. L., and Dulin, W. E. (1974b). Effect of diet limitation on the development of diabetes in prediabetic Chinese hamsters. *Diabetologia* **10,** 559–565.

Gerritsen, G. C., Blanks, M. C., Schmidt, F. L., and Dulin, W. E. (1976). Environmental influences on the manifestation of diabetes mellitus in Chinese hamsters. *In* "The Genetics of Diabetes Mellitus" (W. Creutzfeldt, J. Kobberling, and J. V. Neel, eds.), pp. 165–187. Springer-Verlag, Berlin and New York.

Gerritsen, G. C., Connell, M. A., and Blanks, M. C. (1981). Effect of environmental factors including nutrition on genetically determined diabetes of Chinese hamsters. *Proc. Nutr. Soc.* **40,** 237–245.

Gerritsen, G. C., Dulin, W.E., Connell, M. A., Davis, V. I., Miller, R. L., and Butler, L. (1984). Epidemiology and genetics of the spontaneously diabetic Chinese hamster. *In* "Lessons from Animal Diabetes" (E. Shafrir and A. E. Renold, eds.), pp. 81–92. John Libbey, London and Paris.

Ghosh, M., Hausler, H. R., Basu, P. K., and Stachowska, B. (1970). An electron microscopic study of retinal capillaries of normal and spontaneously diabetic Chinese hamsters. *Can. J. Ophthalmol.* **5,** 187–193.

Ginsberg, L. C., Wyse, B. M., and Chang. A. (1981). Analysis of glycosaminoglycan from diabetic and normal Chinese hamster cells. *Diabetes* **30,** 393–395.

Gishizky, M. L., Frankel, B. J., and Grodsky, G. M. (1984). Cyclophosphamide treatment of prediabetic Chinese hamsters. *Acta Physiol. Scand.* **121,** 81–84.

Goetz, P., Sram, R. J., Kodykova, I., Dohnalova, J. Dostalova, O., and Bartova, J. (1976). Relationship between experimental results in mammals and man. II. Cytogenetic analysis of bone marrow cells after treatment of cytembena and cyclophosphamide–cytembena combination. *Mutat. Res.* **41,** 143–152.

Grodsky, G. M., and Frankel, B. J. (1981). Diabetes mellitus in the Chinese hamster. *In* "Etiology and Pathogenesis of Insulin-Dependent Diabetes Mellitus" (J. M. Martin, R. M. Ehrlich, and F. J. Holland, eds.), pp. 239–250. Raven Press, New York.

Grodsky, G. M., Frankel, B. J., Gerich, K. E., and Gerritsen, G. C. (1974). The diabetic Chinese hamster: *In vitro* insulin and glucagon release; the "chemical diabetic"; and the effect of diet on ketonuria. *Diabetologia* **10,** 521–528.

Hackett, N. A. (1979). Proliferation of lung and airway cells induced by nitrogen dioxide. *J. Toxicol. Environ. Health* **5,** 917–928.

Hansmann, I., Neher, J., and Rohrborn, G. (1974). Chromosome aberrations in metaphase II oocytes of Chinese hamster (*Cricetulus griseus*). I. The sensitivity of the preovulatory phase to triaziquone. *Mutat. Res.* **25,** 347–359.

Heinonen, T., and Vainio, H. (1980). Vinyl toluene-induced changes in xenobiotic-metabolizing enzyme activities and tissue glutathione content in various rodent species. *Biochem. Pharmacol.* **29,** 2675–2680.

Heisig, N., and Schall, J. (1971). Investigations on pregnancy and offspring of the spontaneously diabetic Chinese hamster. *Arch. Gynaekol.* **210,** 21–28.

Hepp, K. D., Langley, J., von Funcke, H. J., Renner, R., and Kemmler, W. (1975). Increased insulin-binding capacity of liver membranes from diabetic Chinese hamsters. *Nature (London)* **258,** 154.

Herbold, B. A., Machemer, L., and Rohrborn, G. (1982). Mutagenicity studies with 2,4,5-T on bacteria and mammalian germ cells. *Teratog., Carcinog., Mutagen.* **2,** 91–101.

Holcomb, G. N., Klemm, L. A., and Dulin, W. E. (1974). The polyol pathway for glucose metabolism in tissues from normal diabetic and ketotic Chinese hamsters. *Diabetologia* **10,** 549–555.

Hunt, C. E., Lindsey, S. R., and Walkley, S. U. (1976). Animal models of diabetes and obesity including the PBB/L1 mouse. *Fed. Proc., Fed. Am. Soc. Ex. Biol.* **35,** 1206–1217.

Janes, D. E., Leach, W. M., Mills, W. A., Moore, R. T., and Shore, M. L. (1969). Effects of 2450-MHz microwaves on protein synthesis and on chromosomes in Chinese hamsters. *Non-Ioniz. Radiat.* **1,** 125–130.

Jhanwar, S. C., and Chaganti, R. S. K. (1981). Pachytene chrommere maps of Chinese hamster autosomes. *Cytogenet. Cell Genet.* **31,** 70–76.

Jhanwar, S. C., Prensky, W., and Chaganti, R. S. K. (1981). Localization and metabolic activity of ribosomal genes in Chinese hamster meiotic and mitotic chromsomes. *Cytogenet. Cell Genet.* **30,** 39–46.

Kakati, S., and Sinha, A. K. (1972). Banding patterns of Chinese hamster chromosomes. *Genetics* **72,** 357–362.

Kankaanpaa, J. T. J., Elovaara, E.. Hemminki, K., and Vainio, H. (1980). Effect of maternally inhaled styrene on embryonal and fetal development in mice and Chinese hamsters. *Acta Pharmacol. Toxicol.* **47,** 127–129.

Kato, R. (1968). The chromosomes of forty-two primary Rous sarcomas of the Chinese hamster. *Hereditas* **59,** 63–119.

Kato, R., Bruze, M., and Tegner, Y. (1969). Chromosome breakage induced *in vivo* by a carcinogenic hydrocarbon in bone marrow cells of the Chinese hamster. *Hereditas* **61,** 1–8.

Kennedy, W. R., Quick, D. C., Miyoshi, T., and Gerritsen, G. C. (1982). Peripheral neurology of the diabetic Chinese hamster. *Diabetologia* **23,** 445–451.

Koishi, T. (1983). Chromosomal radiosensitivity in *meiotic oocytes* of Chinese hamsters. *Nippon Sanka Fujinka Gakkai Zasshi* **35,** 351–358.

Komeda, K., Yokote, M., and Oki, Y. (1980). Diabetic syndrome in the Chinese hamster induced with monosodium glutamate. *Experientia* **36,** 232–234.

Korte, A. (1980). Chromosomal analysis in bone marrow cells of Chinese hamsters after treatment with mycotoxins. *Mutat. Res.* **78,** 41–49.

Korte, A., Wagner, H. M., and Obe, G. (1981). Simultaneous exposure of Chinese hamsters to ethanol and cigarette smoke: Cytogenetic aspects. *Toxicology* **20,** 237–246.

Lappenbusch, W. L. (1972). Effect of circadian rhythm on the radiation response of the Chinese hamster (*Cricetulus griseus*). *Radiat. Res.* **50,** 600–610.

Lappenbusch, W. L., Gillespie, L. J., Leach, W. M., and Anderson, G. E. (1973). Effect of 2450-MHz microwaves on the radiation response of x-irradiated Chinese hamsters. *Radiat. Res.* **54,** 294–303.

Like, A. A., Gerritsen, G. C., Dulin, W. E., and Gaudreau, P. (1974a). Studies in the diabetic Chinese hamster: Light microscopy and autoradiography of pancreatic islets. *Diabetologia* **10,** 501–508.

Like, A. A., Gerritsen, G. C., Dulin, W. E., and Gaudreau, P. (1974b). Studies in the diabetic Chinese hamster: Electron microscopy of pancreatic islets. *Diabetologia* **10,** 509–520.

Lin, M. S., Takabayashi, T., Wilson, M. G., and Marchese, C. A. (1984). An *in vitro* and *in vivo* study of a bromodeoxyuridine-sensitive fragile site in the Chinese hamster. *Cytogenet. Cell Genet.* **38,** 211–215.

Losert, W., Rilke, A., Loge, O., and Richter, K. D. (1971). Comparative biochemical studies on the diabetogenic action of streptozotocin in mice, rats, Chinese hamsters and guinea pigs. *Arzneim.-Forsch.* **21,** 1643–1653.

Luse, S. A., Caramia, F., Gerritsen, G., and Dulin, W. E. (1967). Spontaneous diabetes mellitus in the Chinese hamster: An electron microscopic study of the islets of Langerhans. *Diabetologia* **3,** 97–108.

Luse, S. A., Gerritsen, G. C., and Dulin, W. E. (1970). Cerebral abnormalities in diabetes mellitus: An ultrastructural study of the brain in early-onset diabetes mellitus in the Chinese hamster. *Diabetologia* **6,** 192–198.

McCombs, H. L., Gerritsen, G. C., Dulin, W. E., and Chobanian, A. V. (1974). Morphologic changes in the aorta of the diabetic Chinese hamster. *Diabetologia* **10,** 601–606.

Machemer, L., and Lorke, D. (1975). Method for testing mutagenic effects of chemicals on spermatogonia of the Chinese hamster: Results obtained with cyclophosphamide, saccharin, and cyclamate. *Arzneim.-Forsch.* **25,** 1889–1896.

Machemer, L., and Lorke, D. (1976). Evaluation of the mutagenic potential of cyclohexylamine on spermatogonia of the Chinese hamster. *Mutat. Res.* **40,** 243–250.

Machemer, L., and Lorke, D. (1978). Mutagenicity studies with praziquantel, a new anthelmintic drug, in mammalian systems. *Arch. Toxicol.* **39,** 187–197.

McKay, L. R., Brooks, A. L., and McClellan, R. O. (1969). Metabolism and toxicity of ^{241}Am in the Chinese hamster. *U.S. A.E.C. Lovelace Found.*, pp. 203–207.

McKay, L. R., Brooks, A. L., and McClellan, R. O. (1972). The retention, distribution, dose and cytogenetic effects of [^{241}Am]citrate in the Chinese hamster. *Health Phys.* **22,** 633–640.

Madle, S., Korte, A., and Obe, G. (1981). Cytogenetic effects of cigarette smoke condensates *in vitro* and *in vivo*. *Hum. Genet.* **59,** 349–352.

Malaisse, W., Malaisse-Lagae, F., Gerritsen, G. C., and Dulin, W. E. (1967). Insulin secretion *in vitro* by the pancreas of the Chinese hamster. *Diabetologia* **3,** 109–114.

Meier, H., and Yerganian, G. A. (1959). Spontaneous hereditary diabetes mellitus in Chinese hamster (*Cricetulus griseus*). I. Pathological findings. *Proc. Soc. Exp. Biol. Med.* **100,** 810–815.

Meier, H., and Yerganian, G. A. (1961). Spontaneous diabetes mellitus in the Chinese hamster. II. Findings in the offspring of diabetic parents. *Diabetes* **10,** 12–18.

Merkle, J., and Zeller, H. (1979). Absence of povidone iodine-induced mutagenicity in mice and hamsters. *J. Pharm. Sci.* **68,** 100–102.

Mikamo, K., and Kamiguchi, Y. (1983). Primary incidences of spontaneous chromosomal anomalies and their origins and causal mechanisms in the Chinese hamster. *Mutat. Res.* **108,** 265–78.

Mikamo, K., and Sugawara, S. (1980). Colchicine-induced abnormal meiotic chromosomal segregation in primary oocytes of the Chinese hamster 2. Anaphase lagging. *Jpn. J. Hum. Genet.* **25,** 241–248.

Mikamo, K., Kamiguchi, Y., Funaki, K., Sugawara, S., and Tateno, H. (1981). Stage-dependent changes of chromosomal radiosensitivity in primary oocytes of the Chinese hamster. *Cytogenet. Cell Genet.* **30,** 174–178.

Millar, B. C., Millar, J. L., Clutterbuck, R., and Jinks, S. (1983). Studies on the toxicity of cyclophosphamide in combination with mesna *in vitro* and *in vivo*. *Cancer Treat Rev.* **10,** Suppl. A, 63–72.

Milligan, A. J., Whittington, R., and Leeper, D. B. (1982). Lucanthone modification of cyclophosphamide toxicity in the Chinese hamster. *Int. J. Radiat. Oncol., Biol. Phys.* **8,** 667–670.

Miltenburger, H. G., and Korte, A. (1976). On the simultaneous effects of ionising radiation and chemical agents on animal cells. 1st communication. Cytogenetic studies on X-rays and isonicotinic acid hydrazide in the Chinese hamster, *Cricetulus griseus*. *Arzneim.-Forsch.* **26,** 1303–1307.

Miltenburger, H. G., Metzger, P., and Krause, C. (1980). Busulphan-induced chromosomal aberrations in intestinal cells of Chinese hamster. *Mutat. Res.* **79,** 257–62.

Miltenburger, H. G., Engelhardt, G., and Rohrborn, G. (1981). Differential chromosomal damage in Chinese hamster bone marrow cells and in spermatogonia after mutagenic treatment. *Mutat. Res.* **81,** 117–122.

Mitchell, C. E. (1980). Induction of aryl hydrocarbon hydroxylase in Chinese hamsters and mice following intratracheal instillation of benzo[*a*]pyrene. *Res. Commun. Chem. Pathol. Pharmacol.* **28,** 65–78.

Moore, W., and Colvin, M. (1968a). Chromosomal changes in the Chinese hamster thyroid following x-irradiation *in vivo*. *Int. J. Radiat. Biol.* **14,** 161–167.

Moore, W., and Colvin, M. (1968b). Persistence of chromosomal aberrations in Chinese hamster thyroid following administration of ^{131}I. *J. Nucl. Med.* **9,** 165–167.

Moorthy, A. S., and Mitra, J. (1978). Effects of hallucinogenic drugs on Chinese hamster chromosomes 1. Effects of a moderate dose of LSD administered orally and intraperitoneally. *Nucleus (Calcutta)* **21,** 206–211.

Muller, D., and Strasser, F. F. (1971). Comparative studies on the Chinese hamster bone marrow after treatment with phenylbutazone and cyclophosphamide. *Mutat. Res.* **13,** 377–382.

Muller, D., Fritz, H., Langauer, M., and Strasser, F. F. (1975). Nucleus anomaly test and chromosomal analysis of bone marrow cells of the Chinese hamster and dominant lethal test in male mice after treatment with fluorescent whitening agents. *Environ. Qual. Saf.* **4,** 247–263.

Munzner, R., and Renner, H. W. (1983). Mutagenicity testing of quinine with submammalian and mammalian systems. *Toxicology* **26,** 173–178.

Murkherjee, B. B., and Ghosal, S. K. (1969). Replicative differentiation of mammalian sex chromosomes during spermatogenesis. *Exp. Cell Res.* **54,** 101–106.

Natarajan, A. T., Darroudi, F., Bussman, C. J. M., Van Kesteren-Van Leeuwen, A. C. (1983). Evaluation of the mutagenicity of formaldehyde in mammalian cytogenetic assays *in vivo* and *in vitro*. *Mutat. Res.* **122,** 355–360.

Norppa, H., Westermarack, T., and Knuutila, S. (1980a). Chromosomal effects of sodium selenite *in vivo*. III. Aberrations and sister chromatid exchanges in Chinese hamster bone marrow. *Hereditas* **93,** 101–105.

Norppa, H., Sorsa, M., and Vainio, H. (1980b). Chromosomal aberrations in bone marrow of Chinese hamsters exposed to styrene and ethanol. *Toxicol. Lett.* **5,** 241–244.

Norppa, H., Penttila, M., Sorsa, M., Hintikka, E. L., and Llus, T. (1980c). Mycotoxin T-2 of *Fusarium tricinctum* and chromosome changes in Chinese hamster bone marrow. *Hereditas* **93,** 329–332.

Oesch, F., Protic-Sabljic, M., Friedberg, T., Klimisch, H. J., and Glatt, H. R. (1983). Vinylidene chloride changes in drug-metabolizing enzymes: Mutagenicity and relation to its targets for carcinogenesis. *Carcinogenesis (London)* **4,** 1031–1038.

Orci, L., Stauffacher, W., Dulin, W. E., Renold, A. E., and Rouiller, C. (1970). Ultrastructural changes in A cells exposed to diabetic hyperglycemia: Observations made on pancreas of Chinese hamsters. *Diabetologia* **6,** 199–206.

Orci, L., Amherdt, M., Malaisse-Lagae, F., Perrelet, A., Dulin, W. E., Gerritsen, G. C., Malaisse, W. J., and Renold, A. E. (1974). Morphological characterization of membrane systems in A- and B-cells of the Chinese hamster. *Diabetologia* **10,** 529–539.

Parkinson, T. M. (1973). Enhanced intestinal fat absorption in diabetic Chinese hamsters. *Diabetologia* **9,** 505–508.

Pathak, S., Hsu, T. C., and Markvong, A. (1976). Pachytene mapping of the male Chinese hamster. *Cytogenet. Cell Genet.* **17,** 1–8.

Pereira, M. A., Sabharwal, P. S., Kaur, P., Ross, C. B., Choi, A., and Dixon, J. (1981). *In vivo* detection of mutagenic effects of diesel exhaust by short-term mammalian bioassays. *Environ. Int.* **5,** 439–444.

Pericin, C. (1981). Effect of seasonal variation on the acute toxicity of cyclophosphamide in the Chinese hamster (*Cricetulus griseus*) and the mouse under laboratory conditions. *Experientia* **37,** 401–402.

Petersson, B., Elde, R., Efendic, S., Hokfelt, T., Johansson, O., Luft, R., Cerasi, E., and Hellerström, C. (1977). Somatostatin in the pancreas, stomach and hypothalamus of the diabetic Chinese hamster. *Diabetologia* **13,** 463–466.

Pool, B. L., Renner, H. W., and Baltes, W. (1983). Genotoxicity of brown-colored polymerization products formed in smoke flavors. *Nutr. Cancer* **5,** 26–33.

Rabinovitch, A., Renold, A. E., and Cerasi, E. (1976). Decreased cyclic AMP and insulin responses to glucose in pancreatic islets of diabetic Chinese hamsters. *Diabetologia* **12,** 581–587.

Raicu, P., Nicolaescu, M., and Kirillova, M. (1970). The sex chromatin in 5 species of hamsters. *Chromosoma* **31,** 61–67.

Ray, M., and Mohandas, T. (1976). Proposed banding nomenclature for the Chinese hamster chromosomes (*Cricetulus griseus*). *Cytogenet. Cell Genet.* **16,** 83–91.

Renner, H. W. (1979). Possible mutagenic activity of saccharin. *Experientia* **35,** 1364–1365.

Renner, H. W. (1984). Antimutagenic effect of an antioxidant in mammals. *Mutat. Res.* **135,** 125–129.

Renner, H. W., and Munzner, R. (1978). Mutagenic evaluation of single cell protein with various mammalian test systems. *Toxicology* **10,** 141–150.

Renner, H. W., and Munzner, R. (1980). Mutagenicity of sulphonylureas. *Mutat. Res.* **77,** 349–355.

Renner, H. W., and Munzner, R. (1982). Genotoxicity of cocoa examined by microbial and mammalian systems. *Mutat. Res.* **103,** 275–281.

Reznik, G., Mohr, U., and Kmoch, N. (1976a). Carcinogenic effects of different nitroso-compounds in Chinese hamsters. I. Dimethylnitrosamine and *N*-diethylnitrosamine. *Br. J. Cancer* **33,** 411–418.

Reznik, G., Mohr, U., and Kmoch, N. (1976b). Carcinogenic effects of different nitroso-compounds in Chinese hamsters: *N*-Dibutylnitrosamine and *N*-nitrosomethylurea. *Cancer Lett.* **1,** 183–188.

Reznik, G., Mohr, U., and Kmoch, N. (1976c). Carcinogenic effects of different nitroso compounds in Chinese hamsters. Part 2, *N*-Nitroso morpholine and *N*-nitroso piperdine. *Z. Krebsforch. Klin. Onkol.* **86,** 95–102.

Richter, K. D., Loge, O., and Losert, W. (1971). Comparative morphological studies on the diabetogenic activity of streptozotocin in rats, Chinese hamsters, guinea-pigs and rabbits. *Arzneim.-Forsch.* **21,** 1654–1656.

Rohrborn, G., Miltenburger, H. G., Basler, A., and Strobel, R. (1980). Testing propafeone hydrochloride in the Ames *Salmonella* microsome system and in mammalian cytogenetic systems (bone marrow, spermatogonia). *Arzneim.-Forsch.* **30,** 2084–2087.

Sakai, Y., Watanabe, K., Takebe, T., and Ishii, K. (1981). Effect of synthetic trypsin inhibitor on monosodium glutamate-induced hyperglycemia in newborn Chinese hamsters. *J. Jpn. Diabetic Soc.* **24,** 77–79.

Sala, M., Gu, Z. G., Moens, G., and Chouroulinkov, I. (1982). *In vivo* and *in vitro* biological effects of the flame retardants tris(2,3-dibromopropyl)-phosphate and tris(2-chloroethyl)orthophosphate. *Eur. J. Cancer Clin. Oncol.* **18,** 1337–1344.

Schillinger, E., and Loge, O. (1976). Metabolic effects and mortality rate in diabetic Chinese hamsters after long-term treatment with 5-methoxyindole-2-carboxylic acid (MICA). *Arzneim.-Forsch.* **26,** 554–556.

Schlaepfer, W. W., Gerritsen, G. C., and Dulin, W. E. (1974). Segmental demyelination in the distal peripheral nerves of chronically diabetic Chinese hamsters. *Diabetologia* **10,** 541–554.

Schmid, W., and Staiger, G. R. (1969). Chromosome studies on bone marrow from Chinese hamsters treated with benzodiazepine tranquillizers and cyclophosphamide. *Mutat. Res.* **7,** 99–108.

Schmidt, F. L., Leslie, L. G., Schultz, J. R., and Gerritsen, G. C. (1970). Epidemiological studies of the Chinese hamster. *Diabetologia* **6,** 154–157.

Schmidt, F. L., Miller, R. L., Peterson, T., and Soret, M. G. (1976). A microsurgical technique for orthotopic ovarian transplantation in the Chinese hamster. *Surgery (St. Louis)* **80,** 595–600.

Schoffling, K., Federlin, K., Schmitt, W., and Pfeiffer, E. F. (1967). Histometric investigations on the testicular tissue of rats with alloxan diabetes and Chinese hamsters with spontaneous diabetes. *Acta Endocrinol. (Copenhagen)* **54,** 335–346.

Shirai, T., Welsh, G. W., and Sims, E. A. (1967). Diabetes mellitus in the Chinese hamster. II. The evolution of renal glomerulopathy. *Diabetologia* **3,** 266–286.

Siegel, E. G., Wollheim, C. B., Sharp, G. W., Herberg, L., and Renold, A. E. (1979). Defective calcium handling and insulin release in islets from diabetic Chinese hamsters. *Biochem. J.* **180,** 233–236.

Siegel, E. G., Wollheim, C. B., Herberg, L., Sharp, G. W., and Renold, A. E. (1981). Defective Ca^{++} handling and insulin release in islets of diabetic hamsters and mice. *Acta Biol. Med. Ger.* **40,** 19–22.

Silberberg, R., and Gerritsen, G. (1976). Aging changes in intervertebral discs and spondylosis in Chinese hamsters. *Diabetes* **25,** 477–483.

Silberberg, R., Gerritsen, G., and Hasler, M. (1976). Articular cartilage of diabetic Chinese hamsters. *Arch. Pathol. Lab. Med.* **100,** 50–54.

Sims, E. A., and Landau, B. R. (1967). Diabetes mellitus in the Chinese hamster. I. Metabolic and morphologic studies. *Diabetologia* **3,** 115–123.

Siou, G., Conan, L., and el-Haitem, M. (1981). Evaluation of the clastogenic action of benzene by oral administration with 2 cytogenetic techniques in mouse and Chinese hamster. *Mutat. Res.* **90**, 273–278.

Sirek, O. V., and Sirek, A. (1967). The colony of Chinese hamsters of the C. H. Best Institute. A review of experimental work. *Diabetologia* **3**, 65–73.

Sirek, O. V. and Sirek, A. (1971). Spontaneous diabetes in animals. *In* "Diabetes" (M. Ellenberg and H. Aifkin, eds.), pp. 256–260. McGraw-Hill, New York.

Soret, M. G., Dulin, W. E., and Gerritsen, G. C. (1973). Microangiopathy in animals with spontaneous diabetes. *In* "Vascular and Neurological Changes in Early Diabetes" (R. A. Camerini-Davalos and H. Cole, eds.), pp. 291–298. Academic Press, New York.

Soret, M. G., Dulin, W. E., Mathews, J., and Gerritsen, G. C. (1974). Morphologic abnormalities observed in retina, pancreas and kidney of diabetic Chinese hamsters. *Diabetologia* **10**, 567–579.

Soret, M. G., Blanks, M. C., Gerritsen, G. C., Day, C. E., and Block, E. M. (1976). Diet-induced hypercholesterolemia in the diabetic and nondiabetic Chinese hamster. *Adv. Exp. Med. Biol.* **67**, 329–343.

Speit, G., Wolf, M., and Vogel, W. (1980). The SCE-inducing capacity of vitamin C: Investigations *in vitro* and *in vivo*. *Mutat. Res.* **78**, 273–278.

Sram, R. J. (1975). Proceedings: Cytogenetic analysis of 3-chinuclidylbenzilate, 3-chinuclidinol and benactyzine in bone marrow of Chinese hamster. *Act. Nerv. Super.* **17**, 253–254.

Stahl, K. W., and Bayer, U. (1983). Bone marrow genotoxicity of N-methyl, N-nitrosourea (NMU): N-alkanols as sister chromatid exchange (SCE) anti-inducers. *Experientia* **39**, 757–759.

Sütterlin, U., Thies, W. G., Haffner, H., and Seidel, A. (1984). Comparative studies on the lysosomal association of monomeric ^{239}Pu and ^{241}Am in rat and Chinese hamster liver: Analysis with sucrose, metrizamide, and Percoll density gradients of subcellular binding as dependent on time. *Radiat. Res.* **98**, 293–306.

Takanari, H., Pathak, S., and Hsu, T. C. (1982). Dense bodies in silver-stained spermatocytes of the Chinese hamster: Behavior and cytochemical nature. *Chromosoma* **86**, 359–373.

Tateno, H., and Mikamo, K. (1984). Neonatal oocyte development and selective oocyte-killing by X-rays in the Chinese hamster, *Cricetulus griseus*. *Int. J. Radiat. Biol.* **45**, 139–149.

Utakoji, T. (1966). On the homology between the X and the Y chromosomes of the Chinese hamster. *Chromosoma* **18**, 449–454.

Utakoji, T., and Hsu, T. C. (1965). DNA replication patterns in somatic and germ-line cells of the male Chinese hamster. *Cytogenetics* **4**, 295–315.

Vainio, H., Nickels, J., and Linnainmaa, K. (1982). Phenoxyacid herbicides cause peroxisome proliferation in Chinese hamsters. *Scand. J. Work, Environ. Health* **8**, 70–73.

van Sickle, W., Gerritsen, G., and Beuving, L. (1981). A comparison of circadian rhythms in feeding, plasma insulin, glucose and glucagon between normal and diabetic Chinese hamsters. *Chronobiologia* **8**, 1–9.

van Went-deVries, G. F., and Kragten, M. C. T. (1975). Saccharin: Lack of chromosome-damaging activity in Chinese hamsters *in vivo*. *Food Cosmet. Toxicol.* **13**, 177–183.

van Went-deVries, G. F., Freudenthal, J., Hogendoorn, A. M., Kragten, M. C., and Gramberg, L. G. (1975). *In vivo* chromosome damaging effect of cyclohexylamine in the Chinese hamster. *Food Cosmet. Toxicol.* **13**, 415–418.

Vig, B. K., and Miltenburger, H. G. (1976). Sequence of centromere separation of mitotic chromosomes in Chinese hamster. *Chromosoma* **55**, 75–80.

Vinegar, A., Carson, A., and Pepelko, W. E. (1981). Pulmonary function changes in Chinese hamsters exposed 6 months to diesel exhaust. *Environ. Int.* **5**, 369–372.

Vistorin, G., Gamperl, R., and Rosenkranz, W. (1977). Studies on sex chromosomes of four hamster species: *Cricetus cricetus*, *Cricetulus griseus*, *Mesocricetus auratus*, and *Phodopus sungorus*. *Cytogenet. Cell Genet.* **18**, 24–32.

Wappler, E., and Fiedler, H. (1972). The Chinese hamster *Cricetulus griseus*, a spontaneously diabetic animal: A review. *Z. Versuchstierkd.* **14**, 1–16.

Ward, B. C., Childress, J. R., Jessup, G. L., and Lappenbusch, W. L. (1972). Radiation mortality in the Chinese hamster, *Cricetulus griseus*, in relation to age. *Radiat. Res.* **51**, 599–607.

Weathersbee, P. S., Ax, R. L., and Lodge, J. R. (1975). Caffeine-mediated changes of sex ratio in Chinese hamsters, *Cricetulus griseus*. *J. Reprod. Fertil.* **43**, 141–143.

Weringer, E. J., and Arquilla, E. R. (1981). Wound healing in normal and diabetic Chinese hamsters. *Diabetologia* **21**, 394–401.

Wilander, E. (1974). Streptozotocin diabetes in the Chinese hamster. *Acta Pathol. Microbiol. Scand., Sect. A* **82A**, 767–776.

Wilander, E. (1975). Streptozotocin diabetes in the Chinese hamster: Volumetric quantitation of the pancreatic islets and inhibition of diabetes with nicotinamide. *Horm. Metab. Res.* **7**, 15–19.

Wilander, E., and Boquist, L. (1972). Streptozotocin diabetes in the Chinese hamster: Blood glucose and structural changes during the 1st 24 hours. *Horm. Metab. Res.* **4**, 426–433.

Wilander, E., and Gunnarsson, R. (1975). Diabetogenic effects of N-nitrosomethyl urea in the Chinese hamster. *Acta Pathol. Microbiol. Scand., Sect. A* **83A**, 206–212.

Wilander, E., and Tjalve, H. (1975a). Diabetogenic effects of N-nitrosomethylurea with special regard to species variations. *Exp. Pathol.* **11**, 133–141.

Wilander, E., and Tjalve, H. (1975b). Uptake of labeled N-nitrosomethyl urea in the pancreatic islets. *Virchows Arch. A: Pathol. Anat. Histol.* **367**, 27–34.

Wild, D., King, M. T., and Eckhardt, K. (1980). Cytogenetic effect of orthophenylenediamine in the mouse, Chinese hamster, and guinea pig and of derivatives, evaluated by the micronucleus test. *Arch. Toxicol.* **43**, 249–255.

Yao, K. T. (1978). Microwave radiation-induced chromosomal aberrations in corneal epithelium of Chinese hamsters. *J. Hered.* **69**, 409–412.

Yerganian, G. A. (1964). Spontaneous diabetes mellitus in the Chinese hamster, *Cricetulus griseus*. IV. Genetics aspects. *In* "Aetiology of Diabetes Mellitus and Its Complications" (M. P. Cameron and M. O'Connor, eds.), pp. 25–41. Churchill, London.

Yerganian, G. A. (1972). History and cytogenetics of hamsters. *Prog. Exp. Tumor Res.* 2–34.

Yerganian, G. A. Paika, I., Gagnon, H. J., and Battaglino, A. (1979). Rapid induction of epithelial hyperplasia and lymphoplasmacytosis in the Chinese hamster (*Cricetulus griseus*) by mineral and pristane oil. *Prog. Exp. Tumor Res.* **24**, 424–434.

Part III

The European Hamster

Chapter 20

Biology, Care, and Use in Research

Ulrich Mohr and Heinrich Ernst

I.	History and Taxonomy	351
II.	Care and Management	352
III.	Anatomical and Histological Characteristics	353
IV.	Hematological and Clinicochemical Parameters	356
V.	Hibernation	357
VI.	Diseases	360
	A. Spontaneous Infections	360
	B. Spontaneous Tumors	360
	C. Miscellaneous Disorders	360
VII.	Use in Research	361
	References	364

I. HISTORY AND TAXONOMY

The European or field hamster, *Cricetus cricetus* L., is a member of the family Cricetidae and of the subfamily Cricetinae. Thriving in areas with a continental climate and a clay or loess soil where it can burrow, this animal is found throughout the lowlands of central and eastern Europe. This species is particularly abundant in rural areas bordering highly industrialized regions, especially those where both grain and root crops are cultivated (Reznik *et al.*, 1979).

The European hamster is considerably larger than its Syrian and Chinese relatives (Fig. 1). On average, the adult male measures 27–32 cm in length (excluding the tail) and the female 22–25 cm. When kept under laboratory conditions, the average body weight is 450 gm for males and 350 gm for females (Reznik *et al.*, 1973). The animal has a stocky body and a short tail. Its coat is usually brown on the dorsal surface and white on the ventral surface. Throughout the summer, however, the fur in the dorsal areas tends to be a lighter color than in winter. The snout and paws are whitish and the border between dorsal and ventral fur is marked with small white patches. In its natural habitat, the European hamster is a solitary animal and each adult lives in its own burrow. These burrows vary in depth from 30 to 60 cm in summer to over 2 m in winter and consist of one or more nests, food stores, and hibernating chambers. From mid-October to the end of March, the burrows are sealed and the occupants hibernate.

Although the biology and behavior of the wild European hamster had been observed and described in some detail by Petzsch (1937) and Eibl-Eibesfeldt (1953), the animal was little used for scientific research purposes, principally because its marked aggression both toward its own kind and toward animal caretakers made maintenance and breeding in the laboratory extremely difficult. In the early 1970s, however, a breeding colony was established by us from captured European hamsters and the animal began to be used in a variety of investigations.

Fig. 1. Comparison of adult male European (left) and Syrian hamster (right).

II. CARE AND MANAGEMENT

Early attempts at maintaining and breeding captured European hamsters under controlled laboratory conditions have been described in detail by Mohr *et al.* (1973b) and Reznik-Schüller *et al.* (1974). Fifty wild European hamsters were captured in the Hannover–Braunschweig area of Germany and served as the initial stock for the Hannover breeding colony (strain MHH: EPH). The animals were housed individually in Makrolon cages (type III, 800 cm²) and were kept under standard laboratory conditions (temperature 20° ± 2°C; relative humidity 55 ± 5%; air change eight times per hour). They received a pelleted diet (RMH-B, Hope Farms, Woerden, The Netherlands) and water *ad libitum*. Under such conditions, it was found that the captured animals did not hibernate during the winter months, although they did become drowsy and hypothermic. This state persisted for anything from several hours to 2 days. Such relatively short, "winter sleeplike states" have been observed in other small rodents kept in the laboratory (Aszodi, 1921). The first generation of laboratory-bred European hamsters also showed such intermediate hibernating behavior, while in animals born later this pattern appears to have been lost.

The European hamster's marked natural aggression proved to be the principal obstacle when trying to breed the animal in the laboratory. Even in the wild, a female will only allow a male to approach when she is in estrus; at all other times, fierce aggression is shown (Fig. 2). Even when the female is in estrus, lengthy foreplay, as described by Petzsch (1937) and Eibl-Eibesfeldt (1953), is required to reduce the aggression to a level where willingness to mate is increased. Laboratory studies of the breeding of European hamsters have revealed that females in captivity have a regular four-stage estrus cycle (Fig. 2) (Reznik-Schüller *et al.*, 1974). Proestrus lasts only a few hours, estrus 1–2 days, and metestrus about 6 hr. The longest stage of the cycle is diestrus, which lasts 2–4 days. In early breeding studies, determination of the estrus cycle by daily examination of Giemsa-stained vaginal smears had to be abandoned because the procedure caused considerable stress to the females. Estrus was then established by observing the be-

Fig. 2. Variations in the aggressiveness of female hamsters in relation to their menstrual cycle.

havior of the animals and by noting when the female and male showed mutual acceptance.

For mating, a special box (150 × 100 × 50 cm) with stainless-steel-lined walls and floor was designed. This box is divided into two halves by a steel fence with a trapgate (12 × 12 cm). With the gate closed, a male and female are placed on either side of the fence, through which they can communicate without being able to bite one another. If the female tries to attack the male, this is a sign that she is not in estrus. The animals are returned to their own cages and are tested in the same manner over the next few days. When the female is in estrus, the aggression usually abates and the animals show considerable interest in one another; then the gate is lifted. Foreplay follows a distinct pattern and requires a considerable amount of space, hence the size of the mating box. The female runs in a figure of eight and the male follows closely behind uttering his mating call, a loud "sniff" which increases in volume with the female's readiness to mate. Finally, the hamsters copulate several times before mating is completed.

Pregnancy, which normally lasts for about 20 days in the wild (Asdell, 1946), has been reduced in the laboratory to 15–17 days. While in its natural habitat, the mating season of the European hamster lasts from April until August (Petzsch, 1937), laboratory-bred females are able to deliver litters, averaging nine young, two to five times throughout the entire year. Although wild hamsters are known to live only solitarily, laboratory-bred animals show considerably reduced aggression and, when caged in groups from the day of weaning, integrate well and develop a social order with the biggest male dominant. Although in their natural habitat, European hamsters usually have a life span of up to 8 years (Gaffrey, 1961), hamsters bred in the laboratory live for about 5 years. This reduction is due to the lack of hibernation.

III. ANATOMICAL AND HISTOLOGICAL CHARACTERISTICS

Some of the major anatomical and histological parameters of the laboratory-bred European hamster are summarized below. The lymphatic and endocrine organs and the spleen have been omitted because clinically relevant parameters have not yet been established fully.

Nasal cavity. The initial part of the nasal cavity (Fig. 3), the nasal vestibule, is approximately 4 mm long and is lined with stratified, less-keratinized squamous epithelium. The nasal

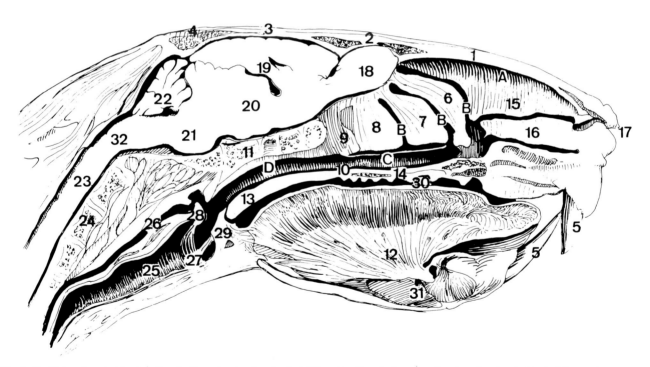

Fig. 3. Sagittal section sketch of skull and adjacent parts of neck of adult hamster at level of nasal septum (nasal septum lacking). 1,Os nasale; 2,os frontale; 3,os parietale; 4,os interparietale; 5,dens incisivum; 6,endoturbinale I; 7,endoturbinale II; 8,endoturbinale III; 9,endoturbinale IV; 10,choanae; 11,os sphenoidale; 12,lingua; 13,palatum molle; 14,palatum durum; 15,concha nasalis dorsalis; 16,concha nasalis ventralis; 17,naris; 18,bulbus olfactorius; 19,cerebrum; 20,adhesio interthalamica; 21,pons; 22,cerebellum; 23,medulla spinalis; 24,vertebrae cervicales; 25,trachea; 26,esophagus; 27,larynx; 28,cartilago arytenoidea; 29,epiglottis; 30,cavum oris; 31,mandibula; 32,medulla oblongata. (From Reznik, *et al.*, 1979; with permission of National Cancer Institute.)

cavity proper, the nasal conchae, the nasopharynx, and nasal septum are lined with respiratory mucosa consisting of ciliated, pseudostratified, columnar epithelium and subepithelial mucus-producing glands. The nasal cavity is formed by the ethmoid, nasal, premaxillary, maxillary, and palatine bones and is bisected by a cartilaginous, centrally located septum.

Paranasal cavity. The European hamster has a paired maxillary sinus with a small rostrodorsal portion (2.5 × 1 mm) and a caudoventral portion (5 × 1 mm). The maxillary sinus joins the nasal cavity via the nasomaxillary opening into the middle nasal meatus (Fig. 3). The paranasal sinus is lined by respiratory epithelium, and the glands beneath are composed of both serous and mucous parts.

Vomeronasal organs. The vomeronasal organs (5 × 1 mm) are situated at the ventral margin of the anterior nasal septum. The ventral margins of the glands are covered by epithelium similar to that found in the olfactory region of the nose. Olfactory sensory cells are interspersed throughout the supporting columnar epithelium. The dorsal edges are a continuation of the respiratory epithelium of the nasal cavity.

Cheeks. Like other members of the family Cricetidae, the European hamster has cheek pouches. These are generally 60–70 mm long and 12–15 mm wide, and have a carrying capacity of 20–30 gm. The mucosa of the empty pouch is marked by deep folds. The lining epithelium, like that of the entire oral cavity, is of the nonkeratinizing cell type.

Teeth. The incisors (Fig. 3) are the first teeth to appear (on days 4 and 5 postpartum), and teething is complete within 35 days of birth. The adult European hamster has only one set of permanent teeth, which consists of 4 continuously growing incisors and 12 molars. The lower incisors pass through almost the entire length of the mandible and terminate caudally at the roots of the last molar.

Tongue. The European hamster has a fairly mobile, spoon-shaped tongue (Fig. 3) which is 30–35 mm long and 10–20 mm wide. Sublingual veins are visible on its ventral surface, while the epithelium forms numerous papillae on the dorsal surface.

Salivary glands. The salivary system consists mainly of parotid, mandibular, sublingual, and zygomatic glands which do not differ greatly from those of other rodents in their histology and anatomical topography.

Larynx. The larynx (Fig. 3) of the European hamster is about 6 mm in diameter and can be up to 9 mm long. It contains three single median cartilages (thyroid, cricoid, and epiglottis) and three paired cartilages (arytenoid, corniculate, and cuneiform). The interior of the larynx is divided into the vestibule, glottis, and infraglottis. A median ventricle is formed at the base of the epiglottis (more pronounced in old hamsters). The dorsal parts of the epiglottis, the vestibule of the larynx, and the vocal cords are covered by stratified squamous epithelium, while other parts have a pseudostratified, ciliated columnar epithelium.

Trachea. In the adult European hamster, the cervical trachea (Figs. 3, 4), which divides at the sixth rib, is about 37 mm long and consists of 15 hyaline, cartilaginous rings, the first being the largest and the last the smallest (luminal diameters of ~3.9 mm and ~2.7 mm, respectively). The tracheal lumen is lined by pseudostratified, ciliated columnar epithelium, interspersed with a few submucosal glands.

Lungs. The lungs of the adult European hamster (Fig. 4) have an average weight of 1.8–2.0 gm and a volume (measured by water displacement) of about 2.0–2.6 ml. The right lung is divided into five lobes (cranial, middle, diaphragmatic, accessory, and intermediate accessory) by deep interlobular

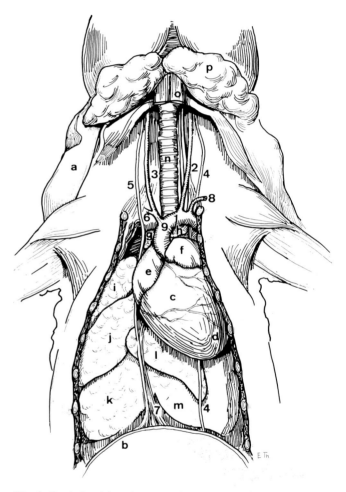

Fig. 4. Cervical and thoracic organs of 1-year-old male European hamster. Sternum, clavicle, and parts of ribs and cervical muscles removed; mandibular gland folded back. a,Bursa buccalis; b,diaphragm; c,ventriculus dexter; d,ventriculus sinister; e,auricula dextra; f,auricula sinistra; g,thymus dexter; h,thymus sinister; i-1,pulmo dexter; i,lobus cranialis; j,lobus medius; k,lobus caudalis; l,lobus accessorius; m,pulmo sinister; n,trachea; o,mm. sternohyoid et sternothyroid; p,gl. mandibularis; 1,arcus aortae; 2,a. carotis communis sinistra; 3,a. carotis communis dextra; 4,n. vagus sinister; 5,n. vagus dexter; 6,a. subclavia dextra; 7,v. cava caudalis; 8,a. subclavia sinistra; 9,a. brachiocephalica. (From Reznik *et al.*, 1979; with permission of National Cancer Institute.)

fissures which extend from the main bronchi. The left lung has only one lobe and extends from the diaphragm to the midpoint of the first rib. The bronchi have a typical epithelium composed of both ciliated and mucus-producing cells. There is no cartilage supporting the muscular walls, and subepithelial glands are lacking.

Heart. The heart (Fig. 4) measures approximately 19 mm from base to apex, is about 10–11 mm wide, and weighs 1.3–1.5 gm. The base lies mainly in the right half of the thorax, while the apex is situated more to the left. There is no contact with the diaphragm. The atria are recognizable by the triangular auricles (right usually larger than left). The left ventricle projects into the fourth intercostal space.

Large vessels. About 7 mm distal to the origin of the ascending aorta is the base of the brachiocephalic trunk, which divides into the right subclavian and right common carotid arteries (Fig. 4). The left common carotid is a branch of the aortic arch and originates at the level of the second thoracic vertebra. The two pulmonary veins form a single venous trunk before joining the left auricle of the heart. The caudal vena cava runs into the right auricle along with the cranial vena cava.

Pharynx. The pharynx (Fig. 3) is lined by nonkeratinizing, squamous-cell epithelium and is made up of three parts: the nasal portion, the oral portion, and the laryngeal portion.

Esophagus. About 45 mm in length, the esophagus (Fig. 3) forms the connection between pharynx and forestomach. It enters the stomach at the margin between the glandular stomach and forestomach.

Stomach. The entire stomach (Fig. 5) is situated in the left hypochondrium and is divided by a distinct constriction into a larger, blind forestomach and a smaller glandular stomach. When full, the stomach extends into the right side of the abdominal cavity and weighs 10–11 gm (the empty weight is ~3 gm). The forestomach has a squamous-cell epithelial lining, and the glandular stomach a typical gastric mucosa.

Duodenum. In an animal weighing about 400 gm, the duodenum (Fig. 5) is approximately 120 mm long and 3.5 mm wide. The common duct (a combination of bile and major pancreatic ducts) enters the duodenum within its cranial part about 30 mm from the pylorus. Histologically, the duodenal wall does not differ from that of other rodents.

Jejunum. The longest portion of the intestine is the jejunum (Fig. 5) which is about 510 mm long and 4 mm wide. Its contorted loops lie predominantly in the dorsal region of the mesogastrium. Histologically, the jejunum is similar to the duodenum but the villi are fewer and smaller (average height 0.35 mm, diameter 0.15 mm).

Ileum. The length of the ileum (Fig. 5) ranges between 25 and 35 mm, and it can be up to 3 mm wide. It joins the cecum via the cone-shaped ostium ileocecocolicum. The villi have a broad base and an average height of 24 mm. Goblet cells occur frequently and are found mostly at the base of the villi.

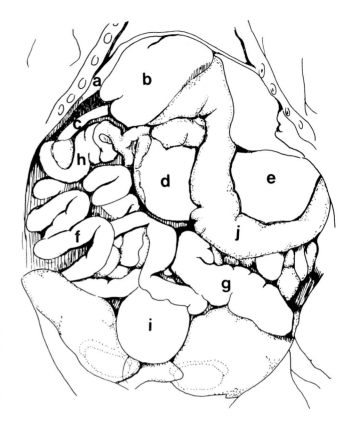

Fig. 5. Diagram of thoracic and abdominal regions of male hamster, with positions of various organs. a,Diaphragm; b,hepar; c,duodenum; d,ventriculus; e,proventriculus; f,jejunum; g,cecum; h,colon ascendens; i,vesica urinaria; j,adipose tissue. (From Reznik *et al.,* 1979; with permission of National Cancer Institute.)

Cecum. In a well-nourished European hamster, the cecum (Fig. 5) may measure up to 120 mm in length and 15 mm in width. It has a distinct apex, body, and freely mobile base. Macroscopically, the folds of the cecum, the plicae semilunares, can be seen from the outside.

Colon. Depending on the size and weight of the hamster, the colon (Fig. 5) has an average length of about 370 mm (ascending 100 mm, transverse 60 mm, descending 210 mm) and an average width of 4 mm. Histologically, the crypts are deeper than those of the small intestine, and the crypt epithelium consists mainly of goblet cells. The folds of the colon occur at 1 to 2-mm intervals are are formed by tunica mucosa and submucosa, as well as by the inner circular layers of the tunica muscularis. Lymphatic tissue is diffusely distributed in follicular patterns.

Rectum. The rectum of the male European hamster is considerably longer (40–60 mm) than that of the female (35–45 mm). Many lymph nodes are found in the lamina propria and submucosa, and there is a considerable increase in the thickness of the tunica muscularis. Toward the end of the rectum, the mucosa consists of several longitudinal folds which contain large veins in the lamina propria. The rectum has fewer goblet cells than the colon. At the anorectal line, the columnar epithelium of the rectum changes into stratified squamous, nonkeratinized epithelium, whereas the zona cutanea of the anus is covered by a keratinized squamous epithelium and contains modified sebaceous glands.

Liver. Although the size of the European hamster liver (Fig. 5) may vary greatly, it weighs on average 15 gm. Usually brown, the color of the organ depends on factors such as blood content, age, and nutritional state of the animal. Deep indentations divide the liver into six parts (bipartite left portion, left lateral lobe, caudate and papillary processes of the caudal lobe, quadrate lobe, and right medial lobe). Impressions of the forestomach, intestine, and kidneys can be seen at the visceral surface. The lobular structure of the liver is just visible macroscopically.

Pancreas. The pancreas consists of three well-defined elongated lobes (gastric, duodenal, and splenic). The organ is light in color and difficult to distinguish from mesenteric fat tissue. The main pancreatic duct forms a common duct with the bile duct 15 mm before it reaches the duodenum. Histologically, the exocrine pancreas has a tubuloalveolar structure and the acini form lobules containing intralobular ducts. These flow together into interlobular ducts and form the main duct. The acini reach their maximum size when the animal is about 1 year old. Ductal diameter and epithelial height vary according to age. The islets of Langerhans contain three cell types: α (red in Azan staining), β (blue in Azan staining), and δ (clear in Azan staining).

Kidneys. The kidneys weigh about 930 mg each and lie retroperitoneally in the lumbar region, projecting between the first and third lumbar vertebrae. They have a smooth surface and are reddish brown in color. At the cranial pole of each kidney is an adrenal gland which is separated from the organ by fatty and connective tissue. Each kidney contains one papilla. The cortex and medulla have no distinct border, and the glomerula and tubules are not visible macroscopically.

Ureters. The ureters are lined by transitional epithelium and are surrounded by rich retroperitoneal fatty tissue.

Urinary bladder. Depending on the degree of distention, the urinary bladder (Fig. 5) can measure from 4 to 20 mm in diameter. Its walls are lined by transitional epithelium and are thin enough to permit easy viewing of the contents. The urethra discharges the urine separately from the vagina and anus in females and via the penis in males.

Male genital organs. The testes of the adult hamster vary in size but, during the sexually active season, are about 20 mm in length and 12 mm in diameter. In the autumn and winter, however, the testes and accessory glands involute and are at their smallest during hibernation. Various cell types are microscopically visible within the seminiferous tubules of the adult male during the sexually active period. Sertoli cells and spermatogonia are seen in the basal portion, while the following layers consist of resting spermatocytes, spermatocytes, and spermatids at different stages of maturity. The interstitial tissue of the testes consists of Leydig and connective tissue cells, together with blood vessels and nerves.

The epididymis, the reservoir and site where sperm maturation is completed, is firmly attached to the testis. Microscopically, the different ductules are lined with ciliated columnar epithelium and contain circularly arranged muscle fibers in their walls. The epithelium of the ductus deferens is of the pseudostratified, columnar type and the superficial cells have long, regular microvilli.

The paired vesicular glands are normally about 22 mm long and 9 mm wide. During the sexually active period, they weigh up to 10 times more than in winter (450 and 45 mg, respectively). Microscopically, the epithelium of the vesicular glands is columnar or flat (depending on its functional state) and has goblet cells.

The prostate gland is about 16 mm long and 10 mm wide, but is barely visible during the winter months. The glandular tissue is composed of numerous follicles with internal papillary projections lined by simple columnar epithelium. The bulbourethral glands are about 5 mm in diameter. Both lobules discharge their secretion though a separate duct into the urethra. The acini of the lobules are lined with columnar epithelial cells.

Female genital organs. Each ovary measures about 5 mm in diameter and weighs approximately 26 mg; a small variance in size occurs with estrus and the seasonal cycle. The cortex contains follicles at different stages of maturity, intact or involuting corpora lutea, albicantia, and atretica. The central medulla is composed of vascularized connective tissue.

The bipartite uterus is an undivided body with separate cylindrical uterine horns which have a diameter of 2–3 mm and are up to 59 mm long. The cervix, a 10-mm-long tube, has a firm consistency, distinguishing it from the vagina and uterus. From the cervix to the vagina, the mucosal lining of the uterus is composed of a single columnar epithelium which is arranged in high, longitudinal folds.

The vagina is a fibromuscular tube, the orifice of which lies at the base of the clitoris. Caudally, it is lined with a keratinized squamous epithelium; cranially, the mucosa is of the nonkeratinized epithelial type. Since the European hamster has a short sexual cycle, the epithelium of the vaginal mucosa is in a constant state of restoration, transformation, and disintegration. Young female hamsters reach sexual maturity when they weigh about 200 gm; at this time, in the spring, the epithelium closing the vagina disappears. The orifice is almost completely round and, depending on the age of the female, is about 3 mm in diameter.

IV. HEMATOLOGICAL AND CLINICOCHEMICAL PARAMETERS

Normal blood values of laboratory-bred European hamsters show variations in increasing age and also during hibernation.

There are no significant differences between the sexes for most hematological values. The number of erythrocytes shows the greatest variation among the different age groups (Tables I and II); the lowest count is at 13 days old and the highest at 120 days, with numbers thereafter gradually decreasing. The erythrocyte diameters are found to become smaller with age. The number of leukocytes is low in 13-day-old hamsters (males 3830, females 3400), highest in 65-day-old males (8300), and subsequently decreases (5840 in 900-day-old males). The highest rates of thrombocytes are found in 13-day-old hamsters, while the lowest values are seen in 95-day-old animals. Thereafter, the number of thrombocytes increases again.

During hibernation, the entire metabolism of the European hamster is reduced to a minimum (Kayser, 1961). At this time, there is a reduction in the number of some blood cells (Table III). Whereas Raths (1953, 1957) found a slight but insignificant increase in blood cells, we found few obvious variations. However, leukopenia was found to occur during hibernation. The number of leukocytes decreased from an average of 5350 to 740. The lowest counts were observed after 16 weeks of hibernation. Thrombocytes decreased during hibernation from 226,000 to 24,170. However, 1 day after awakening from hibernation, the numbers of thrombocytes and leukocytes returned to the average levels recorded during the summer. Hamsters kept in a nonhibernating state during the winter show no changes in blood cell numbers.

The clinicochemical data in Table IV (average and standard deviations) are listed according to the age of the animals. At 120 days of age, a slight decrease of the liver transaminases and alkaline phosphatase is seen. Such changes are physiological and have also been observed in the growing and developing young of other species.

Table V gives additional data for the European hamster, including a hemogram and serum chemistry values.

V. HIBERNATION

In its natural habitat, the mean body temperature of the European hamster was found to be 34.8°C (Eisentraut, 1956). However, environmental temperatures of more than 28°C are compensated by a rise in body temperature (outside 35°C, body 42°C) (Kayser, 1939). Likewise, when outside temperatures drop, the animal's body temperature decreases accordingly. As the weather becomes colder and winter approaches, the European hamster becomes less active and eventually hibernates. The hypothermic body temperature re-

Table I

Male European Hamsters: Mean Number of Blood Cells and Diameters[a]

Cell type	Age of hamsters (days)							
	13	25	30	65	95	120	360	900
Erythrocytes ($\times 10^6/mm^3$)								
$\bar{x}$	3.41	4.87	5.06	7.45	7.98	8.37	7.97	7.30
s	0.52	0.44	0.56	0.49	0.69	0.49	0.63	1.00
Erythrocyte diameter (μm)								
$\bar{x}$	7.59	7.31	7.17	6.72	6.78	6.80	6.87	6.74
s	0.80	0.64	0.53	0.59	0.56	0.56	0.08	0.53
Thrombocytes ($\times 10^3/mm^3$)								
$\bar{x}$	295	221	200	210	208	202	258	229
s	76	40	71	32	32	17	64	53
Leukocytes ($\times 10^3/mm^3$)								
$\bar{x}$	3.83	7.09	6.01	8.30	7.91	7.53	6.74	5.84
s	1.18	1.51	2.44	2.16	1.45	1.28	1.58	2.51
	Differential blood counts (%)							
Lymphocytes	64.71	78.67	72.43	60.73	66.57	69.25	76.60	65.60
Juveniles	0.46	0.67	0.57	0.63	0.93	1.25	0.71	0.79
Stab cells	0.62	0.83	0.71	1.38	0.93	1.25	0.26	1.29
Neutrophils	31.54	16.67	24.71	34.63	29.79	26.00	21.30	38.29
Eosinophils	0.77	1.33	0.57	1.13	1.00	1.75	0.34	0.71
Monocytes	1.85	0.83	1.00	1.00	1.50	0.75	1.46	1.43

[a] $\bar{x}$, Mean value; s, standard deviation. Number of hamsters: 20 males per age group.

Table II

Female European Hamsters: Mean Number of Blood Cells and Diameters[a]

Cell type	Age of hamsters (days)							
	13	25	30	65	95	120	360	900
Erythrocytes ($\times 10^6/mm^3$)								
$\bar{x}$	3.71	4.88	5.57	7.07	7.79	8.40	7.84	6.52
s	0.44	0.50	0.47	0.97	1.09	0.30	0.54	0.84
Erythrocyte diameter (μm)								
$\bar{x}$	7.59	7.31	7.17	6.71	6.75	6.80	6.87	6.74
s	0.80	0.64	0.53	0.60	0.58	0.56	0.08	0.53
Thrombocytes ($\times 10^3/mm^3$)								
$\bar{x}$	265	241	201	232	204	244	260	234
s	59	28	62	64	31	45	28	15
Leukocytes ($\times 10^3/mm^3$)								
$\bar{x}$	3.40	6.87	5.54	8.11	7.64	7.60	6.70	6.25
s	1.19	1.46	2.85	1.63	1.88	0.79	1.60	1.68
Differential blood counts (%)								
Lymphocytes	68.71	72.83	71.70	68.09	62.70	69.17	73.80	60.40
Juveniles	0.29	0.67	0.40	0.92	1.00	0.50	0.78	1.00
Stab cells	0.71	0.83	0.60	1.18	1.00	1.50	0.11	2.00
Neutrophils	27.86	24.00	24.50	25.64	28.30	26.67	23.80	35.00
Eosinophils	0.79	0.83	0.80	1.64	1.36	1.00	0.26	0.40
Monocytes	1.14	0.67	0.70	1.36	0.90	1.17	2.11	1.33

[a] $\bar{x}$, Mean value; s, standard deviation. Number of hamsters: 20 females per age group.

Table III

Influence of Hibernation on Hematological Parameters (Mean Values) of European Hamsters

Cell type	Nonhibernating (in winter)	Weeks of hibernating			24 hr after awakening
		6	14	16	
Male ($N = 20$)					
Erythrocytes ($\times 10^6/mm^3$)	7.1	7.8	7.9	8.0	7.4
Leukocytes ($\times 10^3/mm^3$)	5.35	0.931	1.066	0.74	5.83
Thrombocytes ($\times 10^3/mm^3$)	226.0	24.17	59.5	40.2	220.0
Lymphocytes (%)	69.2	74.3	70.3	67.0	76.2
Neutrophils (%)	29.3	24.3	28.2	28.4	26.1
Stab cells (%)	0.2	0.0	1.3	0.6	0.6
Monocytes (%)	0.0	0.3	4.7	1.3	1.2
Female ($N = 20$)					
Erythrocytes ($\times 10^6/mm^3$)	7.2	8.6	8.1	7.8	7.4
Leukocytes ($\times 10^3/mm^3$)	5.62	1.508	0.725	0.73	5.62
Thrombocytes ($\times 10^3/mm^3$)	251.6	32.67	35.0	45.4	214.0
Lymphocytes (%)	71.6	84.7	75.7	65.0	75.2
Neutrophils (%)	25.2	14.7	17.8	32.8	25.2
Stab cells (%)	0.2	0.3	1.2	0.4	0.0
Monocytes (%)	1.2	0.3	4.3	1.4	1.4

Table IV

Clinicochemical Values of European Hamsters[a]

Sex and age (days)	GOT (μ/liter)	GPT (μ/liter)	AP (μ/liter)	Ca (mmol/liter)	Mg (mmol/liter)	Na (mmol/liter)	K (mmol/liter)
Male							
30							
$\bar{x}$	46.7	30.3	2415	2.47	1.40	155	10.02
s	6.2	5.2	355	0.10	0.13	12	4.94
60							
$\bar{x}$	46.2	34.2	1155	2.56	1.33	143	8.89
s	10.2	4.6	297	0.20	0.14	2	1.43
120							
$\bar{x}$	30.0	23.9	279	2.62	1.18	152	9.27
s	3.7	4.2	47	0.07	0.07	4	1.26
Female							
30							
$\bar{x}$	45.7	27.5	2352	2.45	1.33	157	7.20
s	3.4	6.3	541	0.19	0.12	5	0.62
60							
$\bar{x}$	43.5	36.0	1228	2.47	1.33	137	8.61
s	6.5	6.7	252	0.17	0.13	27	0.97
120							
$\bar{x}$	30.4	24.3	303	2.60	1.15	147	7.58
s	6.7	9.3	58	0.07	0.08	5	1.40

[a]$\bar{x}$, Mean value; s, standard deviation. Number of hamsters: 10 males and females per group.

Table V

Hemogram and Serum Chemistry Values of the European Hamster[a]

Parameter	Mean ± SD	Range	Ref.
Erythrocytes ($\times 10^6/\mu$l)	7.6 ± 0.4	7.1–8.4	b
Packed-cell volume (%)	49.2 ± 1.6	47–52	b
Hemoglobin (gm/dl)	18.0 ± 0.7	17–19	b
Total leukocytes ($\times 10^3/\mu$l)	7.4 ± 2.6	4.8–12.8	b
Neutrophils (%)	23.2 ± 2.5		b
Lymphocytes (%)	74.0 ± 2.3		b
Monocytes (%)	2.6 ± 0.6		b
Eosinophils (%)	0.07 ± 0.1		b
Basophils (%)	0.02 ± 0.1		b
Urea nitrogen (mg/dl)	29.0 ± 4.3	19–44	b
Alanine aminotransferase (SGPT) (Reitman–Frankel units)	27.7 ± 5.2	20–56	b
Cholesterol (mg/dl)	110.9 ± 17.4	80–140	b
Lipase (Cherry–Crandall)	0.4 ± 0.3	0–1	b
Total bilirubin (mg/dl)	0.4 ± 0.2	0.2–0.8	b
Total protein (gm/dl)	6.5 ± 0.3	6.0–7.3	b
Thyroxin (μg/dl)	2.4 ± 0.7	1.6–3.3	b
Glucose (mg/dl) (nonfasted)	57.8 ± 22	15–115	b
Blood pH	7.40 ± 0.02	—	c

[a]Table prepared by Drs. Farol Tomson and K. Jane Wardrop.
[b]Silverman and Chavannes (1977).
[c]Malan and Arens (1973).

sults in reduced energy expenditure, so that whereas in summer heat production is approximately 30 cal/24 hr, this value drops to 0.75 cal/24 hr during hibernation (Kayser, 1959). The heartbeat frequency also decreases to about 10 beats/min compared to the normal 175 beats/min. Hibernation, however, is not continuous, and zoologists have found that the longest phases of uninterrupted sleep last 4 days (Kayser, 1961). The animal will then interrupt its hibernation spontaneously and usually remains awake for about 8 hr, during which time it eats, drinks, and defecates.

A hibernating experiment was conducted using captured European hamsters kept under cold laboratory conditions (room temperature 4° ± 1°C; relative humidity 90 ± 5%; no light). In December, at the start of this experiment, the males had an average body weight of 295 ± 29 gm and the females 179 ± 42 gm. During the first 2 weeks of hibernation, the female hamsters showed a body weight gain of 9 gm, while the males lost about 11 gm. Apparently, females do not hibernate so deeply at this stage, since 60% were often awake during the first 14 days. During the same period, 71–100% of the males were found to be in deep hibernation.

When held in a warm hand for more than 5 min, deeply hibernating hamsters began to wake up and their rectal temperature increased to 16°C. It has been reported that, upon wakening from hibernation, the rectal temperature of the hamster increases from 6°C to 27°C within 100 min (Kayser et al., 1954).

In March, after 11 weeks of hibernation, the males became more active and revealed a body weight loss of around 26%. Females, however, were found to awaken after just 9 weeks and had lost only about 15% of their body weight. In contrast to these artificial laboratory conditions, Kayser (1961) reported that in their natural environment, male European hamsters came out of hibernation 15 days earlier than the females.

The onset of hibernation and awakening in spring are thought to depend on climatic factors such as temperature, hours of daylight, and humidity. In experiments conducted under cold laboratory conditions, it was found that all hibernating European hamsters were awake by April despite the existing cold temperatures. On the other hand, hamsters were always found to be in hibernationlike states during December and January, even when the laboratory temperature was 22°C. The exact reasons for such behavior are unknown, but it would appear that the hibernation cycle is also controlled by factors which are not related to climate.

VI. DISEASES

A. Spontaneous Infections

1. Viruses

So far only two groups of viruses as a possible cause of disease have been identified in European hamsters. Rabies virus was recovered from 5 out of 266 free-living hamster carcasses, which had been collected in Eastern Slovakia in 1972 and 1973 (Svrcek et al., 1980). Hannoun et al. (1974) described a spontaneous and clinically unapparent papovavirus infection in European hamsters captured in France.

2. Bacteria

Severe, occasionally fistulating head abscesses were observed in both captured and bred European hamsters and associated with corynebacteria, staphylococci, *Pasteurella pneumotropica*, and *Pasteurella multocida* as causative agents (Kunstyr et al., 1976).

3. Protozoan and Metazoan Parasites

Spironucleus muris has been reported to be a facultative pathogen in an outbreak of chronic enteritis in seven newly captured European hamsters as a result of stress exposure (Matthiesen et al., 1976). The main pathological findings were swelling of the liver and edema of the intestinal wall, the pancreas, and the mesenteric lymph nodes. Histologically, the liver showed areas of necrosis, cystic changes, and bile duct proliferation. The intestinal lesions were characterized by shortened villi and dilated Lieberkühn's crypts, containing necrotic debris and inflammatory cells. *Spironucleus muris* was found in large numbers both intra- and extracellularly in the small intestine. Some hamsters were also heavily infested with *Hymenolepis diminuta* and trichomonads.

Other spontaneous parasitic infestations of the European hamster including sarcocystosis, trypanosomiasis, taeniasis, cysticercosis, metastrongylosis, and capillariosis have been observed and described as clinically unapparent (Tenora and Murai, 1970; Murai, 1970; Kunstyr et al., 1976; Meszaros, 1977).

B. Spontaneous Tumors

Spontaneous benign and malignant tumors in old European hamsters are very common. The tumor incidence is approximately 70% in animals reaching more than 2 years of age (Ernst et al., 1987b). Neoplasms of the lymphatic system were the most common tumors found at necropsy of European hamsters which have been bred at the Hannover Medical School (strain MHH: EPH). In males, these were followed in a decreasing order of incidence by pheochromocytomas, malignant soft tissue tumors of various histological types and locations, and adenomas of the prostate. In females, ovarian granulosa cell tumors were the second commonest tumor type, followed by malignant soft tissue tumors and pheochromocytomas. Other tumors had a low incidence (Ernst et al., 1987b).

Bronchogenic squamous-cell carcinomas, similar to those in humans have been found in several wild hamsters which were captured in an industrialized area of West Germany. However, the possibility could not be ruled out that these tumors were not truly spontaneous but due to environmental pollutants (Mohr et al., 1973b).

C. Miscellaneous Disorders

Old European hamsters are affected with a high incidence of dental anomalies. According to Althoff et al. (1986), up to 80% of the animals which were more than 2 years old spontaneously developed a nodular cementofibrous dysplasia of the periodontal membrane of the maxillary incisors. Histological changes consisted of an irregularly oriented connective tissue which included abundant denticlelike structures. These usually multifocal and bilateral lesions frequently caused atrophy of the neighboring alveolar bone due to compression.

A second, age-related dysplastic process of the maxillary incisors, characterized by irregular proliferation of dentinumlike material, was observed in up to 65% of conventionally kept European hamsters and was classified as "odontogenic dysplasia (Althoff et al., 1985a).

Although the exact etiology and pathogenesis of both types of dental dysplasia are unknown, chronic microtrauma to the incisors (due to food collection, type of pelleted diet and housing, etc.), as well as probable genetic disposition, may lead to these non-neoplastic processes (Althoff et al., 1986).

Kunstyr et al. (1987) postulated a connection between these abnormalities and overt purulent osteomyelitic processes, which appeared regularly in the region of the lower jaw incisors of old European hamsters.

VII. USE IN RESEARCH

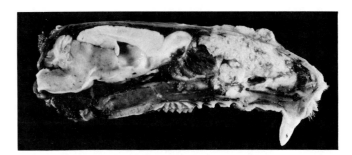

Fig. 7. Squamous-cell carcinoma invading the os nasale and the ductus nasopharyngicus after 35 weeks of treatment with 0.20 N-dibutylnitrosamine (DBN) LD_{50}.

Throughout the 1970s, besides using the European hamster for general experimental research work, Mohr, Reznik, and their colleagues tried in particular to prove that the animal represented a new sensitive model for carcinogenesis studies. The three basic requirements of such a laboratory model have been said to be (1) that lesions are induced by chemicals within a surveyable period during which no spontaneous tumors develop; (2) that tumors develop in those organs which are also affected in humans and are similar in type to those seen in humans; and (3) that a minimum of infectious diseases occur which could interfere with experimental data (Reznik et al., 1979). Like the Syrian golden hamster, the European hamster not only appears to fulfill these criteria but also has been found to have a high sensitivity to carcinogens. With known nitroso compounds, for example, well-differentiated respiratory tract tumors have been seen to develop within 13 weeks (Mohr et al., 1972). Wide-ranging carcinogenesis research has therefore been conducted using the European hamster, and many different substances have been tested (see Figs. 6–12) for their dose-related carcinogenic effects and threshold dose levels (Table VI).

The longevity and size of the European hamster as compared to other conventional laboratory rodents provide this animal with considerable advantages for various areas of research. It is suitable, for example, for a variety of additional studies during the course of an investigation, including periodic examination of hematological parameters and clinical biochemistry (Emminger et al., 1975; Reznik et al., 1975c). The comparative longevity of the European hamster also means that information may be gained on the biological effects of less toxic or less carcinogenic substances, or of low doses of highly effective compounds given over long periods (Ketkar et al., 1978; Richter-Reichhelm et al., 1978). The size of the animal facilitates surgical procedures and means, for example, that

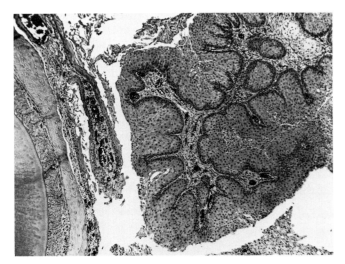

Fig. 6. Segment of squamous-cell papilloma of nasal cavity after decalcification after 20 weeks of treatment with N-diethylnitrosamine (DEN) at a dose of 20 mg/kg body weight. Hematoxylin–eosin stain. Magnification: ×60.

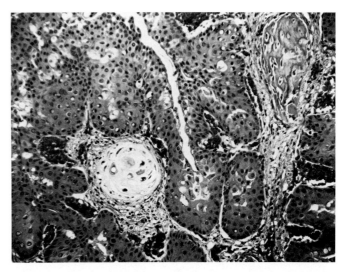

Fig. 8. Histological aspect of a nonkeratinizing bronchogenic squamous-cell carcinoma after DBN treatment. Hematoxylin–eosin stain. Magnification: ×270.

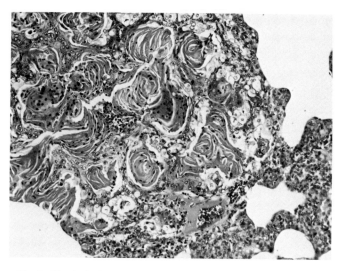

Fig. 9. Histological aspect of a keratinizing squamous-cell carcinoma of the lung after DBN treatment. Hematoxylin–eosin stain. Magnification: ×120.

Fig. 11. A 10 × 8 mm papilloma of the forestomach after 38 weeks of treatment with 0.05 nitrosopiperidine (NP) LD_{50}.

early alterations in the carcinogenic process may be detected by radiology and bronchography (Freyschmidt *et al.*, 1975; Reznik *et al.*, 1975b; Eckel *et al.*, 1973, 1974a,b, 1975).

The European hamster has also proved useful in inhalation studies, because its tidal volume has been found to be larger than that of other laboratory rodents (Kmoch *et al.*, 1976). Inhalation experiments have demonstrated that 30% more particulate matter is deposited in the lungs of the European hamster than in the Syrian hamster (Kmoch *et al.*, 1975; Reznik *et al.*, 1975a).

As the European hamster is a hibernator in its natural habitat and also shows considerably reduced metabolism when kept under cold laboratory conditions, this animal has been used not only for general research into hibernation (Reznik-Schüller and Reznik, 1973, 1974; Reznik *et al.*, 1975c; Hilfrich *et al.*, 1977), but also as a model for estimating the significance of metabolic activation of compounds in carcinogenic and toxicological studies (Mohr *et al.*, 1973a; Reznik and Mohr, 1977; Reznik *et al.*, 1977).

Fig. 10. Numerous mixed carcinomas in both lungs after 35 weeks of treatment with 0.20 *N*-dibutylnitrosamine (DBN) LD_{50}. Magnification: ×1.5.

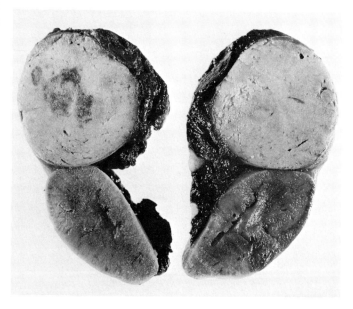

Fig. 12. Cut surface of adrenal pheochromocytoma 85 weeks after beginning of treatment with 0.20 urethane LD_{50}. Grey-white tumor masses are surrounded by adrenal capsule and are larger than adjacent kidney on *right*. Magnification: ×1.24.

Table VI
Organ Distribution of Experimentally Induced Tumors in European Hamsters[a]

Compound LD$_{50}$/route of administration[b]	TBA/ENA[c]	Respiratory system	Fore-stomach	Liver	Other organs	Type of tumor	Reference
Diisopropanolnitrosamine (DIPN)— s.c.	97/144	78(Nc) 32(Tr) 11(Lu)			14(Pa), 1(Co) 1(Ut)	Squamous-cell papilloma Squamous-cell carcinoma Adenocarcinoma	Reznik and Mohr (1977)
Diethylnitrosamine (DEN)	90/108	83(Nc), 5(La), 35(Tr), 8(Lu) 57(Nc), 1(La), 10(Lu)	50	63		Cholangiocellular adenoma Cholangiocellular carcinoma Squamous-cell papilloma Squamous-cell carcinoma	(Fig. 6) Mohr et al. (1972)
246 (♂) / 293 (♀) s.c.		7(Nc), 40(Lu) 7(Nc), 12(Lu)		2	1(Ki)	Adenoma Adenocarcinoma Cholangiocellularcarcinoma	Reznik et al. (1977)
351 (♂) / 413 (♀) s.c.				6 1	1(Br) 1(Nt)	Hepatocellular carcinoma Hemangioendothelioma Glioblastoma	
Dimethylnitrosamine (DMN) 28 (♂) / 43 (♀) s.c.	77/108	2(Nc)	13	31 3		Squamous-cell papilloma Squamous-cell carcinoma Hepatocellular carcinoma Cholangiocellular carcinoma	Mohr et al. (1974a); Richter-Reichhelm et al. (1978)
		2(Lu)		30	24(Ki) 4 3 1 15(Ov) 2(Pr) 2(Ut) 1(Pa) 2(Ad)	Hemangioendothelioma Sarcoma Lymphoblastic leukemia Malignant schwannoma Adenoma Granulosa cell tumor Adenocarcinoma Leiomyoma Islet cell carcinoma Neuroblastoma	
Dibutylnitrosamine (DBN) 2462 (♂) / 1866 (♀) s.c.	48/55	12(Nc) 33(Nc), 3(Tr), 2(Lu) 15(Nc), 10(Lu) 1(Nc), 10(Lu) 38(Lu)	21		3(Pt), 3(Ub) 12(Ub), 1(Ur) 9(Ub)	Papillary polyp Squamous-cell papilloma Squamous-cell carcinoma Adenocarcinoma Mixed carcinoma Transitional-cell papilloma Transitional-cell carcinoma	Althoff et al. (1974) (Figs. 7–9) (Fig. 10)
N-Nitrosopiperidine (NP) 226 (♂, ♀)/s.c.	54/60	18(Nc), 3(La), 9(Tr) 28(Nc), 1(Lu)	38		1(Pt) 8(Mo,Ch,Po, Pt,Es)	Squamous-cell papilloma Squamous-cell carcinoma	(Fig. 11) Mohr et al. (1974b)
N-Nitrosomorpholine (NM) 429 (♂, ♀)/s.c.	58/60	14(Lu) 4(Nc), 1(Lu) 7(Nc) 10(Nc), 8(La), 20(Tr), 3(Lu) 29(Nc), 3(La), 4(Tr), 1(Lu)		23	5(Pt), 13(Mo, Ch,Po,Es) 3(Pt), 8(Mo, Ch,Po,Es)	Adenoma Adenocarcinoma Papillary polyp Squamous-cell papilloma Squamous-cell carcinoma	Mohr et al. (1974b)
N-Nitroso-2,6-dimethylmorpholine (NDMM) 850 (♂, ♀)/s.c.	109/128	6(Lu) 12(Nc), 1(Lu) 11(La), 20(Tr) 92(Nc) 1(La), 2(Tr)			1(Mo) 1(Ph)	Adenoma Adenocarcinoma Squamous-cell papilloma Squamous-cell carcinoma	Althoff et al. (1985b)

(continued)

Table VI (Continued)
Organ Distribution of Experimentally Induced Tumors in European Hamsters[a]

Compound LD_{50}/route of administration[b]	TBA/ENA[c]	Respiratory system	Fore-stomach	Liver	Other organs	Type of tumor	Reference
1120 (♂, ♀)/i.g.		46(Lu)	13		1(Co), 3(Ha)	Adenoma	
		2(Nc)			2(Pi), 2(Ma)		
					1(Co), 1(Ma)	Adenocarcinoma	
						Mixed carcinoma	
				20		Hemangioendothelioma	
				2		Hepatocellular adenoma	
				1		Hepatocellular carcinoma	
				2		Cholangiocellular adenoma	
				5		Cholangiocellular carcinoma	
				1	1(Sk)	Fibrosarcoma	
					9(Ki,Ub)	Transitional-cell carcinoma	
					2	Lymphoblastic leukemia	
					1(Br)	Medulloblastoma	
Urethane 2463 (♂, ♀)/i.p.	39/60	6(Nc)	4			Squamous-cell papilloma	Mohr et al. (1974c) (Fig. 12)
		2(Lu)			4(Ki)	Adenoma	
		2(Lu)				Adenocarcinoma	
				3		Hepatocellular adenoma	
					6(Ad)	Pheochromocytoma	
					22(St,Pe)	Fibrosarcoma	
N-Methyl-N-nitrosourea (MNU) 113 (♂, ♀)/s.c.	39/60		4			Squamous-cell papilloma	Mohr et al. (1974a)
					9(Sk)	Squamous-cell carcinoma	
50.5 (♂, ♀)/i.v.					5(Sk)	Carcinosarcoma	Ketkar et al. (1977)
					24(St)	Fibrosarcoma	
					3(He,St)	Neurofibrosarcoma	
					2	Lymphoblastic leukemia	
N-Methyl-N-nitro-N-nitrosoguanidine (MNNG) 1378 (♂) / i.g. 1070 (♀)	11/20	2(Nc), 1(Tr)	11		2(Ph)	Squamous-cell papilloma	Ketkar et al. (1978)
		1(Nc), 2(La)	4			Squamous-cell carcinoma	
			1		1(Gs)	Fibrosarcoma	
1,1-Dimethylhydrazine (UDMH) 373 (♂) / s.c. 325 (♀)	19/29			2		Hepatocellular carcinoma	Ernst et al. (1987)
				1		Hepatocellular adenoma	
					10(St)	Neurofibrosarcoma	
					7(St)	Malignant schwannoma	
					2(Sk)	Malignant melanoma	
					2(Gs), 1(Sg,Ki)	Adenocarcinoma	
					1(Sp)	Reticulum cell sarcoma	
					1	Lymphoblastic leukemia	
					1(Ad)	Malignant pheochromocytoma	

[a]Dose–response relationships, sex differences, the effect of hibernation on tumor development, and the time of tumor development have not been taken into consideration and are referred to in the cited literature.
[b]LD_{50} (mg/kg body weight). s.c., Subcutaneous; i.g., intragastric; i.p., intraperitoneal; i.v., intravenous.
[c]TBA, Tumor-bearing animals; ENA, effective number of animals.
[d]Ad, Adrenal gland; Br, brain; Ch, cheek; Co, colon; Es, esophagus; Gs, glandular stomach; Ha, Harder's gland; He, heart; Ki, kidney; La, larynx; Lu, lung; Ma, mammary gland; Mo, mouth; Nc, nasal cavity; Nt, nose tip; Ov, ovary; Pa, pancreas; Pe, peritoneum; Ph, pharynx; Pi, pituitary gland; Po, pouch; Pr, prostate; Pt, palate; Sg, salivary gland; Sk, skin; Sp, spleen; St, subcutaneous tissue; Tr, trachea; Ub, urinary bladder; Ur, ureter; Ut, uterus.

REFERENCES

Althoff, J., Mohr, U., Page, N., and Reznik, G. (1974). Carcinogenic effect of dibutylnitrosamine in European hamsters (Cricetus cricetus). J. Natl. Cancer Inst. (U.S.) 53, 795–800.

Althoff, J., Koch, W., and Sommer, N. (1985a). Zahnanomalien beim Feldhamster und anderen Versuchsnagern. DTW, Dtsch. Tieraertzl. Wochenschr. 92, 21–22.

Althoff, J., Mohr, U., and Lijinsky, W. (1985b). Comparative study on the carcinogenicity of N-nitroso-2,6-dimethylmorpholine in the European hamster. J. Cancer Res. Clin. Oncol. 109, 183–187.

Althoff, J., Koch, W., and Reichart, P. (1986). Cemento-fibrous dysplasia of the periodontal membrane (studies of the European hamster maxillary incisors). Oral Pathol. 15, 11–15.

Asdell, A. B. (1946). "Patterns of Mammalian Reproduction." Cornell Univ. Press (Comstock), Ithaca, New York.

Aszodi, Z. (1921). Über künstlich erzeugte winterschlafähnliche Zustände an Mäusen. Biochem. Z. 113, 70–88.

Eckel, H., Reznik-Schüller, H., Reznik, G., Hilfrich, J., and Mohr, U. (1973). Demonstration of nitrosamine-induced tumors in the respiratory tract of the European hamster (*Cricetus cricetu* L.) by X-ray examination. *Strahlentherapie* **145**, 600–603.

Eckel, H., Reznik, G., Reznik-Schüller, H., and Mohr, U. (1974a). Bronchographic studies of the European hamster (*Cricetus cricetus* L.), the Syrian golden hamster (*Mesocricetus auratus* W.) and the Chinese hamster (*Cricetus griseus* M.). *Z. Versuchstierkd.* **16**, 322–326.

Eckel, H., Reznik-Schüller, H., Ohse, B., and Mohr, U. (1974b). Radiological detection and sequential observation of bladder tumors in the European hamster. *Br. J. Cancer* **30**, 496–502.

Eckel, H., Reznik-Schüller, H., Reznik, G., Ohse, B., and Mohr, U. (1975). Diagnosis of experimentally-induced bronchogenic tumours in the European hamster with bronchographs. *Z. Krebsforsch.* **83**, 207–212.

Eibl-Eibesfeldt, I. (1953). Zur Ethologie des Hamsters (*Cricetus cricetus* L.). *Z. Tierpsychiatr.* **58**, 204–252.

Eisentraut, M. (1956). "Der Winterschlaf mit seinen ökologischen und physiologischen Begleiterscheinungen." Fischer, Jena.

Emminger, A., Reznik, G., Reznik-Schüller, H., and Mohr, U. (1975). Differences in blood values depending on age in laboratory-bred European hamsters (*Cricetus cricetus* L.). *Lab. Anim.* **9**, 33–42.

Ernst, H., Rittinghausen, S., Wahnschaffe, U., and Mohr, U. (1987a). Induction of malignant peripheral nerve sheath tumours in European hamsters with 1,1-dimethylhydrazine (UDMH). *Cancer Lett. (Shannon, Irel).* in press.

Ernst, H., Kunstyr, J., and Mohr, U. (1987b). Spontaneous tumours of the European hamster (*Cricetus cricetus* L.). *Z. Versuchstierkd.* (submitted for publication).

Freyschmidt, J., Reznik, G., Reznik-Schüller, H., and Rippel, W. (1975). X-Ray enlargement modified for use in experimental animal science. *Lab. Anim.* **9**, 305–311.

Gaffrey, G. (1961). "Merkmale der wildlebenden Säugetiere Mitteleuropas." Akad. Verlagges. Leipzig.

Hannoun, C., Guillon, J.-C., and Chatelain, J. (1974). Infection spontanée latente à papovavirus chez le hamster européen. I. Spontaneous latent papovavirus infection of the European hamster. I. Isolation of the virus in golden hamsters and in newborn mice. *Ann. Microbiol. (Paris)* **125A**, 215–226.

Hilfrich, J., Züchner, H., and Reznik-Schüller, H. (1977). Studies of the ovaries of hibernating and non-hibernating European hamsters (*Cricetus cricetus*). *Z. Versuchstierkd.* **19**, 304–308.

Kayser, C. (1939). Les changes respiratoires des hibernants. *Ann. Physiol. Physicochim. Biol.* **15**, 1087–1219.

Kayser, C. (1959). Les échanges respiratoires du hamster ordinaire et du lerot en hibernation. *C. R. Seances Soc. Biol. Ses Fil.* **153**, 167–170.

Kayser, C. (1961). The physiology of natural hibernation. *Int. Ser. Monogr. Pure Appl. Biol., Div. Zool.* **23**.

Kayser, C., Pietsch, M. L., and Lucot, M. A. (1954). Les échanges respiratoires et la fréquence cardiaque des hibernants au cours du réveil de leur sommeil hibernal. Recherches physiologiques sur l'incrément thermique critique. *Arch Sci. Physiol.* **8**, 155–193.

Ketkar, M., Reznik, G., Haas, H., Hilfrich, J., and Mohr, U. (1977). Tumours of the heart and stomach induced in European hamsters by intravenous administration of *N*-methyl-*N*-nitrosourea. *J. Natl. Cancer. Inst. (U.S.)* **58**, 1695–1699.

Ketkar, M., Reznik, G., and Green, U. (1978). Carcinogenic effect of *N*-methyl-*N'*-nitro-*N*-nitrosoguanidine (MNNG) in European hamsters. *Cancer Lett.* **4**, 241–244.

Kmoch, N., Reznik, G., and Mohr, U. (1975). Inhalation experiments with ^{14}C-labelled cigarette smoke. II. The distribution of cigarette smoke particles in the hamster respiratory tract after exposure in two different smoking systems. *Toxicology* **4**, 373–383.

Kmoch, N., Reznik, G., and Schleicher, A. (1976). Inhalation experiments with ^{14}C-labelled cigarette-smoke. III. Body size-dependent distribution of particulate matter in small rodents during cigarette smoke inhalation. *Toxicology* **6**, 219–223.

Kunstyr, I., Ernst, H., Merkt, M., and Reichart, P. (1987). Spontaneous pathology of the European hamster (*Cricetus cricetus*). Malocclussion, dysplastic and inflammatory processes on the jaws. *Z. Versuchstierkd.* **29**, 171–181.

Kunstyr, I., Reznik, G., Matthiesen, T., and Friedhoff, K. (1976). Spontaneous infectious diseases of European hamster. *Z. Versuchstierkd.* **18**, 166.

Malan, A., and Arens, H. (1973). Pulmonary respiration and acid–base state in hibernating marmots and hamsters. *Respir. Physiol.* **17**, 45–61.

Matthiesen, T., Kunstyr, I., and Tuch, K. (1976). *Hexamita-muris* infection in mice and European hamsters in a laboratory animal colony. *Z. Versuchstierkd.* **18**, 113–120.

Meszaros, F. (1977). Parasitic nematodes of the hamster (*Cricetus cricetus* L.) in Hungary. *Acta Zool. Acad. Sci. Hung.* **23**, 133–138.

Mohr, U., Althoff, J., and Page, N. (1972). Brief communication: Tumours of the respiratory system induced in the common European hamster by *N*-diethylnitrosamine. *J. Natl. Cancer Inst. (U.S.)* **49**, 595–597.

Mohr, U., Althoff, J., Spielhoff, R., and Bresch, H. (1973a). The influence of hibernation upon the carcinogenic effect of *N*-diethylnitrosamine in European hamsters. *Z. Krebsforsch.* **80**, 285–288.

Mohr, U., Schüller, H., Reznik, G., Althoff, J., and Page, N. (1973b). Breeding of European hamsters. *Lab. Anim. Sci.* **23**, 799–802.

Mohr, U., Haas, H., and Hilfrich, J. (1974a). The carcinogenic effects of dimethylnitrosamine and nitrosomethylurea in European hamsters (*Cricetus cricetus* L.). *Br. J. Cancer* **29**, 359–365.

Mohr, U., Reznik, G., and Reznik-Schüller, H. (1974b). Carcinogenic effects of *N*-nitrosomorpholine and *N*-nitrosopiperidine on European hamster (*Cricetus cricetus*). *J. Natl. Cancer Inst. (U.S.)* **53**, 231–237.

Murai, E. (1970). The hamster *Cricetus cricetus*: A new host of paranoplocephala-omphalodes cestoda anoplocephalidae. *Parasitol. Hung.* **3**, 43–50.

Petzsch, H. (1937). Die Fortpflanzungsbiologie des Hamsters *Cricetus cricetus* L.). *Naturforscher* **13**, 337–340.

Raths, P. (1953). Untersuchungen über die Blutzusammensetzung und ihre Beziehungen zur vegetativen Tonuslage beim Hamster (*Cricetus cricetus* L.). *Z. Biol.* **106**, 109–123.

Raths, P. (1957). Über die Abhängigkeit der Blutzusammensetzung von der allgemeinen Aktivitätslage beim Hamster. *Zool. Anz.* **159**, 139–152.

Reznik, G., and Mohr, U. (1977). Effect of di-isopropanol-nitrosamine in European hamsters. *Br. J. Cancer* **36**, 479–486.

Reznik, G., Reznik-Schüller, H., and Mohr, U. (1973). Comparative studies of organs in the European hamster (*Cricetus cricetus* L.), the Syrian golden hamster (*Mesocricetus auratus* W.) and the Chinese hamster (*Cricetus griseus* M.). *Z. Versuchstierkd.* **15**, 272–282.

Reznik, G., Kmoch, N., and Mohr, U. (1975a). Inhalation experiments with ^{14}C-labelled cigarette smoke. I. Determination of the effectiveness of two different smoking systems with labelled cigarettes. *Toxicology* **4**, 363–371.

Reznik, G., Eckel, H., Freyschmidt, J., and Reznik-Schüller, H. (1975b). Age-dependent skeletal development in the European hamster—radiological investigations. *Z. Versuchstierkd.* **17**, 233–239.

Reznik, G., Reznik-Schüller, H., Emminger, A., and Mohr, U. (1975c). Comparative studies of blood from hibernating and non-hibernating hamsters (*Cricetus cricetus* L.). *Lab. Anim. Sci.* **25**, 210–215.

Reznik, G., Reznik-Schüller, H., and Mohr, U. (1977). Carcinogenicity of *N*-nitroso-diethylamine in hibernating and non-hibernating European hamsters. *J. Natl. Cancer Inst. (U.S.)* **58**, 673–680.

Reznik, G., Reznik-Schüller, H., and Mohr, U. (1979). "Clinical Anatomy of the European hamster (*Cricetus cricetus* L.)." U.S. Govt. Printing office, Washington, D.C.

Reznik-Schüller, H., and Reznik, G. (1973). Comparative histometric investigations of the testicular function of European hamsters (*Cricetus cricetus*) with and without hibernation. *Fertil. Steril.* **24,** 698–705.

Reznik-Schüller, H., and Reznik, G. (1974). The influence of hibernation upon the ultra-structure of the Leydig cells and spermatids of the European hamster. *Fertil. Steril.* **25,** 621–635.

Reznik-Schüller, H., Reznik, G., and Mohr, U. (1974). The European hamster (*Cricetus cricetus* L.) as an experimental animal: Breeding methods and observations of their behavior in the laboratory. *Z. Versuchstierkd.* **16,** 48–58.

Richter-Reichhelm, H.-B., Green, U., Ketkar, M. B., and Mohr, U. (1978). The carcinogenic effect of dimethylnitrosamine in laboratory-bred European hamsters (*Cricetus cricetus*). *Cancer Lett.* **4,** 1–4.

Silverman, J., and Chavannes, J. M. (1977). Biological values of the European hamster (*Cricetus cricetus*). *Lab. Anim. Sci.* **27,** 641–645.

Svrcek, S., Alexander, R., Vrtiak, O. J., Ondrejka, R., Macicka, O., and Grulich, I. (1980). Izolyatsiya virusa beshenstva iz khomyaka obbyknovennogo (*Cricetus cricetus*) v Chekhoslavakii. Isolation of rabies virus from the common hamster (*Cricetus cricetus*) in Czechoslovakia. *Acta Vet. Acad. Sci. Hung.* **28,** 21–26.

Tenora, F., and Murai, E. (1970). *Hymenolepis straminea* cestoda hymenolepididae, parasite of *Cricetus cricetus* in Hungary. *Parasitol. Hung.* **3,** 33–42.

Part IV

Other Hamsters

Chapter 21

Biology, Care, and Use in Research

Connie A. Cantrell and Dennis Padovan

Introduction	369
Phodopus sungorus (Dzungarian hamster)	370
I. Taxonomy, History, and Genetics	370
II. Care and Management	370
III. Anatomy, Physiology, and Reproduction	371
IV. Diseases	371
V. Research Uses	372
References	374
Mystromys albicaudatus (South African hamster)	376
I. Taxonomy, History, and Genetics	376
II. Care and Management	377
III. Anatomy, Physiology, and Reproduction	377
IV. Diseases	379
V. Research Uses	379
References	380
Mesocricetus brandti (Turkish hamster)	382
I. Taxonomy, History, and Genetics	382
II. Care and Management	382
III. Anatomy, Physiology, and Reproduction	382
IV. Diseases	383
V. Research Uses	383
References	383
Mesocricetus newtoni (Rumanian hamster)	384
Text	384
References	385
Cricetulus migratorius (Armenian hamster)	385
Text	385
References	386

INTRODUCTION

The scope of this chapter is limited to those hamsters which have been utilized as laboratory animals and are not covered in other chapters of this book. The Syrian hamster (*Mesocricetus auratus*) is the subject of Chapters 1–16, the Chinese hamster (called *Cricetulus griseus* by some authorities and *Cricetulus barabensis* by others) is described in Chapters 17–19, and the European hamster (*Cricetus cricetus*) is covered in Chapter 20. This chapter will describe the laboratory use of the Dzungarian hamster (*Phodopus sungorus*), the South African hamster (*My-*

stromys albicaudatus), the Turkish hamster (*Mesocricetus brandti*), the Rumanian hamster (*Mesocricetus newtoni*), and the Armenian hamster (*Cricetulus migratorius*). For more information on the taxonomic classification of hamsters, please refer to the Foreword of this book and the tabulation accompanying it.

Phodopus sungorus (Dzungarian Hamster)

I. TAXONOMY, HISTORY, AND GENETICS

References in the literature to *Phodopus sungorus* may actually include two distinct species, *P. sungorus* and *P. campbelli*. *Phodopus campbelli* has been considered a subspecies of *P. sungorus* in the biomedical literature, but many authorities consider *P. campbelli* a separate species (Honacki *et al.*, 1982). Laboratory colonies in the Soviet Union and Great Britain have been referred to as *P. sungorus campbelli,* and colonies in Germany have been identified as *P. sungorus sungorus.* Because of the ambiguity of some references and to conform with the literature, they will be considered here as a single species, *P. sungorus.* If care is not taken when new animals are introduced to an established breeding colony, hybrids between these two species may be developed. The common names which have been applied to this hamster are the Dzungarian (or Djungarian or Zungarian) hamster, striped hairy-footed hamster, and Siberian hamster. Karyological studies suggest that *Phodopus* are more closely related to *Cricetus* and *Cricetulus* than to *Mesocricetus* and *Mystromys* (Gamperl *et al.*, 1978).

The first laboratory colony of *Phodopus* was established by M. N. Meier at the Zoological Institute of the Academy of Sciences of the Soviet Union in Leningrad from animals trapped in Tuva, Siberia and identified as *P. sungorus campbelli*. One female and two males from this colony were used to establish another colony at the Institute of Experimental and Clinical Oncology in Moscow (Pogosianz and Sokova, 1967; Yerganian, 1972), and animals from this colony were subsequently used to start laboratory colonies in various locations in Europe. Another colony was established at the Max-Planck Institute of Comparative Physiology in West Germany from four animals trapped near Omsk in western Siberia and identified as *P. sungorus sungorus* (Hoffmann, 1973, 1978a). Animals from this colony have also been distributed to other laboratories.

Phodopus have 28 chromosomes, including 5 pairs of large, 3 pairs of medium, and 5 pairs of small autosomes (Pogosianz and Brujako, 1971; Pogosianz, 1975; Gamperl *et al.*, 1977; Spyropoulos *et al.*, 1982). The X chromosome is a medium-sized metacentric chromosome, and the Y chromosome is a medium-sized acrocentric chromosome (Das and Savage, 1978). The chromosome banding patterns have been described by Thust (1974), Bigger and Savage (1976), Vistorin *et al.* (1977), and Gamperl *et al.* (1977, 1978). The low chromosome number and the fact that all but the smallest chromosomes can be recognized with routine staining methods make the cells of this hamster useful in cytogenetic studies. The pattern of meiosis in male *Phodopus* has been described by Pogosianz (1970), and polyploid spermatocytes occur commonly (Pogosianz and Brujako, 1971).

A coat and eye color mutation designated pink-eyed dilute after the similar condition in mice (*Mus musculus*) has been described by Pogosianz and Sokova (1981). The trait is due to an autosomal-recessive gene. The hair is a yellowish color and the eyes are pink. There do not appear to be any differences in viability, fertility, or behavior associated with this gene. The trait can be detected at birth because the normally dark eye color is not visible beneath the closed eyelids.

II. CARE AND MANAGEMENT

Phodopus are generally described as being tame and nonaggressive (Pogosianz and Sokova, 1967; Jordan, 1971; Hoffmann, 1981), but Pilborough (1971) found them to be aggressive, prone to fighting, and more difficult to handle than Syrian hamsters. In most colonies, fighting and cannibalism have been infrequent and *Phodopus* can be kept in groups or colony bred. Cannibalism occurred in 30% of the litters when the parents were handled daily for urine collections during the first few days after birth of the litter (Herberg *et al.*, 1980). Mothers with young are usually left with the fathers, and both parents participate in care of the young (Gibber *et al.*, 1984; Pogosianz and Sokova, 1967).

Various types of solid-floored caging with sawdust litter have been used for *Phodopus*. Breeding groups have been provided with nest boxes or hay for nest building. Hoarding of food occurs when nest boxes are provided. The usual animal room temperatures (20°–25°C) and relative humidity (50%) are apparently satisfactory for this species (Herberg *et al.*, 1980; Pogosianz and Sokova, 1967; Pilborough, 1971; Jordan, 1971; Duncan *et al.*, 1985).

The reproductive activity and body weights of *Phodopus* are influenced by daily light cycles (see Section V, Research Uses). Constant photoperiods in laboratory colonies may cause year-round breeding, but they may also cause obesity in older animals as reported in some colonies (Jordan, 1971). Dim red light has been used during manipulations of *Phodopus* during the dark phase of the light cycle without apparent alteration of photoperiodic responses (Yellon *et al.*, 1982).

III. ANATOMY, PHYSIOLOGY, AND REPRODUCTION

Phodopus (Fig. 1) are small, colonial, nocturnal hamsters. Adult males are about 11 cm long and weigh about 40 to 50 gm. Adult females are about 9 cm long and weigh about 30 gm. The short tail is about 1 cm long and usually hidden in the fur. The fur on the dorsal surfaces is gray with a dark-brown or black stripe along the midline from the nape of the neck to the base of the tail. The ventral fur is white. When exposed to natural light cycles, some may develop a white winter coat with gray on the head and shoulders. The color change is enhanced by exposure to colder temperatures (Figala *et al.*, 1973). Annual body weight cycles also occur with exposure to natural photoperiods. Weights are greater during the summer months than during the winter months. The footpads are covered with fur, which has led to the appellation "hairy-footed hamster." Like the other hamsters, except *Mystromys*, *Phodopus* have internal cheek pouches. Females have four pairs of mammary glands. Males possess a unique network of small auxiliary tubules branching from the middle segment of the epididymis (Nagy *et al.*, 1982). A macroscopic description of the digestive tract with a detailed microscopic description of the cecum has been published by Snipes (1979). Suzuki *et al.* (1983, 1984) have published ultrastructural studies of the parotid and mandibular salivary glands. The urine-concentrating ability of *Phodopus* is similar to that of *Cricetulus griseus* and greater than *Mesocricetus auratus* and *Cricetus cricetus* (Trojan, 1977). *Phodopus* have lateral scent-marking glands and roll on their side or back to rub the flank glands on the substrate (Daly, 1976). They also have a large midventral sebaceous gland that is larger in males than females (Figala *et al.*, 1973). The normal rectal temperature is 36.1°–37°C (Heldmaier, 1975; Heldmaier and Steinlechner, 1981a,b). *Phodopus* exhibit a circadian rhythm in testosterone and luteinizing hormone (LH) levels that does not occur in Syrian hamsters (Hoffmann, 1981).

Newborn *Phodopus* are hairless except for vibrissae and weigh about 1.8 gm. The eyes and ears are closed, but the incisors have erupted at birth (Pogosianz and Sokova, 1967; Pilborough, 1971). At 3 days the dorsal skin has darkened and the midline stripe is visible. They gain a righting response on the second to fourth day (Daly, 1976). The ears open on the third or fourth day, and the body hairs become apparent at the sixth to ninth day. At about the tenth day, the eyes open and the young begin to leave the nest and chew on solid food. They can be weaned at 16–18 days. At this age they begin scent-marking behavior (Daly, 1976). The mean weight at 20 days is 17.4 gm, with a range of 11.7–21.5 gm. (Jordan, 1971).

Phodopus breed throughout the year in laboratories, but a decreased reproductive rate has been reported during the summer and winter months (Pogosianz and Sokova, 1967; Jordan, 1971). With natural light cycles there is testicular regression during midwinter to less than 10% of the size in the summer, and breeding may cease completely in midwinter (Hoffmann, 1972, 1973, 1978b; Figala *et al.*, 1973). The gestation period is about 18 days (Daly, 1976; Figala *et al.*, 1973). Litter sizes average about four and vary from one to nine (Pogosianz and Sokova, 1967; Jordan, 1971; Herberg *et al.*, 1980; Figala *et al.*, 1973). The mean age of females at the first parturition has been reported as 90 days by Herberg *et al.* (1980) and 139 days by Jordan (1971), but may be influenced by the seasonal depression of breeding. The onset of puberty in males depends on the photoperiod. Males raised with 16 hr of light per day have well-developed testes at 35–40 days of age. Males raised with short photoperiods of 8 hr/day do not reach puberty until about 150 days of age (Hoffmann, 1978a, 1979a, 1981). The reproductive life of most females was reported by Pogosianz and Sokova (1967) to be about 1 year. The average life span was reported as 1 year by Herberg *et al.* (1980) and 2 years by Heldmaier and Steinlechner (1981b).

A 4.5% incidence (8 of 179 females) of abdominal pregnancies was reported in one colony during a 12-month period (Buckley and Caine, 1979). All the affected animals had previously produced normal litters. Abdominal fetuses, which numbered from one to five, were mummified and were at a similar stage of development with good skeletal and limb formation. A few had placental attachments to the abdominal wall, but most were free in the peritoneal cavity. Some of the mothers had an associated peritonitis. No evidence of rupture of the uterus or oviduct was found in any of the hamsters.

IV. DISEASES

Very few reports of diseases in *Phodopus* have been published. A high incidence of dermatophytosis caused by *Tri-*

Fig. 1. Phodopus sungorus. (Photograph courtesy of Dr. Robert E. Johnston, Ithaca, New York.)

chophyton mentagrophytes was reported in one colony. Signs included alopecia and hyperkeratosis of the limbs, ear pinna, and ventral surface of the body (Young, 1973a,b). Occasional subcutaneous abscesses resulting from fight wounds, thymomas, and cystic ovarian follicles were also reported in this colony as well as infestation with *Myocoptes musculinus*. Another colony had a 66% incidence of thyroid parafollicular-cell adenocarcinomas in *Phodopus* over 1 year old (Quimby *et al.*, 1982). There was a 10% incidence of pulmonary metastasis, and the tumors were associated with obesity and endocrine alopecia. A 30% incidence of spontaneous neoplasia was reported in animals over 2 years old by Pogosianz *et al.* (1970). Spontaneous neoplasms reported included mammary carcinomas, cutaneous papillomas, and squamous-cell carcinomas, pulmonary adenomas and carcinomas, liver adenomas and carcinomas, lymphosarcomas, an ovarian carcinoma, and an angiosarcoma (Pogosianz, 1975; Pogosianz *et al.*, 1970; Sokova, 1971).

V. RESEARCH USES

Marked seasonal changes in reproductive activity has made *Phodopus* a popular subject for studies of photoperiodism and the pineal gland. Male *Phodopus* exhibit large active testes, enlarged accessory glands, and greater body weight when exposed to photoperiods longer than 13 hr/day. Shorter photoperiods cause testicular regression to less than 10% of the active testicle size, decreases in body weight of about 30%, and a change to a whitish winter pelage (Duncan *et al.*, 1985; Hoffmann, 1972, 1973, 1978a, 1979a,b, 1981, 1982; Hoffmann and Kuderling, 1975; Logan and Weatherhead, 1978; Figala *et al.*, 1973). Weight reduction results mainly from decreased body fat though brown fat remains relatively constant (Hoffmann, 1981; Wade and Bartness, 1984). Continuous short photoperiods do not maintain regressed testes and winter pelage indefinitely, however, and after several months there is spontaneous testicular recrudescence and molt to the summer pelage (Hoffmann, 1979b, 1981). The photoperiodic weight reduction also occurs in orchidectomized *Phodopus* (Hoffmann, 1978b. 1981; Wade and Bartness, 1984) but not in ovariectomized *Phodopus* (Wade and Bartness, 1984). Changes in pelage color also occur in orchidectomized *Phodopus* (Hoffmann, 1981). Orchidectomized *Phodopus* fail to gain as much weight during long photoperiods as intact males (Hoffmann, 1978b). A 1- or 5-min light pulse in the middle of the dark phase of an 8-hr daily photoperiod cycle was found by Hoffmann (1979a, 1982) to have the same stimulatory effect on testicular development as a 16-hr daily photoperiod. This marked regulatory effect of brief light exposures may not be unexpected in a nocturnal burrowing species which would not normally expose itself to prolonged periods in the open during daylight hours. The photostimulatory effect on testicular growth is also associated with increased levels of plasma and pituitary follicle-stimulating hormone (FSH) and a slower elevation of plasma and pituitary LH (Simpson *et al.*, 1982).

Phodopus exhibit a seasonal variation in pituitary melanocyte-stimulating hormone (MSH) content, which is responsive to the duration of daily photoperiods (Logan and Weatherhead, 1980a). The change to the white winter pelage is accompanied by decreased levels of MSH in the pituitary (Logan and Weatherhead, 1979), and the seasonal change has been investigated as a possible model for normal melanin formation (Logan and Weatherhead, 1978). Melanogenesis in *Phodopus* hair follicles has been shown to be stimulated by MSH (Logan *et al.*, 1981) and inhibited by melatonin *in vitro* (Logan and Weatherhead, 1980b). However, *in vivo* studies (Duncan and Goldman, 1984a,b) indicate that prolactin mediates pigment changes in the pelage. MSH did not influence pelage color *in vivo* and may be inhibited by circulating levels of melatonin.

The summer pelage develops up to 30 days of age regardless of the photoperiod. The summer pelage is maintained with a long photoperiod but changes to the winter pelage with short photoperiods and then spontaneously back to the summer pelage at 110–150 days of age (Hoffmann, 1981).

The photoperiod also influences the development of puberty in *Phodopus*. Young, male *Phodopus* exposed to a 16-hr daily photoperiod develop large testicles with fully active spermatogenesis at about 35 days of age. Puberty is delayed in male *Phodopus* raised with a short (8-hr) daily photoperiod. Photoperiodic sensitivity begins between 7 and 14 days of age (Yellon and Goldman, 1980; Hoffmann, 1981), and the testes of those exposed to short photoperiods stop growing until about 130 days of age before continuing to full development at 5–7 months (Hoffmann, 1978a, 1979b, 1981). The slow testicular development is correlated with lower blood FSH, LH, and prolactin levels in juvenile male *Phodopus* raised with short daily photoperiods compared to those raised with longer daily photoperiods (Yellon and Goldman, 1980, 1984). Similar inhibitions in ovarian and uterine development were observed in female *Phodopus* raised with an 8-hr daily photoperiod compared to those raised with a 16-hr photoperiod (Hoffmann, 1981).

The pineal gland controls the photic gonadal, weight, and pelage color responses in *Phodopus*. Pinealectomy prevents molt to the winter pelage and weight loss in *Phodopus* exposed to short photoperiods. Adult male *Phodopus* pinealectomized while the testes were in the regressed winter condition and then exposed to long photoperiods (16 hr of light daily) did not show the testicular development expected with the long photoperiod (Hoffmann and Kuderling, 1975). Conversely, pinealectomy prevented the regression of hypertrophic testes after exposure to short photoperiods (Hoffmann, 1979b). The pineal also regulates testicular development in juvenile *Phodopus*. Pinealectomy of juvenile male *Phodopus* raised with an

8-hr daily photoperiod decreased the inhibitory effect of the short photoperiod, and the testes developed more rapidly than in intact animals raised in the short photoperiod. Pinealectomy also slightly retarded the development of the testes in juvenile *Phodopus* raised in a long (16-hr) daily photoperiod (Brackmann and Hoffmann, 1977). Morphological studies of the pineal gland have been conducted to study postnatal development (van Veen *et al.*, 1978) and ultrastructure (Karasek *et al.*, 1982, 1983). Ultrastructural alterations of the pituitary related to photoperiod have also been reported (Wittkowski *et al.*, 1984).

Studies in *Phodopus* have shown that melatonin is at least one of the pineal hormones involved in photoperiodic responses. Natural circadian levels of melatonin or *N*-acetyltransferase, which is involved in the formation of melatonin, in the pineal gland were correlated with photoperiod length (Yellon *et al.*, 1982; Hoffmann *et al.*, 1980, 1981; Goldman *et al.*, 1981; Illnerova *et al.*, 1984). Pineal melatonin levels were increased during the dark phase of the light cycle, and longer dark phases caused more extended melatonin elevations. Constant darkness produced a regular 14-hr elevation of melatonin associated with the time of activity (Yellon *et al.*, 1982). Continuous chronic melatonin treatment of adult male *Phodopus* in the winter state with involuted testes and winter pelage transferred to a long (16-hr) daily photoperiod resulted in suppression of testicular development, inhibition of body weight gains, and delay in molting to the summer pelage (Hoffmann, 1972, 1973, 1981). Chronic melatonin treatment also delayed testicular and accessory gland development in juvenile *Phodopus* (Brackmann, 1977; Hoffmann, 1979b, 1981; Goldman *et al.*, 1984), accelerated testicular regression in late summer (Hoffmann and Kuderling, 1977; Hoffmann, 1981), and caused an increase in brown adipose tissue (Heldmaier and Hoffmann, 1974). The unnatural continuous chronic melatonin exposure produced in these experiments did not simulate the normal circadian cycle. An attempt to simulate a circadian rhythm of melatonin levels revealed that short 4–6 hr/day melatonin exposures in young pinealectomized *Phodopus* stimulated testicular growth while long (8–12 hr/day) melatonin exposures suppressed testicular development (Carter and Goldman, 1981, 1983a,b), even though the total daily melatonin doses were the same in both cases.

The demands on *Phodopus* caused by its small body size in the severe climate of its native Siberia has stimulated interest in studying its thermoregulatory ability. *Phodopus* apparently remain active throughout the winter and do not hibernate. However, periodic torpor lasting 4–8 hr with body temperature declines to 17°–23°C were observed in animals which had molted to the winter pelage (Heldmaier and Steinlechner, 1981b; Figala *et al.*, 1973). Torpor was a response to photoperiod rather than ambient temperature during the first winter of life, but exposure to lower ambient temperatures was required to produce torpor during the second winter.

Phodopus are poorly adapted to resist heat and may become hyperthermic at an ambient temperature of 34°C and die at 36°C (Heldmaier, 1975), but they are much better able to withstand low temperatures. The ability of *Phodopus* to withstand cold is mainly accounted for by an efficient norepinephrine-mediated, nonshivering thermogenesis (Weiner and Gorecki, 1981; Bockler *et al.*, 1982; Heldmaier, 1975; Heldmaier *et al.*, 1982a; Buchberger *et al.*, 1983). Thermal insulation of the hair coat remains essentially constant all year (Heldmaier *et al.*, 1982b), although the winter hair coat is about 2 mm longer than the summer hair coat (Heldmaier and Steinlechner, 1981a). The basal metabolic rate and shivering thermogenesis contribute little to seasonally improved cold tolerance (Heldmaier *et al.*, 1982a). Their thermogenic abilities exhibit an annual cycle regulated by photoperiod and cold exposure (Steinlechner and Heldmaier, 1982; Heldmaier *et al.*, 1981, 1982a,b; Bockler and Heldmaier, 1983; Buchberger *et al.*, 1983). Photoperiod adaptations appear to be mediated at least partially by melatonin, which enhanced nonshivering thermogenesis and increased mitochondrial activity in brown adipose tissue (Steinlechner and Heldmaier, 1982; Heldmaier *et al.*, 1981). The experimental thermogenic cold limit of *Phodopus* during the summer is about $-22°C$, while the winter cold limit is about $-50°C$ (Heldmaier *et al.*, 1982a,b). Neither acute food deprivation (Puchalski *et al.*, 1983) nor surgical removal of brown fat, which may comprise up to 6% of the body weight of *Phodopus* (Rafael and Vsiansky, 1983), had a detectable effect on thermogenic ability.

The small chromosome number and relatively easily distinguished chromosomes have encouraged the use of *Phodopus* in cytogenetic and carcinogenesis studies (Pogosianz, 1970, 1975; Das and Savage, 1978; Bigger and Savage, 1976; Kopnin and Lucas, 1982; Kopnin, 1982; Kopnin and Gudkov, 1982; Kakpakova *et al.*, 1981). Neoplasms have been produced with the chemical carcinogens 7,12-dimethylbenz[*a*]anthracene (DMBA), 3-methylcholanthrene (MCA), methylnitrosourea (MNU) (Pogosianz, 1975), and diethylnitrosamine (DEN) (Warzok and Thust, 1977), and with the oncogenic viruses Rous sarcoma virus, simian adenovirus 7, and human adenovirus 12 (Pogosianz, 1975). A transplantable mammary gland carcinoma, which displayed a high rate of metastasis, has been studied at the Institute of Experimental and Clinical Oncology in Moscow (Kiseleva, 1972).

The animals in one colony of *Phodopus* with glycosuria and ketonuria have been studied as a possible model of diabetes mellitus. Glycosuria and ketonuria developed independently and occurred together or separately. The average age at which glycosuria was first detected was 76 days. Glycosuria was either constant or intermittent. Glycosuric animals had high blood levels of glucose, insulin, and triglycerides. Glycosuria was associated with β cell degranulation and glycogen deposition in the islets of Langerhans. Ketonuria was more frequent and occurred as early as 3 days of age and was usually detected by 21 days of age. Blood and urine acetoacetate and β-hydroxybutyrate concentrations were normal, and the ketonuria was due to increased urinary acetone (Herberg *et al.*, 1980; Herberg, 1979; Vesely and Herberg, 1981).

Phodopus have been the subject of limited behavioral studies (Crawley, 1984a,b; Gibber *et al.*, 1984). Following the separation of mated pairs, males gain weight and both sexes become less active. These changes are reversed when the pairs are reunited. This separation syndrome has been suggested as a possible animal model of depression (Crawley, 1984a,b).

REFERENCES

Bigger, T. R. L., and Savage, J. R. K. (1976). Location of nucleolar organizing regions on the chromosomes of the Syrian hamster (*Mesocricetus auratus*) and the Djungarian hamster (*Phodopus sungorus*). Cytogenet. Cell Genet. **16**, 495–504.

Bockler, H., and Heldmaier, G. (1983). Interaction of shivering and non-shivering thermogenesis during cold exposure in seasonally acclimatized Djungarian hamsters (*Phodopus sungorus*). J. Therm. Biol. **8**, 97–98.

Bockler, H., Steinlechner, S., and Heldmaier, G. (1982). Complete cold substitution of noradrenaline-induced thermogenesis in the Djungarian hamster, *Phodopus sungorus*. Experientia **38**, 261–262.

Brackmann, M. (1977). Melatonin delays puberty in the Djungarian hamster. Naturwissenschaften **64**, 642–643.

Brackmann, M., and Hoffmann, K. (1977). Pinealectomy and photoperiod influence testicular development in the Djungarian hamster, *Naturwissenschaften* **64**, 341–342.

Buchberger, A., Heldmaier, G., Steinlechner, S., and Latteier, B. (1983). Photoperiod and temperature effects on adrenal tyrosine hydroxylase and its relation to non-shivering thermogenesis. Pfluegers Arch. Gesamte Physiol. Menschen Tiere **399**, 79–82.

Buckley, P., and Caine A. (1979). A high incidence of abdominal pregnancy in the Djungarian hamster (*Phodopus sungorus*). J. Reprod. Fertil. **56**, 679–682.

Caine, A., and Lyon, M. F. (1979). Reproductive capacity and dominant lethal mutations in female guinea pigs and Djungarian hamsters following X-rays or chemical mutagens. Mutat. Res. **59**, 231–244.

Carter, D. S., and Goldman, B. D. (1981). Antigonadal and progonadal effects of programmed melatonin infusion into juvenile Djungarian hamsters (*Phodopus sungorus*). Biol. Reprod. **24**, 23A (abstr.).

Carter, D. S., and Goldman, B. D. (1983a). Antigonadal effects of timed melatonin infusion in pinealectomized male Djungarian hamsters (*Phodopus sungorus sungorus*): Duration is the critical parameter. Endocrinology (Baltimore) **113**, 1261–1267.

Carter, D. S., and Goldman, B. D. (1983b). Progonadal role of the pineal in the Djungarian hamster (*Phodopus sungorus sungorus*): Mediation by melatonin. Endocrinology (Baltimore) **113**, 1268–1273.

Coe, J. E. (1981). Comparative immunology of Old World hamsters–Cricetinae. Adv. Exp. Med. Biol. **134**, 95–102.

Crawley, J. N. (1984a). Evaluation of a proposed hamster separation model of depression. Psychiatry Res. **11**, 35–47.

Crawley, J. N. (1984b). Preliminary report of a new rodent separation model of depression. Prog. Neuro-Psychopharmacol. Biol. Psychiatry **8**, 447–457.

Daly, M. (1976). Behavioral development in three hamster species. Dev. Psychobiol. **9**, 315–323.

Das, R. K., and Savage, J. R. K. (1978). Chromosome replication patterns in the Djungarian hamster (*Phodopus sungorus*). Chromosoma **67**, 165–176.

Duncan, M. J., and Goldman, B. D. (1984a). Hormonal regulation of the annual pelage color cycle in the Djungarian hamster, *Phodopus sungorus*, I. Role of the gonads and the pituitary. J. Exp. Zool. **230**, 89–95.

Duncan, M. J., and Goldman, B. D. (1984b). Hormonal regulation of the annual pelage color cycle in the Djungarian hamster, *Phodopus sungorus*. II. Role of prolactin. J. Exp. Zool. **230**, 97–103.

Duncan, M. J., Goldman, B. D., DiPinto, M. N., and Stetson, M. H. (1985). Testicular function and pelage color have different critical daylengths in the Djungarian hamster, *Phodopus sungorus sungorus*. Endocrinology (Baltimore) **116**, 424–430.

Figala, J. (1972). Biorytmiska fenomen hos den sibiriska hamstern (*Phodopus sungorus* Pallas) vid polcirkeln. Fauna Flora **67**, 207–210.

Figala, J., Hoffmann, K., and Goldau, G. (1973). Zur Jahresperiodik beim dsungarischen Zwerghamster *Phodopus sungorus* Pallas. Oecologia **12**, 89–118.

Gamperl, R., Vistorin, G., and Rosenkranz, W. (1977). New observations on the karyotype of the Djungarian hamster, *Phodopus sungorus*. Experientia **33**, 1020–1021.

Gamperl, R., Vistorin, G., and Rosenkranz, W. (1978). Comparison of chromosome banding patterns in five members of Cricetinae with comments on possible relationships. Caryologia **31**, 343–353.

Gibber, J. R., Piontkewitz, Y., and Terkel, J. (1984). Response of male and female Siberian hamsters towards pups. Behav. Neural. Biol. **42**, 177–182.

Goldman, B. D., Hall, V., Hollister, C., Reppert, S., Roychoudhury, P., Yellon, S., and Tamarkin, L. (1981). Diurnal changes in pineal melatonin content in four rodent species: Relationship to photoperiodism. Biol. Reprod. **24**, 778–783.

Goldman, B. D., Darrow, J. M., and Yogev, L. (1984). Effects of timed melatonin infusions on reproductive development in the Djungarian hamster (*Phodopus sungorus*). Endocrinology (Baltimore) **114**, 2074–2083.

Heldmaier, G. (1975). Metabolic and thermoregulatory responses to heat and cold in the Djungarian hamster, *Phodopus sungorus*. J. Comp. Physiol. **102**, 115–122.

Heldmaier, G., and Hoffmann, K. (1974). Melatonin stimulates growth of brown adipose tissue. Nature (London) **247**, 224–225.

Heldmaier, G., and Steinlechner, S. (1981a). Seasonal control of energy requirements for thermoregulation in the Djungarian hamster (*Phodopus sungorus*), living in natural photoperiod. J. Comp. Physiol. **142**, 429–437.

Heldmaier, G., and Steinlechner, S. (1981b). Seasonal pattern and energetics of short daily torpor in the Djungarian hamster, *Phodopus sungorus*. Oecologia **48**, 265–270.

Heldmaier, G., and Steinlechner, S. (1981c). Seasonal control of thermogenesis by photoperiod and ambient temperature in the Djungarian hamster. Cryobiology **18**, 96–97 (abstr.).

Heldmaier, G., Steinlechner, S., Rafael, J., and Vsiansky, P. (1981). Photoperiodic control and effects of melatonin on nonshivering thermogenesis and brown adipose tissue. Science **212**, 917–919.

Heldmaier, G., Steinlechner, S., and Rafael, J. (1982a). Nonshivering thermogenesis and cold resistance during seasonal acclimatization in the Djungarian hamster. J. Comp. Physiol. **149**, 1–9.

Heldmaier, G., Steinlechner, S., Rafael, J., and Latteier, B. (1982b). Photoperiod and ambient temperature as environmental cues for seasonal thermogenic adaptation in the Djungarian hamster, *Phodopus sungorus*. Int. J. Biometeorol. **26**, 339–345.

Herberg, L. (1979). Spontaneously hyperglycemic laboratory animals—models of human diabetes syndrome? Horm. Metab. Res. **11**, 323–331.

Herberg, L. (1981). Decreased tissue guanylate-cyclase activity in glycosuric Djungarian hamsters (*Phodopus sungorus*) that is correctible with insulin. Horm. Metab. Res. **13**, 422–426.

Herberg, L. (1982). Spontaneously hyperglycemic animals—models of human diabetes? Z. Versuchstierkd. **24**, 3–15.

Herberg, L., and Coleman, D. L. (1977). Laboratory animals exhibiting obesity and diabetes syndromes. Metab. Clin. Exp. **26**, 59–99.

Herberg, L., Buchanan, K. D., Herbertz, L. M., Kern, H. F., and Kley, H.

K. (1980). The Djungarian hamster, a laboratory animal with inappropriate hyperglycemia. *Comp. Biochem. Physiol. A* **65A**, 35–60.

Hoffmann, K. (1972). Melatonin inhibits photoperiodically induced testes development in a dwarf hamster. *Naturwissenschaften* **59**, 218–219.

Hoffmann, K. (1973). The influence of photoperiod and melatonin on testis size, body weight, and pelage colour in the Djungarian hamster (*Phodopus sungorus*). *J. Comp. Physiol.* **85**, 267–282.

Hoffmann, K. (1978a). Effects of short photoperiods on puberty, growth and moult in the Djungarian hamster (*Phodopus sungorus*). *J. Reprod. Fertil.* **54**, 29–35.

Hoffmann, K. (1978b). Effect of castration on photoperiodically induced weight gain in the Djungarian hamster. *Naturwissenschaften* **65**, 494.

Hoffmann, K. (1979a). Photoperiodic effects in the Djungarian hamster: One minute of light during darktime mimics influence of long photoperiods on testicular recrudescence, body weight and pelage colour. *Experientia* **35**, 1529–1530.

Hoffmann, K. (1979b). Photoperiod, pineal, melatonin and reproduction in hamsters. *Prog. Brain Res.* **52**, 397–415.

Hoffmann, K. (1981). Pineal involvement in the photoperiodic control of reproduction and other functions in the Djungarian hamster *Phodopus sungorus*. *In* "The Pineal Gland" (R. J. Reiter, ed.), Vol. 2, pp. 83–102. CRC Press, Boca Raton, Florida.

Hoffmann, K. (1982). The effect of brief light pulses on the photoperiodic reaction in the Djungarian hamster *Phodopus sungorus*. *J. Comp. Physiol.* **148**, 529–534.

Hoffmann, K., and Kuderling, I. (1975). Pinealectomy inhibits stimulation of testicular development by long photoperiods in a hamster (*Phodopus sungorus*). Experientia **31**, 122–123.

Hoffmann, K., and Kuderling, I. (1977). Antigonadal effects of melatonin in pinealectomized Djungarian hamsters. *Naturwissenschaften* **64**, 339–340.

Hoffmann, K., and Nieschlas, E. (1977). Circadian rhythm of plasma testosterone in the male Djungarian hamster (*Phodopus sungorus*). *Acta Endocrinol. (Copenhagen)* **86**, 193–199.

Hoffmann, K., Illnerova, H., and Vanecek, J. (1980). Pineal N-acetyltransferase activity in the Djungarian hamster: Effect of one minute light at night. *Naturwissenschaften* **67**, 408–409.

Hoffmann, K., Illnerova, H., and Vanecek, J. (1981). Effect of photoperiod and of one minute light at nighttime on the pineal rhythm of N-acetyltransferase activity in the Djungarian hamster, *Phodopus sungorus*. *Biol. Reprod.* **24**, 551–556.

Honacki, J. H., Kinman, K. E., and Koeppl, J. W. (1982). "Mammal Species of the World." Allen Press and Association of Systematics Collections, Lawrence, Kansas.

Horst, H. J. (1979). Photoperiodic control of androgen metabolism and binding in androgen target organs of hamsters (*Phodopus sungorus*). *J. Steroid Biochem.* **11**, 945–950.

Illnerova, H., Vanecek, J., and Hoffmann, K. (1983). Regulation of the pineal melatonin concentration in the rat (*Rattus norvegicus*) and in the Djungarian hamster (*Phodopus sungorus*). *Comp. Biochem. Physiol.* **74**, 155–159.

Illnerova, H., Hoffmann, K., and Vanecek, J. (1984). Adjustment of pineal melatonin and N-acetyltransferase rhythms to change from long to short photperiod in the Djungarian hamster *Phodopus sungorus*. Neuroendocrinology **38**, 226–231.

Jordan, J. (1971). The establishment of a colony of Djungarian hamsters (*Phodopus sungorus*) in the United Kingdom. *J. Inst. Anim. Technicians* **22**, 56–61.

Kakpakova, E. S., Malakhova, E. M., Massino, Y. S., and Pogosianz, H. E. (1976). Karyological peculiarities and malignancy of hybrids of normal and tumor tissues of the dwarf hamster. *Dokl. Biol. Sci. (Engl. Transl.)* **230**, 392–394.

Kakpakova, E. S., Malakhova, E. M. Massino, Y. S., and Pogosianz, H. E. (1980). Somatic cell hybrids of the Djungarian hamster malignancy and evolution of the karyotype. *Carcinogenesis (N.Y.)* **1**, 539–546.

Kakpakova, E. S., Massino, Y. S., and Malakhova, E. M. (1981). Dzhungarian hamster cell lines resistance to actinomycin D and 6-mercaptopurine: Karotype, morphology, and malignancy. *Sov. Genet. (Engl. Transl.)* **17**, 322–328.

Karasek, M., King, T. S., Hansen, J. T., and Reiter, R. J. (1982). Quantitative changes in the numbers of dense-core vesicles and "synaptic" ribbons in pinealocytes of the Djungarian hamster (*Phodopus sungorus*) following sympathectomy. *Cytobios* **35**, 157–162.

Karasek, M., King, T. S., Brokaw, J., Hansen, J. T., Petterborg, L. J., and Reiter, R. J. (1983). Inverse correlation between "synaptic" ribbon number and density of adrenergic nerve endings in the pineal gland of various mammals. *Anat. Rec.* **205**, 93–99.

Kiseleva, N. S. (1972). Metastasization of transplantable mammary gland carcinoma of *Phodopus sungorus* Pall. on intramuscular and intraperitoneal inoculation. *Bull. Exp. Biol. Med. (Engl. Transl.)* **72**, 1187–1190.

Kopnin, B. P. (1982). Amplification of portions of the genome in mammalian somatic cells resistant to colchicine. I. Trisomy of chromosome 4 in the presence of gene amplification and colchicine resistance in dwarf hamster cells. *Sov. Genet. (Engl. Transl.)* **18**, 1118–1126.

Kopnin, B. P., and Gudkov, A. V. (1982). Amplification of portions of the genome in mammalian somatic cells resistant to colchicine. II. Marker chromosomes with long homogeneously staining regions in colchicine-resistant dwarf hamster cells. *Sov. Genet. (Engl. Transl.)* **18**, 1127–1135.

Kopnin, B. P., and Lucas, J. J. (1982). New dwarf hamster cell lines with selective cytoplasmic and nuclear genetic markers. *Sov. Genet. (Engl. Transl.)* **18**, 987–993.

Logan, A., and Weatherhead, B. (1978). Pelage color cycles and hair follicle tyrosinase activity in the Siberian hamster. *J. Invest. Dermatol.* **71**, 295–298.

Logan, A., and Weatherhead, B. (1979). Photoperiodic dependence of seasonal variations in melanocyte-stimulating hormone content of the pituitary gland in the Siberian hamster (*Phodopus sungorus*). *J. Endocrinol.* **83**, 41P. (Abstract)

Logan, A., and Weatherhead, B. (1980a). Photoperiodic dependence of seasonal changes in pituitary content of melanocyte-stimulating hormone. *Neuro-endocrinology* **30**, 309–312.

Logan, A., and Weatherhead, B. (1980b). Post-tyrosinase inhibition of melanogenesis by melatonin in hair follicles *in vitro*. *J. Invest. Dermatol.* **74**, 47–50.

Logan, A., Carter, R. J., Shuster, S., Thody, A. J., and Weatherhead, B. (1981). Melanotrophin-potentiating factor (MPF) potentiates MSH-induced melanogenesis in hair follicle melanocytes. *Peptides (Fayetteville, N.Y.)* **2**, 121–123.

Malakhova, E. M. (1978). Malignancy in hybrid cells of the Djungarian hamster and characteristics of their karyotype in *in vitro* and *in vivo* cultures. *Dokl. Biol. Sci. (Engl. Transl.)* **242**, 438–440.

Markaryan, D. S., and Avdzhian, M. V. (1972). A cytogenetic study of Rous sarcoma virus in Jungarian hamsters and monkeys. *Sov. Genet. (Engl. Transl.)* **8**, 992–995.

Mordes, J. P., and Rossini, A. A. (1981). Animal models of diabetes. *Am. J. Med.* **70**, 353–360.

Nagy, F., Pendergrass, P. B., and Scott, J. N. (1982). Structural features of a specialized region of the epididymis of the Siberian hamster (*Phodopus sungorus*). *J. Submicrosc. Cytol.* **14**, 673–682.

Pilborough, G. S. (1971). An introduction to the Djungarian hamster (*Phodopus sungorus*). *J. Inst. Anim. Technicians* **22**, 50–55.

Pogosianz, H. E. (1970). Meiosis in the Djungarian hamster. I. General pattern of male meiosis. *Chromosoma* **31**, 392–403.

Pogosianz, H. E. (1975). Djungarian hamster—A suitable tool for cancer research and cytogenetic studies. *J. Natl. Cancer Inst. (U.S.)* **54**, 659–664.

Pogosianz, H. E., and Brujako, E. T. (1971). Meiosis in the Djungarian hamster. II. Polyploid spermatocytes. *Cytogenetics* **10**, 70–76.

Pogosianz, H. E., and Sokova, O. I. (1967). Maintaining and breeding of the Djungarian hamster under laboratory conditions. *Z. Versuchstierkd.* **9**, 292–297.

Pogosianz, H. E., and Sokova, O. I. (1981). Pink-eyed dilution mutation in Djungarian hamster. *Z. Versuchstierkd.* **23**, 294–295.

Pogosianz, H. E., and Sokova, O. I. (1982). Tumors of the Djungarian hamster. *IARC Sci. Publ.* **34**, 451–455.

Pogosianz, H. E., Sokova, O. I., and Prigozhina, E. L. (1970). Djungarian hamster: A new animal for experimental-oncological research. *Biol. Abstr.* **51**, 12356 (abstr.).

Puchalski, W., Bockler, H., and Heldmaier, G. (1983). Effect of food deprivation on thermogenic capacity in the Djungarian dwarf hamster *Phodopus sungorus*. *J. Therm. Biol.* **8**, 99–101.

Quimby, F., Nunez, E., Finlay, B., and Lok, B. (1982). Adenocarcinomas of the thyroid parafollicular (C) cells in Djungarian hamsters (*Phodopus sungorus*). *Lab. Anim. Sci.* **32**, 413 (abstr.).

Rafael, J., and Vsiansky, P. (1981). Adaptive changes in brown adipose tissue of Dsungarian hamsters. *Cryobiology* **18**, 97 (abstr.).

Rafael, J., and Vsiansky, P. (1983). Non-shivering thermogenesis in Djungarian hamsters (*Phodopus sungorus*): Studies on the role of brown adipose tissue in relation to other calorigenic sites. *J. Therm. Biol.* **8**, 103–105.

Simpson, S. M., Follett, B. K., and Ellis, D. H. (1982). Modulation by photoperiod of gonadotrophin secretion of intact and castrated Djungarian hamsters. *J. Reprod. Fertil.* **66**, 243–250.

Snipes, R. L. (1979). Anatomy of the cecum of the dwarf hamster (*Phodopus sungorus*). *Anat. Embryol.* **157**, 329–346.

Sokova, O. I. (1971). Transplantable mammary tumors of the Djungarian hamster. *Biol. Abstr.* **52**, 9105 (abstr.).

Sonnenschein, C., Roberts, P. D., and Yerganian, G. (1969). Karyotypic and enzymatic characteristics of a somatic hybrid cell line originating from dwarf hamsters. *Genetics* **62**, 379–392.

Spyropoulos, B., Ross, P. D., Moens, P. B., and Cameron, D. M. (1982). The synaptonemal complex karyotypes of palearctic hamsters *Phodopus roborovskii* Satunin and *Phodopus sungorus* Pallas. *Chromosoma* **86**, 397–408.

Steinlechner, S., and Heldmaier, G. (1981). Photoperiodic and melatonin-induced changes in cold resistance and NST in the Djungarian hamster at different times of the year. *Cryobiology* **18**, 97 (abstr.).

Steinlechner, S., and Heldmaier, G. (1982). Role of photoperiod and melatonin in seasonal acclimatization of the Djungarian hamster, *Phodopus sungorus*. *Int. J. Biometeorol.* **26**, 329–337.

Suzuki, S., Ago, A., Mohri, S., Nishinakagawa, H., and Otsuka, J. (1983). Fine structure of the parotid gland of Djungarian hamster (*Phodopus sungorus*). *Exp. Anim.* **32**, 175–184.

Suzuki, S., Ago, A., Mohri, S., Nishinakagawa, H., and Otsuka, J. (1984). Fine structure of the mandibular gland of the Djungarian hamster (*Phodopus sungorus*). *Exp. Anim.* **33**, 487–496.

Tamarkin, L. Reppert, S. M., Orloff, D. J., Klein, D. C., Yellon, S. M., and Goldman, B. D. (1980). Ontogeny of the pineal melatonin rhythm in the Syrian (*Mesocricetus auratus*) and Siberian (*Phodopus sungorus*) hamsters and the rat. *Endocrinology (Baltimore)* **107**, 1061–1064.

Thust, R. (1974). G-banding and late replication of the Djungarian dwarf hamster chromosomes. *Exp. Pathol.* **9**, 153–156.

Trojan, M. (1977). Water balance and renal adaptations in four Palearctic hamsters. *Naturwissenschaften* **64**, 591–592.

van Veen, T., Brackmann, M., and Moghimzadeh, E. (1978). Post-natal development of the pineal organ in the hamsters *Phodopus sungorus* and *Mesocricetus auratus*. A fluorescence microscopic and microspectrofluorometric investigation. *Cell Tissue Res.* **189**, 241–250.

Vesely, D. L., and Herberg, L. (1981). Decreased tissue guanylate cyclase activity in glycosuric Djungarian hamsters (*Phodopus sungorus*) that is correctable with insulin. *Horm. Metab. Res.* **13**, 422–426.

Vistorin, G., Gamperl, R., and Rosenkranz, W. (1977). Studies on sex chromosomes of four hamster species: *Cricetus cricetus, Cricetulus griseus, Mesocricetus auratus,* and *Phodopus sungorus*. *Cytogenet. Cell Genet.* **18**, 24–32.

Voss, K. M., Herberg, L., and Kern, H. F. (1978). Fine structural studies of the islets of Langerhans in the Djungarian hamster (*Phodopus sungorus*). *Cell Tissue Res.* **191**, 333–342.

Wade, G. N., and Bartness, T. J. (1984). Effects of photoperiod and gonadectomy on food intake, body weight, and body composition in Siberian hamsters. *Am. J. Physiol.* **246**, R26–R30.

Warzok, R., and Thust, R. (1977). Morphology of diethylnitrosamine-induced lung tumours in Dzungarian dwarf hamsters. *Exp. Pathol.* **13**, 44–51.

Weiner, J., and Gorecki, A. (1981). Standard metabolic rate and thermoregulation in five species of Mongolian small mammals. *J. Comp. Physiol. B* **145B**, 127–132.

Wittkowski, W. Hewing, M., Hoffmann, K., Bergmann, M., and Fechner, J. (1984). Influence of photoperiod on the ultrastructure of the hypophysial pars tuberalis of the Djungarian hamster, *Phodopus sungorus*. *Cell Tissue Res.* **238**, 213–216.

Yellon, S. M., and Goldman, B. D. (1980), Short days inhibit the developmental pattern of FSH and prolactin in the male Djungarian hamster, *Phodopus sungorus*. *Biol. Reprod.* **22**, 88A (abstr.).

Yellon, S. M., and Goldman, B. D. (1984). Photoperiod control of reproductive development in the male Djungarian hamster (*Phodopus sungorus*). *Endocrinology (Baltimore)* **114**, 664–670.

Yellon, S. M., Tamarkin, L., Pratt, B. L., and Goldman, B. D. (1982). Pineal melatonin in the Djungarian hamster: Photoperiodic regulation of a circadian rhythm. *Endocrinology (Baltimore)* **111**, 488–492.

Yerganian, G. (1972). History and cytogenetics in hamsters. *Prog. Exp. Tumor Res.* **16**, 2–41.

Yosida, T. H. (1976). Spontaneous diabetes in Djungarian hamster, *Phodopus sungorus*. *Annu. Rep. Nat. Inst. Genet. (Jpn.)* **26**, 46.

Young, C. M. (1973a). *Trichophyton mentagrophytes* in the Djungarian hamster (*Phodopus sungorus*). *Z. Versuchstierkd.* **15**, 378 (abstr.).

Young, C. M. (1973b). *Trichophyton mentagrophytes* in the Djungarian hamster (*Phodopus sungorus*). *Vet. Rec.* **94**, 287–289.

Mystromys albicaudatus (South African Hamster)

I. TAXONOMY, HISTORY, AND GENETICS

Mystromys albicaudatus is the sole member of its genus and the only hamster native to Africa, where it occurs in the dry grasslands of South Africa. Wild populations of *Mystromys* have not been well studied, but some mammalogists believe they are disappearing from South Africa (Dean, 1978). Chromosome banding patterns do not indicate a close relationship to *Cricetus, Cricetulus, Phodopus,* or *Mesocricetus* (Gamperl *et al.*, 1978).

The first laboratory colony of *Mystromys* was established at the South African Institute of Medical Research in Johannesburg, by D. H. S. Davis in 1941 (Hall *et al.*, 1967; Hallett and Meester, 1971). Twenty-four animals from this colony were sent to Chapman H. Binford at the Armed Forces Institute of Pathology (AFIP) in Washington, D.C., in 1962. The

21. BIOLOGY, CARE, AND USE IN RESEARCH

AFIP breeding colony was started with 6 breeding pairs from this group, and 10 more breeding pairs were imported from South Africa and added to the colony in 1969 (Hall *et al.*, 1967; Hallett and Politzer, 1972; Howell and Hall, 1977). Animals from the AFIP colony have been distributed to several other laboratories in the United States.

Mystromys have 32 chromosomes (Taitz, 1954), including 7 metacentric pairs, 3 submetacentric pairs, and 5 acrocentric pairs of autosomes. The X chromosome is the longest submetacentric chromosome. The Y chromosome is acrocentric. There are G-banding similarities of *Mystromys* chromosome number 6 with *Cricetus cricetus* chromosome number 4, *Cricetulus griseus* chromosome number 4, part of *Phodopus sungorus* chromosome number 1, and part of *Mesocricetus auratus* chromosomes number 2 and 3 (Gamperl *et al.*, 1978).

II. CARE AND MANAGEMENT

Mystromys are nocturnal, live in burrows, and are reported to estivate during dry periods in the wild (Hall *et al.*, 1967). They are continuously active during the night with a slight tendency for a lull around midnight, but there is no evidence of a crepuscular activity pattern (Perrin, 1981). Walker (1975) reported the belief that they may secrete some protective substance making them distasteful to predators.

Mystromys are normally docile but may fight if crowded or when unfamiliar adults are mixed. Most breeding colonies utilize lifelong monogamous pairings. When accustomed to handling, they do not usually attempt to bite, although certain individuals and mothers with infants are more prone to biting, and protective gloves are advisable. *Mystromys* should be grasped around the body, although smaller individuals can be lifted by the base of the tail. They should not be lifted by the tip of the tail, since the skin may tear off.

Mystromys can be maintained in standard rodent cages. Solid-floored cages are required for mothers with infants. They do well on standard commercial rodent diet without supplementation, water *ad libitum,* a 12-hr light cycle, and the usual temperature and relative humidity recommended for rodent colonies.

III. ANATOMY, PHYSIOLOGY, AND REPRODUCTION

Mystromys (Fig. 2) have a head and body length of 136–185 mm with a 50- to 82-mm tail. The fur is gray to brown on the dorsal surfaces and white on the ventral surfaces, feet, and tail. The gray dorsal pelage of the younger animals becomes pro-

Fig. 2. Mystromys albicaudatus, adult male.

gressively more brown with age. Unlike other hamsters, *Mystromys* do not have cheek pouches. The liver has five lobes and lacks a gallbladder (Perrin and Curtis, 1980). Adult males average 145 gm (range 105–185 gm), and females average 95 gm (range 75–125 gm) (Hall *et al.*, 1967). Organ weights and organ/body weight ratios have been published by Becker and Middleton (1979). Females have two pair of inguinal mammary glands and a rudimentary prostate gland (Hall *et al.*, 1967). The eyes are similar to other rodents, and the retina lacks cones (Rodrigues *et al.*, 1971a, 1972).

According to Perrin and Curtis (1980), the lengths of the small and large intestines are typical of an omnivorous diet. The small intestine makes up 53% of the intestinal length, the colon 40%, and the cecum 7%. The stomach (Fig. 3) is divided into glandular and nonglandular parts. The esophagus empties into an enlarged, nonglandular compartment with numerous mucosal papillae supporting a heavy bacterial growth (Hall *et al.*, 1967; Davis, 1963). This region is separated from the glandular compartment by a small, nonglandular, nonpapillated area. The papillae are composed of keratin and lack a connective tissue core. They are absent at birth and begin to form during the third postnatal week, when the infant *My-*

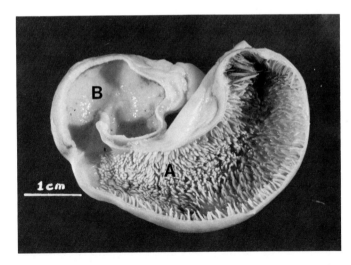

Fig. 3. Stomach of adult *Mystromys albicaudatus*. (A) Nonglandular region with mucosal papillae; (B) glandular region.

stromys begin ingesting solid food. During the second postnatal week the nonglandular mucosa becomes colonized by cocci and coccobacilli. After the papillae form, they become heavily colonized by anaerobic bacilli, and the cocci remain only between the papillae and in the nonpapillated, nonglandular region (Maddock and Perrin, 1981, 1983). In the adult the papillae are about 1.8 mm long and occur at a density of about $550/cm.^2$ The pH in the nonglandular compartment is about 4.6, and the pH in the glandular compartment is about 2.7. The function of the nonglandular compartment and its papillae with their heavy bacterial growth is unknown. It does not appear to be an adaptation for fermentation but may provide a compartment for amylase activity (Perrin and Maddock, 1983).

Normal serum chemistry values for *Mystromys* are listed in Table I. There appears to be little difference between sexes, and the values have been combined. Males had higher serum sodium and lower chloride values than females (Becker *et al.*, 1979), but the differences were slight. Becker *et al.* (1979) found serum glucose levels slightly higher in females, Streett and Highman (1971) found slightly higher values in males, and a larger study by Stuhlman *et al.* (1972a) with 736 determinations found no difference in serum glucose levels related to sex. Serum alkaline phosphatase levels appear to be higher in females (mean of 0.70 Sigma units) than in males (mean of 0.58 Sigma units) (Streett and Highman, 1971). The results of urinalyses from *Mystromys* were unremarkable except for a high rate of glycosuria (>90%) and mild ketonuria (82% in males) in a colony thought not to be diabetic. Urine production over 24 hr was about 0.0456 ml/gm of body weight (Becker *et al.*, 1979).

Neutrophils are the most frequent leukocyte in the peripheral blood and contain eosinophilic, slightly elongated granules up to 1 μm in length and about 0.4 μm in width (Prieur *et al.*, 1979).

Mystromys are born hairless except for a few vibrissae on the snout, have a head and body length of about 54 mm, and weigh about 6.5 gm (range 5.0–7.8 gm) (Meester and Hallett, 1970; Hallett and Meester, 1971). Infant *Mystromys* remain attached to the nipples and are dragged around by the mother until they are 15–20 days old (Fig. 4). A sparse growth of hair erupts at 4–6 days, incisors erupt at 3–5 days, and the eyes open at 16–25 days of age. The external auditory meatus opens at 3 or 4 days, but response to sound is not apparent until 13–15 days. They begin eating solid food at about 21 days of age (Meester and Hallett, 1970; Hallett and Meester, 1971; Maddock and Perrin, 1983). *Mystromys* have an average life span of 889 days (2.4 years) and a maximum reported life span of 2198 days (6 years) (Davis, 1963).

Mystromys breed throughout the year in the laboratory, but there is some indication that they may not breed during the winter in the wild (Meester and Hallett, 1970). Gestation is about 38 days (range 36–39 days) and litter size averages 3 (range 1–6). The average age of the females at the birth of the first litter is 4.8 months. The sex ratio of offspring seems to favor males by about 10%. There is a fertile postpartum estrus, and breeding females are normally continuously pregnant, with births occurring at about 39-day intervals. Natural weaning occurs with the birth of the following litter, although weanlings may compete with newborns for nipples if they are not removed from the cage. Weaning at 25 days or 25 gm of body

Table I

Blood Chemistry Values of *Mystromys albicaudatus*

Serum component	Mean value	Reference
Chloride	105 mEq/liter	Becker *et al.* (1979)
Sodium	146 mEq/liter	Becker *et al.* (1979)
Potassium	5.4 mEq/liter	Becker *et al.* (1979)
Calcium	9.4 mg/dl	Becker *et al.* (1979)
Glucose	95 mg/dl	Becker *et al.* (1979)
	87 mg/dl	Streett and Highman (1971)
	117 mg/dl	Stuhlman *et al.* (1972a)
Urea nitrogen	19 mg/dl	Becker *et al.* (1979)
	25 mg/dl	Streett and Highman (1971)
Total protein	6.9 gm/dl	Becker *et al.* (1979)
Aspartate aminotransferase (glutamic oxaloacetic transaminase, SGOT)	50 Sigma–Frankel units	Streett and Highman (1971)
Alanine aminotransferase (pyruvic transaminase, SGPT)	41 Sigma–Frankel units	Streett and Highman (1971)
Alkaline Phosphatase	0.64 Sigma units	Streett and Highman (1971)
Aldolase	8.4 milliunits/ml	Streett and Highman (1971)

Fig. 4. Female *Mystromys albicaudatus* with 12-day-old young attached to nipples.

weight (whichever comes later) has been recommended (Hall et al., 1967; Meester and Hallett, 1970). Since *Mystromys* only have four nipples, births in excess of four may not survive unless there is more than one lactating female in the cage and fostering occurs. With harem mating, litters are often mixed and readily accepted by foster mothers. Infants which become dislodged from the nipples are usually able to reattach. The maximum age reported at the birth of the last litter was 1642 days (4.5 years) (Davis, 1963).

IV. DISEASES

Few spontaneous diseases have been reported in *Mystromys*. Hall et al. (1967) reported a few cases of pneumonitis, enteritis, and septicemia, but the etiologies were not determined. They also reported a peculiar loss of the upper incisors, but the cause was unknown and it has not been reported by others. A report of spontaneous tumors by Rantanen and Highman (1970) included a squamous-cell carcinoma, a skin adnexal tumor, an osteosarcoma, a uterine leiomyosarcoma, two hypophyseal adenomas and an hepatic adenocarcinoma. All the affected animals were over 3 years of age except an 11-month-old female with an adenoma of the hypophysis. An outbreak of ringtail was reported among suckling *Mystromys* in one colony when a mechanical malfunction resulted in an excessively low relative humidity in the facility (Stuhlman and Wagner, 1971). Severe cases developed annular constrictions of the tail with necrosis and autoamputation. A fatal antibiotic toxicity was reported from the same colony after animals ingested topical antibiotic ointment applied to surgical sites following muscle biopsies (LaRegina et al., 1978). Affected animals died with a hemorrhagic cecitis and colitis.

Mystromys are susceptible to Tyzzer's disease, although no naturally occurring cases have been reported. Oral inoculation with *Bacillus piliformis* resulted in necrotic lesions in the liver, myocardium, cerebellum, brain stem, and tunica muscularis of the intestines (Waggie et al., 1985). Clusters of *B. piliformis* were present in cells bordering necrotic foci. Brain lesions have not been reported in other species with Tyzzer's disease.

In a survey of a laboratory colony of *Mystromys* for murine viral antibodies, titers were found to pneumonia virus of mice and Sendai virus. No antibodies were detected to minute virus of mice, K virus, reovirus type 3, mouse encephalomyelitis virus (GD VII), mouse adenovirus, mouse hepatitis virus, or lymphocytic choriomeningitis virus.

A hereditary partial albinism of the eyes and head has been described in one colony. The eyes appear red and the skin and hair of the head, especially of the ear pinnae, are paler than normal. There was also retinal degeneration which seemed to be independent of the degree of hypopigmentation. Because of the gross similarity of the hypopigmentation to Chediak–Higashi syndrome, this condition has been investigated as a possible animal model but was found to be unrelated (Rodrigues et al., 1971b, 1972; Prieur et al., 1979).

V. RESEARCH USES

Mystromys have been experimentally inoculated with several different viruses, bacteria, and parasites. Following experimental inoculation with arboviruses, no viremia was detected with Wesselsbron, Spondweni, Pongola, or West Nile viruses. Viremia lasted 5 days or less with Bunyamera, Chikungunya, Rift Valley fever, Middleburg and Simbu viruses, and was not accompanied by clinical illness (McIntosh, 1961). *Mystromys* was also used in an early, partially successful attempt to produce an attenuated poliomyelitis virus (Gear, 1952) and were reported to be susceptible to infection with Coxsackie virus (Davis, 1963). They have been shown to be resistant to infection with *Mycobacterium leprae* and *M. microti* (Grasset et al., 1946). They are susceptible to plague (Davis, 1963) and were shown to be susceptible to oral infection, as are rats and mice, resulting from cannibalism of cagemates which died from *Yersinia pestis* infection (Rust et al., 1972). *Mystromys* have been found to provide a good model for American cutaneous leishmaniasis caused by *Leishmania braziliensis* (Beacham et al., 1982; McKinney and Hendricks, 1980; Sayles et al., 1981) and to maintain heavy chronic infections with *L. donovani* for at least a year (Mikhail and Mansour, 1973). Golden hamsters (*Mesocricetus auratus*) develop a more rapidly progressing disease and die about 40 days after infection with *L. donovani*. *Mystromys* have also been used as experimental hosts for *Schistosoma mansoni* (Pitchford and Visser, 1960) and *S. haematobium* (Pitchford and Visser, 1962).

There has been a limited amount of research reported on periodontal disease and dental caries in *Mystromys*. Initial reports by Ockerse (1953, 1956) indicated that dental caries were only produced when a high-sugar diet was fed to the mothers throughout gestation and lactation and continuously to the offspring until they were 8–11 months old. Adult *Mystromys* did not develop caries when fed the same diet for 20 months. Larson and Fitzgerald (1968) found a high rate of caries in *Mystromys* fed a high-sucrose diet combined with oral inoculation of a caries-associated strain of *Streptococcus* of human origin. They speculated that the chronic feeding of high-sugar diets required to produce caries in *Mystromys* by Ockerse may have selected for a caries-active oral flora which was transferred from the mother to the young.

Since *Mystromys* appear to have a low rate of spontaneous tumors, they have been considered as possible subjects for carcinogen testing but were found to be more resistant to the known carcinogenic effects of diethylnitrosamine (Yamamoto *et al.*, 1972) and azaserine (Roebuck and Longnecker, 1979) than other laboratory rodents.

The observation of excessively wet cages resulting from polydipsia and polyuria among some of the *Mystromys* in the colony at the University of Missouri Medical Center in Columbia in 1969 led to the recognition of diabetes mellitus in these animals. About 22% of the animals were found to be hyperglycemic. This colony originated from six adults and six preweanlings obtained from the Armed Forces Institute of Pathology the previous year (Stuhlman *et al.*, 1974). Hyperglycemia has not been reported in the AFIP colony and a survey of the original South African colony at the South African Institute of Medical Research in Johannesburg revealed less than 3% incidence of hyperglycemia (Hallett and Politzer, 1972). Researchers at the University of Missouri developed a technique for repeated blood sampling for glucose determinations (Stuhlman *et al.*, 1972b) and have published several reports describing diabetes mellitus in *Mystromys*. Glycosylated hemoglobin levels may be used to recognize diabetic *Mystromys* (Little *et al.*, 1982). The condition appears to be inherited as a non-sex-linked, polygenic trait. It is characterized by hyperglycemia, glycosuria, and ketonuria. Severe cases may show polydipsia, polyuria, and polyphagia with severe weight loss. Ketoacidosis may lead to coma and death. Obesity is not a feature of the disease, and the age of onset, degree of severity, and rate of progression are variable and signs may be intermittent (Packer *et al.*, 1970; Stuhlman *et al.*, 1972a, 1974). Pancreatic lesions consisting of β-cell vacuolization, degranulation, glycogen accumulation, and degeneration are associated with hyperglycemia (Stuhlman *et al.*, 1975; Goeken *et al.*, 1972). There is thickening of capillary basement membranes and mesangial proliferation in the renal glomeruli (Riley *et al.*, 1975; Schmidt *et al.*, 1980). Capillary basement membrane thickening is also present in skeletal muscle (Yesus *et al.*, 1976), and 10–15% of diabetic *Mystromys* develop cataracts (Stuhlman, 1979). Structural and functional alterations in isolated hepatic mitochrondria from diabetic *Mystromys* have also been described (Schmidt *et al.*, 1974).

A study of lymphocyte mitogen response in *Mystromys* showed a better response to T-lymphocyte mitogens (phytohemagglutinin-P, concanavalin A, and pokeweed mitogen) at a lower cell concentration than required for inbred mice. The larger body size and lower cell concentration required could provide enough lymphocytes from a single *Mystromys* to test several variables without the necessity of pooling cells from several inbred mice. No differences in maximal mitogen response due to sex or age were observed in *Mystromys* between 2 and 24 months of age. The maximal response in inbred mice is between 4 and 26 weeks of age. *Mystromys* exhibited a poor response to B-lymphocyte mitogens (*Escherichia coli* lipopolysaccharide and PPD tuberculin) (Howell and Hall, 1977).

Mystromys were also used in a comparative study of body tin depositions (Furchner and Drake, 1976), which were not different from the other species examined (mouse, rat, rhesus monkey, and dog).

REFERENCES

Beacham, B. E., Romito, R., and Kay, H. D. (1982). Vaccination of the African white-tailed rat, *Mystromys albicaudatus*, with sonicated *Leishmania braziliensis panamensis* promastigotes. *Am. J. Trop. Med. Hyg.* **31**, 252–258.

Becker, S. V., and Middleton, C. C. (1979). Organ weights and organ:body weight ratios of the African white-tailed rat (*Mystromys albicaudatus*). *Lab. Anim. Sci.* **29**, 44–47.

Becker, S. V., Schmidt, D. A., and Middleton, C. C. (1979). Selected biological values of the African white-tailed sand rat (*Mystromys albicaudatus*). *Lab Anim. Sci.* **29**, 479–481.

Davis, D. H. S. (1963). Wild rodents as laboratory animals and their contribution to medical research in South Africa. *S. Afr. J. Med. Sci.* **28**, 53–69.

Dean, W. R. J. (1978). Conservation of the white-tailed rat in South Africa. *Biol. Conserv.* **13**, 133–140.

Furchner, J. E., and Drake, G. A. (1976). Comparative metabolism of radionuclides in mammals. XI. Retention of ^{113}Sn in the mouse, rat, monkey and dog. *Health Phys.* **31**, 219–224.

Gamperl, R., Vistorin, G., and Rosenkranz, W. (1978). Comparison of chromosome banding patterns in five members of Cricetinae with comments on possible relationships. *Caryologia* **31**, 343–353.

Gear, J. (1952). Immunity to poliomyelitis. *Ann. Intern. Med.* **37**, 1–22.

Goeken, J. A., Packer, J. T., and Rose, S. D. (1970). Ultrastructure of pancreatic islets in diabetic and nondiabetic *Mystromys albicaudatus*. *Fed. Proc., Fed. Am. Soc. Exp. Biol.,* **29**, 357 (abstr.).

Goeken, J. A., Packer, J. T., Rose, S. D., and Stuhlman, R. A. (1972). Structure of the islet of Langerhans: Pathological studies in normal and diabetic *Mystromys albicaudatus*. *Arch. Pathol.* **93**, 123–129.

Grasset, E., Murray, J. F., and Davis, D. H. S. (1946). Vole bacillus: Susceptibility of South African wild rodents to the vole strain of acid-fast bacillus and to other acid-fast bacilli; preliminary report. *Am. Rev. Tuberc.* **53**, 427–439.

Hall, A., Persing, R. L., White, D. C., and Ricketts, R. T. (1967). *Mystromys albicaudatus* (the African white-tailed rat) as a laboratory species. *Lab. Anim. Care* **17**, 180–188.

Hallett, A. F., and Meester, J. (1971). Early postnatal development of the South African hamster *Mystromys albicaudatus*. *Zool. Afr.* **6**, 221–228.

Hallett, A. F., and Politzer, W. M. (1972). Comment on Packer *et al. Arch. Pathol.* **93**, 178 (lett.).

Herberg, L. (1979). Spontaneously hyperglycemic laboratory animals—models of human diabetes-syndrome? *Horm. Metab. Res.* **11**, 323–331.

Herberg, L. (1982). Spontaneously hyperglycemic animals—models of human diabetes? *Z. Versuchstierkd.* **24**, 3–15.

Honacki, J. H., Kinman, K. E., and Koeppl, J. W. (1982). "Mammal Species of the World." Allen Press and Association of Systematics Collections, Lawrence, Kansas.

Howell, H. M., and Hall, J. L. (1977). African hamster lymphocytes as a model for immunologic studies. *Res. Vet. Sci.* **23**, 293–297.

Hunt, C. E., Lindsey, J. R., and Walkley, S. U. (1976). Animal models of diabetes and obesity, including the PBB/Ld mouse. *Fed. Proc., Fed. Am. Soc. Exp. Biol.* **35**, 1206–1217.

Keogh, H. J., and Issacson, M. (1978). Wild rodents as laboratory models and their part in the study of diseases. *J. S. Afr. Vet. Assoc.* **49**, 229–231.

LaRegina, M., Kier, A. B., and Wagner, J. E. (1978). A fatal enteric syndrome in *Mystromys albicaudatus* (white-tailed rat) following topical antibiotic treatment. *Lab. Anim. Sci.* **28**, 587–590.

Larson, R. H., and Fitzgerald, R. J. (1968). Caries development in the African white-tailed rat (*Mystromys albicaudatus*) infected with a streptococcus of human origin. *J. Dent. Res.* **47**, 746–749.

Little, R. R., Parker, K. M., England, J. D., and Goldstein, D. E. (1982). Glycosylated hemoglobin in *Mystromys albicaudatus*: A diabetic animal model. *Lab. Anim. Sci.* **32**, 44–47.

McIntosh, B. M. (1961). Susceptibility of some African wild rodents to infection with various arthropod-borne viruses. *Trans. R. Soc. Trop. Med. Hyg.* **55**, 63–68.

McKinney, L., and Hendricks, L. D. (1980). Experimental infection of *Mystromys albicaudatus* with *Leishmania brasiliensis*: Pathology. *Am. J. Trop. Med. Hyg.* **29**, 753–760.

Maddock, A. H., and Perrin, M. R. (1981). A microscopical examination of the gastric morphology of the white-tailed rat *Mystromys albicaudatus* (Smith, 1834). *S. Afr. J. Zool.* **16**, 237–247.

Maddock, A. H., and Perrin, M. R. (1983). Development of the gastric morphology and fornical bacterial/epithelial association in the white-tailed rat *Mystromys albicaudatus* (Smith, 1834). *S. Afr. J. Zool.* **18**, 115–127.

Meester, J., and Hallett, A. F. (1970). Notes on early postnatal development in certain southern African Muridae and Cricetidae. *J. Mammal.* **51**, 703–711.

Mikhail, J. W., and Mansour, N. S. (1973). *Mystromys albicaudatus*, the African white-tailed rat, as an experimental host for *Leishmania donovani*. *J. Parasitol.* **59**, 1085–1087.

Mordes, J. P., and Rossini, A. A. (1981). Animal models of diabetes. *Am. J. Med.* **70**, 353–360.

Ockerse, T. (1953). Experimental dental caries in the white-tailed rat in South Africa. *J. Dent. Res.* **32**, 74–77.

Ockerse, T. (1956). Experimental periodontal lesions in the white-tailed rat in South Africa. *J. Dent. Res.* **35**, 9–15.

Packer, J. T., Kraner, K. L., Rose, S. D., Stuhlman, R. A., and Nelson, L. R. (1970). Diabetes mellitus in *Mystromys albicaudatus*. *Arch. Pathol.* **89**, 410–415.

Perrin, M. R. (1981). Notes on the activity patterns of 12 species of southern African rodents and a new design of activity monitor. *S. Afr. J. Zool.* **16**, 248–258.

Perrin, M. R., and Curtis, B. A. (1980). Comparative morphology of the digestive system of 19 species of southern African myomorph rodents in relation to diet and evolution. *S. Afr. J. Zool.* **15**, 22–33.

Perrin, M. R., and Maddock, A. H. (1983). Preliminary investigations of digestive processes of the white-tailed rat *Mystromys albicaudatus* (Smith, 1834). *S. Afr. J. Zool.* **18**, 128–133.

Pitchford, R. J., and Visser, P. S. (1960). Some observations on *Schistosoma mansoni* in rodents in the Transvaal. *Ann. Trop. Med. Parasitol.* **54**, 247–249.

Pitchford, R. J., and Visser, P. S. (1962). Maintenance of *Schistosoma haematobium* in the laboratory. *Trans. R. Soc. Trop. Med. Hyg.* **56**, 173 (lett.).

Prieur, D. J., Olson, H. M., and Young, D. M. (1979). Partial oculocutaneous albinism in *Mystromys albicaudatus*: Nonhomology with the Chediak-Higashi syndrome. *Lab. Anim. Sci.* **29**, 40–43.

Rantanen, N. W., and Highman, B. (1970). Spontaneous tumors in a colony of *Mystromys albicaudatus* (African white-tailed rat). *Lab. Anim. Care* **20**, 114–119.

Riley, T., Stuhlman, R. A., Van Peenen, H. J., Esterly, J. A., and Townsend, J. F. (1975). Glomerular lesions of diabetes mellitus in *Mystromys albicaudatus*. *Arch Pathol.* **99**, 167–169.

Rodrigues, M., Streett, R. P., Jr., and Highman, B. (1971a). Partial ocular albinism in *Mystromys albicaudatus* (African white-tailed rat). *Arch. Pathol.* **92**, 212–218.

Rodrigues, M., Streett, R. P., Jr., Highman, B., and Fine, B. S. (1971b). An entity simulating the Chediak-Higashi syndrome in *Mystromys albicaudatus* (African white-tailed rat). *Lab. Invest.* **24**, 444–445 (abstr.).

Rodrigues, M., Fine, B. S., Highman, B., Streett, R. P., Jr. (1972). Partial ocular albinism in *Mystromys albicaudatus* (the African white-tailed rat). *Arch. Ophthalmol.* **87**, 337–346.

Roebuck, B. D., and Longnecker, D. S. (1979). Response of two rodents, *Mastomys natalensis* and *Mystromys albicaudatus*, to the pancreatic carcinogen aza serine. *J. Natl. Cancer Inst.* **62**, 1269–1271.

Rust, J. H., Harrison, D. N., and Marshall, J. D. (1972). Susceptibility of rodents to oral plague infection: A mechanism for the persistence of plague in inter-epidemic periods. *J. Wildl. Dis.* **8**, 127–133.

Sayles, P. C., Hunter, K. W., Stafford, E. E., and Hendricks, L. D. (1981). Antibody response to *Leishmania mexicana* in African white-tailed rats (*Mystromys albicaudatus*). *J. Parasitol.* **67**, 585–586.

Schmidt, G., Townsend, J. T., and Vorbeck, M. L. (1973). Evaluation of hepatic mitochondrial function in the spontaneously diabetic *Mystromys albicaudatus*. *Tex. Rep. Biol. Med.* **31**, 595–596 (abstr.).

Schmidt, G., Martin, A. P., Stuhlman, R. A., Townsend, J. F., Lucas, F. V., and Vorbeck M. L. (1974). Evaluation of hepatic mitochondrial function in the spontaneously diabetic *Mystromys albicaudatus*. *Lab. Invest.* **30**, 451–457.

Schmidt, G. E. Martin, A. P., Townsend, J. F., and Vorbeck, M. L. (1980). Basement membrane synthesis in spontaneously diabetic *Mystromys albicaudatus*. *Lab. Invest.* **43**, 217–224.

Srivastava, P. K., Townsend, J. F., Stuhlman, R. A., and Lucas, F. V. (1974). Somatic chromosomes of the African white-tailed rat. *J. Hered.* **65**, 223–226.

Streett, R. P., Jr., and Highman, B. (1971). Blood chemistry values in normal *Mystromys albicaudatus* and Osborne-Mendel rats, *Lab. Anim. Sci.* **21**, 394–398.

Stuhlman, R. A. (1971). The genetic mode of transmission of spontaneous diabetes mellitus in *Mystromys albicaudatus*. M. S. Thesis, University of Missouri, Columbia.

Stuhlman, R. A. (1979). Animal model: Spontaneous diabetes mellitus in *Mystromys albicaudatus*. *Am. J. Pathol.* **94**, 685–688.

Stuhlman, R. A., and Wagner, J. E. (1971). Ringtail in *Mystromys albecaudatus*: A case report. *Lab. Anim. Sci.* **21**, 585–587.

Stuhlman, R. A., Packer, J. T., and Doyle, R. E. (1972a). Spontaneous diabetes mellitus in *Mystromys albicaudatus*: Repeated glucose values from 620 animals. *Diabetes* **21**, 715–721.

Stuhlman, R. A., Packer, J. T., and Rose, S. D. (1972b). Repeated blood sampling of *Mystromys albicaudatus* (white-tailed rat). *Lab. Anim. Sci.* **22**, 268–270.

Stuhlman, R. A., Srivastava, P.K., Schmidt, G., Vorbeck, M. L., and Townsend, J. F. (1974). Characterization of diabetes mellitus in South African hamsters (*Mystromys albicaudatus*). *Diabetologia* **10**, 685–690.

Stuhlman, R. A., Packer, J. T., Doyle, R. E., Brown, R. V., and Townsend, J. F. (1975). Relationship between pancreatic lesions and serum glucose values in *Mystromys albicaudatus*. *Lab Anim. Sci.* **25**, 168–174.

Taitz, L. S. (1954). Chromosomes of the white-tailed rat (*Mystromys albicaudatus*, Wagner, 1841). *S. Afr. J. Sci.* **51**, 143–148.

Waggie, K. S., Thornburg, L. P., and Wagner, J. E. (1986). Experimentally induced Tyzzer's disease in the African white-tailed rat (*Mystromys albicaudatus*). *Lab. Anim. Sci.* **36**, 492–495.

Wagner, J. C., and Bokkenheuser, V. (1961). The mycobacterium isolated from the dassie *Procavia capensis* (Pallus). *Tubercle* **42**, 47–56.

Walker, E. P. (1975). "Mammals of the World," 3rd ed., Vol. 2. Johns Hopkins Univ. Press, Baltimore, Maryland.

Yamamoto, R. S., Kroes, R., and Weisburger, J. H. (1972). Carcinogenicity of diethylnitrosamine in *Mystromys albicaudatus* (African white-tailed rat). *Proc. Soc. Exp. Biol. Med.* **140**, 890–892.

Yesus, Y. W., Esterly, J. A., Stuhlman, R. A., and Townsend, J. F. (1976). Significant muscle capillary basement membrane thickening in spontaneously diabetic *Mystromys albicaudatus*. *Diabetes* **25**, 444–449.

Mesocricetus brandti (Turkish Hamster)

I. TAXONOMY, HISTORY, AND GENETICS

Mesocricetus brandti is larger and has a wider geographic distribution than the closely related species *M. auratus.* (Syrian hamster) and *M. newtoni* (Rumanian hamster). The serological similarities of their immunoglobulins (Coe, 1981) and the G-banding homologies of their chromosomes (Popescu and DiPaolo, 1980) support the close taxonomic relationship of these three species. Colonies of *M. brandti* have been established in laboratories in the United States and Rumania from animals trapped in several areas of Turkey and Iran. *Mesocricetus brandti* has a polymorphic karyotype. Animals from some populations have a diploid number of 44, and other populations have a diploid number of 42. These types are phenotypically indistinguishable and interbreed readily to produce fertile offspring with a diploid number of 43. In laboratory hibernation experiments, animals with 42 chromosomes hibernated less than animals from the population with 44 chromosomes (Lyman and O'Brien, 1977; Popescu and DiPaolo, 1980; Lyman *et al.*, 1983).

Some researchers (Lyman and O'Brien, 1977; Todd *et al.*, 1972) have reported successful experimental crosses between male *M. brandti* and female *M. newtoni* with no offspring resulting from the reciprocal cross of *M. brandti* females and *M. newtoni* males, while other investigators (Raicu *et al.*, 1972) have reported the opposite results. In either case the hybrids were sterile. The offspring of male *M. brandti* and female *M. newtoni* crosses grew larger than either parent, and some of the females developed diabetes mellitus (Lyman and O'Brien, 1977).

II. CARE AND MANAGEMENT

Laboratory care of *M. brandti* is the same as for *M. auratus.* Like *M. auratus*, *M. brandti* is normally a solitary species, with social contacts between individuals only during the breeding season. Although litters may be left together until they are 6 weeks old, individual caging is required to prevent severe fighting. They are nocturnal and eat mainly at night, and they store food and hibernate during the winter when exposed to natural light cycles. Cannibalism of the young is common, and partial darkness and nesting material are required to prevent it (Lyman and O'Brien, 1977; Murphy, 1977).

III. ANATOMY, PHYSIOLOGY, AND REPRODUCTION

Mesocricetus brandti (Fig. 5). is similar in appearance to *M. newtoni* and has dark chest and subauricular patches, although they are not as dark as *M. newtoni*. They are larger than either *M. auratus* or *M. newtoni* and have an average body weight of about 150 gm. They have a median life span in the laboratory of about 670 days with a maximum of approximately 4 years. The young are born blind and hairless. The eyes open in about 12–13 days, when the young begin eating solid food and drinking from a water bottle. They are weaned by 20 days of age and average about 90 gm at 6 weeks of age. When exposed to natural light cycles, animals born in late summer do not reach sexual maturity until the following spring (Lyman and O'Brien, 1977).

Mesocricetus brandti is reported to be harder to breed in the laboratory and have a lower fecundity than *M. auratus*. Litter sizes range from 1 to 13, with an average of 6. *Mesocricetus brandti* may not produce more than three litters per year in the laboratory and probably only produce two litters per year in the wild (Lyman and O'Brien, 1977; Murphy, 1977). Females may spontaneously quit cycling and remain acyclic for 5–6 months (Hall and Goldman, 1982). Most animals reach puberty within 7 or 8 weeks (Lyman and O'Brien, 1977), but occasional males do not reach sexual maturity until they are 5–6 months old. Others undergo spontaneous testicular regression soon after reaching sexual maturity (Carter *et al.*, 1982). Estrous cycles are less regular and breeding less successful after the first year of life. Exposure to natural light cycles usually causes anestrus in females and testicular regression in males during the winter months (Lyman and O'Brien, 1977).

Females have a 4- or 5-day estrous cycle, and estrus is char-

Fig. 5. Mesocricetus brandti. (Photograph courtesy of Dr. Michael R. Murphy, San Antonio, Texas.)

acterized by the appearance of large numbers of nucleated epithelial cells in vaginal smears. The gestation period is 14 or 15 days. Females may produce their first litters at 7 or 8 weeks of age, although they have not attained their full adult size at this age (Lyman and O'Brien, 1977; Murphy, 1977; Frank and Johnston, 1981; Hall and Goldman, 1982).

IV. DISEASES

Very little has been published about the diseases of *M. brandti*, although they are probably susceptible to the same conditions as *M. auratus*. Diarrhea ("wet-tail"), osteomalacia, and malocclusion were reported in one colony. Obesity was also a problem with weights exceeding 200 gm (Lyman and O'Brien, 1977).

V. RESEARCH USES

The close relationship of *M. brandti* to the well-known *M. auratus* has led to their use in many comparative studies. These have included work in genetics (Popescu and DiPaolo, 1980; Raicu *et al.*, 1970), tissue transplantation (Shaffer *et al.*, 1971; Palm *et al.*, 1967), immunology (Coe, 1981), and behavior (Albers *et al.*, 1983). Most of the behavioral research has been related to reproductive activities (Murphy, 1977, 1978; Frank and Johnston, 1981; Johnston and Brenner, 1982).

Mesocricetus brandti has been a popular subject for hibernation studies because it is easier to maintain and breed in the laboratory than most other hibernating species, and it hibernates more readily and remains in continuous hibernation for longer periods than *M. auratus* (Pivorun, 1977; Lyman and O'Brien, 1974). Unlike ground squirrels and marmots, which acquire large fat stores and hibernate continuously through the winter, hamsters store food which they consume during interruptions in hibernation. Hibernation in *M. brandti* lasts about 5–6 months and is usually characterized by 2–7 days of torpor alternating with 1 or 2 days of activity and euthermic body temperatures. Hibernation may continue as long as 10 months in some individuals, and some may remain in continuous torpor for 20–30 consecutive days (Lyman and O'Brien, 1977; Hall and Goldman, 1980; Hall *et al.*, 1982).

Hibernation is initiated by exposure to colder temperatures, which is augmented by shorter daily photoperiods. Hibernation has been experimentally induced by exposure to 5° or 10°C ambient temperatures and 8 hr of light per day (Lyman and O'Brien, 1977; Hall *et al.*, 1982; Lyman *et al.*, 1983; Hall and Goldman, 1982). Short photoperiods result in testicular regression within 60–90 days (Hall *et al.*, 1982), and high testosterone levels prevent or terminate hibernation (Hall and Goldman, 1980). Short photoperiods accompanied by cold exposure cause testicular regression in 15–30 days, and most males begin hibernating within 2 weeks. Females exposed to these conditions stop ovulating within 30 days, and most begin hibernating within 60–90 days (Hall and Goldman, 1982). Testicular size spontaneously increases after 180–210 days, and this is apparently important in terminating hibernation. Castrated males hibernate longer than intact males, and some castrated males remained in hibernation for more than 2.5 years (Hall *et al.*, 1982). Pineal melatonin levels may modulate photoperiod influences on gonadal function and hibernation (Goldman *et al.*, 1981; Carter *et al.*, 1982). Estradiol has less effect in early hibernation than testosterone but may play a role in terminating hibernation in females in the spring (Hall and Goldman, 1980). Hibernation in *M. brandti* has been shown to increase longevity (Lyman *et al.*, 1981) and slow the accumulation of lipofuscin in the brain and heart (Papafrangos and Lyman, 1982).

Oncological research in *M. brandti* has been limited. Experimental inoculation of Rous sarcoma virus caused an 89% incidence of spindle cell sarcomas in muscles after an average latent period of 32 days (Zil'Fyan *et al.*, 1969).

REFERENCES

Albers, H. E., Carter, D. S., Darrow, J. M., and Goldman, B. D. (1983). Circadian organization of locomotor activity in the Turkish hamster (*Mesocricetus brandti*). *Behav. Neural. Biol.* **37,** 362–366.

Carter, D. S., Hall, V. D., Tamarkin, L., and Goldman, B. D. (1982). Pineal is required for testicular maintenance in the Turkish hamster (*Mesocricetus brandti*). *Endocrinology (Baltimore)* **111,** 863–871.

Coe, J. E. (1981). Comparative immunology of Old World hamsters—Cricetinae. *Ad. Exp. Med. Biol.* **134,** 95–102.

Frank, D. H., and Johnston, R. E. (1981). Determinants of scent marking and ultrasonic calling by female Turkish hamsters, *Mesocricetus brandti*. *Behav. Neural Biol.* **33,** 514–518.

Goldman, B., Hall, V., Hollister, C., Reppert, S., Roychoudhury, P., Yellon, S., and Tamarkin, L. (1981). Diurnal changes in pineal melatonin content in four rodent species: Relationship to photoperiodism. *Biol. Reprod.* **24,** 778–783.

Hall, V. D., and Goldman, B. D. (1980). Effects of gonadal steriod hormones on hibernation in the Turkish hamster (*Mesocricetus brandti*). *J. Comp. Physiol.* **135,** 107–114.

Hall, V. D., and Goldman, B. D. (1982). Hibernation in the female Turkish hamster (*Mesocricetus brandti*): An investigation of the role of the ovaries and of photoperiod. *Biol. Reprod.* **27,** 811–815.

Hall, V. D., Bartke, A., and Goldman, B. D. (1982). Role of the testis in regulating the duration of hibernation in the Turkish hamster, *Mesocricetus brandti*. *Biol. Reprod.* **27,** 802–810.

Johnston, R. E., and Brenner, D. (1982). Species-specificity of scent marking in hamsters. *Behav. Neural Biol.* **35,** 46–55.

Lyman, C. P., and O'Brien, R. C. (1974). A comparison of temperature regulation in hibernating rodents. *Am. J. Physiol.* **227,** 218–223.

Lyman, C. P., and O'Brien, R. C. (1977). A laboratory study of the Turkish hamster *Mesocricetus brandti*. *Breviora* **442**, 1–27.

Lyman, C. P., O'Brien, R. C., Greene, G. C., and Papafrangos, E. D. (1981). Hibernation and longevity in the Turkish hamster *Mesocricetus brandti*. *Science* **212**, 668–670.

Lyman, C. P., O'Brien, R. C., and Bossert, W. H. (1983). Differences in tendency to hibernate among groups of Turkish hamsters (*Mesocricetus brandti*). *J. Therm. Biol.* **8**, 255–257.

Murphy, M. R. (1977). Intraspecific sexual preferences of female hamsters. *J. Comp. Physiol. Psychol.* **91**, 1337–1346.

Murphy, M. R. (1978). Oestrous Turkish hamsters display lordosis toward conspecific males but attack heterospecific males. *Anim. Behav.* **26**, 311–312.

Palm, J., Silvers, W. K., and Billingham, R. E. (1967). The problem of histocompatibility in wild hamsters. *J. Hered.* **58**, 40–44.

Papafrangos, E. D., and Lyman, C. P. (1982). Lipofuscin accumulation and hibernation in the Turkish hamster, *Mesocricetus brandti*. *J. Gerontol.* **37**, 417–421.

Pivorun, E. (1977). Mammalian hibernation. *Comp. Biochem. Physiol. A.* **58A**, 125–131.

Popescu, N. C., and DiPaolo, J. A. (1980). Chromosomal interrelationship of hamster species of the genus *Mesocricetus*. *Cytogenet. Cell Genet.* **28**, 10–23.

Raicu, P., Nicolaescu, M., and Kirillova, M. (1970). The sex chromatin in five species of hamsters. *Chromosoma* **31**, 61–67.

Raicu, P., Ionescu-Varo, M., Nicolaescu, M., and Kirillova, M. (1972). Interspecific hybrids between Romanian and Kurdistan hamsters. *Genetica* **43**, 223–230.

Shaffer, C. F., Streilein, J. W., Freedberg, P. S., and Sherman, J. (1971). Studies on antilymphocyte serum and transplantation immunity in Syrian hamsters. *Transplantation* **11**, 396–403.

Todd, N. B., Nixon, C. W., Mulvaney, D. A., and Connelly, M. E. (1972). Karyotypes of *Mesocricetus brandti* and hybridization within the genus. *J. Hered.* **63**, 73–77.

Zil'fyan, V. N., Fichidzhyan, B. S., and Kumkumadzhyan, V. A. (1969). Pathogenicity of Rous sarcoma virus in *Mesocricetus brandti*. *Biol. Abstr.* **50**, 1365 (abstr.).

Mesocricetus newtoni (Rumanian Hamster)

Mesocricetus newtoni, which occurs in eastern Bulgaria and Rumania, is isolated geographically from both *M. auratus* (Syrian hamster) and *M. brandti* (Turkish hamster) (Raicu and Bratosin, 1968). *Mesocricetus newtoni* is serologically similar to *M. auratus* with identical serum IgG_1, IgA, and IgM but lacks some antigens found in *M. auratus* IgG_2 (Coe, 1981). Colonies of *M. newtoni* have been established at the University of Bucharest and the Agricultural Institute of Bucharest in Rumania (Raicu *et al.*, 1968).

The close relationship of *M. newtoni* to the well-studied *M. auratus* has resulted in comparative chromosome studies (Raicu *et al.*, 1970a; Voiculescu, 1974; Popescu and DiPaolo, 1980). The karyotypic differences which are present between *M. auratus* and *M. newtoni* primarily occur as a result of deletion of sex chromosome heterochromatin and autosomal translocations which indicate that *M. auratus* was the ancestor of *M. newtoni* (Popescu and DiPaolo, 1980).

Mesocricetus newtoni has 38 chromosomes (Raicu and Bratosin, 1968; Todd *et al.*, 1972), with 18 pairs of autosomes consisting of 2 metacentrics, 5 submetacentrics, and 11 subtelocentrics. In females the pair of X chromosomes are the longest subtelocentric ones, whereas in males the Y chromosome is the shortest submetacentric one. The arms of the X chromosome in *M. newtoni* are much longer than those of *M. auratus* (Raicu *et al.*, 1968). Chromosomal replication and the distribution of chromosomes during metaphase have been described for *M. newtoni* (Raicu *et al.*, 1970c, 1972b), and Voiculescu (1973) has reported the occurrence and location of constitutive heterochromatin zones on *M. newtoni* chromosomes.

Raicu and Bratosin (1968) unsuccessfully attempted crosses between *M. newtoni* and agouti-type *M. auratus*, but successful crosses were obtained when a partial albinotic mutant ($c^d c^d$) of *M. auratus* was utilized (Raicu and Bratosin, 1968; Raicu *et al.*, 1969). The hybrids resulting from crosses between *M. newtoni* females ($2n = 38$) and *M. auratus* males ($2n = 44$) had an intermediary number of chromosomes ($2n = 41$) and an average yield of young per litter that was less than 50% of the average litter size of each parental species. The *M. newtoni* female and *M. auratus* male hybrids and the offspring of the reciprocal crosses have been unable to produce viable young (Raicu and Bratosin, 1968; Raicu *et al.*, 1969; Todd *et al.*, 1972). These hybrids have been used to study the occurrence of mixed mitochondrial populations in the cells of hybrids (Raicu *et al.*, 1970b).

Some investigators (Todd *et al.*, 1972; Lyman and O'Brien, 1977) have reported successful crosses between female *M. newtoni* and male *M. brandti* with infertile results from the reciprocal cross of *M. newtoni* males and *M. brandti* females, while other researchers (Raicu *et al.*, 1972a) have reported the opposite results. All of the reports indicated that the hybrids are sterile. The offspring of female *M. newtoni* and male *M. brandti* crosses grew larger than either parent, and some of the females developed diabetes mellitus (Lyman and O'Brien, 1977). The hybrids resulting from the crosses between female *M. brandti* ($2n = 42$) and male *M. newtoni* ($2n = 38$) had chromosome numbers ($2n = 40$) which were intermediate between the two parent species, and the karyotype represented two distinct chromosome sets corresponding to those in the two parent species. The number of offspring per litter for the female *M. brandti* and male *M. newtoni* crosses averaged about 50% of the number expected in nonhybrid litters (Raicu *et al.*, 1972b).

Mesocricetus newtoni (Fig. 6) is similar in appearance to *M. brandti*, with both having dark chest and subauricular patches, although these areas are darker in *M. newtoni*. *Mesocricetus newtoni* is approximately the same size as *M. auratus* with an

Fig. 6. Mesocricetus newtoni. (Photograph courtesy of Dr. Robert E. Johnston, Ithaca, New York.)

average adult weight of about 100 gm (Murphy, 1977). The face of *M. newtoni* appears more pointed and ratlike than either *M. auratus* or *M. brandti*, although the nasal portion of the skull is not narrower (Lyman and O'Brien, 1977).

Mesocricetus newtoni has a low fecundity and is considered to be harder to breed in the laboratory than either *M. auratus* or *M. brandti*. The gestation period of *M. newtoni* is 16 days (Murphy, 1977).

REFERENCES

Coe, J. E. (1981). Comparative immunology of Old World hamsters—Cricetinae. *Adv. Exp. Med. Biol.* **134,** 95–102.

Lyman, C. P., and O'Brien, R. C. (1977). A laboratory study of the Turkish hamster *Mesocricetus brandti*. *Breviora* **442,** 1–27.

Murphy, M. R. (1977). Intraspecific sexual preferences of female hamsters. *J. Comp. Physiol. Psychol.* **91,** 1337–1346.

Popescu, N. C., and DiPaolo, J. A. (1980). Chromosomal interrelationship of hamster species of the genus *Mesocricetus*. *Cytogenet. Cell Genet.* **28,** 10–23.

Raicu, P., and Bratosin, S. (1968). Interspecific reciprocal hybrids between *Mesocricetus auratus* and *M. newtoni*. *Genet. Res.* **11,** 113–114.

Raicu, P., Hamar, M., Bratosin, S., and Borsan, I. (1968). Cytogenetical and biochemical researches in the Rumanian hamster (*Mesocricetus newtoni*). *Z. Sauegetierkd.* **33,** 186–192.

Raicu, P., Ionescu-Varo, M., and Duma, D. (1969). Interspecific crosses between the Rumanian and Syrian hamster: Cytogenetic and histological studies. *J. Hered.* **60,** 149–152.

Raicu, P., Nicolaescu, M., and Kirillova, M. (1970a). The sex chromatin in five species of hamsters. *Chromosoma* **31,** 61–67.

Raicu, P., Vladescu, B., Borsan, I., and Staicu, S. (1970b). Heterogeneity of mitochondria in the interspecific hybrid *Mesocricetus newtoni* and *M. auratus*. *Heredity* **25,** 465–469.

Raicu, P., Vladescu, B., and Kirillova, M. (1970c). Distribution of chromosomes in metaphase plates of *Mesocricetus newtoni*. *Genet. Res.* **15,** 1–6.

Raicu, P., Ionescu-Varo, M., Nicolaescu, M., and Kirillova, M. (1972a). Interspecific hybrids between Romanian and Kurdistan hamsters. *Genetica* **43,** 223–230.

Raicu, P., Popescu, N. C., Cioloca, L., Nicolaescu, M., and Kirillova, M. (1972b). Replication of the chromosomal complement and idiogram in the Romanian hamster (*Mesocricetus newtoni*). *Caryologia* **25,** 283–294.

Todd, N. B., Nixon, C. W., Mulvaney, D. A., and Connelly, M. E. (1972). Karyotypes of *Mesocricetus brandti* and hybridization within the genus. *J. Hered.* **63,** 73–77.

Voiculescu, I. (1973). On the constitutive heterochromatin in several rodent species. *Rev. Roum. Biol. Ser. Zool.* **18,** 163–169.

Voiculescu, I. (1974). A comparative study of the chromosome banding patterns of *Mesocricetus newtoni* and *Mesocricetus auratus*. *Z. Sauegetierkd.* **39,** 211–219.

Cricetulus migratorius (Armenian Hamster)

Laboratory colonies of *Cricetulus migratorius* have been established in the United States and in the Soviet Union. Laboratory care is the same as for the Chinese hamster (*C. griseus*). These hamsters (Fig. 7) weigh 40–80 gm and produce litters averaging 6 or 7 after an 18- or 19-day gestation period. They are born hairless, with the eyes and ears closed. The eyes and ears open by 2 weeks of age, when the young begin to nibble solid food. They may be weaned at 18 days (Yerganian *et al.*, 1967; Lavappa and Yerganian, 1970; Lavappa, 1977; Sviridenko, 1969). An immunoglobulin comparison revealed an IgM similarity with *Mesocricetus*, suggesting that this species is more closely related to *Mesocricetus* than is *C. griseus*, which do not show this immunoglobulin similarity (Coe, 1981). A taxonomic, morphometric description of the teeth and mandible of the Syrian subspecies has been published by

Fig. 7. Cricetulus migratorius. (Photograph courtesy of Dr. Michael R. Murphy, San Antonio, Texas.)

Pradel (1981). Akbarzadeh and Arbabi (1979) reported the occurrence of three dwarfs with axial skeleton malformations in a litter of five from a laboratory colony in Iran, but none of the dwarfs survived longer than a month and the etiology was not determined.

Cricetulus migratorius has been used primarily for cytogenetic and oncology studies. They have a small chromosome number (2n = 22), but the identity of individual chromosomes is difficult to distinguish with routine staining procedures and the X and Y chromosomes are isomorphic (Yerganian and Papoyan, 1965; Yerganian *et al.*, 1967; Sonnenschein and Yerganian, 1969; Solari, 1974; Lavappa, 1977; Papoyan *et al.*, 1975). The commencement of meiosis in males displays a high degree of synchrony beginning at 14 days of age and reaching completion at 27 days. Studies of meiotic chromosomes have been published by Lavappa and Yerganian (1970), Solari (1974), Allen (1979), and Pathak *et al.* (1979). Chromosome replication has also been studied in hybridized cells of *C. migratorius* and *C. griseus* by Sonnenschein *et al.* (1969), Sonnenschein (1970), and Yerganian and Nell (1966). In contrast to most other species, cell cultures from *C. migratorius* and *C. griseus* often retain euploidy (Yerganian *et al.*, 1967) and have been used for *in vitro* chemical and radiation mutagenesis studies (Yerganian and Lavappa, 1971; Lavappa and Yerganian, 1971; Lavappa, 1974; Papoyan *et al.*, 1975). *Cricetulus migratorius* is susceptible to radiation (Papoyan *et al.*, 1975) and chemical (Zil'Fyan *et al.*, 1970)-induced lymphosarcoma, and neonates are susceptible to human adenovirus type 12-induced tumors (Kang and Hahn, 1974). Squamous-cell carcinomas of the lip have been produced with repeated applications of dimethylbenzanthracene (Zil'Fyan *et al.*, 1972).

Four inbred strains of *C. migratorius* were developed by Yerganian (1979). Both strains IV and VIII have been reported to develop angioimmunoblastic lymphadenopathy and have been suggested as an animal model of this human disease (Yerganian *et al.*, 1978). *Cricetulus migratorius* is also susceptible to a dose-dependent hyperbilirubinemia caused by diethylstilbestrol and have been considered as a model for estrogen hepatic toxicity (Coe *et al.*, 1983). Immunized *C. migratorius* spleen cells have been hybridized with mouse myeloma cells to produce stable monoclonal antibody-secreting hybridomas (Sanchez-Madrid *et al.*, 1983). This system provides a valuable potential source of monoclonal antibodies to mouse antigens.

REFERENCES

Akbarzadeh, J., and Arbabi, E. (1979). Malformed vertebrate in *Cricetulus migratorius* (grey hamster): A case report. *Lab. Anim.* **13,** 299–300.

Allen, J. W. (1979). UrdU-dye characterization of late replication and meiotic recombination in Armenian hamster germ cells. *Chromosoma* **74,** 189–207.

Coe, J. E. (1981). Comparative immunology of Old World hamsters—Cricetinae. *Adv. Exp. Med. Biol.* **134,** 95–102.

Coe, J. E., Ishak, K. G., and Ross, M. J. (1983). Diethylstilbestrol-induced jaundice in the Chinese and Armenian hamster. *Hepatology* **3,** 489–496.

Honacki, J. H., Kinman, K. E., and Koeppl, J. W. (1982). "Mammal Species of the World." Allen Press and Association of Systematics Collections, Lawrence Kansas.

Kang, Y. S., and Hahn, S. (1974). Spontaneous morphological transformation in adenovirus type 12-induced tumor cells of Armenian and Chinese hamster. *Korean J. Zool.* **17,** 51–56.

Kato, H. (1979). Preferential occurrence of sister chromatid exchanges at heterochromatin–euchromatin junctions in the Wallaby and hamster chromosomes. *Chromosoma* **74,** 307–316.

Lavappa, K. S. (1974). Induction of reciprocal translocations in the Armenian hamster. *Lab Anim. Sci.* **24,** 62–65.

Lavappa, K. S. (1977). Chromosome-banding patterns and idiogram of the Armenian hamster, *Cricetulus migratorius*. *Cytologia* **42,** 65–72.

Lavappa, K. S., and Yerganian, G. (1970). Spermatogonial and meiotic chromosomes of the Armenian hamster, *Cricetulus migratorius*. *Exp. Cell Res.* **61,** 159–172.

Lavappa, K. S., and Yerganian, G. (1971). Latent meiotic anomalies related to an ancestral exposure to a mutagenic agent. *Science* **172,** 171–174.

Loukashkin, A. S. (1944). The giant rat-headed hamster, *Cricetulus triton nestor* Thomas of Manchuria. *J. Mammal.* **25,** 170–177.

Markaryan, D. S., Martirosyan, D. M., and Avdzhian, M. V. (1972). Cytogenetic characteristics of 2 new transplantable cell lines of gray hamsters *Cricetulus-Migratorius* transformed by Rous sarcoma virus. *Bull. Exp. Biol. Med. (Engl. Transl.)* **72,** 1315–1317.

Papoyan, S., Zilfian, V., Yerganian, G., Kumkumadjian, V., and Fichidjian, B. (1975). *Cricetulus migratorius,* Armenian grey hamster as a new object for investigation of certain radiobiological effects. *Stud. Biophys.* **53,** 81–83.

Pathak, S., Lau, Y. F., and Drwinga, H. L. (1979). Observations on the synaptonemal complex in Armenian hamster spermatocytes by light microscopy. *Chromosoma* **73,** 53–60.

Pradel, A. (1981). Biometrical remarks on the hamster *Cricetulus migratorius* (Pallas 1773) (Rodentia, Mammalia) from Krak des Chevaliers (Syria). *Acta Zool. Cracov.* **25,** 271–292.

Sanchez-Madrid, F., Szklut, P., and Springer, T. A. (1983). Stable hamster–mouse hybridomas producing IgG and IgM hamster monoclonal antibodies of defined specificity. *J. Immunol.* **130,** 309–312.

Solari, A. J. (1974). The relationship between chromosomes and axes in the chiasmatic XY pair of the Armenian hamster (*Cricetulus migratorius*). *Chromosoma* **48,** 89–106.

Sonnenschein, C. (1970). Somatic cell hybridization: Pattern of chromosome replication in viable Chinese hamster × Armenian hamster hybrids. *Exp. Cell Res.* **63,** 195–199.

Sonnenschein, C., and Yerganian, G. (1969). Autoradiographic patterns of chromosome replication in male and female cell derivatives of the Armenian hamster (*Cricetulus migratorius*). *Exp. Cell Res.* **57,** 13–18.

Sonnenschein, C., Roberts, D., and Yerganian, G. (1969). Karyotypic and enzymatic characteristics of a somatic hybrid cell line originating from dwarf hamsters. *Genetics* **62,** 379–392.

Sviridenko, P. A. (1969). Growth and development of grey hamsters (*Cricetulus migratorius* Pall.). *Biol. Abstr.* **50,** 12521 (abstr.).

Yerganian, G. (1967). The Chinese hamster. *In* "The U.F.A.W. Handbook on the Care and Management of Laboratory Animals" (A. N. Worden, ed.). 3rd ed., pp. 340–352. Williams & Wilkins, Baltimore, Maryland.

Yerganian, G. (1972). History and cytogenetics in hamsters. *Prog. Exp. Tumor Res.* **16,** 2–41.

Yerganian, G. (1979). Chinese and Armenian hamsters. *In* "Inbred and Genetically Defined Strains of Laboratory Animals" (P. L. Altman and D. D. Katz, eds.), Vol. 3, Part 2, pp. 500–504. Fed. Am. Soc. Exp. Biol., Bethesda, Maryland.

Yerganian, G., and Lavappa, K. S. (1971). Procedures for culturing diploid cells and preparation of meiotic chromosomes from dwarf species of hamsters. *Chem. Mutagens* **2**, 387–410.

Yerganian, G., and Nell, M. B. (1966). Hybridization of dwarf hamster cells by UV-inactivated Sendai virus. *Proc. Natl. Acad. Sci. U.S.A.* **55**, 1066–1073.

Yerganian, G., and Papoyan, S. (1965). Isomorphic sex chromosomes, autosomal heteromorphism and telomeric associations in the grey hamster of Armenia, *Cricetulus migratorius,* Pall. *Hereditas* **52**, 307–319.

Yerganian, G., Cho, S. S., Ho, T., and Nell, M. N. (1967). Euploidy and chromosome alterations in normal, malignant, and tumor-virus transformed cells of the Chinese and Armenian hamsters. *In* "Genetic Variations in Somatic Cells" (J. Klein, M. Vojtiskova, and V. Zeleny, eds.), pp. 349–360. Academia, Prague.

Yerganian, G., Paika, I., Gagnon, H., and Murthy, A. S. K. (1977). Angioimmunoblastic lymphadenopathy (AIL) and immunoblastic sarcoma of B-cells of the Armenian hamster. *Proc. Am. Assoc. Cancer Res.* **18**, 205 (abstr.).

Yerganian, G., Gagnon, H. J., and Battaglino, A. (1978). Animal model: Autoimmune-prone inbred Armenian hamster. *Am. J. Pathol.* **91**, 209–212.

Zil'Fyan, V. N., Fichidzhyan, B. S., and Kumkumadzhyan, V. A. (1970). Induction of leukemia in hamsters with 9,10-dimethyl,1,2-benzanthracene. *Biol. Abstr.* **51**, 8888 (abstr.).

Zil'Fyan, V. N., Fichidzhyan, B. S., and Kumkumadzhyan, V. A. (1972). Experimental cancer of the lip in grey hamsters. *Biol. Abstr.* **53**, 1486 (abstr.).

ACKNOWLEDGMENTS

The authors would like to recognize the significant contribution of Mr. James Paysse, editorial assistant—Science Information Services at the Delta Regional Primate Research Center. The clerical assistance of Ms. Carolyn Boudreaux is appreciated.

This work was supported in part by the National Institutes of Health Grant No. RR00164-23 Support for Regional Primate Research Center.

Index

A

Abdominal cavity, 12
Abscess, 130, 131, 360, 372
Acinar cells, 15
Actinomyces bovis, 130
Actinomyces viscosus, 288
Activity patterns, 268
Acyclovir, 232
Addison's disease, 240
Adenocarcinoma, 372
Adenohypophysis, 311
Adenoma, 159, 164, 324
  of adrenal cortex, 164
    Chinese hamster, 324
    intestinal 159
Adenovirus type 2, induction of tumors by, 204–205
Adrenal demedullation, 82
Adrenal gland, 25, 32, 88–89, 164–165, 170, 240–241
  amyloid, 170
  blood supply to, 25
  endocrinology of, 32
  experimental infections of, 240–241
  neoplasia of, 164–165
  perfusion of, 88–89
Adrenal hormones, 53–54
Adrenal necrosis, 240
Adrenalectomy, 82, 265–266
  effect on sodium intake, 265–266
Aflatoxin B, 277, 286, 340
  carcinogenicity of, 286
  chromosomal damage from, 340
  teratogenic effects of, 277
Aging, 24, 29, 30, 169–175, 253, 254, 306–307, 326, 327, 357, 360
  amyloidosis and, 254
  blood cell numbers and, 357
  blood pressure and, 24
  chromosome abnormalities associated with, 306–307
  diseases associated with, 169–175, 327

female reproductive system, 30
fertility and, Chinese hamster, 326, 327
incidence of thrombosis and, 253, 254
teeth and, European hamster, 360
testis, Syrian hamster, 29
thyroid, Syrian hamster, 34
Aggression, 175, 270, 306, 314, 325, 352, 370
  Chinese hamster, 306, 314, 325
  Dzungarian hamster, 370
  European hamster, 352
  prevention of, 175
  sexual differences in, 270
Airways, measurements of, Syrian hamster, 21
Alanine aminotransferase, 53
Albinism, 379
Albumin, serum, 48
Alimentary system, parasites of, 137
Alkaline phosphatase, 50–51
Alloxan, 336
Alopecia, 149–152, 372
Amebiasis, 237, *see also* specific organisms
Amino acids, 302
Amoebae, 142–143
Amylase, 52
Amyloidosis, 34, 169–171, 254–255, 286
Analgesics, teratogenic effects of, 276
Analytical techniques for clinical chemistry, 45–46
Anatomy, 10–36, 309–313, 353–356, 371, 377
  Chinese hamster, 309–313
  Dzungarian hamster, 371
  European hamster, 353–356
  South African hamster, 377
  Syrian hamster, 10–36
Androgen, 12, 270, 274
Anesthesia, methods for, 77–82
Anesthetics, 44, 77–82
Angioimmunoblastic lymphadenopathy, 386
Animal model, *see* Model
Annual cycle of hibernation, 274

Annual reproductive cycle, 271
Antibiotic toxicity, 181–186, 238
Antibiotic–associated enterocolitis, 180–196
Antibiotic–associated pseudomembranous enterocolitis, *see* Antibiotic-associated enterocolitis
Antibiotics, 120, 181–186, *see also* specific diseases
Antidotes for radioactive contamination, 281
Antimetabolic agents, 186
Antimicrobial agents, 181–186
Antimicrobial–associated colitis, hamster model of, 180, 238
Aorta, 311, 332
   Chinese hamster, 311
   diabetic Chinese hamster, 332
Appetite control, diabetic Chinese hamster, 334
Arboviruses, 231–232
Arenavirus, *see* Lymphocytic choriomeningitis virus
Armenian hamster, 385–386
Arterial blood gases, 88
Arteriolar nephrosclerosis, 173
Arthropod parasites, 149–153, *see also* specific organisms
Artificial sweeteners, toxicity of, 339–340
Aspartate aminotransferase, 53
Aspicularis tetraptera, 144–146
Atrial thrombosis, 253, 254
Atypical ileal hyperplasia, *see* Proliferative ileitis
Auricular thrombosis, 253
Autonomic nerves, microscopic anatomy of, 311
Autonomic nervous system, diabetic Chinese hamster, 332

B

B lymphocytes, 220
Babesia microti, 240, 323
Babesia rodhaini, 240
Bacillus Calmette Guerin vaccine, 231
Bacillus piliformis, 323, 379, *see also* Tyzzer's disease
Bacterial diseases, 112–131, 322–323, 360, *see also* specific bacteria and bacterial diseases
   Chinese hamster, 322–323
   European hamster, 360
   Syrian hamster, 112–131
Bacterial enteritis, 238
Baculum, Syrian hamster, 29
Balantidium coli, 142
Basophils, 301
Bdellonyssus bacoti, *see* Ornithonyssus bacoti
Behavior, 264–268, 272, 306, 352–353, 374, 377, 382
   aggression, European hamster, 352
   Chinese hamster, 306
   drinking, 264–265
   Dzungarian hamster, 374
   feeding, 266–267
   hoarding, 267
   sexual, 267–268, 353
   South African hamster, 377
   thermoregulatory, 272
   Turkish hamster, 382
Behavioral research, support requirements, 269
Besnoitia jellisoni, 235, 240–241
Bile, collection of, 77
Bilirubin, Syrian hamster, 48–49
Biomethodology, 70–93
Birth weight, 301

Bladder, Syrian hamster, 27
Blood, 13, 24, 26, 44–45, 54, 55, 75–76, 88, 301, 314
   coagulation time, Syrian hamster, 26
   collection techniques, 44–45, 75
   diabetic Chinese hamsters, 314
   gases, 55, 301
   neonatal, 54
   normative data, 54, 301
   pH, 24, 55
   pressure, 24, 88, 301
   sample preparation, 45
   smears, methods for, 76
   vessels, 24
   volume, 13, 24, 54, 301
Blood cells, 34, 301, 357, 358
Blood chemistry, *see* Serum chemistry
Bluthaare, *see* Vibrissae
Body conformation, Syrian hamster, 10
Body temperature, 13, 301, 357
   European hamster, 357
   Syrian hamster, 301
Bone marrow, 58
Bone marrow cell proliferation, Chinese hamster, 313
Brain, 310–311, 331–332, 337
   Chinese hamster, 310 311
   diabetic Chinese hamster, 331–332
   effects of radiation on, Chinese hamster, 337
Breeding, 13, 62–64, 314–315, 352–353, 371, 378–379, 382
   age and, 13
   diabetic Chinese hamsters, 314–315
   Dzungarian hamster, 371
   European hamster, 352–353
   light cycle and, 62
   South African hamster, 378–379
   Syrian hamster, 63–64
   Turkish hamster, 382
Bronchopulmonary lavage, methods for, 76
Brown adipose tissue, 11–12, 275

C

Caging, *see* Housing
Campylobacter, 114, 115, 117, 121
Campylobacter jejuni, 114, 127–128
Cancer, animal model of, 259
Cannibalism, 115, 175, 315–316, 370, 382
   Chinese hamster, 315–316
   Dzungarian hamster, 370
   Syrian hamster, 115, 175
   Turkish hamster, 382
Carbohydrate metabolism, Chinese hamster, 307
Carcinogen susceptibility, 259–260, 281–287
Carcinogenesis, 21, 259–260, 281–285, 361
   cheek pouch, 281–285
   European hamster, 361
   genetics and, 259–260
   kinetics of, 283
   lung, 21
Cardiac output, Syrian hamster, 24
Cardiomyopathy, 259
Cardiovascular system, 24–27, 173
   noninfectious disease of, 173

Caries, 13, 288–289, 379
  South African hamster, 379
  Syrian hamster, 13
Castration, 85
Cataenotaenia pusilla, 149
Caviomonas mobilis, 142
Cecal mucosal hyperplasia, 121
Cecum, Syrian hamster, 18
Cell lines, derived from Chinese hamster, 341
Cell–mediated immunity, 205, 243
Cellular alterations in hibernation, 275
Central nervous system, 35, 174, 233–235, 278
  effects of radiation on, 278
  experimental infections of, 233–235
  spontaneous lesions of, 174
  Syrian hamster, 35
Cerebral hemorrhage, 327
Cervical ganglionectomy, 86
Cesarean derivation, 85
Cestodes, 136, 146–149, *see also* specific organisms
  of the alimentary system, 146–149
  parenteral, 136
Cheek pouch, 16–17, 25, 202, 281–285, 354
  anatomy of, 16, 354
  blood supply, Syrian hamster, 25
  chemical carcinogenesis in, 281–285
  immune system of, 17
  tumor transplantation into, 202
Chemical carcinogenesis, 281–287
Chemoprophylactic agents, toxicity of, 284
Chilomastix bettencourti, 142
Chinese hamster, 305–316
Cholangiocarcinoma, 237
Cholangioma, of liver, 160
Cholecystoduodenostomy, 86
Cholecystotomy, 85
Cholesterol, 29, 50, 307
Chromosomal abnormalities, 203–204, 306–307, 325–326
  Chinese hamster, 306–307, 325–326
  of transformed cell lines, 203–204
Chromosome number,
  Chinese hamster, 306
  Dzungarian hamster, 370
  Rumanian hamster, 384
  South African hamster, 377
  Syrian hamster, 13
  Turkish hamster, 382
Chromosomes, *see* Cytogenetics
Chromosome toxicology, 338–341
Cigarette smoke, inhalation studies with, 260, 279, 280, 283
Ciliates, of the alimentary system, 142
Circadian rhythm, 268, 315, 370, 371
  of activity, 268
  of feeding, Chinese hamster, 315
  of reproductive activity, 370, 371
Circadian timing of hormone secretion, 269–270
Circulatory system, neoplasia of, 163
Clara cell, Syrian hamster, 21
Clindamycin, 182, 187, 191, 238
Clindamycin–induced colitis, 238
Clinical chemistry, 44, 46
Clostridium difficile, 180–196, 238
Clotting, *see* Coagulation

Coagulation, 57–58
Coagulopathy, 253
Coat color, *see* Pigmentation
Complement, 217
Contact hypersensitivity, 221
Corticosteriods, 32, 53–54, 192–193, 240–242, 276
  effect on immunity, 240–241, 242
  sexual differences in secretion, 32
  teratogenic effects of, 276
Cortisone, 125, 192, 202, 284, 336
  diabetogenic effects of, 336
  effect on carcinogenesis in cheek pouch, 284
  tumor transplantation and, 202
  Tyzzer's disease and, 125
Costovertebral gland, see Flank organ
Creatine phosphokinase, 52–53
Creatinine, Syrian hamster, 47
Creutzfeldt-Jacob disease, hamster model of, 234
Cricetulus griseus, *see* Chinese hamster
Cricetulus migratorius, *see* Armenian hamster
Cricetus cricetus, *see* European hamster
Cryptococcus, 131, 231
Cryptosporidium, 142
Cutaneous application of chemicals, 89
Cyclic nucleotides, 313
Cystotomy, 83
Cytogenetics, 4–5, 306, 341, 370, 377, 382, 384
  Chinese hamster, 306, 341
  Dzungarian hamster, 370
  Rumanian hamster, 384
  South African hamster, 377
  Syrian hamster, 4–5
  Turkish hamster, 382
Cytomegalovirus, 106

D

Degenerative renal disease, 327
Demodex, 323
Demodex aurati, 149–152, 241
Demodex criceti, 149, 152, 241
Demyelinizing neuropathies, animal model of, 256–257
Dendritic cells, 221
Dental anomalies, 173, 360
  European hamster, 360
  Syrian hamster, 173
Dental formula, Syrian hamster, 13
Dental therapy, animal testing of, 288
Dermatophytosis, 372
Description, 3, 306, 351, 371, 377, 382, 384, 385
  Armenian hamster, 385
  Chinese hamster, 306
  Dzungarian hamster, 371
  European hamster, 351
  Rumanian hamster, 384
  South African hamster, 377
  Syrian hamster, 3
  Turkish hamster, 382
Development, 270, 378
Diabetes mellitus, 325, 327, 328, 330–336, 373, 380
  biochemistry of, 332–333
  chemically induced, Chinese hamster, 335–336
  disease associated with, 327, 328

Diabetes mellitus (*cont.*)
  genetics of, 333–334
  immunology of, 334
  morphology, 331–332
  physiology of, 330–332
  South African hamster, 380
  spontaneous, Chinese hamster 330–335
  therapy, 334–335
Diabetic animals, breeding of, 314
Diabetogenic agents, Chinese hamster, 335–336
Diarrhea, 112, 115, 121, 126, 127, 193–196, 238, 323
Diet, 13, 63, 158, 174, 175, 277, 288–289, 315–316, 334–335, 336, 352
  caries–conducive, 13, 288–289
  Chinese hamster, 315–316
  diabetic hamster, 334–335
  effect on microsomal drug metabolism, 277
  effect on tumor incidence, 158
  European hamster, 352
  lactating females, 316
Digestive system, *see also* Gastrointestinal system and specific organs
  noninfectious disease of, 173–174
9,10-dimethyl-1,2-benzanthracene (DMBA), 282
Dipetalonema vitae, 233
Diseases, infectious, 95–153, 228–243, 321–324, 360, 371–372, 379, 383, *see also* specific diseases and organisms
  Chinese hamster, 321–324
  Dzungarian hamster, 371–372
  European hamster, 360
  South African hamster, 379
  Syrian hamster, 95–153
  Turkish hamster, 383
Djungarian hamster, *see* Dzungarian hamster
DNA, Chinese hamster, 307
Drinking behavior, physiological regulation of, 264–265
Drugs, 179–199, 277, 337–338, 339, 340
  inducing diabetes, 335–336
  metabolism of, 277
  therapy with 179–180, 181–186, 188–191, 192, 193, 194, 195
  toxicity of, 181–194, 337–338, 339, 340
Dystocia, 326
Dzungarian hamster, 370–374

E

Early tumor antigens, 204–205
Ejaculate, collection of, 76
Electrocardiogram, 88
Electrolytes, Syrian hamster, 49–50, 51
Encephalitozoon cuniculi, 136, 235
Endocrine glands, noninfectious disease of, 172
Endocrine system, 32, 164–166, 240–241, *see also* specific glands and hormones
  experimental infections of, 240–241
  neoplasia of, 164–166
  Syrian hamster, 32
Endogenous Chinese hamster virus, 322
Endometrial neoplasia, 324–325
Entamoeba histolytica, 237
Entamoeba muris, 142
Enteritis, 112–121, 122, 164, 360
Enterocolitis, 180–196
Environmental pollution studies, 337, 339
Environmental requirements, 62–63, 314

Enzootic intestinal adenocarcinoma, *see* Proliferative ileitis
Enzymes, 18, 50–53, 308, *see also* specific enzymes
  Chinese hamster, 308
  small intestine, 18
  Syrian hamster, 50–53
Eosinophils, 301
Epidermoid carcinoma of cheek pouch, 281–285
Epidermoid sarcoma, 282
Epilepsy, 255–256
Epizootic leukemia, 163–164, 202–203, 238
Equine rhinopneumonitis, 232, 237
Erythrocytes, 55–56, 301
Escherichia coli, 112–115, 131
Esophageal constriction, 86
Esophageal decontamination, 89
Estradiol, secretion of, 270
Estrous cycle, 13, 30–31, 64, 269, 314, 352, 269, 383
  Chinese hamster, 314
  European hamster, 352
  Syrian hamster, 13, 30–31, 64
  Turkish hamster, 383
Ethanol intake, 266, 283
  carcinogenesis and, 283
Ether, 78
European hamster, 351–366
Euthanasia, 89–90
Examination, 70
Eye, diabetic Chinese hamster, 331

F

Fasting, 266–267
Feeding, 18, 37, 63, 128, 266–267, 315–316
  behavior, 266–267
  Chinese hamster, 315–316
  effect on clinical chemistry, 37
  pattern, Syrian hamster, 18
Female protein, 217, 255
Female reproductive system, 29–30
Filters of infection, hamsters as, 242
Flagellates, alimentary system, 137–142
Flank organ, Syrian hamster, 12
Fluanisone, 81
Follicle stimulating hormone, 269, 308, 372
Food consumption, 266–267, 301, 306, 334
  Chinese hamster, 306
  diabetic hamster, 334
  regulation of, 266–267
Food requirements, Syrian hamster, 13
Fungal infections, *see* Mycoses
Fungus, *see* Mycoses
Fur, *see* Hair

G

Gall bladder, 19
Gamma-glutamyltransferase (GGT), 283
Gastric intubation, 73
Gastrointestinal system, 17–18, 159–161, 238–239, 312, 332, 355, 377
  diabetic Chinese hamster, 332
  European hamster, 355
  experimental infections of, 238–239
  microscopic anatomy, Chinese hamster, 312

neoplasia of, 159–161
South African hamster, 377
Syrian hamster, 17–18
Genetic toxicology, Chinese hamster, 338
Genetics, 216, 251–252
    immunobiology and, 216
Geographical distribution, Syrian hamster, 3
Gestation period, 13, 64, 270, 301, 385
    Armenian hamster, 385
    Rumanian hamster, 385
    Syrian hamster, 13, 270, 301
Giardia, 139–140
Giardia muris, 139–140
    treatment for, 135
Globulins, Syrian hamster, 49
Glomerular thrombosis, 237
Glucose, 46, 267, 307
    food intake and, 267
    metabolism of, Chinese hamster, 307
    serum, Syrian hamster, 46
Golden hamster, *see* Syrian hamster
Golden Syrian hamster, *see* Syrian hamster
Gray hamster, *see* Chinese hamster

H

Hair, 11, *see also* Pigmentation
Halothane, 79
Hamster enteritis, *see* Proliferative ileitis
Handling, 64–65
Heart, 24, 355
    European hamster, 355
    Syrian hamster, 24
Heartbeat, rate, 301
Heat, 13, *see also* Estrous cycle
    duration of, 13
Heavy metals, teratogenic effects of, 276
Helminths, experimental infection with, 233, 235, 276
Hematocrit, 54, 56
    adult, 56
    neonatal, 54
Hematology, 54–58
Hematopoietic system, 163–164, 175, 278, 325
    neoplasia of, 163–164, 325
    noninfectious disease of, 175
    radiation resistance of, 278
Hemoglobin, 56, 301
Hemogram, 301, 359
    European hamster, 359
    Syrian hamster, 301
Hepatitis, 121, 237
Hepatoma, 324
Herbicides, teratogenic effects of, 276–277
Hereditary traits, 252
Heredity, diabetic Chinese hamster, 333–334
Herpesvirus, 232, 237, 240
    effect on placenta, 240
    experimental infection, 232, 237
Hexamastix, 142
Hexamita muris, *see* *Spironucleus muris*
Hibernation, 23, 48, 62, 273–275, 357–366, 383
    blood pH, 23
    environmental conditions for, 62

    European hamster, 357–360
    serum proteins and, 48
    Turkish hamster, 383
Hindleg paralysis, inherited, 257
Histocompatibility, 218
Histology, European hamster, 353–356
Histoplasma, 231
Histoplasma capsulatum, 181, 236, 239–240
    drug treatment for, 181
History of use, 5–6, 305–306, 352, 370, 376
    Chinese hamster, 305–306
    Dzungarian hamster, 370
    European hamster, 352
    South African hamster, 376
    Syrian hamster, 5–6
Hoarding behavior, 18, 63, 267
Hormones, 29, 32–33, 53–54, 268–270, 274, 308–309, *see also* specific hormones
    adrenal, 32–33
    Chinese hamster, 308–309
    circadian rhythm, 269–270
    control of behavior, 268
    control of hibernation, 274
    regulation, of pigmentation, 272–273
    reproductive, 29, 269–270
    sexual differentiation and, 270
    thyroid, Syrian hamster, 33
Housing, 61–62, 77, 152, 269, 313–314, 326, 352, 370, 377, 382
    Chinese hamster, 313–314
    Dzungarian hamster, 370
    European hamster, 352
    infertility and, Chinese hamster, 326
    light–dark cycle, 269
    prevention of mite infestations, 152
    South African hamster, 377
    Turkish hamster, 382
Human tumors, transplantation into conditioned hamsters, 202
Husbandry, 61–66, 175, 313–316, 370, 377, 382
    Chinese hamster, 313–316
    Dzungarian hamster, 370
    prevention of aggression, 175
    South African hamster, 377
    Syrian hamster, 61–66
    Turkish hamster, 382
Hydrocephalus, 234, 256
Hymenolepis diminuta, 146–149, 360
Hymenolepis nana, 146–149
Hyperkeratosis, 372
Hyperphagia, diabetic Chinese hamster, 334
Hypertension, experimentally induced, 83
Hypophysectomy, 84
Hypothalamus, effects of monosodium glutamate on, 336
Hypothermia, 82, 273, 276
    as anesthesia, 82
    effect on fetal development, 276

I

Ibuprofen, inhibition of cheek pouch carcinogenesis by, 285
Identification, 65
Immune system, 101, 216–217, 242–243, 334
    diabetic Chinese hamster, 334
    effects of lymphocytic choriomeningitis virus on, 101

Immune system (*cont.*)
  effects of corticosteroids on, 242–243, 284
  ontogeny of, 216–217
Immunity, 221, 284
  neoplasia and, 284
  pulmonary, 221
Immunogenetics, 216, 218–219
Immunoglobulins, 217–218
Impetigo, hamster model of, 241
Inbred lines, carcinogen susceptibility of, 260
Infectious disease, *see* Disease, infectious
Infertility, Chinese hamster, 326
Inflammatory disease, of the lung, 222
Influenza A virus, 236
Inhalant anesthetics, 78–79
Inhalation studies, 21, 278–281, 337, 362
  Chinese hamster, 337
  European hamster, 362
  radioactive compounds, 281
  Syrian hamster, 21
Injectable anesthetics, 79–80
Injections, 72–75
Insulin, 307, 330–332
Integument, 149–153, 161–162, 175
  neoplasia of, 161–162
  noninfectious disease of, 175
  parasites of, 149–153
Intestinal microflora, 238, 322
  Chinese hamster, 322
  South African hamster, 378
  Syrian hamster, 238
Intestine, 17–18, 25, 159, 174, 278
  blood supply to, 25
  effects of radiation on, 278
  morphology of, 17–18
  neoplasia of, 159
  spontaneous lesions of, 174
Intracerebral injections, 75
Intranasal injections, 72
Intrapulmonary instillations, 74
Intrarectal instillations, 74
Intraspinal injections, 75
Intratracheal instillation studies, 278, 279–280
Intratracheal instillations, 74
Intubations, methods for, 72–75
Irradiation, *see* Radiation
Islets of Langerhans,
  Chinese hamster, 309–310
  development of, 19
  diabetic Chinese hamster, 331
  Syrian hamster, 33

J

Jacobson's organ, *see* Vomeronasal organ

K

Ketamine, 79, 80
Kidney, 25, 27, 161, 170–171, 310, 327, 331, 333, 356
  amyloid, 170
  blood supply, Syrian hamster, 25
  Chinese hamster, 310
  degenerative disease of, Chinese hamster, 327
  diabetic Chinese hamster, 331, 333
  European hamster, 356
  experimental infections of, 236–237
  neoplasia of, 161
  noninfectious disease of, 171
  Syrian hamster, 27
Kilman's rat virus, 322

L

Lactate dehydrogenase, 53
Lactation, 13, 316
  diet and, 316
Lactic acid, 51
Langerhans cells, 221
Laparoscopy, 89
Large intestine, 18
Larynx, European hamster, 354
Lateral vein of tarsus, 72, 76
  blood collection from, 76
  injection into, 72
Legionella pneumophila, 236
Leishmania, 239
Leprosy, *see Mycobacterium leprae*
Leptospira, 236–237
Leukemia,
  Chinese hamster, 325
  epizootic, 163–164, 202–203
Leukocytes, 56, 57, 301
Life span, 13, 286, 301
Light–dark cycle, 62, *see also* Photoperiod
Lipid metabolism, Chinese hamster, 307
Lipids, circulating, Syrian hamster, 49, 50
Liponyssus bacoti, *see Ornithonyssus bacoti*
Litter size, 13, 64, 301, 314, 315
  Chinese hamster, 314, 315
  Syrian hamster, 13, 64
Liver, 19, 25, 89, 159, 170, 172, 237–238, 324, 332–333, 337, 356
  amyloid, 170
  biochemistry, 332–333
  blood supply, 25
  effects of radiation on, 337
  European hamster, 356
  experimental infections of, 237–238
  neoplasia of, 159, 324
  noninfectious disease of, 170, 172
  perfusion, 89
Lung, 20, 21, 159, 174, 222, 354
  European hamster, 354
  immunity in, 222
  neoplasia of, 159
  spontaneous lesions of, 174
  Syrian hamster, 20, 21
Luteinizing hormone, 269–270, 308, 372
Lymph nodes, 26–27
Lymphatic system, 26–27, 175
  spontaneous lesions of, 175
  Syrian hamster, 26–27
Lymphocyte mitogen response, South African hamster, 380
Lymphocytes, 301
Lymphocytic choriomeningitis virus 96, 100–104
Lymphoma, horizontally transmissible, *see* Epizootic leukemia

Lymphoma, infectious, *see* Epizootic leukemia
Lymphoreticular cells, 219–221

M

Macrophages, 221
Major histocompatibility complex, 218–219
Malaria, 235
Male reproductive system, Syrian hamster, 28–29
Malleomyces pseudomallei, 231
Mammary glands, Syrian hamster, 29
Mange, 149–152
Marking behavior, Chinese hamster, 306
Masculinization, 270
Mastitis, 129
Maternal mortality, 172
Measles virus, 233
Melanocyte-stimulating hormone, 372
Melanomas, spontaneous, 161–162
Melatonin, 35, 271–273, 373
Melioidosis, 231
Mesocricetus auratus, *see* Syrian hamster
Mesocricetus brandti, *see* Turkish hamster
Mesocricetus newtoni, *see* Rumanian hamster
Metabolism, 307–308, 332–333, 357
    diabetic hamster, 332–333
    effects of hibernation on, 357
Metal ion metabolism, Chinese hamster, 308
Methohexital, 80, 81
Methoxyflurane, 78–79
Microbial flora, 17, 180, 322, 378
    intestinal, Chinese hamster, 322
    intestinal, South African hamster, 378
    of stomach, Syrian hamster, 17
Microsporum, 131
Microwave radiation, 337, 338–339
Milk, collection of, 77
Mineral deficiency, 176
Minerals, 302
Mites, 149–153
    of the integument, 149–152
    nasal, 152–153
Models of human disease, 196, 228, 252–261, 330–335, 386
    angioimmunoblastic lymphadenopathy, 386
    antibiotic-associated enterocolitis, 196
    spontaneous diabetes mellitus, 330–335
Monocytes, 301
Monosodium glutamate, 336
Morphology of organs, Chinese hamster, 309–313
Morphophysiology, 10–36
Mouse encephalomyelitis virus, 106
Mouth, spontaneous lesions of, 173
Muscles, 10, 16, 313, 333
    of cheek pouch, 16
    diabetic hamster, 333
    microscopic anatomy, 313
    Syrian hamster, 10
Muscular dystrophy, animal model of, 258–259
Musculoskeletal system, 10, 161, 174
    neoplasia of 161
    spontaneous lesions of, 174
Mycobacterium bovis, 231
Mycobacterium kansasii, 229, 231, 240–241
    corticosteriods and, 240–241
Mycobacterium leprae, 231
Mycoplasma, 322–323
Mycoplasma pneumoniae, 235–236
Mycoplasma pulmonis, 130
Mycoses, 131, 231, *see also* specific organisms
    experimentally induced, 231
    spontaneously occurring, 131
Mycotoxins, 340
Myeloproliferative disease, 325
Myocoptes musculinus, 372
Mystromys albicaudatus, *see* South African hamster

N

N-acetylation, 260–261
Nasal cavity, 19, 152–153, 159, 353–354
    European hamster, 353–354
    mites, 152–153
    neoplasia of, 159
    Syrian hamster, 19
Natural habitat, 3, 272, 351
    Dzungarian hamster, 272
    European hamster, 351
    Syrian hamster, 3
Natural killer cells, 220–221
Necropsy, 90–91
Nematodes, 136, 143–146, 233, *see also* specific organisms
    of the alimentary system, 143–146
    experimental infections, 233
    parenteral, 136
    spontaneous infections, 136, 143–146
Neonatal enucleation, 86
Neoplasia, 106, 157–167, 201–214, 222–223, 259–260, 281–285, 287, 324–326, 337, 338, 360, 363–364, 372, 373, 379, *see also* specific neoplasms
    of central nervous system, 167
    of cheek pouch, 281–285
    chemical induction of, 259–260, 281–285
    Chinese hamster, 324–326, 337
    of connective tissue, 160, 161
    Dzungarian hamster, 372
    endocrine system, 164–166
    European hamster, 360, 363–364
    experimentally induced, 373
    of gastrointestinal system, 159–160
    of hematopoietic system, 163–164
    immunity to 222–223
    incidence of, 287
    of integument, 161–163
    of liver, 337
    of musculoskeletal system, 161
    nitroso compounds and, 338
    radiation induced, 214, 281
    of reproductive system, 166–167
    of respiratory system, 158–159
    South African hamster, 379
    spontaneous 157–167, 207–209
    of urinary system, 161
    virally induced, 106, 203, 204–205, 210–213
Nephrectomy, 83
Nephrosclerosis, 327
Nephrosis, 171

Nerves, peripheral, Chinese hamster, 311
Nervous system, 35–36, 332
  diabetic hamster, 332
  Syrian hamster, 35–36
Nesting behavior, Chinese hamster, 306
Neurobiology, use in, 268–269
Neuromuscular terminals, Chinese hamster, 312
Neuropathy, sex–linked, 257
Neutrophils, 301
Nitrosamines, 286
Nitroso compounds, 338, 340
N–nitrosomethylurea, 336
Nocardia, 231
Nomenclature, 4
Norepinephrine, 35
Normative data, 301–302
Notoedres, 152
Nutrient requirements, 302
Nutrition, 63, 175
Nutritional disease, 175–176

O

Obesity, 257
Octomitus, 142
Octomitus muris, *see Spironucleus muris*
Olfaction, 19, 268
  male sexual behavior and, 268
Oncogenes, 204
Oncological research, 201–205
Ontogeny, of immunity, 216
Oocytes, glycogen content of, 312
Ophthalmic venous sinus, 26, 73, 75
  blood collection from, 75
  injection into, 73
Opisthorchis, 237–238
Oral cancer, animal model of, 281–285
Oral cavity, Syrian hamster, 12–13
Orbital venous sinus, *see* Ophthalmic venous sinus
Orchidectomy, effects of, 372
Ornithonyssus bacoti, 152–153
Ornithonyssus sylvarum, 153
Os penis, *see* Baculum
Osteosarcoma, 161
Ovariectomy, 84
Ovary, 26, 29–30, 269
  blood supply to, 26
  hormone secretion by, 269
Ovulation, 13, 30, 269–270
Oxygen consumption, 87
Oxyuris obvelata, *see Syphacia obvelata*

P

Pancreas, 18, 25, 160, 166, 173, 331, 356
  blood supply to, 25
  diabetic Chinese hamster, 331
  European hamster, 356
  neoplasia of, 160, 166
  noninfectious disease of, 173
  Syrian hamster, 18
Pancreatic duct ligation, 86

Pancreatic islets, 308, 309–310
  metabolism, 308
  microscopic anatomy of, 309–310
Pancreaticocolostomy, 86
Pancreatitis, 86
Papilloma, 159, 161
  of the stomach, 159
  of urinary bladder, 161
Papovavirus, 106
Parainfluenza 1, *see* Sendai virus
Paramyxovirus, *see* Sendai virus and Simian virus 5
Parasites, 14, 135–153, 188–189, 229, 233, 235–243, 323–324, 360, 371,
    see also specific organisms
  Chinese hamster, 323–324
  Dzungarian hamster, 371
  European hamster, 360
  experimental infections, 229, 233, 235–243
  parenteral, 135
  spontaneous infections, 135–153
  treatment for, 188–191
Parathyroid gland, 33, 165–166, 172
  neoplasia of, 165–166, 172
  noninfectious disease of, 172
  Syrian hamster, 33
Parenteral parasites, 136
Parkinson's disease, 255
Parotid gland, 15
Parvovirus, 106, 232–233
Pasteurella, 130, 131
Pasteurella multocida, 130, 360
Pasteurella pneumotropica, 129, 130, 131, 360
Pathogen-free hamsters, production of, 135
Pentatrichomonas hominis, 142
Pentobarbital, 79, 80, 81
Perfusions, methods for, 88–89
Periodontal diseases, 288–289, 379
  South African hamster, 379
Periodontitis, 173, 328
  Chinese hamster, 328
Peripheral blood, 55
Peripheral nervous system, 35
Pesticides, 276–277, 338, 339
  teratogenic effects of, 276–277
  toxicity of, Chinese hamster, 338, 339
pH, of blood, 24, 301
Pharmacological compounds, 276–278, 284
  cheek pouch neoplasia, 284
  teratogenic effects of, 276
  toxicology of, 277–278
Pharyngeal decontamination, 89
Phodupus sungorus, *see* Dzungarian hamster
Photoperiod, 11, 28, 30, 35, 272, 274, 308–309, 311, 314, 370, 372–373,
    383
  Chinese hamster, 308–309, 311, 314
  control of pelage color, 11
  Dzungarian hamster, 370, 372–373
  effect on ovulation, 30
  effect on reproductive systems, 28, 270–273
  effect on thermoregulatory behavior, 272
  hibernation, 274
  pineal gland and, 35
  reproductive cycle and, 308–309

INDEX

Syrian hamster, 11, 28, 35
Turkish hamster, 383
Photoperiodicity, of activity, 268
Physiological data, Syrian hamster, 13, 87–89
Physiological measurements, methods for, 87–89
Pigmentation, 11, 272–273, 372–374
　effect of photoperiod on, 11, 372–374
　hormonal regulation of, 272–273
Pineal, 34–35, 271–272, 274, 311, 372–373
　Chinese hamster, 311
　Dzungarian hamster, 372–373
　hibernation and, 274
　mediation of photoperiodism by, 271–272
　Syrian hamster, 34–35
Pinealectomy, 84
Pink-eyed dilute mutant, 370
Pinworm, see *Syphacia*
Pituitary gland, 34, 311
　Chinese hamster, 311
　neoplasia of, 166
　noninfectious disease of, 172
　Syrian hamster, 34
Pituitary hormones, 272, see also specific hormones
　measurement of, 272
Placenta, 240, 276
　infections of, 240
　permeability of, 276
Plant-borne compounds, teratogenic effects of, 276–277
Plasma, 54
Platelets, 301
Pneumocystis carnii, 136
Pneumonia, 96, 99, 129, 130
Pneumonia virus of mice (PVM), 96, 98–100, 322
Pneumonitis, 136
Pneumovirus, see Pneumonia virus of mice
Poliomyelitis, 32
Polycystic disease, 175
Prediabetic Chinese hamster, 330–331
Pregnancy, 19, 48, 172, 314, 315, 353
　determination of, 314, 315
　European hamster, 353
　liver and, 19
　maternal mortality, 172
　serum proteins and, 48
Prenatal mortality, 175
Progesterone, 270, 308
　Chinese hamster, 308
　secretion of, 270
Prolactin, 32–33, 272–273, 372
　fur pigmentation and, 272–273
　measurement of, 272
Proliferative ileitis, 112–121
　clinical signs of, 115
　control of, 120
　diagnosis of, 120
　epizootiology of, 115
　etiology of, 113–114
　history of, 112–113
　pathogenesis of, 114–115
　pathology of, 116–120
　treatment for, 120
Prostatic hyperplasia, 257–258

Protein, 63, 171
　dietary requirements, 63
　kidney pathology and, 171
Proteins, in serum, 47–48
Proteus, 131
Protozoa, 135–143, 360, see also specific organisms
　of the alimentary system 137–143
　European hamster, 360
　parenteral, 136
　spontaneous infections, 136–143
　treatment for, 135
Protozoa–free colonies, 135
Pseudomembranous enterocolitis, see Antibiotic–associated enterocolitis
Pseudomonas, 131
Pseudotuberculosis, 129
Puberty, 13, 301
Pulmonary granulomas, Chinese hamster, 327
Pulmonary immunity, 221–222
Pulmonary measurements, Syrian hamster 20–21, 23–24
Pulse rate, Syrian hamster, 13

Q

Quaking, 255

R

Rabies, 234, 360
Radiation, 125, 131, 278–281, 324, 337
　induction of neoplasia, 214, 324
　infections following, 125, 131
　toxicity of, 337
　Tyzzer's disease and, 125
Radiation syndromes, 278, 281
Radioactive contamination, antidotes for, 281
Radiographic techniques, 71–72
Radioisotopes, effect on chromosomes, 338
Red blood cells, see Erythrocytes
Red mite, see *Ornithonyssus bacoti*
Regional enteritis, see Proliferative ileitis
Renal papillectomy, 83
Reovirus type 3, 96, 105–106, 322
Reproduction, 64, 176, 270, 370–371, 372, 378, 385
　Dzungarian hamster, 370–371, 372
　regulation by photoperiod, 270
　Rumanian hamster, 385
　senescence, 176
　South African hamster, 378
　Syrian hamster, 64
Reproductive capacity, 64
Reproductive cycle, annual, 35, 270, 271
　photoperiod and, 35, 270
Reproductive disorders, 326–327
Reproductive hormones, 29, 269–270, 308–309, see also specific hormones
　Chinese hamster, 308–309
　Syrian hamster, 29
Reproductive life, 13, 301
Reproductive maturity, female Syrian hamster, 30
Reproductive physiology, 269–273
Reproductive system, 28–31, 166, 172, 356
　European hamster, 356

Reproductive system (*cont.*)
   noninfectious disease of, 172
   Syrian hamster, 28–31
Reproductive tract, cells of, 312
Respiration, normative data, 301
Respiratory functions, measurement of, 87
Respiratory infections, bacterial, 129
Respiratory rate, 13, 301
Respiratory research, 21–23
Respiratory system, 19–24, 158–159, 174
   comparison to human, 21
   experimentally induced infections of, 235–236
   measurements of, 21, 23
   morphology of 19–24
   neoplasia of, 158–159
   noninfectious disease of, 174
      Syrian hamster, 19–24
Respiratory tract,
   experimental infections of, 235–236
   technique for inoculations into, 235
Respiratory volume, 87
Restraint, 65
Reticuloendothelial system, experimental infections of, 239–240
Retina, 310
Retinoic acid, 276
Retinoids, 284
Ringtail, 379
Ringworm, *see Trichophyton mentagrophytes*
Rumanian hamster, 384–385

S

Saliva, collection of, 76
Salivary glands, 15, 312, 354
   Chinese hamster, 312
   European hamster, 354
   Syrian hamster, 15
Salmonella, 125–127, 130
Salmonella paratyphi, association with *Schistosoma*, 127, 233
Salt balance, 264, 265–266
Sanitation procedures, 62
Sarcoidlike granuloma, 231
Sarcoma, of cheek pouch, 282–283
Schaumann bodies, 231
Schistosoma, 127, 233, 237
   association with *Salmonella*, 127, 233
Scrapie, 234
Seizures, 256
Sendai virus, 96–98
Sentinels of infection, 242
Serotonin, 275
Serum albumin, 48
Serum chemistry, 46–53
   diabetic Chinese hamster, 333
   European hamster, 356–357, 359
   South African hamster, 378
   Syrian hamster, 46–53
Serum glucose, 46
Serum proteins, 47–48
Sex determination, 10, 28, 65–66
   of adults 10, 28, 65–66
   of neonates, 28
Sexual behavior, 267–268, 306

Sexual differentiation, 270
Sexual maturity, 28, 30, 63, 356
   determination of, Syrian hamster, 63
   female European hamster, 356
   female Syrian hamster, 30
   male Syrian hamster, 28
Sexual receptivity, 270
Sialectomy, 85
Simian virus 5, 40, 96, 104–105, 204–205
Simian virus 40–induced fibrosarcoma, 222–223
Skeletal system, diabetic Chinese hamster, 332
Skin, *see* Integument
Slow acetylation, 260
Small intestine, 17
Sodium intake, 265–266
Sorbitol dehydrogenase, 53
South African hamster, 376–380
Specimen collection, 44–45
Specimen handling, 44–46
Spermatozoa, 312
Spironucleus muris, 137–139, 360
   European hamster, 360
Spiroplasma, 235
Spleen, neoplasia of, 325
Splenectomy, 85
Spondylosis, 328
Spontaneous animal model of human disease, definition, 252
Spontaneous disease model, 253–261
Sporozoa, 142
Staphylococcus aureus, 130, 131, 241
Staphylococcus epidermidis, 241
Starvation polydipsia, 265
Stomach, 17, 159, 174, 355
   European hamster, 355
   neoplasia of, 159
   spontaneous lesions of, 174
   Syrian hamster, 17
Streptococcus, 129, 130, 131
Streptococcus mutans, 288
Streptozotocin, 335–336
Striped-back hamster, *see* Chinese hamster
Subacute sclerosing panencephalitis, 233–234
Sublingual gland, 15
Submaxillary gland, 15
Suprachiasmatic nuclei, 269, 271
Surface area, of body, 301
Surgical techniques, 82–86
Syndoyomita muris, *see Spironucleus muris*
Syphacia mesocriceti, 144
Syphacia muris, 144, 145
Syphacia obvelata, 143–144
Syphilis, *see Treponema pallidum*
Syrian hamster, description of, 3

T

T lymphocytes, 219–220
Taenia, 136
Tapeworms, *see* Cestodes
Taste, 14–15
Taste buds, 14
Taxonomy, 4, 305, 370
   Chinese hamster, 305

INDEX

Dzungarian hamster, 370
Syrian hamster, 4
T-cells, 217
Teeth, 13–14, 173, 312, 354, 360
  abnormalities of, 173, 360
  Chinese hamster, 312
  European hamster, 354
  Syrian hamster, 13–14
Temperature, 13, 71, 276
  body, 13, 71
  effect on fetal development, 276
  method of measuring, 71
Teratogenesis, diabetic Chinese hamster, 334
Teratology, 275–277
Terminal ileitis, see Proliferative ileitis
Testes, Syrian hamster, 28–29
Testicular development, photoperiod and, 28, 308, 372, 373
Testicular tissue, diabetic Chinese hamster, 332
Tetratrichomonas microti, 141
Thermoregulation, 273–275, 373 see also Brown adipose tissue
  Dzungarian hamster, 373
  Syrian hamster, 273–275
Thoracic cavity, 12
Thrombosis, 173, 253–254
Thymectomy, 83
Thymus, ontogeny of, 217
Thyroid gland, 33, 165, 172, 275
  hibernation and, 275
  neoplasia of, 165
  noninfectious disease of, 172
  Syrian hamster, 33
Thyroid hormones, 33–34, 53
Thyroparathyroidectomy, 84
Tidal volume, 301
Tiletamine, 79, 81
Tobacco, 283
Tongue, 14, 354
  European hamster, 354
  Syrian hamster, 14
Toolan's H-1 virus, 106, 322
Toxicology, 181–194, 277–278, 338–341,
  Chinese hamster, 336–341
  Syrian hamster, 181–194, 277–278
Toxoplasma, 229, 230, 235, 240
Trachea, 20, 174, 354
  European hamster, 354
  spontaneous lesions of, 174
  Syrian hamster, 20
Transmissible enterocolitis, see Tyzzer's disease
Transmissible leukemia, see Epizootic leukemia
Transplantation, 88, 125
  into anterior chamber of eye, 89
  Tyzzer's disease and, 125
Transplantation antigens, 218
Trauma, 175, 326
Trematodes, 237–238, see also specific organisms
"Trembler," 255
Treponema pallidum, 229
Treponema pertenue, 229
Trichinella spiralis, 235
Trichomas, 323
Trichomonas wenyoni, 141
Trichophyton mentagrophytes, 131, 372

Trichosomoides, 136
Tritrichomonas, 140
Tropical rat mite, see *Ornithonyssus bacoti*
Tryptophan, 275, 276
  induction of hibernation by, 275
  teratogenic effect of, 276
Tularemia, 128–129
Tumors, see Neoplasia
Turkish hamster, 271, 273, 382–383
  effects of pinealectomy on, 271
  prolactin, 273
  thermoregulation and hibernation, 273
Typhlohepatitis, see Tyzzer's disease
Tyzzer's disease, 121–125, 323, 379
  Chinese hamster, 323
  clinical signs of, 123
  control of, 124
  diagnosis of, 124
  epizootiology of, 123
  etiology of, 122
  history of, 121–122
  pathogenesis of, 122
  South African hamster, 379

U

Urea nitrogen, serum, 46–47
Urinalysis, 54
Urinary bladder, 27, 161
  neoplasia of 161
Urinary system, 27–28, 356
  European hamster, 356
  Syrian hamster, 27–28
Urine, 54
Urogenital system, male Syrian hamster, 28
Urography, intravenous, 71
Uses in research, 6, 201–289, 361–364, 386, 372–374, 379, 383
  Armenian hamster, 386
  Dzungarian hamster, 372–374
  European hamster, 361–364
  South African hamster, 379
  Syrian hamster, 6, 201–289
  Turkish hamster, 383
Uterine fistula, 326–327
Uterus, 26, 324–325
  blood supply, 26
  neoplasia of, 324–325

V

Vagina, 30
Vaginal mucus, collection of, 76
Vaginitis, 131
Vancomycin, 185, 187
Vascular system, 24–27, 239–240, 355
  European hamster, 355
  experimental infections of 239–240
  Syrian hamster, 24–27
Vasectomy, 85
Vena cava, blood collection from, 76
Vena plantaris lateralis, see Lateral vein of tarsus
Venezuelan equine encephalitis, 231–232
Vertebral formula, 10

Vibrissae, 11
Viroid, 163
Viruses, 231–234, *see also* specific viruses and viral diseases
    Chinese hamster, 322
    European hamster, 360
    experimental infections, 231–234.
    induction of tumors by, 203, 204–205, 210–213
    South African hamster, 379
Vision, 36, 268
Vitamin A, 276, 302
    teratogenic effect of, 276
Vitamin deficiency, 176
Vitamin E deficiency, 63
Vitamins, 302
Vomeronasal organ, 35, 354
    European hamster, 354
    Syrian hamster, 35

W

Wasting disease, 102
Water balance, regulation of, 32–33, 264–265
Water consumption, 63, 264–265, 302, 306, 308
    Chinese hamster, 306, 308
    Syrian hamster, 63, 302
Water delivery, 316

Water deprivation, effect on blood parameters, 54
Water regulation, 32–33
Weaning, 64, 301
Weight, 301, 306
    Chinese hamster, 306
    Syrian hamster, 301
Wet-tail, *see* Proliferative ileitis
Whiskers, *see* Vibrissae
White mutant, 309

X

X-irradiation, 337, 338

Y

Yaws, *see Treponema pertenue*

Z

Zinc deficiency, 336
Zolazepam, 79, 81
Zona fasiculata, 32
Zona glomerulosa, 32
Zona reticularis, 32

10/28/14

Norris Medical Library
BOOKSALE
Price: $2.00
Please Pay at Loan Desk